Environmental Regulation and Impact Assessment

Environmental Regulation and Impact Assessment

Leonard Ortolano
Stanford University
Stanford, California

John Wiley & Sons, Inc.
New York • Chichester • Brisbane • Toronto • Singapore • Weinheim

Acquisitions Editors	Cliff Robichaud, Sharon Smith
Marketing Manager	Jay Kirsch
Senior Production Manager	Lucille Buonocore
Senior Production Editor	Nancy Prinz
Manufacturing Manager	Mark Cirillo
Illustration Editor	Jamie Perea

This book was set in Times Roman by BiComp, Inc., and printed and bound by Hamilton Printing Company. The cover was printed by Phoenix Color Corp.

Recognizing the importance of preserving what has been written, it is a policy of John Wiley & Sons, Inc. to have books of enduring value published in the United States printed on acid-free paper, and we exert our best efforts to that end.

The paper in this book was manufactured by a mill whose forest management programs include sustained yield harvesting of its timberlands. Sustained yield harvesting principles ensure that the number of trees cut each year does not exceed the amount of new growth.

Library of Congress Cagaloging-in-Publication Data:

Ortolano, Leonard.

Environmental regulation and impact assessment / Leonard Ortolano.

p. cm.

Includes bibliographical references.

ISBN 0-471-31004-2 (cloth)

1. Environmental impact analysis—United States. 2. Environmental laws—United States. 3. Environmental management—United States —Case studies. I. Title.

TD194.65.078 1996 96-26400

333.7′14′0973—dc20 CIP

Printed in the United States of America

10 9 8 7 6 5 4 3 2 1

To Patti Walters

Preface

Books on environmental regulation typically give little attention to environmental impact assessment (EIA) and vice versa. This book differs from others by emphasizing that EIA and environmental regulations are closely coupled. EIA and environmental regulations are implemented using many of the same forecasting and evaluation procedures. Moreover, EIAs often make explicit use of regulations, such as ambient air and water quality standards.

This book emphasizes both *methods* of environmental analysis and the *process* of environmental planning. Case studies are used to highlight the following attributes of environmental decision processes: multiple decision makers with conflicting objectives, uncertainties introduced by incomplete information, and coordination among government agencies. While much of the case study material is from the United States, several examples highlight experiences in other countries.

A five part organizational scheme is employed. Part One introduces categories of participants typically involved in making decisions affecting the environment. It also examines decision criteria, such as productive efficiency and intergenerational equity, that play important roles in decision making. In addition, Part One characterizes the two main decision contexts treated in the text: (1) the design and implementation of environmental regulations, and (2) the assessment of environmental impacts of proposed projects and programs.

The focus of Part Two is the theory and practice of environmental regulation. In addition to analyzing traditional command and control strategies, Part Two investigates incentive-based regulatory schemes: pollutant discharge fees, subsidies for waste management and tradable pollution rights. These market oriented approaches have attracted attention during the past few decades because of their potential to cut the cost of meeting environmental quality goals. The analysis of regulatory strategies uses techniques from environmental economics, including procedures for placing monetary values on environmental damages and for conducting benefit-cost analyses. A comparative evaluation of alternative regulatory schemes highlights practical implementation problems.

Part Three clarifies policy implementation issues by examining several environmental management programs in the United States. Emphasis is given to federal programs for managing air and water quality and hazardous waste. Pollution prevention measures adopted by industry as an alternative to waste treatment are also considered. In addition, Part Three analyzes the U.S. environmental impact assessment program and highlights its significance in spawning EIA activities throughout the world.

Parts Four and Five concentrate on methods used in formulating regulations and conducting impact assessments. Part Four surveys techniques for predicting and evaluating environmental impacts, conducting risk assessments, involving citizens in planning, and resolving environmental disputes. Part Five details concepts and procedures for assessing environmental changes associated with biological and visual resources, noise, air quality, and water resources. An appendix introduces geographic information systems and highlights their utility in environmental planning.

A preliminary version of this book was used in four classes, two at Georgia Institute of Technology and two at Stanford University. Students in these classes were majoring in a variety of subjects, including city planning, civil engineering, public policy, economics, biology, and applied earth sciences. Students and instructors in these classes made good use of teaching aids at the ends of chapters: lists of key concepts and terms and sets of discussion questions.

Readers are assumed to have a knowledge of elementary algebra, but no prior background in environmental science or economics is required. Calculus is used occasionally in footnotes that contain supplementary information.

Instructors who have used my previous textbook, *Environmental Planning and Decision Making* (New York: Wiley, 1984) may be interested in the relationship between this book and the earlier one. The linkage is strong, since the work that resulted in this book began as an effort to revise the 1984 text. As my work on revisions progressed, I became convinced that a more deliberate synthesis of EIA and traditional environmental regulation was called for.

In developing this synthesis, some things in the 1984 textbook were dropped, and many others were added. The excised material concerns land use-environment relationships, a topic whose importance remains undiminished. Practical considerations played a role here: the new book was growing in length and instructors using the 1984 book often omitted the land use chapters because the subject was covered in land use planning courses. The material added to the 1984 text includes nearly all of Parts One and Two. Although traces of the 1984 book may be found throughout, they are clearly discernable in the second half of this book. The seven chapters in Parts Three and Four grew from Chapter 5 through 8 of the 1984 text. And except for the treatment of geographic information systems, the chapters in Part Five are updated, expanded versions of the last five chapters in the 1984 book.

Stanford, California
September, 1996

Acknowledgments

Many people helped in preparing individual sections of chapters, and I acknowledge their specific contributions in notes throughout the text. I use this space to express my thanks to students, faculty and staff who contributed to entire chapters. The following students at Stanford University helped me prepare first drafts of chapters: Greg Browder, Katherine Kao Cushing, Alnoor Ebrahim, Melissa Geeslin and Samibhar Sankar.

The penultimate draft was pretested in classes taught by Professors William Patton and Anne Shepherd of the Department of City Planning at the Georgia Institute of Technology. Comments by students in these classes were most helpful, as were those provided by Professors Patton and Shepherd. In addition, Christi Bowler, a research assistant to Professor Shepherd, offered numerous suggestions. Two faculty members at other schools also conducted extensive reviews of the penultimate draft: Peter Flachsbart, University of Hawaii at Manoa; and Connie Ozawa, Portland State University. I am grateful to all of these faculty members and students for the careful attention they devoted to the review process.

A draft of the textbook was also pretested in two classes at Stanford: Environmental Planning Methods and Environmental Policy Design and Implementation. The following students from those classes participated in extensive meetings conducted to improve the book's readability: Kevin Armstrong, Alyssa Cobb, Rachel Daniel, Jeffrey Deason, John Hicks, Alexandra Knight, Jennifer Nachbaur, Gilbert Serrano, Anna Steding, Richard Strubbe, Bliss Temple, Maya Trotz, Stephanie Wien, Sarah Young and Jianyu Zhang.

My editor at John Wiley and Sons, Cliff Robichaud, and his assistants Catherine Beckham and Sharon Smith, helped improve the quality of the text by coordinating external reviews at different stages of manuscript preparation. My thanks to the following faculty members for their participation in an early survey used to determine the contents of the text: Craig Chase, Slippery Rock University; Cheryl Contant, University of Iowa; Timothy Duane, University of California at Berkeley; Peter Flachsbart, University of Hawaii; Steven Gordon, Ohio State University; Allen Harrison, College of Santa Fe; Valentine James, University of Virginia; B. Thomas

Lowe, Ball State University; William Marsh, University of Michigan at Flint; Burrell Montz, State University of New York at Binghamton; Stanley Specht, University of Colorado at Denver; and Bruce Stiftel, Florida State University. I also thank Walter Lynn of the University of Pennsylvania for his suggestions about what the book should include.

It is impossible to fully acknowledge the contribution of my assistant, Duc Wong, whose secretarial skills, organizational abilities and unfailing good spirits kept the manuscript production process moving forward. She was assisted in this by several Stanford students including Jean Han Chung, Melissa Geeslin, Toby Goldberg and Kirsten Rhodes. I owe special thanks to Patti Walters for her help with numerous graphic design issues, and to Greg and Monica Lam-Niemeyer and Patti Walters for their efforts in designing the cover.

Contents

Part One

Participants and Criteria in Environmental Decision Making

Contemporary environmentalism was born in the 1960s. During that decade, public concern over environmental degradation increased and efforts to maintain environmental quality intensified. Measures to control water and air pollution were expanded in many countries. New laws and regulations required government agencies to account for the environmental impacts of their decisions. Increased attention to the environmental effects of human actions led to the development of a new field of study: *environmental planning and management.*

The first of the four chapters in Part One puts contemporary environmentalism in perspective. It draws upon the history of American environmentalism to illustrate both traditional and present-day goals of environmental management. The second chapter builds on this historical account by describing criteria commonly used in making decisions that affect the environment. Some of these criteria concern whether monetary benefits of environmental protection exceed their costs and whether the distribution of benefits and costs is equitable. Other decision-making criteria relate to legal and moral obligations to maintain the earth's habitability.

The third chapter introduces typical participants in environmental regulation and the institutional setting within which those participants operate. Although the third chapter emphasizes the United States, it also examines situations in other countries, and institutions created to solve international environmental problems. The fourth and final chapter in Part One demonstrates how environmental regulations affect the planning of major development projects such as dams and chemical refineries. Case studies in the fourth chapter demonstrate how public and private organizations take part in implementing air and water quality laws and environmental impact assessment requirements.

Collectively, the first four chapters introduce organizations that participate in environmental regulation and impact assessment, and they analyze the criteria those organizations use in making decisions. Part One provides a foundation for examining the political, economic, and technical issues treated in subsequent chapters.

Chapter 1

Themes in American Environmentalism

Why do people want to safeguard the environment?[1] Is it because environmental protection is a good economic investment? Or are environmental controls implemented because of moral and legal obligations to preserve species and habitats? This chapter analyzes why environmental protection measures are undertaken by examining the historical roots of contemporary environmentalism in the United States.

The introduction to American environmentalism in this chapter demonstrates that *environmentalism* and *environmental planning and management* have existed, under different names, for over a century. Although terms such as environmental management have recent origins,[2] there is a long history of citizens organizing to influence environmental quality, and of technical specialists using both theory and empirical knowledge to manage the environment. This chapter's analysis of themes in the history of American environmentalism sets the stage for a systematic consideration of alternative bases for environmental decision making, the subject of Chapter 2. Although the historical details that follow concern the situation in the United States, the main themes parallel those in many other countries.

This chapter contains four sections, three of which provide a basis for understanding the historical and likely future goals of environmental planning and management. The first concerns perennial *anthropocentric* (human-centered) motives for protecting the environment: safeguarding public health, using natural resources efficiently, maintaining the earth's ability to support human life, and preserving places prized for their aesthetic and spirit-renewing qualities. The second section introduces the contemporary movement for *environmental justice.* Although the movement grew out of a protest against locating hazardous waste disposal facilities in communities of ethnic minorities, it has forced decision makers to focus attention on how the costs and gains of environmental protection are distributed.

The third section of the chapter shifts the perspective by introducing *biocentrism,* the concept that nonhuman species have inherent value in addition to any uses they may have in supporting and enhancing human life. The discussion introduces philosophical

[1] I owe thanks to Elizabeth Maynard Schaefer, a writer and editor based in Oakland, CA. She provided helpful comments on an early draft of this chapter.

[2] The term *environmentalist* has a long history. Currently the term refers to a person concerned with how human actions affect the natural environment. However, a century ago, an environmentalist was a person interested in how the physical environment influenced the way societies functioned and developed.

theories that support granting legal protection to nonhuman species and natural objects. The fourth and final section digresses from the chapter's focus on the goals of environmental protection and traces the emergence of environmental planning and management as a recognized professional field.

ANTHROPOCENTRIC BASES FOR ENVIRONMENTAL MANAGEMENT

Motivations for environmental protection often have an anthropocentric basis. In the United States, environmental management programs have been developed largely because of concerns about how a deteriorating environment affects people. Most people do not view the environment as having value in and of itself. The history of American environmentalism reflects four enduring anthropocentric bases for environmental concern: human health, efficiency of resource use, life support, and personal renewal (see Table 1.1).

Public Health in the Emerging Industrial City

Environmental quality in the industrial cities that emerged in nineteenth-century America was unimaginably low by contemporary standards. People pushed garbage, cinders, and other unwanted litter into streets to join animal manure in various states of decay. Human wastes frequently went to overflowing privies or cesspools or to open lots or fields. Wastewater from industrial establishments was freely discharged into streets. When wastewater was collected in pipes, it was released to waterways without treatment. Offensive odors emanated from belching smokestacks, open sewers, and decaying animal manure. As a result of poor sanitary conditions, by the middle of the nineteenth century, America's larger cities experienced recurring epidemics of typhoid, cholera, yellow fever, smallpox, and typhus.

TABLE 1.1 Anthropocentric Bases for Environmental Management

- *Protection of public health*
  Provide engineering works to meet people's needs for safe water and basic sanitation.
- *Efficient use of resources*
  Avoid waste by using natural resources productively.
- *Maintenance of natural systems*
  Safeguard the habitability of the planet for human and nonhuman species valued by humans.
- *Preservation of wilderness*
  Preserve areas valued by people as sources of beauty and renewal.

Limited knowledge about the causes of disease hampered efforts to deal with epidemics. The main sources of knowledge were studies documenting associations between the existence of particular environmental conditions and the occurrence of disease. This approach, which is a key element of the modern science of *epidemiology,* was the basis for a monumental study in 1842 by Edwin Chadwick on how the living conditions of English workers correlated with health.[3] The Chadwick report, which has been called "the most influential environmental health document of the nineteenth century," had a significant impact on American public health practice.[4] The report provided a basis for similar studies in New York City and Boston.

Chadwick's studies led him to recommend a disease prevention program based on improved urban sanitation through engineering works. He advocated the "great preventives" of "drainage, street and house cleansing by means of supplies of water and improved sewerage, and especially the introduction of cheaper and more efficient modes of removing all noxious refuse from the towns . . . [all] operations for which aid must be sought from the science of the Civil Engineer, not from the physician. . . ."[5] Although Chadwick was motivated by humanitarianism, his report demonstrated "that a program of envi-

[3] The report of Chadwick's investigations, *On an Inquiry into the Sanitary Conditions of the Labouring Population of Great Britain,* detailed the sanitary deficiencies existing in English slums and analyzed their effects.

[4] The quotation is from Petulla (1987, p. 17). Epidemology is concerned with the incidence, distribution, and control of disease within a population.

[5] Chadwick's remarks are as they appear in Petulla (1987, p. 18); the quotation following in the same paragraph is from Eisenbud (1978, p. 43).

ronmental rehabilitation would pay for itself because the health of the people would improve, family structure would be strengthened, and the thousands of citizens who had been trapped by squalor would begin to enter the mainstream of economic life. . . ." In short, improving urban environmental quality could be justified on economic grounds.

During the time of Chadwick's study, the medical profession was torn by internal disputes over how best to explain the outbreaks of epidemics. Some held that disease was spread by direct contact, and thus the way to prevent disease transmission was by setting up quarantine restrictions to isolate the sick. An alternate, widely held explanation was that disease-causing microorganisms could generate spontaneously in filth.[6] Following this reasoning, the way to secure the health of the public was to clean up putrid alleys and streets, and provide ample pure water, fresh air, and unadulterated food.

Scientific breakthroughs during the latter half of the nineteenth century went far toward explaining how diseases are transmitted. The work of Louis Pasteur and others firmly established that microbes arise from other microbes. They cannot generate spontaneously in a sterile medium. By the turn of the century, the emerging profession of *sanitary engineering* had the beginnings of a scientific foundation.[7]

Two aspects of the early history of public health are notable because of their recurring presence in the history of environmental protection. One concerns the role of citizens' groups, illustrated by efforts of what would now be termed *environmental nongovernmental organizations* to improve conditions in the slums and tenements of nineteenth-century New York City. Two citizen's groups that played major roles as advocates of sanitary reform were the Association for Improving the Conditions of the Poor (formed in 1843) and the Council of Hygiene and Public Health of the Citizen's Association (formed in 1864). These groups lobbied politicians, waged public education campaigns, relied on favorable treatment from the press, and, more generally, used many of the same strategies employed by contemporary environmental organizations.[8] Later in the century, the above-mentioned organizations were joined by citizens' groups made up of educated, middle class women who undertook major campaigns to reduce air and water pollution. Among the most effective was the Ladies' Protective Association of New York City, formed in 1884. Other large American cities had similar groups.

The power of citizens' groups to improve urban sanitary conditions is further demonstrated by the activities of *settlement workers,* such as those at Hull House settlement, founded in Chicago in 1888.[9] The Hull House charter called on members to investigate and improve the conditions of the industrial districts of Chicago. A sample of accomplishments linked to Hull House includes the work of the following:

- Jane Addams, who, in addition to founding Hull House, contributed to the reform of Chicago's woefully inadequate system of garbage collection;
- Alice Hamilton, whose investigations of a typhoid epidemic helped show a direct relationship between the discharge of sewage and the outbreak of disease in several Chicago neighborhoods; and
- Florence Kelly, whose work led to improvements in the sanitary conditions within both tenement residences and the workshops housed in tenement buildings.

Hull House settlement workers were also active in putting environmental issues on the agendas of trade unions.

A second notable aspect of early public health efforts is the debate that took place on whether or not to treat wastewater prior to discharge. In the

[6] For details on *spontaneous generation* and other early theories of how diseases were spread, see Duffy's (1990) history of public health in the United States.

[7] During the 1970s, what was once called sanitary (or public health) engineering in the United States came to be called *environmental engineering.*

[8] Duffey (1968) provides an account of the influence of these and other citizens' groups in improving public health in nineteenth-century New York City. His analysis reinforces the view that improving sanitary conditions of the urban poor was sometimes urged on economic grounds. In tracing the mob violence and pillage connected with the Draft Riots of 1863 in New York City, he cites reports linking the riots to the frustration of slum dwellers living in squalid conditions. Duffey (1968, p. 553) observes: "The 1863 uprising awakened the compassion of a good many decent citizens and helped to convince those with little feeling for their fellow men that social reform was sound economics." The property damage and casualties (estimated at about 1000) "may have provided a far greater impetus to the health and social welfare movement than any single event" (Duffey, 1968, p. 552).

[9] This discussion of Hull House settlement is based on Gottlieb (1993, Chap. 2).

early twentieth century, many engineers viewed wastewater treatment as an unnecessary expense. It was cheaper, they argued, to rely on natural waterways to dilute sewage discharges and then use filtration and other methods to treat public drinking water prior to distributing it. Much sanitary engineering practice in the period was based on the aphorism "dilution is the solution to pollution." Engineers developed procedures for calculating how much streamflow was needed to dilute waste discharges. They also investigated how streams "purify" themselves by recovering oxygen from the atmosphere to offset dissolved oxygen used by microbes in metabolizing waterborne organic waste.

In contrast to sanitary engineers, physicians and those concerned with preserving streams for recreational use urged treatment prior to discharge.[10] There is a present-day version of this debate, but it does not concern whether wastes should be treated at all; most people agree that some waste treatment is essential. Current disputes focus on whether the appropriate level of treatment should be based on economic feasibility or on protection of public health.

Conservation as the Efficient Use of Resources

Early in the twentieth century, America witnessed the arrival of a conservation movement bent on using advances in science and economics to manage forests, water, and other natural resources more efficiently. Advocates of improved management were not opposed to exploiting the environment to satisfy material needs. Indeed, they favored development. Their main concern was to avoid wasting natural resources. The idea of equating conservation with efficient development was a key theme of Theodore Roosevelt's presidential administration, and it was advanced by Gifford Pinchot, Roosevelt's principal advisor on natural resources.

Pinchot was the first head of the U.S. Forest Service and a leading advocate of scientific management of forests. His attitudes toward the environment are reflected in his "three principles." For Pinchot, the "first principle of conservation is development, the use of natural resources now existing on this continent for the benefit of the people who live here now."[11] Pinchot's emphasis on using resources *now* is, in part, a response to critics who argued that Roosevelt and Pinchot intended to withhold resources from development. The second of Pinchot's principles is that "conservation stands for the prevention of waste." Thus, although resources are to be used, their use is to be characterized by wise management and careful stewardship. The third of Pinchot's principles reflects the ideological overtones of the early conservation movement: "The natural resources must be developed and preserved for the benefit of the many, and not merely for the profit of a few." This is consistent with the antimonopoly sentiment that was widespread in America during Theodore Roosevelt's administration.

Leaders of the conservation movement in America argued that experts in fields such as geology, forestry, and hydrology should play key roles in managing the use of natural resources.[12] Highly trained specialists were needed to carry out resource management activities in a scientific and efficient manner. During the first half of the twentieth century, many resource development agencies in the United States hired technically trained specialists to manage multiple demands for land and water.

Many contemporary economists hold views reflecting Pinchot's concerns over the efficient use of natural resources. Economists are trained to identify combinations of labor, natural resources, and other *factor inputs* that can be used to produce goods and services efficiently. The economists' goal of maximizing the difference between benefits and costs (measured in monetary units) is termed the *productive efficiency objective.*[13] For many years, economists involved in resource management have been associated with specialties such as agricultural economics and

[10] For more on the early debate between sanitary engineers and physicians, see Petulla (1987).

[11] The quotations in this paragraph are from portions of Pinchot's book, *The Fight for Conservation,* reprinted in Nash (1968, p. 58–62).

[12] The relationship between the conservation movement and the scientific management of resources is analyzed by Hays (1959).

[13] Some economists use the term *economic efficiency* (as opposed to productive efficiency) when referring to the objective of maximizing monetary benefits minus costs. However, other economists use economic efficiency in a broader sense based on Pareto's criterion, a topic treated in Chapter 5. This broader usage of efficiency is also termed *Pareto efficiency.* Because the term economic efficiency has two meanings in economics (productive efficiency and Pareto efficiency), the term is not used in this textbook unless the context makes its meaning unambiguous.

water resources economics. In the past few decades a new specialty has evolved: *environmental economics.* One of its assumptions is that the acceptance of wastes from industries and municipalities is an appropriate (and unavoidable) use of the environment. Many environmental economists believe the amount of waste reduction should be determined by analyzing the economic benefits and costs involved.[14] Economists' concerns with productive efficiency are consistent with Pinchot's ideas about how the environment should be used.

Maintenance of Natural Systems

As in the case of economists, natural scientists have long been concerned with how resources are used. Decades before ecology became an established science, George Perkins Marsh (1864) argued persuasively for the need to keep "harmonies of nature" from being turned into discords. His widely cited book, *Man and Nature; or, Physical Geography as Modified by Human Action,* is a synthesis of numerous scientific works and personal observations. The firsthand accounts that formed a key part of *Man and Nature* were made possible by Marsh's posts as U.S. minister to Turkey (1849–54) and Italy (1861–62). These assignments allowed Marsh to observe the adverse environmental effects of plant and animal domestication, forest clearing, irrigation system development, and other human activities in the Mediterranean basin.

Marsh examined the ability of natural systems to withstand disturbances and to restore themselves following major perturbations such as those caused by "geologic convulsions." He was distressed with the way human actions could "interfere with the spontaneous arrangements of the organic or inorganic world" and cause instabilities and irreversible changes in nature. Based on his studies of the adverse environmental impacts of human actions, Marsh warned that continued "human improvidence" could threaten the earth's ability to support human life.

As the science of ecology developed in the early twentieth century, others began articulating Marsh's concerns using a more rigorous scientific basis. Aldo Leopold's ecological studies led him to the following position (Leopold 1933, p. 635):

> A harmonious relation to land is more intricate, and of more consequence to civilization, than the historians of its progress seem to realize. Civilization is not, as they often assume, the enslavement of a stable and constant earth. It is a state of mutual and interdependent cooperation between human animals, other animals, plants, and soils which may be disrupted at any moment by the failure of any of them. Land despoliation has evicted nations, and can on occasion do it again.

The ability of humans to inadvertently destroy natural systems increased dramatically in the 1940s. The post–World War II period witnessed the emergence of an array of new substances, including radioactive materials and synthetic organic chemicals. Many of these substances are persistent. They do not decay or decompose rapidly into simpler, less-harmful materials. Some scientists have responded to the recent increase in people's ability to disturb natural systems with demands for additional controls on human actions.

Contemporary scientists' calls for restraint are often based on the same concerns that motivated George Perkins Marsh and Aldo Leopold: the human ability to irreversibly destroy natural systems. The sense of urgency that often accompanies these calls is exemplified by Rachel Carson's *Silent Spring* (1962). The title of this widely read book refers to a hypothetical future springtime in which birds and other animals are silenced inadvertently. They are destroyed by synthetic chemicals transmitted from one life-form to another as part of the normal production and consumption activities in nature. Pesticides, such as *n*[dichlorodiphenyltrichloroethane] (DDT), are the chemicals of principal interest in *Silent Spring.* Carson (1962, p. 7) described the unintended effects of pesticide use:

> These sprays, dusts, and aerosols are now applied almost universally to farms, gardens, forests and

[14] As discussed in Chapters 5 and 7, much of *environmental economics* is concerned with how to get public and private decision makers to account for the social costs of their actions. Social costs include unpriced environmental damages as well as private costs. In recent years, the field has gone beyond the traditional concerns of efficiency by considering how costs and gains are distributed. In the late 1980s, economists who felt constrained by the scope of conventional environmental economics helped found a new field, *ecological economics,* which integrates aspects of ecology and economics by taking a broad-scoped, long-term view of environment-economy interactions. For an introduction to this new field, see Costanza (1991).

> homes—nonselective chemicals that have the power to kill every insect, the "good" and the "bad," to still the song of birds and the leaping of fish in the streams, to coat the leaves with a deadly film, and to linger on in soil—all this though the intended target may be only a few weeds or insects. Can anyone believe it is possible to lay down such a barrage of poisons on the earth without making it unfit for all life?

Silent Spring is just one of many books written since World War II to alert the public to the seriousness of unintentional environmental disruptions.

Renewal and Spiritual Values

The preceding perspectives on the improvement of public health, efficiency in natural resource use, and the maintenance of natural systems are tied to environmental engineering, economics, and ecology, respectively. Each of these viewpoints is pragmatic. Edwin Chadwick and other advocates of public health reform maintained that disease prevention via improved urban sanitation would save governments money. Gifford Pinchot and others involved in the conservation movement focused on eliminating inefficiencies in resource use. Ecologists, such as Aldo Leopold and Rachel Carson, emphasized the importance of maintaining the earth as a suitable habitat for human life. In contrast, the ideas introduced below concerning religion and spiritual renewal have a less tangible quality. These ideas, however, are as consequential as the more practical concerns of engineers, economists, and ecologists. Indeed, some would argue that these spiritual and philosophical matters are of the utmost significance to the long-term habitability of the planet.

Several of the issues examined in the following pages arose initially during efforts to preserve parts of the American wilderness. Some observations about this *preservationist movement* are therefore in order. Until the eighteenth century, the untamed wilderness was generally considered to be mysterious and dangerous. It was a place to be conquered and put to productive use. Romantic poets, such as William Wordsworth and Samuel Coleridge, played an important early role in putting a positive value on the remote, mysterious, and solitary nature of wild places. The cause of wilderness preservation gained momentum in the United States during the late nineteenth century, when only a small fraction of the original wilderness remained.[15]

Many reasons for preserving wilderness have economic and ecological bases. For example, wilderness provides scientists with an opportunity to study natural processes, and the numerous species present constitute a pool of diverse genetic patterns that may eventually prove useful in agriculture, industry and medicine. In addition, wilderness offers opportunities to hunters, anglers, and others engaged in outdoor recreation. Such practical reasons for preserving wilderness are important, and they provide foundations for many environmental management programs, but they are not considered here. Rather, the concern is with spiritual and philosophical arguments for wilderness preservation.

American Transcendentalism

The religious fervor that has sometimes characterized the cause of wilderness preservation is frequently based on ideas of the American transcendentalists of the nineteenth century, especially Ralph Waldo Emerson and Henry David Thoreau. A central tenet in their philosophy is that in possessing a soul, a person has the potential to transcend beyond the material portion of the universe and thereby perceive higher, spiritual truths. According to historian Roderick Nash (1973, p. 68), transcendentalists believed "one's chances of attaining moral perfection and knowing God were *maximized* by entering wilderness."

Emerson is generally acknowledged as the intellectual leader of the American transcendentalists. His views on the communion with God obtainable in wilderness are summarized in a passage from his essay, *Nature:*

> In the woods, we return to reason and faith. . . . Standing on the bare ground—my head bathed by the blithe air and uplifted into infinite space—all mean egotism vanishes. I become a transparent eyeball; I am nothing; I see all; the currents of the Universal Being circulate through me; I am part or parcel of God.[16]

---

[15] This discussion of the American preservationist movement relies heavily on Nash (1968 and 1973).

[16] The Emerson and Thoreau quotations in this discussion are from the volume of readings edited by Nash (1968).

Arguments for wilderness preservation at the turn of the century frequently relied on transcendentalists' views on how spiritual values are reflected in nature. This is illustrated by John Muir's valiant but unsuccessful effort to prevent the damming of Hetch Hetchy Valley in the Sierra Nevada of California. Muir had studied Emerson's work and had similar, strong feelings about the presence of God in nature. Roderick Nash, quoting from several of Muir's writings, observes, "At one point Muir described nature as 'a window opening into heaven, a mirror reflecting the Creator.' Leaves, rocks, and bodies of water became 'sparks of the Divine Soul.' "[17] In defending Hetch Hetchy Valley from those who would inundate it with a reservoir, Muir proclaimed, "Dam Hetch Hetchy! As well dam for water-tanks the people's cathedrals and churches, for no holier temple has ever been consecrated by the heart of man."

Although religious sentiment influenced Muir's attempts to preserve Hetch Hetchy Valley and other wild areas, it was not his only motivation. A dominant theme in Muir's work, and that of many other preservationists at the turn of the century, was the importance of providing people with opportunities for spiritual and emotional renewal. For Muir and others, the tribulations of the mechanized, technological life ushered in by the industrial revolution required the kind of regeneration that only wilderness could provide.

Spiritual Renewal and Re-creation

People who view the wilderness as a source of physical and spiritual "re-creation" follow in the footsteps of Henry David Thoreau. For him, forest and wilderness provided "the tonics and barks which brace mankind." In joining the political cause for preservation, Thoreau argued that the loss of wilderness would lead people to become weak and dull and lacking in creativity.

The importance of wilderness as a source of spiritual renewal is a theme that has been sounded many times since Thoreau wrote. For some, the very idea that wild places still exist is a source of strength. In 1960, the American writer Wallace Stegner articulated the importance of the wilderness idea:

> The reminder and the reassurance that [wilderness] is still here is good for our spiritual health even if we never once in ten years set foot in it. It is good for us when we are young, because of the incomparable sanity it can bring briefly, as vacation and rest, into our insane lives. It is important to us when we are old simply because it is there—important, that is, simply as idea.[18]

Sometimes the spectacular beauty of a wild place is what makes it a "tonic." The aesthetic features of wild areas are often cited in arguments for wilderness preservation, especially when these features are unique. This was the case in the 1960s when citizens' groups successfully opposed federal attempts to inundate portions of the Grand Canyon in Arizona as part of a large water project.

The struggle to preserve wild and scenic areas has generally been waged by individuals with sufficient income to spend their leisure time in the wilderness. In contrast, the emergent *movement for environmental justice* reflects the concerns of people who often have modest financial resources.

MOVEMENT FOR ENVIRONMENTAL JUSTICE

Environmental programs often impose unequal costs on the people and groups they affect. This raises questions about the fairness of the resulting distribution of costs and gains. For example, government requirements to install expensive emission control devices on old motor vehicles are often criticized for imposing disproportionate burdens on the poor.

Before the 1980s, little systematic consideration had been given to the *distribution* of costs and gains of environmental protection. Prior to that period, researchers had attempted to assemble statistics to establish whether one or another environmental program benefited the rich more than the poor (after adjusting for differences in taxes paid) or vice versa. However, the pre-1980 studies were plagued by a paucity of both data and methods for analyzing distributional issues. Although many of these early studies pioneered in treating a difficult analytic problem,

[17] From Nash (1973, p. 125). The famous quote in defense of Hetch Hetchy Valley that follows appeared originally in Muir's *The Yosemite* and is cited by Nash (1973, p. 168).

[18] The quotation is from Stegner's essay, "The Wilderness Idea," reprinted in Nash (1968, pp. 192–97).

they often did not attract attention, except among researchers.[19]

There are other explanations for the lack of attention to whether the costs and gains of environmental protection are distributed fairly. Legal scholar Richard Lazarus (1993, p. 842) points to the U.S. Environmental Protection Agency's historical resistance to "embracing a distributional mandate in its enforcement of [environmental] laws. The agency has consistently viewed 'sociological' concerns such as distributional impacts as outside the purview of its purely 'technical' mandate of establishing technically effective and economically efficient pollution control standards."[20] Still another reason for the inattention to distributional issues is the professional bias of environmental policy analysts, many of whom have economics backgrounds. Economists have given far more attention to productive efficiency than to distributional issues, and this emphasis on efficiency was reflected in the advice economists gave to U.S. environmental policymakers before the 1980s.[21]

Public attention was at last drawn to distributional questions related to environmental protection in 1982 (see Table 1.2). The impetus was a national protest by African-Americans "after the mostly black Warren County, North Carolina, was selected as the burial site for 32,000 cubic yards of soil contaminated with highly toxic polychlorinated biphenyls, or PCBs."[22] Among the 400 people arrested during the unsuccessful protest was William Fauntroy, the congressional representative from the District of Columbia. In the aftermath of his arrest, Fauntroy asked the U.S. General Accounting Office (GAO) to investigate whether race has a relationship to where hazardous waste disposal facilities are located.

TABLE 1.2 Key Events in the Movement for Environmental Justice

1982	National protest takes place over proposed disposal of PCBs in Warren County, North Carolina.
1983	U.S. General Accounting Office study documents that race is correlated with the location of commercial hazardous waste facilities in the South.
1987	United Church of Christ Commission for Racial Justice provides further documentation of statistical associations between race and the location of hazardous disposal facilities.
1990	Conference at the University of Michigan brings together researchers, activists, and U.S. Environmental Protection Agency officials concerned with environmental justice issues.
1992	U.S. Environmental Protection Agency summarizes evidence showing disproportionate environmental costs borne by racial minorities and low-income groups.
1994	President Clinton issues Executive Order 12898 to address environmental justice issues.

The GAO surveyed all off-site commercial hazardous waste landfills in the eight southeastern states comprising Region IV of the U.S. Environmental Protection Agency. (An off-site landfill is one that is not part of or adjacent to an industrial facility.) The GAO study identified four such sites and found, "Blacks make up the majority of the population in three of the four communities where the landfills are located."[23] In the fourth community, 38% of the population was black.

In 1987, a more comprehensive study exploring whether race influences the location of hazardous waste facilities was conducted by the United Church

[19] For a review of the literature on the distribution of costs and benefits of environmental protection, see Lazarus (1993); this literature is considered briefly in Chapter 2.

[20] Lazarus uses economic efficiency to mean productive efficiency. In many circumstances, national laws require that benefits and costs be compared in setting pollution control standards. Exceptions occur, for example, when Congress asks the U.S. Environmental Protection Agency to set standards to protect the public health, without regard to a benefit-cost comparison.

[21] Examples are given by Sagoff (1988). For more on why environmental equity issues received relatively little attention before the 1980s, see Bullard (1990) and Lazarus (1993).

[22] This quotation is from Bullard and Wright (1992, p. 41). Facts in this paragraph and the one that follows are from Bullard and Wright (1992), and Lazarus (1993).

[23] This is from the GAO report as quoted by Lazarus (1993, p. 801).

of Christ Commission for Racial Justice (CRJ). Charles Lee, director of the CRJ study, highlighted the study's most important findings.[24]

- Across the United States, race was consistently the variable, among those variables tested, that correlated most significantly with the location of commercial hazardous waste facilities.
- Communities with the greatest number of commercial hazardous waste facilities had the highest composition of minority residents. In communities with either two or more facilities *or* one of the nation's five largest landfills, the average percentage of minority residents was more than three times that of communities without facilities (38% vs. 12%).
- In communities with one commercial hazardous waste facility, the average minority percentage of the population was twice the average minority percentage of the population in communities without such facilities (24% vs. 12%).
- Although socio-economic status appeared to play an important role in the location of commercial hazardous waste facilities, race still proved to be the most significant factor. This remained true after the study controlled for urbanization and regional differences. Incomes and property values were substantially lower in communities with commercial disposal facilities compared to communities in the surrounding counties without facilities.
- Three out of the five largest commercial hazardous waste landfills in the United States were located in predominantly black or Hispanic communities. These three landfills accounted for 40 percent of the total estimated commercial landfill capacity in the nation.

The dramatic results of the studies by the GAO and CRJ provided impetus for investigations of related issues. For example, was race associated with other environmental inequities, such as disproportionate exposure of minorities to lead, pesticides, and contaminated fish? In 1990, researchers investigating such questions met at the University of Michigan. In addition to examining data on the unfair exposure of minorities to pollution, the researchers pondered the contention that (despite the findings of the CRJ and others) poverty, not race or ethnicity, explained why racial and ethnic minorities shouldered the disproportionate share of environmental hazards reported in earlier studies. The organizers of the meeting provided the following explanation of why race, independent of income, is associated with the unfair distribution of environmental hazards.

> Classic economic theory would predict that poverty plays a role. Because of limited income and wealth, poor people do not have the means to buy their way out of polluted neighborhoods. Also, land values tend to be lower in poor neighborhoods, and the neighborhoods attract polluting industries seeking to reduce the costs of doing business. However, the mobility of minorities is additionally restricted by housing discrimination, amply demonstrated by researchers to be no insignificant factor. Then, because noxious sites are unwanted (the "NIMBY," or not-in-my-backyard syndrome) and because industries tend to take the path of least resistance, communities with little political clout are often targeted for such facilities. The residents tend to be unaware of policy decisions affecting them; they are not organized; and they lack the resources (time, money, contacts, knowledge of the political system) for taking political action. Minority communities are at a disadvantage not only in terms of resources, but also because of underrepresentation on governing bodies. When location decisions are made, this underrepresentation translates into limited access to policy makers and lack of advocates for minority interests.[25]

As a result of the well-publicized studies that followed the Warren County protest, the movement for environmental justice came of age in the United States in the early 1990s. By then, environmental justice had become a mainstream issue. The U.S. Environmental Protection Agency (EPA 1992) issued a report summarizing data showing that disproportionate pollution burdens fall on low-income groups and racial minorities. And in 1994, President Clinton issued Executive Order 12898, which re-

[24] The findings listed are paraphrased from Lee (1993, pp. 48–49).

[25] This quotation is from Bryant and Mohai (1992, p. 7).

quires each federal agency to establish a strategy addressing "disproportionately high and adverse human health . . . effects of its programs, policies, and activities on minority populations and low-income populations." The executive order contains numerous other provisions, including support of community participation in human health research and data collection.[26]

The early 1990s also witnessed the emergence of environmental justice as an issue for environmental nongovernmental organizations. Large and powerful environmental organizations such as the Sierra Club began turning their formidable energies to environmental justice issues. For example, in 1990 the executive director of the Sierra Club called for "a friendly takeover of the Sierra Club by people of color." He went on to add, "The struggle for environmental justice in this country and around the globe must be the primary goal of the Sierra Club during its second century."[27] Additionally—and what may be of even greater significance than the attention of mainstream institutions—there has been widespread, multiracial grassroots support for the environmental justice movement.[28]

BIOCENTRIC PERSPECTIVES ON THE ENVIRONMENT

The preceding sections describe themes arising from the history of environmentalism that have an anthropocentric basis. They rest on the idea that only humans have *intrinsic* value; only people are valuable in and of themselves. Nonhumans, plants, and natural objects have only *instrumental* value; they have value only as means (or instruments) to meet human ends.

According to the anthropocentric view, only human beings have *moral standing*. In other words, only human beings "count" or are valuable in themselves, and *moral obligations* should extend only to humans.[29] Thus, for example, maintaining the population of a particular species of animal or plant is justified only in terms of its value to humans. Perhaps the gene structure of the species will be important in medicine or industry some day. Alternatively, as in the case of the panda bear or the Labrador seal, support for species preservation may be based on the sentiments of the general public. The key point, in this context, is that the rationale for preservation is not based on the intrinsic value of the species. There is no arguing that the species has a moral right to life.

Anthropocentrism has played a dominant role in the history of environmentalism. It is, for example, reflected in the views of Gifford Pinchot and other prominent conservationists of the early 1900s. For Pinchot, conservation meant the wise and efficient *use* of natural resources for the benefit of *people.* Anthropocentrism is also reflected in the wilderness preservation movement in the United States. For example, some preservationists argue that wilderness provides an important source of renewal for people who, increasingly, find themselves trapped in a humdrum existence in crowded cities. Although the sentiment reflected in the wilderness preservation example is different from the starkly utilitarian views of Pinchot, the interests of humans remain at the center of the argument.

Biocentrism reflects an alternative perspective. The biocentric view assumes that some (or for some advocates of biocentrism, all) nonhuman species have intrinsic value.[30] Thus, for example, a dog has a value that is not derived from its use as a household pet for humans. Because a dog is a living organism, a dog is valuable in itself.

---

[26] For details on Executive Order 12898 and its relationship to other federal actions to enhance environmental justice, see Bryant (1995, pp. 217–19).

[27] From a speech celebrating the Sierra Club's centennial by then-executive director Michael Fisher, as reported in *Sierra, The Magazine of the Sierra Club,* vol. 78, no. 3 (May/June 1993): 51.

[28] For details on the history of grassroots activism for environmental justice, see Bullard and Wright (1992), and Bullard (1993). Robert Bullard, who is a professor of sociology at the University of California-Los Angeles, has been described as "the most prolific writer and advocate on the subject of 'environmental injustice' " (Lazarus, 1993, p. 803).

[29] *Moral standing* and *moral obligations* are technical terms used by philosophers specializing in ethics. A description of technical definitions of these terms is beyond the scope of this text. Nontechnical definitions are given in Chapter 2. For a detailed discussion of moral standing and moral obligations, see Van De Veer and Pierce (1994, Chap. 1).

[30] Some authors use the term *ecocentrism* instead of biocentrism. For example, Armstrong and Botzler (1993, p. 369) introduce the following definition: "Ecocentrism is based on the philosophic premise that the natural world has inherent or intrinsic value."

The roots of modern biocentric thought can be traced to the beginning of the twentieth century.[31] Among the most influential advocates of what is now termed biocentrism was Albert Schweitzer, whose medical work in Africa is widely known. Schweitzer developed an ethical system based on the idea that "right conduct" requires a human to give every living being the same "reverence for life that he gives his own."[32] While Schweitzer's ethical perspective approves of taking the lives of nonhumans, those actions are to be taken only when necessary and with the maximum alleviation of pain.

Biocentrism is also reflected in the writings of John Muir. Consider, for example, Muir's answer to the question, What are rattlesnakes good for? His response was that rattlesnakes are "good for themselves, and we need not begrudge them their share of life." In a different context, Muir indicated that "the universe would be incomplete without man; but it would also be incomplete without the smallest transmicroscopic creature that dwells beyond our conceitful eyes and knowledge." According to Muir, in the wilderness people learn a need to respect "the rights of all the rest of creation."[33]

In recent times, the view that humans have moral obligations toward plants and animals has been linked closely to an ethical perspective advocated by the ecologist Aldo Leopold. The last portion of his book, *A Sand County Almanac,* contains the basis for what Leopold referred to as a *land ethic.* For Leopold, the community to which ethics applies should not be restricted to people. It needs to be enlarged to include other members of the *biotic community:* soils, waters, plants, and animals. Leopold felt that decisions about the use of land should consider, in addition to economic factors, "questions of what is ethically and esthetically right." Leopold's land ethic provides a criterion for answering these questions. For him, a "thing is right when it tends to preserve the integrity, stability and beauty of the biotic community. It is wrong when it tends otherwise" (Leopold 1949, p. 262).

The 1970s represent a watershed in the history of biocentric thought: *environmental ethics* emerged as a discipline.[34] Ethics, the branch of philosophy concerned with what ought or ought not to be done, had traditionally considered only humans as having inherent value. Environmental ethics expanded the scope of ethics by ascribing inherent value to species other than humans. The growth in the 1970s of environmental ethics as a field of study is reflected by the creation of scholarly periodicals such as *Environmental Ethics* and *Ecophilosophy.* Moreover, established philosophical journals began publishing articles on the moral rights of nonhuman species. In addition, university courses and textbooks on environmental ethics began to appear.[35]

Philosophers do not agree among themselves about which, if any, nonhuman species (or ecosystems) ought to be valued in and of themselves regardless of their benefit to humans. This dispute among philosophers is important because only species having intrinsic value receive consideration in a theory of moral philosophy. Some philosophers believe that only humans have inherent value. Others believe that moral consideration should be extended to all *sentient beings,* species capable of experiencing pain and suffering.[36] Meanwhile, other philosophers have argued that granting moral consideration *only* to sentient beings is fallacious. They believe there is a philosophical justification for granting rights to all animal *and* plant species. Still other philosophers believe that moral obligations extend even to habitats.[37]

---

[31] The view that nonhuman species have intrinsic value is not a new idea. Before the scientific advances of the 1700s, many cultures viewed the earth as a living creature. According to Mander (1991, pp. 215–18), for example, people as diverse as the Navajo in the southwestern United States and the Aborigines in Australia believed in the earth as a living planet. Mander argues that many "native peoples" viewed humans as a part of a web of life that included plants, animals, people, and rocks.

[32] The remarks by Schweitzer in this paragraph are as quoted by Nash (1989, p. 60).

[33] The quotations of Muir in this paragraph are cited in Nash (1973, pp. 128–29).

[34] The 1970s was also a decade in which feminist theory was applied to problems of environmental resource use. The resultant perspective, which has been termed *ecofeminism,* has grown significantly since its origins in the early 1970s. Central theories concern relationships between male dominance, the subjugation of women, and the misuse of environmental resources. According to Adams (1993, p. 1), "Ecofeminism argues that the connections between the oppression of women and the rest of nature must be recognized to understand adequately both oppressions."

[35] For a listing of over 60 textbooks and anthologies on environmental ethics, see Ralston (1996, pp. 222–29).

[36] For more on this point, see Regan (1983) and Singer (1993).

[37] A historical account of these alternative perspectives is given by Nash (1989).

TABLE 1.3 Types of Environmental Ethics[a]

- An anthropocentric ethic, in which humans are valued intrinsically and the environment is valued instrumentally. This ethical system applies traditional ethics to the environment.
- An animal ethic, in which the welfare of animals counts morally. Some philosophers distinguish between sentient and non-sentient animals and argue that only sentient beings should count morally.
- An ethic based on respect for all life, plant as well as animal.
- An ethic that counts endangered species as morally relevant. This position reflects a respect for life at the species level.
- A land ethic, in which the moral concern is for the biotic community.

[a] Based on material in Ralston (1996, pp. 206–10).

The absence of agreement on which species (or ecosystems) are worthy of moral consideration has led philosophers to distinguish several different types of environmental ethics. For example, Holmer Ralston III, a professor of philosophy at Colorado State University, has identified five possible ways of dividing the subject of environmental ethics. His division of the field is given in Table 1.3.

Deep ecology is an example of a widely discussed system of environmental ethics. While the term deep ecology has its roots in a philosophical system advanced by the Norwegian philosopher Arne Naess, it is most widely known by its basic principles. The principles of deep ecology listed here are from Fox's (1990, pp. 114–15) interpretation of the work of Naess and Naess's colleague George Sessions:

1. The well-being and flourishing of human and nonhuman life on Earth have value in themselves. . . . These values are independent of the usefulness of the nonhuman world for human purposes.
2. Richness and diversity of life-forms contribute to the realization of these values and are also values in themselves.
3. Humans have no right to reduce this richness and diversity except to satisfy *vital* needs.
4. The flourishing of human life and cultures is compatible with a substantial decrease of the human population. The flourishing of nonhuman life requires such a decrease.
5. Present human interference with the nonhuman world is excessive, and the situation is rapidly worsening.
6. Policies must therefore be changed. These policies affect basic economic, technological, and ideological structures. The resulting state of affairs will be deeply different from the present.
7. The ideological change is mainly that of appreciating *life quality* (dwelling in situations of inherent value) rather than adhering to an increasingly higher standard of living. There will be a profound awareness of the difference between big and great.
8. Those who subscribe to the foregoing points have an obligation directly or indirectly to try to implement the necessary changes.

The principles of deep ecology constitute a critique of the anthropocentric rationale for conservation and preservation.

Although biocentric ethical perspectives have attracted much attention, biocentrism is still far from the mainstream of Western moral philosophy. Not surprisingly, some have argued that granting moral consideration to nonhuman species, when followed to its logical end, leads to an absurdity. Passmore (1974, p. 126) put it this way: "If men were ever to decide that they ought to treat plants, animals, and landscapes precisely as if they were *persons,* if they were to think of them as forming with men a moral community in the strict sense, that would make it impossible to civilize the world—or, one might add, to act at all or even to continue living."

Although moral obligations to nonhuman animals and to plants and natural objects are not widely accepted, the idea that such obligations exist has played a role in environmental decision making since at least the 1970s. For example, U.S. endangered species legislation passed in that decade was based, in the view of some at least, on the "legal idea that a listed nonhuman resident of the United States is guaranteed, in a special sense, life and liberty. . . ."[38]

[38] This quotation is from Petulla (1980, p. 51); the second quotation in this paragraph is from Hunter as it appears in Nash (1989, p. 179).

Biocentric ideas have also affected the policies of influential environmental groups, such as Greenpeace. According to Robert Hunter, a founding member of that organization, in 1974 Greenpeace widened its concern "from the concept of sanctity of human life to the sanctity of all life."[39]

The question of whether plants, animals, and ecosystems are worthy of moral consideration must be faced by those making decisions influencing the environment. To the extent that such moral consideration is granted, it is relevant in judging whether one action affecting the environment is preferred to another. Criteria used in making such moral judgments are introduced in the next chapter.

A NEW PROFESSION: ENVIRONMENTAL PLANNING AND MANAGEMENT

Up to this point, the chapter has provided a historically based analysis of what motivates environmental protection. Anthropocentric concerns, such as safeguarding public health and using natural resources efficiently, have been dominant goals in environmental management programs. However, since the 1970s, two new issues have influenced debates on how much environmental protection to provide: environmental justice, and moral obligations to nonhuman species.

This section digresses from the chapter's main focus on goals motivating environmental protection. Instead of discussing goals, this section traces the emergence (in the 1970s) of new environmental problem-solving institutions and a new professional field, *environmental planning and management.* The process of satisfying new environmental regulations, particularly requirements to produce *environmental impact statements* for proposed development projects, required that professionals from different disciplines work together.[40] Calls for interdisciplinary environmental problem solving were commonplace during the 1970s, and the need to create institutions that would facilitate interdisciplinary problem solving became increasingly evident.[41]

Several types of organizations emerged to build bridges among environment-related disciplines. Government agencies, such as the U.S. Army Corps of Engineers, created environmental units in their field offices and staffed them with specialists in environmental engineering, urban planning, biology, and other disciplines that could contribute to improving the agencies' compliance with environmental requirements. Regional and local governments responsible for implementing new environmental policies and programs also created environmental units. In addition, many private businesses, particularly those with a major financial stake in improving their environmental performance, formed new organizational units to meet environmental requirements.

The 1970s was also a time when many new environmental consulting firms were created and staff capabilities of established environmental engineering consulting companies were expanded. The new environmental regulations of the 1970s, together with increasing public expectations about acceptable environmental performance by firms and government agencies, increased the demand for consultants to provide environmental services.

Along with the growth in organizations where environmental problems were being approached by interdisciplinary teams came incentives to create new forums for information exchange. In the United States, the need to bring together professionals from different environment-related disciplines led to creation of the *National Association of Environmental Professionals* (*NAEP*) in 1975.[42] In addition to creat-

[39] For an account of the relationship between deep ecology and the work of environmental nongovernmental organizations, see Devall (1992).

[40] The National Environmental Policy Act (NEPA) of 1969 made these statements mandatory for many types of actions by U.S. federal agencies. Chapter 15 considers the implementation of NEPA.

[41] Dorney (1989, p. 92) provides a definition attributed to Guy Berger: "Interdisciplinary . . . describes the [relationships between] two or more different disciplines which may range from simple communication of ideas to mutual integration of organizing concepts, methodology, . . . terminology, data [and] organization of research. . . ."

[42] Arnold (1979) reports that similar (but distinct) associations of environmental professionals were formed in the states of California, Illinois, Minnesota, and New York. These new professional organizations were also formed in other countries; see, for example, Dorney's (1989) account of the *Ontario Society for Environmental Management* (formed in the mid-1970s). International associations of environmental professionals from different disciplines were also formed. An example is the *International Association for Impact Assessment* (*IAIA*), which was founded in 1980. Many of these new associations produce their own journals and newsletters; see, for example, *Impact Assessment,* the journal of the IAIA.

ing new forums for the exchange of ideas, NAEP also established a "code of ethics and standards of practice for environmental professionals." Adherence to the code became a condition of membership in NAEP. Some years later, NAEP created a process that allowed experienced and highly qualified members to become "Certified Environmental Professionals."[43]

Although there is wide agreement that a new profession has been created, there is no consensus on how to characterize it and what to call it. A monograph, *The Professional Practice of Environmental Management,* by Robert Dorney, who was an environmental consultant and a professor at the University of Waterloo, characterizes the lack of consistent terminology to describe the new profession that NAEP's founders saw developing.

Dorney uses the term *environmental manager* to refer to a new type of environmental problem solver. He describes an environmental manager (Dorney 1989, p. 15):

> . . . a systems-oriented professional with a natural science, social science, or less commonly, an engineering, law or design background, tackling problems of the human-altered environment on an interdisciplinary basis from a quantitative and/or futuristic viewpoint . . . those identified as environmental managers commonly hold professional credentials in planning, landscape architecture, engineering or law.

For Dorney, environmental planning is a component of a broader management function that also includes implementation. He describes an *environmental planner* as a person who

> . . . puts his or her knowledge of the ecosystem into a planning process or predictive frame of reference to effect a better fit between the works of humans and nature. . . .

The dilemma, as Dorney recognized, is that the kind of person Dorney describes as an environmental manager would be called an environmental planner by others. The lack of agreement on definitions results, in part, from the wide range of educational backgrounds among persons who call themselves environmental planners and environmental managers. The absence of clear definitions also stems from the recent origins of interdisciplinary environmental problem solving. For example, university-level teaching and research programs in environmental planning and management are largely a product of the past 30 years. Because there is not yet agreement on definitions, this book will avoid making fine distinctions between environmental planning and environmental management.

Environmental planners and managers work in a broad range of contexts, and the following list provides an overview of typical work settings and activities.[44]

- New facility development, including work on assessing impacts of proposed projects, mitigating environmental impacts during construction, and monitoring post-project environmental impacts. Facilities include public works and private development projects.
- Old facility decommissioning; for example, dismantling obsolete nuclear power plants, cleaning up toxic wastes at military bases scheduled for closure, and rehabilitating "ecologically derelict" lands such as strip-mined areas.
- Facility operations, including monitoring compliance with pollution control requirements, reducing waste generation by recycling materials, and conducting environmental audits of facility performance.
- Government policy analysis and environmental program implementation, as illustrated by studies evaluating the effectiveness of environmental laws and policies.
- Urban and regional development, including analyses of the suitability of land for different uses, and the development of policies to control land use in ways that conserve resources and minimize the public's exposure to natural hazards such as landslides and floods.

This listing is illustrative, not comprehensive. Subsequent chapters of this book detail many activities

[43] Requirements for certification relate to academic training and professional experience; in addition, a written examination must be taken and applicants for certification must have their credentials and experience approved by a board of peers.

[44] This overview is based on Dorney's (1989) treatment of the subject.

that are part of environmental planning and management.[45]

Key Concepts and Terms

Anthropocentric Bases for Environmental Management

Chadwick's influence on public health
Epidemiology
Waste treatment vs. dilution
Nongovernmental organizations
Settlement workers
American conservation movement
Pinchot's principles of conservation
Productive efficiency
Ecological basis for environmental management
American transcendentalism
Preservationist views on wilderness

Movement for Environmental Justice

Associations between race and landfill location
NIMBY syndrome
Distribution of environmental hazards

Biocentric Perspectives on the Environment

Biocentrism vs. anthropocentrism
Schweitzer's reverence for life
Leopold's land ethic
Animal rights
Obligations toward nonhuman species
Deep ecology

A New Profession: Environmental Planning and Management

Environmental planner
Environmental manager
National Association of Environmental Professionals (NAEP)
Interdisciplinary teams

Discussion Questions

1-1 Examine your own reasons for caring about how people affect the environment. To what extent do your personal motives overlap with the economic, ecological, spiritual, and philosophical viewpoints presented in this chapter?

1-2 Criticize the use of an anthropocentric framework for making decisions. In light of this critique, do you feel that anthropocentrism provides a reasonable approach to evaluating the environmental consequences of human actions?

1-3 Some supporters of improved public health and sanitation in nineteenth-century American industrial cities felt those improvements would reduce government spending on health care. What arguments could be advanced to support this position? Could similar arguments be made today?

1-4 The chapter supposes that the anthropocentric motivations for improving environmental quality manifested up through the early twentieth century are applicable today. Do you agree? Think of some contemporary motivations for improving environmental quality that did not exist before the middle of the twentieth century.

1-5 Which, if any, of Gifford Pinchot's principles of conservation are applicable today? How would you modify Pinchot's principles to reflect the current context of natural resource use?

1-6 Provide an example (from your own experience or the literature) of an instance in which one or more racial or ethnic minority groups have appeared to shoulder an unfair burden of environmental hazard. To what degree does your example allow you to comment on whether poverty or race and ethnicity was the source of the injustice?

1-7 Analyze racial and ethnic factors that may contribute to environmental injustice in the United States. In your analysis, consider the possible roles of racism and the relative economic power and political leverage that minority groups possess. Consider also the degree to which environmental groups with strong lobbying capabilities in Washington have represented minority interests, and the extent to which minority groups have, in the past, shown an interest in environmental issues. What does

[45] Because the subject is vast and will expand as new environmental and resource management problems become apparent, no single book can provide a full treatment.

your analysis suggest about ways to eliminate environmental injustice based on race and ethnicity?

1-8 Tom Regan, a contemporary American philosopher and a leading advocate of animal rights, has dubbed Leopold's land ethic a form of "environmental fascism." Regan (1983, p. 361) believes the "implications [of the land ethic] include the clear prospect that the individual [animal or plant] may be sacrificed for the greater biotic good, in the name of the integrity, stability, and beauty of the biotic community." Provide an example illustrating the kind of individual sacrifice Regan has in mind. Do you feel such sacrifices are "right" if they serve the biotic community?

1-9 Which of the eight principles of deep ecology articulated by Naess and Sessions do you agree with? If you agree with all eight, what sorts of obligations to "implement the necessary changes" (a phrase from the eight principles) would you be prepared to assume?

References

Adams, C., ed. 1993. *Ecofeminism and the Sacred.* New York: Continuum.

Armstrong, S. J., and R. G. Botzler, eds. 1993. *Environmental Ethics: Divergence and Convergence.* New York: McGraw-Hill.

Arnold, N. 1979. NAEP: Past, Present and Future. *The Environmental Professional* 1: 3–5.

Barbour, I. G. 1980. *Technology, Environment, and Human Values.* New York: Praeger.

Bryant, B. I., ed. 1995. *Environmental Justice: Issues, Policies and Solutions.* Washington, DC: Island Press.

Bryant, B. I., and P. Mohai, eds. 1992. *Race and the Incidence of Environmental Hazards.* Boulder, CO: Westview Press.

Bullard, R. D. 1990. *Dumping in Dixie: Race, Class and Environmental Quality.* Boulder, CO: Westview Press.

Bullard, R. D., ed. 1993. *Confronting Environmental Racism.* Boston: South End Press.

Bullard, R. D., and B. H. Wright. 1992. "The Quest for Environmental Equity: Mobilizing the African-American Community for Social Change." In *American Environmentalism,* eds. R. E. Dunlap and A. G. Mertig, 39–49. Philadelphia: Taylor & Frances.

Carson, R. 1962. *Silent Spring.* Boston: Houghton Mifflin.

Chadwick, E. H. 1842. *Report Into the Sanitary Conditions of the Laboring Population of Britain.* Reprint, 1965, ed. M. W. Flinn. Edinburgh, U.K.: Edinburgh University Press.

Costanza, R., ed. 1991. *Ecological Economics: The Science and Management of Sustainability.* New York: Columbia University Press.

Devall, B. 1992. "Deep Ecology and Radical Environmentalism." In *American Environmentalism,* ed. R. E. Dunlap and A. G. Mertig, 51–62, 90–91. Philadelphia: Taylor & Frances.

Dorney, R. S. 1989. *The Professional Practice of Environmental Management,* ed. L. C. Dorney. New York: Springer-Verlag.

Duffey, J. 1968. *A History of Public Health in New York City, 1625–1866.* New York: Russell Sage Foundation.

Duffey, J. 1990. *The Sanitarians: A History of American Public Health.* Urbana, IL: University of Illinois Press.

Eisenbud, M. 1978. *Environment Technology and Health.* New York: New York University Press.

EPA (U.S. Environmental Protection Agency). 1992. *Environmental Equity, Reducing Risk for All Communities.* Report No. EPA 230-R-92-008. Washington, DC.

Fox, W. 1990. *Toward a Transpersonal Ecology.* Boston: Shambala.

Gottlieb, R. 1993. *Forcing the Spring: the Transformation of the American Environmental Movement.* Washington, DC: Island Press.

Hays, S. P. 1959. *Conservation and the Gospel of Efficiency: The Progressive Conservation Movement, 1890–1920.* Cambridge, MA: Harvard University Press.

Lazarus, R. J. 1993. Pursuing Environmental Justice: the Distributional Effects of Environmental Protection. *Northwestern University Law Review 87* (3): 787–857.

Lee, C. 1993. "Beyond Toxic Waste and Race." In *Confronting Environmental Racism,* ed. R. D. Bullard, 41–52. Boston: South End Press.

Leopold, A. 1933. The Conservation Ethic. *Journal of Forestry* 31 (6): 634–43.

Leopold, A. 1949. *A Sand County Almanac.* Oxford: Oxford University Press. Reissued in 1970 as a Sierra Club/Ballantine Book by Ballatine Books, New York.

Mander, J. 1991. *In the Absence of the Sacred: The Failure of Technology and the Survival of the Indian Nations.* San Francisco: Sierra Club Books.

Marsh, G. P. 1864. *Man and Nature; or Physical Geography as Modified by Human Action.* New York: Charles Scribner. Reprint, 1965 (ed. D. Lowenthal), Belknap Press of the Harvard University Press, Cambridge, MA.

Nash, R., ed. 1968. *The American Environment: Readings in the History of Conservation.* Reading, MA: Addison-Wesley.

Nash, R. 1973. *Wilderness and the American Mind,* rev. ed. New Haven: Yale University Press.

Nash, R. F. 1989. *The Rights of Nature.* Madison: University of Wisconsin Press.

Passmore, J. 1974. *Man's Responsibility for Nature, Ecological Problems and Western Traditions.* London: Duckworth.

Petulla, J. M. 1980. *American Environmentalism.* College Station, TX: Texas A&M University Press.

Petulla, J. M. 1987. *Environmental Protection in the United States.* San Francisco: San Francisco Study Center.

Ralston, H., III. 1996. "Philosophy." In *Greening the College Curriculum,* eds. J. Collett and S. Karakashian. Washington, DC: Island Press.

Regan, T. 1983. *The Case for Animal Rights.* Berkeley: University of California Press.

Sagoff, M. 1988. *The Economy of the Earth.* Cambridge: Cambridge University Press.

Singer, P. 1993. *Practical Ethics,* 2d ed. Cambridge: Cambridge University Press.

Chapter 2

Decision Making Based on Efficiency, Equity, and Rights

Many different concerns motivate individuals and governments to protect the environment: a desire to use resources efficiently, an obligation to provide safe drinking water, and so forth. How are these concerns translated into criteria used in making decisions? For any particular development project (or environmental regulation), which decision-making criteria should be used and how should conflicts among criteria be resolved?

This chapter investigates criteria government agencies employ in making decisions that influence the environment. Many of the same criteria are used by private organizations whose development projects must satisfy air and water quality requirements and various other environmental regulations.

The chapter first considers the *productive efficiency* objective and the analytic procedure most commonly used to gauge advances in efficiency: benefit-cost analysis. It then sets the stage for considering other criteria by distinguishing between legal and moral rights. The chapter continues by examining two sets of questions related to environmental justice. One set concerns people living now: How fairly are the costs and benefits of environmental protection distributed within a nation and among nations? A second set of environmental-justice questions concerns future generations: Should proposed projects be undertaken if they compromise the ability of future generations to satisfy their own needs? The chapter concludes by examining moral and legal rights. Some of these rights involve humans; for example, the rights of citizens to a habitable environment. Other rights concern animals. Some people believe certain classes of nonhuman animals have a right to be treated decently.

PRODUCTIVE EFFICIENCY AND BENEFIT-COST ANALYSIS

During the 1930s, the ideas of Gifford Pinchot and other early conservationists about using natural resources efficiently were supported by the emergence of a decision-making procedure known as *benefit-cost analysis* (BCA). Although fundamental elements of benefit-cost analysis had existed for decades, the approach came into widespread use only after the U.S. Congress passed the Flood Control Act of 1936. The act marked the beginning of an enormous federal public works program, and Congress worried about whether agencies would spend the massive funds involved wisely. To ensure fiscal responsibility, the act required that the flood control projects that it called for should be undertaken only "if the benefits to whomsoever they may accrue are in excess of the estimated costs." Following passage of the 1936 act, procedures for calculating monetary benefits and costs were developed, and they have been applied in many public-sector decision-making contexts, including the formulation of environmental policy.

Benefit-cost analysis can be justified as an aid to public decision making in several ways, but the most common rationalization is based on *productive efficiency,* a concept developed by economists.[1] In common language, productive efficiency is increased when more of a desired output is obtained with a given set of inputs. Economists have a more specialized definition, which is described in Chapter 7.

Because productive efficiency is treated extensively in Chapters 5–7, the concept is introduced only in summary form here. A development project (or environmental regulation) is productively efficient if it yields the largest possible value for *net benefits,* defined as benefits minus costs. In this context, benefits and costs are expressed in monetary units, and the terms *benefit* and *cost* have specialized meanings. The technical definitions of these terms are given in Chapters 5 and 6.[2]

In practice, efficiency maximization is an elusive goal.[3] Typically, benefit-cost analysis is used in the more limited sense of identifying whether a policy or plan will increase efficiency (not necessarily maximize it). Common applications of BCA involve determining if a proposed policy or project will yield monetary benefits greater than costs. It is often supposed, when using this type of analysis, that a proposed action should not be undertaken unless it yields monetary benefits greater than costs.[4]

The use of productive efficiency as a goal for public decisions is a dominant theme in the prescriptive literature on environmental policy and natural resources management. Advocates of the productive efficiency objective (many of whom are economists) argue that BCA ought to be used more frequently. Sometimes proponents of BCA have been able to persuade decision makers to include productive efficiency as a criterion for judging the acceptability of environmental and natural resources management plans and programs.

There are countless examples of benefit-cost analysis applications in resources management, and one is mentioned here to demonstrate that the productive efficiency objective has practical impacts on environmental policy. In February 1981, U.S. President Ronald Reagan issued Executive Order 12291, which requires that every major new federal regulation be subject to a benefit-cost analysis. An agency proposing a new regulation must demonstrate that the monetary benefits of its proposal outweigh the costs. Even though this benefit-cost requirement was widely criticized as being a mechanism to slow down the process of environmental regulation, BCA continues to be applied in the process of formulating federal environmental regulations in the United States.

Productive efficiency (and its implementation using benefit-cost analysis) cannot be effectively defended as the sole basis for making public decisions regarding environmental protection and natural resources management. Its shortcomings as a decision-making criterion have been documented extensively, and technical limitations of the approach are detailed in later chapters.[5] The remainder of this section ignores the technical shortcomings of BCA and introduces more fundamental limitations: the method's inability to account for either equity issues or rights to a habitable environment.

[1] For a review of alternative bases for rationalizing BCA, see Copp (1985).

[2] Although these terms are elaborated in Chapter 5, a preview of the definitions is given here. Economic benefits are measured as the amount of money that people would be "willing to pay" to obtain the beneficial consequences. Costs are measured in terms of the opportunities that a society gives up when its resources are used to implement a proposed plan.

[3] Difficulties in estimating monetary benefits of environmental programs contribute to the problem of using BCA to maximize productive efficiency. A special case of productive efficiency—cost effectiveness—is easier to apply. In a *cost-effectiveness* study, the analyst assumes that a particular goal must be met (for example, the goal of having a minimum level of bacterial concentration in a polluted river). A cost-effectiveness analysis attempts to find the least costly way to meet the selected goal.

[4] As a practical matter, many benefits of environmental improvement cannot be measured in monetary units. In such circumstances, the existence of benefits less than costs is *not* interpreted to mean that the proposed action should not be taken.

[5] Technical aspects of BCA are treated in Chapters 6 and 7. For examples of recent critiques of BCA, see Campen (1986) and Sagoff (1988).

A major limitation of productive efficiency as a basis for public decision making is that the *distribution* of costs and gains is not considered. Only aggregate net benefits count, not who gains and who loses. If only productive efficiency were considered in making decisions, unfair outcomes could easily result.[6] For example, hazardous waste disposal facilities might be located in low-income neighborhoods because land prices are low and residents do not have the means to resist. Similar injustices could occur if only productive efficiency were used in deciding how much hazardous waste to export from industrialized nations to developing countries. Indeed, the inequities that have resulted from using productive efficiency to make such choices have recently been a subject of international concern.[7]

The productive efficiency objective is also limited because it fails to consider fundamental legal or moral rights that may make *productively inefficient* solutions desirable. A decision by the U.S. Supreme Court on workplace health standards for cotton dust illuminates conflicts between productive efficiency and rights to a healthful environment. The Court ruled on a case in which the textile industry argued that a government health standard for cotton dust was inappropriate because costs exceeded benefits. The significance of the court's decisions in this case, *American Textile Manufacturers Institute v. Donovan,* was summarized by Karpf (1982, p. 89):

> [T]he Supreme Court . . . concluded that the Secretary of Labor is not required to perform a cost-benefit analysis before promulgating permissible exposure levels for toxic materials and harmful physical agents. Indeed, the Court implied that the Secretary may not limit the stringency of such standards solely because the anticipated costs exceed the anticipated benefits.

The Court's ruling suggests that worker health and safety supersede productive efficiency in this case.

[6] In addition to BCA's inability to consider issues of fairness for those alive now, BCA yields incongruous results when applied to decisions affecting future generations. Perplexing outcomes result because of a process called *discounting* (detailed in Chapter 7), which is used to put cash flows in different years on a comparable basis. A critical analysis of how benefit-cost analysis handles effects on future generations is given by Campen (1986, pp. 59–60).

[7] Concerns over the international trade in toxic waste are examined in Chapter 3.

The following sections introduce additional criteria for making decisions about environmental protection and resources management. These criteria concern factors, such as fairness and legal rights, that may lead decision makers to prefer projects or regulations that are not productively efficient.

DISTINGUISHING MORAL AND LEGAL RIGHTS

In discussions of environmental policy, decision criteria other than maximization of productive efficiency often involve the concept of *rights.* For example, policies to preserve wetlands frequently constrain the rights of property owners. And proponents of species preservation policies sometimes argue that nonhuman species have rights. This section lays a foundation for considering additional decision criteria by introducing the concept of rights and distinguishing between legal rights and moral rights.

Legal rights are set out in laws, constitutions, and comparable institutions. Consequently, they vary in time and place. For example, in the United States, men and women over the age of 18 have a legal right to vote in national elections. Not long ago, individuals age 18 through 20 did not enjoy this right. And before 1920, American women could not vote regardless of their age.

In general, legal rights are granted to persons or entities, such as corporations or vessels, that can be treated as "legal persons." For a legal right to be meaningful, there must be a public body, such as a court, that has the authority to review actions of those who threaten a person's (or entity's) legal rights. In addition, the bearer of a legal right must be able to institute legal actions on its own. These and other criteria to be satisfied in establishing a system of legal rights are set out by Christopher Stone (1974) in his argument (elaborated later in this chapter) for extending legal standing to guardians of natural systems.

In contrast to legal rights, a *moral right* is independent of any particular legal system.[8] Saying that

[8] Some go further and say that moral rights are universal in that they apply to all individuals who are the same in *all relevant respects* (Regan 1983, pp. 267–68). Examples of characteristics that are not relevant to the applicability of moral rights are gender and religion. In contrast to this universalist position, some argue that "the truth or falsity of moral judgments are relative to either the individual making the judgment or the culture of which he or she is a part" (Armstrong and Botzler 1993, p. 54).

moral rights exist is the same as saying that morally responsible individuals ought to behave in certain ways. These individuals, referred to by philosophers as *moral agents,* include all persons capable of understanding and using moral principles in deciding how to act. Individuals incapable of reasoning, such as infants, or persons who have suffered incurable brain damage, are not moral agents, but they are nonetheless granted moral consideration in the sense that moral agents are obliged to treat them in morally responsible ways.

Moral principles, which provide premises in logical arguments about what constitutes morally acceptable behavior, are claims thought to be worthy of belief. They are prescriptive (or normative), not descriptive (or empirical). Not surprisingly, moral claims are often the subject of controversy. For example, consider the moral claim that "people ought not destroy the habitats of endangered species." This premise, combined with the empirical facts that the northern spotted owl is an endangered species and the forests of Oregon contain habitats of the northern spotted owl, yield a moral claim: "people ought not destroy the forests of Oregon." This moral argument against logging certain areas of Oregon has been the subject of much recent controversy.

Moral principles provide a basis for statements about who (or what) is the bearer of a moral right; the latter has been a source of dispute among philosophers. The nature of the controversy is clarified by an informal analysis of the phrase, "B has a moral right against M to be treated with respect."[9] In this phrase, B is the bearer of the right, and M is a moral agent. Thus the phrase can be interpreted as follows: M (moral agent) is not permitted to treat B (bearer of the right) disrespectfully.

A key source of disagreement among philosophers is whether B can be anything other than one or more moral agents. Some philosophers argue that B can only include moral agents, because only moral agents are capable of exercising moral judgment and responding to moral claims, such as the claim to be treated with respect.[10] Others extend the circle of rights' holders to all persons, including infants and others who do not qualify as moral agents. Still other philosophers argue that the circle of beings that bear rights includes humans and animals capable of experiencing pain and suffering (*sentient beings*).

Although discussions of legal and moral rights often employ the same terms—for example, words such as "entitlement," "duty," and "obligation"—the terms are used differently in law and philosophy. Legal requirements may or may not be founded on moral theory, and some laws are unjust and morally fallible. However, moral constraints frequently influence the process of lawmaking. Indeed, laws are often passed to protect rights that are, according to some people, based on valid moral claims.

How can ideas about moral principles and moral obligations help in making decisions affecting the environment? One application of these ideas is in defining what it means to say that a development project (or environmental regulation) yields a just distribution of monetary benefits and costs.

ENVIRONMENTAL JUSTICE AND INTRAGENERATIONAL EQUITY

People have debated alternative meanings of justice since ancient times. Although some conceptions of justice have many advocates, there is no general support for a single definition or theory of justice. Despite the lack of agreement, the subject of justice must be explored because it often plays a key role in how decisions affecting the environment are made. Indeed, some philosophers believe the concept of justice is central to environmental problem solving. For example, Robert Wenz (1988, p. 24), a contemporary American philosopher, argues that "people often differ with one another over questions of environmental policy largely because they are employing different principles of justice and/or accord different weights to some of the same principles."

Distributive justice is about who gets what. In the context of decisions affecting the environment, the "what" involves both good things, such as the advantages of improved air quality, and bad things, as illustrated by the loss of jobs when a factory is closed because it violates environmental regulations. The "who" in "who gets what" includes people living now either within a country or in different countries, and people not yet born. Obligations to the unborn

[9] This analysis is adapted from Van De Veer and Pierce (1994, p. 19).

[10] This position is taken by Carl Cohen, a professor of philosophy at the University of Michigan at Ann Arbor. His views on this subject, as well as opposing views of the type noted further along, are presented in a volume edited by Mappes and Zembaty (1992, pp. 433–75).

are generally considered under the label of *intergenerational equity,* a subject treated in the next section.

Principles of justice often involve patterned phrases, such as "to each according to his needs" or "to each according to her contributions."[11] However, applications of different principles of justice may yield conflicting outcomes. To resolve conflicts, principles of justice are often combined to yield *theories of justice.* In his analysis of the suitability of alternative theories of justice for making decisions affecting the environment, Wenz (1988, p. 43) offers the following observations:

> The best or ideal theory [of justice] would provide an underlying rationale that everyone considers convincing. It would find a place for all reasonable principles of justice, and would convincingly explain why other principles that people might appeal to are not reasonable. It would weight or prioritize competing, reasonable principles of justice. It would thus provide the moral basis for the resolution of all disputes to the mutual satisfaction of all the parties involved.

Unfortunately, there is no widely accepted ideal theory of justice. Two philosophical approaches to distributive justice are introduced for illustrative purposes: one approach is based on utilitarian philosophy, and the second (entitlement theory) is associated with libertarianism (see Table 2.1). The following discussion demonstrates how each of these philosophical approaches could be used in resolving environmental conflicts.

TABLE 2.1 Two Philosophical Approaches to Distributive Justice

Utilitarianism—focus is on consequences

Fairness of a government decision that redistributes wealth is gauged by the collective gain in *utility* (happiness) for all members of a society. Some members may be burdened by the government decision, but it is the overall net change in utility that counts.

Entitlement theory—focus is on procedures

Fairness of a government decision that redistributes wealth is based on whether the present distribution was acquired by force or fraud. If the present distribution of wealth was not acquired by force or fraud, then fair changes in the distribution are those that do *not* involve force, fraud, theft, or enslavement.

Utilitarianism and Distributive Justice

Utilitarianism is not a theory of justice, but its central tenets are often used to rationalize particular principles of justice. The classical form of utilitarianism is often associated with Jeremy Bentham, an eighteenth-century British philosopher concerned with the basis for developing social policy and legislation. According to Peter Singer (1993, p. 3), a contemporary utilitarian philosopher and animal rights activist, the "classical utilitarian regards an action as right if it produces as much or more of an increase in happiness of all affected by it than any alternative action, and wrong if it does not." In this context, *utility* is an ambiguous term used synonymously with happiness and satisfaction. The classical utilitarian is concerned only with the sum total of satisfaction that results when a fixed quantity of goods and services is distributed. Only final consequences matter. Another important principle of utilitarianism concerns the equality of interests: everyone's interests count, and like interests count equally. In other words, the like interests of individuals receive the same weight in calculating changes in happiness resulting from an action, regardless of gender, race, age, and so forth.[12]

Numerous court actions illustrate how utilitarian principles are used in deciding the distribution of costs and gains of environmental decisions. A well-known example is the case of *Reserve Mining v. United States.*[13] The dispute centered on the manner in which wastes from the Reserve Mining Company's operation at Silver Bay, Minnesota, influenced the health of citizens of Duluth, Minnesota, and other communities that used Lake Superior for drinking

[11] For other examples of principles of justice, see Nozick (1974, pp. 156–57).

[12] For more on this principle of equal consideration of interests, see Singer (1993, p. 21).

[13] *Reserve Mining v. United States* is detailed in Stewart and Krier (1978, pp. 262–79).

water. Plaintiffs argued that the company's wastewater discharges contained materials similar to a form of asbestos known to be carcinogenic in some occupational settings. The company maintained that its discharge posed no serious health hazard.

The U.S. district court that heard the case concluded that "the Court has no other alternative but to order an immediate halt to the discharge which threatens the lives of thousands."[14] The decision was based on a "balancing of the equities," in which the gains and costs of alternative court actions were compared. The court determined that the danger to public health (which would continue because Reserve Mining did not have a plan to modify its discharge) outweighed the adverse economic and social impacts of a plant closure. At the time, the plant employed 3300 people and had an annual payroll of over $30 million; in addition, it produced 12 percent of the iron ore in the United States.[15]

Reserve Mining Company appealed the district court's decision and gained a reversal. In reaching its conclusion, the appellate court struggled with the uncertainties surrounding the alleged adverse health effects. It concluded that there might be some risk to health, but that "it cannot be said that the probability of harm is more likely than not."[16] The court gave considerable weight to the societal gains—measured in terms of revenues, employment and ore production—from the Reserve Mining Company's operations. The court was also impressed with the waste cleanup plan the company had devised during the course of its appeal. In the end, the court concluded "that an immediate injunction cannot be justified in striking a balance between unpredictable health effects and the clearly predictable social and economic consequences that would follow the plant closing." In this case, any rights that the citizens of Duluth may have had to a completely safe drinking water supply were overshadowed by the economic importance of the plant's continued operations. The citizens of Duluth were asked to shoulder risks to their health.

[14] The quote is from the presiding judge, as it appears in Stewart and Krier (1978, p. 265).

[15] Facts about Reserve Mining Company are from Stewart and Krier (1978, p. 262).

[16] Materials quoted are from the opinion of the Eighth Circuit Court as they appear in Stewart and Krier (1978, pp. 276–77).

Both the district court and the appellate court used utilitarian ideas in deciding how the costs and gains should be distributed. Each court viewed a fair distribution as one that yielded the greatest net gain for society. *Reserve Mining v. United States* demonstrates the problems that arise in applying utilitarian concepts to outcomes, such as health risks, that are highly uncertain.

Making calculations of "happiness" or "utility" in the face of uncertainty is part of a larger measurement problem faced by utilitarians. Some philosophers doubt whether utility can be measured in quantitative terms at all. They also question the validity of making comparisons of utility among different persons. Some critics challenge the very idea that maximizing the sum of individual utilities *ought* to be a guiding principle.[17]

Libertarianism and the Entitlement Theory

The term *libertarianism,* as currently used, refers to the doctrine that views individual liberty as the central consideration in political matters.[18] Libertarians believe that individuals should be free to do (or not do) as they please, provided they do not infringe on the liberties of others. Because individual liberty often decreases when government involvement in the lives of individuals increases, the size of the state should be minimal. Unlike anarchists, many libertarians believe a minimal state is necessary to protect people against fraud and the use of force.

For many libertarians, private property is crucial to maintaining individual liberty. It follows that if the function of the state is to safeguard life and liberty, then the state should guarantee the rights of private property.[19]

In 1974, Robert Nozick articulated what he termed *entitlement theory,* which has come to be viewed as a libertarian theory of justice.[20] In contrast to utilitarianism, entitlement theory is not concerned with outcomes. Rather, the central concern is with procedures that are just. In entitlement theory, a distribution of goods and services (or, using Nozick's term, "holdings") is just if it arises from another

[17] See, for example, Narveson (1988, pp. 150–53).

[18] This conception of libertarianism is from Narveson (1988, p. 7).

[19] This paragraph is based on an analysis of libertarianism by Wenz (1988).

[20] The original presentation is in Nozick (1974).

just distribution by legitimate means. A transfer of holdings is legitimate if it does not involve force, fraud, theft, or enslavement. Nozick introduces additional principles of justice to analyze distributions of holdings that came about by unjust means, such as by a forceful takeover of property. Entitlement theory is consistent with the libertarian view that people ought to be free to do as they please with what they own, as long as their actions don't violate the ability of others to behave in a like manner.

Based on Nozick's entitlement theory, a just distribution of holdings can be a very unequal distribution of holdings. Indeed, whether or not a distribution of holdings satisfies any conception of social need is irrelevant. Unless the present distribution of holdings was acquired by force or fraud, there would be no justification for redistributing resources to reduce the gap in standard of living between rich and poor. If wealthy individuals were inclined to give some of their holdings to the poor voluntarily, the resulting distribution would be just because the transfer was voluntary. However, the distribution would not be just if the state forced the wealthy to transfer part of their holdings by means of a tax on income, for example. This consequence of entitlement theory challenges government actions that redistribute wealth from rich to poor. However, the libertarian argument against government redistribution of wealth is not entirely convincing. Entitlement theory requires that in order to be viewed as just, the current distribution of holdings cannot have resulted from a prior distribution that was unjust because it was based on force, fraud, theft, or enslavement. As Wenz (1988, p. 75) observes, "there are few, if any, parts of the earth whose environmental resources have reached their current owners without force or fraud."

The libertarian conception of justice is applicable in resolving some environmental conflicts. Wenz demonstrates this by analyzing an environmental dispute involving adjacent property owners in seventeenth-century England.[21] In 1611, William Aldrich brought suit against his neighbor, Thomas Benton, arguing that the latter's actions had deprived him of use of his property. Prior to the suit, Benton had built a hogsty in his orchard, which was adjacent to Aldrich's house. The "stench and unhealthy odors emanating from the pigs" being raised by Benton constituted a nuisance that interfered with Aldrich's ability to use his house. Moreover, the hogsty blocked what was an attractive view of the countryside from a window in Aldrich's house. The actions of Benton, who made his living by selling pigs, were defended on grounds that "the building of the house for hogs was necessary for the sustenance of man" and that "one ought not to have so delicate a nose, that he cannot bear the smell of hogs."

The court ruled that Aldrich was entitled to the 40 pounds he claimed as damages because he had to endure the odor. His right to use his property had, in fact, been diminished. However, the judge also ruled that Benton could keep his hogsty and that no compensation was required for Aldrich's diminished view. According to the court, the loss of view involved a matter of "delight" but not "necessity." For Benton, the compensation he was required to pay Aldrich became a cost of staying in the business of raising pigs.

In Wenz's view, private property rights do not provide a complete basis for arriving at just solutions for some environmental problems. In the case involving Aldrich and Benton, for example, the libertarian emphasis on the need to protect rights to property played the central role. However, even in this simple case, additional principles of justice were involved because both parties claimed rights to use their property. Benton wanted to use his land to make a living, and Aldrich wanted to live without the noxious odors from the hogsty. The judge had to balance these conflicting arguments about rights to property.

Libertarian principles are often not helpful in determining fair solutions to environmental problems involving common property. An example is pollution of the atmosphere. In principle, the *common law of nuisance* (as illustrated in the case involving Aldrich and Benton) could be applied, but complexities arise if multiple polluters cause damages to large numbers of persons.[22] The problem for anyone contemplating

[21] Wenz's (1988) analysis draws from details in the case given by Stewart and Krier (1978). The quotes in this paragraph are from the judge's opinion as presented by Stewart and Krier (1978, pp. 117–18).

[22] The legal concept of *nuisance* encompasses many things, including obstruction to the reasonable use of one's property. A nuisance may offend a particular person or it may interfere with the health, safety, and comfort of the community at large. The *common law* is a system of jurisprudence based on judicial precedents rather then legislative enactments. Common law originated in England and was later applied in the United States.

legal action is in establishing which polluter caused the harm. Proving harm may involve complex scientific issues and could require much research. In addition, if many individuals are adversely affected by the pollution, and each individual suffers only a modest harm, it may be impossible to organize a legal action. Thus, even though the damage from pollution may be substantial when summed over a large number of affected parties, legal action based on nuisance law may be too difficult to organize.

As with utilitarianism, principles of justice derivable from libertarianism are useful in determining fair solutions to *some* environmental disputes. However, no single theory of justice is adequate to deal with all environmental problems.

Evidence of Environmental Inequity

The empirical claims that distributions of costs and gains of environmental protection are unfair typically involve implicit use of a *rule of equity*: "the fair share of any one person is the same as that for anyone else who is the same in all relevant respects" (Wenz, 1988, p. 22). Consider, for example, two communities that are the same in all respects from the point of view of technical suitability as a site for a commercial hazardous waste disposal facility. Suppose one community is composed primarily of low-income Latinos and the second consists largely of middle-class white people. If the decision was made to locate the disposal facility in the Latino community because residents had less political and economic ability to oppose it, the outcome would clearly be discriminatory on the basis of both ethnicity and income. Neither of these factors is relevant. They have no bearing on the technical suitability of the two communities as locations for the waste disposal plant. Bullard (1993a, p. 197) cites a case where a consulting firm used the lack of likely opposition from low-income populations as a basis for its recommendation for where to locate incinerators. He calls the consultant's report "a 'smoking gun' of environmental injustice."

Two questions have played a central role in studies of environmental inequities. First, are benefits and costs of environmental protection programs distributed unfairly? Second, do ethnic minorities and poor people shoulder a disproportionately high share of environmental risks? Table 2.2 contains *hypotheses* that have been studied in response to these two questions.

TABLE 2.2 Hypotheses Related to Environmental Inequities in the United States

- How are benefits and costs of environmental management programs distributed?

  The poor often carry a disproportionate share of the cost of pollution control.

  Benefits of nonurban public recreation areas are enjoyed disproportionately by middle- and upper-income families.

- Do ethnic minorities and poor people shoulder a disproportionately high share of environmental risks?

  Compared to non-Hispanic whites, ethnic minorities have disproportionately high incidences of adverse health effects linked to pollution.

  Racial minorities and low-income populations experience higher than average exposure to selected pollutants, such as pesticides.

Inequity of Environmental Protection Programs

Some studies of environmental inequities investigate how the costs of compliance with U.S. environmental regulations are distributed.[23] If compliance costs make up a larger fraction of total income for poor households than they do for rich households, the cost burden is *regressive*.

There are substantial difficulties in investigating whether costs of compliance with environmental protection programs are regressive. It is hard to determine how much of the cost of pollution abatement results in higher prices and how much results in lower profits. Moreover, many environmental programs are funded using federal and state income taxes, and the

[23] The hypotheses on the distributions of costs and benefits of environmental programs introduced in this section are discussed in detail by Barbour (1980), Lazarus (1993), and Bryant (1995).

poor pay a smaller percentage of income as taxes than middle- and upper-income classes. However, one instance in which distributional impacts appear unambiguous concerns emissions of pollutants from auto exhausts. Costs of reducing auto emissions are believed to fall disproportionately on the poor because lower-income families with autos spend a relatively high fraction of their disposable incomes on auto maintenance. It is sometimes also argued that the poor carry a disproportionate burden when environmental regulations cause plant closures and unemployment.

Next, consider studies examining whether environmental management programs benefit higher-income groups disproportionately. An example is Collins's (1977) analysis of how U.S. government subsidies to build publicly owned wastewater treatment plants affected the distribution of wealth.[24] Collins characterized the distributional consequences of the subsidy program as "perverse" because most of the net monetary benefits of the program went to the highest-income class.[25] He also found that while the poor received some net benefits from the subsidies, the "main redistribution of wealth appears to be from the middle-income classes to the highest-income class."

A study of air pollution in the Boston metropolitan area provides another example of research on distributional consequences of environmental protection. Harrison and Rubinfeld (1978) used standard procedures for estimating monetary benefits of air pollution reduction to calculate the gains attributable to air quality management in the Boston area.[26] The researchers found that "the absolute level of benefits, measured in dollars . . . rises consistently and substantially with income. Only when expressed as a percentage of income are air quality benefits pro-poor."[27]

Although few statistical studies have been done to support it, the following hypothesis has been advanced: Benefits of nonurban public recreation areas, such as wilderness areas and national parks, are enjoyed disproportionately by middle- and upper-income families. In proportion to their total numbers, middle- and upper-income families use these national parks and wilderness areas more frequently than lower-income families. This occurs, in part, because people need cars and good incomes to travel to these areas and take advantage of the recreational opportunities they provide.

Inequity in Distribution of Environmental Risks

Many researchers have investigated whether racial minorities and lower-income classes accept a disproportionately high share of environmental risks. A systematic study of this issue began in the 1980s in the context of hazardous waste landfill siting decisions in the United States. Although their conclusions have been challenged, the early studies found that both race and income appeared to influence where hazardous waste facilities were located.[28] Some studies argued that racial discrimination played a central role in the siting of these facilities.[29]

Once charges of racial discrimination became widely known, additional studies were conducted and spokespersons for minority communities called for action by the U.S. Environmental Protection Agency (EPA). Although EPA did not initiate new data gathering as an immediate response, it did conduct a rigorous examination of earlier reports.

The EPA (1992) study presented only one definitive conclusion regarding disproportionate adverse *health effects*: for lead poisoning, there is a disproportionate impact by race. "A significantly higher percentage of Black children compared to White children have unacceptably high blood [lead] levels."[30]

[24] The federal program examined by Collins was established by the Federal Water Pollution Control Act Amendments of 1972 (see Chapter 12).

[25] This characterization (and the quotation below) is from Collins (1977, pp. 352–53). Collins calculated "net financial incidence" by computing, for each income class, the total monetary value of the subsidy minus taxes paid.

[26] Monetary benefits were estimated using the methods described in Chapter 6.

[27] This quotation is from Harrison and Rubinfeld (1978, p. 314). Their study concerned only monetary benefits, not costs.

[28] Been (1994) provides a critical analysis of evidence supporting the linkages between hazardous-waste-facility siting, race, and income class. She also provides an alternative explanation for the empirical results used in earlier studies of environmental inequity, such as the study by the U.S. General Accounting Office (1983).

[29] Two studies had an especially significant impact: (1) a report prepared by the U.S. General Accounting Office (1983), and (2) a study by the United Church of Christ Commission of Racial Justice (1987). Findings from these reports were summarized in Chapter 1. See, also, Bullard (1990).

[30] The study conclusions are reported in EPA (1992, p. 3). The last quotation in this paragraph is from EPA (1992, p. 16).

EPA reached broader conclusions regarding *exposure to pollutants*: "Racial minorities and low-income populations experience higher than average exposures to selected air pollutants, hazardous waste facilities, contaminated fish, and agricultural pesticides in the workplace." On this last point, EPA indicated that "80–90% of the approximately two million farm workers . . . are racial minorities," and that workplace exposure to chemicals in agriculture is one of the areas of greatest human health risks. Although the EPA study was criticized for ignoring some relevant research, the agency's main conclusions were consistent with results from prior studies showing environmental inequity based on race and socioeconomic status.[31]

Another aspect of the distribution of environmental risks concerns whether unequal enforcement of environmental laws discriminates against racial minorities. Is environmental quality lower in minority neighborhoods because relatively few resources are used to enforce environmental regulations in those neighborhoods? Richard Lazarus, a law professor at Washington University in St. Louis, reviewed prior research on this question. He cites investigations showing that in U.S. minority communities, fines for violating environmental laws are lower, pollution cleanups are slower, and violations of pollution control laws are more frequent.[32]

The selective enforcement of environmental laws has led some scholars to define environmental equity in terms of equal environmental protection under the law. This position is taken by Bunyan Bryant of the University of Michigan School of Natural Resources and the Environment. Bryant's conception of environmental equity, which he distinguishes from environmental justice, is given in Table 2.3. Based on the attention given to environmental inequities in the early 1990s, environmental justice issues will undoubtedly receive increased weight in environmental decision making in the United States.

[31] For details on the earlier studies that EPA allegedly ignored, see Bullard (1993a). For a more general review of the controversy surrounding EPA's 1992 report, see Lazarus (1993).

[32] These facts are reported by Lazarus (1993, pp. 818–19). Lazarus also reports that less-rigorous remedies were used in cleaning up Superfund sites in minority communities. As explained in Chapter 14, Superfund sites are hazardous waste disposal facilities subject to cleanup requirements under the U.S. "Superfund law."

TABLE 2.3 Defining Environmental Equity in Terms of Equal Protection[a]

Environmental Equity

Environmental equity refers to the equal protection of environmental laws. For example, under the [U.S. hazardous waste] cleanup program it has been shown that abandoned hazardous waste sites in minority areas take 20 percent longer to be placed on the national priority action list than those in white areas. It has also been shown that the government's fines are six times greater for companies in violation of RCRA [the Resource Conservation and Recovery Act] in white than in black communities. This is unequal protection.

Environmental Justice

Environmental justice . . . is broader in scope than environmental equity. It refers to those cultural norms and values, rules, regulations, behaviors, policies, and decisions to support sustainable communities, where people can interact with confidence that their environment is safe, nurturing, and productive. . . . Environmental justice is supported by . . . communities where both cultural and biological diversity are respected and highly revered and where distributed justice prevails.

[a] Materials in the table are from Bryant (1995). Reprinted with permission. The definitions in the original are longer and should be consulted for a full description of Bryant's views.

International Environmental Justice Issues

The increased importance of environmental justice as a concern in decision making is not confined to the United States. In their review of equity issues, Alston and Brown (1993) identify the following subjects of international significance:[33]

- *War and Weapons Testing.* The environmental impacts of war are illustrated by the extensive use of

[33] The review of Alston and Brown contains a few items that are not listed here because they do not involve significant environmental consequences.

chemical weapons and defoliants during the Vietnam War, and the deliberate oil spills and burning of oil fields during the Persian Gulf War. A related matter concerns the adverse effects of nuclear weapons testing on indigenous peoples, such as the Aborigines of Australia and the Western Shoshones of Nevada.

- *International Lending.* The influence of international development banks is seen in support of environmentally destructive projects and practices in developing countries. This is illustrated by the role of the World Bank in funding projects damaging the integrity of tropical forests in Brazil. A related aspect of international lending concerns *debt-for-nature swaps,* which may involve actions, such as establishing nature preserves, that affect indigenous people in the Third World. Environmental groups in industrialized countries have organized many of these exchanges. For example, $650,000 of Bolivia's foreign debt was purchased by Conservation International for 15 cents on the dollar. In return, the Bolivian government agreed to set aside the value of the original debt for conservation purposes. A number of debt-for-nature swaps have been described as causing "brazen disregard for the rights of the indigenous inhabitants. . . ."[34]
- *Costs of Solving International Environmental Problems.* Solutions to international environmental problems (such as global climate change) may affect the consumption patterns and development possibilities of Third World countries. The latter are understandably reluctant to shoulder a heavy burden of costs to solve global problems caused by what they view as excessive materialism and frivolous consumption by affluent residents of industrialized countries.
- *Toxic Waste Trade.* Hazardous wastes from industrialized countries are disposed of, often illegally, in developing countries, especially those without the environmental specialists needed to manage them. This trading, which has been referred to as "toxic terrorism," has led to numerous environmental problems and caused some developing countries to ban hazardous-waste imports.

An international environmental equity issue, which was not highlighted by Alston and Brown (1993), concerns factory location decisions of transnational corporations. The equity question turns on whether transnational corporations behave unjustly by locating in Third World countries to take advantage of unenforced or lax environmental regulations. A widely known example involves the more than 1900 *maquiladoras,* assembly plants owned by companies headquartered in the United States, Japan, and elsewhere, that operate along the 2000-mile border between the United States and Mexico. Bullard (1993b, p. 19) sums up the problem as follows:

> These plants use cheap Mexican labor to assemble products from imported components and raw materials, and then ship them back to the United States. . . . Nearly half a million Mexicans work in the maquiladoras. They earn an average of $3.75 a day. While these plants bring jobs, albeit low-paying ones, they exacerbate local pollution by overcrowding the border towns, straining sewage and water systems, and reducing air quality. All this compromises the health of workers and nearby community residents. The Mexican environmental regulatory agency is understaffed and ill-equipped to adequately enforce the country's law. . . .

The problems of international environmental equity summarized here are vexing. Challenging international equity problems will persist until effective international institutions are created to deal with transnational pollution.

INTERGENERATIONAL EQUITY AND SUSTAINABLE DEVELOPMENT

The discussion of equity can be extended to consider the obligations of those living now to future generations. Intergenerational equity has been treated analytically from time to time, but philosophers have only recently tried to deal with it rigorously in the context of environmental management. Since the early 1970s, many people have recognized the unique capabilities of post–World War II generations to adversely affect those born thousands of years from now

[34] This quotation is from an open letter to the environmental and conservation community in 1990 as it appears in Alston and Brown (1993, p. 189). In 1991, a meeting of 220 representatives of Latin American nongovernmental organizations called for a temporary suspension of all debt-for-nature swaps while national policies were being elaborated.

(for example, by inadequate disposal of radioactive wastes from nuclear power plants).[35] This recognition has led to a growing literature on obligations to future generations.[36]

Moral Obligations to Posterity

As in the case of distributive justice among members of the same generation, philosophers disagree on what constitutes a just distribution of burdens and benefits between generations. Although it is tempting to argue that future generations have a moral right to an environment that is habitable and not depleted of natural resources, some philosophers dispute this. They reason that because future generations do not exist, except in the imagination, they cannot possibly have rights. For example, Thomas Thompson (1980, p. 201) a contemporary American philosopher, argues that "[f]uture beings, being non-beings, can have no demands upon me. Even if they did, they have no reality, save potential reality, sufficient to establish an emotive bond of identification with me." Other philosophers contend that the obligations of the present generation are, at best, ambiguous because the present generation cannot know what should be saved for the future. Technology and tastes change rapidly. Arguing along these lines, Passmore (1974, p. 98) observed, "We cannot be certain that posterity will need what we save. . . . There is always the risk, too, that our well-intentioned sacrifices will have the long-term effect of making the situation for posterity worse than it would otherwise be."

Notwithstanding these objections, many philosophers have developed moral arguments to demonstrate the existence of obligations to posterity. A particularly interesting approach was introduced by John Rawls, a Harvard philosophy professor, in a section of his influential treatise, *A Theory of Justice.* Because Rawls's work has provided a starting point for many philosophical analyses of obligations to future generations, a few of his ideas on the subject are sketched below.

Rawls's analysis centers around a hypothetical contract between rational, self-interested persons. His approach requires the contractors to develop "principles which are to assign basic rights and duties and determine the division of social benefits" (Rawls 1971, p. 11). Rawls places the hypothetical contractors behind a "veil of ignorance" in which individuals do not know what strengths or weaknesses, and advantages or disadvantages, they will have in the society.[37] The exercise Rawls designed requires the contractors to think through the various situations they might find themselves in once the veil of ignorance is lifted.

Although Rawls didn't examine justice for generations in the distant future, he did consider justice between successive generations. In this context, Rawls asks the hypothetical contractors in a particular generation to put themselves in the position of not knowing which generation they belong to. Then they should

> try to piece together a just savings schedule by balancing how much at each stage [of history] they would be willing to save for their immediate descendants against what they would feel entitled to claim of their immediate predecessors. Thus, imagining themselves to be fathers, say, they are to ascertain how much they would set aside for their sons by noting what they would believe themselves entitled to claim of their [own] fathers.[38]

The "savings" that Rawls refers to can take the form of a net investment in machinery, or investments in research for new technology, and so forth. Although Rawls did not consider environment quality issues, others have extended his work to treat environmental problems.

Consider, for example, the analysis of Ronald Green, a former student of Rawls. Green applied the social contract analysis envisioned by Rawls to the problem of obligations to distant generations. Green's results include the following "axioms" that pertain directly to environmental quality:

- "The lives of future persons ought ideally to be 'better' than our own and certainly no worse."[39]

[35] As of 1970, few philosophical papers on the subject of duties to posterity existed (Partridge 1980, p. 10).

[36] For an introduction to the literature on obligations to future generations, see Partridge (1980) and Brown Weiss (1989).

[37] Because the principles of justice are drawn up in an initial situation that is fair, Rawls uses the term "justice as fairness" to describe his approach.

[38] This quotation is from Rawls (1971, p. 289).

[39] This axiom and the one following are from Green (1980, pp. 94 and 97, respectively).

In interpreting what is "better," Green accounts for environmental deterioration and resource depletion.

- "Sacrifices on behalf of the future must be distributed equitably in the present, with special regard for those presently least advantaged." In this context, Green highlights what has come to be regarded as a central issue in discussions of environmental protection: If current consumption is to be sacrificed to preserve environmental and natural resources for future generations, how, precisely, should that sacrifice be distributed among the affluent and the less economically developed nations of the world?

The assertion that the present generation has obligations to protect the environment for future generations does not rely solely on theories of moral philosophers. As Edith Brown Weiss (1989, p. 18) points out in advancing her theory of intergenerational equity, the "thesis that we have obligations to conserve the planet for future generations and rights to have access to its benefits is deeply rooted in the diverse legal traditions of the international community. There are roots in the common and the civil law traditions, in Islamic law, in African customary law, and in Asian nontheistic traditions [for example, the reverence for nature that is a characteristic of Shintoism in Japan]." Obligations to future generations are even explicit in the constitutions of several countries (for example, Guyana, Iran, and Papua New Guinea), and in the constitutions of Hawaii, Illinois, and Montana in the United States.[40]

Sustainable Development and Distributive Justice

The idea that the current generation has obligations to restrict its use of the environment in order to preserve it for future generations has emerged as a norm in environmental policy making.[41] One manifestation of how duties to future generations have influenced environmental policy is the way intergenerational equity fits into the many definitions of *sustainable development.* A widely cited definition of sustainable development calls it "development that meets the needs of the present without compromising the ability of future generations to meet their own needs." This conception of sustainable development is from the World Commission on Environment and Development (1987, p. 43), also known as the Brundtland Commission.[42]

The Brundtland Commission's report was not the first to use the term sustainable development, but earlier conceptions of sustainable development focused on conserving species and ecosystems.[43] The Brundtland Commission broadened the notion of sustainable development by focusing on the need to balance obligations to future generations against the needs of those living now. The commission stimulated numerous efforts to define sustainable development in ways that accounted for both intergenerational and intragenerational equity.

Consistency among definitions of sustainable development has not yet been achieved. However, many researchers have moved the discussion of sustainable development beyond vague generalities to make the notion of sustainability useful in planning and policy making.[44] An example is the "almost practical step toward sustainability" proposed by Robert Solow, a professor at the Massachusetts Institute of Technology, and a Nobel Laureate in Economics.[45] Solow views a sustainable path to development as

[40] Brown Weiss (1989) includes an appendix detailing the obligations to future generations in several national and state constitutions.

[41] For evidence for this assertion, see Brown Weiss (1989).

[42] Gio Brundtland, former Prime Minister of Norway, chaired the World Commission on Environment and Development.

[43] An earlier conception of sustainable development was set out in a report produced by the International Union for the Conservation of Nature and Natural Resources (IUCN 1980) in consultation with several United Nations organizations. The World Conservation Strategy suggested that "sustainable economic development" requires

- "the maintenance of essential ecological processes and life-support systems,
- the [preservation] of genetic diversity, [and]
- measures to ensure the sustainable utilization of species and ecosystems."

This summary of the IUCN position is from Tisdell (1993, p. 164).

[44] For a review of efforts to define sustainable development carefully, see Tisdell (1993).

[45] The quotations attributed to Robert Solow are from "An Almost Practical Step Toward Sustainability," An Invited Lecture on the Occasion of the Fortieth Anniversary of Resources for the Future, Washington, DC, October 8, 1992. Similar views are given in a paper by Solow reproduced in Dorfman and Dorfman (1993, pp. 179–87).

> one that allows every future generation the option of being as well off as its predecessors. The duty imposed by sustainability is to bequeath to posterity not any particular thing . . . [except for certain unique and irreplaceable assets such as Yosemite National Park in California, for example] but rather to endow them with whatever it takes to achieve a standard of living at least as good as our own and to look after their next generation similarly. We are not to consume humanity's capital in the broadest sense.

Solow's conception of sustainable development centers on depletion and investments of total capital. In this context, *capital* refers to assets capable of producing goods and services. Total capital includes not only *natural capital* (such as forests and wetlands), but also human capital and human-made capital. Human capital (or social capital) includes "people, their capacity levels, institutions, cultural cohesion, education, information [and] knowledge.[46] Human-made capital includes roads, factories, equipment, and so forth.

Solow argues that sustainable development requires a commitment to a specifiable amount of productive investment. The minimum required investment is one that offsets withdrawals from the inherited stock of total capital. At a lower level of investment, the current generation would be consuming capital at the expense of its descendants.[47]

This conception of sustainability might be termed *weak sustainability* because it makes no significant distinction between human-made capital and natural capital.[48] Solow lumps natural capital together with machines, factories, and so forth in computing humanity's total capital.

Many ecologists and some economists do not agree with Solow that natural capital can be lumped together with human-made capital in tallying up total capital at any particular time. The source of disagreement concerns the ability to make substitutions. When thinking of capital in the form of machines and technical know-how, one can conceive of producing a particular output level by using less labor and more machines, and irreversibility is of no concern. In contrast, what are the substitutes for used-up natural capital, such as species that go extinct?

Goodland and Daly (1995) use *strong sustainability* to mean development that involves reinvestment of returns from economic development in separate categories for natural capital and human-made capital (see Table 2.4). Thus, for example, if an industrial facility is developed on filled wetlands, some of the economic returns from the facility would go toward rehabilitating previously destroyed wetlands. As another example, part of the income from "depleting oil should be reinvested in sustainable energy production, rather than in any asset" (Goodland and Daley 1995, p. 305).

Complexities related to the uneven distribution of wealth have led some economists to propose definitions of sustainable development that are more complicated than those based on weak and strong

TABLE 2.4 Weak and Strong Sustainability[a]

Weak sustainability—maintain total capital intact.

Example: Oil may be depleted as long as part of the net income is invested to replenish total capital (for example, invest in scientific research and education).

Assumption: Human-made capital and natural capital are substitutes.

Strong sustainability—maintain human-made capital and natural capital intact separately.

Example: Oil may be depleted as long as part of the net income is invested to replenish natural capital (for example, invest in solar energy) as well as total capital.

[a] Material in this table is based on Goodland and Daly (1995, pp. 305–306).

[46] This definition of human capital is from Goodland and Daly (1995, p. 304).

[47] Solow's analysis does not provide numerical values that policy makers can use now to determine whether a particular development path is sustainable. However, it highlights the need to revise the national accounts employed in characterizing economic development, such as gross national product (GNP), so that depletions of natural resources and environmental quality can be included in those accounts. For a summary of efforts to revise national accounting procedures to include depletions of natural resources and environmental "capital," see Ahmad, Serafy, and Lutz (1989).

[48] This usage of weak sustainability is from Goodland and Daly (1995).

sustainability. For example, Pearce, Barbier, and Markandya (1990, p. 2) define *development* as a vector of desirable social outcomes, such as the following:

- increases in real income per capita
- improvements in health and nutritional status
- educational achievement
- access to resources
- a "fairer" distribution of income
- increases in basic freedoms

They then define *sustainable* development as a "situation in which the development vector . . . does not decrease over time" (Pearce, Barbier, and Markandya 1990, p. 2). They complement their definition by also requiring that the stock of natural capital be maintained.

Current debates on the need for sustainable development raise a difficult question: If current consumption must be intentionally curtailed in fairness to future generations, how is the burden of this reduced consumption to be distributed among the rich and poor? Can China and India, for example, fairly be expected to avoid burning their plentiful coal reserves because of the threats of global warming faced by future generations while so many Chinese and Indians have living standards well below those considered average in Europe and North America?

Sustainable development thus involves a confounding interplay of two sets of distributive justice questions: (1) How much should the current generation sacrifice in consumption in order to be fair to future generations? and (2) How should the sacrifice in consumption be shared among those living now? The complexity of these questions, coupled with difficulties in measuring sustainability, confounds attempts to describe, in practical terms, "obligations to future generations."

MORAL AND LEGAL RIGHTS TO A HABITABLE ENVIRONMENT

Until recently, only humans were thought to have rights, moral or otherwise. As the environmental historian Roderick Nash (1989, p. 17) points out, after the advent of Christianity, it was increasingly assumed that "nature, animals included, had no rights and that nonhuman beings existed to serve human beings." Since the 1970s, many scholars have argued that the boundary of moral philosophy should be extended to include certain classes of nonhuman animals. A smaller number of scholars has argued that moral consideration should be extended beyond animals to include plants and natural objects as well.

The discussion that follows takes up the issue of moral and legal rights to a habitable environment. The initial concern is with the legal rights of people in some political jurisdictions to a healthful environment. This is followed by a consideration of rights involving animals capable of experiencing pain and suffering. The discussion concludes by examining arguments supporting the legal and moral obligations owed to animals, plants, natural objects, and ecosystems.

Legal Rights

Some people feel that individuals have a moral right to a livable environment and that this right needs to be preserved at the expense of limiting people's freedom to degrade the environment. Restraints on individual freedom are considered necessary to preserve the human habitability of the planet.

Some states within the United States have made the right to a healthful environment a part of their constitutions. For example, in 1971, Article I of the Constitution of the State of Pennsylvania was amended to read,

> The people have a right to clean air, pure water, and to the preservation of the natural, scenic, historic, and aesthetic values of the environment. Pennsylvania's public natural resources are the common property of all the people, including the generations yet to come. As trustee of these resources, the commonwealth shall preserve and maintain them for the benefit of all the people.[49]

The notion that "each person has a fundamental and inalienable right to a healthful environment" almost made its way into federal law in the United States during the late 1960s. Such a right was specified in the Senate version of the bill that eventually be-

[49] This is Section 27 of Article I of the Constitution of Pennsylvania (as amended on May 18, 1971) as quoted by Brown Weiss (1989, p. 324). Appendix B of that source includes similar sections from the constitutions of several other states.

came the National Environmental Policy Act of 1969, but it was deleted by the House-Senate conference committee that prepared the final bill.[50] Although it does not guarantee an individual's right to a healthful environment, the National Environmental Policy Act states that one of its purposes is to "assure for all Americans safe, healthful, productive, and aesthetically and culturally pleasing surroundings."

Velasquez (1982) argues that the moral right to a healthful environment was a key factor in determining the form of several U.S. environmental programs established during the 1970s. Some provisions of the Federal Water Pollution Control Act of 1972 illustrate his view. The act proclaims, as a national goal, the elimination of the "discharge of pollutants into navigable waters" by 1985. To meet this goal, the U.S. Environmental Protection Agency was required to issue permits to all municipalities and firms that made significant wastewater discharges to streams and other surface waters. In deciding on levels of wastewater discharge to permit by 1983, Congress required EPA to insist on the "best available technology economically achievable" for reducing waste loads at various types of facilities. Water pollution control requirements for a particular firm or municipality were to be based on the most stringent levels of wastewater reduction that, in EPA's view, could be obtained economically with available technology. The utilitarian concern of whether the social benefits exceeded the costs of such exacting controls was not to affect EPA's decision making. The question of who would benefit and who would bear the cost was not to influence EPA's decision process either. Velasquez interprets this federal imposition of demanding controls as being based, implicitly at least, on a recognition of the moral rights of citizens to pure water.

Legal rights to a healthful environment have been included in constitutions outside of the United States, particularly in Latin America. Brown Weiss (1989) includes relevant excerpts from the constitutions of countries in which rights of citizens to a healthful environment is unambiguous. These include Brazil, Chile, Ecuador, Korea, Nicaragua, Peru, and Spain.[51]

Rights to environmental quality are further illustrated by Brazil's constitution. It was promulgated in 1988 and includes an entire section, Article 225, dedicated to the environment. Article 225, which spells out a host of environmental rights and responsibilities, begins as follows:

> Everyone has the right to an ecologically balanced environment, an asset for the common use of the people and essential to the wholesome quality of life. This imposes upon the Public Authorities and the community the obligation to defend and preserve it for present and future generations.[52]

Article 225 lists duties which the government is to perform in order to assure the "effectiveness of this right." These include a variety of actions concerning protection of flora and fauna, preservation of "genetic patrimony," and management of ecosystems. Later in Article 225, the Brazilian constitution stipulates,

> Conduct and activity considered illegal and harmful to the environment will subject violators, whether individuals or juridical entities, to penal and administrative sanctions, independent of the obligations to repair the damage caused.

Article 225 reinforces the "diffuse right to the just preservation of the environment, established by the 1985 law creating the 'public civil action of responsibility for damages caused to the environment, to the consumer, to property and rights of artistic, aesthetic, historic, touristic and scenic value.' "[53] The 1985 law led to a wave of court actions in defense of the specified rights, and the 1988 constitution strengthened those rights by giving them a constitutional foundation. With the 1988 constitution, "the right to protect the environment moves from the government to the entire people." The 1985 environmen-

[50] A similar outcome occured recently in the province of Ontario, Canada. In 1994, the Ontario government proclaimed an Environmental Bill of Rights that improved access to the courts for Ontario residents seeking to protect the environment. Unlike an earlier, related bill that was not enacted, the Environmental Bill of Rights did not grant citizens a right to a healthful environment. For an analysis of the 1994 law, see Walker (1995).

[51] The 1991 constitution promulgated by Colombia also includes such rights for citizens. Brown Weiss's account was prepared in 1989.

[52] From Article 225 of Brazil's 1988 constitution as quoted in Brown Weiss (1989, p. 298). The quotation below is from the same source.

[53] This quotation and the one below are from Chang (1990, p. 405). She also observes that between 1985 and mid-1988, there were over 100 court actions based on provisions of the 1985 law in the state of Sao Paulo alone.

tal law together with the 1988 constitution make it possible for Brazilian citizens, acting as private associations or through state attorneys, to use the courts to enforce their legal right to a healthful environment.

Obligations to Nonhuman Animals

People have rights to a habitable environment, but what about nonhuman animals? The claim that animals should not be mistreated has been advanced for centuries.[54] The discussion that follows summarizes three arguments commonly used to support the view that animals are to be protected to one degree or another. The first, and oldest, of these arguments is anthropocentric. Until the 1970s, the dominant theme in discussions of human obligations to animals was that animals had only *instrumental* value. They were, like inanimate property, available to serve the interests of humans, and they were not of value in and of themselves. From this perspective, there is only one reason that humans should temper their treatment of animals: how animals are treated affects humans. For example, it is wrong for you to kick your neighbor's cat, but not because the cat is wronged. It is wrong because your neighbor is wronged. A related view is that it is wrong for a person to treat animals with cruelty because that person might thereby be more inclined to treat people cruelly.

A second argument is that the human treatment of animals should be tempered because some animals are *sentient beings* (that is, they are capable of experiencing pleasure, pain, suffering, and other feelings). This position was not widely embraced in 1789 when it was advanced by Jeremy Bentham.

> The day *may* come when the rest of the animal creation may acquire those rights which never could have been withheld from them but by the hand of tyranny. The French have already discovered that the blackness of the skin is no reason why a human being should be abandoned without redress to the caprice of a tormentor. It may one day come to be recognized that the number of legs, the velocity of the skin, or the termination of the *os sacrum,* are reasons equally insufficient for abandoning a sensitive being to the same fate. What else is it that should trace the insuperable line? Is it the faculty of reason, or perhaps the faculty of discourse? But a full grown horse or dog is beyond comparison a more rational, as well as a more convertible animal, than an infant of a day or a week, or even a month, old. But suppose they were otherwise, what would it avail? The question is not, Can they reason? nor Can they *talk*? but, *Can they suffer?*[55]

Some contemporary utilitarian philosophers have taken up Bentham's position and used sentience as a basis for determining the appropriate treatment of animals. Utilitarian philosopher Peter Singer (1993, p. 57) advances an argument that uses traditional utilitarian principles together with the proposition that the "capacity for suffering and enjoying things is a prerequisite for having interests. . . ." According to this view, *all* sentient animals have interests, and thus the moral worth of any particular act must involve a consideration of its effects on all sentient beings, not just people.

Singer's application of the utilitarian calculus involves a summing of the benefits and burdens of an action, where the benefits and burdens accrue to both persons and sentient animals. No extra weight is given to effects on humans; each individual (person or animal) counts equally. A morally acceptable act is one that maximizes the sum of individual satisfactions minus the sum of individual burdens associated with the act.

The positions of utilitarians who include all sentient beings in the formula for calculating benefits and costs have been attacked on grounds similar to those used in attacking traditional utilitarians. Typical objections concern practical problems in measuring benefits and costs, and the apparent injustice that results if a small number of people are required to shoulder extraordinary burdens to maximize the aggregate utility of others. In including nonhuman sentient beings in the utilitarian calculus, Singer and others expose themselves to additional criticisms, especially because the logical consequence of this position challenges traditional views concerning how animals are used in medical research and agriculture. Attacks against these utilitarian views (as articulated by Singer) and the usual counterarguments are widely discussed in the literature on animal welfare.[56]

[54] For simplicity, the term "animals" is used here as a shorthand for "nonhuman animals."

[55] Bentham's quotation is as it appears in Regan (1983, p. 95).

[56] See, for example, Attfield (1991), Regan (1983), and Singer (1993).

A third argument commonly used by advocates of animal welfare concerns *animal rights.* The discussion illustrating this position relies on Tom Regan, a contemporary American philosopher who has presented a detailed, widely cited defense of animal rights. Regan uses the concept of "subjects of a life" as the centerpiece of his theory. In one description of this concept, Regan (1985, p. 22) observes that while individuals differ widely in their attributes and abilities, these differences are not (and should not be) used to argue that some individuals "have less inherent value, less of a right to be treated with respect than do others." Regan goes on to say:

> It is the *similarities* between those human beings who most clearly, most non-controversially have such value (the people reading this, for example), not our differences, that matter most. And the really crucial, the basic similarity is simply this: we are each of us the experiencing subject of a life, a conscious creature having an individual welfare that has importance to us whatever our usefulness to others. We want and prefer things, believe and feel things, recall and expect things. And all these dimensions of our life, including our pleasure and pain, our enjoyment and suffering, our satisfaction and frustration, our continued existence or our untimely death—all make a difference to the quality of our life as lived, as experienced, by us as individuals. As the same is true of those animals that concern us (the ones that are eaten and trapped, for example), they too must be viewed as the experiencing subjects of a life, with inherent value of their own.[57]

Regan argues that just as disrespectful treatment of human beings in the name of social good is not sanctioned, so it is that disrespectful treatment of animals that qualify as subjects of a life should not be allowed.

In defending his views, Regan systematically presents a series of opposing arguments and shows them (in his view at least) to be deficient.[58] For example, to the objection that animals have less inherent value than human beings because of their "lack of reason, or autonomy or intellect," Regan responds by examining how we treat humans, such as infants and certain mentally handicapped persons, who lack those attributes.

> [I]t is not true that such humans . . . have less inherent value than you or I. Neither, then, can we rationally sustain the view that animals like them in being the experiencing subjects of a life have less inherent value. *All* who have inherent value have it *equally,* whether they be human animals or not.

The theories of animals rights advocates and those who expand the traditional utilitarian calculus to include sentient beings have had a notable impact on private and public decisions related to animal welfare. Examples include consumer boycotts of fur products and laws governing the treatment of animals used in laboratory experiments. Those theories consider individual animals, not entire species or ecosystems. Political efforts to protect species and ecosystems have relied, to a large degree, on arguments unrelated to moral rights of nonhumans.

Toward Legal "Rights of Nature"

In recent years, legal scholars and others have devoted significant attention to the idea of granting some form of legal consideration to animals and trees as well as ecosystems and species. Although the term "rights" often comes up in such contexts, granting a right is only one way to provide legal recognition and protection. Moreover, the idea of granting an explicit legal right to a nonhuman runs into linguistic and other confusions because of difficulties in determining the "interests" of nonhumans. For these reasons, it is instructive to consider how legal protection can be afforded without introducing the notion of rights explicitly. An example demonstrating how protection is granted to nonhumans without granting rights is given by laws preventing cruelty to animals.[59] If a law prohibits cruel treatment of dogs, for example, it can do so without creating any explicit legal rights that dogs enjoy. For example, a law could provide that certain behaviors of humans towards dogs are prohib-

[57] This quotation and the ones in the following paragraph are from Regan (1985, pp. 22 and 23, respectively).

[58] For this systematic presentation of arguments and counterarguments, see Regan (1983).

[59] The ideas and examples in this paragraph and the next are from Stone (1988, p. 44); the quote in the next paragraph is from the same source.

ited and punishable. Legal scholars might say that such a law created duties toward (or "to") dogs.

Another example concerns a U.S. federal regulation that requires "fishermen who accidentally land sea turtles on their decks to give them artificial respiration." In this case, turtles are granted a measure of legal protection by imposing duties on fishermen that can be enforced (in principle at least) by federal authorities. This further illustrates how it is possible to protect nonhumans without granting them legal rights.

Notwithstanding that nonhuman beings can be protected without granting them rights, some people have argued that particular environmental laws have granted *de facto* legal rights to nonhumans.[60] An example is the Endangered Species Act (ESA), a U.S. law that protects animal species listed as "endangered" or "threatened," and plant species listed as endangered. Based on an analysis of the act, its implementation, and its legislative history, Varner (1987, p. 68) finds it "more natural to understand the intent behind ESA as a grant of legal rights to species themselves than as a grant of any weaker status which does not reflect species' membership in the moral and legal communities." Even if Varner is correct, there have been so many exemptions to ESA's provisions that many observers are skeptical about the strength of its guarantees.[61] Also, most justifications for species preservation laws have an anthropocentric basis and do not involve rights of species.

Christopher Stone, a professor of law at the University of Southern California, developed a provocative analysis of legal obligations to nonhumans. In his widely cited essay *Should Trees Have Standing?,* Stone advocated what he termed the "unthinkable" proposition that "we give legal rights to forests, oceans, rivers and other so-called 'natural objects' in the environment. . . ."[62] Anticipating the immediate reaction of many, Stone added that "to say the environment should have rights, is not to say that it should have every right we can imagine, or even the same body of rights as human beings have." Rather, Stone argued, it was time to explore deliberately the ways in which legal rights for natural objects could be integrated into the legal system, and the implications of doing so.

Stone's ideas for how a system of rights for natural objects might work are modeled after the approach used to provide legal representation for "legal incompetents," such as senile persons or corporations that have gone bankrupt. Thus he calls for "a system in which, when a friend of a natural object perceives it to be endangered, he can apply to a court for creation of a guardianship" (Stone 1974, p. 17). Environmental groups, such as Friends of the Earth and the Izaak Walton League among others, are suggested as possible guardians. In describing advantages of the guardian approach, Stone (1974, p. 24) points out that the guardian of the natural object would have

> a continuous supervision over a period of time, with a consequent deeper understanding of the wards' problems. . . . [This approach] would thus assure the courts that the plaintiff [that is, the guardian representing the natural object] has . . . expertise and genuine adversity in pressing a claim. . . .

The practical effect of Stone's essay cannot be appraised without appreciating the context in which it was prepared. At the time he wrote *Should Trees Have Standing?,* the U.S. Supreme Court was about to consider *Sierra Club v. Morton.*[63] The central question before the Court was whether the Sierra Club had legal standing to bring suit to halt development of a proposed ski resort in Mineral King Valley in Sequoia National Forest in California. A federal circuit court had refused to hear the case on appeal because the Sierra Club, in not demonstrating that it would be injured economically if the ski resort were built, did not have *standing to sue.* In this context, establishing standing required that the plaintiff show that it would suffer injury, economic or otherwise. Because the Sierra Club had not demonstrated that

---

[60] As Varner (1987, p. 66) points out, "[a] de facto legal right is one which exists in a community whose legal system does not explicitly acknowledge it." Varner goes on to argue that animal protection legislation provides some de facto legal rights for animals.

[61] For examples of how ESA has been modified to make it easier to obtain exemptions from it, see Nash (1989, pp. 177–79).

[62] This quotation is from Stone (1974, p. 9). For his later views on the same subject, see Stone (1988).

[63] For details on the case, see the U.S. Supreme Court's opinion as reprinted in Stone (1974).

either the club or its members would be harmed, the federal circuit court held that the Sierra Club lacked standing to sue.

Stone had long been contemplating arguments concerning the rights of natural objects, and the pending Supreme Court review of *Sierra Club v. Morton* provided an opportunity to see if his ideas could be persuasive.[64] In *Should Trees Have Standing?,* Stone made explicit mention of *Sierra Club v. Morton* and highlighted the advantages of viewing Mineral King Valley as the plaintiff and the Sierra Club as the guardian for the area. Although Stone's ideas were not embraced by the majority of the Court, they formed the foundation for the dissent of Justice William Douglas. In his very first paragraph, Douglas proclaimed,

> The critical question of "standing" would be simplified and also put neatly in focus if we . . . allowed environmental issues to be litigated before federal agencies or federal courts in the name of the inanimate object about to be despoiled, defaced, or invaded by roads and bulldozers and where injury is the subject of public outrage. . . . This suit would . . . be more properly named *Mineral King v. Morton.*[65]

Later in his analysis, Douglas identified the "legitimate spokesmen" for Mineral King as

> [t]hose who hike it, fish it, hunt it, camp in it, or frequent it, or visit it merely to sit in solitude and wonderment. . . . Those who have that intimate relation with the inanimate object about to be injured, polluted or otherwise despoiled are its legitimate spokesmen.

The majority of the Court found against the Sierra Club, but it gave the organization an indication of how its standing could have been established. The Sierra Club eventually amended its suit and the case was heard.[66]

[64] Stone's motivation for writing *Should Trees Have Standing?* are articulated in Stone (1974, pp. xii–xiv).

[65] This quote and the one that follows are from Justice Douglas's opinion in *Sierra Club v. Morton* as reprinted in Stone (1974, pp. 73 and 76, respectively).

[66] The Sierra Club was able to establish standing by redrafting its complaint to accentuate how the development would infringe its "associational interests," as well as prove detrimental to some club members (Stone 1988, p. 8).

In commenting on the lack of action by U.S. courts on issues associated with the guardianship concept articulated in *Should Trees Have Standing?,* Stone (1988) pointed to changes in standing requirements in the 1970s. As that decade unfolded, federal courts became increasingly liberal in interpreting the conditions plaintiffs needed to satisfy to establish standing to sue in environmental cases. In addition, many of the U.S. federal environmental laws of the 1970s included explicit provisions granting citizens the right to sue to enforce environmental regulations.

Another development that diminished the immediate practical impact of the guardianship concept is, according to Stone (1988, p. 9), recent legislation that "fortified the government's right to sue private environment despoilers through a revival and expansion of the ancient public trust doctrine. . . ." This doctrine holds that certain interests—particularly the air, the sea, and submerged and submersible lands—are common property that should be made available to the entire citizenry. Governmental bodies act as trustees of public trust resources. The expanded use of public trust doctrine is recent and its scope is still evolving.[67]

This section completes the introduction to the principal criteria used in making decisions affecting environmental quality: productive efficiency, equity, and moral and legal rights to a habitable environment. Of the various criteria introduced, legal rights to a habitable environment and productive efficiency (as implemented using benefit-cost analysis) are the ones applied most often in making environmental decisions.

Practical and philosphical difficulties often stand in the way of formulaic implementation of the other criteria, particularly criteria related to moral obligations and moral rights. Beatley (1994, p. 17) offers a perspective on the difficulty in using moral arguments in the context of environmental problem solving. In his analysis of "ethical land-use planning," he suggests that one should not look to ethics for detailed operational standards to resolve conflicts. Rather, he suggests, it is more appropriate to fashion an ethical orientation that provides direction as to how moral

[67] For more on the evolution of public trust doctrine in the United States, see Plater, Abrams, and Goldfarb (1992).

reasoning might influence decisions affecting the environment.

Key Concepts and Terms

Productive Efficiency and Benefit-Cost Analysis (BCA)

Justifications for BCA
BCA and distributional consequences
Value of a human life
Discounting to present value
BCA and rights to a quality environment

Distinguishing Moral and Legal Rights

Moral agents
Moral principles
Prescriptive vs. descriptive claims
Sentient beings
Bearers of rights

Environmental Justice and Intragenerational Equity

Distributive justice
Principles of justice
Jeremy Bentham and utilitarianism
Balancing of the equities
Robert Nozick and entitlement theory
Libertarianism
Common law of nuisance
Rule of equity
Regressive regulatory programs
Environmental inequity in the United States
International environmental injustice
Debt-for-nature swaps
Toxic waste trade
Maquiladoras

Intergenerational Equity and Sustainable Development

John Rawls and social contract theory
Moral obligations to posterity
Alternative definitions of sustainable development
Sustainable development and equity
Natural capital vs. human-made capital
Total capital
Strong vs. weak sustainability

Moral and Legal Rights to a Habitable Environment

Legal rights to a habitable environment
Animal rights
Christopher Stone's use of guardianship concepts
Rights of species and ecosystems
Standing to sue
Public trust doctrine

Discussion Questions

2-1 Provide an example demonstrating that an exclusive reliance on utilitarianism could lead to an environmental program that either violates people's rights or distributes costs and benefits unfairly, or both.

2-2 Economists argue that the inability of benefit-cost analysis to treat distributional consequences is not a major limitation of the technique. Productive efficiency maximizes the "size of the economic pie"; if the resulting distribution of costs and benefits is unsatisfactory, it can be changed by employing income transfers, such as those possible using income tax provisions. What is your response to this defense of benefit-cost analysis?

2-3 Would you support use of President Reagan's Executive Order 12291 for regulations limiting concentrations of suspected human carcinogens in drinking water supplies? What are the advantages and disadvantages of using benefit-cost analysis in that context?

2-4 What does it mean to say "the costs of implementing emission control requirements for automobiles is regressive?"

2-5 Winona La Duke, who has worked with Native Americans on environmental justice issues for over 15 years, offered the following at a Sierra Club "Roundtable on Race, Justice and the Environment" (see *Sierra,* 78(3), May/June 1993: 55):

> My reservation is in northern Minnesota, 36 miles by 36 miles, located between Bemidji and Fargo, one of seven Ojibwa reservations in the North. We were ceded a huge area under a treaty of 1867, a land we call the White Earth. It's a wealthy land

full of lakes, pinelands, farmlands, prairie, and most of the medicines our people have used for centuries. That's why we don't have it today. By 1920, 90 percent of our reservation was in non-Indian hands, seized by a bunch of illegal land transactions. Most of our people were forced off the reservation and into poverty. Three-quarters of them are refugees.

Use this example to comment on the suitability of Robert Nozick's entitlement theory as a framework for analyzing environmental justice issues.

2-6 Lazarus (1993, p. 800) reports, "The Office of Management and Budget and some federal judges . . . have recently suggested that environmental laws actually undermine public health concerns because they make people poorer, and 'richer is safer.' In other words, an individual with more economic resources (i.e., wealth) is likely to be more healthy than an individual with fewer such resources. Hence, because environmental laws decrease economic wealth (or so proponents of this theory assume), they simultaneously decrease public health." Prepare an analysis either defending or attacking this line of reasoning.

2-7 Explain the rudiments of John Rawls's use of a "social contract theory" to derive principles of justice, and describe the way others have extended Rawls's approach to conclude that we have obligations to protect the environment and conserve natural resources for future generations. Characterize your personal views regarding obligations to future generations.

2-8 Consider a situation in which individuals living near a factory are exposed involuntarily to low concentrations of carcinogens emitted from the factory's smokestack. The residents press for strict emission controls, but those controls would impose a heavy financial burden on the factory. It is impossible to estimate the likelihood that any one individual would contract cancer, which in this hypothetical example is assumed to be a cancer that leads to death soon after it appears. How would you characterize the rights of the residents and the rights of the factory? How do the rights conflict? What moral theories could you use to help you reason to a solution of this conflict? What resolution would you consider just? In what ways is this hypothetical situation similar to the case in *Reserve Mining v. United States?* To what extent does your approach to resolving the conflict differ from the court's "balancing of the equities" in resolving the Reserve Mining Company case?

2-9 The discussion about granting rights to nonhumans can be couched in terms of where the line is drawn for granting rights. Should it be drawn at sentient beings? All flora and fauna? Ecosystems? Where do you draw the line? Explain how and why you made your choice. If the position you took involved granting rights to nonsentient beings, how might controversies involving conflicting rights be settled?

References

Armstrong, S. J., and R. G. Botzler. 1993. *Environmental Ethics: Divergence and Convergence.* New York: McGraw-Hill.

Ahmad, Y. J., S. E. Serafy, and E. Lutz. 1989. *Environmental Accounting for Sustainable Development.* Washington, DC: The World Bank.

Alston, D., and N. Brown. 1993. "Global Threats to People of Color." In *Confronting Environmental Racism,* ed. R. D. Bullard, 179–94. Boston: South End Press.

Attfield, R. 1991. *The Ethics of Environmental Concern.* 2d ed. Athens: University of Georgia Press.

Barbour, I. G. 1980. *Technology, Environment, and Human Values.* New York: Praeger.

Beatley, T. 1994. *Ethical Land Use: Principles of Policy and Planning.* Baltimore: The Johns Hopkins University Press.

Been, V. 1994. Locally Undesirable Land Uses in Minority Neighborhoods: Disproportionate Siting or Market Dynamics? *The Yale Law Journal* 103(6): 1383–1422.

Brown Weiss, E. 1989. *In Fairness to Future Generations: International Law, Common Patrimony and Intergenerational Equity.* Tokyo: The United Nations University, and Dobbs Ferry, NY: Transnational Publishers.

Bryant, B., ed. 1995. *Environmental Justice: Issues, Policies and Solutions.* Washington, DC: Island Press.

Bullard, R. D. 1990. *Dumping in Dixie: Race, Class and Environmental Quality.* Boulder, CO: Westview Press.

Bullard, R. D. 1993a. "Conclusion: Environmentalism with Justice." In *Confronting Environmental Racism,* ed. R. D. Bullard, 195–206. Boston: South End Press.

Bullard, R. D. 1993b. "Anatomy of Environmental Racism and the Environmental Justice Movement." In *Confronting Environmental Racism,* ed. R. D. Bullard, 15–39. Boston: South End Press.

Campen, J. T. 1986. *Benefit, Cost and Beyond.* Cambridge, MA: Ballinger.

Chang, L. 1990. The New Emerald Hunters: Brazilian Environmental Jurisprudence, 1988–1989. *Georgetown International Environmental Law Review* III(2): 395–416.

Collins, R. A. 1977. The Distributive Effects of Public Law 92-500. *Journal of Environmental Economics and Management* 4(4): 344–54.

Copp, D. 1985. "Morality, Reason and Management Science: The Rationale of Cost-Benefit Analysis." In *Ethics and Economics,* eds. E. F. Paul, F. D. Miller, Jr., and J. Paul 128–51. Oxford: Basil Blackwell.

Dorfman, R., and N. S. Dorfman, eds. 1993. *Economics of the Environment: Selected Readings,* 3d ed. New York: W. W. Norton.

EPA (U.S. Environmental Protection Agency). 1992. *Environmental Equity, Reducing Risk for All Communities.* Vol. 1, Report No. EPA 230-R-92-008. Washington, DC.

Goodland, R., and H. Daly. 1995. "Environmental Sustainability. In *Environmental and Social Impact Assessment,* eds. F. Vanclay and D. A. Bronstein, 303–22. New York: Wiley.

Green, R. M. 1980. "Intergenerational Distributive Justice and Environmental Responsibility." In *Responsibilities to Future Generations,* ed. E. Partridge, 91–102. Buffalo, NY: Prometheus Books.

Harrison, D., Jr., and D. L. Rubinfeld. 1978. The Distribution of Benefits from Improvements in Urban Air Quality. *Journal of Environmental Economics and Management* 5: 313–32.

IUCN. 1980. *The World Conservation Strategy: Living Resource Conservation for Sustainable Development.* Glaud, Switzerland: IUCN.

Karpf, B. 1982. American Textile Manufacturers Institutes v. Donovan. *Ecology Law Quarterly* 10(1): 87–96.

Lazarus, R. J. 1993. Pursuing Environmental Justice: the Distributional Effects of Environmental Protection. *Northwestern University Law Review* 87(3): 787–857.

Mappes, T. A., and J. S. Zembaty. 1992. *Social Ethics: Morality and Social Policy.* 4th ed. New York: McGraw-Hill.

Narveson, J. 1988. *The Libertarian Idea.* Philadelphia: Temple University Press.

Nash, R. 1989. *The Rights of Nature.* Madison: University of Wisconsin Press.

Nozick, R. 1974. *Anarchy, State, and Utopia.* New York: Basic Books.

Partridge, E., ed. 1980. *Responsibilities to Future Generations.* Buffalo, NY: Prometheus Books.

Passmore, J. 1974. *Man's Responsibility for Nature, Ecological Problems and Western Traditions.* London: Duckworth.

Pearce, D., E. Barbier, and A. Markandya. 1990. *Sustainable Development: Economics and Environment in the Third World.* Hants, U.K.: Edward Elgar.

Plater, Z. J. B., R. H. Abrams, and W. Goldfarb. 1992. *Environmental Law and Policy.* St. Paul, MN: Westview.

Rawls, J. 1971. *A Theory of Justice.* Cambridge, MA: Harvard University Press.

Regan, T. 1983. *The Case for Animal Rights.* Berkeley: University of California Press.

Regan, T. 1985. "The Case for Animal Rights." In *In Defense of Animals,* ed. P. Singer. Oxford: Basil Blackwell.

Sagoff, M. 1988. *The Economy of the Earth.* Cambridge, U.K.: Cambridge University Press.

Singer, P. 1993. *Practical Ethics.* 2d ed. Cambridge, U.K.: Cambridge, University Press.

Stewart, R. B., and J. E. Krier. 1978. *Environmental Law and Policy.* 2d ed. Indianapolis: Bobbs-Merrill.

Stone, C. D. 1974. *Should Trees Have Standing?* Los Altos, CA: William Kaufmann.

Stone, C. 1988. *Earth and Other Ethics.* New York: Harper & Row.

Thompson, T. 1980. "Are We Obligated to Future Others?" In *Responsibilities to Future Generations,* ed. E. Partridge, 195–202. Buffalo, NY: Prometheus Books.

Tisdell, C. 1993. *Environmental Economics: Policies for Environmental Management and Sustainable Development.* Hants, U.K.: Edward Elgar.

U.S. General Accounting Office. 1983. *Siting of Hazardous Waste Landfills and Their Correlation with Racial and Economic Status of Surrounding Communities.* Washington, DC.

United Church of Christ Commission for Racial Justice. 1992. *Toxic Waste and Race in the United States.* New York: United Church of Christ.

Van De Veer, D., and C. Pierce. 1994. *The Environmental Ethics and Policy Book.* Belmont, CA: Wadsworth.

Varner, G. E. 1987. Do Species Have Standing? *Environmental Ethics* 9(1): 57–72.

Velasquez, M. G. 1982. *Business Ethics: Concepts and Cases.* Englewood Cliffs, NJ: Prentice-Hall.

Walker, S. 1995. "The Ontario Environmental Bill of Rights," In *Environmental Rights: Law, Litigation and Access to Justice,* eds. S. Deimann and B. Dyssli. London: Cameron May Ltd.

Wenz, P. S. 1988. *Environmental Justice.* Albany: State University of New York Press.

World Commission on Environment and Development. 1987. *Our Common Future.* Oxford: Oxford University Press.

Chapter 3

Participants in Environmental Regulation

The process of implementing environmental policy varies among and within countries. Notwithstanding the variations, there are commonalities in the forces affecting agencies that implement environmental regulations. These agencies, which are sometimes called *environmental protection agencies,* have short histories.[1] In most countries, the agencies did not exist prior to 1970.

An environmental protection agency operates in a highly constrained context. Typically, such an agency is drawn in opposite directions by those who believe it is either too strict or not strict enough in implementing regulations. Judicial and legislative bodies often oversee the environmental agency's activities, and other agencies and business interests try to keep the environmental regulatory body at bay, lest it interfere with attainment of their own goals. Citizens, acting individually or as members of *nongovernmental organizations* (NGOs), can also influence an environmental agency's behavior.

This chapter characterizes the forces influencing agencies that carry out environmental policies. The first section introduces the U.S. Environmental Protection Agency (EPA) and the many public and private bodies that constitute its *organizational environment.* The second section examines EPA's *rulemaking,* the formal process EPA uses in developing regulations to implement environmental statutes. These first two sections highlight the increasingly important roles played by NGOs in policy design and implementation.

The discussion of EPA is put in a broader context in the third section, which outlines the growth of national-level environmental protection agencies outside of the United States. The fourth and final section continues the international theme. It introduces the United Nations Environmental Programme (UNEP) and analyzes the roles of UNEP and NGOs in helping to solve global environmental problems.

U.S. ENVIRONMENTAL PROTECTION AGENCY AND ITS INSTITUTIONAL SETTING

The principal groups affecting the U.S. Environmental Protection Agency include the president and the president's Executive Office, Congress, the federal courts, state governments, the regulated community,

[1] Names of these governmental bodies vary from country to country. Other common names are "ministry of the environment" and "environment department."

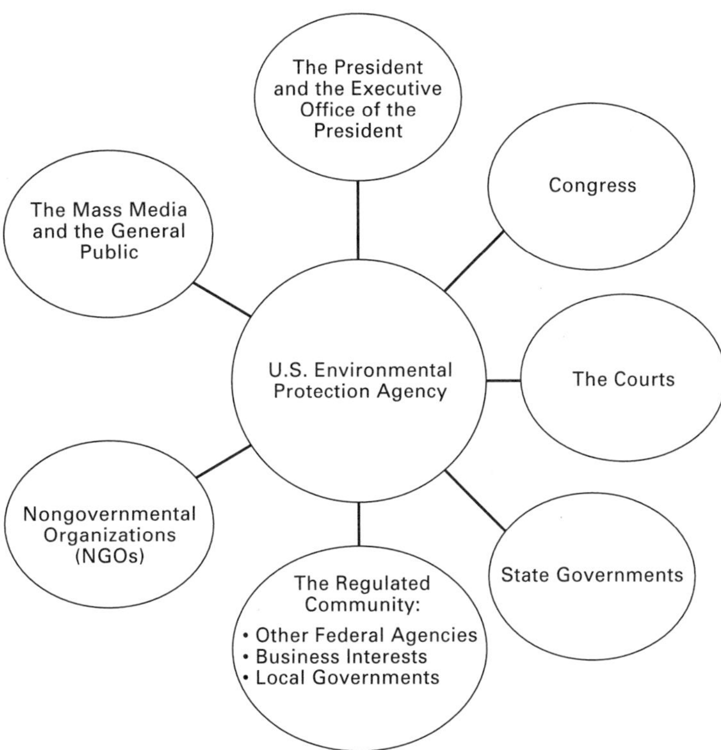

Figure 3.1 The U.S. Environmental Protection Agency and its organizational environment.

NGOs, the mass media, and the general public (see Figure 3.1)[2]. The *regulated community* (also referred to as *regulatees*) includes those subject to EPA's regulations: firms and other private organizations, state and local governments, and other federal agencies. The discussion begins with an introduction to EPA itself.

U.S. Environmental Protection Agency

By the late 1960s, environmental quality degradation had emerged as a politically significant issue. Richard Nixon, who was president at that time, was eager to develop an institution that could effectively deal with environmental problems in a way that was compatible with other policy objectives, particularly economic growth. In 1969, Nixon called for the first of several studies on how the executive branch could best be organized to manage the environment.

The culmination of studies of alternative organizational arrangements was a proposal to create a single environmental agency by bringing together environmental programs that already existed in several federal departments.[3] The new agency was intended to have high visibility, direct access to the president, and the ability to take a comprehensive and systematic approach to pollution.

On July 9, 1970, Nixon used presidential authority to reorganize the federal government and submitted a proposal to Congress to create the Environmental Protection Agency. Under the Reorganization Act of 1939, Congress had 60 days to disapprove of the proposed changes. Since neither house opposed

[2] Helpful critiques of the first two sections of this chapter were provided by Allison Moore, while she was a student at Stanford University, and Patti Collins of the Region 9 office of the U.S. Environmental Protection Agency.

[3] For accounts of the origins and history of EPA, see Landy, Roberts, and Thomas (1994), and Fiorino (1995).

Nixon's proposal, the plan to create EPA went into effect on September 9, 1970.

Ten existing programs were transferred to the newly formed EPA. The largest of these included the water quality management program within the Department of the Interior, and the air quality and solid waste management programs within the Department of Health, Education, and Welfare. Other programs that were part of the reorganization concerned pesticides, radiological health, and water hygiene. In total, EPA "inherited" a staff of about 5200 people and a 1971 (fiscal year) budget of about $1.1 billion.[4]

EPA's principal role is to carry out environmental laws that Congress asks it to implement. In doing so, the agency develops policies and regulations, conducts research and monitoring activities, imposes sanctions, and engages in numerous other activities. In addition to responding to legislative mandates, EPA can influence legislation by proposing new programs to Congress and by informing Congress of actions that could be taken to avoid future environmental problems.

The agency maintains headquarters in Washington, DC, which is where its policy design, rulemaking, and program evaluation activities are centered. EPA's daily interactions with state-level environmental agencies and with firms and other entities subject to EPA rules take place at the agency's regional offices. EPA has ten regional offices distributed across the country, with each regional office having responsibility for a particular group of states.

By the end of its first decade (in 1980) EPA had a staff of about 12,000 and an annual budget of approximately $5 billion.[5] Budget figures for EPA can be misleading because they are greatly influenced by subsidy programs administered by the agency. For example, of the more than $5 billion budgeted for EPA in 1980, $1.5 billion was spent to operate the agency itself. The remainder was passed through EPA as grants for state and local governments (for example, grants for construction of municipal wastewater treatment plants).

[4] These details on EPA's initial staff and budget are from Marcus (1980, p. 36).

[5] Facts in this paragraph are from Portney (1990, p. 10). During the 1980s, EPA acquired many new responsibilities, particularly in cleaning up hazardous waste disposal facilities. Although EPA lost staff during the first Reagan administration (1980–84), by 1994 the agency's staff numbered about 18,500. The agency's budget for fiscal year 1994 was $6.6 billion (Fiorino, 1995, p. 43).

EPA's budget represents only a small fraction of the costs of complying with its environmental quality requirements. Much larger sums are spent by firms and governments to buy cleanup equipment, run wastewater treatment plants, remediate contaminated landfills, and so forth. Table 3.1 shows the costs of complying with EPA's regulatory programs. It indicates that from 1972 to 1990, annual compliance costs (in 1990 dollars) jumped from $30 billion to $115 billion. The latter represented about 2.1% of the gross national product for 1990.

Details on how funds to comply with EPA's programs were spent in 1987 are given in Table 3.2. The two largest expenses were for controlling point sources of water pollution and stationary sources of air pollution (see Figure 3.2). The next largest cost was for managing solid wastes. the costs for dealing with hazardous waste (as a fraction of total cost) is expected to increase, because of increasing expenditures for complying with programs under the Comprehensive Environmental Response, Compensation and Liability Act (CERCLA, also known as *Superfund*) and the Resource Conservation and Recovery Act (RCRA).

Like all bureaucracies, EPA is concerned about the size of its budget and staff. However, the agency's resources and well-being are not guaranteed if it simply carries out vigorously the mandates assigned to it by Congress. It must be careful not to go too far in promulgating and enforcing its regulations, lest it alienate business interests and other regulated groups. If regulated organizations feel they are being treated unreasonably, they typically respond by lobbying Congress to have EPA relax its grip, and this can lead to reduced budget and power for the agency. Moreover, EPA cannot be effective if a large segment of the regulated community chooses to be uncooperative—for example, by violating permit conditions or not submitting required self-monitoring reports. While it has sanctions to deal with failure to comply with its rules, EPA has neither the staff nor the budget to apply sanctions vigorously against the thousands of organizations subject to its regulations. Moreover, EPA lacks the political clout to implement an enforcement effort that makes no compromises. Because EPA must gain the cooperation of the large majority of those regulated in order to be effective, it relies heavily on negotiation and bargaining in implementing its programs. The agency also tries to

TABLE 3.1 Cost of All Environmental Pollution Control Activities in the United States[a]

Total Annualized Costs	1972	1987	1990	2000[b]
In billions of 1986 dollars	26	85	100	160
In billions of estimated 1990 dollars	30	98	115	185
As percent of GNP	0.9	1.9	2.1	2.8

[a] From U.S. Environmental Protection Agency (1991). Capital expenditures were converted to an equivalent stream of annual costs using an interest rate of 7 percent and a useful life of capital facility that varies with facility type (e.g., 30 years for wastewater treatment plants).

[b] Figures for the year 2000 are estimates based on full compliance with EPA requirements based on programs established by 1991.

educate regulated organizations on what they can do to comply with requirements.

Of course, EPA is not free to err on the side of being too lax, because that would provoke criticism (and possibly lawsuits) from environmental groups and counter-reactions from the congressional committees responsible for drafting environmental statutes. When environmental groups feel EPA is not carrying out laws properly, they lobby Congress, and they frequently use the federal courts to sue the EPA

TABLE 3.2 Cost to Comply with EPA Programs in 1987[a]

	Cost in Billions of 1986 Dollars
Water:	
Point sources of pollution	33.6
Nonpoint sources of pollution	0.8
Drinking water	3.1
Air:	
Stationary sources of pollution	19.0
Mobile sources of pollution	7.5
Undesignated	0.3
Solid Waste (Resource Conservation and Recovery Act, RCRA):	16.7
Hazardous Waste:	
RCRA	1.7
Superfund	0.7
Pesticides:	0.5
Toxic chemicals:	0.4
Other (including otherwise unallocated EPA administrative costs):	0.8
Radiation:	0.3
TOTAL:	85.4[b]

[a] Based on U.S. Environmental Protection Agency (1991).

[b] The EPA's total cost for 1987 is reported as $85.29 million; the total shown in the table differs slightly as a result of rounding off all costs to three significant figures.

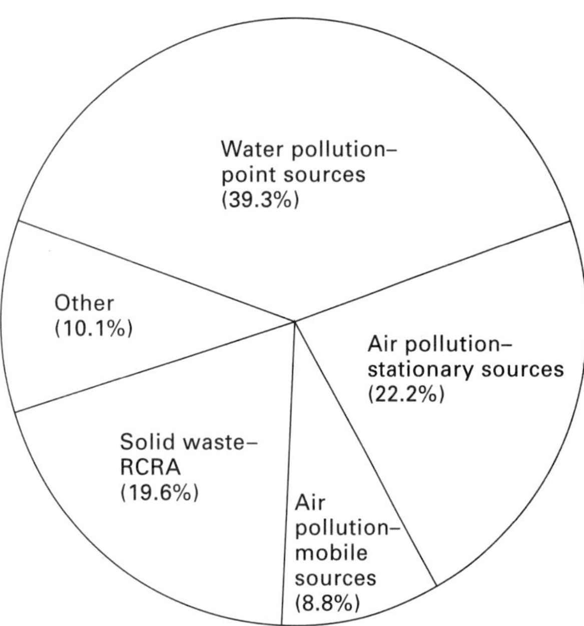

Figure 3.2 Distribution of costs to comply with EPA programs in 1987. Based on material in U.S. Environmental Protection Agency (1991, pp. 2-2–3-5).

administrator. Congress has many options in dealing with a lax EPA. For example, it can amend laws to reduce the degree of discretion that the EPA administrator has in carrying out legislative mandates. Congress can also be vigorous in its oversight, especially when it feels EPA is in danger of being "captured" by the regulated community. For example, in the 1980s, when EPA officials allegedly made sweetheart deals with firms responsible for hazardous waste site cleanups under the Superfund program, Congress stepped up its oversight activities. Thus EPA must balance between being hard on those it regulates and caving in to pressures for regulatory relief.

Congress tries to control EPA by including in statutes various policies and criteria to guide the agency's actions. However, there are several reasons why statutory language may be ambiguous and therefore in need of interpretation by EPA. Sometimes statutes are vague because of compromises struck during the lawmaking process; ambiguity permits flexibility in interpretation and may allow legislative opponents to reach a consensus on a bill. Often the complexities of environmental problem solving lead Congress to enact statutes with ambiguous provisions. Legislators and their staffs typically lack the time and expertise to fully clarify scientific and technical issues, and thus the task of working out details during implementation is left to EPA. For example, in the 1970s, when Congress called on wastewater discharges to install the "best available technology" for abating water pollution, it gave EPA the authority to define what was meant by "best available." In addition, Congress typically leaves it to EPA to detail the strategy and tactics of implementation. Those decisions are reflected in court-enforceable regulations issued by EPA. The agency is also granted latitude in deciding when to impose sanctions available to enforce compliance with environmental laws. Because of its limited resources, EPA must apply sanctions selectively, and this ability to choose is a source of power.

The subsections that follow introduce, in a more systematic way, the principal elements of EPA's institutional setting. They provide an overview of the forces at work in the design and implementation of environmental management programs. Although the discussion concerns the U.S. Environmental Protection Agency, similar forces often affect state and local environmental agencies and environmental agencies in other countries.

The President and the Executive Office

Even though EPA is part of the executive branch, the president has only limited means to control it. One way is through the selection of the EPA's administrator and top-level staff. Indeed, the choice of administrator often sends a clear message. Thus, when President Ronald Reagan appointed Anne Gorsuch as his first EPA administrator, he signaled his intention of reducing EPA's control over business. Gorsuch was selected, in part, because of her loyalty to the president and to his vision of reduced federal interference with the private sector. Similarly, when Reagan appointed William Ruckelshaus (who was respected by many environmentalists) as his second EPA administrator, the message was that EPA would get back on track in enforcing environmental laws. The well-chronicled ineffectiveness of EPA during the Gorsuch period had become a major political liability.[6]

In addition to selecting top-level staff at EPA, the White House frequently attempts to influence both policy direction and individual decisions. Often it is successful, and numerous examples are given in insider accounts such as those by Quarles (1976) and Burford (1986). However, some EPA administrators have managed to resist White House pressure. An example is Russell Train, an EPA administrator during the Nixon administration, who was able to obtain from the White House "written confirmation that the EPA administrator had final authority over the substance of all EPA regulations."[7]

Another way the White House tries to control EPA is through preparation of the president's budget proposals.[8] Budget setting is guided by the Office of Management and Budget (OMB), which is part of the Executive Office of the President. The influence of budgetary control is illustrated by the early years of the Reagan administration. Between 1981 and 1983, EPA's budget had been slashed by nearly one-third, and it was forced to cut back significantly on staff.[9] This greatly curtailed EPA's ability to carry out its monitoring and enforcement work.

In addition to influencing the president's budget, OMB also affects EPA's rulemaking, a process (described in the next section) that yields regulations having the force of law.[10] Since 1971, attempts had been made to require an OMB review of EPA's proposed regulations to ensure that economic and fiscal concerns (such as inflation) were considered in rulemaking. This review was made mandatory in February 1981 when President Reagan issued Executive Order 12291 requiring *all* regulatory bodies to prepare regulatory impact analyses (RIAs) for new (or revised) regulations expected to have an impact exceeding $100 million.

A key provision of Executive Order 12291 was that "regulatory action shall not be undertaken unless the potential benefits to society from the regulation outweigh the costs," and OMB was given responsibility for enforcing this requirement. According to several analyses of OMB performance, Executive Order 12291 served largely as a basis for promoting deregulation. Limitations in both scientific data and methods for evaluating the monetary benefits associated with improved environmental quality made it easy to misuse the RIAs to advance a political agenda centered on providing business interests with relief from regulations.[11]

Congress

Congress provides EPA's reason for being by enacting legislation directing the president to carry out the programs EPA implements. In addition, Congress controls EPA's budget and provides the agency with an important source of power: the authority delegated to the EPA administrator to interpret details of statutes. Congress is also a key source of criticism of EPA's performance.[12]

One way Congress can control EPA is to give it clear implementation standards and policy guidance when passing laws.[13] This includes the specification of *nondiscretionary* duties for the EPA's administra-

[6] For an account of events associated with the departure of Anne Burford (whose name before marriage was Anne Gorsuch) and the arrival of William Ruckelshaus, see Landy, Roberts, and Thomas (1994), and Burford (1986).

[7] This quotation is from Landy, Roberts and Thomas (1994, p. 38).

[8] The president's budget is submitted to Congress in the beginning of each calendar year for appropriations in the next fiscal year (which starts in October). Although Congress makes final budget decisions, the president's budget indicates the executive branch's priorities. For more on how the administration's budgetary process affects EPA, see Fiorino (1995, pp. 70–71).

[9] For details on EPA's budget and staff between 1981 and 1983, see Conservation Foundation (1984, pp. 388–90, and 1987, p. 27).

[10] Note the following distinctions: EPA promulgates *regulations,* which are a form of administrative law, to implement *statutes.* Regulations are binding. The EPA also issues nonbinding *guidance* to assist regulated organizations in complying with statutory requirements. As this discussion demonstrates, the ability of the president to issue executive orders provides another source of White House control over agencies.

[11] For documentation of the misuse of RIAs, see Rosenbaum (1991, p. 133) and studies cited therein.

[12] The discussion that follows on Congress's relations with EPA is limited to congressional checks on EPA performance. For an introduction to Congress's policy development activities concerning environmental matters, see Rosenbaum (1991, chap. 3). A more complete account of congressional oversight of environmental policy is given by Fiorino (1995, pp. 63–69).

[13] Another form of congressional control is the requirement that the Senate approve the president's choice of top-level officials, including the EPA administrator.

tor.[14] As detailed in the following pages, environmental groups have often used the courts to force EPA to carry out statutory provisions that are not subject to the administrator's discretion. Congress has also forced action by EPA using *hammer provisions,* portions of a statute that spell out what amounts to a default regulation. If EPA fails to issue a rule by a stipulated date, the "hammer" falls and the default regulation in the statute applies.

Hammer provisions were first introduced in the 1984 amendments to the Resource Conservation and Recovery Act, in which Congress specified minimum regulatory controls that would apply if EPA failed to issue regulations by specified deadlines. For example, before the 1984 amendments were enacted, *small-quantity generators,* defined as entities generating between 100 kg/month and 1000 kg/month of hazardous waste, were not regulated. In the 1984 amendments to RCRA, Congress required EPA to issue regulations on small-quantity generators by March 31, 1986. It also specified regulatory controls, such as requirements for keeping track of wastes, that would apply for small-quantity generators if EPA failed to meet the deadline. This hammer provision ensured that there would be some level of control over small-quantity generators by March 31, 1986; it did not relieve EPA of its obligation to promulgate more comprehensive regulations later. In this instance, EPA issued its rule on time and the hammer provisions were not invoked. Fortuna and Lennet (1987, p. 116) observe, "EPA's timely behavior was, in large part, due to the existence of the minimum regulatory control provision, [and] thus the provision accomplished its intended purpose."[15]

Congress is able to monitor EPA's implementation of environmental statutes by calling for reports on progress and requesting appraisals of performance from the General Accounting Office. Moreover, congressional committees and subcommittees frequently hold hearings that allow Congress to monitor EPA's implementation of a statute or to amend a statute. These hearings give Congress a chance to hear from all interested parties, including those regulated by its laws.

Although Congress frequently criticizes inadequacies in EPA's performance, many analysts believe shortcomings are inevitable because Congress asks EPA to do more than it possibly can with the information, resources, and time at its disposal. The result is a consistent record of missed deadlines in issuing regulations and a failure to meet cleanup goals. Moreover, Congress itself has routinely amended its own laws to postpone deadlines for compliance with environmental requirements.

Why does Congress appear to have a habit of extending its own deadlines? Reasons include Congress's concern over economic hardships caused by strict implementation of its mandates, and recognition of the many administrative and judicial delays that often plague implementation of complex environmental laws. Rosenbaum (1991, pp. 116–17) adds that an equally important reason is the "lack of political will to face the consequences of the commitments the Congress has made to enforce the law." In effect, Congress shifts to EPA the responsibility for the politically unpopular decisions that must often be made in curbing pollution. Rosenbaum (1991, p. 114) elaborates on this perspective:

> Environmental legislation is often vague and contradictory because Congress cannot or will not resolve major political conflicts entailed in the law. Instead, Congress often papers over the conflict with silence, confusion, or deliberate obscurity in the statutory language. This results in a steady flow of political "hot potatoes" to the bureaucracy, which must untangle and clarify this legal language—often to the accompaniment of political conflict and legislative criticism—or leave the job to the courts. The EPA thus becomes enmeshed in protracted litigation and political bargaining, and program regulations essential to implementing the law often are held hostage to these procedures. Moreover, many regulations can be competently formulated only after research that might require years or decades.

[14] The question of what constitutes a nondiscretionary duty is frequently debated and sometimes the subject of lawsuits. Difficulties involved in deciding which acts are discretionary are apparent from a definition in *Black's Law Dictionary* (1990, 6th ed., St. Paul, NM: West Publishing Co.): *discretionary acts* are those "wherein there is no hard and fast rule as the course of conduct that one must or must not take . . . ." Unless a statute includes unambiguous rules dictating actions the EPA administrator must take, there may be grounds for dispute.

[15] The 1984 amendments to RCRA were designed, in part, to bring EPA back in line after a period in which it had "missed deadlines, proposed inadequate regulations and even exacerbated the hazardous waste problem by suspending certain regulations." Senator George J. Mitchell as quoted by Harris, Want, and Ward (1987, p. 87).

The Courts

In the course of settling disputes among litigants, the courts have influenced both the formulation and implementation of policy by EPA. Court actions are brought against the EPAs administrator frequently; plaintiffs are generally either environmental groups or business interests. When environmental groups bring suit, they frequently allege that EPA either failed to carry out nondiscretionary provisions of an environmental statute or failed to implement a provision properly. Common examples of the latter are suits that claim a particular discharge standard was set too low to achieve statutory goals. In contrast, when business interests bring suit, they often charge that a standard was set inappropriately high, considering the costs and benefits involved in attaining the standard.

Sometimes the courts' actions in interpreting a statute have effects that go well beyond a particular standard. An example is given by the 1977 decision of a federal district court in *Sierra Club v. Ruckelshaus,* which concerned regions that had air cleaner than required by the national ambient air quality standards.[16] Plaintiffs argued that the Clean Air Act of 1970 did not permit air quality in such regions to be degraded all the way down to the national standards. The court found in favor of the Sierra Club, and its decision forced EPA to develop a policy for the *prevention of significant deterioration* (PSD) of air quality in regions having air quality contaminant levels lower than national standards. The policy allows deterioration but limits the amount. In reviewing the courts' role in interpreting the Clean Air Act, Melnick (1983, p. 73) argued that the EPA policy on PSD "was born in the courtroom and has resided there almost constantly for the past decade."

Another example of how court actions influence EPA concerns use of controls on transportation and land use in attaining ambient air quality standards. In 1972, the state of California submitted to EPA its plan to meet Clean Air Act requirements. EPA rejected the plan because it would not meet national ambient air quality standards in major metropolitan areas such as Los Angeles and San Francisco. However, EPA did not issue a substitute plan as required by the Clean Air Act.

A lawsuit was brought in response to EPA's failure to carry out the letter of the Clean Air Act. In *City of Riverside v. Ruckelshaus,*[17] plaintiffs argued that because the state's plan would not meet applicable standards for photochemical oxidants, the EPA administrator was obligated by law to issue land use and transportation control plans that would lead to attainment of the standards. The court found in favor of the plaintiffs and ordered the administrator to issue the required regulations, including those necessary to control land use and transportation. The administrator's response to the Court's decision included a plan that contemplated such extreme measures as gasoline rationing and the imposition of fees on parking spaces to provide an 82-percent reduction in the quantity of gasoline sold in Los Angeles.

EPA's proposed plan would have led to radical changes in lifestyle for the citizens of Los Angeles. Criticism of the plan was intense. The mayor of Los Angeles called the plan "asinine" and "silly," and a senator from California said it would lead to "economic and social chaos."[18] The public outcry was followed by a court action challenging EPA's land use and transportation control regulations.[19] The court found against EPA and spared the agency from having to go forward with the unpopular program. EPA recognized, well before *City of Riverside v. Ruckelshaus,* that it would be thoroughly criticized if it attempted to impose its will on local land use and transportation decision makers. Indeed, in 1972, when EPA unveiled its proposed transportation controls for Los Angeles, it explained publicly that the court and the statute left it no choice.[20]

[16] *Sierra Club v. Ruckelshaus* was first heard in May 1972 in the federal District Court for the District of Columbia. The court's decision survived appeals all the way to the Supreme Court. For an analysis of the court actions and EPA's follow-up activities, see Melnick (1983, chap. 4).

[17] The case was heard in the District Court of the Central District of California (4 E.R.C. 1728, C.D. Cal 1972); see Melnick (1983, chap. 9) for a full assessment.

[18] Mayor Sam Yorty and Senator John Tunney as quoted by Marcus (1980, pp. 133–34).

[19] This case, *Brown v. EPA,* was heard in the United States Court of Appeals, Ninth Circuit (1975), 521 F. 2d. 827. The court decided that EPA could not force the state of California to carry out a plan that EPA itself had written. EPA's position represented a serious threat to traditional state's rights. For more on land use and transportation regulations, see Melnick (1983, chap. 9).

[20] For an elaboration of EPA's views on the matter, see Quarles (1976, p. 202).

Courts have been used frequently for tactical purposes by both environmental groups and regulated parties. When EPA issues a regulation to implement a statute, a court challenge often follows, and if the agency loses it might be forced to change its initial regulation. The amended regulation might lead to additional lawsuits brought by interests who lost ground as a result of the first court's decision. Wenner (1982) analyzed 1900 environmental court actions brought in the United States during the 1970s and found that protracted legal battles were common.

State Governments

Many of the regulatory activities carried out by EPA were once the exclusive domain of state governments, and even as the federal role expanded, states carried out many of their own programs. The federal government had almost no role in environmental protection before enactment of the Federal Water Pollution Control Act of 1948. From that point on, the federal government began to assume an increasingly significant role in regulating air and water pollution. By the early 1970s, the traditional influence of state governments in these regulatory arenas had been overshadowed. Under the Reagan administration, the pendulum began to swing back. Reagan's *new federalism* aimed to return to the states the authority for a variety of domestic programs, including those concerned with managing environmental quality.[21]

Several justifications have been offered for placing increased authority for environmental policy with the states.[22] First, state government is seen as more flexible and responsive to local conditions and needs. In addition, states, acting independently, are likely to develop innovative policy tools and experiment with new approaches to environmental management. Finally, returning authority for environmental programs to the states would reduce what some felt was an imbalance in the traditional sharing of environmental authorities between federal and state governments.

The influence of the Reagan years is seen in the sharp increase in the number of federal air and water quality programs that came to be administered by state agencies.[23] Federal air and water pollution control legislation generally permitted the possibility of state environmental agencies implementing many federal programs. (For example, qualified states could issue wastewater discharge permits under the *National Pollution Discharge Elimination System.*) When states assumed this role, EPA and its counterpart at the state level became partners in program implementation. EPA retained a policy formulation and oversight responsibility for what constituted a national air or water quality management program, and state environmental agencies took on responsibility for implementation. National laws permitted this division of responsibility, but only if a state environmental agency could demonstrate that it had the statutory authority, staff, monitoring capabilities, and other resources required for program implementation.

During the 1980s, many states took the initiative by formulating innovative environmental policies that were often more rigorous than their federal counterparts. This is illustrated by California's tough requirements for clean fuels for motor vehicles, which were later adopted by eight northeastern states. These clean fuels provisions provided the basis for reformulated fuel requirements in the federal clean air legislation of 1990. As demonstrated by Ringquist (1993), many states have become innovators of environmental programs, and some have been at the forefront in efforts to protect the environment. Moreover, policy experiments at the state level have yielded innovative management strategies that have diffused widely among the states and influenced federal environmental policy design.

---

[21] *Federalism* refers to the distribution of power among different levels of government. In a federalist nation, power is distributed among a number of constituent territorial units (typically states or provinces). For case studies demonstrating how federalism works in the context of environmental policy in the United States, see John (1994).

[22] Material in this paragraph is from Ringquist (1993, pp. 44–45). Persuasive arguments can also be made for a strong federal role in environmental regulation. For example, without national regulations, states might try to attract new industry by adopting relatively weak environmental standards. For an analysis of arguments for and against a strong federal role in environmental protection, see Stewart (1977).

[23] According to Ringquist (1993, p. 61), during the first three years of the Reagan administration (1981–83), "the delegation of [federal] environmental programs to the states doubled from 33 percent to 66 percent of all eligible programs." Ringquist (1993, pp. 62–65) also points out that between 1979 and 1988, federal support for state air and water pollution control programs fell by 54 and 68 percent, respectively. States responded by introducing new sources of revenue.

The Regulated Community

Rules issued by EPA influence a wide range of organizations. In addition to affecting private firms, EPA's actions also impact state and local governments and a host of federal agencies. Each of these groups is considered here.

State and Local Governments

Although EPA has numerous occasions to form alliances with state environmental agencies, EPA's relationships with state governments are not always harmonious. In addition to issuing regulations that affect the ways state agencies implement federal environmental programs, EPA also issues regulations that must be complied with by state agencies themselves.

The tension that can exist between EPA and both state and local authorities is demonstrated by events associated with EPA's transportation and land use control plans. As noted above, EPA tried to avoid issuing the plans even though the Clean Air Act required the agency to do so. EPA officials recognized that control over land use and transportation was exercised traditionally by state and local governments. It took a court decision to force EPA to issue land use and transportation control plans. Moreover, even as it issued the plans, EPA announced that it would ask Congress to amend the Clean Air Act so that EPA could release itself from what, not surprisingly, had added much tension to its relationships with state and local authorities.[24]

EPA imposes regulatory burdens on municipal governments and other public entities that operate waste management facilities. For example, under national clean water legislation, all publicly owned wastewater treatment works must obtain discharge permits as part of the National Pollution Discharge Elimination System. In states that have not assumed responsibility for issuing such permits, EPA issues the permits and thereby regulates local public organizations directly.

Other Federal Agencies

Although EPA has numerous programs in which it joins with other federal agencies in cooperative efforts, it also has programs that place it in potential conflict with those agencies. Foremost among these are application of pollution control requirements for federal facilities such as military bases. Although sanctions can be imposed against federal agencies that fail to comply with EPA's regulations, interagency procedures designed to promote compliance without confrontation are a preferred approach.[25] The Environmental Protection Agency's interests are not well served by antagonizing other federal agencies, especially those with the political clout to adversely affect EPA's relations with the White House or Congress. EPA staff often have no alternative to negotiating settlements to disputes with other federal agencies on implementation details because it is not practical to refer them to the top levels of agencies for resolution.

The federal environmental impact statement (EIS) process established by the National Environmental Policy Act of 1969 gives EPA the potential for influencing virtually all federal agencies. Section 309 of the Clean Air Act of 1970 requires EPA to evaluate every impact statement produced under the federal EIS program. In exercising this responsibility, EPA occasionally finds itself embroiled in conflict. Although EPA makes efforts to negotiate differences with other federal agencies and avoid highly charged confrontations, it is not always able to do so. However, critics of EPA argue that such confrontations occur so infrequently that EPA's authority under Section 309 does not provide a significant constraint on the behavior of federal agencies.[26]

Business Interests

The Environmental Protection Agency's approach to the business community fluctuates widely, depending on who is in the White House and who heads the agency. Some EPA administrators—for example, Ruckelshaus during the Nixon administration—have taken aggressive positions in enforcing industry compliance with regulations. Ruckelshaus's efforts bolstered EPA's images as an agency that would live up to its name and protect the environment. Other EPA administrators, such as Anne Gorsuch, the first EPA head under Reagan, have taken a much more concil-

[24] Additional details are given by Quarles (1976).

[25] In addition, some legal experts argue that it would be unconstitutional for EPA to use court actions against other federal agencies to force them to comply with environmental regulations. For a contrary view, see Steinberg (1990).

[26] See, for example, Hill's (1978) analysis of EPA's reviews under Section 309.

iatory position and emphasized the agency's interest in negotiating and cooperating with industry.

EPA administrators in the Bush and Clinton administrations (William Reilly and Carol Browner, respectively) developed nonregulatory programs representing "partnerships" with industry.[27] Consider, for example, EPA's *Green Lights Program.* The several hundred voluntary participants in the program agreed to survey their facilities and, where possible, improve lighting efficiency in more than 90 percent of their square footage within five years. EPA gave participants detailed information on the costs and availability of efficient lighting. According to EPA, the program has saved participants several million dollars annually.[28]

Business interests, for their part, have taken advantage of what Lindblom (1980, p. 73) has characterized as a "special relationship" with government. He observes that "a democratically elected government cannot expect to survive in the face of widespread or prolonged [economic] distress." This accounts for the access that business interests have to politicians. In addition, business interests can often organize the financial resources and legal talent needed for long-term lobbying efforts and extended court battles.

A combination of significant resources and easy access to politicians allows industry to make its views known in a timely and forceful way when it is in conflict with EPA or other regulatory bodies. Business interests often lobby Congress and the White House to reduce the burden imposed by EPA's regulations. Private sector interests are a dominant presence at Congressional hearings on environmental legislation, and, as previously noted, they are vigorous in challenging EPA's regulations in the courts.

The 1970s witnessed a notable addition to the traditional strategies used by business to reduce the burden of environmental regulations: a new type of "public interest group," one supported by private firms and applying legal strategies similar to those used by environmental groups. The oldest of these, The Pacific Legal Foundation, exemplifies the group type. Among the suits it brought were one to prevent California from blocking nuclear development in the state, and another to prevent EPA from requiring Los Angeles to treat its wastewater before discharging it into the ocean.[29] Other groups of this type include the Mountain States Legal Foundation and the Northeast Legal Foundation. According to Rosenbaum (1991, p. 94), "business public interest groups are financed principally by organizations such as the Adolph Coors Company and the Scaife Foundation, who have fought vigorously against most of the major environmental regulatory programs that have been passed during the last three decades."

Since the late 1980s, the number of industry sponsored groups opposing environmental regulation has proliferated as part of the *wise use movement,* a diverse collection of organizations committed to preserving jobs and protecting private property rights. Many advocates of "wise use" believe environmental regulations, such as those protecting wetlands, deprive landowners of constitutionally protected rights to use their property. Not surprisingly, wise use groups and environmental NGOs have engaged in disputes (as illustrated by recent controversies over logging vs. habitat preservation in the Pacific Northwest).[30]

Environmental Nongovernmental Organizations

Organizations concerned with protecting the environment have had an influence on U.S. environmental policy since the Sierra Club was founded in 1892.[31] Between that year and the end of World War II at least five other national-level environmental groups formed (see Table 3.3). The principal concerns of these groups, prior to about 1950, was the protection of public land (especially national parks and other scenic resources) and particular species of wildlife.

Since the 1960s, there has been a sharp increase in the number and size of environmental nongovernmental organizations and the variety of strategies they employ. Table 3.3 shows membership and bud-

[27] For examples of industry efforts to reduce pollution beyond reductions required by regulations, see the discussion of pollution prevention in Chapter 14.

[28] This summary of EPA's Green Lights Program is from Kling and Schaeffer (1993, p. 27). They also describe several other partnerships between EPA and industry.

[29] For information on these suits, see Wenner (1982, pp. 54–55).

[30] For more on the wise use movement, see Echeverria and Eby (1995). They also give many examples of pro-development organizations having environmentally friendly names. For example, the National Wetlands Coalition sponsored by oil and gas companies and developers is a leading opponent of wetlands preservation.

[31] Environmental NGOs in many other countries have played roles similar to those described in the U.S. context. For information on environmental NGOs in countries other than the United States, see Tolba et al. (1992), World Resources Institute (1992), and Yap (1990).

TABLE 3.3 National Environmental Lobbying Organizations in the United States[a]

Organization	Year Founded	Membership (Thousands) 1960	Membership (Thousands) 1990	1990 Budget ($ million)
Sierra Club	1892	15	560	35.2
National Audubon Society	1905	32	600	35.0
National Parks & Conservation Association	1919	15	100	3.4
Izaak Walton League	1922	51	50	1.4
The Wilderness Society	1935	10[b]	370	17.3
National Wildlife Federation	1936	—	975	87.2
Defenders of Wildlife	1947	—	80	4.6
Environmental Defense Fund	1967	—	150	12.9
Friends of the Earth	1969	—	30	3.1
Natural Resources Defense Council	1970	—	168	16.0
Environmental Action	1970	—	20	1.2
Environmental Policy Institute	1972	Not a membership group		
TOTAL		123	3103	217.3

[a] Adapted from "Twenty Years of Environmental Mobilization: Trends Among National Environmental Organization," Mitchell, Mertig, and Dunlap in *American Environmentalism,* Dunlap and Mertig, eds., 1992, p. 13, Taylor & Frances, Philadelphia. Reproduced with permission.

[b] Mitchell and Mertig report this figure as an "estimate."

get figures for 12 environmental NGOs that dominated the environmental movement's presence in Washington as of 1990.[32] These organizations are distinguished from other environmental groups by the importance they attach to lobbying for environmental laws and their effective implementation.

Table 3.3 includes two public interest environmental law groups: the Environmental Defense Fund (EDF) and the Natural Resources Defense Council (NRDC). These organizations are staffed largely by attorneys, scientists, and engineers, and they have built their reputations by uncovering environmental abuses of both public and private organizations and bringing them to the attention of the courts. EDF and NRDC are not the only organizations listed in Table 3.3 that rely heavily on litigation. During the 1970s, the Sierra Club initiated more court actions on environmental grounds than any other environmental group.[33] In addition to lobbying and legal maneuvering, groups included in Table 3.3 also rely on public education campaigns to meet their objectives.

Although the groups listed in Table 3.3 are prominent in Washington, hundreds of other environmental NGOs exist in the United States, and some have budgets and memberships in excess of those listed in the table. For example, in 1990 Greenpeace USA had a budget of $50 million and more than two million members. Some environmental groups not listed in Table 3.3 have relied heavily on direct action to attain their objectives. This has included both civil disobedience and, in the case of Earth First! and the Sea Shepherd Society, various forms of "ecotage"; for example, sinking whaling ships and spiking trees to interfere with timber cutting.[34] In some contexts, the reliance on ecotage has been linked to the principles associated with "deep ecology."[35]

[32] For an analysis of the influence of these national environmental groups, see Cameron, Mertig, and Dunlap (1992).

[33] This fact is from Wenner (1982, p. 47).

[34] For a review of the history of environmental groups and the strategies they employ, see Dunlap and Mertig (1992).

[35] See Chapter 1 for a listing of these principles. For an analysis of the influence of deep ecology on radical environmental groups such as Earth First!, see Devall (1992).

Environmental groups often work as outsiders and rely on litigation, lobbying, direct action, and mass media campaigns. However, during some presidential administrations, staff from environmental NGOs have been recruited into high positions in EPA and other agencies, and they have made their influence felt directly. A recent example is William Reilly, the EPA administrator during the Bush administration; Reilly had previously headed the World Wildlife Fund/Conservation Foundation.

National-level environmental NGOs are not the only ones that have affected environmental policy in the United States. Since the 1970s, an enormous number of grassroots environmental groups have been formed. Their influence is generally felt locally as they struggle to force the cleanup of toxic waste contamination in their communities and block the location of new sources of pollution. Actions of these local groups have also influenced EPA. One example concerns EPA's programs to aid community participation in decisions involving cleanup of toxic waste contamination from landfills and leaking underground storage tanks. Grassroots environmental groups lobbied heavily for these programs, which include the provision of grants to local organizations so they can hire consultants to advise them on technical issues in toxic waste site remediation under Superfund.

A second example of the influence of grassroots environmentalism on EPA involves the environmental justice movement (see Chapter 1). As a consequence of pressures exerted by leaders in this movement, EPA has given increased attention to how its programs and policies influence racial and ethnic minorities.

The Mass Media and the General Public

The influence of the environmental movement is tied closely to the high level of support it has received from the general public.[36] Sympathetic mass media coverage of environmental issues has played an important role in generating this support. Strong public approval is essential to environmental groups, because it legitimizes their demands for new environmental laws and effective implementation of existing laws.

Environmental groups have used the mass media to influence public policy on the environment since the turn of the century. The campaign in the early 1900s to save Hetch Hetchy Valley in Yosemite National Park from being inundated by a water supply reservoir is a classic example. Newspaper editorials, especially in the East, did much to arouse public opposition to the reservoir. By the late 1960s, extensive media coverage of environmental issues had become common. Examples include the media attention given to the Santa Barbara oil spill (1969), the chemical waste problems at Love Canal (1978), and the rupture of the oil tanker Exxon Valdez (1989).

The influence of media coverage and public opinion on the Environmental Protection Agency takes many forms. For example, it can contribute to shifts in the leadership of the agency. This was demonstrated in the early 1980s when TV and newspapers publicized charges that the EPA's administrator and her top-level staff had mismanaged the implementation of the Superfund program. EPA's alleged use of funds to meet partisan political objectives was a long-running media story. The public was constantly told of measures taken in the battle between the White House, EPA, and Congress over access to information required to verify allegations of wrongdoing on EPA's part. This controversy, which became a major political liability for the Reagan administration, led to the administrator's resignation.[37]

Both the mass media and public opinion can also influence EPA's position on particular regulations. An example concerns EPA's action on the safety of the chemical pesticide daminozide, commonly known by the trade name *Alar.* McClosky (1992, p. 86) describes the influence of public opinion:

> The question about the safety of Alar had been debated for 15 years within EPA, which could not make up its mind about whether to force its withdrawal. [When the Natural Resources Defense Council went on national television to present its risk assessment results and to denounce the use of Alar on apples], no customer wanted to buy apples anymore. No supermarkets wanted to sell

[36] For details on public support for the environmental movement, see the various essays in Dunlap and Mertig (1992).

[37] For that EPA administrator's account of these events, see Burford (1986).

> them. The apple growers pledged to stop using Alar, and the manufacturers of it announced the production was ceasing because the market had collapsed. . . . [The] issue had been decided in the marketplace. It did not matter whether EPA had banned it; society had decided through other means that its use was unacceptable.

In the end, the manufacturer of Alar decided to pull the product off the market.

Public opinion and mass media have an influence on Congress that eventually affects how EPA prioritizes its efforts. This was demonstrated by a widely publicized assessment conducted by EPA's environmental experts in the late 1980s. EPA's staff ranked a variety of environmental problems according to their view of the seriousness of the risks involved, and they compared their rankings with those of the general public (EPA, 1987). Results showed many inconsistencies. For some risks, such as stratospheric ozone depletion and global warming, public perceptions of the risks were much lower than those of EPA's experts. The opposite occurred for risks of chemical plant accidents and disposal of toxic wastes. For these, the public perception of risk was high, while EPA's staff found the risk to be medium or low. The rankings offered by EPA's staff were generally supported by a group of outside experts brought together by the agency's administrator to follow up on the staff's assessment (EPA, 1990).

Results from this risk-prioritizing operation demonstrated that EPA was devoting enormous resources to environmental problems (for example, cleaning toxic waste sites) that EPA's experts considered to be of low importance compared to other problems. This inconsistency occurred because EPA's agenda is set by elected officials, who are often responsive to public opinion. In order for EPA to deal with what it perceives as high-priority problems, it must educate both the general public and elected officials on different forms of environmental risk.

EPA'S RULEMAKING PROCESS

When the U.S. Congress passes an environmental statute, it gives the Environmental Protection Agency responsibility for devising detailed implementation procedures and requirements. And when EPA's requirements are issued as rules, they have the force of law. In the United States, formal rules of an agency are the administrative equivalent of legislation. Frequently, the terms *regulation* and *rule* are used interchangeably.

This section examines EPA's rulemaking process. It begins by introducing the Administrative Procedure Act (APA), which details procedural aspects of federal rulemaking. The section then provides an example of the rulemaking process: EPA's regulations defining treatment technologies and discharge standards for industrial wastewater. These regulations provided a basis for writing discharge permits required by the Federal Water Pollution Control Act (FWPCA) Amendments of 1972.

Administrative Procedure Act

A legislature's delegation of implementation responsibilities to an agency puts enormous power in the hands of nonelected officials in the agency. During the 1930s, the potential for administrative abuse of power grew dramatically as the Franklin D. Roosevelt administration tried to work the country out of an economic crisis by implementing new social programs. This period witnessed a sharp increase in the number, size, and authority of federal agencies, which gave rise to debates on how to make the agencies more accountable to citizens.

Finally, in 1946, Congress passed the Administrative Procedure Act to limit the exercise of power by federal agencies. (Many states have since adopted their own versions of APA.) Although APA has many elements, one of its main components (and the only one treated herein) concerns procedures that agencies follow in promulgating rules to implement statutes.[38]

Unless otherwise directed by a statute, a federal agency issuing rules must meet the following *notice-and-comment* steps required by the Administrative Procedure Act:

1. The agency publishes a *notice of proposed rulemaking* in the *Federal Register,* the component of the federal register system used to record execu-

[38] Other important elements of APA concern requirements that agencies must follow in conducting hearings and adjudicating disputes. The law also spells out circumstances under which citizens can have agency decisions reviewed by federal courts.

tive acts such as presidential proclamations and notices of rulemaking.[39] A notice of proposed rulemaking must include the substance of the proposed rule as well as information describing the rulemaking procedure. The notice must also detail the legal authority under which the rule is proposed.

2. Interested parties must be given an opportunity to participate in rulemaking by submitting data and written comments.[40]
3. After considering the comments of interested parties, the agency must publish a *notice of rulemaking* in the *Federal Register.* The notice includes the final rule and a general statement of the rule's basis and purpose.
4. The agency cannot implement the final rule until at least 30 days after it has been published.

These steps are generally called *informal rulemaking* under APA.[41]

The four requirements listed provide safeguards by allowing interested parties to find out about a proposed rule, offer comments on the proposal, and learn about the final outcome. In addition, the 30-day waiting period after publication of the final rule gives affected parties an opportunity to challenge the final rule in federal court.[42]

Agencies such as the Environmental Protection Agency tend to augment APA's informal rulemaking by adding provisions for keeping a detailed written record of all activities undertaken in developing a rule. EPA compiles a thorough administrative record because the opinions of many federal judges in deciding suits brought under APA have emphasized the need for agencies to maintain records of rulemaking activities.

The rulemaking processes followed by EPA and other agencies vary somewhat from case to case. Sometimes individual statutes impose special requirements on agency rulemaking, and administrative changes in procedures occur from time to time. Even though details of the rulemaking procedure differ from one case to the next, the following example illustrates key features of a rulemaking exercise.

Procedures Followed in Issuing Effluent Guidelines

The process used by EPA to issue rules that established *effluent guidelines* under the FWPCA Amendments of 1972 has been studied extensively.[43] (*Effluent* is a synonym for wastewater discharge.) As detailed in Chapter 12, the 1972 amendments introduced a new element in the federal strategy to control water pollution: the National Pollution Discharge Elimination System (NPDES). Under this system, EPA (or states that received authority to implement NPDES) had to regulate about 47,000 industrial wastewater dischargers by issuing permits specifying the quantity and quality of allowable releases. Congress called upon EPA to issue effluent guidelines by October 1973 and to write the permits by December 1974. The effluent guidelines were to give meaning to criteria that the 1972 amendments included as a basis for issuing permits. For example, the amendments called for industry to use the "best practicable technology" (BPT) for controlling water pollution by 1977 and "best available technology" (BAT) by 1983.

Marcus (1980, pp. 141–42) describes the intent of Congress as follows:

> The theory . . . was that the general rules, contained in the effluent guidelines, were to be applied to the particular permit cases. The effluent guidelines would state the BPT and BAT standard that an entire industry category, such as grain mills, glass manufacturers, or petroleum producers had to achieve, while the permits would specify the pollution allowable for a particular plant.

Arguably, EPA experienced more than the usual difficulties in issuing regulations that spelled out the effluent guidelines. The task required surveying wastewater reduction practices in more than 250 groupings of industries to accertain such things as

[39] The *Federal Register* is also used to record amendments to the *Code of Federal Regulations* (CFR) and presidential executive orders. The *Federal Register* is issued daily except Sundays, Mondays and days following holidays.

[40] The APA does not specify a particular time period for public comment. Interested parties often have 30 days or more to comment.

[41] If a statute does not include directives to the contrary, agencies are required to follow APA's informal rulemaking procedure. If a statute requires an agency hearing as part of rulemaking, then APA's more-rigorous, "formal" rulemaking procedure is used.

[42] For details on the rulemaking procedures established by APA, see Davis (1972, chap. 6).

[43] Unless otherwise noted, the facts in this subsection are from Magat, Krupnik, and Harrington (1986).

what pollution control technologies were the best practicable and best available.[44] This was an enormous job and the schedule mandated by Congress was tight. EPA's original goal was to issue effluent guidelines for 22 industries in the second half of 1973, but it issued none in that period. These missed deadlines led the Natural Resources Defense Council to sue EPA, and the settlement took the form of a *consent decree* detailing a rigid timetable for EPA to follow.[45] The decree eliminated opportunities the Office of Management and Budget normally would have had to review EPA's proposed rules. (The more typical role of OMB is elaborated below.)

For any particular industry, the rulemaking procedure followed by EPA consisted of three main parts (see Table 3.4). During the first part, EPA relied on a "technical contractor" to survey industry practices and develop a working conception of BPT and BAT. For each subcategory of an industry, the contractor determined how various production processes affected pollution parameters in a typical plant, what wastewater management procedures were followed, and the costs and energy required to carry out various wastewater reduction schemes. The contractor's product was a draft "development document" containing numerical limits on effluent quality that EPA should use in writing NPDES permits. The EPA staff also engaged an "economic contractor" to study the economic effects of adopting proposed effluent guidelines (for example, how many plants would be forced to close if particular effluent standards were imposed).

During this first phase of rulemaking, EPA staff conducted internal reviews of the contractor's progress and solicited information from other organizations, including firms, industry trade associations, agencies, and environmental groups. The draft development document produced in the first stage was circulated to a wide audience. Typically, a 30-day comment period was provided, and EPA published in the *Federal Register* summaries of comments on the document and its responses to the comments.[46]

TABLE 3.4 EPA's Rulemaking Procedures for Defining BPT and BAT[a]

Contractor Study Stage
Select industry to regulate
Choose EPA technical and economic project officers
Choose technical and economic contractors
Produce draft "development document" and circulate for informal external review[b]
Proposed Rule Stage
Analyze comments from informal external reviews
Conduct EPA internal reviews
Produce proposed development document and rule
Publish *notice of proposed rulemaking* in *Federal Register*[c]
Provide at least 30 days for formal external review
Promulgated Rule Stage
Analyze comments from formal external reviews
Produce final development document and rule
Publish *notice of rulemaking* in *Federal Register*[d]
Provide minimum of 90 days before implementing rule

[a] Adapted from Magat, Krupnik, and Harrington (1986, p. 32). Copyright © 1986 by Resources for the Future.

[b] EPA's informal external review was not required by the Administrative Procedure Act.

[c] Normally, OMB would review the proposed rule prior to publication of this notice in the *Federal Register.*

[d] Normally, OMB would review the final rule prior to publication of this notice in the *Federal Register.*

During the second stage of rulemaking, EPA formed an internal working group to formulate a proposed rule. This is when additional analyses and trade-offs were made in response to comments on the draft development document. The result of EPA's internal assessment was a *proposed rule* that was published in the *Federal Register.* Interested parties had at least 30 days to comment. Typically, comments consisted of formal restatements of criticisms made during reviews of the draft development document.

After the period of formal comment, EPA entered the third stage of its process: rule promulgation. During this stage, there was no contact with outsiders.

[44] EPA initiated its rulemaking by dividing the nation's water polluters into more than 30 industrial categories and over 250 subcategories.

[45] A consent decree is a statement issued by a court spelling out details of a settlement reached by parties in a lawsuit.

[46] The Administrative Procedure Act does not require public comment at this early stage of rulemaking. Public comment is required only when a rule is proposed.

EPA reworked its proposed rule based on its own reactions to comments and published its *final rule* in the *Federal Register.*

Following publication of the final rule, those who wished to contest it had 90 days to register a legal challenge. The rules establishing effluent guidelines were contested frequently. Over 150 lawsuits were brought challenging different effluent guidelines. In many cases, the process of appeal and counterappeal was a lengthy one. Delays in settling these cases resulted in even longer delays before required treatment works were installed.

OMB's Role in Rulemaking

The example just cited does not demonstrate OMB's usual role in rulemaking. During the 1980s, two executive orders were issued that substantially increased the influence of the Office of Management and Budget. One of them, Executive Order 12291, calls for a benefit-cost analysis on proposed major rules, and "authorizes OMB to review virtually all proposed rules for consistency with the substantive aims of the Executive Order before an agency can even ask for public comment. . . ."[47]

If the rulemaking process for effluent guidelines described above had followed normal procedures, OMB could have conducted detailed reviews of each rule at two points: (1) prior to publication of the notice of proposed rulemaking in the *Federal Register,* and (2) prior to publication of the notice of rulemaking in the *Federal Register.*

In extreme cases, OMB's reviews of rules proposed by EPA have taken more than three and a half years. More typically, OMB reviews have added several months to the time required for rulemaking.[48]

Executive Order 12498, issued by President Reagan in January 1985, increased OMB's influence by requiring an agency to notify OMB before it even starts to develop a new rule. The order also requires the agency to submit written justification for the research and information gathering it must do to determine whether or not a regulation is called for. Critics charge that the 1985 order, together with Executive Order 12291, "places the ultimate rulemaking decisions in the hands of OMB personnel who are neither competent in the substantive areas of regulations, nor accountable to Congress or the electorate in any meaningful sense."[49] Although criticisms of OMB's competence are debatable, there is little question that OMB's perspective in reviewing proposed regulations is often restricted, with an emphasis on environmental costs and gains that can be described in monetary units.

The preceding discussion clarifies why implementation of a new environmental program generally takes years. It also demonstrates some of the ways in which the White House, Congress, the regulated community, and environmental groups influence how environmental policies are carried out.

NATIONAL ENVIRONMENTAL AGENCIES OUTSIDE THE UNITED STATES

This section puts the discussion of the U.S. Environmental Protection Agency in perspective by considering how similar agencies outside the United States have developed. Although details about the evolution of national environmental agencies or departments vary from country to country, there are some common patterns. One is that before the 1960s, environmental regulations covering a variety of areas, such as air pollution, water pollution, and solid waste, were not generally administered by a single, national agency. Typically, several such regulatory programs existed, but they were administered by lower-level units within departments (or ministries) of health, natural resources, and so forth. As in the United States, fragmentation of responsibility was the norm.

Another prevalent outcome in the evolution of national environmental agencies outside the United States is that oversight of environmental agency performance has been provided, to varying degrees, by the legislative and executive branches rather than the judiciary. This contrasts with the situation in the United States, where the courts, responding to suits brought by citizens and nongovernmental organizations, have played a dominant role in overseeing EPA's activities. Several explanations exist for why

[47] This quotation is from Morrison (1986, p. 1062).

[48] The slow downs reported here are for rules expecting to have an impact exceeding $100 million. Fiorino (1995, p. 72) provides numerous examples of EPA rules that were delayed by OMB.

[49] This quotation is from Morrison (1986, p. 1064).

ordinary citizens and interest groups in other countries rely less on courts to influence performance of an environmental agency; among them, (1) statutes limit the circumstances under which an agency can be sued by citizens (for example, citizens may need to demonstrate that personal harm results from an agency's action); (2) it is not customary for citizens to sue agencies; and (3) potential plaintiffs cannot gain access to either the information or resources needed to wage court battles.

Another common theme is that nonenvironmental governmental units often work actively to limit the way a national environmental agency carries out its tasks. In the United States, this role has been played by the Office of Management and Budget. In other countries, an office of national economic planning or its equivalent may try to reduce the influence of an environmental agency. Sometimes resource development agencies try to control the environmental agency's influence, especially where environmental regulations could affect development agency projects.

Not surprisingly, another common feature is that national environmental agencies outside the United States frequently engage in the same balancing act as the U.S. EPA: demands of environmentalists for increased regulation are reconciled with pressure to prevent environmental requirements from impeding economic development. The discussion in this section expands on these general points.

As noted, national-level environmental agencies didn't begin to appear until the 1960s.[50] National environmental agencies in industrialized countries generally were formed in the late 1960s and early 1970s. Examples are shown in Table 3.5. The forces leading to creation of these agencies were typically the same as those that led to establishment of the U.S. EPA. Regulatory programs for domestic water supply, wastewater management, air quality, and so forth had typically grown piecemeal and were administered independently by different organizational units. A unified perspective that accounted for interdependencies across regulatory programs was lacking, and the creation of an environmental agency was seen as a way to eliminate fragmentation and enhance coordination of regulatory activities. In addition, in the late 1960s, governments in industrialized countries were under increasing pressure from their citizens to curb the increasing environmental decay that accompanied rapid economic growth.

National-level environmental protection institutions generally did not emerge in developing countries until after the United Nations Conference on the Human Environment (UNCHE), which was held in Stockholm in 1972. The UNCHE (or Stockholm Conference), together with many international meetings held to prepare for the conference, provided occasions for industrialized countries and developing countries to gain perspectives on each other's problems and their differing perceptions of the link between environment and development. According to Tolba and his colleagues (1992, p. 742), the industrialized countries "came [to Stockholm] to discuss international solutions to pollution problems which resulted from growth of industrial activity. They were also concerned with nature conservation. . . . The developed countries expressed little interest in those problems as compared with 'the pollution of poverty' and the inefficiency of resource use caused by underdevelopment." Runnals (1986, p. 194) comments on the concerns that developing countries had about Stockholm:

> They feared that environmental rules and regulations would slow down their development, that stringent health regulations aimed at carcinogens or toxic substances would be used to block their exports, and that their attempts to exploit their forest and mineral wealth would be curtailed. Finally, many of them feared that the newly found environmental concerns of the World Bank and the United States Agency for International Development (USAID), among others, would add substantially to the costs of their development projects, and prolong their execution without making additional financial resources available to cover these costs.

One consequence of the Stockholm Conference was an increase in the priority given to management of environmental quality by developing countries. The case of Kenya, which in 1974 created the first national-level environmental protection unit in Africa, is illustrative. In 1972, the government of Kenya formed a Working Committee on the Human Environment to develop a report for an international

[50] Details on the formation of these agencies are given by Tolba et al. (1992, p. 702).

TABLE 3.5 Examples of National Environmental Agencies Created in Industrialized Countries before 1972

Year	Country	Organization
1969	Sweden	National Environmental Protection Board
1970	United States	Environmental Protection Agency
1970	United Kingdom	Department of Environment
1970	Canada	Department of Environment
1971	Japan	Environmental Agency
1971	France	Ministry for the Environment and the Protection of Nature

meeting at Founex, Switzerland, that was convened to prepare for the Stockholm Conference. Following the meeting at Founex, the Kenyan government created the National Environment Secretariat (within the Ministry of Natural Resources) to represent Kenya at the UNCHE. Two years after the Conference, the National Environment Secretariat was elevated in stature and made part of the Office of the President. Although the National Environment Secretariat has sometimes been stymied by industries and development agencies in meeting its regulatory goals, it has been effective in keeping environmental quality management on the national agenda.[51] Like environmental agencies in other developing countries, it has received assistance from various UN organizations and development assistance agencies (such as the U.S. Agency for International Development).

As illustrated by Table 3.6, many developing countries created national-level units soon after the UNCHE. By 1992, on the twentieth anniversary of the Stockholm Conference, national-level environmental agencies existed in virtually all industrialized countries and a majority of developing countries. For example, as of 1992, 37 out of the 51 countries in Africa had established "Ministries of Environment, Natural Resources, or Nature Conservation/Protection."[52]

Although many developing countries have created national environmental management agencies, these agencies have often had difficulty establishing their legitimacy and meeting their regulatory goals. Environmental agencies in many developing nations have had to struggle with low budgets and a shortage of experienced staff. Perhaps more significantly, the agencies often lack political clout. The combined effect of these factors is that, in many countries, environmental statutes and policies are not implemented effectively.

INTERNATIONAL RESPONSES TO GLOBAL ENVIRONMENTAL PROBLEMS

The 1970s witnessed the emergence of global environmental problems that dwarfed in significance the international environmental concerns of earlier periods. Before the 1970s, the main international environmental issues concerned the protection of migratory wildlife and the quality and use of international waters. Since 1970, international environmental lawmaking has covered an increasingly complex and expanding range of topics.

Of the many environmental issues that have a global dimension,[53] the ones treated in this text include transboundary air pollution leading to deposition of acidic substances (a phenomenon commonly

[51] For an account of strategies used by the National Environmental Secretariat to accomplish its regulatory goals, see Hirji and Ortolano (1991).

[52] This fact is from Tolba et al. (1992, pp. 724–25).

[53] A comprehensive review of international environmental issues is given by Tolba et al. (1992). Typical listings of important environmental issues having an international dimension include overpopulation; loss of biodiversity; devegetation, including deforestation and overgrazing; global climate change; depletion of stratospheric ozone; acid deposition; transboundary water pollution; and the global trade in toxic wastes.

TABLE 3.6 Examples of Environmental Agencies Created in Developing Countries after The Stockholm Conference

Year	Country	Organization
1973	Brazil	Special Secretariat of Environment
1974	Kenya	National Environment Secretariat
1975	Thailand	National Environment Board
1977	Philippines	National Environmental Protection Council
1980	Saudi Arabia	Monitoring and Environmental Protection Agency

termed *acid rain*), depletion of ozone in the stratosphere, global climate change, loss of biodiversity, and the international trade in hazardous waste. Technical aspects of these problems are introduced in later chapters. The remainder of this chapter introduces key participants involved in developing international environmental policies. Two global problems are considered: stratospheric ozone depletion and the hazardous waste trade. The former is used to clarify the roles played by the United Nations Environment Programme and environmental NGOs in fostering international agreements. A discussion of trade in hazardous waste demonstrates the increasingly important influence of domestic environmental policies on international politics and trade.

Negotiating International Agreements: The Case of Ozone Depletion

Ozone depletion first surfaced as an issue in the 1970s, but it was elevated in importance in 1984 when British scientists discovered a hole in the layer of stratospheric ozone over Antarctica.[54] Later evidence showed that the ozone layer was thinning in the high latitudes of both hemispheres. By the late 1980s, empirical evidence linked chlorinated and brominated chemicals to the destruction of ozone. Chlorofluorocarbons (CFCs) were identified as being especially significant. Effects of stratospheric ozone depletion on humans include skin cancers, eye disorders (such as cataracts), and suppression of the immune response system. Negative impacts on ecosystems, including reductions in agricultural and fisheries productivity, have also been identified.

During the 1980s, scientists predicted what would happen if no steps were taken to halt the destruction of the ozone layer. The scenarios were alarming and the international response was expeditious.

UNEP's Role in Managing Negotiations

Since the Stockholm Conference in 1972, the United Nations Environment Programme (UNEP) has been a major player in the international environmental arena.[55] The organization was created by the UN General Assembly in response to a UNCHE call for a new UN institution that would inspire and coordinate environmental activities primarily (but not exclusively) through the UN system. UNEP serves the international community as a clearinghouse for environmental monitoring data and as a facilitator of international agreements.[56]

The process UNEP uses to foster international agreements is well illustrated by events leading up to the Montreal Protocol on Substances that Deplete the Ozone Layer. In response to growing scientific concern over the harm that could result from a thinning layer of ozone in the stratosphere, UNEP's Governing Council decided in 1976 to make ozone depletion one of its five top-priority problem areas. Soon

[54] The *stratosphere* is an upper portion of the atmosphere. As explained in Chapter 22, too *much* ozone is a source of problems in the *lower* atmosphere.

[55] There are other international organizations that have played important roles in helping solve global environmental problems. For an account of these organizations and their influence, see Young (1994).

[56] The agenda followed by UNEP is dictated by its Governing Council, which consists of 58 UN member nations elected for three-year terms. UNEP is headquartered in Nairobi, Kenya, and led by an executive director.

thereafter, UNEP organized the first of a series of international conferences to bring together scientists, NGOs, and government representatives to discuss the ozone depletion problem. This UNEP conference, which took place in 1977, adopted a World Plan of Action on the Ozone Layer. Although the conference suggested international actions to regulate use of CFCs, the ozone depletion problem was not considered urgent at that time.

In 1978, UNEP organized The Coordinating Committee for the Ozone Layer to further delineate the ozone depletion problem. The committee, which consisted of NGOs and representatives of government agencies, played a major role in establishing priorities for scientific research.

In 1981, UNEP formed the Ad Hoc Working Group of Legal and Technical Experts for the Elaboration of a Global Framework Convention for the Protection of the Ozone Layer. The working group began meeting in January 1982, and the process of international bargaining began in earnest. (The framework convention stage of international lawmaking is a key step in formulating agreements. A *framework convention* establishes principles, but it does not include binding agreements.)[57]

The key nations in the negotiations to reduce ozone depletion included the principal CFC-producing countries: the United States, Great Britain, France, West Germany, and Italy. These five countries accounted for 75 percent of worldwide CFC production. U.S. producers faced domestic pressures forcing them to cut back on CFCs, and thus the United States was eager to get an international agreement that would "level the playing field" by having other nations cut back CFC production as well. European producers resisted cutbacks because they wanted to preserve their large overseas markets for CFCs and avoid the cost of adopting substitutes.

As of 1985, key participants could only agree on a framework convention: the 1985 Vienna Convention for the Protection of the Ozone Layer. The Vienna Convention set out an agenda for research and information exchanges, but it did not include any specific obligations. By the late 1980s, scientists had made clear the negative results of continued production of CFCs. This provided incentive for the CFC-producing countries to resume negotiations.

Agreements reached during negotiations following the Vienna Convention were formalized in 1987 by the Montreal Protocol on Substances that Deplete the Ozone Layer. In contrast to the Vienna Convention, the Montreal Protocol included binding obligations to make CFC reductions. The protocol required, among other things, that signatory governments freeze production of selected CFCs at their 1986 levels and cut back production by 50 percent (of 1986 levels) in the year 2000.[58]

After the meetings at Montreal, UNEP organized additional meetings to strengthen the Montreal Protocol. Throughout this process, UNEP's executive director at the time, Mostafa Tolba, participated directly in the negotiations and had an important influence on the outcome. Porter and Brown (1991, p. 49) describe Tolba's role as follows: "During the negotiations on the Montreal Protocol, he was a key player, along with the conference chair, Austria's Winfried Lang, in the process of reaching consensus, lobbying hard for a complete phaseout of CFCs in informal talks with chiefs of [the European Community] delegations." Porter and Brown also cite the important role that UNEP's executive director played in working out compromises and pushing for stringent implementation timetables at several UNEP-sponsored meetings (in the late 1980s) that strengthened the Montreal Protocol. By 1990, the original protocol had been amended to call for a complete phaseout in production of selected CFCs by the year 2000.

Influence of Environmental NGOs

Increasingly, national environmental NGOs, such as the Environmental Defense Fund in the United States, have taken an active role in international environmental problem solving. These national groups work alongside international environmental NGOs in trying to hammer out agreements that reduce

[57] In this context, a *convention* is an international legal instrument used frequently in formalizing general international agreements on environmental issues. Once a convention is in place, *protocols,* which include binding obligations to perform specific actions, can be negotiated.

[58] The Montreal Protocol went into effect in 1989, at which time 23 nations had ratified it. By April 1992, 76 countries had ratified the protocol.

global problems.[59] The significance of environmental NGOs in international contexts is demonstrated by their roles in responses to the stratospheric ozone problem.

In addition to lobbying government negotiators, NGOs brought results from their own analyses to the meeting in Montreal. For example, representatives from the Environmental Defense Fund, which included an attorney, an economist, and an atmospheric physicist, shared data and analytical results with European NGOs. That information helped convince some European governments to accept the Montreal Protocol's call for a 50-percent reduction in selected CFCs by the year 2000.[60] Information exchange among NGOs and between NGOs from one country and negotiators from another is common. One reason NGOs were influential relates to UNEP's policy that NGOs be permitted to present their views at all international meetings in which agreements on ozone depletion could be reached.[61]

Many environmental NGOs felt the Montreal Protocol was weak, and they quickly took steps to strengthen it. For example, just weeks after the meetings in Montreal, the annual meeting of national affiliates of Friends of the Earth International passed a resolution making protection of the ozone layer the subject of a priority campaign. Their goals were to ensure that participating nations enforced provisions of the protocol, and to press for strengthening those provisions at future international meetings.

During the years immediately following announcement of the Montreal Protocol, environmental NGOs around the world waged national-level campaigns to increase awareness of the ozone problem, and to gain public and political support for a reduction in the production and use of CFCs and other ozone-depleting compounds. Among the campaigns were programs to educate voters, schoolchildren, and the press about alternatives to CFCs, and nationwide consumer boycotts against firms producing CFC-based products. In addition to waging campaigns, NGOs maintained a strong presence at international meetings convened to negotiate tougher measures to slow down the depletion of stratospheric ozone. The final positions taken at several of these meetings relied heavily on action statements presented by representatives of NGOs.

Methods NGOs employed in their efforts to control depletion of stratospheric ozone are likely to be used in future attempts to negotiate international solutions to environmental problems. Cook (1990, p. 337) argues that the "strengthened NGO voice [reflected in NGO activities associated with the Montreal Protocol] has changed the international political landscape. Where governments once had to negotiate only with one another, they now must respond to citizens who speak for the preservation of the Earth." The strategies and tactics employed by NGOs in the case of the Montreal Protocol were soon put to use on other, related issues, particularly the international concern over global climate change.[62]

The increasingly severe CFC cutback requirements agreed to by signatories to the Montreal Protocol do not mean the ozone depletion problem is solved. UNEP and many NGOs see a need to strengthen existing agreements, and much work remains in terms of monitoring to determine if accepted cutbacks are made and to resolve whether or not improvement of the ozone depletion problem actually takes place. In addition, efforts to comply with provisions of the Montreal Protocol are being stymied by the lucrative black market for virgin (as

[59] Some international NGOs take the form of loosely associated federations of national organizations; examples include Greenpeace and Friends of the Earth International. Another type of international NGO follows the "think tank" model in which technical and legal expertise resides in the hands of a core staff of professionals. This type is exemplified by the World Resources Institute in Washington, DC, and the International Institute for Environment and Development in London and Buenos Aires. A unique form of international NGO is the International Union for the Conservation of Nature (IUCN). It links members in about 60 countries, more than 100 government agencies, and 400 or so NGOs, and is governed by a general assembly of delegates from its member organizations. The IUCN was established (under a different name) in 1948 following an international meeting on the protection of nature. For more information on international NGOs and their significance, see McCormick (1993) and Caldwell (1990).

[60] The role of the Environmental Defense Fund in influencing the Montreal Protocol is described by Porter and Brown (1991, p. 61).

[61] Information in the remainder of this section is from Cook (1990).

[62] The international policy response to climate change issues has been both slower and more disputable because of the high level of uncertainty over the rate at which climate change might occur. For an account of efforts to formulate policy on global climate change, see Young (1990).

opposed to recycled) CFCs. Illegal imports of CFCs are a problem in several countries, including India, China, Russia and the United States. In the U.S., for example, over 1 million pounds of illegally imported CFCs were seized by federal authorities in 1995, and CFCs were the second largest illegal import (after illegal narcotics).[63]

International Trade in Hazardous Waste

Even though UNEP and NGOs were effective in building an international consensus for CFC cutbacks, they have not attained the same success in dealing with the global waste trade. Although international trade in waste has long been practiced among industrialized countries, such trade began to extend to developing countries during the 1980s. The reaction of some developing countries in response to what they viewed as "toxic colonialism" has been to ban the importation of toxic waste. However, this has increased the pressure on other developing countries to accept such waste. Many actions to resolve the international tensions associated with toxic waste trading have been neither adequate nor enforceable.

Problems associated with hazardous waste transported from industrialized to developing nations have three origins: growth of the chemical industry, increasingly stringent regulations on hazardous waste disposal in industrialized countries, and desperate economic conditions in many developing countries.

The number of commercially available chemicals has increased dramatically since World War II.[64] By 1982, the total number was estimated at 60,000 substances, which was about 350 times the number in 1940. By 1992, about 100,000 chemicals were commercially available. The increase in the number of new substances was accompanied by a corresponding rise in total production of chemicals. Problems in dealing with rapid growth in chemical waste products are compounded by the paucity of information regarding their effects. While about 1000 new chemicals become available each year, the capacity to test these chemicals for carcinogenicity and other biological effects is far from adequate.[65]

A notable international trade in toxic waste emerged in the 1980s, a result, in part, of increasingly strict regulations on the disposal of hazardous waste in industrialized countries. In the United States, for example, the 1984 amendments to the Resource Conservation and Recovery Act imposed stringent requirements on facilities that accepted hazardous waste. Many landfills that had been receiving hazardous materials closed because they found it too costly to meet the new requirements. This increased the demand for facilities that would accept hazardous waste. However, new disposal facilities could not be developed easily because citizens feared the danger of being near these sites. Opposition to siting of hazardous waste facilities came to be characterized by the acronym *NIMBY* (*not in my backyard*).

The resulting shortage of hazardous waste disposal sites drove up fees at existing facilities. The new, stringent regulations in the United States had parallels in other industrialized nations. Entrepreneurs soon discovered that, even accounting for the costs of shipping, companies could profit by transporting hazardous waste to developing countries. By the late 1980s, a lucrative global trading network had evolved, and many developing countries (mainly in Latin America and Africa) were targeted as hazardous waste recipients.[66]

At that time, many developing countries faced a critical need for hard currency. Their massive international debts made it tempting to receive U.S. dollars (or other international currency) for accepting hazardous waste from industrialized nations. However, accepting such waste often proved to be a poor bargain. In the 1980s, most developing countries did not have regulations covering many of the hazardous substances contained in imported waste. In addition, they did not have the staff or facilities to manage and monitor hazardous wastes or to appraise fully the dangers they posed.

During the 1980s, numerous incidents were reported in which wastes from industrialized nations were being dumped in developing countries, often illegally, and causing great harm.[67] For example, 228 fifty-five gallon drums, identified as "pure solvents"

[63] Facts about the illegal import of CFC's are from Chin (1996).

[64] Facts in this paragraph are from Tolba et al. (1992).

[65] Petulla (1987, p. 63) indicates that testing a new chemical takes two to four years and costs up to $1 million. He also notes that more than "10,000 untested separate chemical entities are in widespread use."

[66] For further information on the countries targeted as recipients, see Hilz (1992).

[67] For details on such incidents, see Center for Investigative Reporting and Moyers (1990).

by the shipper, went from the United States to Zimbabwe. Instead of solvents, the drums contained highly toxic waste. In other cases, freighters from the United States and Europe laden with dioxin-laced incinerator ash crossed the oceans looking for developing countries that would accept their cargoes. The media reported the wanderings of these "poison ships," as well as numerous incidents of damage caused by exported wastes when they were improperly stored or treated in developing nations.

The regulatory response to problems caused by the growing international trade in toxic waste has taken many forms, but a common one involves the notion of *prior informed consent:* waste exporters must notify a nation intended as the waste recipient and receive its written consent before any shipment can take place. (Typically, regulations also include provisions for accurate labeling of waste shipments.) The prior informed consent idea underlies the unilateral action taken by the United States to restrict hazardous waste exports. It is also the central theme in the international agreement known as the Basel Convention on the Control of Transboundary Movement of Hazardous Wastes and Their Disposal, which was reached in 1989 and went into effect in May 1992.

Although the Basel Convention has been hailed as a first step in managing the international toxic waste trade problem, it has many critics. Indeed, some have raised the question of whether it merely legalizes "toxic terrorism." The main concern of the convention's detractors is that most developing nations have neither the regulations nor the waste management experience to deal with hazardous waste.

An equally significant shortcoming of the Basel Convention and other prior informed consent schemes is their lack of enforceability. Consider the prior informed consent requirements that were at the heart of the U.S. program to regulate the hazardous waste trade. Fake labeling of toxics as "recyclables" or "fuel substitutes" was commonplace. In addition, the U.S. EPA did not have an effective enforcement strategy, and it failed to coordinate its regulatory program with customs officials. As a consequence, the U.S. prior informed consent requirements were routinely ignored.[68]

[68] This paragraph relies on Hilz's (1992) discussion of the U.S. regulations governing the trade of hazardous waste.

Many African states, which were among the main recipients of waste exports, viewed the trade in toxics as a form of environmental injustice in which poor nations were exploited by industrialized countries and prosperous business enterprises. They lobbied unsuccessfully for a more restrictive version of the Basel Convention, one that would have banned all waste exports to countries that lacked the same levels of technology and waste management facilities as the exporting countries. They also advocated a scheme that would include monitoring of disposal sites by UN inspectors. Their proposals did not win the support of the industrialized nations, many of whom saw an economic need to maintain the trade in hazardous materials.

Finally, in 1995, a Conference of the Parties to the Basel Convention adopted an amendment that would immediately ban all exports of hazardous waste from many industrialized countries to developing countries for *final disposal.*[69] The amendment also bans such exports for *recyclable waste* as of January 1, 1998. However, these changes do not necessarily represent a step forward in slowing the trade of hazardous materials to developing countries. Adoption of the amendment by the Conference of the Parties does not mean the new bans are yet a part of the Basel Convention. Modification of the convention requires ratification by member states, and as of mid-1996, the proposed bans had not been ratified. Moreover, because of ambiguities in the convention's definitions of hazardous and recyclable waste, questions remain about what waste transport would be prohibited by the new bans.[70]

Even if the Basel Convention continues to be strengthened, the problems associated with the global trade in hazardous waste will not be easily solved. Enforceability will remain difficult. Even within the United States, which has some of the most sophisticated hazardous waste management regulations and facilities in the world, illegal disposal and inadequate enforcement are common.[71]

[69] The 1995 amendment bans exports from nations that are members of the Organization for Economic Cooperation and Development.

[70] For information on ambiguities in the definition of hazardous waste in the Basel Convention, see Krummer (1995, p. 50).

[71] Difficulties involved in managing hazardous waste in the United States are detailed in Chapter 14.

Key Concepts and Terms

U.S. Environmental Protection Agency and Its Institutional Setting

EPA administrator
Regulated community
Regulatory impact analysis
Office of Management and Budget
Congressional oversight
"Hammer" provisions
Judicial influence on EPA policy
Federalism
Prevention of significant deterioration
Transportation control plans
Wise use movement
Environmental NGOs
Grassroots environmentalism

EPA's Rulemaking Process

Administrative Procedure Act
Informal rulemaking
Federal Register
Effluent guidelines
Executive Order 12291

National Environmental Agencies Outside the United States

Fragmented regulatory programs
Interagency rivalries
Sources of agency oversight
UN Conference on the Human Environment
Poverty as a cause of pollution

International Responses to Global Environmental Problems

United Nations Environment Programme
International NGOs
Chlorofluorocarbons
Stratospheric ozone depletion
Vienna Convention
Montreal Protocol
Conventions vs. protocols
International trade in toxic waste
NIMBY syndrome
Prior informed consent
Basel Convention

Discussion Questions

3-1 Between 1982 and 1985, Petulla (1987, pp. 87–89) interviewed 92 environmental managers working for various firms in the United States. He reports the following as the "strongest conclusion" of the study:

> [T]he most important indicator of a strong, enlightened environmental management program at any firm is the position of the president or chief executive officer and their attorneys. If that position is committed to environmental compliance and long-term planning to minimize problems of risk and liability, everything else falls into place: a clearly articulated policy, adequate budgets, training programs, augmented environmental divisions, goal-oriented programs, cooperation from plant managers, environmental audits, resource recovery. I found that if the attitude at the top was neutral or negative, the environmental sections tended to muddle along without a plan, "putting out fires," in a state of disorganization, do shoddy work, and suffer morale problems.

How might you use this conclusion to advise the EPA administrator on how private-sector compliance with the U.S. EPA's regulations could be improved?

3-2 A state-level environmental protection agency aims to increase industry compliance with its regulations by directing most of its enforcement effort against large, highly visible polluters that are violating environmental requirements. The agency plans to take these firms to court with the intention of imposing massive fines and, wherever possible, sending corporate officials to jail for knowingly endangering public health and welfare. What are the advantages and disadvantages of this enforcement strategy?

3-3 National environmental NGOs based in Washington, DC, have had a major impact on the formulation and implementation of environmental policies. What reasons can you give to explain the influence these organizations have had? How would your response change if you were describing grassroots environmental groups instead of the environmental NGOs with a major presence in Washington, DC?

3-4 Consider the following "ethical dilemma" as reported by Petulla (1987, p. 148):

An environmental manager found out that toxic waste underground storage drums were leaking, reported it to the company, and asked for money in the budget to do the cleanup. Company officials said they were in no position to perform remediation, and would tell him when they were. The manager believed they would stall indefinitely. He did not consider reporting the company to government agencies but quit a year later because no action was taken. He regretted he did not tell the company why he left.

What would you have done if you had been in the position of the environmental manager?

3-5 A U.S. Environmental Protection Agency (1987) study, *Unfinished Business,* found that EPA was spending enormous resources on cleaning up hazardous waste landfills that were contaminating groundwater; the agency's staff did not view this as a high-risk issue. At the same time, the staff felt that very serious problems, such as indoor air pollution and stratospheric ozone depletion, were not receiving a large share of the agency's resources. Does this outcome surprise you? Why do you think EPA gives high priority to issues the staff believes to be relatively unimportant, while some high-risk environmental problems receive little attention?

3-6 Describe the difficulties the U.S. Environmental Protection Agency experienced in issuing its effluent guidelines under the Federal Water Pollution Control Act Amendments of 1972. In preparing your response, consider the scope of EPA's task and the significance of the outcome for the regulated community. What steps could EPA have taken to speed up the rule-making process in this case?

3-7 Consider a national environmental protection agency that has recently been created in a developing nation. You are hired as a consultant to advise the agency on strategies it should follow to increase its budget and political power. What advice would you offer?

3-8 Residents of a community that is largely populated by low-income families have received word that a regional hazardous waste landfill may be located on county land near the edge of their town. A local nongovernmental organization is formed to represent the citizens in discussions with the private entrepreneurs and government officials who are trying to establish the landfill. Because you are familiar with the experiences of environmental NGOs, you are hired as a consultant to advise the newly formed NGO. What advice would you offer about options they might pursue?

3-9 Although many have charged that the international trade in hazardous waste is immoral and represents a form of environmental injustice, others have argued that there is nothing immoral or unjust about it. Supporters of trade in toxic waste reason that if a developing country knowingly accepts a particular shipment of hazardous waste in return for money, the transaction should be viewed as an ordinary business deal between willing buyers and sellers. How would you respond to this argument?

3-10 The Basel Convention represents a start toward reaching a satisfactory resolution of problems associated with the international trade of hazardous waste. However, many feel the *prior informed consent* scheme that the Convention relies on is unenforceable. What steps can UNEP take to promote a more satisfactory resolution of the problem?

References

Burford, A. M. 1986. *Are You Tough Enough?* New York: McGraw–Hill.

Caldwell, L. K. 1990. *International Environmental Policy.* 2d ed. Durham, NC: Duke University Press.

Cameron, R., A. G. Mertig, and R. E. Dunlap. 1992. "Twenty Years of Environmental Mobilization: Trends Among National Environmental Organizations." In *American Environmentalism,* eds. R. E. Dunlap and A. G. Mertig 11–26. Philadelphia: Taylor & Frances.

Center for Investigative Reporting and B. Moyers. 1990. *Global Dumping Ground.* Washington, DC: Seven Locks Press.

Chin, K. 1996. Customs, EPA Crack Down on CFC Crooks, *Chemical Engineering* 103(4): 43 and 51.

Conservation Foundation. 1984. *State of the Environment 1982.* Washington, DC.

Conservation Foundation. 1987. *State of the Environment: A View Towards the Nineties.* Washington, DC.

Cook, E. 1990. Global Environmental Advocacy: Citizen Activism in Protecting the Ozone Layer. *Ambio* 19(6–7): 334–38.

Davis, K. C. 1972. *Administrative Law Text.* St. Paul, MN: West Publishing.

Devall, B. 1992. "Deep Ecology and Radical Environmentalism." In *American Environmentalism,* eds. R. E. Dunlap and A. G. Mertig, 51–62. Philadelphia: Taylor & Frances.

Dunlap, R. E., and A. G. Mertig, eds. 1992. *American Environmentalism.* Philadelphia: Taylor & Frances.

EPA (U.S. Environmental Protection Agency). 1987. *Unfinished Business: A Comparative Assessment of Environmental Problems.* Vol. I. Washington, DC: Office of Policy Analysis.

———. 1990. *Reducing the Risk: Setting Priorities and Strategies for Environmental Protection.* Report of the Science Advisory Board: Relative Risk Reduction Strategies Committee (Report #SAB-EC-90-021). Washington, DC.

———. 1991. *Environmental Investments, The Cost of a Clean Environment.* Report of the Administrator, Environmental Protection Agency. Washington, DC: Island Press.

Echeverria, J., and R. B. Eby, eds. 1995. *Let the People Judge: Wise Use and the Property Rights Movement.* Washington, DC: Island Press.

Fiorino, D. J. 1995. *Making Environmental Policy.* Berkeley: University of California Press.

Fortuna, R. C., and D. J. Lennett. 1987. *Hazardous Waste Regulation: The New Era.* New York: McGraw–Hill.

Harris, C., W. L. Want, and M. A. Ward. 1987. *Hazardous Waste—Confronting the Challenge.* New York: Quorum Books.

Hill, W. W. 1978. *The National Environmental Policy Act and Federal Water Resources Planning: Effects and Effectiveness in the Corps and SCS.* Report No. IPM-4. Stanford, CA: Department of Civil Engineering, Stanford University.

Hilz, C. 1992. *The International Toxic Waste Trade.* New York: Van Nostrand–Reinhold.

Hirji, R., and L. Ortolano. 1991. Strategies Used by Kenya's National Environment Secretariat to Promote Environmental Protection. *The Environmental Professional* 13: 154–65.

John, D. 1994. *Civic Environmentalism: Alternatives to Regulation in States and Communities.* Washington, DC: CQ Press.

Kling, D. J., and E. Schaeffer, 1993. EPA's Flagship Programs: Existing Programs to Promote Pollution Prevention in Innovative Ways. *EPA Journal* 19(3): 26–27.

Kummer, K. 1995, *International Management of Hazardous Wastes,* Oxford: Clarendon Press.

Landy, M. C., M. J. Roberts, and S. R. Thomas, 1994. *The Environmental Protection Agency: Asking the Wrong Questions from Nixon to Clinton* (expanded edition). New York: Oxford University Press.

Lindblom, C. A. 1980. *The Policy Making Process.* 2d ed. Englewood Cliffs, NJ: Prentice–Hall.

Magat, W. A., A. J. Krupnick, and W. Harrington. 1986. *Rules in the Making.* Washington, DC: Resources for the Future.

Marcus, A. A. 1980. *Promise and Performance, Choosing and Implementing Governmental Policy.* Westport, CT: Greenwood Press.

McClosky, M. 1992. "Twenty Years of Change in the Environmental Movement: An Insider's View." In *American Environmentalism,* eds. R. E. Dunlap and A. G. Mertig, 77–88. Philadelphia: Taylor and Frances.

McCormick, J. 1993. "International Nongovernmental Organizations: Prospects for a Global Environmental Movement." In *Environmental Politics in the International Arena: Movements, Parties, Organizations and Policy,* ed. S. Kamieniecki, 131–43. Albany: State University of New York Press.

Melnick, R. S. 1983. *Regulation and the Courts: The Case of the Clean Air Act.* Washington, DC: The Brookings Institution.

Morrison, A. B. 1986. OMB Interference with Agency Rulemaking: The Wrong Way to Write a Regulation. *Harvard Law Review* 99: 1059–74.

Petulla, J. M. 1987. *Environmental Protection in the United States.* San Francisco: San Francisco Study Center.

Porter, G., and J. W. Brown. 1991. *Global Environmental Politics.* Boulder, CO: Westview Press.

Portney, P. R., ed. 1990. *Public Policy for Environmental Protection.* Washington, DC: Resources for the Future.

Quarles, J. 1976. *Cleaning Up America: An Insider's View of the Environmental Protection Agency.* Boston: Houghton Mifflin.

Ringquist, E. J. 1993. *Environmental Protection at the State Level: Politics and Progress in Controlling Pollution.* Armonk, NY: M. E. Sharpe.

Rosenbaum, W. A. 1991. *Environmental Politics and Policy,* 2d ed. Washington, DC: CQ Press.

Runnals, D. 1986. "Factors Influencing Environmental Policies in International Development Agencies." In *Environmental Planning and Management, Proceedings of the Regional Symposium on Environmental and Natural Resources Planning,* Manila, Phillipines, 19–21 February 1986. Manila: Asian Development Bank.

Steinberg, M. W. 1990. Can EPA Sue Other Federal Agencies? *Ecology Law Quarterly* 17(2): 317–53.

Stewart, R. B. 1977. Pyramids of Sacrifice? Problems of Federalism in Mandating State Implementation of National Environmental Policy. *Yale Law Journal* 86(5): 1196–1272.

Tolba, M. K., O. A. El-Kholy, E. El-Hinnawi, M. W. Holdgate, D. F. McMichael, and R. E. Munn, eds. 1992. *The World Environment 1972–1992.* London: Chapman & Hall.

Wenner, L. M. 1982. *The Environmental Decade in Court.* Bloomington: Indiana University Press.

World Resources Institute. 1992. *World Resources 1992–93.* Oxford: Oxford University Press.

Yap, N. 1990. NGOs and Sustainable Development. *International Journal.* Winter: 75–105.

Young, O. R. 1994. *International Governance: Protecting the Environment in a Stateless Society.* Ithaca, NY: Cornell University Press.

Chapter 4

Project Development: Actors, Processes, and Environmental Factors

Environmental regulations often influence decisions on major development projects such as highways, dams, and steel mills. How do proponents of such projects satisfy environmental requirements without sacrificing their economic objectives and other goals? This question is investigated by analyzing particular cases. Although case studies cannot provide generalizable conclusions, they allow a detailed investigation of how environmental concerns are treated in developing major projects. In analyzing cases, this chapter considers *all* aspects of planning, including politics and economics, to discover how environmental factors influenced project decisions.

The next section introduces a case-study *public* project, the New Melones Dam, and is followed by sections that extend the case by highlighting activities and processes typically associated with public projects. The final section of the chapter analyzes a case involving a petrochemical facility proposed by Dow Chemical Company. The Dow case highlights ways in which the planning of private development projects differs from public sector planning.

MULTIPURPOSE WATER PROJECT: NEW MELONES DAM

The New Melones Dam on the Stanislaus River in California was planned and built by the U.S. Army Corps of Engineers (hereinafter *the Corps*). An analysis of the project, which was a source of major struggles between environmentalists and project supporters, serves several purposes. Categories of actors that often participate in decision making for major public projects are identified. In addition, strategies used by parties in conflict over development objectives are highlighted. The New Melones case also demonstrates ways in which individual citizens and groups can intervene in public sector decision processes in the United States. Finally, the case illustrates difficulties in coordinating the interests of different government agencies operating in the same geographical area. Major events in the New Melones project's history are listed in Table 4.1.

Project Planning Prior to 1966

The New Melones project was controversial well before the start of construction in 1966, but for reasons unrelated to the project's environmental impacts.[1] Early conflicts stemmed from differences between the Corps and the U.S. Bureau of Reclamation (hereinafter *the Bureau*), the two largest water resources development agencies in the United States. Both agencies had plans to build a dam at the same site.

Damage from floods on the Stanislaus River during the 1930s led the U.S. Congress (in 1944) to authorize the Corps of Engineers to determine the feasibility of building a single-purpose dam to control floods. The Corps developed plans for a reservoir that would store 0.45 million acre feet (*maf,* a widely used unit of water volume). The project was to be located at the site of the Melones Dam, a small irrigation project developed in the 1920s by local farmers (see Figure 4.1).

At the same time the Corps was designing its flood control project, the Bureau of Reclamation was planning to expand its program of providing federally subsidized irrigation water to farmers in southern California. Working together with the California Department of Water Resources, the Bureau determined that a dam at the New Melones site could productively store 1.1 maf. The Bureau, with the approval of Congress, successfully negotiated with the Corps to have the latter plan a 1.1-maf project that would provide both flood control and irrigation water.

Although the Corps and The Bureau were able to establish the technical feasibility of a 1.1-maf project in the late 1940s, the Corps could not proceed without funds from Congress. Funding proved difficult. During the Eisenhower administration in the early 1950s, federal water projects had a low priority, and progress on the proposed New Melones project stalled. Moreover, in that same period, a state of California analysis showed that the irrigation water was not needed. In the state's analysis of New Melones, the ratio of economic benefits to costs was only 0.73, which is much less than the value required for federal water projects. Congress does not authorize water projects for construction if monetary benefits are lower than costs.

During the late 1950s, the Bureau of Reclamation refined its plans for the Central Valley Project,

TABLE 4.1 Key Events in the History of New Melones Dam

1944	U.S. Congress authorizes the Army Corps of Engineers (*the Corps*) to plan a 0.45-million-acre-foot (maf) reservoir for flood control.
1962	U.S. Congress authorizes the Corps to plan a 2.4-maf multipurpose project to be operated by the Bureau of Reclamation (*the Bureau*).
1966	Construction of 2.4-maf project begins.
1971	Sierra Club threatens to sue the Corps if construction of the 2.4-maf project continues without an environmental impact statement.
1972	Environmental Defense fund sues the Corps and the Bureau alleging the EIS is inadequate.
1973	State Water Resources Control Board (WRCB) denies the Bureau's application for rights to full use of 2.4 maf; the Bureau appeals in federal district court.
1974	California voters reject a citizen-initiated ballot measure that would block completion of the 2.4-maf project.
1978	Project construction is completed; U.S. Supreme Court establishes procedures to be followed by WRCB and the Bureau regarding reservoir operations.
1980	California governor vetoes state legislature's bill to permit filling reservoir to 2.4 maf.
1983	State Water Resources Control Board reverses earlier position and permits full use of reservoir.

[1] Except where otherwise noted, facts for the New Melones case study are from Palmer (1982). Individual pages in Palmer (1982) are not cited except where direct quotes are involved.

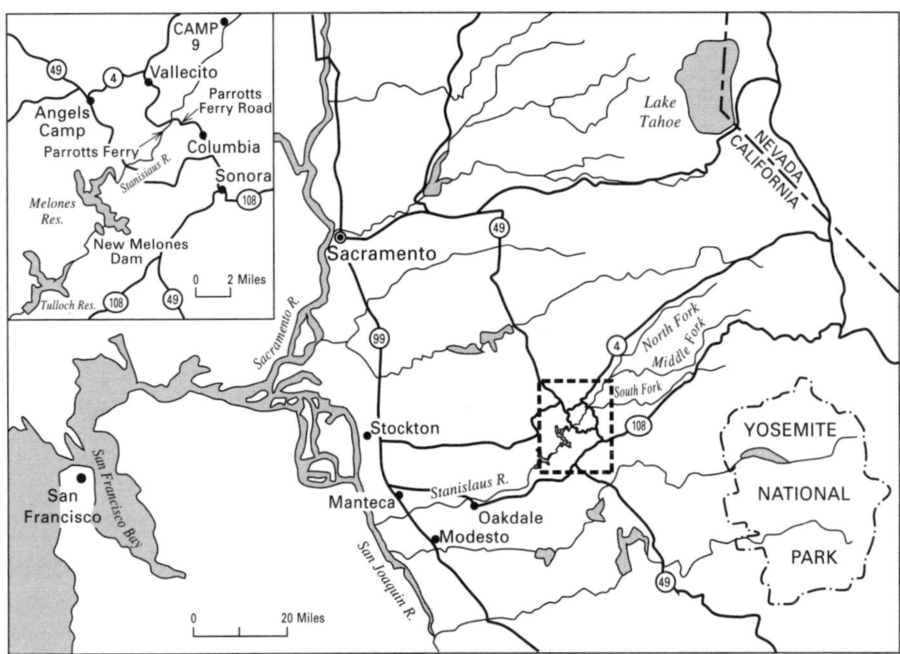

Figure 4.1 Location Map for New Melones Project. Reprinted with permission from Palmer, *Stanislaus: The Struggle for a River, C71* © 1982 by The Regents of the University of California.

a massive network of federal canals and reservoirs to transfer water from northern to southern California. This planning led the Bureau to conclude that irrigation water stored in a 2.4-maf reservoir at the New Melones site could be used productively within its Central Valley Project. Thus the Corps, in cooperation with the Bureau, urged Congress to approve a 2.4-maf project to provide flood control, irrigation water, hydroelectric power, and recreation among other things. In 1962, Congress authorized this greatly expanded version of the original New Melones project. The Corps was to design and construct the project, and the Bureau was to operate it. At the time of Congressional authorization, and for many years thereafter, the Bureau had no definite plans for when and where the irrigation water would be used.

A key factor contributing to the 1962 congressional authorization was a series of major floods on the Stanislaus River in the late 1950s and early 1960s. In 1965, local residents and farmers formed the Stanislaus River Flood Control Association to lobby for rapid completion of the dam.

Initial Opposition: 1962–1973

Not all local interests favored a federal dam.[2] Farmers outside of the flood plain feared the Bureau would export massive quantities of water to farmers in southern California, thereby depriving local farmers of irrigation water. Local opponents to the project formed the Stanislaus River Basin Group and proposed a nonfederal project of 1.1 maf to provide both flood control and irrigation water. Their efforts failed.

During the late 1960s, when the Corps was about to start building the 2.4-maf project, shifts occurred in how people valued the Stanislaus River. Two new issues emerged: the growing popularity of white-water rafting on the Stanislaus, and fears that water quality in the San Francisco Bay-Delta system would be degraded as a result of the proposed project.

In 1962, the year Congress authorized the 2.4-maf project, the Stanislaus River was first scouted

[2] Except where otherwise noted, facts in this section are from Randolph and Ortolano (1975).

as a zone for white-water rafting, a little known sport at the time. Within a decade, white-water rafting was in vogue. By 1971, the nine mile reach from Camp Nine to Parrot's Ferry (see Figure 4.1) was the most heavily used white-water rafting area in California and the second most popular rafting area in the country. The popular nine mile reach would not be inundated if the New Melones project stored 1.1 maf as proposed in the 1940s, instead of 2.4 maf as authorized by Congress in 1962. Rafting enthusiasts argued that the smaller of the two projects should be built.

The possibility that diversions of water from the Stanislaus River to farmers in southern California would degrade water quality led to new sources of opposition in the late 1960s. Environmental groups, especially the Sierra Club and the Environmental Defense Fund (EDF), feared that diversions of water to southern California would significantly diminish the freshwater flows available to flush out the Sacramento-San Joaquin delta and the Suisun and San Francisco bays. Decreased water quality and adverse effects on fish were two likely impacts of inadequate flushing.

Environmental Impact Reporting Requirements

Beginning in 1971, well after preliminary construction activities had started, opponents began taking steps to halt the 2.4-maf project. Their first approach involved use of the National Environmental Policy Act of 1969, with its requirement that federal agencies prepare environmental impacts statements (EIS's) when proposing "major federal actions significantly affecting the quality of the human environment." There was little doubt that the proposed New Melones project would have significant impacts, but as of 1971 an EIS had not been completed. This was not surprising, because national requirements for EIS's were first issued in 1970, and the Corps had been overwhelmed with the task of preparing EIS's for hundreds of projects in various stages of development.

In 1971, while the Corps was working on its EIS for New Melones, it learned that the Sierra Club would file a lawsuit if bids to complete project construction were accepted before the final EIS was issued. The Corps took the threat seriously. In what Corps officials termed an "unprecedented" action, bidding was delayed nearly five months until the final EIS was issued.

Soon after the final EIS came out in May 1972, the Environmental Defense Fund filed suit against the Corps and the Bureau. EDF argued, among other things, that the EIS was inadequate because it contained virtually no analysis of environmental impacts associated with the use of water stored for irrigation. Indeed, as of May 1972, the Bureau still had not specified when and where the irrigation water would be used. EDF felt that, in addition to assessing the impacts of the proposed use of stored water, the EIS should analyze the *uncertainty* associated with projected needs for the water. The position of EDF and other opponents was that a 1.1-maf project could meet needs for flood control and other legitimate project purposes without flooding the highly valued zone between Camp Nine and Parrot's Ferry.

The federal district court judge who heard the EDF case ruled in favor of the environmentalists and required that construction stop until the Bureau prepared a supplemental EIS in response to EDF's concerns about use of the irrigation water. The Bureau prepared the requisite document, which outlined the impacts associated with each of several alternative uses for the stored water. Even though the Bureau still had not specified a definite use for the water, the judge was satisfied and construction was allowed to continue. Although EDF appealed the judge's decision, it was ultimately unsuccessful, and opponents of the 2.4-maf project were forced to rely on other strategies.

Water Rights

Before the Bureau could withdraw water from the Stanislaus River, it needed legal rights to the water. Thus the Bureau made application to the California Water Resources Control Board (WRCB or *the Board*), the agency administering water rights. The application process, which involved public hearings, gave opponents another opportunity to derail the 2.4-maf project. This time the opponents prevailed.

The California Water Resources Control Board acted on the Bureau's application by granting it rights to fill the reservoir to 0.65 maf, except during times of flood, when it could store up to 1.1 maf. While the Bureau's application to use 2.4 maf of water was not granted, the denial was widely recognized as temporary. The WRCB did not grant the water rights requested because the Bureau had no specific plan for using the water. However, the Board's permit gave the Bureau an opportunity to prevail in the future if it could convince the Board "that the benefits that will accrue from a specific proposed [irrigation] use will outweigh any damage that will result to fish, wildlife and recreation in the watershed above New Melones Dam."[3]

Further Struggles to Block the Project: 1973–1983

Opponents of the dam saw WRCB's decision to limit reservoir filling to 1.1 maf as a temporary gain. The Bureau immediately appealed the decision to a federal district court. In addition to the threat of a reversal on appeal, project opponents feared that shifts in the politics of California water might cause WRCB to reverse itself. In addition, the state legislature might pass a law overturning the Board's decision.

Because neither the lawsuit by the Environmental Defense Fund nor the decision by WRCB represented permanent setbacks, the Corps continued construction of the 2.4-maf project. In the meantime, opponents regrouped to plan a new attack, one they hoped would yield a permanent victory.

Ballot Initiative

In late 1973, opponents unveiled a new strategy based on the ability of Californians to pass laws by voting on initiatives proposed by citizens. Many details needed to be worked out: How would the 300,000 signatures of registered voters needed to put a measure on a state ballot be obtained? What form should the measure take? Was there enough grassroots support to wage an effective campaign? Where would campaign funds come from? Through the efforts of a newly created grassroots organization called Friends of the River (FOR), satisfactory answers to all these questions were forthcoming.

In the fall of 1974, California voters had the opportunity to pass Proposition 17, a ballot measure that would have put the Stanislaus in the California Scenic Rivers System and killed the New Melones project. The measure would, however, have allowed for a smaller flood-control dam at the site. A yes vote on the proposition meant a no vote on the New Melones dam. The wording of the ballot measure was a source of confusion. (Did a no vote mean no to the dam?) By the time the heated statewide campaign was over, some citizens opposing the dam didn't know whether to vote yes or no. The proposition was defeated by a 6 percent margin.

Although reasons for the proposition's defeat are debatable, a contributing factor was imbalance in resources available to supporters and opponents. Friends of the River managed on less than $240,000, whereas dam supporters had a campaign fund of over $400,000. The main contributors to the pro-dam campaign were contractors working on the New Melones project, manufacturers of construction equipment, and farming interests.

A Final Outcome

Two years after the statewide ballot to stop the dam had failed, opponents of the project in the state legislature attempted to put parts of the Stanislaus in the California Scenic Rivers System. The bill died in committee.

In July 1978, the U.S. Supreme Court heard an appeal, brought by the state of California, in connection with the Bureau of Reclamation's earlier lawsuit (in a federal district court) to overturn the Water Resources Control Board's decision on water rights. The Supreme Court generally upheld the WRCB's position that the Bureau was required to abide by California water law. However, the Court sent the case back to the district court (in Fresno, California) where the case was originally heard. In doing so, the Supreme Court specified a process that WRCB and the Bureau were to follow. The Bureau was required to submit a reservoir operating plan to the state by the fall of 1979. After the WRCB decided on the acceptability of the plan, the district court would determine if the state's decision on the operating plan was consistent with the intent of Congress when it had authorized the New Melones dam. In sending the case back to the district court, the Supreme Court left the door open to further appeals.

In May 1980, an event that dam opponents long feared took place. Supporters of the 2.4-maf project lobbied the California legislature to pass a bill that allowed use of all of the reservoir's 2.4-maf storage capacity. Although the bill passed, it was vetoed by Jerry Brown, who was then governor of California. At that point, the WRCB's original decision remained in effect.

[3] WRCB Decision 1422 as quoted by Palmer (1982, p. 74).

Finally, in 1983, five years after project construction had been completed, supporters of the dam achieved the result they needed. The Water Resources Control Board, citing new studies that established a need for additional irrigation water, reversed its earlier position and allowed the reservoir to be filled to full capacity.

CHARACTERISTICS OF PUBLIC SECTOR PROJECTS

This section uses the New Melones case to highlight characteristics commonly associated with public sector development projects, especially infrastructure projects. Large public projects often involve a complex web of interactions among federal, state, and local organizations with different goals, and this often makes it difficult to implement major projects without encountering opposition. As demonstrated in the New Melones case, citizens who oppose a project have multiple opportunities to force project developers to consider the environmental consequences of their actions. Many opportunities for citizen intervention were created in the early 1970s in the United States (for example, EIS requirements), and they are available in other countries as well.

Complexity of the Institutional Setting

Figure 4.2 shows the types of organizations that can influence a project developer's decisions. Public development agencies are typically influenced by executive and legislative bodies. In the case of New Melones Dam, these were the U.S. president, who set budget priorities, and the U.S. Congress, which au-

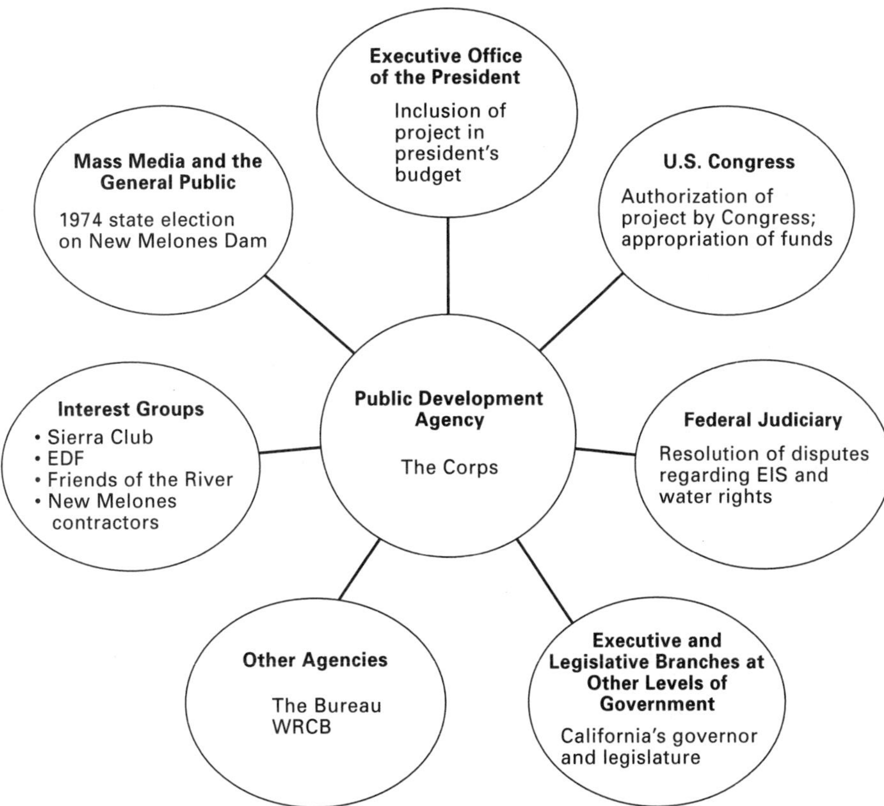

Figure 4.2 Organizations influencing public sector project proponents. Principal actors in the New Melones case.

thorized project planning and construction and appropriated funds to carry out these tasks. For public sector projects undertaken by local agencies, the corresponding executive and legislative bodies are typically a mayor (or city manager) and a board of supervisors (or city council).

In the United States, the courts are often used to resolve disputes over public projects. Two court actions played a role in the New Melones case. The lawsuit brought by the Environmental Defense Fund caused delays in construction and generated new information about project impacts. In addition, environmental studies conducted by the Corps in response to the suit led to modest changes to "mitigate" (or offset) adverse impacts. For example, the EIS process yielded a plan to preserve wildlife habitat to offset habitat that would be lost as a result of inundation.

Executive and legislative bodies at different levels of government may also influence a public project. Although federal development agencies try to gain endorsements from executive and legislative bodies at state and local levels before proceeding, government support sometimes erodes during the long time periods often involved in project planning. Changes in elected officials may lead to opposition where there was once support. For example, although previous California governors had endorsed the New Melones project, Governor Jerry Brown opposed it.[4]

Extensive coordination among agencies at all levels of government is common in planning projects. Following the introduction (in the 1970s) of environmental impact assessment requirements for many types of public projects, interagency coordination became more systematic. EIS requirements are not the only mandates for interagency coordination. For example, federal water resources agencies had been required to coordinate with fish and wildlife agencies for decades before the arrival of EIS requirements in the 1970s.

Often the most contentious interactions are between project proponents and citizens who stand to lose if a project goes forward. This is clearly demonstrated by the Corps' (and the Bureau's) struggles with the Environmental Defense Fund and Friends of the River.

[4] As previously noted, Brown vetoed the California legislature's bill that would have overturned the WRCB's decision to limit reservoir filling to 1.1 maf.

In addition to interactions with citizens groups, project development agencies often communicate with the general public. Newspapers and television typically play an important role in informing citizens of major projects. For controversial projects, development agencies may use a variety of "public involvement techniques" (e.g., public workshops) to keep the public informed, and sometimes these techniques are used to engage citizens in plannning activities. The high degree of involvement of the general public and the mass media in the New Melones project was somewhat exceptional. Voters are not typically asked to decide a project's fate via a ballot measure.

Irreversible Effects of Large Public Projects

Since passage of the U.S. National Environmental Policy Act of 1969, the degree to which a proposed project would cause irreversible change has become a common concern. Analyses of irreversible effects, which are required by the Act, must answer a practical question: irreversible over what time period? One reason the struggle to stop New Melones Dam was so vigorous is that opponents believed the effects of a completed project would never be undone. Chances of taking down a major dam are slim, and even if the project were dismantled, it would take decades before the inundated area resembled its original state.

The effects of a large project are often irreversible for political, not technical, reasons. It is rare, for example, that a major public project is demolished because its adverse effects are recognized and there is enough political support to tear it down. While such project dismantling is unusual, it is not unheard of. Consider, for example, the Pruitt-Igoe public housing project in St. Louis, Missouri. In the 1950s, the St. Louis Public Housing Authority (PHA) spent \$36 million to construct the 33-building, 2,800-unit Pruitt-Igoe housing project. In 1965, ten years after project completion, the PHA spent several millions more trying to change the project to eliminate numerous adverse effects caused by inadequacies in design and construction. For example, uninsulated steampipes led to countless painful burns, and low-hung windows without protective grates were linked to the death of three children from accidental falls.[5] The

[5] For more on the adverse effects of the Pruitt-Igoe project, see Bailey (1965).

project's design and construction deficiencies, and its location in an already badly blighted area, led eventually to its being overtaken by vandals. Crime rates skyrocketed. In 1973, twenty years after its construction, the PHA oversaw the demolition of all 33 buildings.[6]

Decision Making under Uncertainty

Another characteristic of large public projects is that some of their effects are difficult to predict. It is widely recognized that forecasts of project costs are uncertain, but the degree of uncertainty is often higher in predictions of economic benefits and environmental impacts. Estimation of economic benefits typically requires long-term projections of population and economic activity, and many assumptions must be made about how people are likely to behave in the post-project setting.

Although uncertainties in the projections of economic benefits of the New Melones project were not acknowledged by the Corps of Engineers, they were nonetheless present. This is illustrated by the Corps' calculation of the benefits of recreational use of the proposed reservoir. The Corps estimated there would eventually be four million recreational visitor-days per year at the reservoir. Opponents argued that this figure was absurb: the four million visitor-day figure was even higher than usage at nearby Yosemite National Park. Moreover, there were already 20 reservoirs used for recreation within 75 miles of New Melones; many of the sites were underutilized, and 15 were closer to major population centers than New Melones.[7] These challenges to the Corps' estimates of recreational use were significant, because the Corps' high recreational use figures were translated into economic benefits that increased the benefit-cost ratio for the project. During the early 1970s, opponents argued that the Corps' recreational benefit projections were most certainly inflated. Notwithstanding the validity of attacks on the benefit-cost ratio calculations, they did not have much influence because the Corps did not need to formally defend its economic analysis after Congress authorized the 2.4-maf reservoir in 1962.[8]

Arguably the most notable source of uncertainty surrounding the New Melones project concerned the use of stored irrigation water. Critics claimed that the Bureau's projected need for the water was exaggerated. However, strong support from those affected by floods and from farming interests in southern California allowed the Bureau and the Corps to complete the project despite the lack of details on when and where the irrigation water would be used.

PLANNING ACTIVITIES AND PROCESSES

The New Melones case study provides a context for introducing activities typically undertaken in planning public projects. For simplicity, planning activities are presented in a linear order. In practice, however, these activities often overlap, and it is common to redo studies as new information becomes available. Alexander (1979, p. 114) illuminates this point in commenting on the numerous "models" used to describe the process of planning physical works:

> [All] these models have one thing in common. They see planning as a sequential, multi-staged process in which many of the phases are linked to their predecessors in *feedback loops.* The central idea is that feedback causes the planners to rework previous analyses and, sometimes, reach different conclusions.

What are the phases or components of project planning and how are they linked? This section answers these questions and introduces planning terminology. It also indicates how environmental considerations can be integrated into all phases of project planning.

Most models that describe project planning contain the following components: identification of problems and goals, formulation of alternative plans, im-

---

[6] For information on the demolition of the Pruitt-Igoe project, see Kay (1973). Pruitt-Igoe is only one of many U.S. public housing projects demolished in recent years. In 1996, the federal government was "in the middle of an eight-year plan to spend more than $2.5 billion to tear down 100,000 apartments in high-rise buildings nationwide." (MacFarquhar, N. "U.S. Getting Public Housing Closer to the Ground," *New York Times,* June 2, 1996, national ed., pp. 1 and 15.)

[7] Details about recreational reservoir sites are from Randolph and Ortolano (1975, p. 246).

[8] For a critique of the Corps' benefit-cost analysis for the New Melones project, see Parry and Norgaard (1975).

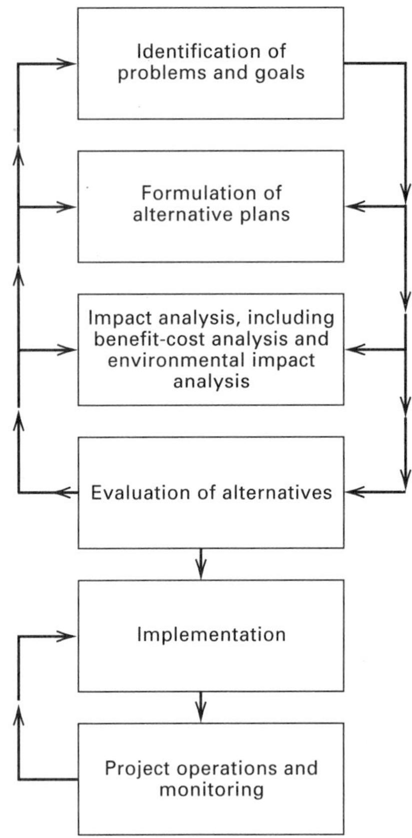

Figure 4.3 Components of Project Planning.

pact analysis, evaluation of alternative plans leading to a recommended course of action, plan implementation, and project operations and monitoring.[9] The lines between when one activity ends and another starts are often blurred. In fact, it can be argued that the first four activities (identification of problems and goals, formulation of alternatives, impact analysis, and evaluation) are carried out simultaneously, *not* sequentially. As suggested by Figure 4.3, during any particular stage of a planning study, information flows in both directions among all four activities. As planning proceeds, the four activities are carried out continually, but with increasing degrees of refinement. At any point, one of the activities may receive more emphasis than the others, and this accounts for the tendency of planners to describe where they are in a study by referring to only one of the activities.[10]

Identifying Problems and Goals

The common usage of the word "problem" involves the notion of an unsatisfactory situation; there is dissatisfaction with the status quo, and "problem identification" involves articulating what is wrong with the current situation. In the context of project planning, problem identification involves much more. Large physical works, especially infrastructure projects, are often built to deal with problems expected to occur in the future because of anticipated growth. Thus the problem-definition activity requires more than looking at what is currently unsatisfactory. It also involves making projections of economic development and population growth to identify anticipated problems so that actions can be taken to deal with them.

In the case of the New Melones project, the initial conception of the problem focused on flood damages. Population was increasing in the floodplain of the Stanislaus River, and projections of future growth indicated that flood damages would increase unless actions were taken either to manage development in the floodplain or to control the river. In the 1940s, when the Corps undertook its first studies for the New Melones project, placing limits on floodplain development was not generally considered as a strategy for reducing flood damages. The emphasis was on flood control.

More generally, problem definition also involves identifying goals, objectives, and constraints that guide the process of delineating and evaluating alternative plans. These goals, objectives, and constraints are sometimes termed *evaluative factors.*

For public projects, the process of identifying goals and objectives is often complex because of the large number of affected parties. In the New Melones case, the main parties in the early 1940s were the Corps, the owners of floodplain land, and the governments of cities subject to flooding below the dam site. To illustrate the nature of evaluative factors, consider the Corps' perspective. While its overall goal was to reduce flood damages, it faced constraints that influenced its view of the problem. The Corps' legisla-

[9] The terminology used in describing elements of planning studies is not standardized, but the terms presented here are commonly used.

[10] For a case example demonstrating points made in this paragraph, see Ortolano (1974).

tive mandate from Congress restricted its options: it could undertake physical works only to control flooding. Thus the Corps did not consider restrictions on land use or flood-proofing of structures as ways of reducing flood damages. In addition, the Corps was not free to implement projects for which monetary benefits did not exceed costs.

Local farmers downstream of the dam site were interested in reducing flood damages, but they wanted a solution they could afford. For a Corps of Engineers project, there are requirements for "local sponsors" to pay for a portion of the project. In the early 1940s, local cost-sharing requirements were minimal, and the bulk of a Corps project's costs were paid from the federal budget. Thus local interests were eager to find a solution to *their* problem that also met constraints on what the Corps could do. For a single-purpose flood control project in the early 1940s, it was not hard for the Corps and local interests to agree on a definition of the problem and on acceptable solutions.

As the scope of the New Melones project expanded beyond flood control to include irrigation and other project outputs, the number of affected interests grew, and there were many different conceptions of "the problem." Even as early as the 1950s, conflicting objectives had surfaced. At that time, the Bureau of Reclamation argued that irrigation water was needed for export out of the basin, but the state of California saw no such need, and many local farmers opposed water export. By the 1970s, the number of interested parties with differing views of the problem had grown to where achieving a consensus was virtually impossible.

Formulation of Alternatives

The conceptualization of alternative plans rests on *assumptions,* either explicit or implicit, about goals, objectives, and constraints. Consider again the New Melones study during the early 1940s. The Corps of Engineers translated goals and objectives about reducing future flood damages into numerical limits on the volume of floodwater that would need to be stored at different locations on the Stanislaus River. It also identified technical constraints related to river flows, soil conditions, and so forth, to eliminate unsuitable dam sites. In the process of conceiving alternative plans, constraints are used in two ways. First, they provide a basis for *designing* a plan. For example, the goal of keeping downstream flood elevations below target levels gives a basis for determing the minimum size of a flood control reservoir at the New Melones site. Second, constraints are used to judge the *feasibility* of a plan. For instance, a reservoir to store a particular volume at the New Melones site would have been unacceptable if projected economic costs exceeded benefits.

Environmental factors can play a key role in designing alternative plans if those factors are introduced early in the planning process. If the original planning exercise for the New Melones project had attached a high value to the nine mile reach between Camp Nine and Parrot's Ferry, the plan for a 2.4-maf reservoir at New Melones might have been ruled out because of environmental and social costs. However, in 1962, when the 2.4-maf project was authorized by Congress, no one could conceive of the high value that would later be associated with the nine mile reach.

The process of conceptualizing alternative plans often involves more art than science. Except for routine problems, the process relies on creativity and innovation. Before the 1970s, much of the "art" in water resources engineering involved use of *professional judgment* to narrow the range of alternatives early in the planning process. Often this narrowing was (and in many cases still is) undertaken by development agency planners who restricted their attention to actions their agencies could implement.

The environmental impact reporting requirements of the 1970s were an attempt to expand the design process. Environmental values were required to be integrated into the process of formulating alternatives. As demonstrated in Chapter 15, the situation in practice has not lived up to the theory articulated in EIS requirement. The conceptualization of alternatives often continues to be driven by technical and economic constraints. Issues related to the environment often are first brought up during the process of impact assessment at a time after a proposed project has been identified. At that point, it is difficult for project proponents to view impact assessment results dispassionately.

Impact Analysis

Impact analysis involves forecasting and describing changes (impacts) expected to result from a proposed action. An important question concerns which impacts to analyze. Large projects bring about many changes, but resources are rarely available to study all potential impacts in depth.

An economic impact analysis is often performed and is legally required for some types of projects. For example, in the United States, virtually all large federal water resources projects must include studies of economic benefits and costs. Another example involves the World Bank, which generally will not fund development projects unless their monetary benefits exceed their costs.

Although economic impact studies have been done routinely in the United States since the 1940s, comprehensive analyses of environmental and social impacts were not common before the National Environmental Policy Act of 1969 was enacted. Assessments of project impacts were undertaken before then, but they were typically limited in scope. For example, before the New Melones project was authorized in 1962, the Corps had commissioned studies of how the project would influence fish and wildlife in the Stanislaus River basin. These analyses were done in response to federal laws requiring the Corps to coordinate its plans with fish and wildlife agencies.

The issue of which impacts to study and how thoroughly to study them cannot be resolved using a formula. Every case is different. If goals and constraints are carefully articulated in the early stages of planning, they provide guidance for an impact assessment. One approach to identifying which impacts to study involves a process known as *scoping,* in which interested parties are brought together to develop a consensus on what to study before an impact analysis begins.[11]

Evaluation of Alternatives

Proposals that meet all constraints are, by definition, feasible. The evaluation stage of planning addresses the question, Which of the feasible proposals should be implemented? If there is only one feasible option, then the question is whether to implement the project or do nothing.

In principle, information from an environmental impact assessment plays a central role in evaluation. In practice, however, project proponents often decide on a project's location and scale before many impact studies are undertaken. This leads to the commonly observed outcome in which environmental impact analyses serve only to *mitigate* (or offset) adverse effects of a project previously selected based on technical and economic criteria.

Many methods have been developed to aid in project evaluation (see Chapter 17). Because project evaluation often rests on learning about the values of affected individuals, citizens are sometimes involved in the process.

Implementation

The time at which a development agency such as the Corps of Engineers recommends a proposal marks the end of a cycle of planning. At that point, those with authority to make final decisions review planning study outputs, especially the recommended plan, and their decisions may lead to a new round of planning. It is not uncommon for decision makers to either reject recommended plans outright (for example, because of controversies about negative impacts) or send them back for modification. For instance, Congress asked the Corps to redo the New Melones planning study more than once, as project purposes other than flood control were added, and as the scale of the project increased.

Before the 1970s, environmental issues did not play a major role in the process of building physical works. That situation has changed. Many construction contracts now specify practices that must be followed to minimize adverse environmental and social impacts during construction.

Project Operations and Monitoring

After a project is built, its operations are sometimes (but not commonly) monitored for their impacts on the environment. For example, some governments, such as the state of California, stipulate that projects subject to environmental impact reporting requirements must be monitored to ensure that agreed-upon impact reduction measures are carried out. Other examples include efforts by environmental groups to monitor operations of facilities to ensure compliance with environmental standards.

[11] The scoping concept grew out of early attempts to implement the environmental impact statement requirements of the U.S. National Environmental Policy Act, and it is discussed further in Chapter 15.

More common examples of monitoring are the measurements taken to comply with air and water quality laws. Many countries have pollution control regulations that require facility operators to *self-monitor* stack emissions and wastewater discharges and report results to environmental protection agencies. The latter often conduct their own monitoring.

Many environmental impact analysts have called for an increase in monitoring of impacts of new projects. One argument for post-project monitoring is that it can reduce adverse environmental effects. If monitoring reveals that impacts are unacceptable, changes can be made by either retrofitting the project or modifying project operations. This is the position advocated by Holling (1978) and his colleagues under the rubric of *adaptive environmental management.*

A second argument supporting an increase in monitoring relates to opportunities to improve the state of the art of impact prediction. Relatively few post-project monitoring studies are conducted. With increased monitoring, it would be possible to evaluate the accuracy of forecasts and identify limitations of existing impact prediction methods.

Citizen Participation in Public Sector Planning

Before the 1960s, citizen involvement in public sector planning was limited. In the United States, citizens were able to access the planning process for public projects through formal public hearings, but these were generally held after the development agency had made its decisions. For the most part, citizens were content to leave matters in the hands of planners, engineers, and elected and appointed officials. Public projects were planned and evaluated by technical specialists who coordinated, as appropriate, with government officials and agencies having an interest in their work.

During the 1960s, citizens called for increased opportunities to influence the planning of public projects. Many citizens claimed that the technical specialists planning large public projects were not accounting for environmental and social values.

By the 1970s, many development agencies observed that projects selected on the basis of a benefit-cost analysis and the judgments of technical specialists were often challenged in court by those concerned about environmental and social impacts. Some agencies responded by revamping their planning procedures to facilitate public involvement.

The planning processes devised in the 1970s by large infrastructure agencies such as the Corps of Engineers recognized that citizens could play a role in identifying problems and project objectives, and they could also contribute to project evaluation.[12] Indeed, some agencies recognized the difficulty in making tradeoffs between environmental losses and monetary gains *without* involving those affected. Agencies even found that citizens could assist in formulating alternative projects and analyzing impacts. Often citizens had detailed knowledge of an area that professionals might overlook and that information could be useful in identifying plans. Still more frequently, agencies recognized that citizens could help identify which impacts to analyze in detail.

The movement to involve citizens directly in the planning of public projects was not restricted to the United States. Citizen participation in planning took hold in many countries that had strong democratic traditions. The growing involvement of citizens in planning was associated with the increased influence of environmental nongovernmental organizations. Often grassroots environmental groups sprang up in response to a proposed public sector project and then exerted pressure on agencies to give citizens a role in planning and decision making.

PRIVATE SECTOR PROJECT PLANNING

Planning a private development project differs from public sector planning in two fundamental ways. First, opportunities for direct citizen participation in decisions made by private developers is limited. For the most part, citizens are restricted to participating in decisions of governmental bodies. An example is the citizen participation in meetings conducted by a city council prior to changing a local zoning ordinance to accommodate a proposed project. Second, a private project's success or failure often turns on the developer's ability to manage and allocate risks. For a developer, risks come in many forms and are present in

[12] For an analysis of how the Corps modified its water resources planning procedures to accommodate public involvement, see Mazmanian and Nienaber (1979).

all phases of a project including operations. For example, developers of office buildings run the risk that adequate amounts of office space will not be leased. As Peiser (1990, p. 500) put it: "A developer's fundamental role in city building is to accept risks that no one else will. Developers' success depends on their ability to unbundle huge risks into manageable proportions that can be alloted among various participants."

Because financial risks faced by private developers are commonly appreciated, this section does not emphasize them. Instead it focuses on *regulatory risk,* the risk that a project will be delayed or blocked because of a developer's failure to obtain approvals needed from government organizations. Regulatory risk is clarified by analyzing the public reviews faced by Dow Chemical Company in its efforts to build a petrochemical facility in California during the 1970s. The analysis also demonstrate how citizens can influence private development decisions.

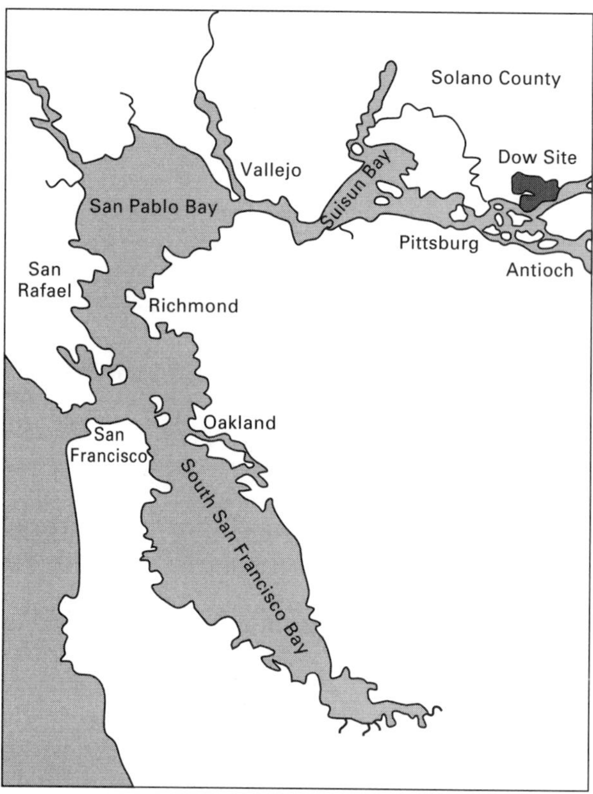

Figure 4.4 Dow Site Location in the San Francisco Bay Area. Reprinted with permission from Duerksen, *Dow vs. California: A Turning Point in the Envirobusiness,* pp. 6–7. © 1982 by the Conservation Foundation. All Conservation Foundation copyrights have been transferred to the World Wildlife Fund.

Regulatory Risk and Dow Chemical in California

In February 1975, the Dow Chemical Company announced plans to build a $500 million petrochemical complex in California, about 35 miles northeast of San Francisco.[13] The project, which included 13 different units, was to be built on a 600-acre site in Solano County, and a 250-acre site one-half mile away in Contra Costa County (see Figure 4.4). The facility was to convert naptha and other inputs into widely used substances including ethylene acetone, and styrene. The project would create about 1000 full-time manufacturing jobs and 1000 construction jobs. Dow's schedule called for initiating construction in 1976, beginning production at the first unit in 1978, and finishing all 13 units by 1982. By any measure, this was a mammoth project.

Soon after Dow released its plans, the Atlantic Richfield Company (ARCO) announced a "world-scale" chemical complex near the proposed Dow project. Not surprisingly, environmental groups in the San Francisco Bay area began to fear a major build up of chemical plants and related industrial facilities with attendant adverse effects on air and water quality and local wetlands. Environmentalists were also concerned that the proposed Dow project would attract a new wave of population to the area, and that the resulting demands for public services would outstrip the ability of local governments to provide them.

The regulatory environment faced by Dow Chemical was extraordinarily complex. Dow needed to obtain a minimum of 65 permits from a host of federal, state, and local agencies (see Table 4.2).

In addition to permit requirements, Dow's project was subject to California's environmental impact reporting requirements. Under the California Environmental Quality Act (CEQA), which was passed in 1970, an *environmental impact report* (EIR) is re-

[13] The facts in this section are from Duerksen (1982); page numbers are cited only for direct quotes.

TABLE 4.2 Minimum Number of Permits Required for Dow's Proposed Petrochemical Project

Federal		
U.S. Army Corps of Engineers		4
U.S. Coast Guard		1
	Subtotal-federal	5
State		
Regional Water Board		7
Water Resources Control Board		1
Department of Water Resources		1
Fish and Game		3
Reclamation Board		1
State Lands Commission		1
Bay Area Air Pollution Control District		26
	Subtotal-state	40
Counties		
Sacramento		1
Solano		11
Contra Costa		8
	Subtotal-counties	20
Grand Total		65

Reprinted with permission from Duerksen, *Dow vs. California: A Turning Point in the Envirobusiness,* pp. 6–7. ©1982 by the Conservation Foundation. All Conservation Foundation copyrights have been transferred to the World Wildlife Fund.

quired for a broad range of decisions by state agencies and local governments. Although several state agencies and county governments made decisions in this case, Solano County took the lead role in the *EIR process* because it had the greatest involvement with the project. The EIR process, which included compiling information on environmental and social effects and responding to public and agency comments on a draft environmental impact report, was completed quickly. The rapid pace led to problems. State agencies and interested citizens later charged that the EIR process was completed so fast they did not have adequate time to review the draft environmental impact report.

The EIR process was also criticized because the environmental impact report was prepared by a consulting company that was paid by Dow itself. Even though this arrangement was legal under CEQA, and the EIR was prepared for Solano County (not Dow), it led to questions about the objectivity of the environmental impact report. One state official claimed the document was "a sweetheart EIR that reeked with what was convenient for Dow."[14]

In December 1975, the Solano County Board of Supervisors accepted the final EIR as adequate and took steps to facilitate the implementation of Dow's project. For example, the county rezoned a portion of the project's 600-acre site from agricultural to industrial use.

Dow faced two problems because of the speed with which the EIR process was completed. One was that soon after the final EIR was accepted by Solano County, a lawsuit was brought in an attempt to stop

[14] Larry King, then of the state attorney general's office, as quoted by Duerksen (1982, p. 47).

the project. The suit, which was initiated by Friends of the Earth, the Sierra Club, and a local environmental organization, People for Open Space, alleged that Solano County violated procedures in implementing the EIR process and in rezoning agricultural lands that were part of the project site. This suit was of no consequence in the long term.

The second response to the rapidity of the EIR process did not occur until several months later, when Dow applied to the U.S. Army Corps of Engineers for permits. Because a decision by the Corps to issue permits to Dow would have major environmental impacts, the federal *environmental impact statement* (EIS) process was set in motion. The EIS process is the federal counterpart to the state-level EIR process.[15] Even though each process calls for written documentation of projected environmental impacts, the processes were not coordinated.

In April 1976, the Corps issued a draft EIS and circulated it to various governmental agencies and interested citizens for review and comment. By this time, several state agencies had recognized that reviews they had conducted during the EIR process orchestrated by Solano County were inadequate because they had been done too quickly. That, and their suspicion that the EIR was incomplete, led state agencies to raise new questions and ask Dow for much additional information.

While Dow's often contentious negotiations with state agencies over information needed to complete the EIS review were proceeding, the company ran into another stumbling block. This one concerned permits Dow needed from the Bay Area Air Pollution Control District (BAAPCD), the agency charged with implementing both state and federal air quality regulations in the San Francisco Bay area. During the summer of 1976, the BAAPCD denied Dow's request for a permit for a proposed styrene plant, one of the 13 units in Dow's proposed complex.

The basis for the BAAPCD's denial was a provision in the U.S. Clean Air Act of 1970 that barred significant new sources of air pollution from *nonattainment areas,* regions failing to meet national ambient air quality standards (NAAQS) for one or more pollutants. For example, if an area failed to meet ambient standards for particulates, no significant new sources of particulates would be permitted.[16]

Dow's problem was that even though it would use air pollution control devices at its plants, the remaining emissions would constitute significant releases of three pollutants for which the NAAQS were already violated: particulates, hydrocarbons, and oxidants. Thus Dow's request for a permit was denied. The company appealed the decision to the BAAPCD's hearing board.

By the fall of 1976, Dow was reaching the limits of its patience with several agencies of the state of California. The stalemate between Dow and the agencies led to an intervention by Bill Press, one of then-governor Jerry Brown's top aides. Press wanted to speed the process of issuing state agency permits because he felt the governor needed to "give some signal of pro-business sympathy."[17] Many within the business community felt that regulatory gridlock was rampant in California. After extensive negotiations with Press, Dow agreed to participate in two full days of hearings with all state agencies that issued permits Dow needed. The agencies delivered over 70 detailed questions to Dow by a mutually agreed-upon deadline: November 5, 1976.[18] About one month later (on December 8) the hearings were held; they generated more than 1000 pages of testimony. The governor's staff promised Dow officials a definitive "yes or no" response reflecting the collective view of all responsible state agencies within 30 days. The deadline passed with no response. On January 19, 1977, after investing \$10 million on the planning and design for the project, Dow dropped its proposal. The governor's office was dumfounded by Dow's decision, which, as one observer put it, "shook the windowsills from Sacramento to San Diego."[19]

After all its efforts, Dow felt it had to cut its losses. As the company's press release put it, "the

[15] The federal process leads to preparation of an EIS as required by the U.S. National Environmental Policy Act, and the state process yields an EIR under the California Environmental Quality Act. The required contents of an EIS and an EIR are similar but not identical.

[16] This provision of the Clean Air Act was modified in the late 1970s to make it possible for projects such as the one proposed by Dow to obtain permits in nonattainment areas under certain conditions (see Chap. 13).

[17] The quotation by Press appears in Duerksen (1982, p. 75).

[18] The complete list of questions is given by Morse (1984, pp. 149–57).

[19] This quotation and the one that follows are both from Duerksen (1982, p. 109).

permitting process for new facilities has proved to be so involved and expensive it is impractical to continue." Dow Chemical Company had allocated about $150 million for the proposed project. The decision to drop the project reflected the view that the company "should not be holding up money that would otherwise be spent somewhere else in the world."[20] The president of Dow put it plainly: "In the final analysis, what killed us was the uncertainty."[21] The regulatory risks were too high to allow the proposed project to go forward.

Private Development and Regulatory Risks

Important sources of regulatory risk faced by a private project proponent include:

- uncertainty about how terms in environmental laws will be interpreted by regulators;
- requests for substantially more environmental impact information than anticipated;
- changes in regulations that take place during project planning;
- miscommunications between regulators and developers; and
- delays caused by legal actions and administrative appeals brought by project opponents.

Uncertainties about outcomes of the regulatory process translate into risks that regulatory compliance costs will be higher than expected or that a project will be delayed or stopped completely. Costs of compliance include funds to construct treatment facilities, mitigate adverse effects, and pay development fees, and these costs can mount quickly. Unanticipated delays can lead to substantial costs in the form of increased interest on outstanding loans taken by developers, or lost opportunities to earn income by diverting tied-up funds into alternative projects. Several of these regulatory risks and costs were present in the Dow case, and they are elaborated in the remainder of this chapter.

The Nature of Regulatory Risks

Dow dropped its plans for the proposed petrochemical project because it believed the regulatory risks were too high. Ray Brubaker, who was head of the Dow division in charge of the proposal, described the company's position succinctly:

> Our real problem was with the statewide permits. We weren't making timely progress. With no positive results to show after spending two and a half years and $4.5 million to get four permits out of 65, I had to cut my losses.[22]

Regulatory risks are often incurred because of ambiguities regarding what actions are permissible. The dispute between Dow and the Bay Area Air Pollution Control District over what constituted "significant emissions" of air pollutants is illustrative. At the time of the BAAPCD denial of Dow's permit application, if expected emissions of a new source of air pollution were deemed *significant,* the source could be barred. The Clean Air Act of 1970 provided no definition of "significant source," and it was left to the BAAPCD to provide one. The BAAPCD interpretation was that if one "could measure the emissions on a monitor at ground level, then it was significant."[23] The Dow staff found this unreasonably stringent and pointed to other regional air pollution control agencies that set much less exacting numerical emission limits (for example, if the pollutant is emitted at or above 150 lbs/hr, then it is significant). The Dow staff also disputed the many "worst case" assumptions regarding wind speed and other meteorological conditions that the BAAPCD used in its air quality models for estimating ground level emissions. Notwithstanding Dow's objections, the BAAPCD held firm and refused to issue the permit.

Dow also faced unanticipated regulatory costs when the data and predictions provided in its environmental impact assessments were deemed incomplete. The risk to a developer that additional information will be called for is unavoidable because there are no standards for minimum levels of detail in an environmental impact analysis. For a large, complex project like the one proposed by Dow, there may be dozens of potential impacts. The dilemma for the project proponent (or its consultants) is in sorting out which impacts to study and at what levels of detail. If impacts that regulators or project opponents

[20] Ray Brubaker, a Dow manager at the time, as quoted by Duerksen (1982, p. 111).

[21] Paul Oreffice, as quoted by Morse (1984, p. 146).

[22] Ray Brubaker, as quoted by Duerksen (1982, p. 111).

[23] Feldstein, then second in command at the BAAPCD, as quoted by Duerksen (1982, p. 60).

deem important are not studied carefully, a developer may be asked to do more. That is what happened to Dow.

Dow's costs in dealing with requests for new information skyrocketed during the federal EIS process. State agencies that had made superficial reviews of the environmental impact report prepared for Solano County under state law came back in force during reviews of the draft EIS developed by the Corps of Engineers. For example, the California Resources Agency wanted new information in four areas it deemed critical: air quality; increased ship traffic; secondary growth, exemplified by the increased development of other industrial facilities that would provide services to Dow's new plants; and chemical spills and water quality. Jack Jones, the Dow official coordinating compliance with environmental regulations at the time, found many of the questions to be "impossible." He observed: "We answered what questions we could. I was spending, at that time, $200,000 per month on outside consultants to answer questions and do studies."[24]

Another source of risk demonstrated by the Dow case involves miscommunications. Here, also, an example involves the Bay Area Air Pollution Control District. At the time of Dow's permit application, if an area failed to attain national ambient air quality standards, a significant new source of air pollution could be denied a permit. The source of the misunderstanding was whether Dow's proposed project was in a nonattainment area. Jack Jones, who was then coordinating Dow's compliance with regulations, argued that in 1974, BAAPCD staff assured Dow that it need not be concerned with the nonattainment requirements of the federal Clean Air Act. According to Jones, "[BAAPCD staff] told us that San Francisco was not classified as a nonattainment area and there would no such designation in the foreseeable future."[25] It is clear that a miscommunication took place. Milton Feldstein, who was then second in command at BAAPCD, said he was "amazed" at Jones's assertion. In Feldstein's view, "the company didn't seem to understand what ambient air standards were all about."[26]

The risks noted above all concern interactions between project developers and regulators, but interactions with project opponents can also be a source of risk. In many places, opponents can delay a project using routine administrative procedures or the courts. Critics of the Dow project exercised their rights by bringing a lawsuit based, among other things, on alleged inadequacies in the environmental impact report. That suit might have been a source of delay had other events not caused Dow to pull out. Developers can lose significant amounts of money in protracted court battles, and they sometimes drop proposed projects in midstream to avoid the risk of delays that opponents could cause by suing them.

Reducing Regulatory Risks

The regulatory hurdles faced by Dow Chemical Company are not unique to Dow. Nor are they unique to California. Since the 1970s, private companies and governments at all levels have become sensitive to the risks that businesses face in dealing with increasingly complex and burdensome environmental regulations. Several strategies have been used in attempting to reduce those risks.

To avoid the stigma of being "anti-business," some state and local governments have tried to reduce the time and uncertainties associated with obtaining permits. For example, in the late 1970s, Georgia implemented a "consolidated one-stop shopping system for environmental permits."[27] Similar one-stop shopping schemes have been used by others. The idea is to simplify the permit process by having developers work with a single agency that coordinates with all other agencies of that government. Thus if a developer needed permits from five state agencies, the developer would fill out only one application form instead of five. Moreover, the developer would not have to interact with all five agencies. In this context, one-stop shopping means that a single state agency coordinates the application and consolidates the concerns of all state agencies involved. In theory, at least, the developer would receive a single, consolidated multi-agency response to a permit application in a timely fashion.

One-stop shopping can reduce regulatory risk, but it is no panacea. Putting the burden of coordination on one agency instead of the project proponent may not necessarily make the process faster. In addi-

[24] Jack Jones, as quoted by Duerksen (1982, p. 69).

[25] Jack Jones, as quoted by Duerksen (1982, p. 55).

[26] This quotation is from Duerksen (1982, p. 57).

[27] The quotation characterizing Georgia's permit system is from Duerksen (1982, p. 128).

tion, it will not eliminate frustrations that developers experience when they are forced to work with regulators who are overworked or inexperienced, neither of which is an uncommon circumstance. Moreover, one-stop shopping will not reduce regulatory uncertainties that developers experience when laws and regulations contain ambiguous terms that must be interpreted by regulators on a case-by-case basis.

Companies have taken their own steps to reduce regulatory risk. One strategy has been to strengthen communications with potential project opponents at an early stage in planning. This includes communications with both individual citizens and environmental groups who may have concerns about a proposal. Some firms have created ad hoc organizational units to work with outside interests on particular projects. An example is the "communication task force" created by Du Pont to facilitate siting of its titanium dioxide plant in Taiwan in the late 1980s. Du Pont's efforts to communicate with project opponents was instrumental in allowing it to build the plant at Kuanyin in northern Taiwan.[28]

When communication barriers between developers and opponents of a project have persisted, some firms have used mediation to resolve disputes on project siting. As described in Chapter 18, mediation often provides an alternative to the court system for resolving disputes over siting of controversial facilities.

In addition to improving communications with potential opponents, many firms are enhancing their environmental performance. As of the late 1980s, "a number of multinationals in environmentally intensive industries . . . had begun to develop a kind of 'corporate environmentalism.'"[29] The experience of Johnson & Johnson is illustrative. During the 1980s, the company set up programs to ensure that all of its units were obeying government and corporate environmental rules. In addition, it made efforts to "build links (some educational, some financial) with local communities and [continue] the environmental education of employees. [It was also] trying to cut packaging and make as much use of recycled and recyclable materials as possible."[30]

Many corporations that aim to be environmentally responsible have focused on *pollution prevention* activities: waste source reduction, recycling, and reuse. As explained in Chapter 14, firms often have strong incentives to pursue pollution prevention strategies vigorously.

Some companies have gone beyond pollution prevention and included environmental responsibility as an element of facility siting and design. An example is Du Pont's effort to in establish a one billion dollar multiproduct industrial facility in the Asturias region in northern Spain. In 1989, Du Pont invited the U.S.-based Wildlife Habitat Enhancement Council (WHEC) to help design its new facilities in ways that would enhance environmental values, particularly wildlife conservation, wetlands restoration, water quality improvement, and the preservation of local history and culture. Working with WHEC, Du Pont has relied heavily on environmental factors in siting, designing, and constructing its plants in the Asturias region. Du Pont commissioned an "independent" environmental impact assessment for the proposed project to gauge the environmental impacts and risks, which were found to be minimal. The company expects to devote 15 percent of its investment for the facility on pollution control, and at least one production unit will be designed for "zero emissions." According to Bill Walker, project director for Du Pont's facility at Asturias, "[w]orking with WHEC has helped to both mitigate some concern in the community prior to construction and demonstrate the compatibility of industry and the environment."[31] This degree of corporate environmental responsibility is not common, and it demonstrates how firms can reduce regulatory risks by working with local communities and integrating environmental factors into facility siting and design.[32]

In summary, firms can reduce regulatory risks by improving communications with local communities and potential opponents of proposed projects,

[28] The task force was created after Du Pont's unsuccessful experience in trying to locate the plant in Lukang, a historically significant city whose citizens vigorously opposed Du Pont's proposal. For information on the Du Pont project in Taiwan, see Ortolano (1993).

[29] This quotation is from Holmberg (1992, p. 186).

[30] This quotation is from Cairncross (1991, p. 283).

[31] Bill Walker as quoted by Schmidheiny (1992, p. 235). All facts about the Du Pont facility in Spain are from Schmidheiny (1992).

[32] Most firms have a long way to go toward engaging in environmentally sensitive plant siting, design, and construction. Impediments that corporations face in changing past practices regarding the environment are analyzed by Holmberg (1992, pp. 171–78).

and by using pollution prevention (waste reduction, recycling, and reuse) to supplement traditional waste treatment. Firms can also reduce risks by analyzing environmental impacts of proposed projects and by using the results to make decisions about facility design, siting, and construction.

Key Concepts and Terms

Multipurpose Water Project: New Melones Dam

Interagency coordination
Interagency competition
Benefit-cost ratio
Environmental impact statement
Water rights

Characteristics of Public Sector Projects

Irreversible impacts
Complex institutional settings
Multiplicity of actors and goals
Decision making under uncertainty
Citizen participation in planning

Planning Activities and Processes

Models of project planning
Identification of evaluative factors
Formulation of alternatives
Impact analysis and evaluation
Implementation and operations
Post-construction monitoring
Public participation techniques

Private Sector Project Planning

Regulatory risk
Financial risk
Regulatory gridlock
One-stop shopping for permits
Corporate environmentalism
Pollution prevention

Discussion Questions

4-1 Much of this chapter rests on the following assertion: It is difficult to understand how environmental factors influence project decisions without considering all aspects of the planning process, particularly those concerned with politics and economics. Use materials from the two case studies in this chapter to argue for or against this position.

4-2 What are the main differences in the *processes* used to plan public as compared to private development projects? Indicate differences in the factors influencing the behavior of public and private project developers.

4-3 This chapter includes one case study of a public sector project and a second case study of private sector development. To what degree are the types of organizations that influenced project outcomes the same in both cases?

4-4 In many jurisdictions, consultants that perform environmental impact assessments for proposed projects are paid directly by project developers. Does this arrangement compromise the ethics of consultants? What can be done to reduce the likelihood that consultants have a conflict of interest in assessing projects when developers pay for assessments?

4-5 Some have argued that interest groups occasionally misuse environmental impact assessment laws in their efforts to stop projects they oppose. Can this argument be supported by the case studies in this chapter?

4-6 Imagine that you are hired to advise the state of California on steps it can take to reduce *regulatory gridlock*. What advice would you offer?

4-7 Suppose you are a consultant to a corporation that wants to locate a large industrial complex in an area that contains citizens groups that may oppose the project because of its environmental impacts. How would you advise the corporation on steps it could take to reduce the risk that its project will be delayed or stopped.

References

Alexander, E. R. 1979. "Planning Theory." In *Introduction to Urban Planning,* ed. A. J. Catanese and J. C. Snyder, 106–119. New York: McGraw–Hill.

Bailey, J. 1965. The Case History of a Failure. *Architectural Forum* 123(5): 22–29.

Cairncross, F. 1992. *Costing the Earth.* Boston: Harvard Business School Press.

Duerksen, C. J. 1982. *Dow vs. California. A Turning Point in the Envirobusiness Struggle.* Washington, DC: Conservation Foundation.

Holling, C. S., ed. 1978. *Adaptive Environmental Assessment and Management.* Chichester, England: Wiley.

Holmberg, J., ed. 1992. *Making Development Sustainable.* Washington, DC: Island Press.

Kay, J. H. 1973. Pruitt-Igoe Housing Project. *The Nation* 217(9): 284–86.

Mazmanian, D., and J. Nienaber. 1979. *Can Organizations Change?* Washington, DC: Brookings Institution.

Morse, C. W. 1984. *Environmental Consultation.* New York: Praeger.

Ortolano, L. 1974. A Process for Federal Water Planning at the Field Level. *Water Resources Bulletin* 10(4): 766–78.

Ortolano, L. 1993. Controls on Project Proponents and Environmental Impact Assessment Effectiveness. *The Environmental Professional* 15(4): 252–363.

Palmer, T. 1982. *Stainislaus: The Struggle for a River.* Berkeley: University of California Press.

Parry, B. T., and R. B. Norgaard. 1975. Wasting a River. *Environment* (Jan./Feb.): 17–27.

Peiser, R. 1990. Who Plans America? Planners or Developers? *Journal of the American Planning Association* 56(4): 496–502.

Randolph, J., and L. Ortolano. 1975. Effect of NEPA on the Corps of Engineers' New Melones Project. *Columbia Journal of Environmental Law* 1(2): 232–73.

Schmidheiny, S. 1992. *Changing Course: A Global Business Perspective on Development and the Environment.* Cambridge, MA: MIT Press.

Part Two

Designing and Implementing Environmental Regulations

Part Two analyzes how concepts from economics and engineering are synthesized in designing and implementing environmental regulations. The first three chapters provide a foundation by introducing topics from environmental economics. Chapter 5 shows how basic economic concepts related to supply, demand, and markets are used to characterize environmental pollution problems. In Chapter 6, those basic economic concepts are applied to the task of placing monetary values on damages caused by pollution. Chapter 7 integrates concepts from Chapters 5 and 6 by introducing benefit-cost analysis and calculating efficient levels of pollution abatement.

The remaining chapters in Part Two concern regulatory strategies. Chapter 8 analyzes the traditional approach to regulation, *command and control.* Using this scheme, an environmental agency issues "commands," typically by specifying particular pollution abatement technologies or by restricting waste discharges. The agency then follows up with control activities, which include checking to see if its requirements have been met and imposing sanctions when regulations are violated. Chapter 9 provides a perspective on difficulties of monitoring and enforcing environmental requirements. Although the chapter emphasizes command and control because that approach is so widespread, the implementation difficulties it describes apply also to other regulatory schemes.

The last two chapters in Part Two examine what are often termed *market-based policy instruments* for environmental regulation. Using these instruments, a regulatory agency doesn't issue detailed requirements telling polluters what they must do. Instead, the agency provides economic incentives that could, in theory at least, lead polluters to cut back their wastes to efficient levels. One form of incentive, *pollution discharge fees,* is analyzed in Chapter 10. A second market-based approach to regulation, *tradable discharge permits,* is considered in Chapter 11. Using this approach, waste discharge permits can be bought and sold by polluters.

All three of the regulatory schemes—command and control, pollution discharge fees, and tradable pollution permits—have been applied extensively in many countries. Case studies in Part Two demonstrate practical applications of the strategies. Part Two ends with a comparative evaluation highlighting strengths and weaknesses of each of the three regulatory approaches.

Chapter 5

Economic Framework for Analyzing Environmental Problems

Economists begin their analysis of environmental protection by recognizing that what is commonly termed *pollution* is an unavoidable by-product of living.[1] The key regulatory question concerns how best to manage pollution by, for example, minimizing waste production, recycling waste materials, and treating wastes to reduce destructive discharges.

This chapter introduces economic concepts used to analyze environmental regulations. It begins by viewing the environment as an economic asset capable of providing a variety of services. Next, the chapter presents two types of economic models: a model of consumer choice, which provides a foundation for placing monetary values on the benefits of environmental regulation; and a model of production, which enables analysts to estimate how firms will react to different regulatory policies. Finally, the chapter introduces several conditions that define a perfectly competitive market. This market form is a theoretical ideal. A model of perfectly competitive markets is used to investigate how real markets fail by not providing consumers and producers with prices that reflect the social costs of their decisions.[2]

THE ENVIRONMENT AS AN ECONOMIC ASSET

Economists often use the term *residuals* to mean the material or energy "left over" from the various consumptive and productive activities carried out by individuals, firms, and governments. For example, the residuals generated by a ride on a bus include the noise, heat, hydrocarbons, particulates, and carbon monoxide that emanate from the bus. Residuals correspond to what are ordinarily called pollutants or

---

[1] Helpful comments on an early version of this chapter were received from Melissa Geeslin and Greg Browder (while they were students at Stanford University); David Lougee of Stanford's Technical Communications Program; and Elizabeth Schaefer, a writer and editor in Oakland, CA.

[2] Readers interested in a more complete introductory treatment of topics in this chapter are referred to textbooks used in a first course on economics—for example, Samuelson and Nordhaus (1992). More advanced treatments are found in textbooks on microeconomics, such as the ones by Mansfield (1988) and Baumol and Blinder (1991).

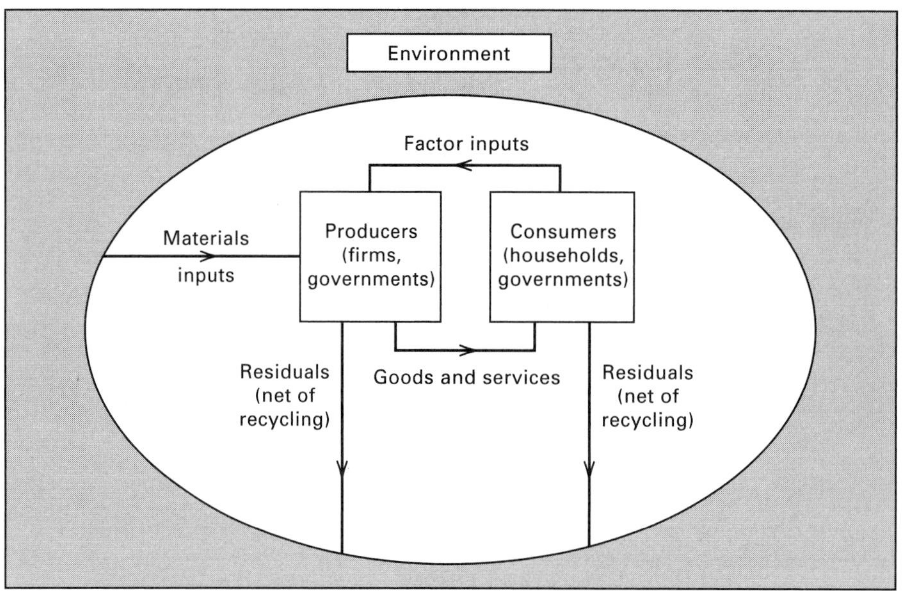

Figure 5.1 Traditional economic activities and the environment.

waste discharges, and this text uses these terms interchangeably.

The natural environment can be viewed as a large shell that surrounds the economic system (see Figure 5.1). Firms and governments produce goods and services, which are sold to consumers. The production of goods and services utilizes *factor inputs;* for example, labor provided by households and raw materials from the environment. (Economists use the term *households* to include consumers of goods and services and owners of factor inputs.) Production also yields residuals, which flow back to the environment. Consumers use the income received from firms and governments to purchase goods and services. The process of consuming these goods and services yields additional residuals, which are ultimately released into the environment. The residuals depicted in Figure 5.1 are discharged to the environment after recycling and waste management activities are carried out by producers and consumers.

Residuals, which are a normal consequence of virtually all productive and consumptive activities, consist of either materials or energy. *Materials residuals* are substances returned to the environment in solid, liquid, and gaseous forms. *Energy residuals* take the form of noise or waste heat; for example, the heat that goes up a chimney when wood is burned in a fireplace. What are generally termed "wastes" can often be recycled (in which case they become factor inputs), or they can be treated using physical, chemical, and biological processes.

Physical laws dictate that residuals cannot be eliminated entirely. Consider, for example, the first two laws of thermodynamics. The first law states that energy cannot be created or destroyed; it can only be transformed from one state to another. Although the second law of thermodynamics cannot be stated in such simple terms, it indicates that any process to transform energy will inevitably produce heat that is not available for doing work.[3] For example, the combustion of coal in a power plant transforms chemical energy into heat that turns water into steam. The

[3] The second law of thermodynamics can be stated in terms of *entropy,* a measure of the amount of energy in a system that is not available for doing useful work. The second law states that entropy increases in a system undergoing change. For example, when coal is burned to produce heat, the amount of useful work that can be done with the heat is less than the useful work that could have been done with the coal if the coal could have been turned into energy with 100 percent efficiency. In the example involving burning coal to produce electricity, each energy transformation produced an energy residual, waste heat.

steam passes through a turbine and thereby generates electricity. During each of these energy transformations, some energy is lost as waste heat.

A Materials Balance Perspective

The law of conservation of mass plays a central role in designing policies for controlling residuals.[4] Policies that fail to consider this physical law sometimes shift the problems caused by residuals from one *medium* (air, land, or water) to another. For example, regulations aimed at improving water quality may end up causing solid waste and air quality problems. When additional matter is removed from wastewater, more solids have to be disposed of in landfills or by incineration; the latter is a source of air pollution.

A *material balance perspective* of the environment recognizes that processes used to treat wastes do not eliminate them. Rather, treatment processes transform wastes into substances that may cause fewer objectionable effects when they are ultimately released into the environment. Consider, for example, the use of screens and other physical devices to remove solids from wastewater at a municipal treatment plant. Solids are taken out of wastewater with the hope that when they are returned to the environment as dried sludge, incinerator ash, or airborne materials, they will cause less of a problem than if they flowed directly into a watercourse.

Although the materials balance idea may seem elementary, governments have sometimes adopted environmental policies that seem to ignore the law of conservation of mass. Such policies may result inadvertently because of the narrow focus of many environmental programs. Often one program exists for air quality, one for water quality, one for noise, and so on, with little coordination among the programs. An example involving air quality management in New York City several decades ago demonstrates what can happen if materials balance concepts are not adequately considered.[5] The city adopted strict controls on airborne emissions from apartment house incinerators. Requirements were so stringent that many apartment house owners found it more economical to stop using their incinerators than to bring them into compliance with the new regulations. The result was a substantial increase in solid waste left for curbside pickup and disposal by the city's sanitation department. The required expansion in solid waste collection and disposal efforts was so great that the rigorous incinerator regulations were not enforced.

[4] The mass conservation law states that although matter can be transformed, it cannot be created or destroyed.

[5] The example is from Freeman, Haveman, and Kneese (1973, p. 29).

TABLE 5.1 Categories of Environmental Services[a]

Materials Inputs
Inputs, such as oil and timber, to various economic activities
Waste Receptor Services
Acceptance of residuals such as wastewater and noise
Life-Support Functions
A hospitable, healthful environment including clean air and pure water
Amenity Services
Beautiful landscapes and pleasant spaces for recreation and personal renewal

[a] Based on Freeman, Haveman, and Kneese (1973, pp. 21–22).

Environmental Outputs

Economists often view the natural environment as *capital,* an asset capable of producing goods and services. *Human-made capital* includes goods such as tractors and turbines used in producing food and electricity, respectively. What goods or services (or *natural capital*) does the environment provide?

Table 5.1 describes four categories of services provided by the environment.[6] One class of services consists of materials inputs, such as oil, timber, and iron ore used in production activities. These are conventionally termed *natural resources.* A distinction is sometimes made between natural resources that are *renewable* and those that are *depletable* (or *nonrenewable*). An example of a renewable resource is groundwater, which can be used indefinitely, provided the usage rate is below the rate of groundwater recharge.

[6] The discussion of services provided by the environment is based on Freeman, Haveman, and Kneese (1973, pp. 21–22).

In contrast, depletable resources, such as oil and coal, exist in stocks that decrease as the resource is used.

Another category of services provided by the environment is termed *waste receptor services.* These services are available because of the environment's ability to dilute and transport residuals, and because natural processes can transform some residuals into harmless substances.

A third type of service provided by the environment, *life-support functions,* is illustrated by the photosynthetic activities of plants. If plants did not transform solar energy into organic matter, humans could not live. Without photosynthesis, there would be no food to serve as a source of human energy.

The last item in Table 5.1 is labeled *amenity services.* Examples include the many pleasant outdoor spots that people go to for rest and recreation, and the beautiful landscapes that attract people seeking to renew their spirits.

When viewed as a capital good, the natural environment is of concern only to the extent that it affects people. This perspective is deliberately anthropocentric. For example, many people like dolphins, and thus the value of preserving dolphins is included in economists' frameworks for analyzing environmental problems. In contrast, far fewer people care about the snail darter, a little known fish that is endangered. In comparison to dolphins, economists would attach almost no value to the snail darter. Nonhuman species and natural objects can be considered when treating the environment as a capital good, but their significance is gauged entirely by the degree to which people care about them. Biocentric arguments play an insignificant role in the economics of the environment.

The entries in Table 5.1 can be viewed as economic services in the following sense: people are *willing to pay* to receive more of them, or to avoid a reduction in their quality or quantity. The categories in Table 5.1 provide a basis for a technical definition of pollution: adverse effects that occur when use of waste receptor services interferes with life support functions, amenity services, or the provision of materials inputs. Using this definition, it is possible to have waste discharges into the environment that would not be called pollution. An example is the disposal of human waste by a lone backpacker in the Alaskan wilderness. If backpacking in the now-isolated sections of Alaska became popular, the discharge of human wastes in these areas might adversely affect someone and satisfy this technical definition of pollution. In most circumstances of concern to environmental regulators, pollution results when waste is discharged.

Another concept related to waste receptor services is the *assimilative capacity of the environment.* Societies often make judgments about which levels of environmental quality are satisfactory. For example, the quality of a stream may be judged acceptable if the concentration of oxygen dissolved in the stream is above a certain level, if the bacterial counts are below some number, and so on. When such judgments are made, it may be possible to calculate the quantity of waste that can be released into the environment without violating established levels of acceptable quality. This amount of residuals is the environment's assimilative capacity.

CONSUMER CHOICE THEORY AND DEMAND CURVES

The economists' theory of consumer choice plays a key role in environmental regulation because it provides a conceptual basis for deriving a demand curve. For an individual consumer, a demand curve for a particular good (or service) is a mathematical relationship indicating the amount of the good the consumer would purchase at different price levels. For most goods, consumption increases as price decreases, and a demand curve represents this relationship between quantity and price. This section shows how demand curves are derived in theory.

Although there are many different ways of describing how consumers behave, the model used in introductory treatments of consumer choice theory has the following general form. An individual consumer has a fixed income and is faced with the question of how much to purchase of each of two different goods.[7] The price of each of the goods is fixed by market forces. The individual consumer is said to be

[7] In a more general treatment of consumer choice theory, a consumer chooses among n commodities (where n is an arbitrary number), and a commodity is either a good or a service. This text introduces the consumer's choice for the case involving only two goods because the theory can be presented using simple graphs. Analogous results can be derived for the n variable case using calculus; see, for example, Baumol and Blinder (1993).

a *price taker* because he purchases goods in quantities too small to influence market prices.

Utility Maximization

A model describing a consumer's behavior is developed by introducing a criterion for decision making. The criterion commonly used by economists since the late nineteenth century rests on the concept of *utility.*[8] For an economist, utility provides a basis for describing the way a consumer would rank alternative *bundles* of goods based on personal preference.[9] Saying that a consumer derives more utility from one bundle of goods than another is equivalent to saying that the bundle with higher utility is preferred. When a consumer maximizes utility, he maximizes his satisfaction of personal preferences in choosing among alternative bundles of goods.

Consider how a consumer chooses among alternative combinations Q_1 and Q_2, where Q_i represents the amount of the ith commodity ($i = 1, 2$). For any particular consumer, the *utility function* $U(Q_1, Q_2)$ is a measure of the consumer's preference for the commodity bundle (Q_1, Q_2). Any curve with $U(Q_1, Q_2) =$ constant is called an *indifference curve.* The consumer has the same level of preference for all combinations of Q_1 and Q_2 on any particular indifference curve.

Indifference curves provide information about preferences. In Figure 5.2, for example, combinations of Q_1 and Q_2 on curves with high values of $U(Q_1, Q_2)$ are preferred to Q_1, Q_2 combinations on curves with lower values. Consider two bundles of goods represented by $(\overline{Q}_1, \overline{Q}_2)$ and $(\overline{\overline{Q}}_1, \overline{\overline{Q}}_2)$. Based on the utility values shown in Figure 5.2, the consumer prefers bundle $(\overline{Q}_1, \overline{Q}_2)$ over bundle $(\overline{\overline{Q}}_1, \overline{\overline{Q}}_2)$. Nothing can be said about the extent of preference, because the utility function indicates preference *ordering,* not the quantitative degree of preference.[10]

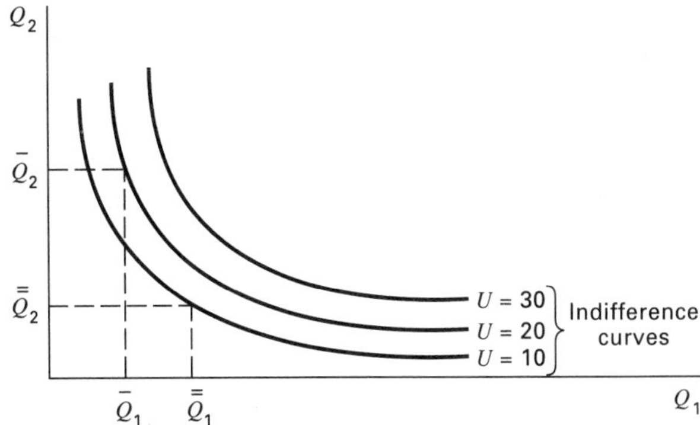

Figure 5.2 Indifference curves as measures of preference.

A standard model of consumer choice assumes that an individual will pick the commodity bundle (Q_1, Q_2) that makes her utility as high as possible, given the limits imposed by the consumer's fixed income (I). In abstract terms, the consumer chooses Q_1 and Q_2 to maximize $U(Q_1, Q_2)$, subject to the constraint that

$$p_1 Q_1 + p_2 Q_2 = I \qquad \textbf{(5-1)}$$

where p_i is the price of the ith good ($i = 1, 2$). The consumer is assumed to have *perfect information:* she knows her preference orderings and the prices of both goods.

Figure 5.3 is the same as Figure 5.2 except it contains a line representing the budget constraint faced by the consumer, as expressed in equation (5-1). The slope of the *budget line* is $-\left(\frac{p_1}{p_2}\right)$.[11] By inspection of Figure 5.3, a utility-maximizing con-

---

[8] The economists' term utility has a historical connection to the utilitarian philosophy developed by Jeremy Bentham and others. However, the term utility is now used differently by economists and philosophers. For a discussion of how philosophers use utility, see Chapter 2.

[9] An older theory of consumer choice treated utility as a quantity that could be measured in units of "utils." Modern treatments of consumer choice rely on the ordinal interpretation of utility used here. Introductions to utility in contemporary economics textbooks have a consumer rank *commodity bundles,* where a bundle consists of a collection of goods and services. For simplicity, this text refers only to goods. For an introduction of consumer choice theory, see standard economics textbooks such as Samuelson and Nordhaus (1992).

[10] The conceptual and practical problems in measuring utility are the subjects of many books; they are not of central importance here. For an introduction to methods for measuring utility, see Keeney and Raiffa (1993).

[11] To obtain this result, rewrite the budget constraint in the standard form for the equation of a straight line: $Q_2 =$ slope (Q_1) + intercept. The result is

$$Q_2 = -\left(\frac{p_1}{p_2}\right)Q_1 + \left(\frac{I}{p_2}\right).$$

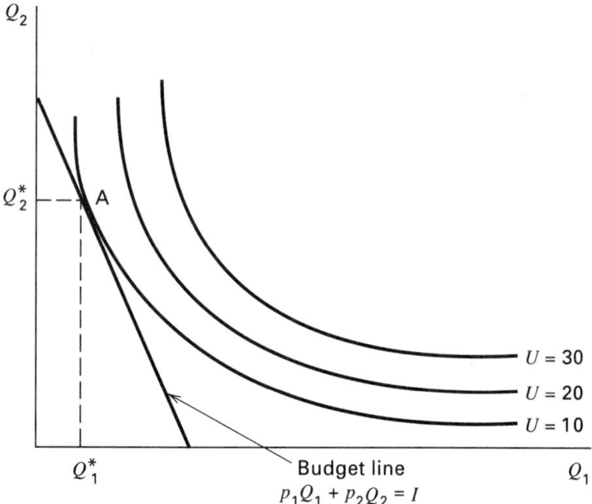

Figure 5.3 Utility-maximizing values of Q_1 and Q_2.

sumer will spend her income (I) on the commodity bundle (Q_1^*, Q_2^*). At the utility-maximizing level of consumption, the budget line is tangent to a contour of $U(Q_1, Q_2)$.

The slope of the budget line is equal to the slope of the indifference curve at the point of tangency (point A in Figure 5.3).[12] This equality of slopes can be interpreted using the concept of marginal utility. Economists use the word *marginal* to mean "additional" or "incremental," and thus the *marginal utility* of a good is the incremental gain in utility associated with consuming one more unit of the good. At the point of tangency, the ratio of the marginal utility of consuming the ith good divided by p_i is a constant (for $i = 1, 2$).[13] In other words, the utility gain from consuming one more unit of the ith good is proportional to p_i. This relationship is the basis for economists' assertions that the price of a good is a measure of its value.[14]

Derived Demand Curves

This summary of consumer choice theory provides a conceptual basis for deriving demand curves, relationships describing how much of a good would be consumed at different prices. Demand curves for an individual consumer can be derived by repeating the utility-maximizing exercise in Figure 5.3 for different values of one of the prices. To derive the demand curve for good i, everything is held constant except the price of good i.

The demand curve for good 1 is derived for illustrative purposes. Proceed by letting the price of commodity 1 decrease. At lower values of p_1, the budget line intersects the horizontal axis at higher values of Q_1 (see Figure 5.4).[15]

[12] For $U(Q_1^*, Q_2^*)$ = constant, the total derivative is zero; i.e.,

$$dU = 0 = \frac{\partial U}{\partial Q_1} dQ_1 + \frac{\partial U}{\partial Q_2} dQ_2,$$

where the derivatives are evaluated at (Q_1^*, Q_2^*). It follows that

$$\text{slope} = \frac{dQ_2}{dQ_1} = -\frac{\dfrac{\partial U}{\partial Q_1}}{\dfrac{\partial U}{\partial Q_2}} = -\frac{p_1}{p_2}.$$

The final equality above results because the slope of the budget line is $-\frac{p_1}{p_2}$. This equality holds for point A in Figure 5.3.

[13] Interpreting the word marginal to mean incremental, the marginal change in utility from increasing consumption of the ith good by ∂Q_i is $\frac{\partial U}{\partial Q_i}$. Rearranging results from the previous note, the marginal utility per dollar spent is the same for each good:

$$\frac{\dfrac{\partial U}{\partial Q_1}}{p_1} = \frac{\dfrac{\partial U}{\partial Q_2}}{p_2}$$

In the more general case involving n goods, calculus methods are used to show an analogous result:

$$\frac{\dfrac{\partial U}{\partial Q_i}}{p_i} = \text{constant} \qquad \text{for } i = 1, \ldots, n.$$

For a derivation of this result, see Baumol and Blinder (1991).

[14] Relationships between market price and value are explored further in chapters 6 and 7.

[15] The horizontal intercept of the budget line is the value of Q_1 when $Q_2 = 0$. Putting $Q_2 = 0$ in

$$p_1Q_1 + p_2Q_2 = I$$

gives

$$Q_1 = \frac{I}{p_1}.$$

If I is constant and p_1 decreases, the value of Q_1 gets larger.

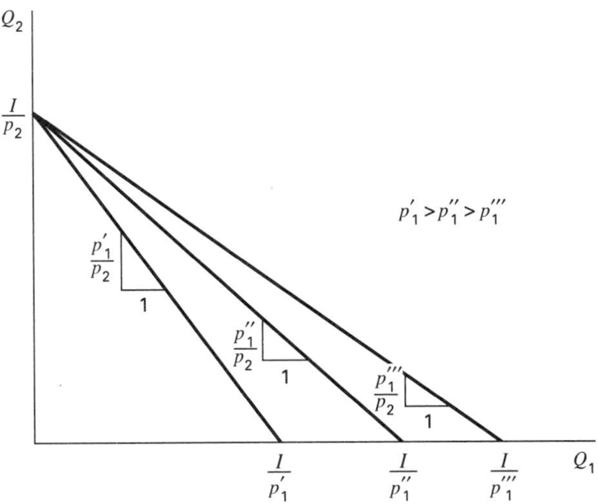

Figure 5.4 Budget lines for different values of p_1 with I and p_2 fixed.

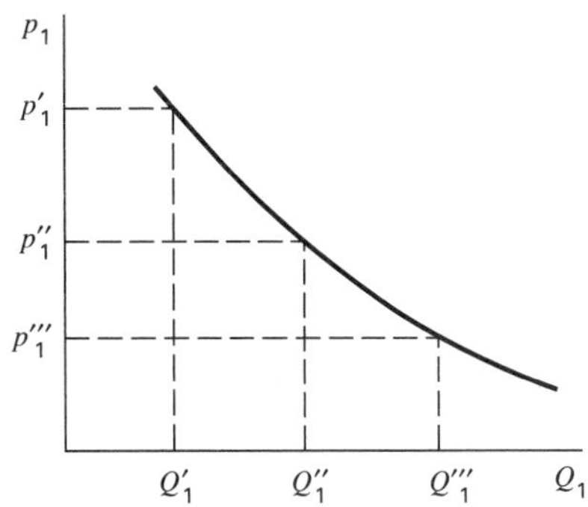

Figure 5.6 Derived individual demand curve for commodity 1.

Now imagine superimposing the indifference curves in Figure 5.2 and the budget lines in Figure 5.4. The result, shown in Figure 5.5, is a set of utility-maximizing commodity bundles, one for each value of p_1. The figure contains all the information needed to find the utility-maximizing values of Q_1 associated with each p_1. These combinations of p_1 and Q_1 are plotted in Figure 5.6, which shows an individual consumer's demand curve for good 1. For most goods, the demand curve slopes downward to the right.[16] In other words, at lower prices, a utility-maximizing consumer will purchase larger quantities of the commodity (all other things being equal).

Conceptually, if curves like those in Figure 5.6 were derived for every member of a community, the curves could be added to yield a *market demand curve* for commodity 1. Figure 5.7 shows the derivation of a market demand curve for a hypothetical community with only two consumers. The process of summing individual demand curves (demonstrated in the figure) is sometimes termed *horizontal addition.*[17] The next chapter demonstrates how market demand curves are used in estimating monetary benefits of pollution reduction.

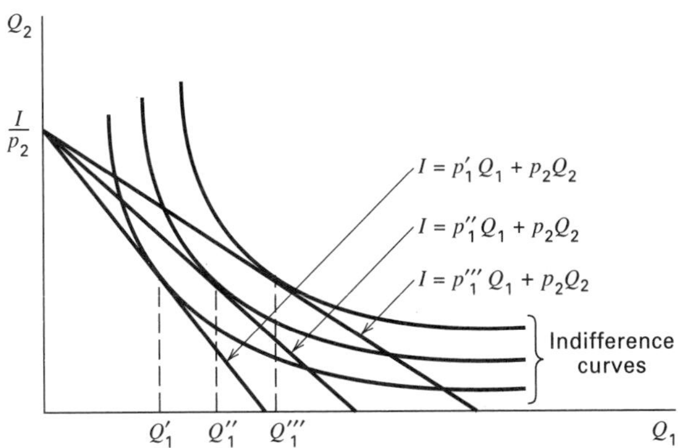

Figure 5.5 Utility-maximizing consumption levels with variations in p_1.

PRODUCTION THEORY AND SUPPLY CURVES

Economists' theory of production also plays a role in environmental policy making. Production theory concerns how firms decide on the amounts of various factor inputs needed to produce a particular quantity of output. This theory is also used to derive supply curves: relationships indicating how much output firms will produce at different output prices. When

[16] Economists use the term *normal good* to describe a good for which demand increases when income increases. For such goods, the demand curve slopes downward to the right.

[17] This process for constructing demand curves using horizontal addition does not apply for *public goods*, a special type of commodity whose characteristics are defined later in this chapter.

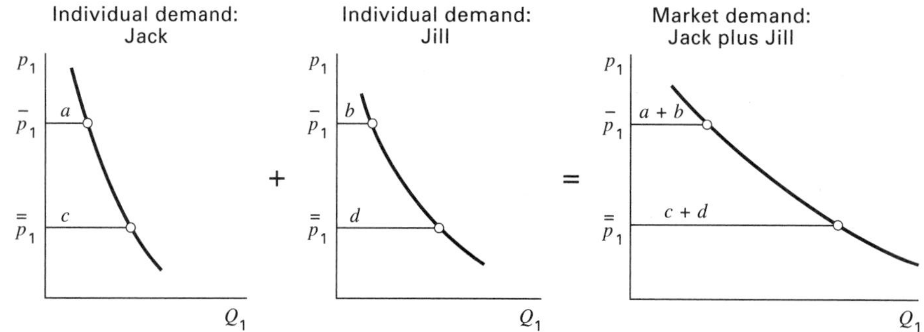

Figure 5.7 Horizontal addition of individual demand curves to obtain a market demand curve.

new environmental regulations are imposed, they often increase the cost of production. Economists use supply curves to gauge how environmental regulations might affect outputs and prices of products.

Production Functions and Isoquants

Theories of how firms make decisions often rest on information about technology used in production. Elementary versions of these theories assume that the relationship between factor inputs and outputs can be expressed mathematically in the form of a *production function.* For simplicity, assume that a firm produces a quantity Q of a single output and only two inputs are used in the production process.[18] In this case, a production function can be represented as

$$Q = F(X_1, X_2), \quad \textbf{(5-2)}$$

where X_i represents the amount of the ith factor input ($i = 1, 2$). The production function $F(X_1, X_2)$ describes the *maximum* amount of output that can be obtained with the particular set of inputs (X_1, X_2). A production function summarizes the technological relationships between inputs and outputs. The function reflects *cost effectiveness* in production because it indicates the maximum output obtainable with a given set of inputs.

Production functions come in many forms, depending on the technologies involved. A form suitable for an introductory discussion has the following characteristics.

- For every set of inputs (X_1, X_2), there is a unique output Q.
- The function $F(X_1, X_2)$ is smooth in the conventional mathematical sense.[19]
- $F(X_1, X_2)$ has the property that, all other things being equal, output increases if any one input increases.

Figure 5.8 shows contours of a hypothetical production function $F(X_1, X_2)$. The contours are called *isoquants* because they represent all combinations of the inputs X_1 and X_2 that yield the same quantity of output, Q.

Firms are able to use more of one input and less of another and still produce the same output level. For any particular combination of X_1 and X_2, the slope of an isoquant indicates how much input 2 must increase to compensate for using one less unit of input 1. Input substitution is a key idea: it is possible to substitute more of one input to compensate for the use of less of another while maintaining a constant total output. The absolute value of the slope of an isoquant is called the *marginal rate of technical substi-*

[18] Although it is possible to present a theory of production for an arbitrary number of factor inputs, this discussion considers only the case of two inputs. The main ideas of production theory can be presented using graphs if only two inputs are considered.

[19] It is assumed that $F(X_1, X_2)$ is a continuous function and has first and second partial derivatives that are also continuous.

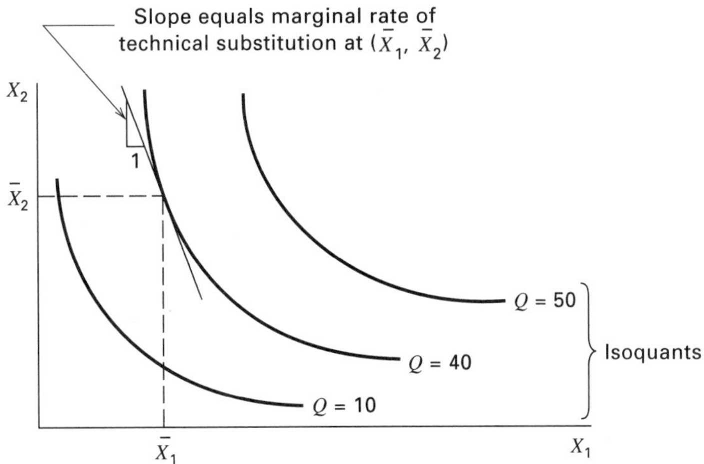

Figure 5.8 Contours of a production function.

tution (see Figure 5.8). The term *technical* is a reminder that $F(X_1, X_2)$ characterizes a production technology.

Least Cost Production

The theory of least cost production involves an exercise in cost effectiveness. The goal is to find the combination of factor inputs, in this case X_1 and X_2, that yields a particular output at a minimum total cost.[20] Production cost is represented by

$$w_1X_1 + w_2X_2 = C, \tag{5-3}$$

where w_i = price of input i ($i = 1, 2$),
X_i = quantity of input i ($i = 1, 2$), and
C = cost of production.

Equation (5-3), like the budget line in the previous section, plots as a straight line on a graph with the quantities X_1 and X_2 as axes.

Figure 5.9 shows the results of placing isoquants and cost lines on the same graph. The combinations of factor inputs that yield a particular output level at minimum cost can be identified graphically. Consider, for example, the production of 40 units of output. It is possible to produce this level of output with many combinations of inputs (for instance, points S and U in Figure 5.9). However, there is only one combination of X_1 and X_2 that yields 40 units of output for a cost of 200 (point S). And there is no set of inputs that can yield 40 units of output if less than 200 is spent. Cost-effective production levels occur where the slope of the cost line is tangent to an isoquant. Points R and T in Figure 5.9 are two other

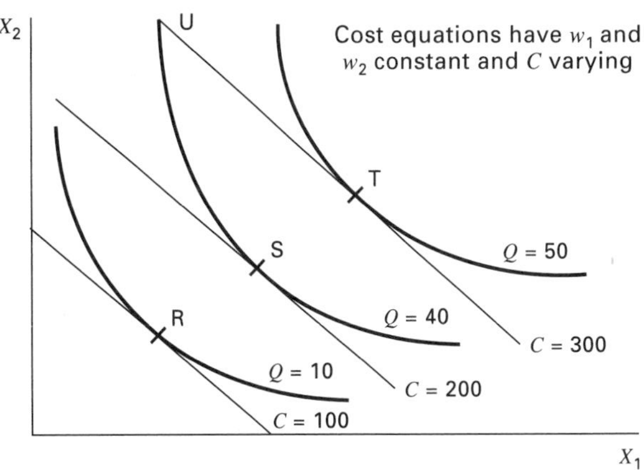

Figure 5.9 Cost-minimizing values of X_1 and X_2 with variations in C.

[20] It is also possible to frame the analysis in the following form: Find the maximum value of output that can be obtained with a particular budget. For "well-behaved functions" such as $F(X_1,X_2)$, this analysis yields the same results as the case of finding factor inputs that minimize the cost of producing a particular output.

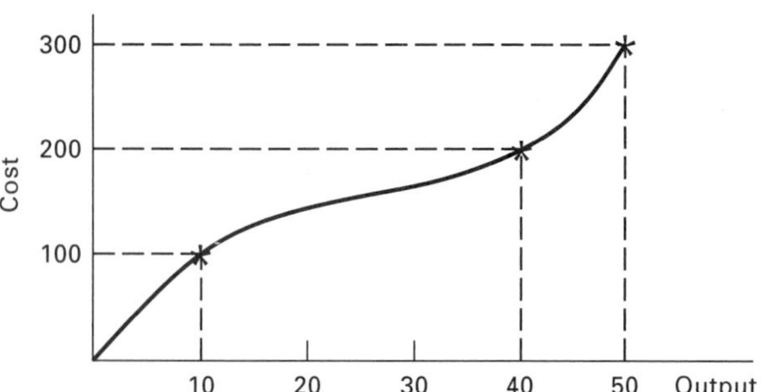

Figure 5.10 Cost curve derived from results in Figure 5.9.

points of tangency between cost equations and isoquants.[21]

The points of tangency provide a basis for identifying the minimum total costs to produce different levels of output. Suppose the combinations of cost and output identified as points of tangency in Figure 5.9 are replotted on a graph of cost versus output. The result, shown in Figure 5.10, is called a *cost curve.*

[21] The condition involving points of tangency has an interesting economic interpretation. The *marginal product* (or output) associated with factor i is, by definition, the increase in total output that results from using one more unit of input i and holding all other inputs constant. This is recognizable as the partial derivative of $F(X_1,X_2)$ with respect to X_i, calculated at a particular value of X_1, X_2. Using calculus, it can be shown that at the points of tangency, the marginal product per dollar spent on each input is the same. The proof involves calculating the total derivative of $F(X_1,X_2)$ = constant:

$$d\{F(X_1,X_2) = \text{constant}\} = 0 = \frac{\partial F}{\partial X_1}\,dX_1 + \frac{\partial F}{\partial X_2}\,dX_2,$$

where the partial derivatives are calculated for a particular X_1, X_2. The slope of an isoquant is

$$\frac{dX_2}{dX_1} = -\frac{\dfrac{\partial F}{\partial X_1}}{\dfrac{\partial F}{\partial X_2}}.$$

Recall the calculation of the slope of a budget line from the previous section. Using the same reasoning, the slope of equation (5-3) is $-(w_1/w_2)$. Combining this with the previous result gives

$$\frac{\dfrac{\partial F}{\partial X_1}}{w_1} = \frac{\dfrac{\partial F}{\partial X_2}}{w_2}.$$

Cost Terminology

The following terms are widely used in production theory.

Fixed costs: Some costs are fixed; they are paid even when no output is produced. Fixed costs include such things as contractual commitments for rent and maintenance. What constitutes fixed costs depends on the time period of analysis. In the *long run* (an ambiguous term), there are no fixed costs.[22]

Variable costs: These costs vary with the quantity of output. Variable costs are illustrated by wages and payments for raw materials and fuel.

Total cost: In the *short run,* a firm incurs both fixed and variable costs. The total cost (TC) of producing a given output is the sum of the two (see Figure 5.11a). In the theory of least cost production, the total cost function "associates with each level of output the *minimum* total expenditure at which the output can be produced and distributed, given factor prices."[23]

Average cost: The average cost (AC) associated with a given output is the cost divided by the output (see Figure 5.11b). If *total cost* is in the numerator, then the resulting ratio is the average total cost. The average fixed cost (or average variable cost) is defined using fixed (or variable) cost in the numerator.

Marginal cost: The marginal cost (MC) associated with a given output is the additional

[22] Distinctions between the terms long run and short run cannot be stated in terms of calendar time because the defining characteristic of the long run is that *all* costs are variable (that is, none of the cost is fixed). For a company that just built a factory for which it plans to pay with a 15-year mortgage, the long run might be 15 years. In contrast, for a company with no physical plant, the long run might be as short as a few months.

[23] This quotation is from Lloyd (1967, p. 111). *Factor prices* are the prices of factor inputs such as land and labor.

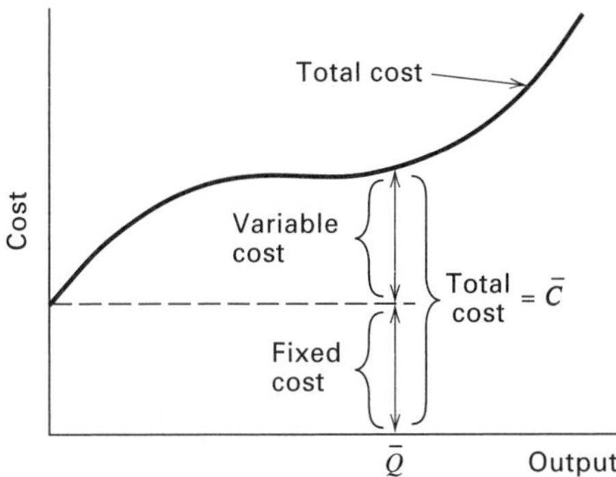

(a) Total cost = variable cost + fixed cost

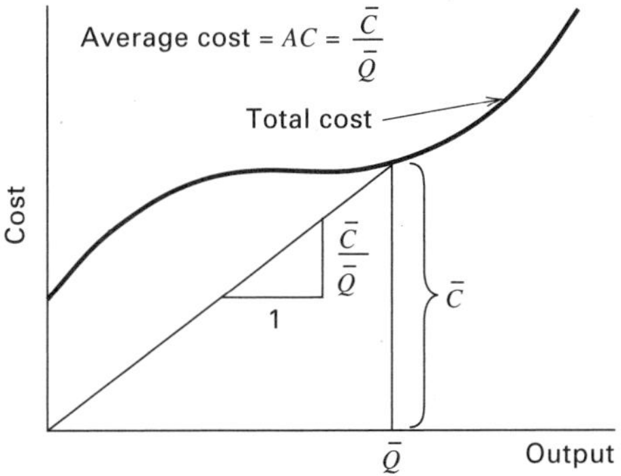

(b) Computing average total cost of producing $\bar{Q}$

Figure 5.11 Total, fixed, variable, and average costs.

cost incurred to produce one more unit of output (when operating at a given level of output). At any point on the total cost curve, the slope is equal to the marginal cost[24] (see Figure 5.12).

Because marginal cost is (by definition) the cost of producing one more unit of output, the total variable cost of producing a particular output level, say $\bar{Q}$, is the sum of each of the marginal costs up to $\bar{Q}$. In other words, the total variable cost of producing $\bar{Q}$ is the area under the marginal cost curve between the origin and $\bar{Q}$.[25]

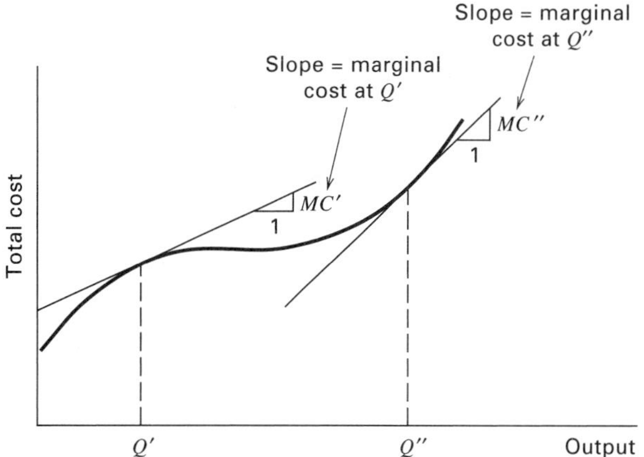

Figure 5.12 Marginal cost as the slope of the total cost curve.

Although the shape of a cost curve depends on the underlying production technology, introductory treatments of production theory employ curves of the form shown in Figure 5.13 to demonstrate relationships between different types of costs. An interesting characteristic of these curves is that the marginal cost curve intersects the average cost curve at the minimum average cost. More generally, when

[24] The marginal cost curve is the first derivative of the total cost curve. Because the total cost (TC) is the sum of total variable cost (TVC) and total fixed cost (TFC), the marginal cost is also equal to the first derivative of total variable cost. This follows because the derivative of TFC, which is a constant, equals zero:

$$MC = \frac{d}{dQ}(TVC + TFC) = \frac{d}{dQ}(TVC).$$

[25] Students unfamiliar with calculus can demonstrate this result by responding to question 5.4 at the end of the chapter. Using calculus, the result is based on definitions of total cost and marginal cost, and the relationships between derivatives and integrals. As previously noted,

$$MC = \frac{d(TVC)}{dQ}.$$

It follows from the definition of a derivative that

$$TVC = \int_0^{\bar{Q}} MC\, dQ,$$

where TVC is for the production of $\bar{Q}$. The integral is the area under the MC curve between 0 and $\bar{Q}$.

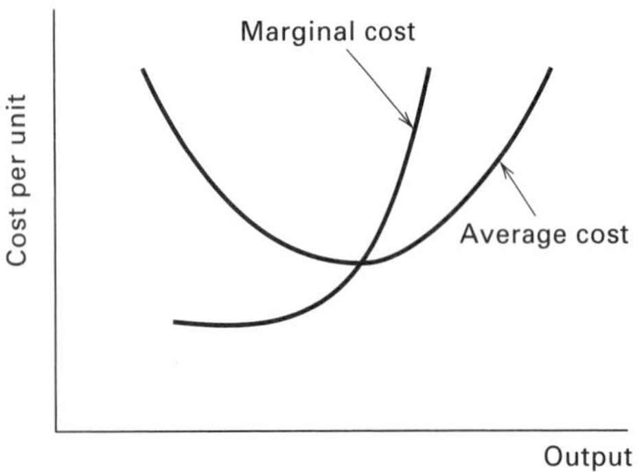

Figure 5.13 Typical shapes of short-run marginal and average cost curves.

average cost is falling, the marginal cost must be lower than the average cost.[26]

Supply Curves

A firm's supply curve for a commodity shows how much of the commodity the firm would produce at different market prices. All other things being equal, the higher the price of the commodity, the greater the quantity produced. Typically, supply curves slope upward to the right.

The previous exercise to determine the minimum cost to produce a given output is extended to derive a firm's short-run supply curve.[27] Consider a firm operating in what economists term a *perfectly competitive market.* One characteristic used to define perfectly competitive markets is that individual firms in those markets are price takers. (Other characteristics are given in the next section). Each firm in a perfectly competitive market is so small that its decisions cannot influence the prices of its output or inputs. If the market price of a firm's output is p, the firm can expect to receive revenues of pQ if it sells Q units of output. Price is constant in the short run.

A key factor in a firm's decision of how much to supply is the relationship between the firm's marginal cost of production and the market price of its output. In determining this relationship, economists often assume that firms decide the level of output to produce based on a goal of profit maximization.[28] In mathematical terms, the firm picks its output level (Q) to maximize profit (π), defined as

$$\pi = pQ - TC, \qquad \textbf{(5-4)}$$

where p = market price per unit of output, and TC = total cost of producing Q. The firm "solves" this profit maximization problem by adjusting its output level to the point where its *marginal cost is equal to the market price* of a unit of output.[29]

[26] To see this result, let

$$C(Q) = \text{total cost to produce } Q.$$

The following is true by definition:

$$MC = \frac{dC(Q)}{dQ}, \text{ and } AC = \frac{C(Q)}{Q},$$

To simplify notation, let $C = C(Q)$. By hypothesis, MC is decreasing, and thus the derivative of MC is negative:

$$\frac{d}{dQ}\left[\frac{C}{Q}\right] = \left\{\frac{Q\dfrac{dC}{dQ} - C\dfrac{dQ}{dQ}}{Q^2}\right\} < 0.$$

Therefore,

$$Q\frac{dC}{dQ} < C \text{ (for } Q > 0\text{), or}$$

$$\frac{dC}{dQ} = MC < \frac{C}{Q} = AC.$$

A similar procedure can be used to establish the following result: when the average cost is rising, the marginal cost must be greater than the average cost.

[27] The cost curves in Figure 5.13 are typical for firms operating in the short run. In more complete introductions to economics, long-run cost curves are derived from sets of short-run curves. For purposes of this discussion, it is unnecessary to introduce long-run cost curves, and the remainder of the theoretical development here focuses on short-run conditions.

[28] This assumption is commonly used in modeling the behavior of firms. However, the assumption is often disputed, even among economists, and alternative models of a firm's behavior have been developed (see, for example, Cyert and March, 1992).

[29] The result that $p = MC$ is the condition associated with a maximum profit can be deduced as follows: Suppose a firm produces output (Q) at a point where $p > MC$. In this setting, the firm is not maximizing profit because it could earn an added profit of ($p - MC$) by producing one more unit of output. Now suppose the firm sets the level Q at a point where $p < MC$. In this case, the firm can increase its profit by ($MC - p$) if it cuts back output by one unit. Thus profits are not maximized at either $p < MC$ or $p > MC$. A rule for maximizing profit is to produce at the point

Footnote continued on opposite page

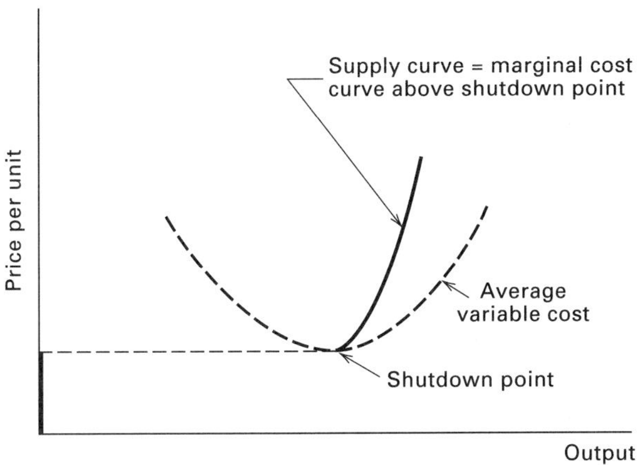

Figure 5.14 Short-run supply curve for a firm.

Above some minimum price, the supply curve for a profit-maximizing firm in a perfectly competitive market is identical to its marginal cost curve (see Figure 5.14). This results because the firm decides its output level by setting price equal to marginal cost. The price that just equals the firm's average variable cost is a limiting value in deriving a supply curve. If the price is so low that the firm cannot cover the average variable cost of making its product, the firm will lose money with each unit produced. A profit-maximizing firm would shut down in this circumstance.

Figure 5.15 uses a firm's supply curve to estimate how the firm would respond to a proposed environmental regulation. The horizontal line represents the market price (p) of the firm's output. The firm's supply curve prior to regulation is labeled S_1. The proposed regulation consists of a "pollution tax" of K per unit of output produced by the firm. This tax reflects the cost of environmental degradation caused by the firm's production process.

The effect of the tax is to cause the firm's supply curve to shift upward by an amount K. The supply curve shifts because the tax increases the firm's marginal cost of production, and the marginal cost curve (above a shutdown point) equals the supply curve. The regulation will cause a shift in the supply curve from S_1 to S_2 and a corresponding decrease in the firm's output from Q_1 to Q_2. Because the firm is a price taker, it has no opportunity to pass the increased production cost along to consumers by raising the price of its product.

These results are for a single firm. A perfectly competitive market includes many firms, and each has its own supply curve. In theory, the *market supply curve* is derived by adding up the supply curves for each of the firms in the market. The summing exercise involves the same type of "horizontal addition" used in Figure 5.7 to determine market demand.

Production theory indicates that a profit-maximizing firm (in a perfectly competitive market) will adjust its output level so that the cost of producing the last unit of output is equal to the market-determined price of that unit. In this sense, the market price reflects the value of resources used by the firm in producing its last unit of output. An economist would say this price represents an *opportunity cost* because it reflects the value of opportunities *foregone* by the producer in not using the factor inputs in an alternative activity. (The notion of opportunity cost is important, and it is described in a subsequent section.)

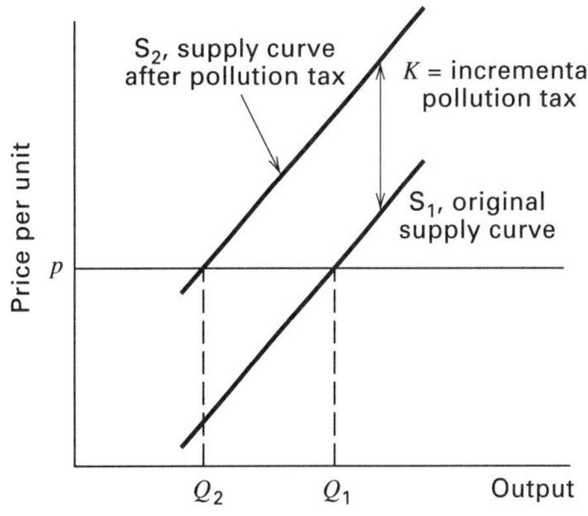

Figure 5.15 Effects of a proposed pollution tax on a firm's output.

where $p = MC$. The same result can be derived using calculus. The condition for maximization of p is

$$\frac{d\pi}{dQ} = 0 = p - \frac{d}{dQ}(TC), \text{ or}$$

$$p = \frac{d}{dQ}(TC).$$

By definition, the first derivative of total cost equals marginal cost, and thus the result is $p = MC$. Conditions on the second derivative of π must be satisfied to ensure that the resulting optimum is a maximum and not a minimum or saddle point.

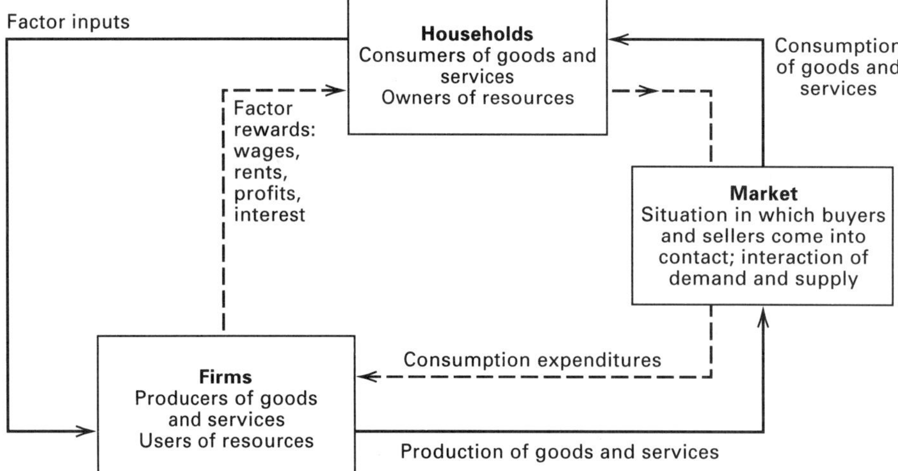

Economics proceeds by making models of society, which are simplified representations of reality. Simplifying assumptions of the model:
(a) no government
(b) all incomes spent, not saved
(c) no international trade
(d) closed self-contained system
Conventional economic analysis can readily deal with complications (a), (b), and (c), but has failed to properly address the implications of (d).

Figure 5.16 Conventional model of an economic system. From: Turner, Pearce, and Bateman. Environmental Economics: An Elementary Introduction, © 1993, p. 16. Reprinted with permission of the Johns Hopkins University Press.

COMPETITIVE MARKETS AND PARETO EFFICIENCY

Producers and consumers come together in *markets,* institutions in which buyers and sellers make exchanges. An idealized market form, perfect competition, is important in environmental policy analysis because it provides a baseline for measuring whether real ("imperfect") markets can be improved by introducing environmental regulations. This section introduces characteristics that define perfect competition. It also explains why economists often use perfectly competitive markets as benchmarks for evaluating market performance.

Figure 5.16, a conventional model of a market economy, includes two sets of exchanges. Consumers receive goods and services from producers in exchange for money. In addition, producers receive factor inputs (for example, land and labor) from consumers in exchange for money.[30]

[30] Many individuals in the economy have dual roles: they are both producers and consumers. For purposes of this discussion, the existence of dual roles is unimportant and can be ignored.

Perfectly Competitive Markets

A *perfectly competitive market* has a large number of small producers of a single homogeneous product. Each of the firms is a price taker, and all participants in the economy have perfect information about prices, technology, and costs of production.[31] Current producers are free to leave the market if they choose, and potential producers are free to enter. No barriers (such as government regulations) impede a firm's entry to or exit from the market.

Imagine that all goods and services and factor inputs in the economy represented in Figure 5.16 are traded in perfectly competitive markets. Imagine, also, that the market reaches an equilibrium in which the many interactions among consumers and producers yield a set of prices that allow all markets to "clear" simultaneously. A market clears when the

[31] In modern economies, these conditions are rarely (if ever) satisfied. The model of perfect competition is largely of interest from a theoretical point of view. It provides a benchmark for examining departures from a theoretical ideal.

quantity of products or services supplied by firms equals the quantity purchased by consumers.

Models of perfectly competitive markets include assumptions about the behaviors of consumers and producers. Individual producers are assumed to maximize profits subject to constraints on technology (as reflected in production functions). Individual consumers are assumed to maximize utility subject to constraints on income. As in the case of producers, the number of individual consumers is large and each is a price taker. No consumer is large enough to affect market prices by his or her individual actions.

Economists use the term *general equilibrium* to describe models in which *all* markets for products and factor inputs reach equilibrium simultaneously. The general equilibrium for a set of perfectly competitive markets has the following characteristics.

- Individual consumers maximize utility subject to constraints on income.
- Individual producers maximize profits subject to constraints on technology.
- All markets clear in the sense that (at equilibrium prices) producers supply exactly as much as consumers want to buy.
- Firms do not make *excess profits* (that is, the revenues of firms just cover their costs, where costs include appropriate returns for entrepreneurial talent and risk taking).

The market clearing condition for a particular commodity is depicted in Figure 5.17. For this commodity, the equilibrium price (p^*) is one where the market supply curve intersects the market demand curve. Any other price leads to adjustments in the behavior of participants in the market. If prices were such that demand exceeded supply, a shortage would exist and competition among buyers would drive up the price. Similarly, a price that yields supply greater than demand cannot be an equilibrium price. In this case, the excess supply exerts a downward pressure on prices. At the equilibrium price, the market clears.

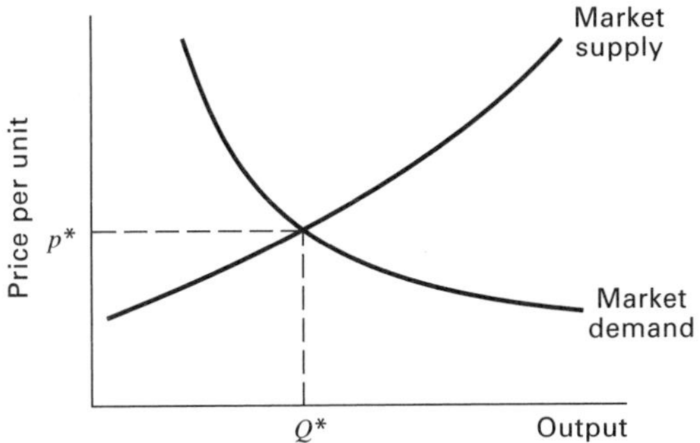

Figure 5.17 Market equilibrium as the intersection of supply and demand curves.

Pareto-Efficient Allocations

One reason economists have given much attention to models of perfectly competitive markets is that the equilibrium conditions correspond to allocations of inputs and distributions of outputs that are Pareto-efficient.[32] A resource allocation is said to be *Pareto-efficient* (or a *Pareto-optimum*) if it is impossible to modify the allocation to make even one person better off (by increasing utility or profit) without making at least one other person worse off. *Resource allocation* is used here in a broad sense and includes the allocation of factor inputs among firms, and the distribution of goods and services among consumers. If a resource allocation is not Pareto-efficient, it may be possible to reallocate resources in a way that makes one or more persons better off (in their own opinion) without making anyone worse off. Such a reallocation would be labeled a *Pareto improvement.*

The notion of a Pareto-efficient allocation can be illustrated by supposing that two consumers hold particular bundles of goods, and each knows what the other has. Suppose they can make an exchange of goods in which each registers an increase in utility and no one else in the economy is influenced by the trade. By definition, the pre-trade economy was not a Pareto-efficient allocation. The post-trade state represents an improvement in that both consumers increased utility and no one was worse off. The trade itself is a Pareto improvement, but the final resource allocation may not be Pareto-efficient. Other reallocations that are Pareto improvements may be possible.

[32] The term Pareto efficiency is named for Vilfredo Pareto, an Italian-born, Swiss economist who at about the turn of the century introduced a criterion for judging alternative resource allocations. *Pareto's criterion* states that a change in the allocation of inputs and outputs is an improvement if it harms no one and improves the lot of some people (in their own opinion). This criterion is not the only basis for appraising changes in a resource allocation. Chapter 7 introduces a concept of efficiency that rests only on the computation of monetary benefits and costs.

Pareto efficiency also applies on the production side. At a Pareto-efficient resource allocation, production is *cost-effective* in that each firm produces a given level of output at a minimum cost. In addition, it is not possible to produce more of one output without producing less of another.

As previously mentioned, perfectly competitive markets yield Pareto-efficient resource allocations. A more complete version of the relevant theorem (known as the *first theorem of welfare economics*) is: If (1) there are "enough markets," (2) consumers and producers operate in perfectly competitive markets, and (3) a general equilibrium set of prices and outputs exist, then the allocation of resources will be Pareto-efficient. Note that Pareto-efficient resource allocations are not unique. There are many such allocations. The theorem says that one such allocation will be attained only if the three conditions are satisfied.

The condition that there be "enough markets" requires elaboration. The underlying idea is that all goods and services valued by participants in the economy must be traded in markets. If there are goods, such as pure air, that are valued but not traded in markets, the theorem's result does not follow. Because many services provided by the environment are neither privately owned nor traded in markets, the practical applicability of the theorem is limited.

SIGNIFICANCE OF PRICES IN COMPETITIVE MARKETS

As previously noted, economists use the term "cost" in a way that differs from cost as monetary outlays. For a firm, the cost of production is found by summing up expenditures for factor inputs. However, the existence of government interventions such as minimum wage requirements and subsidy programs means the firm's monetary expenditures are different from what an economist would reckon as costs. For economists, costs reflect foregone opportunities, and the term *opportunity cost* is used to distinguish the economists' conception of cost from the everyday view of cost.

The notion underlying opportunity cost is sacrifice in the face of limited resources. If capital, labor, and other resources are used to produce a set of outputs, they cannot be used for other purposes. The most highly valued other purpose provides the basis for determining cost. For example, the opportunity cost of farmland used to produce wheat is the *maximum* amount the land would yield if it were used for something else, such as housing or corn production. Even if a farmer producing wheat inherited land "free and clear," using the land to produce wheat involves a cost: the maximum return from using the land for other purposes. The only circumstance in which the opportunity cost of a factor input is zero is when the input has no alternative economic use. For example, a tiny, isolated mid-ocean island used by an air-transport company to refuel planes would have a zero cost if it had no other uses.

The economists' view of costs as opportunities foregone is important in environmental policy analysis because pollution and natural resource use often involve significant (opportunity) costs that do not show up in cost accounts and are easily overlooked. Consider, for example, a wastewater discharge that causes a river in an urban area to give off noxious odors. The discharge can impose significant costs in foregone opportunities for recreational use of the river (among other uses). However, these costs do not show up in accounting statistics. From the waste discharger's point of view, use of the river to transport waste has a zero cost.

Although opportunity costs are often difficult to compute, there is one set of circumstances in which market prices reflect opportunity costs: the general equilibrium conditions for perfectly competitive markets. Under perfect competition, the price of any factor input reflects the "opportunities for the use of that input, and the demand for each of the final goods to whose production that input can contribute."[33] In other words, perfectly competitive market prices of inputs reflect the values of inputs in alternative uses.

If all goods and services were traded in perfectly competitive markets, it would be easy to estimate opportunity costs because they would be given by market prices. However, for some services provided by the environment, such as the provision of breathable air, there are no markets at all. Moreover, many services provided by the environment are provided in markets that are not competitive. In these circumstances, market prices may depart widely from opportunity costs. Consider, for example, market prices of

[33] This quotation is from Walsh (1970, p. 267); see that source for details of the theoretical argument interpreting market prices as opportunity costs of inputs and values of outputs.

irrigation water paid by farmers in California. Some farmers receive an acre-foot of water for as little as $50, while others may pay as much as $400.[34] Government subsidy programs often make the prices that farmers pay useless as measures of the opportunity cost of water.

When market prices are unavailable as a measure of opportunity costs, economists rely on the concept of a shadow price. By definition, *shadow prices* are prices that would exist if goods and services were produced and traded in perfectly competitive markets.[35] Shadow prices reflect opportunity costs.

WHY ENVIRONMENTAL RESOURCES ARE NOT ALLOCATED EFFICIENTLY BY MARKETS

Competitive markets have been celebrated since the days of Adam Smith's description of the *invisible hand* that would guide the actions of independent, profit-seeking individuals to an outcome that serves the public interest.[36] Modern versions of the invisible hand argument take models of perfectly competitive markets as points of departure. The contemporary argument uses mathematical models to demonstrate that competitive markets guide resources into uses that allow consumers to maximize utility and producers to maximize profits. A Pareto-efficient outcome is attained without any deliberate effort to reach one. No central planning or government intervention takes place, yet a Pareto-efficient allocation results.

Unfortunately, the virtues of competition that Adam Smith and many others have described do not apply to the use of many environmental resources. Economists have several ways of describing why this is so, and alternative explanations provide insights into what institutions, other than markets, might be useful in managing the environment. The following concepts play a key role in these alternative explanations: property rights, public goods, common property resources, and external costs. Although these terms overlap in some ways, they are introduced separately.[37]

Property Rights

One explanation of why environmental resources are not properly managed by ordinary market interactions starts with the concept of property rights: "the rights of an owner to use and exchange property as the owner sees fit."[38] Models of perfect competition assume that *all* resources, goods, and services are owned by individuals and that a system of rights to private property exists.

Tietenberg (1992, pp. 45–47) details a structure of property rights that could produce economically efficient outcomes in a well-functioning market economy. For Tietenberg, a system of well-defined property rights has four characteristics:

1. *Universality.* All resources are privately owned, and all entitlements completely specified.
2. *Exclusivity.* All benefits and costs accrued as a result of owning and using the resources should accrue to the owner, and only to the owner, either directly or indirectly by sale to others.
3. *Transferability.* All property rights should be transferable from one owner to another in a voluntary exchange.
4. *Enforceability.* Property rights should be secure from involuntary seizure or encroachment by others.

Because many environmental resources (for example, natural waterways and ambient air) are not privately owned, competitive markets cannot allocate them in Pareto-efficient ways. It might be tempting to argue that environmental resources could be allo-

[34] "Acre-foot" is a commonly used unit of volume for irrigation water in the United States. One acre-foot is the volume of water in a one foot layer of water spread over one acre.

[35] The term shadow price originated in mathematical optimization theory, which provides one basis for estimating these prices. For more on shadow prices, see Sassone and Schafer (1978).

[36] "Every individual endeavors to employ his capital so that its produce may be of greatest value. He generally neither intends to promote the public interest, nor knows how much he is promoting it. He intends only his own security, only his own gain. And he is led by an invisible hand to promote an end which was no part of his intention. By pursuing his own interest he frequently promotes that of society more than when he really intends to promote it" (Adam Smith in *The Wealth of Nations,* as quoted by Gregory and Ruffin, 1986, p. 35).

[37] Economists don't agree among themselves on the definitions of these terms. See Dorfman and Dorfman (1993, p. 77) for a commentary on the lack of consistency in the way economists use the terms "common property resource," "public good," and "externality."

[38] This quotation is from Gregory and Ruffin (1986, p. 34).

cated efficiently if property rights were created to allow those resources to be owned and traded by private parties. However, as the following discussion makes clear, creation of private property rights may not always enhance Pareto efficiency.

Public Goods

For goods and services that satisfy the economists' definition of *public goods,* private ownership is not the key to attaining Pareto-efficient resource allocations. The definition of public goods is unrelated to whether goods or services are provided by public entities. For an economist, a public good (for example, a radio signal or a lighthouse) is defined as having each of the following characteristics:

1. *Nonrival consumption.* The consumption of the good by one person does not reduce its consumption by others. For example, one family's ability to benefit from a dam that provides flood protection is in no way diminished by the existence of neighboring families who are protected by the same dam.
2. *Nonexcludability.* The condition in which it is excessively costly or impossible for a seller to exclude anyone from consuming a good or service once it has been produced.[39] For example, it is virtually impossible to exclude a resident in an area protected by a flood control dam from enjoying the protection provided by the dam.

The market system will generally fail to provide public goods because nonexcludability means producers are unable to collect payments from all those who consume their products. Thus public goods are often provided by governments on the basis of a collective decision process. However, the defining characteristics of a public good lead to the *free rider problem,* which is linked to the provision of public goods at less than Pareto-efficient levels.

A free rider is a person who enjoys the benefits of a good or service without paying, and the possibility of free riding occurs because of the nonrival consumption of public goods. (A good such as toothpaste is characterized by rival consumption; its use by one person means there is less available for others.) Faced with the issue of financing a public good voluntarily, individuals often understate their willingness to pay for it because they gamble on the good being provided to meet the demands of others. Free riders may either consume the good for free or consume the good by making a payment that is less than they would have made if consumption had been rival. Public television provides an example. Many individuals enjoy the benefits of public television without paying any station membership fees. Others make contributions, but at rates lower than what they would be willing to pay if programming were not provided as a public service.

Common Property Resources

The term *common property resource* (or *collective good*) is used to characterize situations in which there is rival consumption for a resource, but it is difficult to exclude users from consuming it. Examples include fisheries, hunting grounds, groundwater basins, and common pools of oil. As in the case of public goods, the difficulty in excluding or controlling users makes it hard to have common property resources managed by private parties, unless potential users constrain their individual consumption (voluntarily or otherwise).[40]

The negative effects of allowing common property resources to be exploited without restraints were popularized in a widely cited essay, *The Tragedy of the Commons,* by the biologist Garrett Hardin (1968).[41] In his essay, Hardin describes events in an imaginary cattle grazing area, "the commons," that is open to all. No constraints exist on how many cattle an individual member of the community may graze. If a limited number of cattle per year are grazed, the commons can renew itself and provide a satisfactory grazing area indefinitely. However, each herdsman seeks to maximize his own gain by adding more animals to his herd. Because there are no restrictions on the size of an individual's

[39] The costs of exclusion include those involved in defining and enforcng property rights in a good (Gregory and Ruffin, 1986, p. 206).

[40] Public goods differ from collective goods in terms of rivalry in consumption. For a public good, such as a radio station's signal, consumption is nonrival; in contrast, consumption of oil from a reservoir, a collective good, is rival.

[41] These negative effects had been detailed more than a decade earlier in a technical article by Gordon (1954).

herd, the commons are used excessively and eventually ruined for grazing.

The "tragedy of the commons" also applies to the waste receptor services provided by the environment. Consider, for example, a stream on which standards of quality have been established. This stream can accept a certain amount of water pollution without violating the standards, and that quantity is its assimilative capacity. Like the grazing area, the assimilative capacity of the stream is common property. In the absence of socially imposed controls, the stream's ability to carry pollutants will be viewed by each discharger as a free good. Thus no waste source will have an incentive to invest to reduce wastewater discharges. Without this incentive, the stream's waste receptor services are likely to be overused, causing a violation of the stream quality standards. The need for incentives to prevent overuse of common property resources is frequently cited in justifying government intervention to manage environmental quality.

Analysts who have probed the problem of common property resources typically agree on the need for an alternative to unregulated private use. Effective management of common property resources requires restraints on individual users to "prevent the resources of the community at large from being destroyed by excessive exploitation."[42] Examples of effective management include limits on hours of allowable pasturing and other rules that regulated use of common pasture in the medieval manorial economy. Another example is the seasonal closure rule developed by the Canadian Atlantic Coast lobster-conservation program.[43]

External Costs

Externalities exist when interactions among producers and consumers are not adequately reflected in market prices. An example of an *external cost of production* is the ecosystem destruction caused by air pollutants from power plants fueled by high-sulfur coal. *External costs of consumption* are illustrated by the case of a cigar smoker in a restaurant. The smoker pays the private cost of the cigar (the purchase price), but the act of consumption imposes an external cost on all those disturbed by the cigar smoke.[44]

Resources are not allocated efficiently when external costs are imposed. For example, producers that impose external costs produce more output than they would if they accounted for the full cost of their production. In this context, the full cost (or *social cost*) consists of the firm's private costs plus the external costs it imposes on others.

The example that was introduced in connection with Figure 5.15 demonstrates how resources are misallocated when external costs are ignored. In that example, the firm produced Q_1 units of output when it was not forced to account for the external costs of its pollution. A regulation forced the firm to pay a pollution tax of $\$K$ per unit of output produced. The amount of the tax was based on the damages (or external costs) imposed. As shown in Figure 5.15, if the firm was faced with a pollution tax its marginal cost of production would increase. The firm in the example is a profit maximizer, and consequently it would reduce its output to keep its profit at a maximum in the face of the pollution tax.

Economists use the term *market failure* to characterize outcomes in which externalities are ignored. In the example in Figure 5.15, the market failed: market interactions alone did not cause the firm to account for both its private costs *and* its external costs. For an economist concerned with Pareto efficiency, the key question is what can be done to force the firm to *internalize* its external costs by accounting for the costs imposed by its waste discharges.

Governments have traditionally attempted to force polluters to internalize the external costs of their wastes by issuing regulations that limit the levels of waste discharge, and by imposing sanctions for noncompliance with the regulations. The phrase *command and control* characterizes this approach.

---

[42] This quotation is from Gordon (1954) as it appears in the collection of essays edited by Dorfman and Dorfman (1993, p. 106).

[43] These and other institutions for managing common property resources are detailed by Gordon (1954).

[44] For completeness, note that not all externalities are adverse. Adverse external effects are called *external diseconomies* to distinguish them from beneficial ones, called *external economies*. An illustration of an external economy is the increase in pollination of apple orchards that occurs when a beekeeper locates hives near an apple grower's orchard. The pollinating activities of the bees increase the yields of the orchard owner, and the latter receives the benefits for free. They are *external benefits*.

Many economists feel there are more efficacious ways to force polluters to internalize external costs. They attempt to remedy market failure using market-based incentives. Two approaches are commonly advocated. In one scheme, regulators impose a pollutant discharge fee (or tax) that forces polluters to account for external costs. A second approach assumes that excessive pollution results because there are no property rights to clean air and water (or rights to discharge wastes). In this second approach, regulators create a market in rights to discharge waste; these rights can be bought and sold. Later chapters describe each of these approaches: command and control, discharge fees, and marketable rights to discharge waste.

Key Concepts and Terms

The Environment as an Economic Asset
- Materials residuals
- Energy residuals
- Materials balance perspective
- Environment as a capital good
- Waste receptor services
- Amenity services
- Assimilative capacity of the environment

Consumer Choice Theory and Demand Curves
- Consumer's preference
- Utility functions
- Indifference curves
- Utility maximization
- Budget constraint
- Marginal utility
- Individual vs. market demand curves

Production Theory and Supply Curves
- Production function
- Isoquants
- Cost effectiveness
- Marginal rate of technical substitution
- Fixed, variable, and total cost
- Average vs. marginal cost
- Supply curves for firms and industries
- Price equal to marginal cost

Competitive Markets and Pareto Efficiency
- General equilibrium
- Equilibrium price
- Pareto improvement
- Pareto-efficient resource allocation
- First theorem of welfare economics

Significance of Prices in Competitive Markets
- Opportunity cost
- Shadow price

Why Environmental Resources Are Not Allocated Efficiently By Markets
- "Invisible hand" concept
- Property rights
- Rival vs. nonrival consumption
- Free riders
- Collective vs. public goods
- Common property resources
- Tragedy of the commons
- External costs of production
- External costs of consumption
- Private vs. social cost
- Market failure
- Internalization of external costs

Discussion Questions

5-1 Policymakers often devise environmental management programs for a specific *environmental medium* such as air, land, or water. Identify some significant cross-media impacts that are ignored by the traditional single-medium approach. (For example, consider how curbing emissions from motor vehicles might improve water quality, or how wastewater treatment can influence air quality.) Explain the following observation by tracing cross-media effects: In some major cities of the world, a cost-effective way to reduce flood damages is to improve garbage collection services.

5-2 Some people feel that pollution occurs only if the ambient environment is degraded to cause damage. What do you think of this definition of pollution? Does it characterize how the word pollution is used in the mass media?

5-3 Although linear demand curves are generally not observed in real situations, this exercise uses them to simplify the computations. Consider a hypothetical market with only two persons. Their individual demands for a single commodity are shown here.

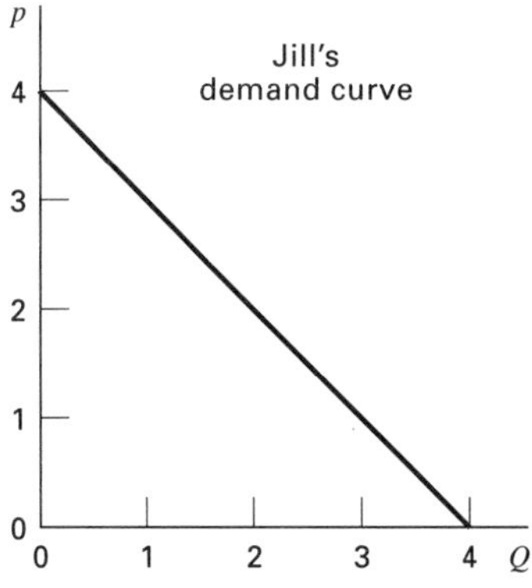

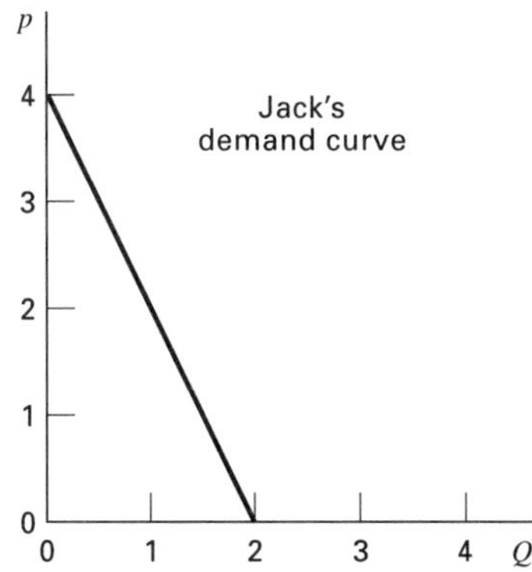

Sketch the market demand curve for the commodity and write an algebraic equation describing market demand in terms of price. What would happen to the market demand curve if a third person entered the market?

5-4 Consider a firm producing a single homogeneous product. The total cost (TC) of producing an output Q is given by

$$TC = Q^2 + 1.$$

(a) What is the total fixed cost? Write the relationship for total variable cost (TVC) in terms of Q.
(b) Calculate the values of TVC and marginal cost (MC) for all integer values of Q between 0 and 6.
(c) Use the following procedure to prepare a plot of MC. In considering the marginal cost of moving from a given integer value of Q, call it $\overline{Q}$, to the next highest integer value of Q, plot the MC value against the corresponding average quantity, namely,

$$[\overline{Q} + (\overline{Q} + 1)]/2.$$

This plotting procedure is used because, in this example, for simplicity, Part B examines only integer values of Q. For the simple function used for total cost in this exercise, the aforementioned plotting procedure yields the exact marginal cost equation

$$MC = 2Q$$

For more complex total cost relationships, the plotting procedure yields only an approximation to MC.
(d) Compute the area under the MC curve between the origin and Q, for $Q = 1, 2, \ldots, 5$. These areas should correspond to values of TVC for the integer values under consideration.

5-5 A firm's supply curve shows the quantities supplied by a firm at different prices, assuming all other things are equal. A single point on a supply curve represents the quantity supplied, and a change in demand for the product leads to a movement along the supply curve. This type of change is different from a shift in the supply curve. A supply curve shift occurs, for example, when the entire curve moves because of changes in prices of factor inputs, or changes in technology. For each of the following, indicate whether there would be a movement along the supply curve, or a shift in the supply curve. (If there is a shift, is it downward to the right or upward to the left?) In each case, assume the firm operates in a perfectly competitive market.

(a) An increase in the market price of the firm's output
(b) A fall in the price of one of the firm's inputs
(c) A technological innovation that decreases production costs
(d) A pollutant discharge fee (for example, a fee of 10 cents per lb of biochemical oxygen

demand discharged in the firm's wastewater.)

5-6 Suppose the government imposes a tax on a product in order to reduce pollution. (Examples include taxes on gasoline or gasoline-guzzling "luxury" cars.) The effect of such a tax is equivalent to shifting the supply curve upward by the amount of the tax per unit. (From a consumer's perspective, the total price per unit of commodity would be the same with either a tax or an equivalent increase in the cost of production.) Whether such a new tax per unit is a burden to producers or consumers depends on the shape of the demand curve. Consider two extreme cases: (A) a vertical demand curve, and (B) a horizontal demand curve. For each case, analyze the influence of a tax ($\$t$ per unit) on the equilibrium price and indicate whether the tax is borne by consumers or producers or both.

5-7 Consider a market you are familiar with (e.g., air transportation in the United States). Which, if any, of the conditions associated with perfectly competitive markets are satisfied?

5-8 It is common to find goods or services that are *partial* public goods in the sense that they have some attributes of public goods. Which of the following have characteristics of public goods: gasoline, flood control dams, libraries, museums, pure air, national defense, and community water supply.

5-9 A wealthy urban dweller from the eastern United States purchases a large ranch in an arid zone in the western United States and finds, to his surprise, that there is groundwater 20 feet below his property. The new ranch owner puts together a consortium of investors to plan an enormous pumping and piping system in order to sell huge quantities of groundwater to a nearby city. One effect will be a significant drawdown in the groundwater table; the aquifer is currently used as irrigation water by many family-sized farms near the ranch. Is this a problem involving public goods? Collective goods? Externalities? What external costs are involved? What actions could be taken to force the new consortium to internalize the external costs it will impose?

References

Baumol, W. J., and A. S. Blinder. 1991. *Microeconomics, Principles and Policy.* Dallas, TX: Harcourt, Brace, Jovanovich.

Cyert, R. M., and J. G. March. 1992. *A Behavioral Theory of the Firm,* 2d ed. Cambridge, MA: Blackwell Business.

Dorfman, R., and N. S. Dorfman, eds. 1993. *Economics of the Environment: Selected Readings,* 3d ed. New York: W. W. Norton.

Field, B. C. 1994. *Environmental Economics: An Introduction.* New York: McGraw–Hill.

Freeman, A. M., III, R. H. Haveman, and A. V. Kneese. 1973. *The Economics of Environmental Policy.* New York: Wiley.

Gordon, H. S. 1954. The Economic Theory of a Common-Property Resource: The Fishery. *The Journal of Political Economy* (April). Reprinted in Dorfman, R., and N. S. Dorfman, eds. 1993. *Economics of the Environment: Selected Readings,* 3d ed. New York: W. W. Norton.

Gregory, P. R., and R. J. Ruffin. 1986. *Essentials of Economics.* Glenview, IL: Scott, Foresman.

Hardin, G. 1968. The Tragedy of the Commons. *Science* 162: 1243–48. Reprinted in Dorfman, R., and N. S. Dorfman, eds. 1993. *Economics of the Environment: Selected Readings,* 3d ed. New York: W. W. Norton.

Keeney, R. L., and H. Raiffa. 1993. *Decisions with Multiple Objectives: Preferences and Value Tradeoffs.* Cambridge, U.K.: Cambridge University Press.

Lloyd, C. 1967. *Microeconomic Analysis.* Homewood, IL: R. D. Irwin

Mansfield, E. 1988. *Microeconomics: Theory Applications,* 6th ed. New York: W. W. Norton.

Samuelson, P. A., and W. P. Nordhaus. 1992. *Economics,* 14th ed. New York: McGraw–Hill.

Sassone, P. G., and W. A. Schafer. 1978. *Cost-Benefit Analysis: A Handbook.* New York: Academic Press.

Tietenberg, T. 1992. *Environmental and Natural Resources Economics,* 3d ed. New York: Harper Collins.

Turner, R. K., D. Pearce, and I. Bateman. 1993. *Environmental Economics: An Introduction.* Baltimore: Johns Hopkins University Press.

Chapter 6

Economic Valuation of Environmental Resources

An improvement in environmental quality may result in economic benefits, and a deterioration may impose economic costs. Estimating the monetary value of changes in environmental quality raises several questions.[1] How are the benefits (or costs) of environmental changes defined? And how can these benefits (or costs) be measured in practice? This chapter explores these questions by surveying the main methods economists use to estimate economic effects of changes in environmental resources.

In theory, the process of estimating benefits of environmental improvements requires construction of the demand curves introduced in Chapter 5. This demand (or benefit) information can be combined with appropriate cost information to help improve the allocation of resources. Increasingly, environmental policymakers are calling for comparisons of monetary costs and benefits, and the process of combining information on costs and benefits is taken up in the next chapter. This chapter focuses exclusively on *valuation,* "the placement of monetary values on environmental goods or services or [on] the impacts of environmental quality changes."[2]

[1] I am grateful to Greg Browder for his work on early drafts of this chapter while he was a student at Stanford University, and to Professor Lawrence Goulder of Stanford's Department of Economics for his comments on the final draft of the chapter.

ENVIRONMENTAL RESOURCES AS CAPITAL ASSETS

Economists view environmental resources as forms of *natural capital* that provide four categories of services: materials inputs, waste receptor services, life-support functions, and amenity services (see Table 5.1). A lake provides an example of how a single environmental resource can yield multiple services. The lake may function to support life by serving as a source of drinking water. It may also provide amenity services in the form of opportunities for recreational boating and lakeside picnicing. The lake may also be a source of materials inputs; for example, by producing fish which are harvested. Finally, the lake may serve as a receptor of wastewater from a nearby city. In estimating the value of the lake, it is necessary to consider not only services provided in the current year, but also services that may be provided in future years.

[2] This definition of valuation is from Dixon et al. (1994, p. 4).

Economists define total economic value of the environment as the sum of active-use and passive-use values (see Figure 6.1). As the name implies, *active-use values* stem from people utilizing a natural asset. In the lake example, active users include anglers, boaters, people who use the lake as a water supply, and those who discharge wastewater into the lake. Active-use values are further classified in terms of direct and indirect use. The illustrations just cited are all direct uses. Indirect use of the lake takes place, for example, when people sit nearby and enjoy the view.

Since the 1960s, economists have formally recognized that individuals have *passive-use values* to the extent that they would be willing to pay to maintain certain environmental resources even though they may never actively use them. Suppose that the previously mentioned lake is Lake Tahoe, which lies on the border of California and Nevada. Many people who will never visit Lake Tahoe would be willing to pay to preserve its beauty. They are passive users, and their willingness to pay reflects the value they place on the lake. Passive-use values are subdivided, depending on whether or not an individual contemplates active use in the future. If such potential use is contemplated, an *option value* is implied. In other words, the person values the option of being able to use the lake in the future. An individual can also value a resource without considering the possibility of using it. For example, a person may take comfort in knowing that a place as beautiful as Lake Tahoe exists and be willing to pay to preserve it, even though he or she never plans to go there. The person values the lake simply because it exists. This is an example of what economists term *existence value.*

The economic value of an environmental resource rests on the concept of *willingness to pay* (WTP). In estimating a monetary value for an environmental change, WTP is the *maximum* amount a person would pay to achieve an improvement (or prevent a decline) in environmental quality. People make trade-offs in choosing among goods and services, and WTP provides a measure of how people value goods and services relative to each other. By convention, the relative value of goods is expressed in units of money.

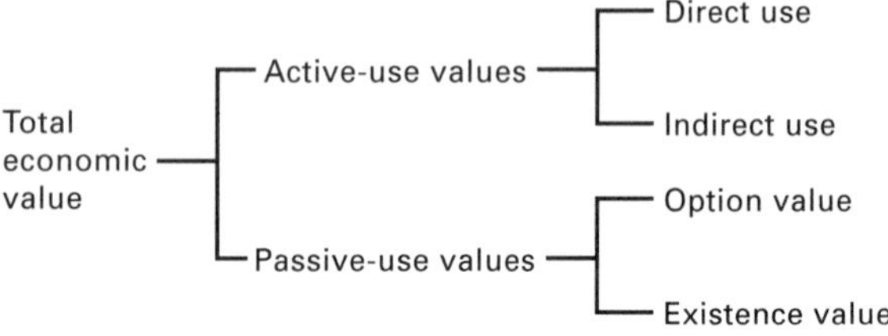

Figure 6.1 Types of values provided by the environment.

Willingness to pay cannot be used to measure the value of goods a person considers nontradable. For example, some people feel that the destruction of a species is not something that can be traded off against any possible gains. This concern with only tradable goods and services is a limitation of economic evaluation methods.

ECONOMIC VALUE AND CONSUMERS' SURPLUS

The centerpiece in the economists' definition of economic value is the demand curve, which is derived by analyzing the behavior of a utility-maximizing consumer (see Chapter 5). The height of an individual's demand curve at any point represents the person's *marginal willingness to pay.* It indicates how much money the consumer would give up in exchange for one more unit of product (or service). As an example, consider Figure 6.2, which represents an individual's weekly demand for jars of juice, a product that can be purchased only in discrete quantities. The figure shows that the consumer would be willing to pay a high price for the first jar of juice, but WTP would decrease as additional jars were purchased. The hypothetical consumer would be willing to pay as much as $2.00 for her first jar of juice, a maximum of $1.50 for her second, and so forth. These numbers represent the consumer's marginal willingness to pay for the first jar, the second jar, and so on. Willingness to pay diminishes as additional jars of juice are purchased, because the incremental satisfaction derived from a good generally decreases with increasing consumption of the good.

The price a person is *willing* to pay for a good is often greater than the price the person *must* pay to purchase it. Recall from Chapter 5 that an individual consumer in a competitive market does not establish the price; rather, the price is set in the marketplace. According to Figure 6.2, if the market price of juice is $1.00, the consumer will buy three jars. The total

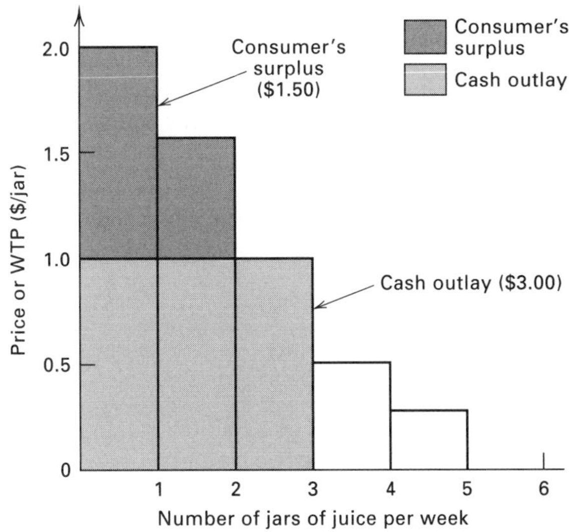

Figure 6.2 An individual's WTP curve for juice.

cost is three times $1.00, or $3.00. However, the total value of three jars of juice to the consumer is the sum of the marginal WTP values up to the third jar, that is, $2.00 + $1.50 + $1.00 or $4.50. Hence, the consumer would be willing to pay up to a maximum of $4.50 for three jars of juice but is only required to pay $3.00. The difference between the total value and the total cost to the consumer, $1.50, is termed the *consumer's surplus.*[3]

The concept of consumer's surplus was introduced by French economist Jules Dupuit (1844) in a treatise on the value of public works. Contemporary treatments of benefit estimation build on Dupuit's ideas by observing that a consumer's cash outlay for a product underestimates the total value of the product because it omits the consumer's surplus. In reckoning total value, economists take the sum of market value (as expressed in price times quantity consumed) and consumer's surplus. In the example represented in Figure 6.2, the total purchase price of $3.00 underestimates the total value that the consumer places on the three jars of juice. The total value is properly reckoned as the area under the consumer's demand curve between the origin and the quantity consumed, (the sum of shaded areas for consumer's surplus and cash outlay in Figure 6.2).

The foregoing example involving jars of juice illustrates the concepts of marginal willingness to pay, total value, and surplus for an *individual* consumer. In analyzing markets, individual preferences are summed across the relevant population in order to form aggregate curves. Figure 6.3 represents concepts identical to those in Figure 6.2 except that a market demand curve is used. The market demand curve is a smooth line instead of step function because it aggregates individual preferences. Not all individuals will have identical preferences, and hence there will be many individualized versions of Figure 6.2, which are added horizontally to construct the smooth market demand curve in Figure 6.3. The sum of the two shaded areas in Figure 6.3—cash outlay plus consumers' surplus—represents the total economic benefit to society of selling jars of juice, and this conception of total benefit is used for estimating the economic value of environmental resources. As in the analogous case for a single consumer, the total benefit

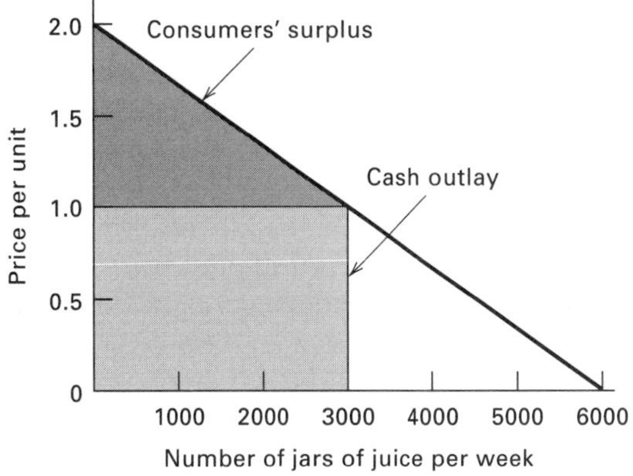

Figure 6.3 Market demand curve for juice.

[3] The discussion that follows concerns consumer's surplus as introduced by Jules Dupuit in 1844 and later elaborated by Alfred Marshall in the early 1900s. As Marshall explained it, the consumer's surplus is measured by the area under an ordinary demand curve and above the horizontal price line. Economists after Marshall carefully articulated the conditions under which Marshall's measure of the surplus corresponded to the "true" utility surplus as measured by indifference curves and utility functions. This later work led to development of more fundamental measures of changes in economic welfare, and controversy over the suitability of using consumer's surplus in the sense elaborated by Marshall. These complexities, which are beyond the scope of this text, are detailed by Freeman (1993, Chap. 3) and Hanley and Spash (1993, Chap. 2).

includes consumers' surplus and is represented by the area under the market demand curve between the origin and the total quantity consumed.

The economic value of environmental change is estimated in two contexts: *ex ante* (before the fact) analysis, and *ex post* (after the fact) analysis.[4] An ex ante analysis provides information for those who must decide between alternative policy actions. Examples include an analysis of the economic impact of different levels of pollution abatement, or an appraisal of the monetary value of maintaining a wetlands area in its natural state. Ex post analyses are performed to assess the consequences of an action already taken. An example is an assessment of economic damages due to an oil spill from a leaking tanker ship.

Economic valuation techniques for ex post and ex ante analyses are similar. One of the primary distinctions is that ex ante analysis may consider a number of different future states, whereas ex post analysis focuses on the differences between the ex post situation and the setting just prior to the action or event.[5]

Two broad categories of approaches to valuing an environmental resource can be distinguished: market-based methods and contingent valuation methods. *Market-based methods* estimate values based on choices people *actually* make. They rely on market prices. *Contingent valuation methods* construct hypothetical situations and ask people what choices they *would* make if they were in those circumstances. Contingent valuation relies on *expressions* of preference, not preferences *revealed* by individuals' purchases.

This chapter introduces each of the procedures listed in Table 6.1, beginning with the market-based methods. Some of the market-based methods—hedonic property value, travel cost, and hedonic wage—involve construction of demand curves using information derived from individuals' market choices. The other market-based methods in the table use market data, but they don't involve construction of demand curves. The last approach in Table 6.1, contingent valuation, is notable because it is the only available procedure for estimating passive-use values. There are many monetary valuation procedures and it is not feasible to review all of them. Table 6.1 includes methods that are commonly used.[6]

[4] This discussion of *ex ante* and *ex post* analysis relies on Freeman (1993, p. 14).

[5] Ex post analyses can also be used to help improve and verify ex ante evaluation techniques. For example, comparing actual damage caused by a waste discharge with an ex ante assessment of damage provides a basis for evaluating the assessment procedure.

HEDONIC PROPERTY VALUE METHOD

The hedonic property value approach infers prices for environmental services that are attributes of housing. The name of the method derives from the word "hedonism," but it is not particularly descriptive.[7] The central idea behind *hedonic pricing* is that the total price of a good (or service) can, in some cases, be disaggregated into inferred prices for attributes of the good. For example, the price of a house depends on characteristics such as lot size and number of bedrooms. Hedonic pricing attempts to identify the contributions of individual attributes to the overall price.

In the context of environmental management, the hedonic property value approach involves holding all characteristics of a house constant except for an environmental characteristic, such as the sulfur dioxide (SO_2) concentration of the ambient air. The contribution of the environmental characteristic to the price of the house is then inferred. For example, if two houses are equivalent in all respects except that one is subject to annoying transportation noise, the difference in price between the two houses reflects the value of the less noisy setting. Hedonic pricing has been used extensively to place monetary

[6] Alternate categorizations of monetary valuation procedures appear in the literature. For example, Turner, Pearce, and Bateman (1993, p. 15) divide methods into categories based on whether or not a demand curve is estimated as part of the evaluation procedure. Using their classification scheme, the following procedures in Table 6.1 are what Turner and his colleagues called *demand curve methods:* hedonic property value, travel cost, hedonic wage, and contingent valuation methods. Turner, Pearce, and Bateman classify all other methods in Table 6.1 as *nondemand curve methods.* For other methods and a more complete account of environmental valuation procedures, see Freeman (1993) and Turner (1993). Books on valuation that contain both theory and detailed case studies include Dixon et al. (1994) and Hanley and Spash (1993).

[7] Hanley and Spash (1993, p. 74) trace the origin of hedonic price analysis to theoretical advances in modeling consumer behavior made in the 1960s. Application of the hedonic price method to environmental attributes dates at least as far back as Ridker and Henning (1967), who used it to estimate the influence of air pollution on residential property values.

TABLE 6.1 Commonly Used Economic Valuation Procedures

Market-Based Methods

- *Hedonic Property Value Method.* Statistical analysis of factors influencing property values is used to derive implicit prices and demand curves for environmental attributes such as improved air quality.
- *Travel Cost Method.* Data on visitation rates and expenditures per visit to public outdoor recreation areas are used as a basis for deriving demand curves for those areas.
- *Defensive Expenditures Procedure.* Benefits of improved environmental quality are calculated as decreases in expenditures on actions taken to avoid effects of environmental degradation.
- *Production Function Approaches.* Economic effects are described in terms of (1) how alterations in environmental quality affect production processes and output levels, and (2) how changes in production influence revenues, profits, and prices.
- *Methods of Valuing Health and Longevity.*
  - *Human Capital Technique.* An individual is viewed as a unit of capital and valued in terms of expected earnings.
  - *Hedonic Wage Method.* Statistical analysis of factors influencing wage rates is used to derive implicit value of a life.

Contingent Valuation Methods

People's willingness to pay for an attribute of an environmental resource is estimated based on responses of individuals to survey questions. Respondents indicate how they would act contingent on being placed in a hypothetical circumstance.

values on environmental attributes related to noise, air, and water quality.

Hedonic Price Function

The first step of any hedonic pricing study involves identifying all variables that influence the price of the good. In the case of housing, the variables would include the following:

1. Characteristics of neighborhoods, for example, median income and school quality, represented by the variables N_i, for $i = 1$ to m neighborhood attributes.
2. Attributes of a house, such as number of rooms and lot size, represented by the variables H_j for $j = 1$ to n house attributes.
3. Measures of environmental quality, such as ambient noise levels and SO_2 concentrations in the ambient air, represented by the variables E_k for $k = 1$ to p environmental attributes.

The hedonic property value approach relies on a large set of data, where each *data record* includes (for a particular house) the price of the house (P) and values of attributes related to neighborhood, house, and environmental quality. The data are for a sample of houses within the geographical area under study. After the data are assembled, a statistical analysis procedure called *regression analysis* is used to estimate a *hedonic price equation*:[8]

$$P = f(N_1, N_2, \ldots N_m, H_1, H_2, \ldots H_n, E_1, E_2, \ldots E_p) \qquad \textbf{(6-1)}$$

where $f(\,)$ is called a hedonic price function.

The hedonic price equation provides a basis for determining the *implicit price* associated with an environmental attribute, for example, E_1. If all other variables are held constant, the slope of a curve of P vs. E_1 gives information on how the price P changes with very small changes in E_1. Using calculus, the slope is

[8] Regression analysis is used to estimate coefficients in an equation, such as $y = a_1x_1 + a_2x_2$. In this case, y is the *dependent variable*, x_1 and x_2 are *independent variables*, and a_1 and a_2 are the coefficients to be estimated based on observations of y, x_1 and x_2. Regression analysis is introduced in elementary statistics textbooks and is discussed further in Chapter 16.

the partial derivative of the function $f(\ )$ with respect to E_1. This derivative is termed the *implicit price of* E_1.

Under special conditions, the relationship between E_1 and the implicit price of E_1 is identical to the demand curve for E_1. This holds only if all households have identical incomes and utility functions. Although this is a restrictive case, it illustrates the basic concepts. The more general case requires an additional set of computations and is beyond the scope of this text.[9]

To examine the hedonic property value approach, consider a hypothetical setting in which the goal is to estimate the monetary value of improved air quality in an urban area. Assume all households have identical incomes and utility functions. Suppose, further, that ambient air quality is represented by an index based on concentrations of SO_2, particulates, and ozone, and that the quality of unpolluted air is indexed at a value of 100. Human activities resulting in decreased air quality are reflected by index values of less than 100. Hence, heavily polluted areas have low index values while areas with little pollution have index values close to 100. Assume that a statistical analysis of housing prices yields the hedonic price equation:

$$P = N^2 + H^2 + 10{,}000 - (100 - E)^2 \qquad \textbf{(6-2)}$$

where P = housing price,
N = neighborhood quality level,
H = house quality level, and
E = air quality index value.

In this case, the implicit price[10] of E is $200 - 2E$.

The plot in Figure 6.4 shows the (implicit) price an average household is willing to pay for incremental improvements in air quality when it buys a house.[11] If all households in the area are assumed to have the same demand for air quality, regardless of whether they buy a house, Figure 6.4 represents an average household's demand curve for air quality.

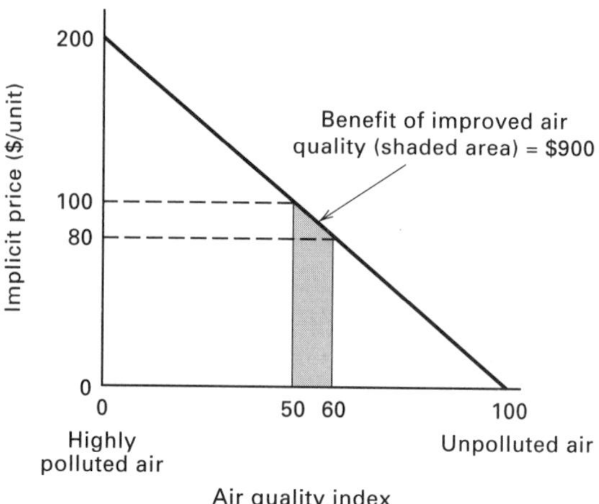

Figure 6.4 Household's demand curve for air quality.

The value of an air quality improvement is estimated by calculating differences in areas under households' demand curves with and without the improvements. For example, a gain in air quality represented by a shift in the air quality index from 50 to 60 results in an economic benefit represented by the shaded area in Figure 6.4, or $900, for a single household. If the air quality were improved and the index moved from 50 to 60, individuals would be better off and willing to pay for the improvement. The change in value for one household, $900, is multiplied by the total number of households in the study area to estimate the aggregate economic benefit of the improvement in air quality. For instance, if there are 100,000 households in the air basin affected by changes in air quality, then the economic benefit of the improvement in air quality is $90 million.[12]

[9] This more complicated case is analyzed by Freeman (1993). He also examines the simpler case presented here.

[10] The implicit price of E is $\frac{\partial P}{\partial E}$, the partial derivative of the hedonic price function, equation (6-2), with respect to E.

[11] This example utilized a simple hedonic price function. Suppose the function took a more complex form:

$$P = NH[10{,}000 - (100 - E)^2]$$

In this instance, the implicit price of E is equal to $NH(200 - 2E)$, and the value of a marginal improvement in air quality depends on the neighborhood, type of house, and air quality.

[12] A question that arises in hedonic price studies is, "With no increase in regionwide income or population due to migration into the region, how can everyone's house increase in value when air quality improves?" Trijonis and colleagues (1985, p. 108) answer the question by observing that the hedonic price "approach is not based on changes in house prices. Therefore, everyone's house price does not have to increase for benefits to exist." Trijonis and colleagues argue further that many people incorrectly believe that the basis for benefits using the hedonic property value technique is a change in demand for housing due to an increase in environmental quality. The hedonic price method is not concerned with

Footnote continued on opposite page

TABLE 6.2 Variables in the Hedonic Price Equation from a Study of Air Quality in the Boston Region[a]

Examples of Variables	Coefficient	Explanation of Variables
RM^2	0.0057	Average number of rooms
AGE	1.26×10^{-4}	Proportion of owner units built prior to 1940
Log RAD	0.017	Index of accessibility to radial highways
TAX	-3.53×10^{-4}	Full value of property tax
PTRATID	−0.030	Pupil teacher ratio by town school district
CRIM	−0.014	Crime rate
NOX^2	−0.0058	Nitrogen oxide concentration, parts per hundred million

[a] Based on Harrison and Rubinfeld (1978). The material contained in this table are only some of the variables they used in developing a hedonic price equation.

The air quality example used here supposes that all households have identical incomes and preferences. In practice, this cannot be assumed, and thus the implicit price function is not the demand curve per se. Application of the hedonic property value method is complex and involves much more than using statistical methods to find a hedonic price function, and calculus to derive an implicit price equation. Specialists in both economic theory and statistical analysis are required in implementing the hedonic price approach.

Illustrative Hedonic Price Analysis

Dozens of hedonic price studies have been performed to estimate the influence of air quality degradation on property value. In summarizing many of the early studies, Freeman (1979) shows that commonly used forms of hedonic price functions include linear equations, and equations that contain logarithmic transforms of measured variables. For example, in studies of how air quality influences housing values in the Boston metropolitan area, Harrison and Rubinfeld (1978) experimented with several functional forms. They finally selected a form that used, as the housing price variable, the log of the median value of owner-occupied housing units in a census tract.[13]

Table 6.2 includes some of the variables used in the Boston study. As shown in the table, the variable used to measure air pollution is the square of the concentration of nitrogen oxide (NOX), in units of parts per hundred million (pphm). The coefficient of NOX has an interesting economic interpretation, but its significance is not obvious because of the nonlinear form of the hedonic price equation. Based on their calculations, Harrison and Rubinfeld found that when NOX and all the other variables in their analysis take on median values, an increase of $1613 in the median housing value is associated with a 1 pphm reduction in NOX.[14]

Because they could not assume that all households earned the same income, Harrison and Rubinfeld had to do additional statistical analyses to estimate values of willingness to pay for improvements in air quality. They experimented with willingness-to-pay equations that depended upon household income, NOX, and persons per dwelling unit. Depending on the form of the willingness-to-pay equation they employed, the average annual benefit (per household) of proposed reductions in NOX by emission controls on automobiles ranged from nearly $60 per year to almost $120 per year.

the demand and supply of houses, but only with the parameters that underly the housing choices that are made. The authors elaborate on this point: "The hedonic approach uses differentials in house price to provide information on how consumers trade air quality [and other environmental attributes] for other goods (living area, location, etc.). With his information, the structure of the consumer's demand for air quality can be logically deduced [for example, see Figure 6.4]. Note that this is not the demand for housing units but for only one aspect of the housing unit, the air quality aspect."

[13] It is common for hedonic property value studies that center on air quality in the United States to use, as a source of basic data, the U.S. Census of Population and Housing.

[14] As Harrison and Rubinfeld (1978) point out, this numerical result changes if a different form of the hedonic price function is used.

Potential Problems with Hedonic Pricing

Problems with the hedonic property value technique can be summarized in the following terms.[15]

- *Omitted Variables.* If an important characteristic that influences the price of the good is omitted in the process of constructing a hedonic price equation, coefficients in the equation will be incorrect. These errors will, in turn, produce an inaccurate implicit price function.
- *Correlated Characteristics.* If two variables representing characteristics are closely correlated, it is not possible to interpret their individual coefficients in the hedonic price equation. For example, an airport may generate both noise and air pollution. An increase in air pollution may be accompanied by an increase in noise pollution, and both may result in a fall in house prices. The hedonic price equation will not be helpful in distinguishing the decrease in house prices attributable to noise pollution from the decrease attributable to air pollution.
- *Choice of Functional Form.* Many functional forms have been used in hedonic price studies, including quadratic and logarithmic functions. However, economic theory does not provide a basis for identifying the "correct" functional form. Experts try to select forms that are easy to work with, fit the data, and have sensible economic interpretations.
- *Market Segmentation.* Housing markets may be segmented in numerous ways, for example, rental vs. owner-occupied housing. If the existence of different market segments is not recognized and all segments are grouped together in the statistical analysis, the coefficients in the hedonic price equation may be inaccurate. Each market segment requires a separate hedonic price equation.
- *Perfect Housing Markets.* The hedonic pricing technique assumes that markets are perfectly competitive and in equilibrium. In the air quality case, for example, the method assumed that all buyers were fully informed of air pollution levels and the associated threats to their welfare, and that buyers accounted for effects of air pollution in making housing purchases. The hedonic property value method also assumes no surplus or shortage of housing: supply and demand are equal and the housing market is in equilibrium. These assumptions are not met in most housing markets.

Notwithstanding the above limitations, the hedonic pricing method can provide useful information on the value of environmental characteristics associated with housing. Based on his review of 15 different air pollution—hedonic property value studies covering 11 cities in the United States and Canada. Freeman (1979, p. 161) made the following observations: "First, the hypothesis that property values are affected by air pollution is generally supported by the evidence. Only 3 studies [yielded results inconsistent with the hypothesis]. . . . Second, the numerical values reported [in the 15 studies] are generally plausible and broadly consistent both within cities as derived from different studies and between cities."

TRAVEL COST METHOD

The travel cost method (TCM) has been used extensively to calculate the value of outdoor recreation areas, such as parks and reservoirs.[16] The method infers prices for visits to recreational areas based on the cost of traveling to and utilizing those areas. The TCM cannot estimate passive-use values because the underlying data reflect active uses.

If entry fees for outdoor recreation areas varied substantially across sites or over time, a demand curve for outdoor recreation areas could be estimated using data on how the number of visits to a recreation site varied with admission price. However, public owners of recreation areas typically charge either no entry fee or a nominal fee that does not vary significantly over time. Thus, admission fees at publicly owned recreation areas do not generally reflect the willingness of people to pay for visits to those areas.

The travel cost method assumes the " 'price' of visiting a park or other recreation area (even one for which entry is free) will vary according to the travel

[15] This summary is based on Hanley and Spash (1993, pp. 78–81). Some of the problems listed here are common in many domains of econometrics, a branch of economics in which statistical analysis methods and economic theory are used to obtain estimates of relationships describing economic variables. For an introduction to econometrics, see Gujarati (1988).

[16] The original idea for the travel cost method is attributed to the economist Harold Hotelling, who mentioned it in a letter to the director of the U.S. Park Service in 1947 (Hanley and Spash, 1993, p. 83). However, it is often associated with Clawson and Knetsch (1966) who elaborated upon the method in a monograph entitled *Economics of Outdoor Recreation.*

TABLE 6.3 Travel Cost and Site Usage Data for Visits to a Hypothetical Recreation Site[a]

Zone	Population	Cost Per Visit	Number of Visits in a Season	Visits per 1000 Population
1	1000	$1	500	500
2	4000	$3	1200	300

[a] Adapted from Clawson and Knetsch, *Economics of Outdoor Recreation,* © 1966, p. 79. Reprinted with permission of the Johns Hopkins University Press.

cost of visitors coming from different places" (Portney 1994, p. 4). The method estimates willingness to pay for use of a recreation area by determining how users of the area would respond to increases in the cost of traveling to the site.[17] In the TCM, *travel costs* include the monetary expenses incurred in getting to the site, the value of the time spent traveling (*time costs*), and admission fees. Monetary travel costs depend on the location of the household, and time costs depend on income levels of household members. However, admission fees are usually the same for all visitors.

The first step in the travel cost method involves dividing the area surrounding the recreation site into zones. A common zoning scheme uses concentric circles with the recreation site at the center. Alternatively, zones may be defined using political jurisdictions such as cities or counties.

The travel cost method requires data on *zonal population* and how many visits to the site are made by people in each zone. The latter is obtained by surveying a sample of visitors to the site to determine their zones of origin. The number of visitors from each zone is then divided by the zonal population to determine the number of visitors per 1000 population for each zone.

TCM Applied to a Hypothetical Site

Table 6.3 shows results from a travel cost survey for a hypothetical reservoir recreation area. There is no entry charge. The area surrounding the recreation site is divided into two zones. Visitors from zone 1, which is closest to the site, have a travel cost per visit of $1. Visitors from zone 2 incur a cost of $3 per trip. In addition to indicating costs per visit, the table shows the population of each zone and the number of visits per thousand people in each zone. Data in the columns for travel cost and visits per thousand population are plotted in Figure 6.5.

The travel cost method for estimating a demand curve for the reservoir assumes that the relationship in Figure 6.5 is applicable under conditions in which travel cost increases because the admission price goes up. Visitors are assumed to respond to a higher entry fee the same way they (according to Figure 6.5) react to a rise in the cost of transportation to reach the site.

The demand curve for a recreational site is developed using the following iterative procedure:

1. Select an entry fee arbitrarily and add it to the travel cost (for each zone) to obtain a new cost to visit the site.

2. Use the relationship between travel cost and visits per 1000 population to obtain (for each zone) a new visitation rate per 1000 population.

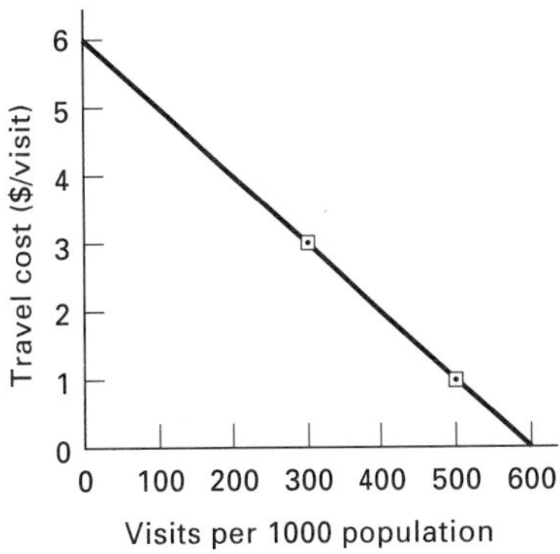

Figure 6.5 Travel cost vs. visits per 1000 population for a hypothetical recreation site. Adapted from Clawson and Knetsch, *Economics of Outdoor Recreation,* © 1966, p. 79. Reprinted with permission of the Johns Hopkins University Press.

[17] The TCM method assumes that "fairly large samples of persons, chosen randomly as far as their interest in outdoor recreation is concerned, will react similarly to changes in recreation costs" (Clawson and Knetsch, 1966, p. 78).

TABLE 6.4 Estimates of Site Visits at Different Entry Prices[a]

	Number of Visits at Entry Fee of:				
Zone	**$0**	**$1**	**$2**	**$3**	**$4**
1	500	400	300	200	100
2	1200	800	400	0	0
Total Visits	1700[b]	1200	700	200	100

[a] Adapted from Clawson and Knetsch, *Economics of Outdoor Recreation*, © 1966, p. 80. Reprinted with permission of the Johns Hopkins University Press.

[b] The number of visits at a zero entry fee is based on survey data (see Table 6.3).

3. Obtain the total number of visits per zone by multiplying the population in each zone by the number of visits per 1000 population.
4. Sum the results in step 3 for each zone to obtain the total number of visits at the entry fee selected in step 1.
5. Return to step 1 and repeat the procedure with a new entry fee.

To illustrate the procedure, observe (in Table 6.4) that if there is no entry fee for the hypothetical recreation area, the total number of visits in a season is 1700. This number is obtained from the survey of visitors, and it provides one point on the demand curve in Figure 6.6. Now suppose that the price of admission was $1 instead of zero. This means the cost per visit from zone 1 would increase from $1 to $2, and (using Figure 6.5) the number of visits per 1000 population from zone 1 would drop from 500 to 400. Since there are 1000 people in zone 1, the estimated number of visits from that zone would be 400 with an entry price of $1. The same reasoning is used in analyzing zone 2. In this case, travel cost would increase from $3 to $4, trips per 1000 population would drop from 300 to 200, and the total number of visits would decrease from 1200 to 800 (that is, 200 visits per 1000 population times 4000, the zone 2 population). Summing the number of visits from each zone yields a grand total of 1200 visits with an entry price of $1.

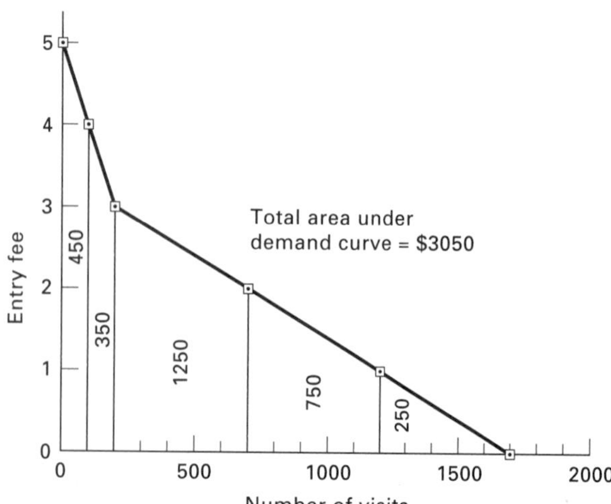

Figure 6.6 Demand curve for a hypothetical reservoir.

Now increase the entry fee, say to $2, and repeat the steps in the iterative procedure. As shown in Table 6.4, with a $2 entry fee there would be 700 visitors (300 from zone 1 and 400 from zone 2). This iterative procedure is repeated for entry fees of $3, $4, and $5. An admission fee of $5 brings the travel cost in zone 1 up to $6, and there would be no visitors at or above this price. Results from the procedure are plotted in Figure 6.6.

The appropriate measure of the economic benefit of the hypothetical reservoir is the maximum amount visitors would be willing to pay to use the site. Because the admission price is zero, the maximum WTP is equal to the total area under the demand curve in Figure 6.6.[18] As indicated in the figure, the first 100 visits to the hypothetical site are valued

[18] Another consequence of the zero entry fee is that the total economic value is equal to the total consumers' surplus.

at between \$4 and \$5, and the economic value, as represented by the total willingness to pay for these 100 visits, is \$450 (the area under the demand curve between 0 and 100). The results of calculating willingness to pay for the next 100 visits, and so on, are indicated in the figure. Adding up the area yields a total of \$3050, which is a measure of the economic value of recreational opportunities provided by the site.

Complications with the Travel Cost Method

The foregoing simplified example highlights the main steps in the TCM. However, actual use of the TCM is more complicated because the analyst must account for the following:[19]

- *Multi-purpose Trips.* If an individual visits more than one site on the same trip, then a question arises: How much of the travel cost should be attributed to the site under investigation? The problem is significant when visits are made by vacationers who journey a long distance and then visit several places in the same region.
- *Substitute Sites.* The value of a site depends not only on the site itself, but also on the availability of nearby, similar sites. Consider the hypothetical case of two lakes that are identical except that one is ten miles closer to the main access road than the other. The lakes are nearly perfect substitutes, but people will generally choose to visit the closer lake. If the presence of a substitute were not accounted for, results would be in error. The economic consequences of losing the closer lake as a recreational area would not be as large as indicated by the TCM because people could simply drive a little farther to experience the same environmental attributes. There is no simple way to incorporate substitute sites into the demand curve derived by the travel cost method.
- *Time Costs.* The economic value of time spent in travel to a site is an important component of overall travel costs, but there is no agreed-upon way to estimate that value. One approach is to use the individual's wage rate, but this may not be accurate because the travel time might be spent on weekends or during a vacation when the person would be unable to substitute work time for travel time. Moreover, even vacation time has an opportunity cost associated with it. If a person was not traveling, she could enjoy her free time in another way. The problem is that time costs vary for individuals and it is difficult to collect data that reflect these variations.[20]
- *Distance Costs.* There are two measures used in calculating cost per mile (or kilometer) in TCM studies: (1) the *incremental* cost of driving, as given by fuel costs, and (2) the *average* cost of driving, which includes an allowance for vehicle depreciation, insurance, and so forth. Models of consumer behavior typically assume that individuals maximize utility by comparing incremental costs to incremental benefits of consumption. However, many people also consider average costs when making choices.[21] The decision to use incremental driving costs versus average driving costs affects the outcome of TCM studies, but there is no nonarbitrary way to decide which to use.

DEFENSIVE EXPENDITURES PROCEDURES

The *defensive expenditures* approach to valuation rests on the following concept: People spend money to avoid some consequences of environmental degradation, and those expenditures reflect how much they value improvements in environmental quality. Examples of defensive expenditures are abundant. Households may react to decreased quality of their tap water by increasing their purchases of bottled water and water filters, or they may spend money on fuel to boil their water before drinking it. And firms requiring high-quality input water may react to increased pollution of their supplies by spending more money on water treatment. Households and firms may respond to increased air pollution due to suspended particulates by spending additional funds to keep things clean. In some cases, air-conditioning

[19] This discussion of problems with the travel cost method is based on Hanley and Spash (1993, pp. 86–91).

[20] For an introduction to several methods used to estimate time costs, see Hanley and Spash (1993, pp. 89–90).

[21] The average per-mile cost of driving can be much higher than the fuel cost per mile. Hanley and Spash (1993, p. 88) report on a TCM study where the economic benefits calculated using average cost per mile of driving was about 2.5 times as high as benefits computed using incremental costs.

devices and air filters may be used to improve indoor air quality.

The central idea in estimating environmental benefits by examining defensive expenditures concerns what economists call *substitute goods.* In effect, bottled water for a household is a substitute for tap water, and a company's use of polluted intake water that it has treated is a substitute for unpolluted intake water that is usable without treatment. Benefits of increased environmental quality are measured by decreased spending on substitutes.

While the defensive expenditure procedure has a certain intuitive appeal and is often used in practice, it is flawed in several ways.[22] One shortcoming of the method is that, in general, substitutes are imperfect in that they typically compensate only partially for the effects of pollution. For example, suppose households install double-glazed windows to offset indoor noise caused by increased traffic on a nearby road. The benefits of quiet are underestimated by this approach because double-glazed windows are an imperfect defense against noise. Moreover, the noise reduction is effective only when the residents are indoors with their windows closed. During periods when residents open their windows or are outdoors in their gardens, the double-glazed windows have no effect.

Another difficulty in using the defensive expenditure technique is that items purchased to offset pollution may yield benefits unrelated to pollution reduction. For example, double-glazed windows decrease indoor heating expenditures in addition to cutting down indoor noise. Similarly, air conditioners that reduce indoor suspended particulate levels also provide temperature control. How much of the expenditure for double-glazed windows or air conditioners should be counted as a defense against pollution?

Defensive expenditures undertaken by firms can be particularly misleading as estimates of environmental benefits.[23] As will be demonstrated, this occurs when defensive expenditures are large enough to affect market supply curves.

[22] According to Freeman (1993, pp. 114–15) the defensive expenditure approach is flawed in a fundamental way when it is used to analyze a "nonmarginal change" in environmental quality. His mathematical analysis shows that even when the market good purchased as a defensive measure is a perfect substitute for environmental quality, use of defensive expenditures to analyze nonmarginal changes in quality underestimates the benefits of a cleaner environment.

PRODUCTION FUNCTION APPROACHES

Production function approaches are based on relationships between environmental quality variables and the output of marketable goods. Links between environmental quality and marketable products exist in many industrial sectors, particularly agriculture, timber, and commercial fisheries, and these links are reflected in empirically derived production functions.[24] As clarified below, the relationships between environmental variables and market outputs often take the form of dose-response curves. Improvements in environmental quality can lower production costs, which can lead to increases in the quantity of marketed goods supplied by producers. Reductions in environmental quality often have opposite effects. Production function approaches to valuation are introduced here using examples.

Prices Unaffected by Productivity Changes

Consider a case in which effects of changes in environmental quality on production costs are so small they do not affect market prices. In this context, the evaluation focuses on first determining how alterations in environmental quality affect the output of products, and then valuing the changes in output in monetary terms.

The approach is illustrated using a study by Hufschmidt and Dixon (1986) to find the economic value of fishery losses in Tokyo Bay, Japan. The fishery was degraded by a land reclamation program: wetlands and contiguous areas of shallow water in the bay were filled with earth, sand, and rubble to create new land for industrial development.

Effects of reclamation included a significant decline in commercial fishing, since the reclaimed zone was the most productive fisheries area within Tokyo

[23] For additional shortcomings of the defensive expenditures approach, see Hanley and Spash (1993, pp. 99–101).

[24] Production functions are introduced in Chapter 5.

TABLE 6.5 Fisheries Output Losses due to Land Reclamation at Tokyo Bay[a]

Year	Marine Product Output (10^3 tons)	Reduction in Output Compared to 1956–62[b] (10^3 tons)	Market Value of Lost Output (10^9 yen)	Reduction in Gross Profits to Fishermen (10^9 yen)
1972	50	116	62.6	46.3
1973	52	114	61.3	43.5
1974	55	111	56.5	42.4
1975	44	122	66.5	47.2
1976	53	113	51.2	36.4
1977	97	69	19.2	12.1
Annual Average	59	108	52.9	38.0

[a] Values are based on Hufschmidt and Dixon (1986, pp. 114–116). All values are calculated based on 1979 prices.
[b] Average annual output in the period prior to reclamation (1956–62) was 166 × 10^3 tons.

Bay. Hufschmidt and Dixon conducted an ex post study to determine the monetary value of observed declines in "marine product output," which included seaweed and shellfish. Data on outputs and market prices existed for periods before and after 1963, the year the reclamation program was assumed to have affected fishing in Tokyo Bay. The average annual output of marine products from the bay in the several years before 1963 was 166 thousand tons, a figure used as a baseline in determining decreased output because of reclamation. The period from 1972 to 1977 was taken as a representative post-reclamation period, and annual outputs for those years are shown in Table 6.5. The third column in the table shows production losses, which are differences between actual outputs and average output in the predevelopment period (1956–62).

The next step in the analysis centered on placing a monetary value on losses in marine product output. Hufschmidt and Dixon calculated two measures: market value of the decrease in output; and loss to fishermen, represented by reduced profits associated with the productivity loss.

The market value of reduced output was computed by multiplying the decreased fisheries output by "unit values" of output (in 10^6 yen per 10^3 tons). The unit values were based on the actual gross revenues to fishermen in different years. Results from these calculations are shown in Table 6.5 in the column labelled "market value of lost output."

A second measure of value used by Hufschmidt and Dixon reflects lost profits to fishermen. This was determined by subtracting capital and operating costs (per 10^3 tons of output) from the market value of lost output. Inadequacies in production cost data forced Hufschmidt and Dixon to make several assumptions about production costs.[25] The resulting losses in profits are given in the last column in Table 6.5.

The Tokyo Bay example introduces the approach to using changes in productivity to place a monetary value on environmental resources. However, because it was an ex post study, the example did not raise questions that arise in an ex ante context, where the valuation is conducted to help decide whether to undertake a proposed action. These questions are considered now using a hypothetical example that introduces complexities that arise when proposed environmental changes cause shifts in market prices.

[25] For example, Hufschsmidt and Dixon (1986, p. 111) assumed that the opportunity cost of the fishermen's labor was zero. They note that "[t]his is a very strong (and probably unrealistic) assumption," but they had no data on labor costs. Hufschmidt and Dixon reasoned that, for traditional reasons, fishermen would not be willing to work in another occupation or a different location.

TABLE 6.6 Production Function Approach to Valuation

Delineate Physical Effects
A dose-response curve or a production function is used to relate changes in environmental quality to changes in output.

Analyze Producers' Responses
Producers take actions in response to changes in environmental quality. This leads to changes in marginal production costs and short-run supply curves.

Investigate Market Responses
When producers' responses are aggregated, they may affect the industry-wide supply curve, and the equilibrium price and quantity of output.

Estimate Economic Changes
The aforementioned changes in output price and quantity are translated into changes in consumers' surplus and a surplus for producers.

Environmental Changes Affect Prices

The example centers on a proposal to reduce ozone levels in a region, and the evaluation problem is to estimate the value of the expected decrease in ozone to farmers in the region. This is a realistic evaluation problem because high levels of ozone adversely influence growth of some crops.[26] Steps in the evaluation are listed in Table 6.6.

The first step in the analysis involves estimating how output of existing crops would be affected by the proposed decrease in ozone. This is a scientific question, and results are sometimes presented in a *dose-damage* (or *dose-response*) curve.[27] For simplicity, suppose the proposed reduction in ozone concentration would increase annual agricultural output by two tons per acre per year, and the market price of the crop is constant at $10 per ton.[28] If productivity gains do not require adjustments in other inputs, such as labor at harvest time, the economic value of ozone reduction would be estimated at $20 per acre per year. Suppose, however, that an additional $2 of labor per ton is required to harvest the increased yield. In this case, the economic benefit is the increase in agriculture output value *net* of the increased labor, or $16 per acre per year.

A key question can be raised at this point: Will producers change their outputs in response to the proposed improvements in environmental quality? Farmers customarily make adaptations in response to changing economic and environmental conditions. In this example, farmers may find they can increase profits by shifting from existing crops to a crop such as spinach, which is sensitive to ozone but has a relatively high market value. In this circumstance, the previous analysis would yield incorrect results.

Market Level Effects

Other questions concern market level effects. Are the changes made by individual producers significant enough, in aggregate, to affect industrywide production costs? For example, if the proposed change in ozone causes a shift in the industrywide supply curve for a crop, then a new market equilibrium price and quantity will result for that crop, and it must be accounted for in estimating monetary benefits. The resulting complexities center on measurements of consumers' surplus, and a related surplus for producers.

For illustrative purposes, suppose the proposed ozone reduction would yield a decrease in the industrywide marginal cost of producing spinach. The expected shift in the supply curve for spinach is as shown in Figure 6.7. The original equilibrium price is OC, and the corresponding equilibrium output is OH. Because of the reduction in ozone concentration, the intersection of the demand curve with the new supply curve yields an equilibrium price of OA and an output of OG. At the new equilibrium, consumers purchase more spinach because the cost of producing spinach has dropped.

[26] This example concerns ozone in the lower atmosphere; high concentrations of ozone (either alone or in combination with oxides of sulfur and nitrogen) have been linked to significant crop losses in the United States.

[27] For details on how *exposure dose* has been defined in economic evaluations of crop loss due to high ozone levels, see Hanley and Spash (1993, pp. 175–77). They also review methods for deriving dose-response curves in ozone-crop loss studies. Dose-response curves are described in the discussion of risk assessment in Chapter 17.

[28] The illustrative case here is from Freeman (1982, pp. 86–88).

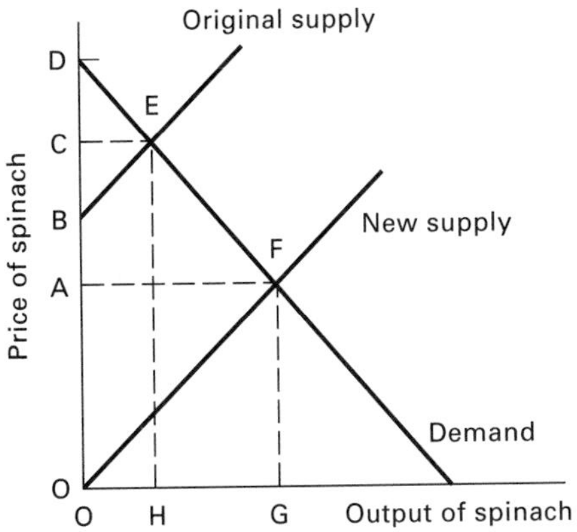

Figure 6.7 Hypothetical market level responses to a reduction in ozone concentration.

The economic gains associated with shifts made by producers and consumers are analyzed in two parts. The gains to consumers consist of increases in consumers' surplus. In the original situation, the consumers' surplus was the triangle CDE. But with the shift in supply, the consumers' surplus expands to ADF. The increase in consumers' surplus is the difference between the two triangles, the area represented by ACEF.

Producers are also affected by changes in the market equilibrium price and quantity. The impact of these market changes on producers is analyzed using *producers' surplus,* a concept introduced by considering a single firm in a perfectly competitive market. Recall from Chapter 5 that in perfect competition, each firm is a *price taker.* A profit-maximizing firm would increase output until its marginal cost of production equaled the market price. The difference between total revenue (price times quantity produced) and total *variable* cost (the area under the marginal cost curve between the origin and the firm's output) equals the producer's surplus.[29]

Producers' surplus for an industry relates to producer's surplus for a firm because the supply curve for an industry consists of the horizontal sum of portions of marginal cost curves of individual firms.[30] Using reasoning similar to that sketched for a single firm, the producers' surplus for an industry is the area below the price line and above the industry supply curve. Triangle BCE in Figure 6.7 represents the industrywide producers' surplus before the decrease in production costs due to reduced ozone pollution. Similarly, triangle OAF shows the producers' surplus after the decline in pollution. The net increase in producers' surplus is the difference between the two triangles: the area OAF minus BCE.

An analysis of the full economic consequences of a change in equilibrium price and quantity accounts for changes in both consumers' surplus and producers' surplus. In the example of Figure 6.7, both consumers and producers experience economic benefits due to decreased costs of production. The magnitude of their economic gain is represented by combining the changes in consumers' surplus (area ACEF) and producers' surplus (area OAF minus area BCE). The sum of these net changes in surplus is represented by the area OBEF.[31]

METHODS OF VALUING HEALTH AND LONGEVITY

For certain improvements in environmental quality, such as reductions of sulfur dioxide emissions to the atmosphere, benefits in the form of improved health and longevity can be substantial. Two different approaches have been taken to evaluate economic benefits of improved health and longevity. One involves calculating foregone earnings associated with illness and premature death. This approach is not tied closely to the fundamental definition of economic benefits (as is willingness to pay), but it has been widely used because it imposes modest data-gathering and analysis burdens. A second method, which has a more solid conceptual foundation, is based on observing what people have been willing to pay to take on high-risk occupations.

Each of the two valuation methods yields monetary values of human life (in a statistical sense), and

[29] This analysis is for short-run conditions, and thus total costs include both fixed and variable costs. Producer's surplus is not the same as profit because producer's surplus does not account for *fixed* costs. A long-run analysis of the response of firms to a reduction in ozone level would consider how affected firms might alter their portfolios of fixed assets.

[30] This relationship between supply curves for individual firms and the market supply curve is described in Chapter 5.

[31] The proof of this assertion involves an exercise in geometry. The term *social surplus* is used to refer to the sum of consumers' surplus and producers' surplus.

this has been a source of controversy. Some people object to placing monetary value on a life. The customary response to such objections is that individuals and governments routinely make choices that imply deliberate trade-offs between increased welfare and risk of death. For example, people who smoke cigarettes are often aware that they are trading off the pleasures smoking brings them against a risk of shortening their lives. Their choice involves an implicit utility-risk trade-off. Methods for placing an economic value on human life aim to make such implicit valuations explicit. The methods do not place a value on a particular individual's life; they concern the lives of individuals in a statistical or actuarial sense.

Valuation Based on Foregone Earnings

A frequently used approach for placing monetary value on the premature loss of life or increased morbidity is the *human capital technique.*[32] The value of a life lost is taken as the present value of the anticipated time stream of future earnings for an individual. A human being is viewed as a unit of capital and is valued (as with other forms of capital) in terms of expected earnings.

A rudimentary version of the human capital procedure requires specification of a time stream of future expected earnings and data on the probability that individuals will live to certain ages. The life expectancy data provide a basis for computing the expected value of future earnings.[33] These earnings are then discounted (using an appropriate discount rate) back to an equivalent present value.[34] Given an estimate of the number of lives saved by an improvement in environmental quality, the human capital approach places a monetary value on those lives. The calculation procedure accounts for variations among individuals, such as differences in age, sex, and education, and the future likelihood of individuals being in the labor force.

Many objections have been raised to the human capital approach, in part because it assumes that a person's values rests entirely on his or her earnings. This is morally indefensible, because it places a value of zero on the lives of retired persons and those who are permanently disabled, among others. In addition, many economists object to the approach because it yields monetary values that are not closely tied to willingness to pay as a basis for defining benefits.

The human capital approach has found a somewhat wider acceptance when applied to monetary valuations of morbidity. Here the foregone earnings are the income lost by being absent from work because of illness caused by environmental pollution. Some studies of morbidity also include computations of increased expenditures on doctors, drugs, hospitalization, and so forth.[35]

Valuation Based on Willingness to Pay

In the past few decades, many economists have argued that the central issue in valuing human life (in the context of environmental benefits) is reduction of the risk of death. Freeman (1982, p. 39) explains the central idea with the following example. Imagine a group of 1000 similar people in which each person would be willing to pay $1000 for government policies that would reduce the probability of her or his death by 0.01. The sum of these individuals' willingness to pay is $1 million, and the average (or expected) number of deaths avoided is 10 (the product of 0.01 and 1000). Dividing the aggregate willingness to pay, $1 million, by the number of deaths avoided yields a *statistical value of a life* of $100,000. If individuals actually expressed their willingness to pay by purchasing policies, the value of a life could be inferred from observed data. In practice, an alternative approach is used to determine willingness to pay. In studies of lives saved by pollution reduction measures, labor market data for workers receiving high wages for accepting occupational risks are customarily used to estimate monetary value of a life.

A standard approach to using data on wage-risk tradeoffs revealed by labor market data is based on hedonic pricing. Recall that the hedonic price method requires estimation of a mathematical relationship linking the price of a good or service to characteristics

[32] According to Landefeld and Seskin (1982), applications of the human capital approach to valuing life can be traced as far back as the seventeenth century.

[33] In statistics, the *expected value* is defined in terms of products of quantitative outcomes (such as earnings) and the probability that those outcomes will be realized.

[34] Procedures for calculating the present value of a time stream of future costs are introduced in the next chapter.

[35] For a survey of the human capital approach to estimating economic benefits of reduced morbidity from air pollution control, see Dickie and Gerking (1989). The authors also demonstrate use of defensive expenditures as a procedure for estimating benefits of reduced morbidity.

of the good or service. In the context of studies to infer a statistical value of a life, the price is a wage rate and the service is a job. The valuation procedure, known as the *hedonic wage method,* requires use of statistical methods to develop an equation of the form

$$w = g(J_1, J_2, \ldots, J_n, \text{L. I}) \qquad \textbf{(6-3)}$$

where w represents wage rate (for example, annual earnings), g ( ) is a hedonic wage function, and the variables $J_i (i = 1, \ldots, n)$ represent worker characteristics and job attributes unrelated to the risk of either loss of life (L) or nonfatal injury (I). Examples of personal characteristics that correlate with wage rate are age, sex, and education. Job-related characteristics linked to wage rate are illustrated by union status, working conditions, and prestige. Hedonic wage studies generally include an extensive list of job attributes to insure that the portion of the wage due to acceptance of risk can be isolated.

The hedonic wage approach involves use of calculus. In the simplest application, the partial derivative of the hedonic wage function with respect to any characteristic ($J_1, \ldots, J_n$, L, or I) gives the *marginal implicit price* of that characteristic.[36] Thus, for example, the partial derivative of the hedonic wage function with respect to loss of life (L) gives what has been termed the *marginal implicit value of life,* and the partial derivative with respect to nonfatal injury (I) represents the (marginal) implicit value of a nonfatal injury. As elaborated by Viscusi (1986, p. 199), "these values represent the risk-dollar trade-off, not the value the person would place on certain death or injury." The implicit prices represent willingness to pay for marginal changes in risk. More extensive analyses are required to estimate willingness to pay when nonmarginal changes are involved.

Freeman (1993) presents marginal willingness to pay for reductions in risk of death from 11 hedonic wage studies. He reports "judgmental best estimates" of statistical value of a life from those studies; the estimates range from \$640,000 to \$8.5 million (in 1986 U.S. dollars). While the estimates vary by more than a factor of ten, some of the variation is due to differences in quality of data and in variables included in the studies. The most recent of the 11 studies, the work of Gegax, Gerking, and Schulze (1991), used what Freeman (1993, p. 428) characterized as "a more detailed description of jobs than is available in most labor-market data sets." The result was a statistical value of life of \$1.9 million (in 1986 U.S. dollars) for workers included in the study.

[36] A number of assumptions must be satisfied for this to be true, and they are elaborated by Freeman (1993).

CONTINGENT VALUATION

Each of the benefit estimation procedures previously discussed estimates value by examining market data. An alternative way to determine people's willingness to pay for a good or service is simply to ask them. The *contingent valuation* method asks people to indicate how much they would be willing to pay for environmental services if they were in a hypothetical situation.[37] Although the contingent valuation approach appears simple, its application requires considerable skill and effort.

Contingent valuation was first applied in the 1960s and was initially greeted with skepticism by economists.[38] Many had doubts about the ability of individuals to know their preferences for goods and services they had no experience in purchasing. How, for example, could an individual place a value on enhanced visibility in the Grand Canyon or the preservation of the bald eagle? In addition, concerns were raised about intentionally biased responses. For example, people might overstate their willingness to pay if they thought it would lead to public expenditures to preserve environmental resources they valued.

Although it is still controversial, contingent valuation has attained a degree of acceptance.[39] Contingent valuation has been used to evaluate a variety of environmental resources, including scenery, wildlife, and wetlands. The method has also been employed

[37] In this context, *contingent* means possible future condition. The method is named contingent valuation because "the values revealed by respondents are contingent upon the constructed or simulated market presented in [a contingent valuation] survey." (Portney, 1994, p. 3).

[38] According to Portney (1994, p. 4), contingent valuation was first proposed by S. V. Ciriacy-Wantrup in 1947 and first applied by Davis (1963). Although Davis did not use the term contingent valuation, he did elaborate on the use of "experimental markets" to obtain information on individuals' willingness to pay for use of outdoor recreation areas.

[39] For an indication of the degree of controversy surrounding contingent valuation, see the symposium on contingent valuation reported in the Journal of Economic Prespectives, Vol. 8, No. 4, Fall, 1994, pp. 3–64.

to assess damages from oil spills and toxic wastes. An important factor contributing to the acceptance of contingent valuation was its use in the late 1980s by U.S. federal courts in setting awards to compensate for environmental damages.[40]

Willingness to Pay and Willingness to Accept

When people are asked to express their values in a contingent valuation study, the questions sometimes focus on *willingness to accept* (WTA) instead of willingness to pay. The WTA concept assumes a person has the right to a given level of environmental quality, and the individual should be paid if this right is violated. Willingness to accept refers to the *minimum* amount a person would be willing to accept to forego an improvement (or to allow a deterioration) in environmental quality. For example, assume individuals participating in a CV study have a right to safe drinking water. A CV approach using willingness to accept asks the participants how much money they would have to receive as compensation to make up for the loss of their safe drinking water. Contrast this with questions that ask individuals how much they would be willing to pay to obtain a safe drinking water supply.

Willingness to pay and willingness to accept need not be equal. Willingness to pay is constrained by an individual's income, but there is no upper limit on what a person would be willing to accept for a change in environmental quality. Dorfman (1993, p. 317) notes that theories of consumer behavior indicate WTP and WTA should be about equal when they are small in proportion to income. However, empirical studies indicate that WTA is invariably greater than WTP, often several times as great. Dorfman offers the following explanation:

> [P]eople form ego attachments to their possessions and entitlements, and psychologically regard and resist decreases as invasions of their established rights. On this ground, the observed discrepancies between WTP and WTA are not errors . . . , but reflections of discontinuities in people's preferences that public decisions should take into account.

Both WTP and WTA are well-established measures of economic value, and use of one over the other may depend on whether the environmental attribute in question is considered to belong to the person (WTA), or whether the person must pay to enjoy the attribute (WTP). The discussion that follows uses the willingness to pay concept because among economists "the general view is that WTP values are more reliable indicators of welfare changes than are the WTA results" (Freeman 1993, p. 170). However, in certain contexts, the WTA concept may be more appropriate.

The complexity of eliciting people's true willingness to pay for an environmental attribute can be appreciated by considering the steps in a contingent valuation analysis.[41] These steps, which often center on use of a survey questionnaire, are highlighted in the following pages.

Conducting a Contingent Valuation Study

Once the environmental quality attribute to be evaluated has been identified, a hypothetical setting is constructed. The setting provides a context for eliciting information related to willingness to pay. Participants in a CV study are not asked directly how much they would be willing to pay for the environmental attribute. The preferred approach is to construct a scenario in which the environmental attribute is threatened and then ask respondents how much they would be willing to pay to protect it. For example, a CV study to evaluate economic benefits of improvements in water quality included a hypothetical scenario in which public funding to restore Atlantic salmon in New England had been curtailed, and extinction of

[40] The 1989 U.S. court opinion in *Ohio v. U.S. Department of the Interior (DOI)* represents the first major court decision in the United States concerning the contingent valuation method and the loss of passive-use values. The court held that DOI's regulations for natural resource damage assessments conducted under the federal Comprehensive Environmental Resource Conservation and Liability Act (CERCLA) must contain provisions for compensation for loss of passive-use values. In addition, the court held that contingent valuation was an acceptable method of estimating lost passive-use values. (See Kopp and Smith (1993) for more information on the Ohio case.) Contingent valuation methods have been utilized in a few hazardous waste cleanup cases since the ruling. Other applications of contingent valuation are discussed in National Oceanic and Atmospheric Administration (NOAA, 1993) and Portney (1994).

[41] This specification of steps in a CV study is adapted from Hanley and Spash (1993, pp. 54–57).

Atlantic salmon in New England rivers would occur.[42] The scenario description, which indicated that a private foundation had proposed to maintain and restore the Atlantic salmon, provided a context for asking questions about how much participants in the CV study would be willing to donate to the imaginary foundation.

Results of a CV study can be affected by the choice of method for raising funds to improve the environment. In the CV literature, these methods are termed *bid vehicles,* and they include property taxes, income taxes, entry fees, utility bills, and voluntary contributions. If a survey respondent feels that the environmental cause is worthy but the bid vehicle is inappropriate, she may submit a bid lower than her actual willingness to pay.

The scenario used to elicit responses is typically incorporated into a survey questionnaire. The questionnaire must be pre-tested to ensure that it does not bias the outcome and that it accurately reflects people's willingness to pay for the environmental good or service in question. Pre-testing can be done by administering trial versions of the survey in small group settings. Survey respondents can then be monitored as they complete the survey, and their reactions can be discussed by group members and survey designers. Suggestions for improvement can be incorporated into the final version of the questionnaire.

The survey population includes all those whose welfare would be significantly affected by the proposed action to modify the environment. Typically, only a representative sample of the affected population completes the questionnaire. Information on characteristics of the respondents, such as age, income, political party, education, and so on are typically collected as part of a CV survey, because this information can help in interpreting results.

WTP Elicitation Techniques

Willingness to pay can be elicited from survey respondents in different ways. Methods include telephone surveys, mail surveys, and face-to-face interviews. For example, in a *bidding game,* which relies on interviews, a respondent is first asked if he will pay a certain amount for the environmental change.[43] If the answer is yes, higher amounts are suggested until the respondent finally declines to pay any more. The highest amount for which the respondent answered yes represents his maximum willingness to pay. An alternative is to use open-ended questions, which simply ask the respondent, "What is the maximum amount you would be willing to pay for this good?"

Another way to elicit WTP involves the *referendum question approach.* The respondent is asked if she would be willing to pay a specified amount of money for the environmental change. If she answers yes, then her actual willingness to pay is greater than or equal to the amount mentioned. Respondents can be randomly assigned to groups, with members of each group being asked to respond to a different amount; these data can be used to construct a demand curve.

Still another approach to eliciting WTP involves *contingent ranking.* Using this method, respondents are provided with a set of alternative actions, one of which is the environmental change in question. Each of the other "actions" has an associated monetary value (for example, a vacation trip to Hawaii with an equivalent cash value of $1000). Respondents are asked to rank the alternative actions in order of preference. The value of the environmental resource is inferred from its position in the ranking of alternatives: its value is higher than the monetary value of the next lowest alternative, and lower than the monetary value of the next highest alternative. In this example, if the results place the trip to Hawaii first ($1000), preservation of the environmental resource in question second, and a weekend skiing trip (valued at $500) third, it can be inferred that the maximum value the respondent places on preservation of the environmental resource is between $500 and $1000.

Processing Data and Analyzing Responses

After the WTP bids have been solicited, results are studied to eliminate unusable responses. For example, bids obviously outside an individual's ability to pay are removed. Survey data are often used to develop statistical relationships between WTP and variables such as income, age, and education level of respondents. The validity of the CV survey can then be assessed by determining if WTP varies in reasonable ways. For example, in many contexts, WTP

[42] This CV study is summarized by Field (1994).

[43] This description of elicitation techniques is adapted from Freeman (1993, pp. 170–76).

would be expected to increase with increases in income and the level of environmental quality.

Appraisals of Contingent Valuation

Although economists generally agree that passive-use values ought to be considered in deciding on levels of environmental protection, they do not all agree that the CV method provides meaningful economic measures of passive-use values. For many economists, the only valid approach to assessing the value of goods and services to individuals is to examine choices people make in the marketplace. This is the essence of market-based procedures such as hedonic pricing. Even though hedonic pricing techniques do not center on direct purchases of environmental resources, they rest on data about real, not hypothetical, decisions. The CV method is suspect because it uses responses to hypothetical questions that do not involve cash outlays. What assures that responses will be meaningful and that respondents will not deliberately misstate their preferences or simply be unable to provide informed responses?

Critics of contingent valuation point to studies showing that the wording of questions in a CV questionnaire, or even the order in which questions are asked, can have significant impacts on the answers. For example, Diamond and Hausman (1993, pp. 15–16) cite CV research in which the value of preserving seals depended on when, in the questionnaire, the respondent was asked about seals. If a question about the preservation of whales preceded the question about preserving seals, then the economic value of preserving seals was lower than when the questions were in reverse order. In contrast, the value of preserving whales was not affected by how the questions were sequenced. Apparently, whales were more popular with respondents than seals. Thus when asked about whales first, respondents implicitly placed an upper bound on how they would respond to the question about the value of seals. Many respondents were apparently reluctant to value seals more than whales. As this case shows, the survey process itself can introduce bias in responses, which raises questions about relationships between survey responses and willingness to pay.

Debates about the validity of CV were largely confined to the environmental economics research literature until the mid-1980s. A turning point occurred in the late 1980s, when U.S. federal courts ruled that CV was suitable method for assessing compensation when losses of passive-use values were at stake. In addition, the U.S. Oil Pollution Control Act of 1990 charged the National Oceanic and Atmospheric Agency (NOAA) with assessing compensation requirements for damages to natural resources from oil spills covered under the act. NOAA maintains that passive-use values should be included among the compensable damages associated with oil spills.

During the early 1990s, NOAA commissioned a panel of five distinguished economists to evaluate the use of CV for determining passive use values. The panel's report (NOAA, 1993) provides an authoritative appraisal of the contingent valuation method. Although the panel confirmed the validity of the contingent valuation concept, it identified several limitations. The method's shortcomings were viewed as especially important because it is impossible to validate fully results of studies to determine passive-use values. According to the panel, prior CV "studies suggested that the CV technique is likely to overstate 'real' willingness to pay" (NOAA 1993, p. 4604). The panel also relied on past studies to register their concerns about CV, a few of which are summarized here.

- *Inconsistency with Rational Choice.* Some results from contingent valuation studies seem inconsistent with models of rational choice. For example, it is often expected that more of something regarded as good is better than less, as long as an individual consumer is not satiated. However, some contingent valuation results show WTP values that do not increase with increasing quantities of the good. The panel cited a study in which the average WTP to prevent 2000 migratory birds from dying in oil-filled ponds was not statistically different from the WTP to prevent 200,000 birds from dying.[44]
- *Absence of Meaningful Budget Constraints.* Even if respondents in contingent valuation surveys provide good-faith estimates of their WTP for environmental resources, they may do so without thinking carefully about limits on their disposable incomes. If respondents actually had to make choices be-

[44] For details on the study to evaluate migratory birds, see Devousges et al (1993); the panel cited an earlier, unpublished version of the paper by Devousges and his colleagues.

tween paying for an environmental change and foregoing other opportunities, they might give different responses. Some CV surveys indicate (implausibly) that individuals are willing to devote substantial fractions of their incomes to solving environmental problems.

- *Information Provision and Acceptance.* Even if detailed information on the hypothetical setting is supplied, CV survey participants have a limited ability to assimilate the information. This results, in part, because participants often have little experience with the issues treated in CV studies. For example, intelligent and well-informed respondents might find it difficult to place a meaningful economic value on a program to prevent oil spills.

In spite of the potential shortcomings of contingent valuation, the NOAA panel concluded that, *if properly conducted,* CV studies convey useful information and can produce "estimates reliable enough to be the starting point of a judicial process of damage assessment, including lost passive-use values."[45] The panel included recommendations for conducting contingent valuation studies:

> [We have identified] a number of stringent guidelines for the conduct of CV studies. These require that respondents be carefully informed about the particular environmental damage to be valued, and about the full extent of substitutes and undamaged alternatives available. In willingness to pay scenarios, the payment vehicles must be presented fully and clearly, with the relevant budget constraint emphasized. The payment scenario should be convincingly described, preferably in a referendum context, because most respondents will have had experience with referendum ballots with less-than-perfect background information. Where choices in formulating the CV instrument can be made, we urge they lean in the conservative direction, as a partial or total offset to the likely tendency to exaggerate willingness to pay.

Notwithstanding its shortcomings, contingent valuation remains the only technique available for estimating passive-use values.

[45] This quotation and the following one are from NOAA (1993, p. 4510).

ENVIRONMENTAL RESOURCE VALUATION IN PRACTICE

Each of the valuation techniques in Table 6.1 has been used in practical settings. However, some are more widely used than others. Which method is most appropriate in any particular setting?

Choosing a Valuation Procedure

Although many books have been written on techniques for valuing environmental resources, they cannot answer questions about which method to use. The choice of a method depends on the time and funds available for performing valuations, as well as the intended uses of the valuation results. Selection of an appropriate technique also depends on the qualifications of the staff available to conduct the evaluation. Some techniques—for example, hedonic pricing methods—require staff with extensive backgrounds in economic theory and statistical analysis. Others, such as the defensive expenditures approaches, can be implemented by less-experienced staff.

Table 6.7 provides a starting point for considering which valuation methods might be useful in a particular setting. The table includes only a fraction of the available valuation methods. Much of Table 6.7 summarizes points previously made regarding the applicability of the valuation procedures introduced in this chapter. However, the last column contains new information: ratings of valuation technique applicability as determined by Dixon et al. (1994).

Dixon and his colleagues at the East-West Center's Environment and Policy Institute conducted an extensive review of valuation techniques used in practice. Their review, which was supported by both the Asian Development Bank and the World Bank, provides practical advice to analysts charged with appraising the environmental costs and benefits of proposed development projects.[46] As indicated in the table, Dixon and his associates find hedonic pricing methods only *potentially applicable.* Their rating is based primarily on the extensive data requirements

[46] Many organizations that fund infrastructure projects (such as the World Bank) require that a formal benefit-cost analysis be conducted as part of project planning. Increasingly, these organizations are encouraging use of studies to monetize environmental impacts of projects they support.

TABLE 6.7 Attributes of Selected Valuation Techniques

Valuation Procedures	Typical Applications	Some Underlying Assumptions	Key Items to be Estimated	Applicability Rating by Dixon et al.[a]
Hedonic Property Value Method	Noise near airports Urban air quality	Housing markets are perfectly competitive	Hedonic property value function	Potentially applicable
Travel Cost Method	Outdoor recreation sites Biological preserves	WTP for recreation visits can be inferred from travel expenditures	Visitation rates and costs of travel	Selectively applicable
Defensive Expenditures	Air pollution Water pollution	Expenditures are for goods that are perfect substitutes for environmental services	Cost of defensive actions	Generally applicable
Production Function Approaches	Crop losses Fisheries losses	Assumptions regarding how productivity changes affect supply curves	Influence of supply curve shifts on market prices and outputs	Generally applicable
Human Capital Technique	Value of a life	A life can be evaluated in terms of foregone earnings	Life expectancy and foregone earnings	Generally applicable
Hedonic Wage Method	Value of a life	Labor markets are perfectly competitive	Hedonic wage function	Potentially applicable
Contingent Valuation Methods	Passive-use values of recreation areas Potable water supplies	Expressions of WTP in CV survey responses measure economic value	Meaningful WTP values using surveys and interviews	Potentially applicable

Source: Based on Dixon et al. (1994, pp. 8, 63).

[a] Dixon and colleagues (1994) distinguish three categories of valuation techniques: (1) *generally applicable*—techniques commonly used in project analysis; (2) *selectively applicable*—techniques that "need greater care in their use, make more demands on data or on other resources, . . . or require stronger assumptions than" methods classified as generally applicable (p. 63); and (3) *potentially applicable*—techniques that have had limited use in project analyses because of their "formidable data requirements" (p. 8).

TABLE 6.8 Air Pollution Control Benefits for the United States in 1978 (in billions of 1978 dollars)[a]

	Range	Most Reasonable Point Estimate of Benefits
Health	3.1–40.6	17.0
Soiling and Cleaning	1.0– 6.0	3.0
Vegetation	0.1– 0.4	0.3
Materials	0.4– 1.4	0.7
Property Values	0.9– 8.9	2.3

[a] Adapted from Freeman Air Pollution Control, copyright © 1982 by John Wiley & Sons, Inc. Reprinted with permission of John Wiley & Sons, Inc.

for applying hedonic pricing methods. Dixon and his colleagues rated the travel cost method and contingent valuation methods as *selectively applicable.* These approaches are considered practical, but only in a limited set of circumstances. Production function approaches, defensive expenditures, and the human capital technique are considered *generally applicable.* However, as noted earlier in this chapter, both the human capital technique and the defensive expenditures method have been subject to significant criticisms.

Benefit Evaluation and Policy Analysis

A use of valuation procedures to analyze regulatory programs is demonstrated by Freeman's (1982) study of the economic benefits of air pollution control programs in the United States. Freeman found that most air pollution reduction benefits were associated with improvements in human health (see Table 6.8). Of the $17 billion he estimated as health-related benefits (for 1978), nearly $14 million was attributed to reductions in mortality due to controls on stationary sources of air pollution. A small portion of the health benefits was attributed to controls on mobile sources. Freeman (1982, p. 130) used his data, together with an analysis of costs of pollution reduction, to conclude that "a fair degree of confidence can be attached to a statement that the control of stationary sources has been worthwhile on benefit-cost grounds." He also suggested that the federal mobile source control program, which was estimated to have cost more than $7 billion per year in 1978, would not pass a benefit-cost test. Freeman used his results to argue for improvements in the cost-effectiveness of federal programs for mobile source control.

In many countries, the high cost of meeting environmental goals has led policymakers to require monetary valuations of the benefits of environmental programs. This is demonstrated by the benefit-cost analyses that the U.S. Environmental Protection Agency (EPA) must perform under Executive Order 12291. The order requires federal regulatory agencies to submit to the Office of Management and Budget a *regulatory impact assessment* for each proposed major regulation. Under this order, agencies are instructed that (to the extent permitted by law) regulations should not be promulgated unless their benefits exceed costs. This has the effect of making monetary valuation a significant element in environmental policy design at the federal level in the United States. Although many have argued that these benefit-cost analyses are badly flawed, they play a significant role in the regulatory process.[47] Frequently, EPA cannot move forward with a regulatory program without justifying it on economic grounds.

Key Concepts and Terms

Environmental Resources as Capital Assets
Active-use vs. passive-use values
Direct vs. indirect use

[47] For more on the implementation of Executive Order 12291, see Smith (1984) and Chapter 3 in this book. Difficulties EPA faces in applying economic theory to conduct benefit-cost analysis in response to the executive order are described by VanderHart (1993).

Renewable vs. depletable resources
Option value
Existence value

Economic Value and Consumers' Surplus
Willingness to pay
Ex ante vs. ex post analysis

Hedonic Property Value Method
Implicit price
Hedonic price equation
Regression analysis
Shifts in demand curves
Correlated characteristics
Perfect housing markets

Travel Cost Method
Time cost of travel
Admission fee as a travel cost
Multipurpose trips
Substitute sites

Defensive Expenditures Procedure
Substitute goods
Benefits unrelated to defensive actions

Production Function Approaches
Dose-response curves
Ex post studies of lost productivity
Firms' adjustments to changes in environmental quality
Effects of productivity changes on prices and outputs
Shifts in supply curves
Producers' surplus

Methods of Valuing Health and Longevity
Human capital technique
Foregone earnings
Statistical value of human life
Utility-risk tradeoffs
Hedonic wage method

Contingent Valuation
Compensation for lost passive-use values
CV survey questionnaire
Hypothetical market
Bid vehicles
Contingent ranking
Bias in survey responses
Absence of budget constraints

Environmental Resource Valuation in Practice
Criteria for selecting valuation techniques
Valuation for project-level decisions
Policy and program evaluations
Benefits of proposed regulations

Discussion Questions

6-1 Consider a park or a wilderness area that you have visited. What environmental services does the area provide? Are there any passive-use values associated with the park or wilderness area? What methods are available to place monetary values on the services provided by the area?

6-2 Construct an example to explain the concept of consumer's surplus. Describe the differences between the consumer's total economic value, total cash outlay, and surplus.

6-3 In what ways are the hedonic wage method and the hedonic property value method similar? How do the methods differ?

6-4 What assumptions form the basis of the travel cost method? Do the assumptions about how consumers respond to travel cost changes describe your behavior?

6-5 The following table provides results from a travel cost survey for a hypothetical recreational reservoir.

City	Population	Cost Per Visit	Number of Visits in a Season
A	1000	$1	400
B	2000	$3	400
C	4000	$4	400

a. Use the travel cost method to construct a demand curve for the reservoir; that is, a curve showing entry fee vs. number of visits.
b. Use the demand curve to estimate total benefits (based on willingness to pay) associated with the provision of recreational services at the reservoir.

c. Describe briefly three shortcomings of the travel cost method as a basis for estimating monetary benefits of a proposed recreational reservoir facility.

6-6 Suppose the demand curve for an existing public reservoir used for recreation is defined by

$$Q = 2000 - 20p$$

where Q is the number of visits and p is the admission price. The current admission price is zero, and the government plans to charge an admission price of \$5. What would be the loss in consumers' surplus if the government went ahead with its plans?

6-7 Have you ever made any defensive expenditures in response to environmental changes such as increased noise pollution? If so, use your personal experience as a basis for illustrating the defensive expenditures approach to monetary valuation of environmental attributes.

6-8 Explain how the method used to monetize the productivity losses resulting from land reclamation in Tokyo Bay could be used to place a value on damages from acid rain to commercial forestry operations.

6-9 Dozens of studies have been conducted to estimate the monetary damages of crops lost due to ozone pollution. The starting point in these studies is a dose-damage (or dose-response) function. Hanley and Spash (1993, p. 176) summarize the variables that should influence the definition of dose:

> The ambient ozone concentration, the length of time a particular concentration persists, and the frequency of occurrences combine to form a measure of the dose of an air pollutant to which a plant is exposed: the "exposure dose." Other characteristics of plant exposure may also be important determinants of the nature and magnitude of the effects of ozone on plants: the length of time between exposures, the time of day of exposure, their sequence and pattern, and the total flux of ozone to the plant as it is affected by canopy characteristics and leaf boundary layers.

How would you construct a measure of dose for use in a study to determine dose-damage relationships for crop loss due to ozone pollution? Sometimes dose in crop damage studies is defined as pollutant concentration. How suitable is this definition of dose?

6-10 Provide a critique of the human capital approach for placing monetary values on decreases in morbidity and mortality.

6-11 Distinguish contingent valuation methods from other valuation techniques? In responding, consider the economists' conception of value as measured by willingness to pay. Also, reflect on the differences between passive-use values and active-use values.

6-12 Which valuation technique would you use if you were asked to place a monetary value on each of the following:

a. a wildlife preserve used for ecotourism
b. noise from a proposed airport
c. improvements in the quality of a village water supply
d. commercial fishery losses caused by water pollution

6-13 Many studies have compared results of applying different valuation techniques to the same environmental resource. Hanley and Spash (1993, p. 120) summarize results from work by researchers studying the monetary value of recreational boating on lakes in eastern Texas:

	Willingness to Pay (\$/visit)	
Lake Site	**Travel Cost Method**	**Contingent Valuation Method**
Conroe	32	−0.87
Livingston	102	6
Somerville	24	11
Houston	13	5

What conclusions would you draw from these results? More generally, what would you have concluded if the numerical results from applying each method to the same lake had been approximately equal?

6-14 Assume that you hold a biocentric (as opposed to an anthropocentric) view of the environment. What would you consider to be the principal shortcomings of all procedures for placing monetary values on changes in the environment?

References

Clawson, M., and J. L. Knetsch. 1966. *Economics of Outdoor Recreation.* Baltimore: Johns Hopkins Press.

Davis, R. K. 1963. Recreation Planning as an Economic Problem. *Natural Resources Journal,* 3 (2): 239–49.

Diamond, P. A., and J. A. Hausman. 1993. "On Contingent Valuation Measurement of Nonuse Values." In *Contingent Valuation: A Critical Assessment,* ed. J. A. Hausman, 3–38. Amsterdam: Elsevier.

Devousges, W. H., F. R. Johnson, R. W. Dunford, K. J. Boyle, S. P. Hudson, and K. N. Wilson. 1993. "Measuring Natural Resource Damages with Contingent Valuation: Tests of Validity and Reliability." In *Contingent Valuation: A Critical Assessment,* ed. J. A. Hausman, 91–159. Amsterdam: Elsevier.

Dickie, M., and S. Gerking. 1989. "Benefits of Reduced Morbidity from Air Pollution Control: A Survey." In *Valuation Methods and Policy Making in Environmental Economics,* ed. H. Folmer and E. Van Ierland, 105–122. Amsterdam: Elsevier.

Dixon, J. A., L. F. Scura, R. A. Carpenter, and P. B. Sherman. 1994. *Economic Analysis of Environmental Impacts.* London: Earthscan Publications.

Dorfman, R. 1993. "An Introduction to Benefit-Cost Analysis," In *Economics of the Environment: Selected Readings,* ed. R. Dorfman and N. S. Dorfman, 297–322. New York: W. W. Norton.

Dupuit, J. 1844. On the Measurement of the Utility of Public Works. *Annales des Ponts et Chaussées,* 2d series, vol. 8. Reprinted (as translated from the original, French version) in Munby, D., ed. 1968. *Transport: Selected Readings.* Harmondsworth, Middlesex, U.K.: Penguin Books.

Field, B. C. 1994. *Environmental Economics.* New York: McGraw–Hill.

Freeman, A. M., III. 1979. *The Benefits of Environmental Improvement.* Baltimore: Johns Hopkins University Press.

———. 1982. *Air and Water Pollution Control.* New York: Wiley.

———. 1993. *The Measurement of Environmental and Resource Values: Theory and Methods.* Washington, DC: Resources for the Future.

Gegax, D., S. Gerking, and W. Schulze. 1991. Perceived Risk and the Marginal Value of Safety, *Review of Economics and Statistics* 73 (4):589–96.

Gujarati, D. N. 1988. *Basic Econometrics.* New York: McGraw–Hill.

Hanley, N., and C. L. Spash. 1993. *Cost-Benefit Analysis and the Environment.* Hants, England: Edward Elgar.

Harrison, D., Jr., and D. L. Rubinfield. 1978. Hedonic Housing Prices and Demand for Clean Air. *Journal of Environmental Economics and Management* 5:81–102.

Hufschmidt, M. M. and J. A. Dixon. 1986. "Valuation of Losses of Marine Product Resources Caused by Coastal Development of Tokyo Bay." In *Economic Valuation Techniques for the Environment: A Case Study Workbook,* eds. J. A. Dixon and M. M. Hufschmidt, Baltimore: Johns Hopkins University Press.

Kopp, R. J., and V. K. Smith. 1993. *Valuing Natural Assets.* Washington, DC: Resources for the Future.

Landefeld, J. S., and E. P. Seskin. 1982. The Economic Value of Life: Linking Theory of Practice. *American Journal of Public Health* 72:555–61.

NOAA (National Oceanographic and Atmospheric Administration). 1993. Natural Resource Damage Assessments Under the Oil Pollution Act of 1990. *Federal Register* 58 (10) (January 15):4601–14.

Portney, P. R., 1994, The Contingent Valuation Debate: Why Economists Should Care, *Journal of Economic Perspectives* 8 (4):3–17.

Ridker, R. G., and J. A. Henning. 1967. The Determinants of Residential Property Values with Special Reference to Air Pollution. *Review of Economics and Statistics* 49:246–57.

Smith, V. K., ed. 1984. *Environmental Policy Under Reagan's Executive Order.* Chapel Hill: University of North Carolina Press.

Trijonis, J., et al. 1985. *Air Quality Analysis for Los Angeles and San Francisco Based on Housing Values and Visibility.* Report by the Santa Fe Research

Corporation to the California Air Resources Board (ARB Contract No. A2-088-32). Sacramento, CA.

Turner, R. K., ed. 1993. *Sustainable Environmental Economics and Management: Principles and Practice.* London: Belhaven Press.

Turner, R. K., D. Pearce, and I. Bateman. 1993. *Environmental Economics: An Elementary Introduction.* Baltimore: Johns Hopkins University Press.

VanderHart, P. G. 1993. An Economic Critique of Several EPA Regulatory Analyses. *The Environmental Professional* 15 (4):406–11.

Viscusi, W. K. 1986. The Valuation of Risks to Life and Health: Guidelines for Policy Analysis. In *Benefit Assessment: The State of the Art,* eds. J. D. Bentkover, V. T. Covello, and J. Mumpower. Dordrecht, Netherlands: D. Reidel.

Chapter 7

Efficient Levels of Pollution Abatement

The application of economic reasoning to environmental problems often involves the following question: Will a proposed environmental regulation or development project lead to an improvement in the way society's resources are allocated?[1] In this context, *resource allocation* includes the use of factor inputs by firms and the distribution of outputs among consumers. One basis for identifying improved resource allocations is *Pareto's criterion:* A change in the use of inputs and distribution of outputs is an improvement if it harms no one and improves the lot of some people (in their own opinion).[2] A change that satisfies Pareto's criterion is termed a *Pareto improvement.* As explained in Chapter 5, if it is *not* possible to modify a resource allocation in a way that satisfies Pareto's criterion, the allocation is *Pareto efficient.* At a Pareto-efficient allocation, any change yielding gains for at least one person will harm someone.

Pareto's criterion is acceptable to most economists because resource allocations can be evaluated from a community's perspective without introducing judgments about whether one person's utility gains offset another person's utility losses. Because a Pareto improvement involves no losses in utility, gains and losses to individual's need not be compared.

The practical significance of Pareto's criterion is minimal because the criterion can never be satisfied if a proposed change in the allocation of a community's resources involves losses. Most development projects and environmental regulations are accompanied by both gains and losses. In such cases, Pareto's criterion provides no basis for judging whether implementation of the project or regulation led to an improvement.

POTENTIAL PARETO IMPROVEMENT CRITERION

During the 1930s, economists introduced *compensation criteria* for identifying improvements in the allocation of a community's resources. A widely used compensation criterion involves the notion of a *potential Pareto improvement.* Consider a proposed development project (or environmental regulation) and assume that those who will gain and those who will lose have been identified, and that beneficial and

[1] Melissa Geeslin provided helpful comments on this chapter while she was a student at Stanford University.

[2] When applying Pareto's criterion, each person is the authoritative judge of whether a reallocation of resources causes beneficial or harmful effects.

harmful effects have been calculated in monetary units. (In this context, beneficial and harmful effects are termed *monetary benefits* and *monetary costs,* respectively.) A proposed project (or regulation) is a potential Pareto improvement if the following condition is satisfied: Those who benefit from the project could, *in principle,* transfer enough funds to those who lose to fully compensate them for their monetary losses and still leave at least one beneficiary with a net monetary gain (after the transfer).[3] Such a reallocation is a *potential* improvement because the actual transfer of funds need not take place.[4]

Shortcomings of the Criterion

Use of potential Pareto improvement as an evaluation criterion recognizes that even though a project or regulation causes harm to some, the overall effect may be an improvement for society. Although the potential Pareto improvement criterion has the advantage that it can be satisfied in circumstances involving both gains and losses, the criterion has been criticized because it does not account for the way projects or regulations affect the distribution of wealth. A project that makes the rich richer at the expense of the poor can be judged a gain for society using the potential Pareto improvement criterion. For example, a project would be judged as improving national welfare if it yielded a gain of a million dollars to the richest person in America, while, at the same time, causing losses of a thousand dollars to 999 of the country's poorest families.

[3] The potential Pareto improvement criterion is often associated with two economists, Lord Nicholas Kaldor and Sir John Hicks. The *Kaldor potential compensation test* takes the form of the potential Pareto improvement criterion introduced here: Can the gainers from a proposed change fully compensate the losers and still leave someone better off? The compensation test introduced by Hicks takes acceptance of the proposed change as a reference point and asks: Can the losers from the proposed action fully compensate the gainers for a decision not to proceed with the change. If the proposed change were rejected and compensation were paid by the losers, those who would have gained would be at least as well off as they would have been with the project. The same could be said for those who would have lost. For an elaboration of the differences between the tests proposed by Kaldor and Hicks, see Freeman (1993, p. 86–88).

[4] In discussing potential Pareto improvements, "it is conventional to assume that making transfers of money between individuals is not costly in itself. . . ." Sugden and Williams (1978, p. 90).

Critics of the potential Pareto improvement criterion challenge the value judgments that lie behind the criterion:

1. A change in an individual's welfare is adequately measured by a change in the person's income.
2. A unit of money gained by any one person counts the same (as a contribution to total community welfare) as a unit of money gained by any other person.
3. A change in community welfare is measured by summing the change in income to individual members of the community.

Critics also observe that use of the potential Pareto improvement criterion assumes implicitly that the existing distribution of wealth is acceptable.[5]

Productive Efficiency of an Economy

Even though the potential Pareto improvement criterion does not account for the distribution of costs and gains, it has been put to practical use. An important use of the criterion results from its connection to a measure of an economy's performance known as *productive efficiency.* For economists, a project (or regulation) improves productive efficiency if it increases the net value of the goods and services that an economy produces.

The concept of productive efficiency is clarified by an example. Consider a hypothetical economy with two factor inputs, land and labor, and two outputs, wheat and timber. Suppose all the available land and labor are used to produce 100 units of each output. Furthermore, suppose the allocation of land and labor to produce outputs could be changed to yield 101 units of wheat and 100 units of timber. Such a change in the original allocation would yield an increase in productive efficiency. There might be still

[5] This criticism is based on economists' conception of demand for goods and services as a reflection of the preferences of individuals participating in markets. If the existing distribution of society's wealth were changed significantly, then demand for goods and services would also change because different individuals would be participating in the markets. Demand affects market prices, and those prices (along with empirical estimates of demand curves) play a key role in calculating monetary benefits and costs. Thus the existing distribution of wealth implicitly affects the computation of benefits and costs.

other shifts in the use of land and labor that would provide more of one output without decreasing the quantity of the second output. If those shifts were not made, the economy would not be at maximum productive efficiency.

Productive efficiency is a narrower concept than Pareto efficiency, because the latter concerns both production and distribution. Maximizing productive efficiency does not necessarily yield a Pareto-efficient resource allocation. In an economy that is productively efficient there might be opportunities to redistribute outputs among consumers in ways that satisfied Pareto's criterion. In contrast, an economy that is Pareto efficient must have maximum productive efficiency. Otherwise it would be possible to make changes in the use of factor inputs that satisfied Pareto's criterion. In the language of mathematics, productive efficiency is a necessary condition for Pareto efficiency, but it is not a sufficient condition.[6]

By definition, productive efficiency is augmented if factor inputs are shifted in a way that increases the net value of goods and services produced. Therefore, a proposed development project (or environmental regulation) increases productive efficiency if it creates goods and services with a monetary value higher than the costs of production. In other words, a project (or regulation) enhances productive efficiency if it satisfies the potential Pareto improvement criterion. As an extension of this reasoning, a project that maximizes its contribution to productive efficiency is one that maximizes the difference between monetary benefits and monetary costs. This is the criterion used in a benefit-cost analysis.

BENEFIT-COST ANALYSIS

Benefit-cost analysis refers to a project (or program) evaluation procedure in which attainment of productive efficiency is the objective, and costs and benefits are computed in monetary terms. Some project analysts use the term benefit-cost analysis more broadly to refer to an evaluation procedure in which a project's advantages and disadvantages are described (often qualitatively) and used as a basis for decision making. This broad usage of the term benefit-cost analysis is avoided in this text.

In a benefit-cost analysis, it is customary to calculate benefits in terms of *willingness to pay* for goods and services and to compute costs in terms of *opportunity costs.* If markets for inputs and outputs exist, and if conditions associated with those markets do not depart excessively from perfect competition, then market prices may be appropriate for use in a benefit-cost study. In perfectly competitive markets, prices of factor inputs measure opportunity costs, and prices of outputs reflect values as measured by consumers' willingness to pay. Because markets often depart widely from the ideal of perfect competition, market information generally needs to be modified in calculating benefits and costs.

In circumstances where markets don't exist or where market prices do not properly reflect costs of inputs and values of outputs, economists utilize *shadow prices.* A description by Mishan (1975, p. 81) is illuminating:

> [A shadow price] is the price the economist attributes to a good or factor on the argument that it is more appropriate for the purpose of economic calculation than its existing price, if any. There is nothing very special about the notion of a shadow price. In evaluating any project, the economist may effectively "correct" a number of market prices and, also, attribute prices to unpriced gains and losses that it is expected to generate. He will, for example, add to the cost of a factor, or subtract from the cost of a good, in making allowance for some external diseconomy.

An example illustrates a shadow price as a correction to a market price. Suppose a benefit-cost analysis is being performed for a project that utilizes unskilled labor as an input. Furthermore, suppose the project is for an area where long-term unemployment for unskilled workers is common. In this case, the wage actually paid to unskilled workers would be higher than the opportunity cost of the labor because if the workers were not engaged to build the project, they would probably be unemployed. The wage for unskilled labor in the benefit-cost study for this example project might be "corrected" by making it zero. The opportunity cost is zero because there are no alternative, economically valued uses of unskilled labor in this case.

---

[6] For an analysis of relationships between productive efficiency and Pareto efficiency, see Dorfman (1993).

Suppose a proposed project would yield losses that are unpriced because they involve air pollution damages and other external costs. In this context, a shadow price is the implicit (or attributed) price of the loss. Examples of methods to compute implicit prices are the valuation techniques described in Chapter 6. For instance, the hedonic property value method can be used to find an implicit price for air pollution in urban areas.

A more complete statement of the benefit-cost analysis approach requires the introduction of two parameters: i, the rate of discount, and T, the time horizon used in the analysis. Suppose a proposed development project or environmental regulation will yield a stream of monetary benefits given by $B_1, B_2, \ldots, B_T$, where B_t is the monetary benefit that occurs during the tth year ($t = 1, \ldots T$). Suppose further that the project costs incurred during each year are $C_1, C_2, \ldots, C_T$, where C_t is the cost in year t.

A benefit-cost analysis proceeds by calculating the present value of each of the future costs and benefits. The discount rate, i, plays a key role in this calculation. (An explanation of the procedure for calculating the present value of future costs and benefits follows.) Next the present value of all costs are summed and subtracted from the present value of all benefits. The result, termed the *present value of net benefits,* is used to gauge contributions to productive efficiency.[7] A proposed project will improve productive efficiency if it yields a positive value for the present value of benefits minus costs. The contribution to productive efficiency is maximized if the present value of net benefits is maximized.

Discounting to Present Value

A process called *discounting* puts costs (or benefits) that occur at different times on a comparable basis. Procedures for discounting are explained by first considering a more familiar idea: compound interest on a bank savings account. If P dollars are deposited in a bank account earning interest at i percent per year, the P dollars will grow to $(1 + i)P$ dollars at the end of the first year. That sum, in turn, will grow to $(1 + i)\,[(1 + i)P]$ dollars at the end of the second year. Continuing in this way, a sum of P dollars invested at an interest rate of i percent will yield a sum of $(1 + i)^tP$ dollars after t years.

The concept of discounting a future cost (or benefit) to its present value rests on the answer to the following question: What is the equivalent value, in current monetary terms, of a sum of $(1 + i)^tP$ dollars t years in the future? The answer is P dollars, since an investment of P dollars at an interest rate of i percent would yield $(1 + i)^tP$ dollars after t years. Dividing the future sum, $(1 + i)^tP$, by the factor $(1 + i)^t$ yields its equivalent present value. Consequently, an arbitrary amount of F dollars at some future time t is said to have a *present value* given by

$$\frac{F}{(1 + i)^t}. \qquad \textbf{(7-1)}$$

The process of calculating this ratio is called *discounting to present value.*

When a benefit-cost analysis is conducted, the key ingredients are the estimates of annual benefits and costs, and the parameters i and T. Monetary benefits and costs are calculated using concepts and procedures introduced in chapters 5 and 6. The perspective adopted in making those computations is typically a broad one, often at the level of a nation. For example, in analyzing federal water resources projects in the United States, the benefits and costs are those that accrue to *all* residents of the United States. It is possible to define monetary benefits and costs from a regional or local perspective, but a national accounting stance is more common.

The time horizon (T) is a year in the future "beyond which all costs and benefits are ignored."[8] Thus using a time horizon of 50 years is equivalent to assuming that, as a result of discounting, whatever benefits and costs occur beyond the fiftieth year will be so small they can be ignored in the analysis. To investigate whether the selection of any particu-

[7] Using the terminology introduced here, the present value of net benefits is

$$\left\{\sum_{t=1}^{T} \frac{Bt}{(1 + i)^t} - \sum_{t=1}^{T} \frac{Ct}{(1 + i)^t}\right\} > 0.$$

As explained below, the expression $1/(1 + i)^t$ in these sums is used to discount future costs and benefits to their equivalent present values.

[8] The quotation is from Sugden and Williams (1978, p. 65).

lar time horizon will yield erroneous results, it is necessary to consider what benefits and costs might occur in the years beyond the time horizon and then examine if their present values are notable.

Selecting a Discount Rate

In conducting benefit-cost analyses for public decisions, the values used for the time horizon and discount rate are generally set by policymakers. Selection of the time horizon depends on the choice of a discount rate. At high values of *i*, the process of discounting to present value makes benefits and costs that occur in the distant future relatively insignificant. This is demonstrated in Table 7.1, which shows the present value of $1 for several future times (t) and discount rates. As indicated in the table, for any particular value of *t*, the present value of $1 falls off sharply as the discount rate increases from 4% to 20%. The table also shows that as the time horizon extends beyond 50 years, the discounted value of $1 can be less than a fraction of a penny, depending on the discount rate.

Table 7.2 demonstrates that long-lived, capital intensive projects look more attractive when evaluated at lower rates of discount. The table shows how the present value of net benefits changes when different discount rates are used to evaluate a hypothetical project having present value of costs = $1000, and constant annual benefits of $100 per year for 50 years. In doing the computations for this example, it is convenient to use the *uniform series present worth factor* (K), defined as:

$$K = \frac{(1 + i)^T - 1}{i(1 + i)^T}. \quad \textbf{(7-2)}$$

When the uniform series present worth factor is multiplied by an amount (A) that represents the constant annual amount paid (or received) at the *end* of each of T years, the product is the present value (P) at a discount rate of i.[9] In other words,

$$P = AK. \quad \textbf{(7-3)}$$

In the example in Table 7.2, it is necessary to find the present value of a uniform series of annual benefits of $100 at the end of each year. Consider the first row in Table 7.2, which employs $i = 4\%$. The present value of the annual benefits of $100 at a 4% discount rate is found using equation 7-1 for each year and summing the results:

$$\frac{\$100}{(1.04)} + \frac{\$100}{(1.04)^2} + \ . \ . \ . + \frac{\$100}{(1.04)^{50}}.$$

This sum need not be calculated. Instead, note that the uniform series present worth factor (equation 7-2) for $i = 4\%$ and $T = 50$ years is

$$K = \frac{(1.04)^{50} - 1}{0.04(1.04)^{50}} = 21.5.$$

Using equation 7-3 with $K = 21.5$ and $A = 100$ provides a present value of benefits of $2150.

Table 7.2 lists values of K for i between 4 and 14%. For any value of *i*, the present value of benefits is found using equation 7-3 with $A = 100$ and the

TABLE 7.1 Present Value of $1 at the End of Year t with a Discount Rate i

	Discount Rate i (%)			
Year t	**4**	**8**	**12**	**20**
10	0.676[a]	0.463	0.332	0.162
25	0.375	0.146	0.059	0.011
50	0.141	0.021	0.004	0.0001

[a] Values in the table are equal to $(1 + i)^{-t}$.

[9] The formula for K is derived using

$$P = A\left[\sum_{t=1}^{T} \frac{1}{(1 + i)^t}\right].$$

The term in brackets is based on writing expression 7-1 once for each year t (where $t = 1, . . . , T$). The bracketed sum can be shown to be equal to K, as given in equation 7-2. The required algebraic manipulations are demonstrated in textbooks on engineering economy such as Collier and Ledbetter (1982, pp. 37 and 62). More generally, T in equation 7-2 can be replaced by n, the number of periods used in the analysis. In the more general case, costs and benefits may be calculated for time periods less than a year.

TABLE 7.2 Influence of Discount Rate on the Present Value of Net Benefits for a Hypothetical Project[a]

Discount Rate i (%)	Uniform Series Present Worth Factor	Present Value of Benefits ($)	Present Value of Net Benefits ($)
4	21.5	2150	1150
6	15.8	1580	580
8	12.2	1220	220
10	9.9	990	−10
12	8.3	830	−170
14	7.1	710	−290

[a] The hypothetical project has present value of costs equal to $1000, and annual benefits of $100 per year for 50 years.

appropriate value of K. The present value of *net* benefits is determined by subtracting the present value of costs ($1000) from the present value of benefits. As indicated in Table 7.2, the hypothetical project would yield positive net benefits only at discount rates below about 10%. This example demonstrates that the selection of i plays a key role in determining whether a proposed project will enhance productive efficiency.

Discount rates used in benefit-cost studies are typically set as a matter of government policy. Debates over how to select the discount rate often center on one or more of the following arguments.

- *Cost of Borrowing Money.* One approach to selecting i is to set it equal to the interest rate that the project proponent would have to pay in obtaining funds. For example, project funds might be obtained by selling bonds or by obtaining loans from a development assistance agency such as the World Bank.
- *Market Rate of Interest.* When government agencies build projects using tax revenues, they use funds that might otherwise be available for private investment projects. Shouldn't projects undertaken by government agencies be at least as financially attractive as comparable private projects? Some economists respond affirmatively to this question and argue that government agencies should use a market rate of interest in performing benefit-cost studies. For any particular government project, the discount rate equals the rate of interest paid by private entrepreneurs for funds to build projects with investment risks similar to risks associated with the government project.
- *Social Rate of Discount.* Some economists believe the interest rates in private markets are too high to be used in benefit-cost studies because they don't reflect social goals, particularly goals concerning future generations. Government infrastructure projects, such as roads and water treatment plants, often involve costly initial outlays and benefits that accrue over long time periods. As demonstrated by the example in Table 7.2, such projects might look unattractive at high discount rates, but attractive at discount rates low enough to reflect a government's judgments about the need to build infrastructure projects that benefit future generations. Following this reasoning, a benefit-cost study should use a social rate of discount established by a political process.[10]

Which discount rate to use in a benefit-cost analysis is a question that has been debated among economists for several decades.[11] It is a politically

[10] See Farber and Hemmersbaugh (1993) for an argument favoring use of a social rate of discount as low as 1 percent in a benefit-cost analysis of environmental regulations.

[11] Information on the alternative bases for identifying a suitable value of i in evaluating development projects is given by Sugden and Williams (1978, Chap. 15), and Dixon and Hufschmidt (1986, pp. 43–45). For an analysis of factors affecting the selection of i in benefit-cost studies of proposed environmental regulations, see Arnold (1995, pp. 177–97).

charged issue, because the higher the discount rate in a benefit-cost analysis, the more difficult it is for costly infrastructure projects to yield positive net benefits. In selecting a value for a discount rate, government policymakers often debate the merits of using market interest rates and the interest rate paid on funds borrowed for project construction. Frequently, these debates also consider the importance to future generations of major infrastructure projects.

OPTIMAL LEVELS OF POLLUTION ABATEMENT

The level of pollution reduction that maximizes the difference between the benefits and costs of cutting back waste release is called the *optimal level of pollution abatement.* At this level, abatement is productively efficient. Defining the productively efficient level of pollution reduction as optimal is a *value judgment* that many economists have made, but often only implicitly. Although this position ignores who benefits from and pays for abatement, the value judgment is defended by arguing that the distribution of costs and gains should be dealt with as a separate issue, after the efficient level of abatement has been identified. This defense of productive efficiency as a basis for defining the optimum abatement level has its limitations, since in practice, costs and gains of abatement are not often redistributed.

An example is introduced here to demonstrate the calculations for identifying the optimal level of pollution abatement. The example involves simplifications, but it is complex enough to illustrate why the optimal level of pollution abatement represents more of a theoretical ideal based on productive efficiency than a practical goal for environmental managers.[12]

The example concerns two firms on the fictitious Cedro River in Figure 7.1. The upstream firm, the Margarita Salt Company, takes water from the river, uses the water in its production process, and then discharges it back to the river. The only change in the water that results from this usage is that the concentration of chlorides, in milligrams per liter (mg/l), is increased. Thus, when the Margarita Salt Company's wastewater is discharged, the level of chlorides in the river is increased. This adversely affects the downstream firm, the Long Shot Brewery, because the brewery must treat the water to remove chlorides before it uses the water to make beer.

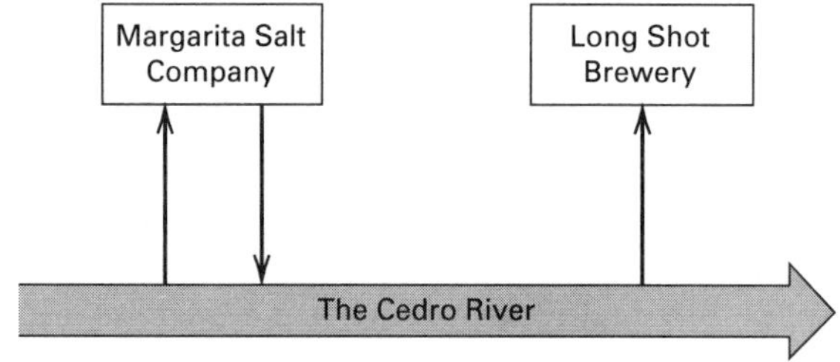

Figure 7.1 Two firms on the Cedro River.

Cost of Chloride Reduction by Margarita Salt Company

The Cedro River Agency to Control Quality (CRACQ) has been established to develop a water quality management plan. Its aim is to get the Margarita Salt Company to account for the damage it is causing the Long Shot Brewery as a result of discharging its chloride load. Each firm has a representative who is a liaison to CRACQ, and both firms are interested in providing CRACQ with

TABLE 7.3 Cost of Chloride Reduction by Margarita Salt Company

Percent Reduction of Chloride Discharge	Cost of Reduction ($\$10^3$)
25	164
50	808
75	2050
100	3980

[12] Although benefit-cost analyses of environmental regulations may not be capable of yielding theoretical optima, those analyses have often been useful in environmental regulatory processes. For case studies of how benefit-cost analyses have assisted U.S. environmental agencies, see Arnold (1995).

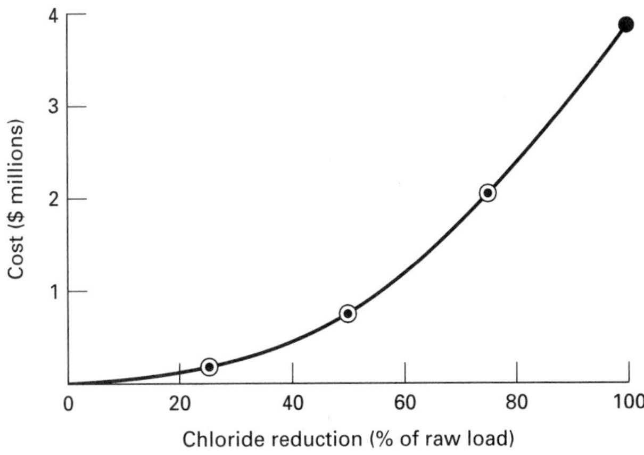

Figure 7.2 Cost of chloride reduction by Margarita Salt.

all the data it needs in making its decisions. (This *is* a hypothetical example.) The chief engineer at CRACQ asks the representative of each firm for some information. From the person representing Margarita Salt, she asks how much it would cost to reduce the existing Margarita Salt Company discharge by 25, 50, 75, and 100%, respectively. (One hundred percent reduction is unrealistic in practice. It is used here only because it simplifies the arithmetic in the example.) The chloride reduction can be accomplished by means of end-of-pipe treatment of wastewater or by a combination of actions involving materials recycling and production process changes.[13] Productive efficiency dictates that any particular chloride reduction must be attained with the least costly combination of actions. Costs provided by the Margarita Salt Company are given in Table 7.3.[14] The chief engineer at CRACQ plots these numbers (see Figure 7.2) and then uses statistical methods to get the equation of a curve that passes closely through all the points. That equation, plotted in Figure 7.2, takes the form

$$C_M(R) = 100(R)^{2.3} \quad \textbf{(7-4)}$$

where R represents the percentage of the "raw" (that is, before reduction) chloride load removed by Margarita Salt, and $C_M(R)$ is the cost of reducing the raw load by R percent. The raw load represents the quantity of chlorides that Margarita Salt releases in the absence of any government intervention to influence its discharge.

Relationship between Chloride Discharge and Cedro River Quality

The CRACQ staff then analyzes how different levels of chloride reduction by Margarita Salt translate into changes in the chloride concentration at the intake of the Long Shot Brewery. First, the staff examines the historic record of flows in the Cedro River. On the basis of these records they decide to perform their analysis for conditions when the river flow at Margarita Salt is approximately 1000 cubic feet per second (cfs), and the flow does not increase significantly between the Margarita Salt Company and the brewery. The CRACQ staff also examines water quality records and finds that under this flow condition, the chloride concentration just above Margarita Salt's point of discharge is about 25 mg/l. The discharge for Margarita Salt is of negligible volume (1 cfs), but it is highly concentrated (100,000 mg/1). The load from Margarita Salt without any chloride reduction is 100,000 cfs · mg/l, the *product* of flow times concentration.[15]

Because chlorides do not degrade over time, the CRACQ staff has an easy job. To find the concentration of chlorides at Long Shot Brewery, they simply conduct a *mass balance* analysis at the Margarita Salt Company's point of discharge. The chloride load downstream of the point is set equal to the load upstream plus the load from the Margarita Salt Company (see Figure 7.3). Once the chloride load downstream from Margarita Salt is found, all that remains is to compute the concentration of chlorides. This concentration is calculated by dividing the total downstream chloride load by the downstream flow. Thus, if Margarita Salt did not reduce its discharge,

[13] Options for reducing pollution, other than traditional waste treatment, are elaborated in Chapter 14.

[14] For simplicity, assume that all costs in this example are given as *present values,* which are computed using a particular discount rate and time horizon.

[15] Although flow and concentration are not given in a consistent unit system, the units indicated are commonly used by water quality specialists in the United States. The product of flow (volume/time) and concentration (mass/volume) is a *mass flowrate* (mass/time).

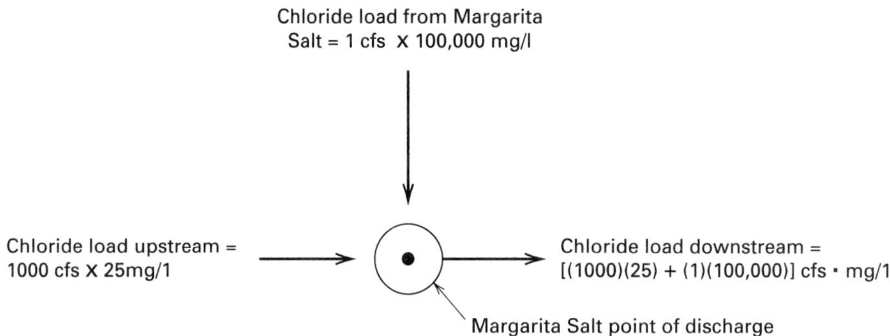

Figure 7.3 Chloride balance for Margarita Salt discharge with no chloride reduction.

the chloride concentration at Long Shot Brewery would be

$$\frac{[(1000)(25) + (1)(100{,}000)]\ \text{cfs} \cdot \text{mg/l}}{1001\ \text{cfs}} = 125\ \text{mg/l}.$$

Using the same reasoning, the staff at CRACQ derives a formula showing the relationship between chloride concentration in the river and discharges at Margarita Salt. If *R* equals the percentage of chloride load reduction, then the diagram for a chloride balance at Margarita Salt is as shown in Figure 7.4. To deduce the relationship between chloride load and concentration, let $[Cl]$ represent the concentration of chlorides (in mg/l) downstream from the discharge point. The value of $[Cl]$ can be found using a mass balance equation like the one just demonstrated. In this instance,

$$[Cl] = \frac{[(1000)(25) + 100{,}000 - 1000\ R]\ \text{cfs} \cdot \text{mg/l}}{1001\ \text{cfs}},$$

or, after some simplifications,

$$[Cl] = 125 - R. \qquad \textbf{(7-5)}$$

In equation 7-5, the result of dividing by 1001 cfs is rounded off to the nearest whole number.

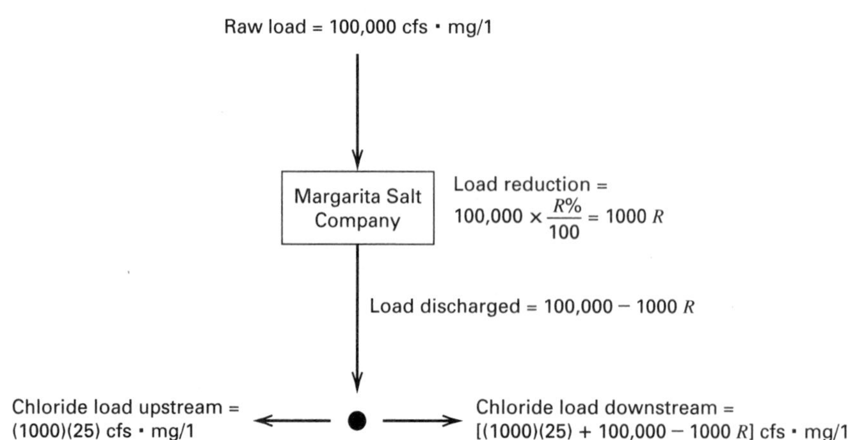

Figure 7.4 Chloride balance for Margarita Salt discharge with *R*% reduction.

TABLE 7.4 Relation between Chloride Reduction and Downstream Chloride Concentration

Percent Reduction of Chloride Discharge by Margarita Salt	Total Chloride Load from Margarita Salt (cfs · mg/l)	Concentration of Chlorides in Cedro River at Long Shot Brewery (mg/l)
0	100,000	125
25	75,000	100
50	50,000	75
75	25,000	50
100	0	25

Cost of Water Treatment at Long Shot Brewery

The chief engineer at CRACQ next turns to the representative from the Long Shot Brewery for information. She asks for an estimate of what the brewery's water treatment costs would be if its intake water had each of several chloride concentrations (see Table 7.4). The engineers at Long Shot Brewery perform the necessary analyses and provide the cost estimates in Table 7.5. Once more the chief engineer derives an equation, this time for a curve that links water treatment cost with the river's chloride concentration. The resulting equation is

$$C([Cl]) = 100[Cl]^2 \qquad \textbf{(7-6)}$$

where $C([Cl])$ is the Long Shot Brewery water treatment cost for any particular value of chloride concentration at the point of water intake. If equations 7-5 and 7-6 are combined, the result can be expressed as

$$C_L(R) = 100(125 - R)^2 \qquad \textbf{(7-7)}$$

where $C_L(R)$ is the Long Shot treatment cost in terms of the percentage of chloride reduction by Margarita Salt. The chief engineer at CRACQ plots this cost equation against both $[Cl]$ and R, and the result is given in Figure 7.5. The horizontal axis in this plot

TABLE 7.5 Cost of Water Treatment at Long Shot Brewery

Chloride Concentration of Intake Water (mg/l)	Cost of Water Treatment for Long Shot Brewery ($\$10^3$)
25	63
50	250
75	563
100	1000
125	1560

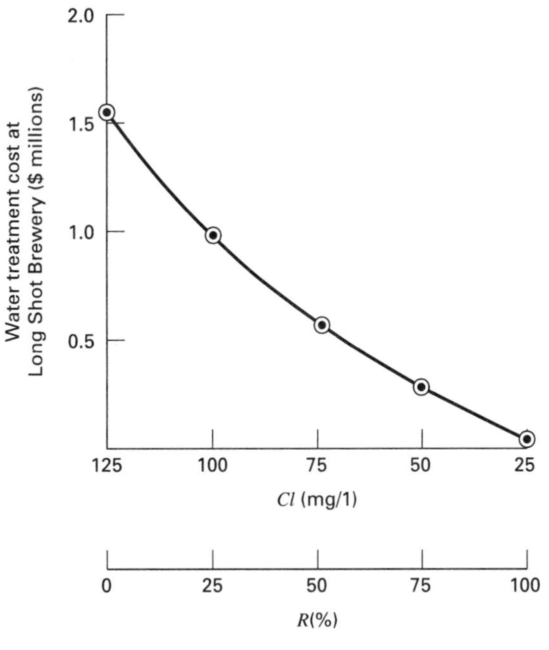

Figure 7.5 Water treatment costs at Long Shot Brewery.

TABLE 7.6 Computation of Benefits to Long Shot Brewery ($\$10^3$)

Percent Reduction of Chloride Discharge by Margarita Salt (*R*)	Cost to Long Shot Brewery if Margarita Salt Does Not Reduce Its Load	Cost to Long Shot Brewery if Margarita Salt Reduces *R*% of Its Load	Benefit to Long Shot Brewery in Terms of Treatment Costs Avoided[a]
0	1560	1560	0
25	1560	1000	560
50	1560	563	997
75	1560	250	1310
100	1560	63	1500

[a] Entries in this column have been rounded off to three significant figures.

is based on the relationship between $[Cl]$ and R in equation 7-5.

Benefits Estimated as Water Treatment Costs Avoided

People who feel that society's resources should be allocated efficiently would advocate that CRACQ set R to maximize the total economic benefit minus the cost of chloride reduction. In this instance, the benefit is defined as the decrease in the brewery's water treatment cost due to chloride removal by the salt company.[16]

The procedure for computing the economic benefit rests on the idea that when Margarita Salt reduces its chlorides by R%, there are water treatment costs (or "damages") avoided by the Long Shot Brewery. If $R = 0$, no damages are avoided and thus the benefit is zero. Alternatively if $R = 100\%$, the damage avoided is the difference between Long Shot's water treatment cost with no upstream chloride reduction (\$1,560,000) and its treatment cost with full chloride reduction (\$62,500). Continuing with this reasoning, the benefits to Long Shot Brewery, in terms of water treatment cost avoided, can be expressed as $B_L(R)$, where

$$B_L(R) = 1{,}560{,}000 - C_L(R). \qquad \textbf{(7-8)}$$

[16] Using terminology of Chapter 6, the cost of water treatment by the brewery is a *defensive expenditure.*

Combining equation 7-8 with equation 7-7 gives

$$B_L(R) = 1{,}560{,}000 - 100(125 - R)^2. \qquad \textbf{(7-9)}$$

Table 7.6 presents benefits for several values of R. The chief engineer now has all the information needed to compute the value of R that maximizes net benefits.[17] To demonstrate the calculation to CRACQ's policymakers, she prepares a table that allows her to select R^*, the net benefit-maximizing percentage of chloride reduction. For any particular R, the equations for $B_L(R)$ and $C_M(R)$ are used

[17] The chief engineer uses calculus to deduce R^*, the value of R that maximizes net benefits. The net benefits, $N(R)$, are defined as

$$N(R) = B_L(R) - C_M(R)$$

or

$$N(R) = 1{,}560{,}000 - 100(125 - R)^2 - 100R^{2.3}.$$

The value of R that maximizes $N(R)$ must satisfy

$$\left.\frac{dN(R)}{dR}\right|_{R = R^*} = 0.$$

Solving

$$\frac{dN(R)}{dR} = 0 = 200(125 - R) - 230R^{1.3}$$

yields $R^* = 30\%$. It can be shown that the second derivative of $N(R)$ is negative for $R > 0$, and this ensures that R^* maximizes $N(R)$.

TABLE 7.7 Summary of Benefits to Long Shot Brewery and Costs to Margarita Salt ($\$10^3$)

Percent Reduction of Chloride Discharge by Margarita Salt	Total Benefits to Long Shot Brewery	Total Costs to Margarita Salt	Difference Between Total Benefits and Total Costs
0	0	0	0
10	240	20	220
20	460	98	362
30	657	250	407
40	838	484	354
50	997	808	189
60	1140	1230	−90
70	1260	1750	−490
80	1360	2380	−1020
90	1440	3120	−1680
100	1500	3980	−2480

to compute the benefits to Long Shot Brewery and the costs to Margarita Salt, respectively. The results for different values of R are given in Table 7.7.

The last column in Table 7.7 indicates that the net benefit-maximizing value of R is 30%. This is confirmed by making a graph of total costs and total benefits and inspecting the differences (see Figure 7.6). The 30% reduction yields a chloride concentration of 95 mg/l at the intake of the Long Shot Brewery. This value of concentration is obtained using equation (7-5) with $R = 30\%$.

Figure 7.7 contains plots of the incremental benefits and costs of chloride reduction. As indicated in the figure, the value of R that maximizes net benefits corresponds to the point at which the incremental benefit equals the incremental cost. At values below $R = 30\%$, it is possible to increase net benefits by doing additional chloride reduction, because the incremental costs of doing so are less than incremental

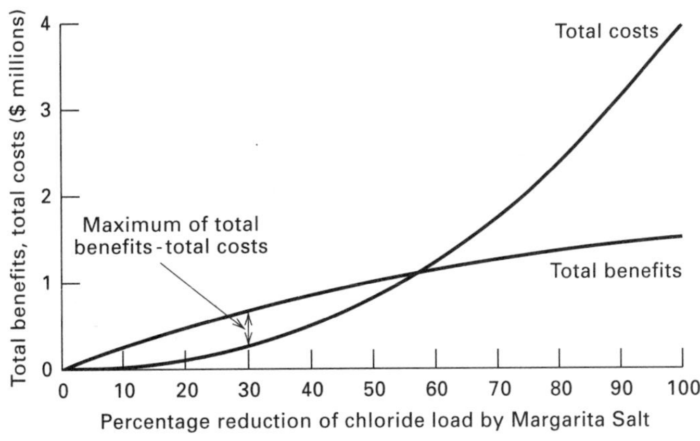

Figure 7.6 Total benefits and total costs.

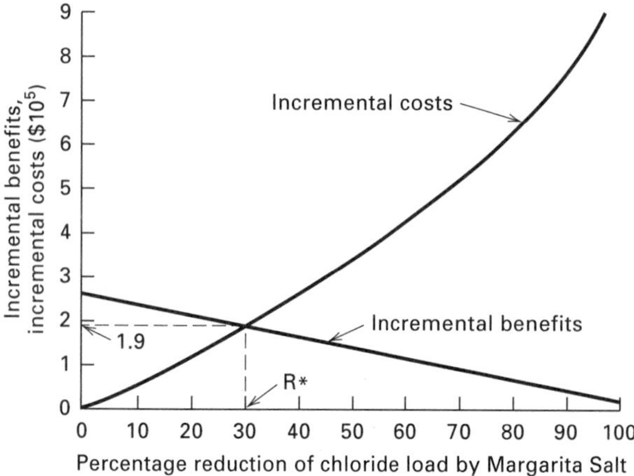

Figure 7.7 Incremental benefits and incremental costs.

benefits. A comparison of incremental benefits and costs in the figure also shows that, compared to $R = 30\%$, net benefits decrease if chloride reduction is increased.[18]

Minimizing the Total Cost of Margarita Salt's Discharge

In this particular example, the definition of pollution abatement benefits is such that the productively efficient level of chloride reduction by Margarita Salt Company also corresponds to the value of R that minimizes the sum of upstream waste reduction cost and downstream water treatment cost. In this context, it is instructive to *imagine* that a single firm owns both the Margarita Salt Company and the Long Shot Brewery. For such a firm, it would make good economic sense to decrease the chloride discharge to the point where Margarita Salt's chloride reduction cost *plus* Long Shot Brewery's water treatment cost is at a minimum. Any other level of chloride reduction would be unnecessarily expensive to a single firm that owned both the salt company and the brewery.

Determination of the chloride reduction that minimizes total cost is based on Table 7.8. For any particular reduction, R, the cost to the Margarita Salt Company is computed using equation 7-4. The corresponding water treatment cost for Long Shot Brewery is determined from equation 7-7. These costs are shown for different values of R in Table 7.8. By inspection of the last column of the table, the combined cost of chloride reduction *and* water treatment is at a minimum when Margarita Salt decreases its chloride load by 30%.[19] This is the same value of R that maximizes net benefits.

How the Cedro River Example Differs from Reality

The Cedro River case demonstrates the benefit-cost analysis approach to environmental quality manage-

[18] The value of R that maximizes net benefits will always have incremental benefits equal to incremental costs, provided that the cost and benefit curves are differentiable and have appropriate concavity properties. Using notation from the previous footnote,

$$N(R) = B_L(R) - C_M(R).$$

The value of R that maximizes $N(R)$ must satisfy

$$\frac{dN(R)}{dR} = 0 = \frac{dB_L(R)}{dR} - \frac{dC_M(R)}{dR},$$

which establishes that incremental benefits are equal to incremental costs when $N(R)$ is at a maximum.

[19] Calculus can also be used to determine the value of R that minimizes $C_T(R)$, the cost of chloride reduction plus water treatment. This total cost is found by adding equations 7-4 and 7-7:

$$C_T(R) = 100R^{2.3} + 100(125 - R)^2.$$

A theorem from calculus requires that $\hat{R}$, the value of R that minimizes $C_T(R)$, satisfy the condition

$$\left.\frac{dC_T(R)}{dR}\right|_{R = \hat{R}} = 0.$$

Solving

$$\frac{dC_T(R)}{dR} = 0 = 230R^{1.3} - 200(125 - R)$$

for R yields $\hat{R} = 30\%$. For this value to provide a minimum total cost, the second derivative of $C_T(R)$ evaluated at $\hat{R}$ must be positive. This can be shown to be the case. Inspection of the first derivatives for $C_T(R)$ above and $N(R)$ in footnote 17 indicates that these derivatives are equal except for their signs. It follows that the R that maximizes net economic benefits (in this particular example) is identical to the R that minimizes the costs of chloride reduction plus water treatment.

TABLE 7.8 Cost of Chloride Reduction and Water Treatment ($10³)

Percent Reduction of Chloride Discharge by Margarita Salt	Chloride Reduction Cost Incurred by Margarita Salt Company	Water Treatment Cost Incurred by Long Shot Brewery	Cost of Chloride Reduction Plus Water Treatment[a]
0	0	1560	1560
10	20	1320	1340
20	98	1100	1200
30	250	903	1150
40	484	723	1210
50	808	563	1370
60	1230	423	1650
70	1750	303	2050
80	2380	203	2580
90	3120	123	3240
100	3980	63	4040

[a] Entries in this column have been rounded off to three significant figures.

ment, but it does not provide a realistic representation of how decisions concerning appropriate levels of discharge are actually made. The example considers only a single contaminant (chlorides), but in real settings, several pollutants are likely to be significant. Moreover, the Cedro River has only a single source of pollution, the Margarita Salt Company. Usually there are many discharges involved.[20]

The linkage between the pollutant released by the Margarita Salt Company and the pollutant concentration downstream is much simpler than corresponding relationships in real rivers. The simplicity in the example results because chlorides are stable, and because both the discharge and the river flow are assumed constant. Not many pollutants behave as simply as chlorides. Complex physical, chemical, and biological processes often transform substances after they are discharged into the environment. In addition, characteristics of the environment itself change over time. These complicating factors often make it difficult to predict accurately the effect that an emission will have on the quality of the ambient environment. When there are numerous discharges, the forecasting task is even more burdensome. In many settings, it is impossible to make meaningful predictions.

[20] Chapter 8 analyzes more complex cases involving multiple sources of pollution.

Another simplification concerns the procedure for estimating the benefits of pollution reduction (or its counterpart, the damages caused by the discharge). The benefit estimation procedure used in the example is straightforward. However, most real situations involve much more than damages in the form of additional treatment costs downstream. Typically, the benefits concern less easily quantified effects such as improvements in aesthetics and reductions in human illness. Many effects of pollution abatement are so difficult to evaluate in monetary terms that the computations are frequently not even attempted. Without such estimates, a net benefit-maximization computation cannot be made.

Despite its many simplifications, the Cedro River case serves several purposes. It demonstrates a rationale that is often advocated in discussions of environmental quality: pollution abatement should be undertaken only if the resulting monetary benefits exceed the costs. In addition, the example illustrates the simplifications and assumptions that must be made to calculate the benefits and costs of a particular environmental management program.

The Cedro River example also assists in understanding the limits of the productive efficiency criterion. Even if a complete benefit-cost analysis could be performed, the numerical results would not be used without examining other factors. Decision makers often consider whether the costs and benefits of pollution reduction are distributed fairly. Issues related to entitlements and rights are also relevant. Here is a sample of questions related to rights and fairness in the Cedro River case.

- Does a discharge by Margarita Salt Company violate the brewery's rights to clean water?
- Should Margarita Salt Company be legally entitled to cause any damages at all?
- If damages are caused, should they be offset by payments from Margarita Salt Company to Long Shot Brewery?
- If forcing Margarita Salt Company to pay for either chloride reduction or downstream damages would cause it to close, should anything be expected of the company?
- Would strong action against Margarita Salt Company be justified if a plant closure would cause widespread unemployment?

These important questions go beyond the boundaries of the productive efficiency approach to environmental management.

NEED FOR GOVERNMENT INTERVENTION IN POLLUTION ABATEMENT

To what extent is government intervention required in managing pollution? In examining this question, consider Margarita Salt's expected behavior if there were no intervention by CRACQ or any other agency. In these circumstances, the Margarita Salt Company would probably discharge its entire raw load. From the salt company's perspective, waste receptor services are provided by the Cedro River free of charge. Any level of chloride reduction costs the company money, but it provides no economic return. Even though substantial downstream costs are associated with the chloride discharge, there is nothing to compel Margarita Salt to account for these costs in its decision making.

It might be argued that the salt company would provide some chloride decrease out of a sense of moral obligation. As a practical matter, the ethical perspectives of Margarita Salt cannot be relied on to yield substantial chloride reductions. This position is supported by evidence that Baumol and Oates (1979) gathered concerning the voluntary activities that firms engage in to eliminate environmental damages. They observed that "moral suasion" is effective in promoting voluntary actions only when brief, acute emergencies are involved. They cite, as an example, emergency conditions associated with intense smog. In such cases, voluntary pollution abatement may be the only way to deal with the emergency. The analysis by Baumol and Oates provides little support for the view that discharges will be cut back voluntarily in routine circumstances.

A cash payment to the Margarita Salt Company would probably provide a more effective incentive for chloride reduction than an appeal to the company's moral obligations. The benefits and costs in Table 7.7 suggest that a suitable payment might be offered without any government intercession. Consider, for example, the chloride reduction costs and the savings in water treatment costs when $R = 10\%$ instead of zero. As shown in the table, Margarita Salt's chloride reduction costs are \$20,000. At $R = 10\%$, the Long Shot Brewery would save \$240,000 in water treatment costs. Clearly there is the potential for a private agreement in which Margarita Salt provides a 10% chloride reduction and receives a payment that more than covers its costs. Long Shot Brewery has an incentive to offer payments to Margarita Salt for chloride reductions up to $R = 30\%$. Beyond 30%, it is cheaper for the brewery to treat its intake water than to pay for Margarita Salt's additional chloride reduction. For example, to attain reductions of 30% and 40%, the costs are \$250,000 and \$484,000, respectively (see Table 7.7). The \$234,000 difference represents the cost of the 10% increment of chloride reduction. The benefits to Long Shot Brewery in going from $R = 30\%$ to $R = 40\%$ are \$657,000 and \$838,000, respectively. The difference, \$181,000, is the additional benefit the brewery receives for the extra 10% reduction. Since this amount is lower than the corresponding increment in costs, Long Shot Brewery has no incentive to pay for the additional chloride decrease.

A theorem formulated by Ronald Coase, a Nobel laureate in economics, indicates conditions under which efficient outcomes could be expected using mutually advantageous trades. In the context of pollution control, the Coase theorem can be paraphrased as follows: If costless negotiation among parties is possible, and if rights to pollute (or rights to an unpolluted environment) are well specified, then the allocation of resources will be productively efficient.[21] Unfortunately, as will be illustrated for the Cedro River example, the conditions of the Coase theorem are not often satisfied for cases involving pollution.

One condition of the Coase theorem is that there is no cost to reach agreements. However, for firms such as Margarita Salt Company and Long Shot Brewery, there may be substantial "transactions costs"; that is, costs to identify and implement mutually advantageous exchanges. There are costs involved in generating information, such as the estimates in Table 7.7, and in negotiating and enforcing agreements. The negotiation process can be complex and lengthy because the firms generally would not know *each other's* treatment costs, and they might rely on tactics such as stalling and bluffing to gain advantages in negotiations.

A second impediment to a private agreement between the firms is an ambiguity in legal rights. Does Margarita Salt have a legally enforceable entitlement to release chlorides, or is its discharge occurring in the absence of laws controlling the quality of the Cedro River? Alternatively, does Long Shot Brewery have a legal right to water that is uncontaminated by Margarita Salt's discharge? Long Shot Brewery would probably not agree to pay for chloride reductions by Margarita Salt if the latter did not have a right to release chlorides. If legal rights are ambiguous, it might be impossible to enforce a private bargain between the two firms.

Arguing in a more general way, Coase (1960) showed that the presence of transaction costs and the lack of well-defined legal rights are often barriers to mutually advantageous exchanges of the type just described. Before such private agreements can be expected to materialize, governments must clarify who has the legal right to the services provided by the environment. Of course, even if rights are clearly specified, the existence of high transaction costs could prevent agreements from being reached. For example, imagine the costs involved in reaching agreements when the sulfur dioxide discharged by a power plant adversely influences the health of an entire community. When many individuals are affected, the costs of bringing them together as a negotiating unit may be greater than the resulting benefits.

[21] This statement of the Coase theorem is adapted from Layard and Walters (1978, p. 192).

FORMS OF GOVERNMENT INTERVENTION

The preceding discussion indicates that without government intervention, pollution reduction is unlikely, except at levels that make economic sense to the firms generating the wastes. The remaining chapters in Part Two emphasize three strategies of actions governments can take to manage pollution: (1) regulations, exemplified by the promulgation of discharge standards; (2) economic incentives in the form of fees (or taxes); and (3) permits to pollute that can be bought and sold in markets.[22] The Cedro River example is used to introduce each of the three strategies.

Using the first strategy, CRACQ could issue *regulations* requiring that Margarita Salt Company cut back its discharge to the productively efficient level. As indicated in the discussion of Figure 7.6, the net benefit-maximizing chloride concentration is 95 mg/l, and this results when $R = 30\%$. Assuming it has the needed authority, CRACQ can promulgate the following requirement: the chloride load released by the Margarita Salt Company can be no greater than 70% of its raw load, which is equivalent to a flow of 1 cfs with a chloride concentration of 100,000 mg/l. Sanctions, such as monetary fines, could be imposed if the company failed to comply with the requirement.

A second strategy rests on the use of a *pollutant discharge fee,* a fee paid per unit of contaminant released. Assume that CRACQ places no standard limiting the quantity of chlorides released by Margarita Salt. Instead, it imposes a fee of $\$Z$ per pound of chlorides discharged. Margarita Salt's response will depend on how the fee compares to the cost of chloride reduction. If Z is set much lower than the incremental cost of chloride reduction, Margarita Salt will have no financial motivation to reduce its discharge. It would release its raw load and pay the consequent

[22] These are not the only possible government strategies. For example, governments sometimes establish programs to subsidize the construction of waste treatment facilities. An example of such a subsidy program is given in Chapter 10.

fee. If, on the other hand, Z is set very high compared to the cost of chloride reduction, Margarita Salt will try to minimize its discharge. If CRACQ knew Margarita Salt's basis for making decisions and its cost of chloride reduction, then the agency could set Z to induce Margarita Salt to decrease its discharge by any amount between 0 and 100%.

Another scheme based on economic incentives involves *marketable pollution permits.* A marketable (or tradable) permit system could be established for the Cedro River as follows. CRACQ first sets an ambient water quality standard for chlorides. Using a mass balance analysis, it then computes the total chloride load the river can accept and still meet this standard. Let Y lb/day represent this total load. Next, CRACQ issues permits to discharge chlorides into the Cedro River. Collectively, the permits allow for only Y lb/day of chlorides. By requiring that each unit of discharge be covered by a permit, the ambient standard is always met. Suppose CRACQ initially allocates *all* of the permits to the Margarita Salt Company.[23] After the initial allocation, the individual permits can be bought and sold. If the price were right, Margarita Salt would sell some of its permits to Long Shot Brewery. As indicated by Table 7.7, the brewery enjoys substantial savings in water treatment when chloride discharges are reduced. Thus the brewery would have an incentive to purchase some permits from the salt company.

Another possibility is for Margarita Salt to sell some permits to a new firm that wants to locate in the area and that needs to dispose of chlorides. This permit trading system would have to be altered for the more realistic case in which the concentration of pollutants in the river depends on the locations of the discharges. The necessary modifications do not change the main idea, which is to encourage private transactions that use the environment's waste receptor services efficiently and that meet environmental quality goals.

The three strategies are presented here as alternatives to each other. As the following chapters make clear, the practical work of environmental management is advanced by using hybrid systems that combine traditional regulatory approaches with innovative economic incentive systems, such as those based on discharge fees and tradable pollution permits.

[23] An alternative is to give all the permits (totaling Y lb/day) to Long Shot Brewery. Another possibility is for CRACQ to auction the permits off to the highest bidders.

Key Concepts and Terms

Potential Pareto Improvement Criterion
- Pareto's criterion
- Interpersonal comparisons of utility
- Compensation tests
- Distribution of costs and gains
- Productive efficiency

Benefit-Cost Analysis
- Willingness to pay
- Opportunity cost
- Shadow price
- Discounting to present value
- Discount rate
- Market vs. social rate of discount
- Time horizon
- Uniform series present worth factor

Optimal Levels of Pollution Abatement
- Mass balance analysis
- Linking discharges and ambient quality
- Costs of pollution reduction
- Benefits as damage costs avoided
- Minimizing pollution reduction plus damage costs
- Maximizing net benefits
- Limitations of benefit-cost approach

Need for Government Intervention in Pollution Abatement
- Voluntary pollution control
- Private agreements to reduce pollution
- The Coase theorem
- Transaction costs
- Legal rights to discharge

Forms of Government Intervention
- Direct regulation using standards
- Effluent versus ambient standards
- Technology-based standards
- Cash subsidies
- Pollutant discharge fees
- Marketable pollution permits

Discussion Questions

7-1 Is there any ethical basis for making interpersonal comparisons of utility? What does your response suggest about the possibilities for objective analysis of how development projects or environmental regulations influence the distribution of income? What interpersonal comparisons of utility are implicit in the use of benefit-cost analysis?

7-2 Construct a hypothetical example to demonstrate the application of the potential Pareto improvement criterion. What value judgments are made in using this criterion?

7-3 Consider a proposed project that involves an initial cost of $10,000 and annual benefits of $2,000 over the next 10 years. Does this project enhance productive efficiency if the discount rate is 5%? How high does the discount rate have to be before the net benefits are negative?

7-4 Consider a market economy of the type that prevails in the United States. What arguments can be used to justify government interventions to influence the way individuals and firms make decisions regarding the generation and disposal of pollutants? In framing your response, consider the differences between social costs and private costs.

7-5 Perform a *sensitivity analysis* for the Cedro River example by determining the net benefit-maximizing levels of chloride reduction if the cost function for Margarita Salt is

$$C_M(R) = 100\ R^2$$

instead of the function in equation 7-4. Suppose CRACQ sets effluent standards to attain economic efficiency. Describe the effects, in terms of differences in total costs to both firms, if CRACQ imposes an effluent standard assuming that Margarita Salt's costs are those in equation 7-4, when in reality, its costs are $100R^2$.

7-6 Consider a stream receiving the wastewater discharge from a cannery. A heavily used park lies downstream of the discharge. Many individuals would be willing to pay a significant amount to rid the park of odors caused by the cannery's waste. Assume the total amount that park users would be willing to pay is enough to pay for pollution abatement to minimize the odors. Discuss the barriers to having the odor reduction measures paid for by private cash transfers from the park users to the cannery.

7-7[24] You are asked to conduct a benefit-cost analysis for determining the productively efficient level of pollution discharged to a particular river. Two firms are located near the river. Let X = the quantity of waste discharged by the upstream firm, and let R = the raw load of the upstream firm.

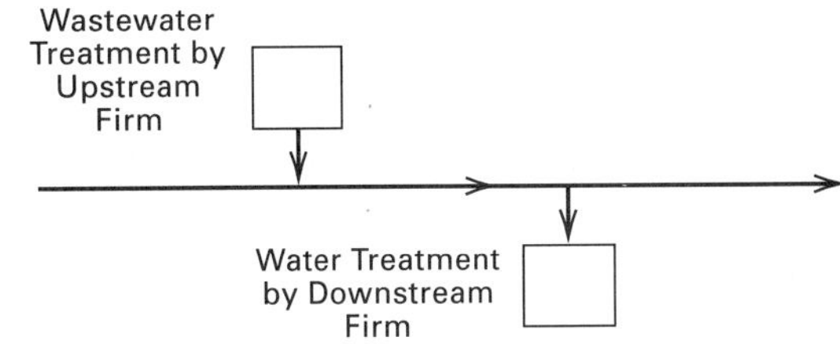

The cost of upstream *wastewater* treatment, as a function of the waste discharge at the upstream firm is

$$m + nX^{-1},$$

where m and n are non-negative parameters, and

$$0 \leq X \leq R.$$

The cost of downstream *water* treatment, as a function of the upstream waste discharge level, is

$$aX^b,$$

where a and b are parameters, with $a > 0$ and $b \geq 1$.

If the upstream firm reduces its discharge to be X, a value less than the raw load R, then the downstream firm enjoys a benefit equal to the water treatment cost that it avoids.

[24] The solution to this problem requires use of calculus.

a. Write an expression for $B(X)$, defined as the benefit associated with waste discharge X, where $X < R$.
b. Find the value of X that maximizes the difference between downstream benefits [given by $B(X)$] and upstream wastewater treatment costs. Check conditions that ensure you have a maximum (and not a minimum or a saddle point). This value of X is the optimal level of pollution.
c. In this context, the optimal level of pollution also can be determined by finding the level of upstream wastewater discharge that minimizes the sum of the upstream wastewater treatment cost plus the downstream water treatment cost. Use this method of finding the optimal level of pollution. Check conditions that ensure that you have found a cost-minimizing value.
d. Compare the results in parts b and c.

References

Arnold, F. S. 1995. *Economic Analysis of Environmental Policy and Regulation.* New York: Wiley.

Baumol, W. J., and W. E. Oates. 1979. *Economics, Environmental Policy and the Quality of Life.* Englewood Cliffs, NJ: Prentice–Hall.

Coase, R. H. 1960. The Problem of Social Cost. *Journal of Law and Economics* 3: 1–44.

Collier, C. C., and N. B. Ledbetter. 1982. *Engineering Cost Analysis.* New York: Harper & Row.

Dorfman, R. 1993. "Some Concepts from Welfare Economics." In *Economics of the Environment: Selected Readings,* 3d ed., R. Dorfman and N. S. Dorfman, eds. New York: W. W. Norton.

Dixon, J. A., and M. M. Hufschmidt, eds. 1986. *Economic Valuation Techniques for the Environment: A Case Study Workbook.* Baltimore: John Hopkins University Press.

Farber, D. A., and P. A. Hemmersbaugh. 1993. The Shadow of the Future: Discount Rates, Later Generations and the Environment. *Vanderbilt Law Review* 46(2): 268–304.

Freeman, A. M., III. 1993. *The Measurement of Environmental and Resource Values.* Washington, DC: Resources for the Future.

Layard, P. R. G., and A. A. Walters. 1978. *Microeconomic Theory.* New York: McGraw–Hill.

Mishan, E. J. 1975. *Cost-Benefit Analysis: An Informal Introduction.* London: George Allen & Unwin Ltd.

Sugden, R., and A. Williams. 1978. *The Principles of Practical Cost-Benefit Analysis.* Oxford: Oxford University Press.

Chapter 8

Environmental Management Based on Command and Control

Command and control is a phrase used to indicate that discharge standards and other requirements, as opposed to economic incentives, are employed to meet environmental quality goals. The idea is straightforward. An environmental agency issues regulations (commands) requiring polluters to take actions to meet environmental goals. The agency then controls by monitoring to see if its requirements are being met; the agency can impose sanctions for noncompliance and offer rewards for compliance.

Different types of programs based on command and control are outlined in this chapter's first section. Most of the chapter concerns a question that is central to the design of many command and control programs: What levels of waste reduction should polluters be required to undertake? An analytic approach to this question is introduced using a hypothetical example involving three waste sources. Application of the analysis procedure is demonstrated using a case study involving firms and municipalities discharging wastewater into the Delaware River estuary. The question of how environmental agencies bring polluters into compliance with regulations is taken up in the next chapter.

TYPES OF ENVIRONMENTAL REQUIREMENTS

Although details of regulatory programs vary, some program elements are widely used. In a typical instance of command and control, one or more of the program elements in Table 8.1 is called for by a statute or administrative order.[1] The environmental

[1] Table 8.1 includes requirements, not incentives used to encourage polluters to meet requirements. Examples of incentives include tax credits and grants to help waste dischargers defray the cost of waste reduction. The discussion in this section presents elements of a regulatory program based on command and control, but it does not evaluate command and control. An evaluation of command and control is conducted at the end of Chapter 11. That evaluation compares command and control with market-based policy instruments for managing environmental quality.

TABLE 8.1 Common Forms of Environmental Requirements

Ambient Standards
Effluent and Emission Standards
Technology-based Effluent and Emission Standards
Performance Standards
Technology-forcing Standards
Technology Standards
Management Practice Standards
Product Bans
Information Generation and Disclosure Requirements

TABLE 8.2 U.S. National Ambient Air Quality Standards[a]

Pollutant	Standards to Protect Public Health
Carbon monoxide (CO)	10 mg/m^3 average over 8 hours[b] 40 mg/m^3 average over 1 hour[b]
Lead (Pb)	1.5 μg/m^3 average over quarter-year[b]
Nitrogen dioxide (NO_2)	100 μg/m^3 annual average
Ozone (O_3)	235 μg/m^3 average over 1 hour[b]
Particulates (measured as PM_{10})[d]	50 μg/m^3 average over a year 150 μg/m^3 average over 24 hours[c]
Sulfur oxides (measured as SO_2)	80 μg/m^3 average over a year 365 μg/m^3 average over 24 hours[c]

[a] From U.S. Environmental Protection Agency.

[b] One day of expected exceedances per year is permitted. Symbols used for units are as follows: mg = milligrams (10^{-3} gm); μg = micrograms (10^{-6} gm); and m^3 = cubic meters.

[c] Not to be exceeded more than once per year.

[d] PM_{10} refers to particulates less than 10 micrometers in diameter.

agency charged with implementation then works out the details based on available policy guidance and scientific information. The discussion that follows describes each of the program elements.

Ambient Standards

An *ambient standard* is a goal set for the quality of the surrounding air or water.[2] Standards are often quantitative and typically expressed in *concentration units* (that is, mass of pollutant per unit volume). Ambient standards are illustrated in Table 8.2, which contains national ambient air quality standards (NAAQS) set by the U.S. Environmental Protection Agency (EPA) under the Clean Air Act.[3] The standards in the table are *health-based* in that Congress required EPA to set standards to protect human health without considering the cost of meeting the standards.[4] Consider, for example, the ambient standard for sulfur oxides (measured as sulfur dioxide, SO_2) for an *averaging time* of one year: 80 micrograms of SO_2 per cubic meter of air (80×10^{-6} gm/m^3) based on readings averaged over a year. The SO_2 standard is less stringent for an averaging time of 24 hours: 365 micrograms per cubic meter. In the United States, one set of ambient air quality standards applies nationwide. In some countries—China, for example—ambient air quality standards vary by region.

Ambient standards are generally set to accommodate certain uses of the environment or to prevent adverse consequences, such as damage to ecosystems. Typically, lawmakers specify the uses to be protected and other criteria to guide development of standards, and environmental agencies use the criteria to set numerical values for environmental parameters. Rulemaking procedures, such as those used by EPA

[2] Ambient standards are sometimes set for noise, particularly in occupational settings. As discussed in Chapter 21, parameters measuring noise exposure are expressed in units of decibels. Although it is not common usage, the term ambient standards can also refer to standards applied to soils in toxic waste site cleanups.

[3] Key provisions of the U.S. Clean Air Act are detailed in Chapter 13.

[4] The Clean Air Act also includes provisions for "secondary" NAAQS, ambient standards to protect "public welfare" (for example, to protect against adverse effects on agriculture). With the following exceptions, the standards in Table 8.1 are also used (currently) as secondary NAAQS: no secondary standard is used for carbon monoxide, and the secondary standard for sulfur dioxide is 1300 μg/m^3 averaged over 3 hours.

and described in Chapter 3, generally allow citizens and interest groups to comment on proposed ambient standards. In some cases, interested parties participate on committees that recommend standards.[5] Environmental agencies are often empowered to modify ambient standards as new scientific evidence becomes available.

Environmental agencies can use violations of ambient standards to force polluters to reduce discharges. For example, in the 1960s, when ambient standards for coliform bacteria in Raritan Bay, New Jersey, were exceeded, the U.S. Public Health Service pinpointed the cause of the violations and ordered the sources responsible to clean up. The Public Health Service took decisive action because the contamination was linked to the spread of disease.[6]

Discharge Standards Based on Ambient Standards

Ambient standards are often used as a basis for setting pollutant discharge standards for individual firms and municipalities. Discharge standards are called *effluent standards* when the discharge is wastewater and *emission standards* when the discharge involves air pollutants. Environmental agencies generally use effluent and emission standards as a basis for issuing permits to discharge waste. When effluent or emission standards are violated, they may trigger an *enforcement response:* an environmental agency's use of sanctions, such as fines and plant shutdown orders, to force polluters to clean up.

For an agency to set discharge standards on the basis of an ambient standard, it must face the following question: How much will ambient water (or air) quality decrease if a particular set of waste discharges are made. If this question cannot be answered, there is little basis for linking release of waste to changes in ambient environmental quality. Mathematical relationships (or "models") are often used to estimate how waste discharges influence ambient quality. For some wastes, such as chlorides in streams, the requisite mathematical relationships are available. For many other pollutants, however, adequate mathematical models don't exist and other approaches must be used to set discharge standards.

Even when mathematical models exist to predict how a waste discharge affects ambient water (or air) quality, using those models to derive discharge standards can be problematic. The predictive accuracy of mathematical models can be challenged. Dischargers can argue that an environmental agency's models are incorrect or poorly constructed. This claim is common. Long delays in cleanup often result because agencies and dischargers debate (in and out of court) the validity of models used to set discharge standards.

In addition to contesting mathematical models, polluters may challenge standards because they appear unfair. If several waste sources influence ambient quality at a particular point, many different combinations of discharge standards may satisfy an ambient standard. Some polluters may be asked to do major cleanups and others may be asked to do little. Which of the alternative sets of discharge standards should be used? Questions related to fairness and model validity arise frequently when environmental agencies set discharge standards on the basis of ambient targets.

Discharge standards are *not* an alternative to ambient standards. The value of using discharge standards and ambient standards in tandem is clarified by considering how ambient environmental quality is influenced by growth. If discharge standards are established in the absence of an ambient standard, what happens when economic development or population growth leads to new pollution sources? If new polluters discharge wastes and existing sources continue to meet their discharge standards, ambient quality must decrease. Sometimes existing sources are regulated stringently to provide a "margin of safety" in meeting ambient standards. In such cases, some new waste sources can be accommodated without violating ambient standards.

---

[5] Flachsbart and Morrow (1987) provide an example of direct citizen participation in setting ambient standards. They describe the work of the Air Advisory Committee of Hawaii's Department of Health in setting statewide ambient standards for hydrogen sulfide (H_2S), a pollutant associated with geothermal energy development. The Committee included representatives from health and environmental groups, energy producers, the construction industry, and state agencies. The committee's recommended standards for H_2S were the subject of public meeting that allowed affected citizens to influence the final standard.

[6] The Public Health Service discovered the violations in ambient standards while investigating an outbreak of infectious hepatitis in the New York metropolitan area. The outbreak originated when people ate contaminated shellfish that were taken from Raritan Bay.

Another way to accommodate growth using discharge standards is to treat new waste dischargers differently from existing polluters. Typically, new polluters must reduce a higher fraction of their wastes than existing sources. This approach to accommodating growth can work provided the environment can be degraded without violating the ambient standard. When no further environmental degradation is possible, new sources may be required to *offset* their proposed discharges by arranging for additional cutbacks by existing sources. As illustrated in Chapter 11, this offsetting approach is used to regulate new sources of air pollution in some metropolitan areas in the United States.

TABLE 8.3 Portion of BPT Effluent Limitations for Metal Finishing Industry[a]

Pollutant	Maximum for Any One Day	Monthly Average Shall Not Exceed
	Milligrams per liter (mg/l)	
Cadmium	0.69	0.26
Chromium	2.77	1.71
Copper	3.38	2.07
Lead	0.69	0.43
Nickel	3.98	2.38
Silver	0.43	0.24
Zinc	2.61	1.48
Cyanide	1.20	0.65

[a] From *Code of Federal Regulations,* Title 40, Part 433, Section 433.13 (revised as of July 1, 1995). The table shows effluent limits for total concentration of metals in discharges. Pollutants other than metals are also regulated under the BPT effluent limitations for metal finishing.

Technology-Based Discharge Standards

Discharge standards can be set without having to judge how much improvement in environmental quality would result if a *particular* source of waste is reduced by a specific amount. Instead of relying on ambient standards, discharge standards can be set based on the technology available for reducing wastes. The resulting *technology-based discharge standards* often take account of the costs to achieve different levels of reduction. Sometimes, however, governments impose requirements calling for the *best available* techniques for reducing pollution. These are usually the most expensive.

Application of technology-based standards is demonstrated by provisions of U.S. water pollution control laws.[7] Prior to the Federal Water Pollution Control Act Amendments (FWPCAA) of 1972, states developed effluent standards based on ambient standards. The 1972 amendments provided a new mandate: all significant pollution sources had to obtain federal wastewater discharge permits that satisfied both ambient standards set by states and technology-based requirements established by the U.S. Environmental Protection Agency.

Technology-based effluent restrictions are illustrated by the 1972 amendments' use of *best practicable control technology* (BPT) currently available. This technology was defined in *effluent guidelines* issued by EPA. When a discharger applied for a permit in the 1970s, effluent restrictions based on guidelines defining BPT were determined. Before a permit was issued, however, a check was made to assure that use of BPT would satisfy applicable ambient water quality standards. Whenever feasible, this check was made using quantitative modeling procedures. If the BPT requirements were not sufficient to meet ambient standards, permit conditions had to be made even more stringent.

Effluent limitations guidelines defining BPT were issued for many industries.[8] For each industry, subcategories were created based on similarities in production processes, effluent characteristics, and the age of facilities. For example, effluent guidelines for the pulp and paper industry included subcategories such as craft mills and sulfite process mills. For any particular subcategory, BPT requirements set allowable limits on the concentrations of pollutants.

Effluent limitations are illustrated in Table 8.3, which shows a portion of BPT requirements for the metal-finishing industry. Activities within the metal-finishing industry include electroplating, chemical

[7] Other examples of technology-based limits include use of *maximum achievable control technology* as a basis for regulating air pollutants in the U.S. Clean Air Act Amendments of 1990.

[8] The process of developing effluent guidelines is described in Chapter 3. BPT requirements were also set for publicly owned treatment works.

etching, and printed circuit board manufacture. Table 8.3 shows limits on the discharge of metals. The complete set of BPT requirements for the metal-finishing industry also includes limits on oil and grease, total suspended solids, pH, and several dozen toxic organic chemicals. The agency issuing a discharge permit for a metal-finishing facility could modify BPT restrictions in the effluent guidelines, but only after the permit applicant had demonstrated that the guidelines did not reflect its particular plant's attributes.

Characteristics of wastewater discharge permits issued in the United States are given in Table 8.4. As shown in the table, virtually all wastewater permits in the United States include restrictions on both concentration (mass of pollutant per volume of wastewater) and mass of pollutant per unit time. Permit conditions on mass/time (or *mass flowrate*) employ different time intervals. More than half the states issued permits limiting mass flowrates in terms of mass per day. Table 8.4 also shows that many permits contain additional performance standards based on mass per unit of input and mass per unit of output.

TABLE 8.4 Types of Limits Included in Wastewater Discharge Permits[a]

Type of Limit	Percentage of States Using the Limit Type (%)
Concentration[b]	100
Mass/unit input[c]	36
Mass/unit output	50
Mass/unit time[d]	
per minute	4
per hour	9
per day	59
per week	14
per month	27
per year	7

[a] Based on a survey of 44 states conducted by Resources for the Future, Inc., and reported by Russell, Harrington, and Vaughan (1986, p. 19).

[b] Concentration is expressed in units of mass of pollutant per volume of wastewater.

[c] Mass refers to mass of pollutant.

[d] Entries below indicate the different time periods used in discharge permits. For example, 4% of the states put restrictions on the mass of pollutant discharged per minute.

Although numerical limits in a wastewater discharge permit are based on available technology, they often leave dischargers the option of deciding which waste reduction procedures to use in meeting requirements. The term *performance standard* is employed when dischargers have this option. Performance standards have an advantage: polluters can select *cost-effective* measures, defined as actions that meet applicable pollution reduction goals at minimum cost. Frequently, waste treatment is not cost-effective. It is often possible to achieve waste reduction goals at lower cost using *pollution prevention,* techniques to reduce the amount of waste generated as well as methods for materials recycling and reuse (see Chapter 14). For example, instead of treating wastewater to remove chromium, an electroplating factory may find it cheaper to change its production process to generate less chromium waste.

It required a massive effort for EPA and other agencies to issue permits based on BPT, but the technology-based strategy had a major advantage. Permits could be written without necessarily predicting how the ambient environment would respond to the discharge reductions called for by BPT requirements. If the scientific knowledge to make predictions of ambient quality was unavailable, a permit could be written using only the effluent guidelines defining BPT.

Technology-based discharge standards have a major shortcoming: if the standards are rigorously enforced, productive efficiency is not attained.[9] This occurs because technology-based standards do not account for differences in discharge *location* or in *costs* of pollution reduction. The discharge limitations apply uniformly to all facilities of a given type. For example, the BPT limitations for metal finishing in Table 8.3 require small factories with high costs per unit of waste removed to meet the same targets for metals as large factories with low unit treatment costs. The *total* cost to meet a particular ambient standard can often be lowered by tailoring discharge requirements to reflect both facility location and pollution abatement costs.

[9] In the context of technology-based water quality standards in the United States, Hunter and Waterman (1996) provide much evidence demonstrating that these standards are not rigorously enforced.

Technology-Forcing Discharge Standards

Regulation of emissions from automobiles in the United States in the 1970s illustrates a case in which standards were promulgated without relating discharges to ambient environmental quality, except in a general way. Emission standards for autos were based on congressional judgments of what reductions were technologically feasible and how well the auto industry could absorb the costs. The standards were *technology forcing* because they could not be met without the development of new technology. Although Congress assumed that benefits of high levels of auto emission control were worth substantial costs, many people challenged this assumption. Congress's positions were also criticized for leading to frequent confrontations between the auto industry and the regulating agency, the U.S. Environmental Protection Agency.

Problems in implementing Congress's policy are illustrated using the emission decreases mandated by the Clean Air Act of 1970. The act called for a 90% reduction in hydrocarbon (HC) and carbon monoxide (CO) emissions (based on 1970 levels) for 1975 model-year cars, and a 90% reduction in oxides of nitrogen (NO_x) emissions (based on 1971 levels) for 1976 model-year cars. These reductions went beyond what was feasible at the time. In passing the act, Congress tried to force auto companies to speed up their research on emission controls. The 1970 act granted the EPA administrator the ability to allow a one-year extension for meeting the 1975–76 standards. The history of how this extension was granted is filled with "eleventh hour" agreements that just barely averted the government's threatened closure of a major auto company's plants, and endless arguments between government and industry experts regarding what levels of emission control were technically feasible.[10]

Ninety percent reductions in CO, HC, and NO_x, which the act required by 1975–76, were not attained until the early 1980s. Moreover, in regulating only emissions from new cars, Congress's policy did not assure that emission standards would be met after new cars had been used.

According to some analysts, the auto emission standards in the Clean Air Act of 1970 provided the U.S. auto industry with incentives against making a major advance in emission control technology.[11] Why would a manufacturer try to decrease emissions beyond those required if the new reductions might eventually be used as a basis for an even more stringent standard? Such an outcome is conceivable because auto emission standards were largely based on technological feasibility, not on practical cost considerations. It has also been argued that the auto industry did not initially pursue its emission control research with intensity because of the industry's many opportunities to press for a relaxation or postponement of the standards; for example, by challenging control requirements in the courts and lobbying Congress.

Some have claimed that the technology-forcing characteristics of the Clean Air Act of 1970 distorted the research activities of U.S. auto manufacturers. In the early 1970s, auto makers had several high-risk research opportunities for reducing vehicle emissions. U.S. manufacturers opted to develop the catalytic-converter technology, since it had a high probability of meeting the federal emission requirements. Mills and White (1978) argued that other riskier technologies, such as the stratified-charge engine, may have been superior. Foreign auto manufacturers, less dependent on meeting the standards by a particular date, explored alternative technologies. They developed some of the earliest and least costly vehicles capable of meeting applicable standards. Another view, offered by Garwin (1978), is that foreign manufacturers felt less capable than their American counterparts of lobbying against the federal standards. This provided foreign manufacturers with substantial motivation to meet the standards by the 1975–76 deadlines specified in the Clean Air Act.

Other Common Forms of Requirements

In addition to ambient standards and discharge standards, command and control schemes can be based on requirements that stipulate use of particular technologies or management practices. Command and

[10] For an insider's view of these events, see Quarles (1976), who was among the EPA officials acting on the auto industry's application for an extension.

[11] The discussion of the influence of federal auto emission regulations on the research efforts of auto manufacturers relies on Mills and White (1978).

control measures also include product bans and requirements to generate and disclose information.

A *technology standard* is similar to a technology-based standard, but it is often more burdensome because it stipulates the particular equipment or process to be used for waste management. A technology standard has one key advantage: it is easy to monitor. The environmental agency only needs to determine if the required equipment (or process) is in place and operating properly. Disadvantages of a technology standard are (1) dischargers have no flexibility in decision making, and (2) less costly approaches may exist for meeting an ambient quality target. Moreover, a technology standard provides no incentive for the discharger to investigate alternative waste reduction methods.

Instead of requiring a waste management technology or performance level, some standards specify waste management practices. For example, the U.S. Environmental Protection Agency has developed a variety of *best management practice* standards to reduce water pollution caused by *nonpoint sources,* which are water pollution sources so diffuse they cannot be easily collected in a single pipe (or channel) and treated. (Dischargers that can be collected in pipes or channels are called *point sources.*) An example of a nonpoint source is a farm that uses chemical pesticides, which are leached by precipitation and irrigation water and transported to groundwater basins and surface waters. Practice standards are used to prescribe behavior that limits pollutant loadings from nonpoint sources.

Another form of regulation is the *product ban,* illustrated by restrictions on pesticides deemed unsafe. The U.S. Environmental Protection Agency has responsibilities for licensing pesticides, and it is capable of banning certain ones by not granting or renewing licenses. (In this context, a license is a permit to manufacturer, test, and sell a product). A widely discussed equivalent of a product ban is the international effort to prohibit production of chlorofluorocarbons.

Since the 1970s, requirements to generate and disclose information have become an increasingly significant form of regulation. An example is the U.S. federal mandate for environmental impact statements, which is detailed in Chapter 15.

Another example of an information disclosure requirement is contained in the Community Right-to-Know Act enacted by the U.S. Congress in the 1980s. Under this law, firms that have hazardous materials on their property are required to inventory those materials and report the results to appropriate agencies of local government. This gives citizens a chance to learn about their exposure to hazards created by local industry. In addition, the U.S. Environmental Protection Agency has created a *Toxic Release Inventory* (TRI), a national database that indicates the types, quantities, and locations of hazardous materials used and stored by firms. After learning (via TRI) which hazardous materials are being used in different places, citizens may pressure companies to change production practices and reduce their use of toxic materials.

THE WASTE LOAD ALLOCATION PROBLEM

A common problem in designing command and control systems to manage air and water quality involves the following question: How should discharge standards be set for different waste sources in order to meet a specified ambient quality goal? This is sometimes called the *waste load allocation problem* because it concerns how the environment's *waste receptor services* are divided among dischargers. The following example involving water quality illustrates how the waste load allocation problem can be analyzed. The same approach can be used in the context of air quality.

Consider a hypothetical stream that receives effluents from three sources. Assume that each wastewater discharge can be described adequately using biochemical oxygen demand (BOD), and that ambient quality is characterized by the concentration of dissolved oxygen (DO) in the stream. A high concentration of BOD in wastewater indicates that the bacterial decomposition (biodegradation) of organic matter will use up a large amount of the oxygen dissolved in the wastewater. After the waste is discharged, the oxygen required during biodegradation comes from the stream. Thus, if the BOD of the effluent increases, the dissolved oxygen concentration immediately downstream of the point of discharge decreases. The concentration of dissolved oxygen in a stream is important because if it drops below about 6 mg/l, some types of fish will not survive.

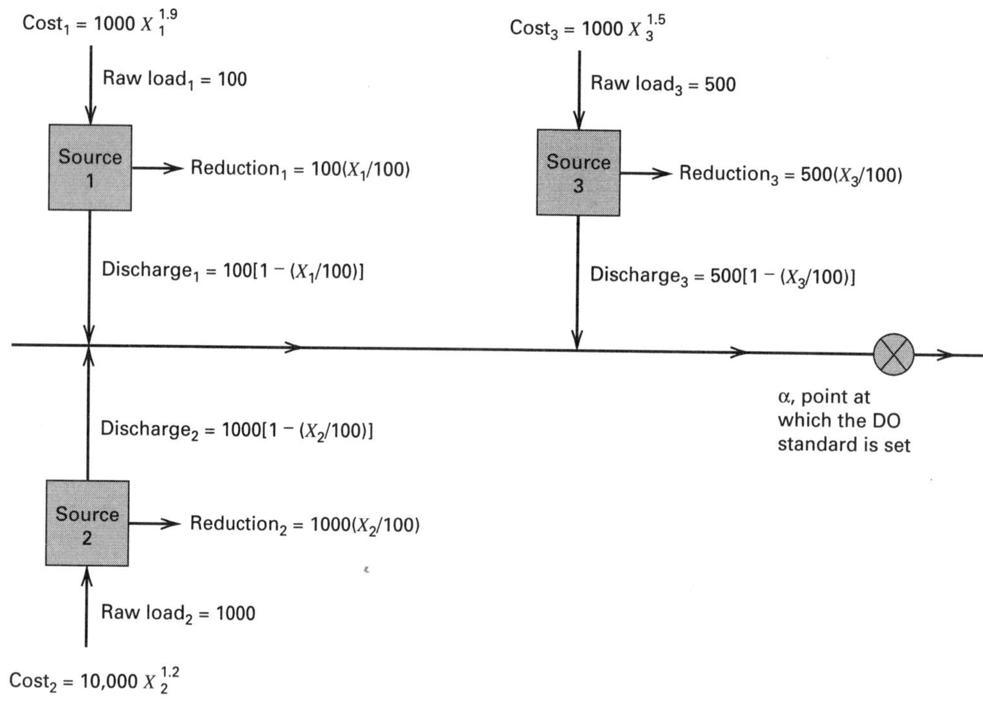

Figure 8.1 The three-waste-source example. Note: Loads, reductions, and discharges are in 1000 lb/day of BOD.

If the DO drops all the way to zero, offensive odors will result from the biodegradation process.

This analysis relies on standard modeling procedures for predicting how much a stream's dissolved oxygen will decrease if a BOD discharge is increased. Mathematical models of BOD and DO are widely used, and these two parameters are employed here to illustrate how effluent standards can be derived to meet ambient standards. Use of only two parameters is a simplification. In real situations, other aspects of water quality, such as turbidity and presence of disease-causing bacteria, are also considered in setting standards.

Assume that an ambient standard for the stream has been set at 6 mg/l of dissolved oxygen under specified low streamflow conditions.[12] Suppose, also, that if no wastes were released into the stream, the dissolved oxygen would be 7 mg/l under these flow conditions. The difference between the DO standard and the DO that would exist in the absence of discharges is 1 mg/l. It represents the portion of the stream's oxygen used to provide waste receptor services.[13]

Figure 8.1 shows the three waste discharges. Sources labeled 1 and 2 are located opposite from each other, and source 3 is farther downstream. The figure also includes a point "α" downstream of source 3. The 6 mg/l DO standard applies only at point α, a location valued for its good fishing. In a real case, standards would be set at many stream locations. However, the single point α is sufficient for illustra-

[12] When setting water quality standards, it is customary to specify how much streamflow is available to dilute wastewater discharges. Typically, periods of low streamflow are selected. If standards can be met when little flow is available for diluting wastes, they will generally be met under other conditions. Velz (1976) describes ways to define low flow conditions.

[13] The notion that the environment provides *waste receptor services* was introduced in the discussion of Table 5.1. The term *assimilative capacity* is sometimes used in this context. For the three waste source example, 1 mg/l of DO is used up in the process of "assimilating" the organic wastewater discharges, and the allowable drop in oxygen sets a limit on the capacity of the stream to assimilate biodegradable organic waste.

tive purposes. Figure 8.1 also shows the *raw loads* of BOD generated at each source. The loads are in units of 1000 lb/day of BOD. Thus, for example, source 1 has a raw load of 100 units, or 100,000 lb/day of BOD.

There are many ways in which the three waste sources can reduce their discharges so that only 1 mg/l of the stream's dissolved oxygen is used up. To analyze the possibilities, let X_1 represent the percentage reduction in the raw BOD load originating at source 1. Define X_2 and X_3 similarly for sources 2 and 3, respectively. Although numerous options exist for waste reduction (for example, materials recycling), this example assumes that only wastewater treatment is used.

Figure 8.1 indicates the BOD reductions and discharges associated with any combination of treatment levels, X_1, X_2, and X_3. It shows, for example, that for source 1, 100 (X_1/100) units are removed by wastewater treatment, and {100 [1 − (X_1/100)]} units are discharged to the stream. These results follow because X_1 is the percent of raw load at source 1 removed by treatment. Enumerating percentage removals, X_1, X_2, and X_3, is equivalent to specifying a set of effluent standards. For any X_i, the corresponding BOD discharge is simply the raw load at source i multiplied by [1 − (X_i/100)]. The corresponding effluent standard requires that the BOD discharge be no greaer than this value.

Costs of Wastewater Treatment

Figure 8.1 also indicates the costs of waste treatment. It shows, for example, that the cost to source 1 of removing X_1 percent of its raw BOD load is 1000 $X_1^{1.9}$. All costs in the figure represent the *present value of costs* to construct and operate treatment facilities. The present value idea is used because the costs to operate facilities occur at different future times.[14]

The total cost of effluent standards requiring removal percentages X_1, X_2, and X_3 is labeled $C_T(X_1, X_2, X_3)$. It is the sum of the costs in Figure 8.1:

$$C_T(X_1,X_2,X_3) = 1000X_1^{1.9} + 10{,}000X_2^{1.2} + 5000X_3^{1.5}. \qquad \textbf{(8-1)}$$

The individual terms in equation 8-1 depart somewhat from typical treatment costs. In general, real costs increase greatly as the percentage reduction approaches 100. This type of increase occurs to some extent for sources 1 and 3. However, the cost expression for source 2 has an exponent close to 1, and it does not show rapidly increasing costs. The costs in this example were chosen intentionally to dramatize the differences between productively efficient standards and those that appear equitable. Such distinctions occur because source 1 has a small quantity of waste that is very expensive to remove. Source 2, which discharges at the same point on the stream, can remove an equivalent quantity of waste for a much lower cost.

Effects of Discharges on Stream Quality

Additional information is required in order to set effluent standards that attain the ambient standard of 6 mg/l of dissolved oxygen. BOD discharges are known to reduce stream DO, but a quantitative description of this effect is needed. The simplest analysis procedures involve *steady-state* conditions in which the streamflow and the waste discharges are all constant. In these circumstances, it is common to assume that a linear relationship exists between waste releases and stream dissolved oxygen at a particular location. For the hypothetical example, ΔDO, the change in DO at point α caused by the BOD discharges, is computed as

$$\Delta\text{DO} = \phi_{1\alpha}\left[100\left(1 - \frac{X_1}{100}\right)\right] + \phi_{2\alpha}\left[1000\left(1 - \frac{X_2}{100}\right)\right] + \phi_{3\alpha}\left[500\left(1 - \frac{X_3}{100}\right)\right]. \qquad \textbf{(8-2)}$$

The expressions in brackets represent the BOD discharges (see Figure 8.1). The value of $\phi_{i\alpha}$, referred to as a *transfer coefficient,* estimates the decrease in DO at point α per unit increase in BOD discharge at source i. The units of $\phi_{i\alpha}$ are milligrams per liter

[14] The present value concept was introduced in Chapter 7. Recall that the concept can be illustrated by considering an ordinary savings account. At 6% interest, it would take only $1.00 today to yield $1.06 one year from now. In this case, $1.06 one year from now has a present value of $1.00.

of DO per 1000 lb/day of BOD. In this example, $\phi_{1\alpha} = \phi_{2\alpha} = 0.002$, and $\phi_{3\alpha} = 0.003$. The values of $\phi_{1\alpha}$ and $\phi_{2\alpha}$ are equal because the effluents from sources 1 and 2 are released at the same stream location. In this example, the value of $\phi_{3\alpha}$ is greater than the other two coefficients because source 3 is closer to point α.[15] The relationship between stream DO and distance from the waste source is such that after a certain distance the stream "regenerates" itself naturally. If no other wastes are added, natural stream reaeration may eventually cause the DO to rise to the concentration that exists upstream of sources 1 and 2.

Before examining alternative effluent standards, consider what the DO at α would be if *no* waste treatment were provided. This is determined using equation 8-2 with appropriate numerical values of $\phi_{i\alpha}$, and with $X_1 = X_2 = X_3 = 0$. Substituting these values in equation 8-2 gives the *change* in DO (in mg/l) as

$$\Delta\text{DO} = (0.002)(100) + (0.002)(1000) + (0.003)(500) = 3.7.$$

With each source emitting its raw load, the stream dissolved oxygen at point α is 3.7 mg/l below the DO value that exists in the absence of any waste loads, 7 mg/l. The difference between the two numbers, 3.3 mg/l, is the resulting DO at point α. This value is significantly lower than the 6 mg/l standard.

If the DO decrease caused by all three effluents is limited to 1 mg/l, the 6 mg/l standard will be met. The DO decrease can be restricted in this way by choosing X_1, X_2, X_3 such that

$$\phi_{1\alpha}\left[100\left(1 - \frac{X_1}{100}\right)\right] + \phi_{2\alpha}\left[1000\left(1 - \frac{X_2}{100}\right)\right] + \phi_{3\alpha}\left[500\left(1 - \frac{X_3}{100}\right)\right] = 1.0. \qquad \textbf{(8-3)}$$

[15] Thomann (1972) presents mass balance analyses for developing the transfer coefficients. The numerical values of $\phi_{i\alpha}$ depend entirely on the particular circumstances. As indicated by Thomann's analysis, it is possible to have a case in which $\phi_{3\alpha} < \phi_{1\alpha}$.

Equation 8-3 sets the ΔDO given by equation 8-2 equal to 1 mg/l. After substituting appropriate values for the transfer coefficients and simplifying, equation 8-3 can be written as

$$0.2X_1 + 2X_2 + 1.5X_3 = 270. \qquad \textbf{(8-4)}$$

Any combination of X_1, X_2, and X_3 that satisfies equation 8-4 can be used to determine effluent standards that meet the 6 mg/l ambient standard. Because many such combinations exist, there are numerous possible sets of effluent standards.

Efficiency and Equity in Setting Effluent Standards

Equal Percent of Waste Removed

One of the simplest approaches to setting standards requires that each source remove the same percentage of its raw load. Using this method, the necessary percentage reduction is found by setting $X_1 = X_2 = X_3 = X$ in equation 8-4 and solving for X. The solution is $X = 73\%$. If each source removes 73% of its raw load, the 6 mg/l standard will be met. The total cost of this equal percentage removal plan is found using equation 8-1.

$$C_T(73,73,73) = 1000(73)^{1.9} + 10{,}000(73)^{1.2} + 5000(73)^{1.5} = \$8{,}310{,}000.$$

Some people argue that an equal percentage removal policy is fair because each waste source is treated in the same way. This position is easily challenged. Dischargers have unequal costs of treatment and are situated at different locations. Consequently, equity is not necessarily attained by imposing an equal percentage reduction requirement.

Inequities in the distribution of costs are demonstrated by considering sources 1 and 2. They are at the *same location,* and a unit of waste from each produces an identical effect on dissolved oxygen. In this sense, the sources appear as equals. Consider, however, what it costs to reduce a unit of BOD at each source when a 73% reduction is imposed. Table 8.5 presents the average per-unit costs: total treatment costs divided by BOD loads removed. The aver-

TABLE 8.5 Standards Based on Equal Percentage Removal Policy

Waste Source	Percent of Waste Removed	BOD Load Removed (10^3 lb/day)	Cost of Treatment ($\$10^3$)	Average Cost per Unit of BOD Load Removed ($\$/10^3$ lb/day)
1	73	73	3470	47.5
2	73	730	1720	2.4
3	73	365	3120	8.5
Total Cost			8310	

age cost of reducing a unit of BOD discharge is much greater for source 1 than for source 2. Is it fair to force them to reduce their raw loads by the same percentage?

Inequities also result because an equal percentage removal policy does not account for differences in the locations of discharges. In this example, the effects of location are represented by the transfer coefficients: $\phi_{1\alpha} = \phi_{2\alpha} = 0.002$, and $\phi_{3\alpha} = 0.003$. A unit of BOD released by source 3 causes 1.5 times as much of a decrease in DO at α as a unit released from the other sources. In these circumstances, would it not be more equitable to have source 3 remove a higher fraction of its raw load?

The most common criticism of the equal percentage removal method is that it fails to utilize society's resources efficiently. Many alternative schemes can meet the 6 mg/l standard at much less than the \$8.31 million cost of the equal percentage reduction arrangement in this example. For instance, a less expensive program can be designed by leaving source 3 at the 73% level and having source 1 treat nothing. To meet the ambient standard, source 2 must remove 73% of its own raw load plus 73% of the raw load at source 1. The percentage of waste removal by source 2 increases to

$$\left[\frac{(0.73)(1000) + (0.73)(100)}{1000}\right] \times 100 = 80.3\%.$$

Using equation 8-1, the total cost of this scheme is

$$C_T(0, 80.3, 73) = 10{,}000(80.3)^{1.2} + 5000(73)^{1.5} = \$5{,}050{,}000.$$

This represents a \$3.26 million cost savings over the equal percentage removal policy. Even greater savings are possible.

Waste Treatment to Minimize Total Costs

Effluent standards can also be based on a *cost-effectiveness criterion:* find the removal percentages that meet the DO target at a minimum total cost. Cost-effectiveness is required if an economy is to be productively efficient.[16] Cost-effective values of X_1, X_2, and X_3 minimize cost while satisfying equation 8-4, the condition requiring a 6 mg/l dissolved oxygen content at point α. In mathematical terms, finding the cost-effective effluent standards is equivalent to finding X_1, X_2, and X_3 to minimize equation 8-1 subject to the requirement represented by equation 8-4.

[16] More generally, cost-effectiveness means a predetermined target or goal is met at minimum cost. A relationship between cost-effectiveness and productive efficiency is established as follows: As noted in Chapter 7, the productive efficiency of an economy is increased if factor inputs are reallocated in a way that increases the net value of goods and services produced. Assume that the social goal of meeting an environmental quality target is *not* met cost-effectively. This means the cost of meeting the environmental goal is not at a minimum. If factor inputs are reallocated to reach the environmental goal at a minimum cost, then some of the inputs originally used to meet the environmental goal will be available to produce other goods and services. Therefore, the original resource allocation (in which the environmental goal was not met cost-effectively) was not productively efficient.

TABLE 8.6 Standards That Minimize Total Cost

Waste Source	Percent of Waste Removed	Percent of Waste Discharged	BOD Load Discharged (10^3 lb/day)	Decrease in DO at α (mg/l)	Cost of Treatment ($\$10^3$)
1	4.12	95.88	95.9	0.192	15
2	100.00	0	0	0	2510
3	46.12	53.88	269	0.808	1566
Total Cost					4091

This cost-minimization problem is solved using standard procedures from calculus.[17] The results are given in Table 8.6.

The cost-effective effluent standards provide about a 50% cost reduction compared to standards in which all sources have the same percentage removal (compare tables 8.5 and 8.6). Because of opportunities for substantial cost savings, one might expect support for the cost-minimizing approach to setting discharge standards. However, support is often low because costs are distributed unevenly. For example, the cost-minimizing standards require source 2 to remove virtually all of its wastes, whereas source 1 removes less than 5%. Since both discharges are at the same stream location, many people would consider this unfair.

Inequities in the distribution of costs can potentially be eliminated by a system of taxes and subsidies. Suppose, for example, it is considered unfair to have source 2 remove its entire BOD load. Instead of disregarding the cost-effective policy, the government could tax sources 1 and 3 and subsidize source 2 so that the final cost distribution is equitable. In this way, a minimum of society's resources are devoted to waste treatment, and the treatment costs are allocated fairly. Although this position seems reasonable, it has not been widely embraced. One reason is the difficulty in getting agreement on what constitutes a fair distribution of costs. Another is that it is often politically infeasible to implement the necessary taxes and subsidies.

Equal Percentage Removal within Zones

Each of the preceding methods for setting effluent standards has shortcomings. The equal percentage removal approach is not productively efficient, and the cost-effective policy often yields an unfair distribution of waste removal costs. The apparent unfairness in the cost-effective scheme results because waste sources are treated differently. Reducing inequity requires answering a hard question: What does it mean to treat polluters in a like manner? If it were possible to define zones such that waste sources in the same zone were alike in all relevant respects, it might be considered fair to require each of the waste sources to remove the same percentage of its raw load. Designing discharge standards based on "equal

[17] One solution procedure begins by solving equation 8-4 for X_3 and back substituting into equation 8-1 to obtain

$$\hat{C}_T(X_1,X_2) = 1000X_1^{1.9} + 10{,}000X_2^{1.2} + 5000(180 - 0.133X_1 - 1.33X_2)^{1.5}.$$

The values of X_1 and X_2 that minimize $\hat{C}_T$ are found by solving $\partial\hat{C}_T/\partial X_1 = 0$, and $\partial\hat{C}_T/\partial X_2 = 0$, simultaneously. This yields $X_2 >$ 100%, which is not physically possible and can be interpreted to mean that source 2 must remove all its wastes if equation 8-1 is to be minimized. Next, the entire optimization problem is reexamined with $X_2 = 100\%$. Equation 8-4 is solved for X_3. Substituting the result into equation 8-1 gives

$$\tilde{C}_T(X_1) = 1000X_1^{1.9} + 10{,}000(100)^{1.2} + 5000(46.7 - 0.133X_1)^{1.5}.$$

The value of X_1 that minimizes $\tilde{C}_T$ is determined by solving $d\tilde{C}_T/dX_1 = 0$. The result, $X_1 = 4.12\%$, together with $X_2 = 100\%$, is used in equation 8-4 to obtain $X_3 = 46.12\%$. The condition assuring that $X_1 = 4.12\%$ provides a minimum (and not a maximum) for $\tilde{C}_T(X_1)$ is

$$\left.\frac{d^2\tilde{C}}{dX_1^2}\right|_{X_1=4.12} > 0.$$

This condition is satisfied.

TABLE 8.7 Standards That Minimize Total Cost with $X_1 = X_2$

Waste Source	Percent of Waste Removed	Percent of Waste Discharged	BOD Load Discharged (10^3 lb/day)	Decrease in DO at α (mg/l)	Cost of Treatment ($\$10^3$)
1	61.3	38.7	38.7	0.077	2490
2	61.3	38.7	387.0	0.774	1400
3	89.9	10.1	50.5	0.152	4260
Total Cost					8150

percentage removal within zones" is based on the (debatable) supposition that a fair outcome results if each waste source in a zone removes the same percentage of its raw load.[18]

Finding effluent standards based on equal percentage removal within zones requires answering another difficult question: How should zones be designated? For illustrative purposes, suppose the three waste sources in the example are placed in two zones: sources 1 and 2 are in one zone, and source 3 is in the other.[19] This zoning recognizes that a unit of BOD from either source 1 or source 2 has the same effect on stream dissolved oxygen. It also recognizes that source 3 should not necessarily remove the same percentage of waste as the others, since a unit of BOD from source 3 causes a greater decrease in DO at point α.

A zoned equal percentage removal program that makes economical use of resources keeps the total cost to a minimum while requiring sources 1 and 2 to reduce their raw loads by the same fraction. Standards based on this approach are determined by finding X_1, X_2, and X_3 to minimize total cost (equation 8-1), while satisfying the conditions that $X_1 = X_2$, and that DO at point α equals 6 mg/l. This latter condition is represented by equation 8-4. The values of X_1, X_2, and X_3 are found using standard procedures from calculus and are shown in Table 8.7.[20]

The characteristics of this example are such that the *zoned* equal percentage removal requirements cost almost as much as the program in which all sources remove 73% of their loads (compare tables 8.5 and 8.7). The high total cost for the zoned scheme results because source 1 has very high average treatment costs, and both sources 1 and 2 must make the same BOD reductions. In other circumstances, the use of zones might be much more economical than the equal percentage removal policy. Imagine a case in which some sources are so far from the point at which an ambient standard is imposed that their discharges have little influence on quality at that point. A zoned equal percentage removal scheme would mandate low levels of treatment for such sources. Consequently, the total cost of meeting the ambient

[18] The supposition is debatable because dischargers in the same zone will typically have unequal costs of waste treatment. If treatment costs are unequal, waste sources in the same zone are not the same in all relevant respects. Because the polluters are not alike, asking them to remove the same fraction of their waste loads is not necessarily fair.

[19] In designating zones, two polar cases can be identified: put all waste sources in one zone, or put each source in its own zone. If all three sources are put into a single zone, the outcome is the equal percentage removal approach. If each source has its own zone, the cost-effective scheme is obtained. Intermediate policies are derived by defining zones that differ from these two extremes.

[20] The solution procedure solves for X_3 using equation 8-4. This result, together with the constraint that $X_1 = X_2$ is substituted into equation 8-1 to yield

$$\overline{C}_T(X_1) = 1000X_1^{1.9} + 10{,}000X_1^{1.2} + 5000(180 - 1.47X_1)^{1.5}$$

The cost-minimizing value of X_1 is found by solving $d\overline{C}/dX_1 = 0$. This value, $X_1 = 61.3\%$, also satisfies the condition required for X_1 to be a minimum:

$$\left.\frac{d^2\overline{C}}{dX_1^2}\right|_{X_1=61.3} > 0.$$

Using $X_1 = X_2 = 61.3\%$ in equation 8-4 yields $X_3 = 89.9\%$.

standard could be much less than when all sources reduce their raw loads by the same proportion.

The zoned equal percentage removal approach may not eliminate all unevenness in the distribution of waste reduction costs. In the hypothetical example, apparent inequities remain even after putting sources 1 and 2 into the same zone. The average cost per unit of BOD removed for source 1 is many times higher than the corresponding cost for source 2.

Zones need not be delineated on a geographical basis. To provide a more equitable outcome, zones may be defined to include waste dischargers of the same type. Suppose, for example, sources 1 and 3 are municipalities and source 2 is a petroleum refinery. It might be argued that sources 1 and 3 should remove the same fraction of their raw loads because both are municipalities and their treatment costs should thus be similar. This approach to zoning does not eliminate all apparent inequities. Because source 3 has a greater impact on DO at point α than source 1, requiring that $X_1 = X_3$ might not be considered fair.

The three-waste-source example demonstrates that no one set of effluent standards will necessarily be productively efficient and provide a fair distribution of costs as well. Compromises between equity and efficiency must be made.

BOD LOAD ALLOCATION FOR THE DELAWARE ESTUARY

Although the preceding example is hypothetical, it has several features in common with an effort to determine effluent standards on the Delaware River estuary during the late 1960s. This effort was led by the Delaware Estuary Comprehensive Study (DECS), a part of what was then the Federal Water Pollution Control Administration.[21] The previous discussion of the waste load allocation problem is extended by examining selected aspects of the DECS analysis.

As in the three-waste-source example, the Delaware Estuary Comprehensive Study focused on BOD and DO. Forty-four sources of BOD were considered. Collectively, the sources accounted for about 95% of the biochemical oxygen demand discharged to the estuary. The DECS obtained estimates of how much it would cost to remove different percentages of the BOD discharge at each source. The costs were represented using straight-line segments as in Figure 8.2. The DECS was thus able to express the total cost of any BOD reduction program as a set of linear equations.

In addition to cost information, the DECS also determined transfer coefficients similar to the ones used previously in this chapter. To do this, the 86-mile Delaware Estuary was divided into 30 sections, each 10,000 or 20,000 ft in length (see Figure 8.3). For each section, two equations representing the law of conservation of mass were written, one for dissolved oxygen and one for biochemical oxygen demand. Steady-state conditions were assumed. The resulting 60 equations were solved simultaneously to yield the transfer coefficients, which provided estimates of how much a decrease in BOD input to any one section of the estuary would increase the DO in each of the 30 sections. Hundreds of transfer coefficients were needed since the ambient quality stan-

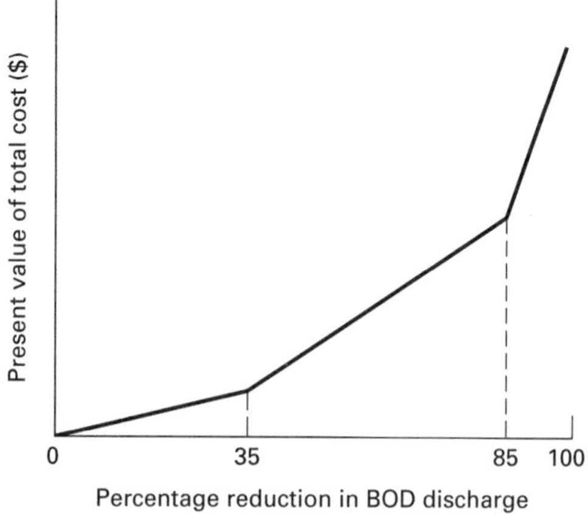

Figure 8.2 Costs of BOD reduction as straight-line segments.

[21] Ackerman et al. (1974) discuss the DECS analyses and how they influenced ambient and effluent standards on the Delaware Estuary. Thomann (1972) and Thomann and Mueller (1987) provide detailed information on the mathematical and engineering aspects of the DECS investigation.

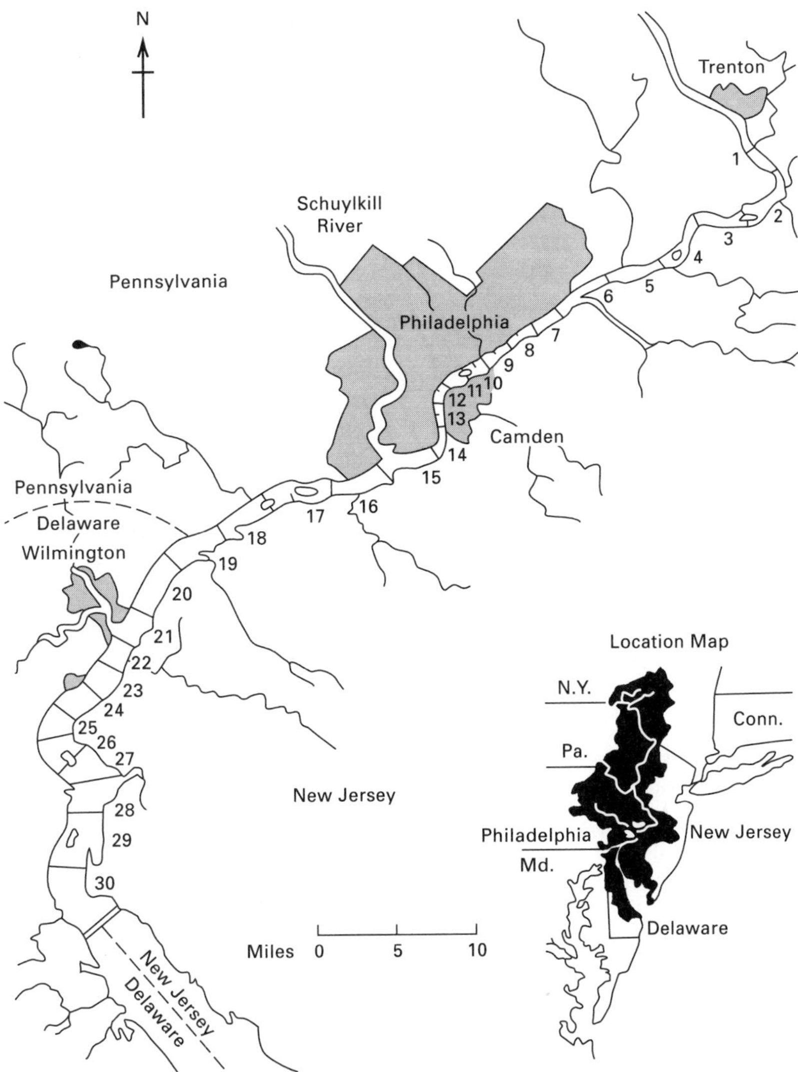

Figure 8.3 Sections used in the DECS analysis of the Delaware Estuary. From *Principles of Surface Water Quality Modeling and Control* by Robert V. Thomann and John A. Mueller. Copyright © 1987 by Addison Wesley Educational Publishers, Inc. Reprinted by permission of Addison Wesley Educational Publishers, Inc.

dards were set for 30 sections, not just a single point. In addition, the influence of BOD releases on DO both downstream *and* upstream of the discharge points was taken into account. Estuaries are coastal waters in which both tides and freshwater flows are present. The tidal influence causes wastes to be transported upstream as well as downstream of the discharge locations.

The Delaware Estuary case differed from the three-waste-source example in which an ambient standard had already been established. In the DECS study, the goal was to provide decision makers with

TABLE 8.8 Alternative Ambient DO Standards used in the Delaware Estuary Analysis[a]

Objective Set Number	Selected Locations (Section Numbers)				
	Trenton (1)	**Philadelphia (14)**	**Chester (18)**	**Wilmington (21)**	**Liston Point (30)**
1	6.5	4.5	5.5	6.5	7.5
2	5.5	4.0	4.0	5.0	6.5
3	5.5	3.0	3.0	3.0	6.5
4	4.0	2.5	2.5	2.5	5.5
5	7.0	1.0	1.0	4.0	7.1

[a] The DO standards, which are for summer average conditions, are given in the table in terms of milligrams per liter. Based on Kneese and Bower (1968, p. 227).

information that would help them select ambient standards. Key information concerned how much it would cost to meet different targets for ambient concentrations of DO.[22]

The Delaware Estuary Comprehensive Study examined five sets of ambient standards. Each was referred to by the DECS as an "objective set." Table 8.8 displays portions of the objective sets for dissolved oxygen in several estuary sections. Objective set 5 represents the 1964 conditions in the estuary. A complete description of the objective sets includes all 30 estuary sections, as well as limits on other variables such as chlorides, turbidity, and pH. Although many water quality indicators were considered by the DECS, dissolved oxygen received the most attention.

Alternative effluent standards that would attain the water quality levels represented by each objective set were determined. The DECS investigated three strategies for allocating BOD discharges: equal percentage removal, minimum cost (or cost-effectiveness), and equal percentage removal within zones. In essence, the analysis used in the three-waste-source example was conducted five times, once for each of the five objective sets.

An effluent limitation for waste source i was represented by X_i, the percentage removal of BOD load by that source. For a set of effluent standards to meet the DO requirements of a particular objective set, the values of $X_1, X_2, \ldots, X_{44}$ had to satisfy various *water quality constraints*. For any section "β" in the estuary, the water quality constraint took the form

$$\phi_{1\beta}W_1\left(\frac{X_1}{100}\right) + \phi_{2\beta}W_2\left(\frac{X_2}{100}\right) + \cdots + \phi_{44\beta}\left(\frac{X_{44}}{100}\right) \geq K_\beta \qquad \text{(8-5)}$$

where,

W_i = total current BOD load discharged by source i,

$\phi_{i\beta}$ = increase in DO in section β resulting from a 1-unit decrease in BOD load discharged by source i, and

K_β = DO increase in section β required to meet the dissolved oxygen standard.

The left side of expression 8-5 is analogous to equation 8-2. For any source i, W_i multiplied by $(X_i/100)$ is the cutback in the existing BOD discharge associated with an X_i percentage load removal. The discharge reduction, $W_i(X_i/100)$, multiplied by the transfer coefficient $\phi_{i\beta}$, indicates the consequent increase in DO in section β. The left side of expression 8-5 aggregates the DO increase in section β caused by each of the load reductions. The inequality assures that the total improvement in dissolved oxygen in

[22] Benefits associated with each of several sets of ambient standards were considered, but most benefits could not be measured in monetary terms. The final selection of ambient standards for the Delaware Estuary reflected the various costs and gains involved, but a rigorous benefit-cost analysis was not used.

section β caused by $X_1, X_2, \ldots, X_{44}$ is at least as great as the improvement required to satisfy the DO requirements of a particular objective set.

The equal percentage removal program for setting effluent standards was determined by trial and error. First a percentage was picked. Then water quality constraints of the type shown in expression 8-5 were checked to see if the DO standards in each of the 30 sections were attained. If the standards were violated in one or more sections, the percentage removal was increased by a small amount, and the water quality constraints were checked again. If the standards were met throughout, then the percentage was decreased by a small amount and the constraints were rechecked. This procedure yielded the single percentage removal that just met the dissolved oxygen requirements.

To find the cost minimizing values of $X_1, X_2, \ldots, X_{44}$, it was necessary to sum the straight-line segments representing BOD reduction costs for each source (see Figure 8.2). The resulting total cost equation was a sum of terms that were *linear* with respect to removal percentages. In other words, each percentage removal was raised to the first power only. The water quality constraints consisted of 30 linear inequalities, one for each section; expression 8-5 illustrates a typical constraint. Because all the equations and inequalities consisted of terms that were linear with respect to percentage removals, the cost-minimizing values of $X_1, X_2, \ldots, X_{44}$ could be calculated using standard mathematical techniques.[23] The cost-minimization problem was easily solved by the DECS.

The third approach to setting effluent standards, the equal percentage removal within zones policy, involved a more sophisticated cost-minimization exercise. To find the appropriate values of $X_1, X_2, \ldots, X_{44}$, the cost-minimization problem was expanded to require all of the sources in any particular zone to remove the same percentage of their discharges. There are many ways to group 44 sources into zones, and there is little theoretical basis for arguing that one grouping makes more sense than another. Several zoning schemes were examined by the DECS. Although the requirements for equal percentage removal within zones increased the mathematical complexity of the cost-minimization problem, the DECS staff devised a special procedure to solve it.

A portion of the results from the Delaware Estuary Comprehensive Study is displayed in Table 8.9. Costs shown include both construction costs and operation and maintenance costs and reflect waste loads that were anticipated for the 1975-80 period. The minimum-cost programs are generally much less expensive than the equal percentage removal schemes. The costs of programs involving equal percentage removal within zones depend on how the zones are defined. The costs in the last column of Table 8.9 are for the zones finally used by the DECS.

The outcome of the standard-setting process for the Delaware Estuary was the adoption of a set of ambient standards representing a compromise between objective sets 2 and 3. The regulatory approach to implement this water quality standard was designed by the Delaware River Basin Commission (DRBC), an interstate agency with broad powers and responsibilities. Their scheme involved four zones, with each waste source within any one zone removing the same percentage of its raw BOD load. These percentages ranged from 86.0 to 89.25.

Aspects of the final outcome illustrate some of the difficulties in implementing effluent standards. Consider the problem of setting limits for the BOD discharges from two petroleum refineries in the same zone. Suppose that at the time of implementing the equal percentage removal within zones policy, one refinery had already implemented costly waste treatment while a neighboring refinery was discharging all of its wastes without treatment. It would be unfair to require each refinery to reduce its existing discharge by the same percentage. To deal with this, the DRBC determined "hypothetical raw loads" on a case-by-case basis. The hypothetical loads were used by DRBC to determine numerical limits ("quotas") on the BOD that could be released by each source. Dischargers often argued that the quotas initially set by DRBC were too low. The final quotas resulted after a series of bargaining sessions involving the dischargers and DRBC. Baumol and Oates (1979, p. 236) observe:

[23] In addition to the water quality constraints, it was necessary to restrict each X_i to be between 0 and 100%. The resulting cost-minimization problem took the form of a "linear program." Such problems are solvable using well-established procedures. A complete formulation of the linear programming problem to find the minimum cost set of $X_1, X_2, \ldots, X_{44}$ is given by Thomann (1972).

TABLE 8.9 Total Costs of Achieving Objective Sets for the Delaware Estuary (in millions of dollars)[a]

	Strategy		
Objective Set	**Equal Percentage Removal**	**Minimum Cost**	**Equal Percentage Removal within Zones**
1[b]	460	460	460
2	315	215	250
3	155	85	120
4	130	65	80

[a] Based on Kneese and Bower (1968, p. 230). All costs are *present values* reported in 1968 dollars.

[b] Cost figures for objective set 1 are the same for each of the three strategies. The high level of DO associated with this objective set required very high levels of treatment and the use of mechanical devices to reoxygenate the estuary artificially.

> The anomalies in the final set of permits are quite striking: for example, the final pollution quotas for petroleum refineries on the estuary ranged from 692 pounds to 14,400 pounds of BOD per day!

The wide variations in the final quotas reflect the difficulties in developing effluent standards that are equitable, especially where judgments must be made to identify hypothetical raw loads.

Another problem in implementing the effluent requirements on the Delaware Estuary arose over challenges to the validity of the transfer coefficients. The DECS investigation assumed that these coefficients provided a suitable basis for estimating how much dissolved oxygen would improve if particular BOD loads were reduced. Several dischargers argued that the coefficients greatly oversimplified the complex functioning of the estuary. They felt that precise quantitative predictions of how dissolved oxygen would change could not be made with available knowledge of estuarine behavior. These dischargers objected to having costly BOD release quotas imposed on the basis of what they viewed as unreliable and overly simple procedures for predicting changes in water quality.

The dispute over the validity of the transfer coefficients illustrates a general problem encountered in setting discharge standards to meet an ambient standard. This method of deriving discharge standards requires an explicit judgment regarding how much ambient quality would improve if the proposed waste reductions were implemented. As previously noted, these judgments are often challenged.

This discussion of the DECS investigation demonstrates how analysts can help decision makers identify trade-offs between equity and productive efficiency, and it introduces some of the practical problems that come up in selecting ambient and discharge standards. The next chapter expands on implementation issues by examining the actions environmental agencies can take when polluters fail to meet environmental requirements.

Key Concepts and Terms

Types of Environmental Requirements

Ambient standards
Effluent and emission standards
Discharge permits
Technology-based standards
Technology-forcing standards
Technology vs. performance standards
Best management practices
Information disclosure requirements

The Waste Load Allocation Problem

Present value of costs
Biochemical oxygen demand
Dissolved oxygen
BOD-DO models
Transfer coefficients
Cost-effectiveness criterion
Productive efficiency
Equal percentage removal

Zoned equal percentage removal
Equity in distributing costs

BOD Load Allocation for the Delaware Estuary
Water quality constraints
Equity-efficiency tradeoffs
Bases for setting ambient standards
Hypothetical raw loads

Discussion Questions

8-1 What criteria would you use to compare alternative types of requirements for meeting an ambient environmental quality target? Examples of criteria include flexibility for regulatees, ease of monitoring, and influence on private sector efforts to investigate new waste reduction technologies.

8-2 Suppose you are on the policy analysis staff of an environmental agency, and you are asked to compare technology standards and performance standards in regulating stationary sources of air pollution to meet a legislatively mandated ambient air quality standard. List criteria you would use to evaluate technology standards and performance standards, and carry out a comparative analysis.

8-3 Effluent standards can be set on the bases of (a) ambient water quality standards, (b) available waste reduction technologies, or (c) a combination of the two. Analyze the advantages and disadvantages of each approach.

8-4 A regional air quality management agency imposes lax regulations on *existing* sources, but it is considering use of stringent performance standards for all *new* stationary air pollution sources that discharge more than 100 tons of sulfur dioxide per year. New sources discharging less than 100 tons per year of SO_2 would face requirements moderately more demanding than those faced by existing sources. Your company owns an existing production facility that discharges 300 tons per year of SO_2, and it is considering opening up a new, more efficient facility with 200 tons per year discharge and closing down the existing plant. Assume that the costs of SO_2 reduction are a significant factor in any decisions the company makes. What options does the company have to avoid the stringent new standards? How would these options affect ambient air quality?

8-5 Of all the different ways of regulating major sources of air pollution (see Table 8.1), which provides the greatest incentive for firms to do research on treatment technologies or production process changes that reduce waste loads?

8-6 Consider a situation in which air pollution is emitted from two firms that are identical in every respect except for location relative to an air quality control point. Firm 1 is closer. The differences in location are reflected in transfer coefficients: $\phi_{i\alpha}$ = change in concentration at air quality control point α per unit change in emission from plant i (i = 1, 2). In this case, $\phi_{1\alpha} = 1.0$ and $\phi_{2\alpha} = 0.5$. The units of $\phi_{i\alpha}$ are mg/l of concentration per 1000 lb/day of discharge. The total cost of emission reduction for *reach* firm is given in the table below. The maximum possible emission from each firm is 7, and control technologies (in this example) are such that only reductions of 1, 2, . . . , 7 are possible (i.e., non-integer values are not possible). Assume that the ambient air quality standard calls for a maximum possible concentration of 3 mg/l at point α. The concentration when there is no air pollution from these two sources is zero.

a. Determine the least costly air pollution control program that satisfies the additional con-

Units of emission reduced (1000 lb/day)	0	1	2	3	4	5	6	7
Total annual cost (in thousands of $)	0	1	3	6	10	16	23	31

dition that both firms remove the same percentage of their uncontrolled discharge.

b. The program in part *a* is not productively efficient. Find the minimum cost control program. (Hint: Since $\phi_{1\alpha} > \phi_{2\alpha}$, calculate the costs of control programs in which firm 1 reduces more of its wastes than firm 2.)

8-7 Consider the information from the three-waste-source example summarized in the table below. Which of these three programs would you advocate and why? Your response should consider both equity and productive efficiency.

8-8 Assume you are the head of a basinwide water quality management agency, and your staff urges you to authorize a study like the one illustrated by the three-waste-source example to determine effluent standards for major sources of pollution in the basin. What questions would you ask of your staff before you made a decision on whether or not to authorize the study?

8-9 Suppose you are a consultant to a company that has just received word that it will be required to reduce its effluent discharge by 80%. The requirement is based on a cost-minimization study like the one in the three-waste-source example. How would you advise your client to proceed?

	Minimum Cost (% reduction)	**Equal Percentage Removal (% reduction)**	**Zoned Equal Percentage Removal (% reduction)**
Waste Source			
1	4	73	61
2	100	73	61
3	46	73	90
Total Cost ($\$10^6$)	4.091	8.31	8.15

References

Ackerman, B. A., S. R. Ackerman, J. W. Sawyer, Jr., and D. W. Henderson. 1974. *The Uncertain Search for Environmental Quality.* New York: Free Press.

Baumol, W. J., and W. E. Oates. 1979. *Economics, Environmental Policy and the Quality of Life.* Englewood Cliffs, NJ: Prentice–Hall.

Flachsbart, P. G., and J. W. Morrow. 1987. "The Development of Hawaii's H_2S Standards and Geothermal Regulations." Paper presented at the 80th Annual Meeting of the Air Pollution Control Association, (June 21-26), New York, NY.

Garwin, R. L. 1978. "Comments on Government Policies Toward Automotive Emissions Control," In *Approaches to Controlling Air Pollution,* ed. A. F. Friedlaender. Cambridge, MA: MIT Press.

Hunter, S., and R. W. Waterman. 1996. *Enforcing the Law: the Case of the Clean Water Acts.* Armonic, NY: M. E. Sharp.

Kneese, A. V., and B. T. Bower. 1968. *Managing Water Quality: Economics, Technology, Institutions.* Baltimore, MD: Johns Hopkins University Press.

Mills, E. S., and L. J. White. 1978. "Government Policies toward Automotive Emissions Control." In *Approaches to Controlling Air Pollution,* ed. A. F. Friedlaender. Cambridge, MA: MIT Press.

Quarles, J. 1976. *Cleaning up America: An Insider's View of the Environmental Protection Agency.* Boston: Houghton–Mifflin.

Russell, C. S., W. Harrington, and W. J. Vaughan. 1986. *Enforcing Pollution Control Laws.* Washington, DC: Resources for the Future.

Thomann, R. V. 1972. *Systems Analysis and Water Quality Management.* New York: Environmental Science Services Division. Reissued by McGraw–Hill, New York.

Thomann, R. V., and J. A. Mueller. 1987. *Principles of Surface Water Quality Modeling and Control.* New York: Harper & Row.

Velz, C. J. 1976. "Stream Analysis—Forecasting Waste Assimilation Capacity." in *Handbook of Water Resources and Pollution Control,* ed. H. W. Gehm and J. I. Bregman, 216–61. New York: Van Nostrand Reinhold.

Chapter 9

Implementing Environmental Regulations

Many environmental regulatory schemes that appear to be sound are not implemented effectively. Even when environmental agencies make enforcement efforts, regulatees may find ways around requirements, and outcomes may be different from those anticipated in policy design. What are the barriers to effective implementation of environmental regulations? And what steps can agencies take to overcome those barriers?

This chapter examines environmental policy implementation problems by evaluating policy outcomes in real settings. The discussion begins with an analysis of pollution control regulations in the People's Republic of China. The chapter then turns to two key implementation issues: monitoring performance and enforcing regulations. The Chinese policy implementation strategy, which rests on cooperation between environmental agencies and polluters, is contrasted with the more coercive approaches used by environmental agencies in the United States. The chapter ends by analyzing factors that influence whether or not polluters comply with regulations.

ENVIRONMENTAL POLICY IMPLEMENTATION IN CHINA[1]

The Chinese regulatory approach is used for illustrative purposes because it contains programs not commonly found elsewhere.[2] In addition, the institutional setting for environmental management is unusual in that most Chinese agencies administering national environmental programs are units of local government. In most countries, national, state, or provincial

[1] This section benefited greatly from my many conversations with three colleagues: Xiaoying Ma, Barbara Sinkule, and Kimberly Warren.

[2] More typical approaches to regulating industrial waste are outlined in chapters 12 and 13, which describe water and air quality management programs in the United States.

authorities carry out national environmental programs.[3]

China's Environmental Protection Programs

China had attempted to control industrial waste discharges using environmental standards before the 1970s, but those standards were often ignored with impunity.[4] After the initiation of Deng Xiaoping's political and economic reforms in 1978, the stage was set for major environmental reforms.

The 1979 Environmental Protection Law of the People's Republic of China authorized the development of standards, and ambient air and water quality standards were issued soon after the law was passed. In addition, a national system of effluent and emission standards was established. Local jurisdictions were granted authority to issue discharge standards that were more stringent than those established for the nation as a whole.

The effluent and emission standards contained a fundamental flaw: only concentration of pollutants was regulated, not *mass flowrate,* the total mass of pollutant discharged per unit time. For example, consider the effluent standard in Guangdong Province for *chemical oxygen demand* (COD), a parameter widely used to characterize the amount of organic matter in liquid waste. Wastewater discharge standards in the province required that the maximum concentration of a factory's COD be less than 110 milligrams per liter.[5] The quality of water receiving the waste discharge depends not only on the concentration of COD discharges, but also on the total mass of COD per unit time. If a factory had a high concentration of COD in its wastewater, it could meet the discharge standard without reducing its mass of COD. The factory could simply add unpolluted water to dilute its wastewater. In many cases, the receiving water would not be much improved by adding dilution water. As will be described, Chinese authorities began to remedy this deficiency in standard setting in 1989.

The 1979 Environmental Protection Law also established three industrial pollution control programs: environmental impact assessment; effluent and emission discharge fees; and the *three synchronizations,* a requirement that the design, construction, and operation of new factories be accompanied by the simultaneous design, construction, and operation of waste management systems. As summarized in Table 9.1, pollution control efforts were expanded during the 1980s, and the 1979 Environmental Protection Law was amended in 1989. The programs introduced during the 1980s are sometimes referred to in China as the "five new programs"; they supplemented the "three old programs" introduced in 1979.

Only one of the five new programs has the appearance of a traditional pollution control policy: the *discharge permit system* (DPS). The DPS is similar to permit schemes commonly used in the United States and elsewhere. The system requires waste dischargers above a certain size to obtain a permit before releasing pollutants. A permit spells out conditions (including maximum pollutant concentrations and mass flowrates) that define legally allowable discharges. Because permits restrict mass flowrates, factories are unable to meet requirements by adding

TABLE 9.1 Pollution Control Programs in China (as of 1993)

The "Three Old Programs"—established by the 1979 Environmental Protection Law

- Environmental impact assessment
- Pollutant discharge fees
- The three synchronizations

The "Five New Programs"—established in the 1980s

- Discharge permit system
- Environmental responsibility system
- Annual assessment of environmental quality in cities
- Limited time treatment orders
- Centralized pollution control

---

[3] The discussion of the U.S. Environmental Protection Agency in chapter 3 illustrates a more common implementation setting, in which federal and state agencies play the main roles in implementing regulations.

[4] This observation is based on Ross (1988, p. 154), who provides an introduction to China's approach to pollution control.

[5] For technical details on COD, see Masters (1991).

water to dilute their wastewater. The DPS constitutes a system of effluent and emission standards promulgated on a source-by-source basis.[6]

The DPS regulations include provisions requiring industries to obtain equipment to monitor the quantity and quality of wastes generated by individual unit processes within an industrial facility. Before the DPS was implemented, many Chinese factories lacked information on how much their individual unit processes and production lines contributed to their total waste loads. The requirement to monitor internal operations helps factory managers identify opportunities for *cleaner production;* for example, by changing production processes to reduce the amount of waste generated. Chinese environmental officials see cleaner production as often being cost-effective and substantially less expensive than waste treatment. (In China and many other countries, the term "cleaner production" is used for what is called "pollution prevention" in the United States.)

In contrast to the discharge permit system, which has counterparts in many countries, the *environmental responsibility system* is uniquely Chinese. Under this system, territorial officials (for example, city mayors), government department heads, and factory directors sign environmental responsibility agreements (or contracts). When leaders sign contracts they commit to meeting specific environmental improvement targets for work under their control. The responsibility system is important because it makes attainment of specific environmental goals a formal part of job performance evaluation for many Chinese leaders.

Another management system that is uniquely Chinese is the *annual assessment of environmental quality in Chinese cities.* Under this system, China's National Environmental Protection Agency conducts an annual, quantitative assessment of environmental quality in major Chinese cities. Parameters used in the assessment include traditional measures of air and water quality and variables that characterize urban infrastructure (for example, the percentage of families provided with trash removal services). Annual assessments are released to the media, and the data have "resulted in acclaim or ridicule for local government officials."[7] Publicity accompanying the annual assessments has caused some government officials to promote environmental quality improvements. No mayor wants to see his or her city at the bottom of the annual assessment rankings.

The last two of the five new programs in Table 9.1 are not programs in the usual sense. *Limited time treatment orders* essentially give environmental officials the ability to order polluters to control their wastes by a set deadline or risk the imposition of significant penalties. Local officials give the order for a target waste cutback and leave it to the polluter to implement needed waste management systems by a fixed date. In some cases, environmental agencies work closely with factories receiving orders to solve pollution problems. Failure to meet the conditions of limited time treatment orders can have serious consequences: fines can be imposed and factories can be shut down until orders are met.

The last item in the table, *centralized pollution control,* is a requirement for cities, countries, and other political entities to actively promote centralized waste management systems. Application of the program in Anyang City in Henan Province is illustrative. The local environmental protection unit in Anyang has created a "Wastewater Management and Treatment Station" to construct and operate centralized treatment works. Facilities include interceptor sewers that collect wastewater discharges from many small factories, and a centralized plant that treats the wastewater.

Institutions for Environmental Management

The 1979 Environmental Protection Law authorized establishment of a complex network of agencies to administer China's new pollution control programs. China's National Environmental Protection Agency provides guidance and regulations for implementing environmental programs authorized by national-level mandates. However, the national programs are carried out primarily by environmental protection bureaus (EPBs), which are units of lower levels of government such as provinces, cities, and counties.

Each EPB is part of a "functional line" which has the National Environmental Protection Agency

[6] During the early 1990s, the DPS was being implemented on a trial basis in more than 100 cities; a final DPS policy is expected to be issued in the late 1990s.

[7] This quote is from Wang and Blomquist (1993, p. 72).

at the top. This functional line is a vertical hierarchy that descends from the national level to provincial EPBs and, eventually, down to city and county EPBs. Consider, for example, the situation in Shenzhen, a city (in Guangdong Province) a city near Hong Kong. The Shenzhen EPB reports vertically upward to the EPB of Guangdong Province, which, in turn, reports to the National Environmental Protection Agency.

Each EPB is also an agency of local government and reports to a local government head.[8] For example, the Shenzhen EPB is a bureau of the Shenzhen city government and is responsible to the mayor of Shenzhen. The mayor's office funds the EPB, and the mayor has influence over the hiring and firing of the head of the EPB. Each environmental protection bureau must work with other agencies reporting to the same local leader. Thus the Shenzhen EPB works with the Shenzhen Economic Commission, the Shenzhen City Industry Office, and other bureaus within the Shenzhen city government.

A local EPB director reports in two directions: vertically up the functional line with the National Environmental Protection Agency at the top, and laterally to the leader of a local government. This dual set of responsibilities and allegiances affects the way EPBs operate, and it makes EPBs fundamentally different from agencies that administer environmental programs in other countries.[9] Because they are units of local governments, EPBs must tailor their pollution control strategies to the economic development goals of their local governments. It would be irrational to do otherwise since the local government leader (for example, the mayor of Shenzhen) influences both the EPB's budget and the cooperation it receives from other government agencies. An EPB would have difficulty enforcing pollution control requirements without the support of its local government leader.

The provisional status of the 1979 Environmental Protection Law impeded the ability of EPBs to implement it. When it was set up, the 1979 law was designated for "trial implementation." Although trial implementation of new laws is common in China, the trial status of the 1979 law worked against the EPBs.[10] Because of the trial status, some factories viewed compliance with the law as a matter of convenience rather than legal obligation.[11] This changed in 1989, when the 1979 Environmental Protection Law was amended and its trial status was removed.

The 1979 Environmental Protection Law contained no provisions diminishing the importance of production targets in China. Thus, whenever there was a conflict between meeting a factory's output targets and environmental protection goals, factory directors tended to ignore the environmental goals. Factory leaders felt that meeting output goals meant good performance evaluations, whereas not meeting pollution control targets would be inconsequential. The situation started to change in 1982, when, for the first time, environmental protection was included as a full chapter in the Sixth Five Year Plan 1981–85. Including environmental protection in the plan assured that assets would be "allocated to environmental protection," and there would be increased "pressure on economic ministries and enterprises to comply with regulations that function as production constraints."[12]

Environmental Responsibility System

The status of environmental protection goals in the eyes of factory directors was elevated significantly by the environmental responsibility system, which was introduced in the 1989 amendments to the Environmental Protection Law. Wang and Blomquist (1993, pp. 61–62) describe the system as follows:

[8] In the Chinese administrative system, the term "local" is interpreted broadly to include provinces, cities and so forth. Local government is distinguished from national government.

[9] In large Chinese cities such as Beijing, the local EPB implements some aspects of national environmental laws jointly with the Environmental Protection Departments (EPDs) of local "industry bureaus," local offices of national-level industry bureaus and ministries. For example, the Chemical Industry Bureau's area office in Beijing is a unit of the Beijing municipal government, and it has a cadre of environmental specialists in its EPD. When the Beijing EPB regulates chemical factories, it works closely with the EPD from the Beijing municipal government's Chemical Industry Bureau.

[10] In 1978, Deng Xiaoping explained why trial implementations were called for: "There is a lot of legislative work to do, and we don't have enough trained people. Therefore, legal provisions will have to be less than perfect to start with, then be gradually improved upon. . . . In short, it is better to have some laws than none, and it is better to have them sooner than later." (Deng Xiaoping as quoted by Ross 1988, p. 140.)

[11] This observation is based on Ross (1988, p. 159).

[12] Quoted material is from Ross (1988, p. 160).

> [L]ocal governments at all levels shall be responsible for the environmental quality within their jurisdictions and shall take measures to improve environmental quality. Thus, individuals such as provincial governors, city mayors, county heads, departmental leaders of local government, and managers of enterprises and other entities assume responsibility for environmental objectives by signing an agreement of environmental protection responsibility. Each agreement sets forth various environmental protection objectives that the entity must satisfy within a given time.

Environmental responsibility agreements are made between an entity and the body that ordinarily oversees its performance. Thus a factory director whose performance is evaluated by a local industry bureau would have the terms of evaluation expanded to include specific environmental targets (for example, reduce COD discharge by 10% by a fixed date). Terms of the agreement are negotiated, and they reflect economic and social factors as well as environmental concerns. Agreements may also contain provisions for incentives (such as monetary rewards for superior performance) and sanctions including fines. Wang and Blomquist (1993, p. 63) describe examples demonstrating that the "reputation of an enterprise, its profit and the welfare of its employees are all at stake under this system."

The environmental responsibility system does not apply only to enterprises. Agreements are also made between governing units. Consider, for example, a city mayor who signs an agreement detailing environmental improvement objectives with the provincial governor. In attempting to meet these objectives, the mayor depends on both city bureaus and the local enterprises under his authority.[13] Each industrial bureau within city government typically has supervisory authority over a number of enterprises within a city. The city's EPB plays a central role, because the EPB helps the mayor select environmental improvement objectives and imposes requirements on enterprises that affect whether or not the mayor's environmental objectives are met. The mayor has a vested interest in ensuring that both city bureaus and local enterprises cooperate with the EPB so that the environmental improvement targets are met.

The Three Synchronizations

Environmental regulation in China proceeds differently than in other places because of the unique institutional context in which regulation takes place. Regulation also proceeds differently because there is a much greater reliance on bargaining and negotiation between environmental regulators and factories. These differences are illustrated by an application of the three synchronizations policy to a new factory in the City of Shenzhen.[14] As previously mentioned, the three synchronizations policy requires that the design, construction, and operation of a new factory be accompanied by the simultaneous design, construction, and operation of the factory's waste management systems.

Factory Design and Construction

In 1986, Chinese and Hong Kong investors organized a joint venture in Shenzhen to create a factory to dye and weave cotton thread. To carry out the first synchronization (the design phase), the investors formally registered with the Shenzhen EPB. As part of registration, the factory owners presented the EPB with information on their proposed facility and their wastewater treatment plant design. In meetings with the EPB, the investors were made aware of local environmental requirements. This first of the three synchronizations makes it difficult for new factory owners to argue later that they were unaware of relevant environmental requirements.

During the first synchronization, the Shenzhen EPB indicated that the proposed maximum wastewater discharge of 210 tons/day for the factory seemed too low, given the expected production level. In response to its concern, the EPB negotiated an agreement whereby the Chinese side of the joint venture accepted responsibility for keeping flows below 210 tons/day. The first synchronization was completed when the EPB gave its stamp of approval. Without this stamp, the Shenzhen municipal government would not authorize construction. Because factory directors know that without EPB approval

[13] In many ways, the mayor of a Chinese city is similar to the director of a "corporation" that includes both city bureaus and the enterprises they supervise. For more on this point, see Sinkule and Ortolano (1995, pp. 20 and 21) and references cited therein.

[14] This subsection summarizes a case study by Sinkule and Ortolano (1995).

they cannot get permission to construct a proposed facility, they generally undertake the first synchronization.

Construction of the weaving and dyeing plant started in October 1986. During the second synchronization (the construction phase), the Shenzhen EPB was authorized to inspect the treatment works as they were being built. The three synchronizations program requires that construction of proposed treatment facilities proceed simultaneously with factory construction. In the case of the weaving and dyeing plant, an inspection of the construction process was not carried out by the EPB. It is not unusual for EPBs to skip the second synchronization. No formal paperwork or stamp of approval is associated with the second synchronization, and an EPB might let the second synchronization pass without an inspection in order to save staff time.[15]

Trial Production at the Factory

Once construction was completed, the weaving and dyeing plant began trial production, including trial operation of its wastewater treatment plant. The trials began in October 1987. In mid-November 1987, the EPB asked the factory leaders to apply for final certification of the factory and the treatment works within two months. This certification, which is given only after an EPB inspects and approves a factory's waste treatment plant operations, ends the third synchronization. The weaving and dyeing plant did not comply with the two-month deadline, and, in what apparently was an oversight, the Shenzhen EPB did not follow up.

Several months after the weaving and dyeing factory had commenced production (and before the final inspection had been performed), local fishermen complained to the Shenzhen EPB about a fish kill downstream of the plant. Although factory authorities acknowledged responsibility for the fish kill, they claimed it was due to a break in a wastewater pipeline. The Shenzhen EPB inspected the factory and found that a so-called "emergency bypass pipeline" was being used routinely to discharge wastewater that bypassed the treatment works.

After learning that the factory was intentionally bypassing its treatment plant, the Shenzhen EPB's leaders brought other officials of the Shenzhen city government to visit the plant. Soon thereafter (in May 1988) the EPB ordered the factory to pay a fine and limit its discharge to 210 tons/day. The plant immediately shut down part of its production process and initiated steps to have its wastewater treatment plant redesigned and rebuilt. The Shenzhen EPB's orders had an immediate effect because the EPB had the leadership of the Shenzhen city government behind it.

Failure to Meet Requirements

During the next 20 months, the weaving and dyeing plant rebuilt its treatment works twice before it satisfied the Shenzhen EPB's requirements for the third synchronization (the operations phase). After the first renovation turned out to be inadequate, the Shenzhen EPB staff intervened by providing technical assistance on measures needed to bring the treatment plant into compliance. Having an EPB expert work together with factory staff to solve waste management problems is not unusual in China. Indeed, during the 1980s there was a shortage of environmental engineering specialists with experience in designing treatment works, and the inexperience of consultants who worked on the first renovation was apparently the reason the initially renovated treatment plant failed to meet the EPB's requirements.

While the weaving and dyeing factory was trying to comply with regulations, the EPB questioned the factory's use of a discharge rate of 210 tons/day. The issue was resolved by means of a three-party negotiation that included representatives of the factory, the Shenzhen EPB, and the Shenzhen City Industry Office. The latter unit, which is part of the Shenzhen city government, was able to negotiate an increase in the factory's allowable discharge from 210 tons/day to 350 tons/day. Three-party negotiations involving factories, EPBs, and local offices of industry bureaus are common in China. Industry bureaus often play a role in "shielding" a factory from the EPB and, once negotiations are complete, ensuring that the factory meets its responsibilities for treatment.

Tools Available to Force Compliance

In the last stages of the Shenzhen EPB's efforts to force the weaving and dyeing factory to reduce pollu-

[15] For other explanations for why an EPB might choose not to inspect construction of a treatment plant, see Sinkule and Ortolano (1995).

tion, the EPB used its authority to issue limited time treatment orders. The EPB insisted (in October 1989) that the factory bring its wastewater treatment plant into full compliance with requirements within three months or risk serious penalties. The factory complied on time, the EPB was satisfied with the plant's performance, and the three synchronizations process was completed.

Although the three synchronizations program is applied throughout China and is generally viewed as one of China's principal environmental protection strategies, it fails to address a major difficulty: the practice of some factories to avoid operating treatment works that have been designed and built to meet requirements of the three synchronizations. This practice appears to be widespread, especially in southern China, where failure to operate treatment works is, in part, attributable to chronic power shortages that result in frequent brownouts. Other environmental programs, such as the use of fines and pollutant discharge fees, are used in dealing with factories that fail to operate their treatment works (see Table 9.2).

TABLE 9.2 Instruments Used by Chinese Environmental Agencies to Promote Environmental Protection (as of 1993)[a]

Sanctions
- Administrative warnings
- Fines
- Limited time treatment orders
- Orders to temporarily shut down or to close enterprises
- Orders to relocate enterprises
- Disciplinary penalties (for example, demotion of factory managers)

Education and Publicity
- Media releases (for example, newspaper articles with data on how cities rate in terms of environmental quality)
- Public "notices of criticism"
- Technical and financial advice to enterprises
- Worker education programs

Additional Economic Incentives[b]
- Pollutant discharge fees
- Grants and loans for construction of waste treatment systems

[a] Based on Sinkule and Ortolano (1995).

[b] Economic incentives to control pollution in China are summarized in Chapter 10.

Negotiation as an Enforcement Style

Chinese environmental policy implementation is characterized by negotiation and consensus building, and it is much influenced by cultural traditions that place importance on maintaining harmonious relationships.[16] Although the negotiation process may be tortuous and time consuming, it is consistent with traditional practices in China. Bargaining also improves the likelihood of policy implementation because potential opponents of a policy are involved in negotiating agreements.

The EPB's reliance on negotiation and the cultivation of harmonious relationships with enterprises is demonstrated by the case of the weaving and dyeing plant. The Shenzhen EPB recognized that the factory was having difficulty obtaining the environmental engineering services it needed to comply with regulations, and the EPB provided the factory with technical advice on how to meet the discharge standards. The weaving and dyeing factory case also demonstrates the way that industry bureaus can assist in consensus building.[17]

[16] The emphasis on bargaining and compromise to reach a consensus is not unique to environmental regulation. As demonstrated by Shirk (1993), negotiating consensus is the dominant style of decision making at all levels of government in China.

[17] Use of a local industry bureau as a third party in negotiations between factories and EPBs has advantages for all sides. Industry bureaus involved in these negotiations have supervisory relationships with the enterprises, and thus they have an interest in the outcome. The advantage to factories is that industry bureaus can use their knowledge of production technology and pollution reduction capabilities of factories in helping to negotiate agreements. The EPBs benefit because their staffs often lack knowledge of industrial production and waste management possibilities; these technical matters vary from one industry to the next. In addition, once a consensus is reached, the industry bureau, as the supervisory unit for the enterprise, can assist the EPB in pressuring the enterprise to live up to its agreement to implement environmental protection measures.

Sinkule (1993, pp. 27–28) characterizes relationships between EPBs and enterprises in terms of traditional Chinese styles of conflict resolution:

> Maintaining harmony and avoiding loss-of-face are central to building consensus and cooperative relationships. A harmonious relationship between individuals or organizations does not mean there is complete agreement on all issues; rather, the relationship between parties includes compromise on the part of one or both parties at times to maintain the overall harmony. Negotiations are usually carried out in a manner that allows the parties to give in on occasion without loss-of-face. In contrast, insisting on following a regulation to the letter and clearly defining right and wrong in a confrontational manner could result in loss-of-face for the party that is forced to admit wrong doing.

The EPBs' emphasis on cooperative as opposed to adversarial relationships with enterprises helps explain the minor role of the judiciary in the implementation of environmental regulations in China.[18]

MONITORING COMPLIANCE WITH REGULATIONS

In China and elsewhere, promulgation of environmental regulations does not ensure compliance. Moreover, while agency records may show high compliance rates based on one-time measurements (for example, tests of new cars' compliance with emission control requirements), *continuing* compliance rates are frequently quite low.

A disparity between one-time compliance and continuing compliance is common in the United States. For example, the U.S. Environmental Protection Agency (EPA) reported (in the early 1980s) that 96% of major private waste sources and 81% of major publicly owned treatment works were in compliance with wastewater discharge permit conditions based on one-time observations.[19] In contrast, a 1983 study by the U.S. General Accounting Office (GAO) reported widespread failure to comply with wastewater discharge limits on a continual basis. Using data covering 531 major wastewater dischargers, the GAO found 31% of the sources had one or more "significant violations" of permit conditions.[20] The GAO also found that permit violations sometimes continued for years before environmental agencies enforced permit conditions.

Why is the rate of continuing (as opposed to one-time) compliance with environmental regulation so low, and why does so much time pass before available sanctions are imposed? These questions do not have simple answers. However, partial explanations are provided by considering how U.S. environmental agencies undertake monitoring and enforcement activities.

Environmental agencies in the United States rely heavily on *self-monitoring:* dischargers measure the quantity and quality of their releases and report the results to environmental agencies. Information from self-monitoring reports is supplemented by *facility inspections,* in which trained agency staff try to "infer compliance status from indirect evidence, such as equipment operating logs, plant and equipment appearance and state of repair, and the visual quality of the current discharges."[21] A more rigorous supplement to self-monitoring comes from *monitoring visits,* where agency staff conduct measurements to determine concentrations and rates of discharge of pollutants. As might be expected, facility inspections cost less and are performed more frequently than

---

[18] As argued by Sinkule and Ortolano (1995, p. 156), the Chinese traditionally have not relied on courts to resolve disputes. The development of legal institutions was disrupted during the Cultural Revolution (1966–76) and few EPBs have lawyers on their staffs. This situation is currently in flux, and some EPB staff see a trend toward increased use of legal institutions in implementing environmental policy.

[19] Data from studies mentioned in this paragraph are from the summary given by Russell, Harrington, and Vaughan (1986, pp. 6–8).

[20] In this context, a significant violation means at least one discharge permit condition was violated for four or more consecutive months.

[21] This quotation is from Russell, Harrington, and Vaughan (1986, p. 22). All facts in this paragraph are from Russell, Harrington, and Vaughan's (1986) summary of several studies of monitoring activities of state environmental agencies. Terms used to describe types of compliance inspections are not standardized. For example, EPA distinguishes 5 types of inspections for checking compliance with wastewater discharge permits. The least rigorous, "reconnaissance inspections" and "compliance evaluation inspections," are similar to what is called a facility inspection here. The most costly site visits involve sampling for toxic substances, which typically requires sophisticated sampling and analysis techniques (Hunter and Waterman, 1996, pp. 41–43).

monitoring visits. The latter are often used at major sources of pollution and at facilities where regulators believe discharge permit conditions are being violated.

Based on their review of existing studies, Russell, Harrington, and Vaughan (1986, pp. 36–37) report that state environmental agencies rely heavily on self-monitoring data to check on compliance with their requirements.[22] Agencies have neither the staffs nor the funds to conduct monitoring visits at all facilities they regulate. The absence of extensive monitoring by agencies makes it difficult to gauge accurately which sources are in compliance with environmental regulations. In addition, agency inspections and monitoring visits are often announced in advance, and this makes it hard to know if agency staff observations reflect typical conditions at facilities.

Self-monitoring activities in the United States may have improved in recent years because of changes in statutory provisions concerning liability. Some U.S. laws require *self-certification* of compliance by company officials as a part of self-monitoring and reporting. Senior managers are personally liable for false reports, and this enhances the attention firms give to the quality of corporate monitoring efforts. It is difficult to quantitatively characterize the influence of self-certification on improving the quality of self-reported monitoring data. However, jail sentences and other criminal law penalties for top management staff in firms that violate environmental requirements have been widely publicized, and the publicity undoubtedly has increased the quality of self-reported data.

The quality of self-monitoring is also enhanced by the common requirement that self-monitoring data sent to environmental agencies be made public.[23] Provisions of some environmental laws allow citizens to use this date to bring lawsuits against polluters. More generally, citizens often play a key role in alerting environmental authorities to violations of environmental laws.

AGENCY ENFORCEMENT OF ENVIRONMENTAL REGULATIONS IN THE UNITED STATES

What happens when an environmental agency believes that a polluter has violated an environmental requirement such as a discharge standard? The answer varies from country to country and even within countries. In some nations, such as China, the environmental agency pursues strategies based on maintaining cooperative long-term relationships with polluters. The agency might provide technical advice and even help the enterprise secure funding to build waste treatment facilities. In other countries, including the United States, a more adversarial strategy is pursued: when initial efforts to encourage compliance fail, environmental agencies attempt to force the waste discharger to clean up. In all cases, a combination of cooperation and coercion is involved. Cultural and institutional factors determine whether the enforcement approach taken by regulators is conciliatory or adversarial. Even within a country, the enforcement style of environmental regulators varies over time depending on factors such as the health of the economy and the political agendas of key public officials. Moreover, geographical differences and other sources of case-to-case diversity in environmental problems contribute to variations in enforcement style within an environmental agency during any particular time period.

In the United States, many environmental agencies view coercion as the essential ingredient in getting polluters to comply with requirements. According to Cheryl Wasserman (1992, p. 41), who has served in key enforcement positions in the U.S. Environmental Protection Agency, "It is now generally recognized that if the polluter expects no consequences from noncompliance . . . [it] has little incentive to undertake any costs of compliance before getting caught." Wasserman underscores the importance of coercion by noting that government programs encouraging compliance by cooperative means (such as offering polluters assistance and in-

[22] According to Russell, Harrington, and Vaughan (1986, pp. 36–37): "Auditing, or checking of self-monitoring sources by the agency, is most commonly conducted on an annual basis, but the reported frequency of audits ranged from once each month to once every five years." They go on to observe that "the cost of audit visits vary as widely as their frequency—from roughly $100 [in 1982 dollars] in some states when discharge measurements are not made to several thousands of dollars in other states when such measurements are made."

[23] For examples of the roles citizens play in monitoring compliance with environmental laws, see Hunter and Waterman (1996, pp. 40–41).

formation) are poorly funded and often viewed as expendable.[24]

A recent study of EPA's enforcement activities since 1970 makes the case that regulatory enforcement by EPA follows the model of a *deterrence system:* "an arms length process whose essential purpose is to detect and punish violators. . . . In a deterrence system the major preoccupation of agency officials is with punishing wrongdoers, both as a means of doing justice in individual cases and as a way of deterring future violations of regulatory standards" (Mintz, 1995, p. 102).

The view that EPA's regulatory style is largely coercive with an emphasis on "deterrence" is common but disputable. A contrasting position is taken by Hunter and Waterman whose analysis of EPA's performance in enforcing clean water laws led them to characterize EPA's enforcement style as more pragmatic than coercive. For Hunter and Waterman (1996, p. 9), a *pragmatic enforcement style* is one that "places results as the primary goal of enforcement personnel, rather than either a strict adherence to the law or negotiation." In employing this pragmatic style, EPA staff exercise considerable discretion. The theory advanced by Hunter and Waterman is that diversity in the regulatory setting—in terms of water quality problems, geographic areas, and participants in the regulatory process—leads to EPA's pragmatic approach to enforcing water quality laws.

Surveys of how state environmental agencies respond to violations of regulations provide a basis for characterizing how U.S. agencies typically enforce environmental laws.[25] Except in emergency situations, the response to noncompliance is often a drawn-out process involving a gradually escalating sequence of communications, commands, threats, and sanctions (see Table 9.3). Initial responses are typically informal: letters, phone calls, meetings, and *notices of violation.* These initial actions advise the discharger of the violation and what is required to correct it. Often a deadline for corrective action is set. If initial informal exchanges fail to produce results, more severe sanctions are threatened. Polluters frequently attempt to negotiate extensions, or even make promises they intend to break, simply as a way of stalling. The longer a polluter stays out of compliance, the longer the expense of pollution abatement can be avoided.

TABLE 9.3 Enforcement Responses Taken by Environmental Agencies[a]

Informal Responses
- Goals:
  - Notify polluters that environmental requirements are being violated
  - Inform polluters of actions polluters must take to comply with requirements
- Techniques:
  - Letters, phone calls, and meetings
  - Official "notices of violation"

Formal Responses
- Goals:
  - Use agency's ability to impose sanctions to force polluters to comply with requirements
- Techniques:
  - Administrative proceedings
  - Civil actions
  - Criminal actions[b]

[a] Table entries describe goals and techniques commonly used by environmental agencies in the United States. The listings are not comprehensive.

[b] Sometimes civil and criminal lawsuits against a polluter proceed in parallel.

If informal measures fail to yield satisfactory results, the environmental agency may invoke coercive measures: formal administrative procedures, and civil and criminal lawsuits. The first formal action typically involves *administrative proceedings:* actions that allow the agency to impose fines or issue *administrative orders* requiring polluters to take particular cleanup actions. Formal administrative measures are similar to actions taken by courts, but they involve the environmental agency's own administrative law judges. Strict procedural rules are followed to protect the rights of those accused of violations. Environmental agencies can often take quasi-judicial actions on their own; these actions account for the bulk of formal enforcement activities in the United States.

Administrative orders issued by an agency are not self-enforcing. If a polluter fails to respond, an environmental agency can force compliance only by relying on the courts. Most suits brought by environ-

[24] For more on this point, see Wasserman (1992, p. 30).

[25] For examples of such surveys, see Russell, Harrington, and Vaughan (1986).

TABLE 9.4 Administrative Proceedings and Criminal and Civil Referrals to the U.S. Department of Justice by the U.S. Environmental Protection Agency.[a]

	Fiscal Year			
	1981		1989	
	Total	**% of Total**	**Total**	**% of Total**
Administrative proceedings	1107	90	4136	91
Civil action	118	10	364	8
Criminal action	0	0	60	1
Total	1225	100	4560	100

[a] Based on Segerson and Tietenberg (1992, p. 68). The original source contains EPA data for the entire period from 1981 to 1989.

mental agencies involve civil law, even though criminal suits may have greater value as a deterrent (especially when corporate officers face the possibility of serving time in jail).[26] Criminal suits are not used frequently because they place heavy demands on an environmental agency. In order to win, the agency may need to prove that an individual or business was knowingly and willfully responsible for violating environmental regulations. In addition, rules governing permissable evidence are more demanding in criminal suits than they are in civil suits.

It is commonly believed that environmental authorities in the United States can simply bring lawsuits against polluters who ignore requirements, but things are not that simple. Court actions are expensive, and they tie up technical staff who must gather evidence and prepare legally defensible cases. In addition, environmental agencies cannot file lawsuits on their own. Typically, environmental agencies must refer cases to public prosecuting bodies, such as the U.S. Department of Justice or the office of a state attorney general. Public prosecutors typically have their own agendas and may not give priority to all the cases environmental agencies think are worthy. Even if an environmental agency prevails in a trial, it may have to endure the process of appeals, which can delay the resolution of a case for several years.

Data on the formal enforcement actions taken by the U.S. Environmental Protection Agency highlight the relative frequency of use of different types of formal measures.[27] Between 1981 and 1989, 87 to 92% of formal enforcement actions in any one year involved administrative proceedings (see Table 9.4). Civil law cases accounted for between 8 and 11% of the formal actions in a year, and criminal law cases were used less than 2% of the time. The number of formal enforcement actions in the United States increased during the 1980s, peaking at 4560 actions in 1989.

Frequently, initiating a court action encourages an out-of-court resolution of disputes. Environmental agencies and those they sue are often eager to avoid the time and expense of a trial. The great majority of environmental suits are resolved by negotiated settlements.

CITIZEN-BASED ENFORCEMENT ACTIVITIES

Individual citizens and nongovernmental organizations (NGOs) influence the enforcement of environmental regulations in several ways, depending on the country involved. In most nations, citizens can affect the enforcement of environmental statutes informally

[26] In the United States, actions brought under civil law protect private rights. In contrast, criminal law concerns offenses against the state. The penal code of a state (or of the United States) defines what constitutes a crime. Environmental statutes such as the U.S. Clean Water Act and the U.S. Clean Air Act contain sanctions that can be imposed under both civil law and criminal law. For more information on the Clean Water Act and the Clean Air Act, see chapters 12 and 13, respectively.

[27] Data in this paragraph are from Segerson and Tietenberg (1992, p. 68). Table 9.4 contains only a portion of the nine years of data summarized by Segerson and Tietenberg.

by reporting suspected violations of environmental regulations to the media and government authorities. Even in China, which does not have a reputation for citizen participation in government, citizen complaints about particular sources of degradation in environmental quality often play a role in the enforcement activities of environmental protection bureaus.

In some countries—for example, Brazil—nongovernmental organizations are able to file lawsuits to enforce provisions of environmental statutes.[28] In these circumstances, NGOs complement the enforcement activities of environmental agencies.

Under what conditions do private parties take advantage of their rights to enforce environmental laws? To examine this question, consider court actions initiated by citizens and NGOs to enforce environmental laws in the United States. Between 1970 and 1987, most cases in which citizens and NGOs sued violators of U.S. environmental laws involved the Clean Water Act.[29] In an analysis of over 1200 citizen-based suits initiated during this period under U.S. environmental laws, Jorgenson and Kimmel (1988, p. 1) found that about 900 actions were brought under the Clean Water Act. The 1972 amendments to the Clean Water Act allow any citizen to initiate a civil suit against any person who is alleged to be in violation of an effluent standard or limitation.[30] Citizens can also bring suit based on alleged violations of an order issued by EPA (or a state) to bring about compliance with an effluent standard. A citizen or NGO must give a 60-day notice to the state before proceeding with a suit. This requirement allows state and federal environmental authorities to take their own enforcement action on the case. The 60-day notice also gives the person alleged to be violating requirements a chance to comply and avoid going to court.

According to Naysnerski and Tietenberg (1992), three features of the Clean Water Act account for its widespread use by citizens:

1. If citizen plaintiffs win in court, their court costs (including attorney fees) are paid by the defendant.
2. Citizen plaintiffs can sue both for injunctions to force compliance with the law *and* for monetary penalties.
3. Citizens can use legally required self-monitoring reports to identify violations of effluent standards, and those reports can be used in court as evidence of noncompliance.

If polluters lose, they face one or more of the following: costs to bring their facilities into compliance, monetary penalties provided for by the statute, court costs incurred by the plaintiff, and the polluter's own costs in preparing a defense and going to court.[31]

Although private enforcement of environmental statutes is not available to citizens in most countries, some scholars underscore its widespread potential. Naysnerski and Tietenberg (1992, p. 131) see private enforcement as having great potential in contexts where improving environmental policy seems a "hopeless venture in the face of an uncooperative, corrupt or merely underfunded government." They believe private enforcement "may have particular relevance in developing countries, since public agencies in those countries are frequently characterized as having neither the administrative apparatus, nor the financial resources, to pursue effective environmental enforcement. Even corrupt enforcement is undermined by the private enforcement process" (Naysnerski and Tietenberg 1992, p. 136).

Weak enforcement of regulations by environmental agencies is, of course, not restricted to developing countries. Consider the United States, for example. During the mid-1980s, many private enforcement actions were taken in response to EPA's weak enforcement record during the Reagan administration. Naysnerski and Tietenberg (1992, p. 114)

[28] General information on citizen suits is available in Naysnerski and Tietenberg (1992); they emphasize the United States and briefly mention the situation in the European Community.

[29] The first U.S. environmental law with provisions for citizen enforcement was the Clean Air Act of 1970. Although the Clean Air Act provisions set an important precedent, they were not widely used, possibly because there was no inexpensive way for citizens to obtain data needed to establish that a waste source violated the Act. This was changed in 1990 when the Clean Air Act was amended to require waste sources to prepare discharge monitoring reports and submit them to environmental agencies. Citizens have access to these reports.

[30] More precisely, the 1972 law was named the Federal Water Pollution Control Act Amendments of 1972. The law was later called the Clean Water Act (see Chapter 12).

[31] Data on cases brought by citizens between 1970 and 1987 indicate that defendants almost always lost. According to Naysnerski and Tietenberg (1992, p. 117), 507 of more than 1200 cases they analyzed were "settled;" of these, "only four involved a decision for the defendant."

argue that many environmental groups organized staff and found donors to enable them to bring citizen suits during the Reagan presidency. Once staff and funding were in place, the private enforcement process became self-sustaining.

FACTORS AFFECTING COMPLIANCE BEHAVIOR

Economists have developed models to explain the degree to which polluters comply with environmental requirements.[32] Waste dischargers are generally assumed to behave as amoral, rational actors with full knowledge of costs. According to economists' theories, the discharger acts to minimize the costs of complying with regulations plus the *expected value* of the costs of noncompliance. This expected value reflects both the penalties associated with noncompliance and the probability that penalties would be imposed.

Political scientists have also developed theories explaining how polluters respond to regulations. Models of political scientists emphasize the limits on "optimizing" behavior placed by incomplete and uncertain information. According to these models, polluters and regulators "muddle through" to find solutions to environmental problems that are satisfactory, but not optimal. Decisions by polluters are explained using both economic and noneconomic factors, such as personal relationships between agency staff and regulatees. These theories often highlight the exercise of discretion by agency staff and the process of negotiating mutually acceptable resolutions to environmental disputes.[33]

Theories developed by both economists and political scientists provide many insights into how polluters respond to environmental regulations. The theories also offer guidance on how environmental agencies can increase polluters' compliance with environmental regulations. However, no one theory has emerged as being definitive.

Compliance behavior refers to actions polluters take in response to environmental regulations. A wide range of actions is possible. At one extreme, waste dischargers may ignore requirements. At the other extreme, dischargers may implement pollution prevention programs that more than satisfy environmental regulations. The factors influencing compliance behavior can be divided into two categories: one category concerns the discharger and the other involves the environmental agency.

Waste Dischargers' Context

Table 9.5 lists attributes of a waste dischargers' context that affect compliance behavior. The first few factors in the table characterize a waste discharger's costs. Factors include *capital costs,* which are the costs of purchasing equipment and building facilities, and the discharger's ability to raise needed funds. Finding resources is a problem for public bodies, especially when voters are asked to pass bond issues to raise funds. Obtaining resources for environmental projects in private organizations is also problematic, since intraorganizational competition for money to launch new projects is often intense. Costs of compliance also include funds to operate and maintain waste reduction systems. In many countries, but not all, polluters calculate their compliance costs by combining initial capital costs with the present value of future operation and maintenance costs.

The economic costs of *not* complying with regulations are more difficult to analyze because they are highly uncertain. A discharger can estimate the maximum size of various penalties on the basis of laws and regulations. However, it is impossible for a polluter to know if the maximum penalties would be imposed. Polluters are also uncertain about *when* penalties would be imposed or if they would be imposed at all. As previously noted, the enforcement response of an environmental agency is typically a drawn-out affair. The agency must first detect violations, and this takes time. Moreover, if the environmental agency has an inadequate budget and staff, many violations of requirements will remain undetected. Even after violations are spotted, there is typically a period of negotiation in which the agency attempts to get the waste discharger to comply "vol-

[32] For an introduction to the literature on compliance behavior, see Tietenberg (1992) and sources cited therein.

[33] For more on the influence of noneconomic factors on compliance with environmental regulations, see, for example, Mazmanian and Sabatier (1981), Sinkule and Ortolano (1995) and Hunter and Waterman (1996). The more general literature on the implementation of government programs—for example, Bardach (1977) and Mitnick (1980)—also provides insights into factors influencing compliance with environmental regulations.

TABLE 9.5 Factors Affecting Compliance Behavior—Waste Dischargers' Context

- *Costs of Compliance*
  - Magnitude of capital and operating costs
  - Ability to raise capital
- *Expected Costs of Noncompliance*
  - Transaction costs
  - Fines and discharge fees
  - Liability for damages and toxic site cleanups
  - Jail sentences
  - Plant closure orders
- *Potential Economic Opportunities*
  - Savings from pollution prevention
  - New product development
- *Knowledge of Requirements and Technology*
  - Understanding of regulations
  - Availability of expertise
  - Availability of technology
- *Cultural Factors and Social Norms*
  - Respect for laws
  - Concern for public image
  - Accountability of organization members
  - Organizational inertia
- *Facility Attributes*
  - Facility type, size, and age
  - Facility location
  - Form of ownership

untarily." If the discharger has clever negotiators and little concern for either its public image or the law, negotiations can continue for years. During this time, the discharger pays *transaction costs,* including costs to negotiate, respond to administrative orders, and wage court battles. At the same time, it avoids the costs of reducing pollution.[34]

Environmental laws in many countries make the costs of noncompliance even more complex to analyze because they allow sanctions under criminal law. Moreover, in some countries, corporations are legally liable to compensate victims of pollution, or liable for the cleanup of contaminated waste disposal sites. Clearly, the calculation of noncompliance costs in monetary terms is not simple when costs involve jail sentences for corporate officials responsible for willful violations of environmental laws. The highly uncertain costs associated with cleaning up toxic waste sites also make it difficult to estimate monetary noncompliance costs.[35]

Another economic factor that influences a discharger's compliance behavior is the opportunity for financial gain by implementing *pollution prevention* (or *cleaner production*) programs. These programs typically include efforts to minimize the generation of waste by recycling and reusing materials and conserving energy and water. Sometimes pollution prevention leads to new outputs promoted as *green products* because they result from materials' recycling and reuse. Many firms have tried to minimize waste generation because it makes financial sense.[36] Stringent environmental regulations and potentially costly liability rules have caused many firms to discover waste minimization measures that would have been profitable even in the absence of environmental regulations.

Compliance behavior is much influenced by the quality of information concerning environmental regulations, and the likelihood that requirements will be enforced or changed. Rapidly changing and difficult-to-understand environmental requirements confound the decision processes of waste dischargers. Frequent regulatory changes and ambiguities in requirements have inspired many firms to minimize the waste they create. The less waste generated, the less a firm is exposed to regulatory risk.[37]

In some contexts, especially in developing countries where environmental engineering services may be in short supply, information about technologies available for waste reduction affects compliance behavior. Polluters' decisions are also influenced by the availability of pollution control equipment and staff capable of operating facilities.

In addition to economic and technical factors, compliance behavior is influenced by social and cul-

[34] Some scholars have modeled interactions between polluters and regulatory agencies using the "theory of games"; see, for example, Russell, Harrington, and Vaughan (1986).

[35] As shown in Chapter 14, high uncertainties in costs of cleanups are a characteristic of toxic waste site cleanups under the Superfund program in the United States.

[36] Examples of firms that have embraced waste minimization as a pollution prevention strategy are given in Chapter 14.

[37] Regulatory risk was introduced in the Dow Chemical Company case study in Chapter 4.

tural variables. In some countries, polluters do not expect environmental agencies to have the political clout and resources needed to enforce regulations.[38] Another sociocultural factor concerns whether citizens view violations of environmental laws as socially irresponsible behavior. The importance of maintaining a good public image has influenced some firms to comply with environmental laws, especially firms that are proud of being solid corporate citizens.

Another set of factors concerns behaviors *within* an organization. During the 1980s, the chief executive officers of many large corporations made commitments to environmentally responsible industrial development. The *greening of corporations* became a common theme in the popular press. However, it takes time for a commitment at the top levels of a large company to filter down and affect the daily behavior of the many corporate workers whose routine decisions affect environmental quality. This lack of quick response to new corporate environmental policies is often characterized as a problem of *organizational inertia.* Corporate leaders committed to making their organizations environmentally responsible face a challenging question: How can environmental factors be integrated into the routine decision making of all workers? Programs that make individuals accountable for how their decisions affect environmental quality have been effective in this regard. The environmental responsibility system in China is one example of how environmental accountability can be institutionalized. Even workers who are not concerned about their organization's environmental goals tend to take those goals seriously when there is an environmental component in their job performance evaluations.

The inertia that impedes an organization's ability to meet environmental goals may be offset by innovative, entrepreneurial staff members. Individuals who know how to get an organization's resources and staff organized to solve environmental problems can play a key role in meeting regulations, particularly when compliance involves innovative pollution prevention methods.

Some factors affecting an industrial waste discharger's compliance with environmental requirements concern the industrial facility itself. Important attributes include plant size, age, and location as well as the kinds of production processes involved. The nature of the owner of a facility can also be influential. For example, a statistical analysis of Chinese enterprises showed that, all other things being equal, state-owned enterprises had a higher degree of waste treatment than enterprises with other forms of ownership.[39] This outcome may have resulted because state-owned firms could exceed budget constraints without risking bankruptcy; thus budget limits were less of an impediment than they were for other types of enterprises.

TABLE 9.6 Factors Affecting Compliance Behavior—Environmental Agency Programs

- *Monitoring (Ability to Detect Violations)*
- *Nature of Enforcement*
  - Ability and will to impose sanctions
  - Swiftness and predictability of sanctions
  - Appropriateness of penalties
- *Degree of Compliance Promotion*
  - Education of regulated community
  - Development of public support
  - Use of economic incentives
  - Improvement of environmental management skills of regulatees

Actions by Environmental Agencies

Waste dischargers' responses to environmental requirements are influenced heavily by the behavior of environmental agencies. In designing programs to implement regulations, environmental agencies typically distinguish among three sets of activities: monitoring, enforcement, and *compliance promotion* (see Table 9.6).

Monitoring is needed to detect violations. Requirements for monitoring are linked to the type of enforcement response an agency makes. If an agency aims to bring about compliance by negotiating, it can often rely on self-reported data and facility inspections. However, if a court must be convinced that a

[38] For an example, see the analysis by Abracosa and Ortolano (1987) of how national environmental impact assessment requirements were ignored in the Philippines under former President Marcos.

[39] This study is reported by Rozelle, Ma, and Ortolano (1994).

discharger did, in fact, violate legal requirements, then the agency's monitoring demands increase.

An environmental agency's monitoring task is particularly difficult if it must demonstrate that a waste discharge causes violations in an ambient standard. The complexities increase if the scientific theories to describe the fate and transport of the relevant pollutants are not well developed, or if there are multiple waste sources. If many waste sources have discharges that affect ambient environmental quality in a given circumstance, any one source can claim its neighbors caused the violation. In these cases, the agency may have to go beyond simple monitoring and construct a mathematical model to describe how each of several sources affect ambient quality. It is expensive to construct models that can withstand legal challenges, which is one reason an agency may spend years pushing for compliance by informal negotiations.

Enforcement Efforts

The credibility of an environmental agency's implementation effort may be destroyed if it does not attempt to detect violations and respond to violations with consistent and appropriate actions. Polluters typically make compliance decisions based on how they think agencies will respond. For their part, agencies may work to build an image of swift and determined enforcement by ensuring that enforcement actions are well known to the regulated community. If agencies publicize successful enforcement actions, regulatees may believe that violations of environmental requirements will not be tolerated.

A U.S. Environmental Protection Agency (EPA 1992, p. 2–3) manual on environmental enforcement describes the importance of polluters' perceptions of agency responses:

> Because perception is so important in creating deterrence, how enforcement actions are taken is just as important as the fact that they are taken.[40] History has many stories of small armies that successfully beat large forces by giving the impression that they were a formidable fighting force. Similarly, enforcement actions can have significant effects far beyond bringing a single violator into compliance if they are well placed and well publicized.

Here is an example of how agencies influence perceptions of regulatees: companies whose officers do jail time for violating environmental laws have been required to publicize their penalties in business sections of major newspapers.

Another notable aspect of agency enforcement concerns the use of "creative settlements" in lawsuits with polluters. Environmental agencies sometimes leverage out-of-court settlements to further agency goals. For example, an agency might propose to reduce a monetary penalty or to extend a compliance deadline as part of a settlement offer. In exchange, the agency might get a commitment by the polluter to restore an ecologically significant area that has been damaged by others. Table 9.7 illustrates creative settlements used by Allegheny County, Pennsylvania, to improve air quality.

Compliance Promotion

In addition to imposing sanctions for noncompliance, environmental agencies can actively encourage compliance. Following are some examples of activities used to promote compliance.[41]

- *Education and Technical Assistance to the Regulated Community.* An example of technical assistance is the previously cited work of an environmental protection bureau in China in helping a weaving and dyeing company comply with wastewater discharge requirements. Many environmental agencies issue technical handbooks and sponsor workshops to provide regulatees with information about pollution control methods and waste minimization practices.
- *Development of Public Support.* Many environmental agencies gain public support by using the media to publicize environmental dangers and opportunities for citizen action. For example, in Taiwan in the 1980s, a power company proposed a dam that would have flooded Taroko Gorge, one of the country's main scenic attractions. Taiwan's

[40] In the context of this quotation, *deterrence* refers to the forces that lead waste dischargers to modify their waste management practices in order to avoid adverse consequences that may be imposed by regulatory agencies. Wasserman (1992, p. 23) summarizes four elements "necessary to create deterence . . . : (i) a credible likelihood of detection of the violation; (ii) swift and sure enforcement response; . . . (iii) appropriately severe sanction; and (iv) that each of these factors be perceived as real [sic]."

[41] The main headings in this list are from EPA (1992).

TABLE 9.7 Creative Out-of-Court Settlements Negotiated by Allegheny County, Pennsylvania (1970–90)[a]

- *Performance Bonds.* Companies posted performance bonds which were forfeited if they failed to carry out settlement terms.
- *Escrow Accounts.* Companies set up special escrow accounts, and funds in those accounts were to be turned over to the county if companies failed to pay penalties incurred for noncompliance.
- *Credit Projects.* Instead of paying penalties, companies reduced emissions beyond required levels.
- *Self-monitoring.*Companies engaged in extensive self-monitoring to increase their awareness of whether they met regulations.
- *Research Requirements.* Facilities performed research studies to identify how they could best meet regulations.

[a] Source: U.S. Environmental Protection Agency (1992, p. 11–10). Entries in this table are examples of settlements Allegheny County courts negotiated with polluters who violated air pollution control regulations.

Environmental Protection Agency relied on the media to alert citizens to the proposed project's effects. Public opposition led project proponents to drop plans for the dam. Eventually a national park was created at Taroko Gorge.

- *Use of Economic Incentives.* Environmental agencies often lobby governments to create incentive programs, such as grants and low-interest loans for construction of treatment works. Many countries promote compliance by subsidizing pollution control projects. Environmental agencies have also promoted "creative financing" schemes, such as the use of bonds with tax-free interest to finance environmental projects.[42]
- *Building Environmental Management Capabilities of Regulatees.* Capabilities of environmental specialists in firms and municipalities have been improved by environmental agencies. In China, for example, environmental protection bureaus sometimes mentor environmental management staff at factories.

[42] For additional examples of creative financing promoted by environmental agencies, see EPA (1992, pp 5–6).

Each environmental agency strikes its own balance between using reactive sanctions and proactive compliance promotion. In the United States, where environmental agencies often have adversarial relations with regulatees, enforcement responses based on gradually escalated sanctions are common. In contrast, China relies much less on litigation. Chinese environmental agencies place more emphasis on promoting compliance and developing cooperative relations with regulatees. The appropriate balance between cooperation and coercion depends heavily on context.

Key Concepts and Terms

Environmental Policy Implementation in China

Local vs. national environmental agencies
Discharge permit system
Three synchronizations policy
Environmental responsibility system
Assessments of urban environmental quality
Centralized pollution control systems
Establishing requirements by negotiating
Cultural factors affecting implementation

Monitoring Compliance with Regulations

Self-monitoring
Self-certification
Inspections and audits
Initial vs. continuing compliance

Agency Enforcement of Environmental Regulations in the United States

Cooperative vs. coercive responses
Deterrence system
Formal vs. informal enforcement
Negotiating voluntary compliance
Administrative orders
Civil vs. criminal law responses
Citizen-based enforcement activities
Citizen and NGO complaints
Citizen-suit provisions in environmental laws
Public vs. private enforcement

Factors Affecting Compliance Behavior
Economic vs. sociocultural factors
Capital and operating and maintenance costs
Costs avoided by noncompliance
Transactions costs
Corporate vs. personal liability
Cost savings via pollution prevention
Legal adequacy of monitoring data
Modeling in support of enforcement
Predictability and credibility of agency action
Creative settlements
Compliance promotion

Discussion Questions

9-1 Chinese environmental officials sometimes describe their pollution control scheme as "environmental protection with Chinese characteristics." How would you interpret this phrase in light of the way the Chinese have (1) adapted widely used forms of environmental regulation to fit into their national context, (2) developed unique programs consistent with established government institutions, and (3) relied on traditional cultural practices and social norms in implementing environmental regulations.

9-2 The Chinese implementation of environmental requirements relies heavily on bargaining between environmental agencies and polluters. In contrast, U.S. environmental agencies often treat waste dischargers as adversaries that need to be coerced into compliance. How would you explain why the approaches to policy implementation in these two countries are so different.

9-3 What lessons can other countries learn from the Chinese approach to environmental management? In framing your response, think about the policies and programs that are uniquely Chinese (for example, the environmental responsibility system).

9-4 During the 1980s, the formal enforcement responses of the U.S. Environmental Protection Agency relied heavily on administrative proceedings. Legal action was taken in only about 10% of the cases where a formal enforcement response was taken. The great majority of court actions relied on civil law as opposed to criminal law. How would you explain this pattern of response?

9-5 Think of a political jurisdiction you are very familiar with. Suppose you were the principal environmental administrator in this jurisdiction. What enforcement approaches would you rely on to bring waste dischargers into compliance with environmental regulations?

9-6 How would the following forms of environmental regulation be ranked and characterized in terms of ease of enforcement by a regulatory agency: technology standards, technology-based discharge standards, and discharge standards based on ambient standards?

9-7 Suppose you are the U.S. EPA administrator and you want to promote voluntary compliance with environmental regulations. Furthermore, suppose you want to emphasize to corporations the economic advantages of preventing pollution instead of trying to manage it using waste treatment and other measures. What actions would you take?

References

Abracosa, R., and L. Ortolano. 1987. Environmental Impact assessment in the Philippines: 1977–1985. *Environmental Impact Assessment Review* 7 (4): 293–310.

Bardach, E. 1977. *The Implementation Game.* Cambridge, MA: MIT Press.

EPA (U.S. Environmental Protection Agency). 1992. *Principles of Environmental Enforcement.* Washington, DC: Office of Enforcement, U.S. EPA.

Jorgenson, L., and J. Kimmel. 1988. *Environmental Citizens Suits: Confronting The Corporation.* Washington, DC: The Bureau of National Affairs, Inc.

Hunter, S., and R. W. Waterman. *Enforcing the Law: The Case of the Clean Water Acts.* Armonk, NY: M.E. Sharpe.

Masters, G. M. 1991. *Introduction to Environmental Engineering and Science.* Englewood Cliffs, NJ: Prentice–Hall.

Mazamanian, D. A., and P. A. Sabatier, eds. 1981. *Effective Policy Implementation.* Lexington, MA: Lexington Books.

Mintz, J. A. 1995. *Enforcement at the EPA: High Stakes and Hard Choices.* Austin: University of Texas Press.

Mitnick, B. M. 1980. *The Political Economy of Regulation.* New York: Columbia University Press.

Naysnerski, W., and T. Tietenberg. 1992. "Private Enforcement." In *Innovation in Environmental Policy,* T. Tietenberg, ed., 109–36. Hants, England: Edward Elgar Publishers.

Ross, L. 1988. *Environmental Policy in China.* Bloomington: Indiana University Press.

Rozelle, S., X. Y. Ma, and L. Ortolano. 1993. Industrial Wastewater Control in Chinese Cities: Determinants of Success in Environmental Policy. *National Resources Modeling* 7 (4):353–78.

Russell, C. S., W. Harrington, and W. J. Vaughan. 1986. *Enforcing Pollution Control Laws.* Washington, DC: Resources for the Future.

Segerson, K., and T. Tietenberg. 1992. "Defining Efficient Sanctions." In *Innovation in Environmental Policy,* T. Tietenberg, ed., 53–73. Hants, England: Edward Elgar Publishers.

Shirk, S. L. 1993. *The Political Logic of Economic Reform in China.* Berkeley: University of California Press.

Sinkule, B. 1993. *Implementation of Industrial Water Pollution Control Policies in the Pearl River Delta Region of China.* Ph.D. dissertation, Dept. of Civil Engineering, Stanford University, Stanford, CA.

Sinkule, B., and L. Ortolano. 1995. *Implementing Environmental Policy in China.* Westport, CT: Praeger Pub.

Tietenberg, T., ed. 1992. *Innovation in Environmental Policy.* Hants, England: Edward Elgar Publishers.

Wang, X., and R. Blomquist. 1993. The Developing Environmental Law and Policy of the People's Republic of China. *The Georgetown International Environmental Law Review* 5:25–75.

Wasserman, C. 1992. "Federal Environment: Theory and Practice." In *Innovation in Environmental Policy,* T. Tietenberg, ed., 21–51. Hants, England: Edward Elgar Publishers.

Chapter 10

Pollutant Discharge Fee Programs

Since the 1970s, many governments have developed pollution control programs based on the *polluter pays principle.* According to this principle, "the polluter should bear the cost of measures to reduce pollution . . . to ensure that the environment is in an acceptable state."[1] In many countries, this principle has been implemented by making polluters pay fees based on the quantity of waste they discharge.

The first half of this chapter introduces the economic theory behind pollutant discharge fees, a tax mechanism used to apply the polluter pays principle. Economic theory shows how numerical values of discharge fees can be selected to yield productively efficient pollution control outcomes. According to the theory, discharge fees can replace discharge standards, a central element in command and control regulatory systems.

The second half of the chapter analyzes how fee systems have been used in Europe and China.[2] As demonstrated by these analyses, practical implementations of pollutant discharge fees bear little relation to theory.[3] In practice, fee systems are linked with traditional command and control schemes, and revenues generated by fees are often used to subsidize pollution control efforts.

[1] As reported by Opschoor and Vos (1989), this definition is used by the Organization for Economic Cooperation and Development (OECD). Opschoor and Vos analyze ambiguities in the definition and show how the polluter pays principle is applied in OECD countries.

[2] Pollutant discharge fees (or taxes) are employed in a number of countries, but they are not used widely in the United States. As Oates (1996, p. 197) observes, the "impact of taxation on the environment in the United States has consisted primarily of the side-effect of taxes . . . introduced with other objectives in mind." Federal and state excise taxes on gasoline provide examples. Brown and Johnson (1984) analyze the feasibility of using effluent charges in the United States.

[3] Readers interested exclusively in practical applications of discharge fees can skip to the section entitled "Effluent Charges and Subsidies in Europe."

EFFLUENT CHARGES AND PRODUCTIVE EFFICIENCY

Several decades ago, an eminent British economist, Arthur Pigou, proposed a tax scheme that would correct resource misallocations resulting from *external costs* of production activities.[4] When external costs exist, a firm's private cost of production is lower than the social cost. What is now called a *Pigouvian tax* makes the cost perceived by the firm equal to the social cost. When a firm accounts for external costs, those costs are said to be *internalized.*

The contemporary theory of pollutant discharge fees is built on Pigou's insights regarding how firms can be made to internalize external costs. The central idea is as follows: A regulatory agency imposes a tax on a firm (or other pollution source) per unit of waste discharged. The tax is not levied on the firm's inputs or outputs, but on the waste discharge itself.[5] When forced to pay a price for each unit of waste released, firms have incentives to reduce the waste they generate.

A common framework to present the theory of effluent charges relies on curves representing the marginal costs and marginal benefits of pollution abatement. Chapters 6 and 7 illustrated how such curves might be derived. The theoretical discussion that follows assumes that the environmental regulatory agency has complete and perfect information about the costs and benefits of pollution abatement by the waste sources it regulates.

Basic Theory: Benefits Independent of Source Location

Figure 10.1 contains a simplified linear representation of marginal cost and marginal benefit curves.[6] The horizontal axis represents the amount of pollution reduction undertaken by a firm, where the maximum possible reduction is $\overline{R}$, the firm's raw load. The *raw load* is the quantity of waste the firm would release if there were no government requirements forcing the firm to reduce its waste discharge. For illustrative purposes, the pollutant can be viewed as chlorides, the substance used in the Cedro River example in Chapter 7. Because the illustration involves a water pollutant, the resulting Pigouvian tax is called an *effluent charge.* If the example involved an air pollutant, the tax would be termed an *emission charge.* The basic argument is unchanged when dealing with either effluents or emissions.[7]

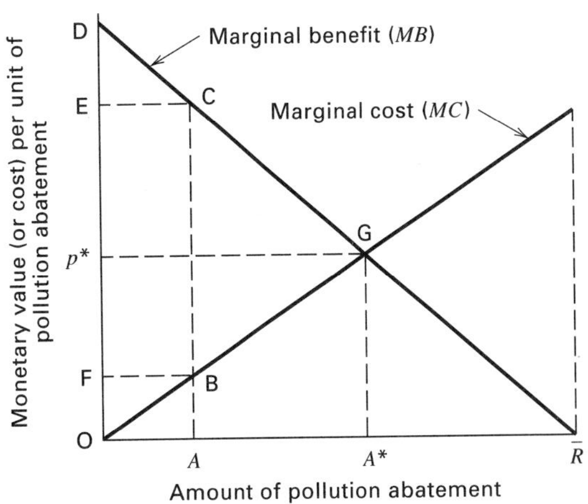

Figure 10.1 Costs and benefits of pollution abatement by a hypothetical firm.

[4] For a statement of the original proposal, see Pigou (1932). External costs were defined in Chapter 5.

[5] An example of an environment-related tax on *inputs* is the U.S. tax on chemical feedstocks used to produce chemical and petroleum products. As explained in Chapter 14, this tax funds cleanups of contamination from hazardous waste management facilities. Illustrations of taxes on *outputs* are the fees some states in the United States impose on batteries and tires. For other examples, see Oates (1996, pp. 195–223).

[6] This presentation of the theory of effluent charges is adapted from an unpublished manuscript by Noll (1994). While nonlinear equations could be used to illustrate the theory of effluent charges, the argument is simpler using linear equations.

[7] Terminology is not standardized in the literature on this subject. Other terms for Pigouvian taxes include effluent (or emission or discharge) fees. In this context, the words *charges, fees,* and *taxes* are often used interchangably. The units used for the charge (or fee or tax) is money per unit of pollutant discharge. To compound the potential for confusion, the term *user fee* is also employed in this literature. However, it has a different meaning. Hahn (1989, p. 11) clarifies the difference: "User fees refer to fees which are designed to cover some or all of the cost of an activity, but which do not necessarily provide appropriate incentives for efficient resource utilization. Examples include water charges, sanitation fees, and airport ticket taxes." For an analysis of how user fees affected firms using municipal sewer services in five U.S. cities, see Hudson, Lake, and Grossman (1981). User fees often serve as an economic incentive for solid waste reduction; see, for example, National Academy of Public Administration (1994, pp. 113–138).

The vertical axis in Figure 10.1 represents cost or value in monetary units. The *marginal cost* (MC) curve represents the amount it costs a firm to reduce its waste discharge by one unit. And the marginal benefit (MB) curve reflects the economic value to society of reducing pollution by one unit. Benefits may be linked, for example, to enhanced recreational opportunities and safer drinking water supplies. For an arbitrary level of pollution abatement, A, the distance OE in Figure 10.1 represents the social benefit of an additional unit of abatement, and the distance OF delineates the firm's cost to cut its pollution by one more unit.

As detailed in Chapter 5, the areas under the marginal cost and marginal benefit curves represent total costs and total benefits, respectively. Thus the area of the triangle OAB is the firm's total cost of reducing A units of pollution, and the area of the quadrilateral OACD is the resulting total benefit of abatement level A.[8] The *productively efficient* level of pollution abatement is, by definition, the value of A that maximizes the difference between total benefits and total costs. The efficient outcome corresponds to the point at which marginal benefit equals marginal cost, the level A^* in Figure 10.1.[9] The corresponding value of net economic benefit is OGD, the difference between total benefit (OA*GD) and total cost (OA*G).

Imposing an effluent charge of p^* dollars per unit of waste discharged provides (in theory, at least) sufficient incentive for the firm to undertake the productively efficient level of abatement (A^*). No discharge standard is required as long as p^* represents the charge at which $MC = MB$. To see why a discharge standard is not needed, consider a scenario in which an environmental regulatory agency has perfect knowledge of the marginal cost and marginal benefit curves in Figure 10.1.[10] Knowing that $MC = MB$ at p^*, the regulatory agency imposes an effluent charge p^*. The firm is free to reduce its raw load by any amount A (including $A = 0$), but it must pay a tax bill of p^* times ($\overline{R}$-A) for the amount discharged.

In analyzing the response of the firm to the regulatory agency's effluent tax p^*, the criterion used by the firm to make its choice of abatement level must be known. The theory of effluent charges *assumes* that the firm sets A to minimize the total cost of dealing with its waste. If an effluent charge p^* is imposed, the firm has two types of costs: waste reduction costs, and effluent charge payments to the environmental agency. For an abatement level A, the charge payments are represented by the area of a rectangle with a base equal to the waste discharge ($\overline{R}$-A) and a height equal to p^*.

A firm acting to minimize its total pollution control expenditures will choose the abatement level at which $p^* = MC$ (see Figure 10.2). Why does a charge of p^* lead the firm to reduce its raw load by A^*? This question is answered in two parts. First, consider the marginal cost of an abatement level greater than A^*. For any $A > A^*$, the cost of reducing an increment of waste is greater than p^*. It would be cheaper for the firm to pay more charges and do less waste reduction. Now consider any $A < A^*$. In this region, the unit charge p^* is higher than the incremental waste reduction cost. If a firm is at an abatement level $A < A^*$, it can reduce its total expenditure by doing more waste reduction and paying less effluent charges. At the point A^*, no further reduction in total cost is possible. A cost-minimizing firm reduces its raw load by A^* and pays effluent charges for the amount of waste ($\overline{R} - A^*$) that it releases.

This theoretical argument can be extended to include multiple waste sources; but complexities occur if the locations of the sources influence the bene-

[8] Strictly speaking, the area OAB represents the firm's total *variable* cost of reducing A units of pollution. Fixed costs are ignored here because they have no influence on the outcome in this analysis.

[9] To establish this result, try any value of $A \neq A^*$ and demonstrate that the area representing net benefits is smaller than OGD, the area representing the net benefits of pollution reduction A^*. The condition that MC = MB at A^* can also be established using standard procedures for optimizing a function of one variable: set up an equation in which net benefits (of reduction level A) equal total benefits minus total costs, and set the first derivative with respect to A equal to zero. As usual, the net benefit function would need to satisfy certain convexity properties for this procedure to yield a maximum (and not a minimum or a saddle point).

[10] Under conditions of perfect knowledge, the agency could use either effluent standards *or* effluent charges to obtain an efficient outcome; this point is taken up in Chapter 11 in an analysis comparing discharge standards, effluent charges, and marketable permits to pollute.

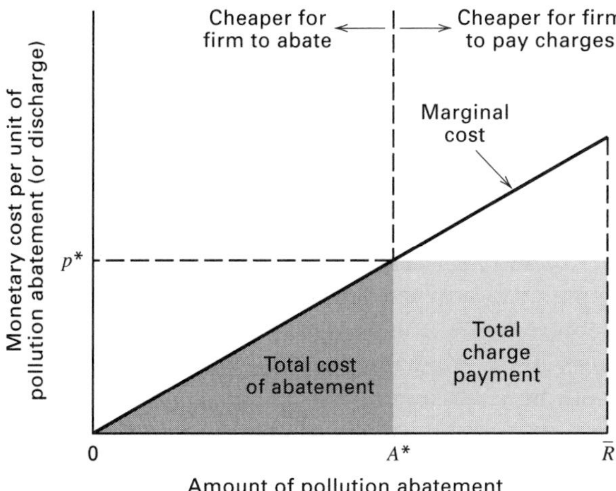

Figure 10.2 Outlays for pollution abatement and effluent charges.

fits of pollution abatement.[11] Multiple sources are considered next. The case where source location affects benefits is treated in a later section.

Example Involving Multiple Sources

To extend the theory presented in connection with Figure 10.2 to multiple waste sources, assume that the incremental benefit of a unit of waste reduction does *not* depend on where the reduction occurs. The U.S. federal program to control acid rain by reducing sulfur dioxide (SO_2) emissions illustrates that marginal benefits can be independent of source location. The program's goal is to reduce total emissions of SO_2 from electric power plants regardless of their locations.[12]

Consider two firms, each discharging the same type of pollutant.[13] Suppose the marginal costs of pollution abatement at firms 1 and 2 (respectively) are given by

$$MC_1 = 2A_1, \qquad \textbf{(10-1)}$$

and

$$MC_2 = \frac{4}{3}A_2, \qquad \textbf{(10-2)}$$

where MC_i = marginal cost of abatement for firm i
i = 1, 2, and
A_i = level of waste reduction by firm i.

The raw loads for firms 1 and 2 are 8 and 12, respectively. The symbol A represents the level of *total* waste reduction, and thus $A = A_1 + A_2$.

The theory is developed in two parts. First, a marginal cost curve is constructed that is analogous to the MC curve for the single waste source case in Figure 10.1. The constructed MC curve represents the *minimum* incremental cost of abatement at an arbitrary level of waste reduction, $A_1 + A_2$. (Abatement costs may be incurred by each firm, and the sum of their costs is at a minimum.) The second part of the analysis introduces a marginal benefit curve and uses the result that net benefits are maximized when $MB = MC$ to determine the value of the effluent fee.

Attaining Abatement at Minimum Cost

Construction of the MC curve corresponding to the least costly approach to abating A units of pollution proceeds by first establishing that $MC_1 = MC_2$ at the point where total abatement costs are minimized. This result is established using a graphical procedure demonstrated in Figure 10.3 for the case in which $A = 10$. (The same procedure can be used for any feasible value of A.) The horizontal axis in Figure 10.3 has a total length of $A = 10$, the abatement level equal to $A_1 + A_2$. The marginal cost curve for firm 1 is plotted in the usual way, with A_1 increasing from the origin to the right, and with marginal cost on the left vertical axis. In contrast, the marginal cost for firm 2 is plotted on a vertical axis on the right side of Figure 10.3, and A_2 increases from right to left on the horizontal axis. This method of arranging the marginal cost curves serves a useful purpose.

[11] For example, for organic pollutants entering a stream, the benefits of waste reduction typically depend on where the waste discharge occurs. The incremental benefits of one less unit of waste discharge often depend on *which* source reduces its waste.

[12] The U.S. policy on acid rain assumes that the incremental benefits, in terms of reductions in acid rain, are independent of where (in the 48 contiguous states) SO_2 reductions occur. The U.S. acid rain policy is analyzed in the next chapter.

[13] An analogous theoretical development is used for cases with more than two waste sources, but calculus is required.

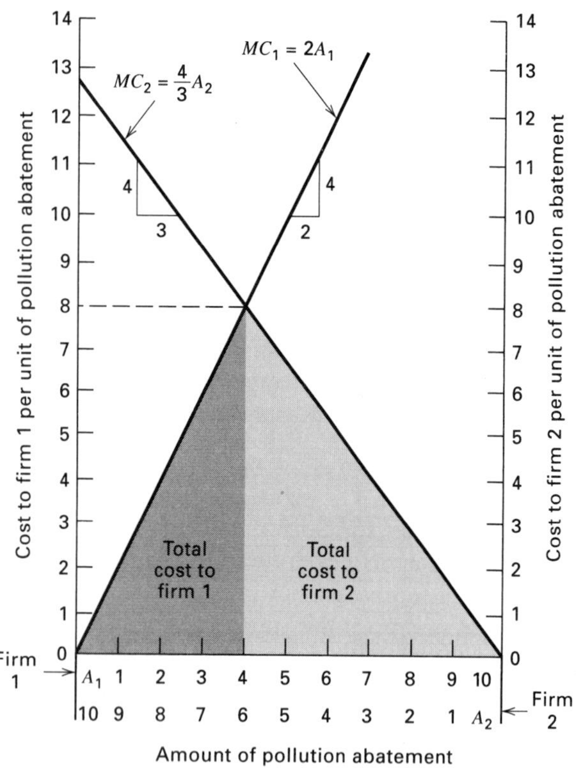

Figure 10.3 Minimum cost of pollution reduction for $A_1 + A_2 = 10$.

Combinations of A_1 and A_2 that correspond to *any* vertical line on Figure 10.3 satisfy the condition that $A_1 + A_2 = A$, which is 10 in this illustration.

Figure 10.3 shows that the sum of the total waste reduction costs for the two firms is at a minimum for values of A_1 and A_2 satisfying the condition that $MC_1 = MC_2$.[14] (Recall that the total cost to firm i of abatement level A_i is the area under the marginal cost curve between 0 and A_i.) In this example, total cost is minimized at $A_1 = 4$ and $A_2 = 6$. The corresponding total costs to firms 1 and 2 are 16 and 24, respectively. An analogous result holds for n firms, where n is arbitrary. To minimize total cost of waste reduction level $A = A_1 + A_2 + \ldots + A_n$, it is necessary that

$$MC_1 = MC_2 = \ldots = MC_n \qquad \textbf{(10-3)}$$

where MC_i is the marginal cost of abatement for firm i (i = 1, . . . , n).[15]

The result that $MC_1 = MC_2$ when total cost is minimized provides a basis for constructing a curve representing the minimum incremental waste reduction cost at abatement level A when the actions of both firms are accounted for. Suppose, for example, that each firm faces the same arbitrary value of an effluent charge ($\overline{p}$). Since each firm responds by setting price equal to marginal cost, the waste reduction of the two firms combined can be found by summing horizontally the individual marginal cost curves. The

[14] At $MC_1 = MC_2$, waste reduction levels are $A_1 = 4$ and $A_2 = 6$. To demonstrate that $A_1 = 4$ and $A_2 = 6$ corresponds to a minimum total cost, pick any other combination of A_1 and A_2 on a vertical line in Figure 10.3 (e.g., $A_1 = 3$ and $A_2 = 7$). Shading in the areas under the two marginal cost curves demonstrates that the total cost is larger than that given in Figure 10.3.

[15] This necessary condition for cost minimization can be established using Lagrange's method of undetermined multipliers, a procedure detailed in textbooks on mathematical optimization theory such as Gue and Thomas (1968). The problem is to find $A_1, A_2, \ldots, A_n$ to minimize

$$\sum_{i=1}^{n} TC_i(A_i),$$

subject to

$$\sum_{i=1}^{n} A_i = A,$$

where $TC_i\,(A_i)$ is the total cost to firm i of reducing its waste discharge by an amount A_i. The Lagrangian function is

$$L(A_1, A_2, \ldots, A_n, \lambda) = \sum_{i=1}^{n} TC_i(A_i) + \lambda\left(A - \sum_{i=1}^{n} A_i\right).$$

A necessary condition for minimizing $\sum_{i=1}^{n} TC_i(A_i)$ is that each of the first partial derivatives of the Lagrangian function be equal to zero:

$$\therefore \quad \frac{\partial L}{\partial A_i} = 0 = \frac{\partial[TC_i(A_i)]}{\partial A_i} - \lambda, \qquad (i = 1, \ldots, n),$$

and

$$\frac{\partial L}{\partial \lambda} = 0 = A - \sum_{i=1}^{n} A_i.$$

By definition of marginal cost, the derivative of $TC_i(A_i)$ with respect to A_i is MC_i. It follows that $MC_i = \lambda$ for all i, and thus $MC_1 = MC_2 = \ldots MC_n$.

procedure is illustrated in Figure 10.4 for the case where $\overline{p} = 8$. As expected from results in Figure 10.3, when $MC_1 = MC_2 = 8$, firms 1 and 2 reduce their waste discharges by 4 and 6, respectively. The total waste reduction is 10. The process of horizontal addition can be repeated for other values of $\overline{p}$, and the results are shown in Figure 10.4. As indicated in the figure, the marginal cost curve (reflecting the actions of *both* firms) that minimizes the total cost of abatement is

$$MC = \frac{4}{5}A. \qquad \textbf{(10-4)}$$

Determining the Effluent Charge

Information needed to complete the theoretical presentation of effluent charges in the two-firm case concerns the incremental benefit of pollution reduction. Recall the simplification used in this example: the benefit associated with a unit of waste reduction is independent of where the reduction occurs. Suppose the marginal benefit of abatement level A is

$$MB = 5.34 - 0.267A. \qquad \textbf{(10-5)}$$

By using the noninteger coefficients in equation 10-5, other values in this hypothetical example will be integers.

Development of the basic theory is completed by finding the effluent charge that an environmental agency needs to impose to bring about the productively efficient level of pollution abatement. As in the single-firm case, the efficient level of abatement yields, by definition, a maximum net benefit. This maximum occurs when $MB = MC$. As indicated in Figure 10.5, the level of abatement corresponding to $MB = MC$ is $A^* = 5$.

Figure 10.5 also shows that if an effluent charge equal to 4 is imposed, the firms will act independently to bring about a total pollution reduction of 5. This outcome results because each cost-minimizing firm reacts to an effluent charge by reducing pollution to the point where the effluent charge equals its marginal cost of abatement. By inspection of Figure 10.4, if firm 1 faced an effluent charge of 4, it would cut back its waste discharge by 2 units; firm 2 would reduce its waste by 3 units. Thus the efficient level of pollution abatement, a total reduction of 5 units, would result from each firm acting independently to minimize its own costs in response to the effluent charge.

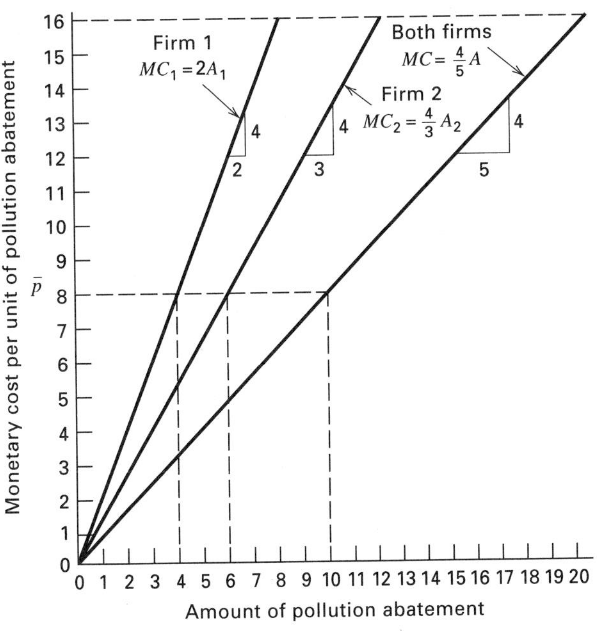

Figure 10.4 Marginal cost of pollution reduction for both firms.

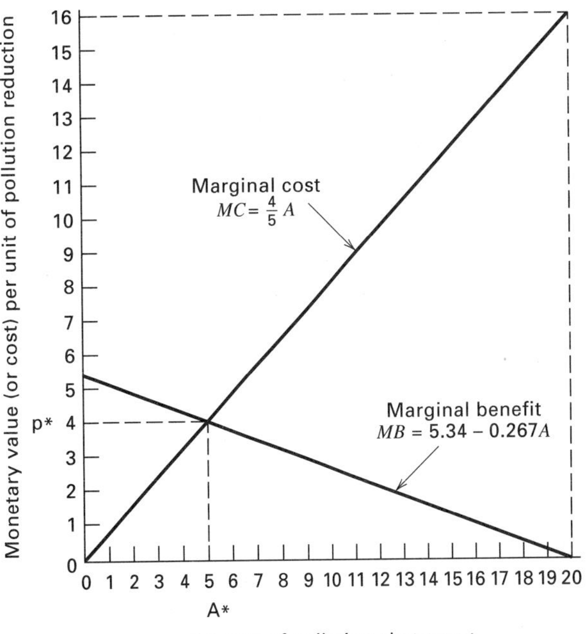

Figure 10.5 Productively efficient level of abatement.

The theory of effluent charges just outlined shows how the effluent charge should be set to attain the productively efficient level of abatement. However, the theory is of limited usefulness because it assumes that a marginal benefit function exists. As shown in Chapter 6, marginal benefit functions are difficult to estimate. A second limiting assumption is that the marginal benefit of pollution abatement is independent of where abatement occurs. The next section demonstrates an extension of the basic theory that eliminates both of these assumptions.

EFFLUENT CHARGES TO MEET AMBIENT STANDARDS

When marginal benefits of pollution reduction depend on where abatement occurs, transfer coefficients are used to extend the theory of effluent charges.[16] In the extension of the theory presented here, transfer coefficients are used to determine how wastewater discharges from different pollution sources degrade ambient water quality at various locations.

Although environmental agencies rarely estimate benefit functions, they often set ambient standards that reflect implicit valuations of the benefits of pollution abatement. Effluent charges can be applied in a context in which ambient water quality standards exist, even if incremental benefit curves cannot be estimated. In this setting, the central economic question involves *cost-effectiveness:* How should effluent charges be set to minimize the cost of attaining an ambient standard?[17]

The following analysis shows how to compute effluent charges when source location must be considered in calculating effects on ambient quality. The analysis builds upon the three-waste-source example presented in Chapter 8. The extension of the three-waste-source example demonstrates numerical computations used in calculating effluent charges based on realistic data.[18]

[16] Transfer coefficients were introduced in Chapter 8.

[17] Cost-effectiveness analysis is widely used when net-benefit maximization is infeasible because benefits cannot be measured in monetary terms.

[18] Recall that the three-waste-source example was modeled on calculations made in determining a pollution control program for waste dischargers on the Delaware Estuary.

Calculating Effluent Charges by Trial and Error

Figure 8.1 (in Chapter 8) summarizes elements of the three-waste-source example. Recall that the example finds the percentage removal of biochemical oxygen demand (bod) needed at each of three sources to meet a 6 mg/l dissolved oxygen (DO) standard at a point α. With this standard, 1 mg/l of dissolved oxygen in the stream provides waste receptor services. The removal percentages that meet the ambient standard at the lowest total treatment cost are $X_1 = 4.12\%$, $X_2 = 100\%$, and $X_3 = 46.12\%$, where X_i represents the reduction in BOD discharge at waste source i. This scheme costs \$4.09 million.[19] Any particular BOD reductions can be implemented using a command and control scheme based on effluent standards. If a polluter does not comply with standards, then penalties involving judicial and administrative procedures can be imposed.

As an alternative to effluent standards, an environmental agency could (in theory) meet the 6 mg/l DO target by charging each source a single, constant price per unit of BOD released. To see how an appropriate price could be set, assume the agency knows the cost of BOD reduction for each source, and it knows how each discharge affects the DO at point α. With this information, the agency can use trial and error to calculate the effluent charge, \$$p$ per 1000 lb of BOD discharged, that leads to attainment of the 6 mg/l DO standard.

To find the value of p, the agency assumes each discharger will decrease its raw load so that the sum of BOD reduction costs and effluent charges is at a minimum. In formulating an equation to represent these costs and charges, it is convenient to use the *capital recovery factor* (C), defined as

$$C = \frac{i(1 + i)^T}{(1 + i)^T - 1} \tag{10-6}$$

where i is the discount rate and T is the number of years over which annual costs are considered in the discharger's analysis. The capital recovery factor converts a single cost of \$$K$ incurred at the present time to an equivalent uniform series of annual costs, \$$CK$,

[19] The minimum-cost pollution control program is detailed in Table 8.6. Alternative programs are shown in Tables 8.5 and 8.7.

incurred over T years.[20] The definition of C is such that the *present value* of these T years of costs is equal to \$$K$. Suppose source 1 uses $i = 0.10$ and $T = 30$ years in its analysis of the costs of pollution abatement. Substituting these values into equation 10-6 gives $C = 0.106$. The present value of BOD reduction cost for source 1 is \$1000 $(X_1)^{1.9}$. This is equivalent to a 30-year series of annual pollution abatement costs equal to[21]

$$0.106\,(1000\,X_1^{\,1.9}). \qquad \textbf{(10-7)}$$

The charge p that leads to attainment of the 6 mg/l dissolved oxygen standard is found by analyzing the waste reduction costs and effluent charges faced by each discharger. Once more, consider source 1. The annual cost of BOD reduction is indicated by expression 10-7. The annual cost of the effluent charge is

$$365p\left[100\left(1 - \frac{X_1}{100}\right)\right]. \qquad \textbf{(10-8)}$$

The term in brackets represents the daily release by source 1 in 1000 lb/day of BOD (see Figure 8.1). For a single day, the effluent charge is equal to the daily BOD discharge times p (\$/1000 lb of BOD). Multiplying this product by 365 days/year gives the annual charge in expression 10-8. During any year, source 1 faces a total cost, $h_1(X_1)$, given by

$$h_1(X_1) = 106X_1^{\,1.9} + 36{,}500p\left(1 - \frac{X_1}{100}\right). \qquad \textbf{(10-9)}$$

Equation 10-9 is a simplified form of the sum of expressions 10-7 and 10-8, representing treatment cost and effluent charge, respectively. Using calculus,[22] the value of X_1 that minimizes the total annual cost is found to be

$$X_1 = (1.81\,p)^{1.11}. \qquad \textbf{(10-10)}$$

Conducting analogous cost-minimization computations for sources 2 and 3, the agency determines the percentages of BOD reduction for these sources as

$$X_2 = (2.86\,p)^5 \qquad \textbf{(10-11)}$$

and

$$X_3 = (2.29\,p)^2. \qquad \textbf{(10-12)}$$

To find the charge needed to meet the 6 mg/l DO standard, a *trial* value of p is selected, and the corresponding values of X_1, X_2, and X_3 are computed using equations 10-10 to 10-12. A check is then made to see if these removal percentages meet the 6 mg/l DO standard. Recall from the previous discussion of the three-waste-source example that the DO decrease (ΔDO) at point α is computed using[23]

$$\begin{aligned} \Delta\text{DO} = {} & \phi_{1\alpha}\left[100\left(1 - \frac{X_1}{100}\right)\right] \\ & + \phi_{2\alpha}\left[1000\left(1 - \frac{X_2}{100}\right)\right] \qquad \textbf{(10-13)} \\ & + \phi_{3\alpha}\left[500\left(1 - \frac{X_3}{100}\right)\right]. \end{aligned}$$

If the DO decrease is greater (less) than 1 mg/l, p is increased (decreased) and the entire calculation is repeated. For any waste source i, if the value of p results in X_i greater than 100%, X_i is simply set at

[20] The capital recovery factor is equal to the inverse of the *uniform series present worth factor* introduced in Chapter 7 (see equation 7-2). For details on how C is derived, see a textbook on engineering economy such as Grant, Ireson, and Leavenworth (1982, pp. 36–37). The derivation of C assumes the cost during any year is incurred at the end of the year. In computing C, i is the discount rate divided by 100.

[21] A more sophisticated analysis of a firm's cash outlays for treatment would include opportunities for tax writeoffs. Johnson's (1967) study of effluent charges on the Delaware Estuary shows how the computations here must be adjusted to account for tax considerations.

[22] The value of X_1 that minimizes $h_1(X_1)$ satisfies

$$\frac{dh_1(X_1)}{dX_1} = 1.9(106)X_1^{\,0.9} - 365p = 0.$$

Solving for X_1 yields the value in equation 10-10. The second derivative of $h_1(X_1)$ is positive for positive values of X_1, and this assures that X_1 found using equation 10-10 is a minimum.

[23] The basis for equation 10-13 is given in Chapter 8; see the discussion of equation 8-2.

TABLE 10.1 Using a Single Effluent Charge to Attain a DO Standard

Trial Value of p (\$/1000 lb)	$X_1 = (1.81\ p)^{1.11}$ (%)	$X_2 = (2.87\ p)^5$ (%)	$X_3 = (2.29\ p)^2$ (%)	Decrease in DO at α (mg/l)
1.0	1.93	100	5.2	1.62
2.0	4.17	100	21.0	1.38
3.0	6.54	100	47.2	0.98
2.96	6.43	100	45.8	1.00

100%. This trial-and-error procedure is completed when a p is found that gives $\Delta DO = 1$ mg/l.

Several steps in the procedure for determining the effluent charge are shown in Table 10.1. The first trial value, $p =$ \$1/1000 lb, is too low. It leads to a DO decrease of 1.62 mg/l, which exceeds the 1.0 mg/l limit. The effluent charge is increased in increments of 1.0 until it yields X_1, X_2, and X_3 that provide more than enough BOD cutback. This occurs at $p =$ \$3/1000 lb, a charge that produces a DO decrease of 0.98 mg/l. The charge that attains the 1 mg/l limit exactly is $p =$ \$2.96/1000 lb.

Now consider the question of cost-effectiveness: Does the final value of p yield removal percentages that meet the 6 mg/l DO standard at a minimum total cost? This question is examined using the cost equations in Figure 8.1 to compute the treatment costs that result when $p =$ \$2.96/1000 lb. These costs, as well as those for the minimum-cost solution calculated previously in Chapter 8, are given in Table 10.2.[24] The total cost using the charge program is greater than the overall minimum cost. Use of p = \$2.96/1000 lb is not cost-effective because a single, uniform effluent charge cannot account for differences in the locations of the sources. Effects of source location are reflected by relationships among the transfer coefficients: $\phi_{3\alpha} = 1.5\phi_{1\alpha} = 1.5\phi_{2\alpha}$.

A program with *nonuniform charges* can yield the cost-minimizing values of X_1, X_2, and X_3 by accounting for the way different sources influence water quality. This involves setting the charge for source 3 at 1.5 times the charge for sources 1 and 2. When this is done, and the computations for finding the charges are repeated, the removal percentages meeting the DO standard are precisely the cost-minimizing values. Some of the trial-and-error calculations are shown in Table 10.3.

Practical Problems in Determining Effluent Charges

The example involving three waste sources demonstrates how, in theory, effluent charges can be set to attain an ambient standard when discharge location influences ambient water quality. The example also shows that a system of nonuniform charges can be designed to meet an ambient standard at minimum cost.

In a real situation, the correct charges could not be determined precisely using the procedure in the foregoing example. One reason is that an environmental agency would have only a rough idea of the costs of waste reduction at each source. Moreover, even in the best circumstances, the agency would have only an approximate estimate of how changes in discharges influence ambient environmental quality. For many water quality indicators, transfer coefficients or equivalent water quality modeling results would not be available.

Another practical complication concerns the assumption that each discharger sets its percentage removal so that the waste reduction cost plus the effluent charge is at a minimum. This assumption is the basis for the trial-and-error process used in computing charges. Other factors, such as the dischargers' inability to raise money for treatment plant construction, may influence the selection of a percentage removal. In such cases, the decisions made by the polluter will not be predicted correctly using the theory.

Because of these complications, only luck would lead an agency to set effluent charges so that ambient

[24] The minimum-cost solution for the three-waste-source example is detailed in Table 8.6.

TABLE 10.2 Treatment Costs Using Effluent Charges and the Minimum-Cost Approach

	Removal Percentages That Minimize Cost		Removal Percentages Using p = $2.96/1000 lb	
Waste Source	**Percentage of Waste Removed**	**Cost ($10³)**	**Percentage of Waste Removed**	**Cost ($10³)**
1	4.12	15	6.43	34
2	100	2510	100	2510
3	46.12	1566	45.82	1551
Total Cost		4091		4095

standards were met precisely. If the charge were set too low, the standard would be violated. If, in the preceding example, a single uniform charge of $1/1000 lb were used (instead of the required $2.96/1000 lb), then $\Delta DO = 1.62$ mg/l instead of the maximum 1.0 mg/l decrease associated with meeting the DO standard (see Table 10.1). If the charge were set too high, the stream DO would be greater than required. More resources would be devoted to BOD reduction than are called for by the ambient standard.

Some economists believe that the theoretically correct effluent charges can be set using an iterative scheme. For instance, suppose the agency's cost estimates are in error. The agency sets the charge at $1/1000 lb, thinking that this would be just enough to meet the 6 mg/l DO standard. After the polluters respond to the effluent charges, the agency observes that the DO decrease at α is 1.62 mg/l instead of the expected 1.0 mg/l. Under these circumstances, the agency could revise the charge upward, wait for the dischargers to respond to the increase, and then check to see if the 6 mg/l standard is met. Continuing with this iterative approach, the charge would eventually increase to a level that just meets the standard.

An iterative process for setting charges confounds the decision making of polluters. They may respond to the initial charges by investing in facilities and equipment. Decisions based on one set of charges may be difficult to modify in response to new, higher charges. Because long-term resource commitments are involved, investments made in reaction to the initial, lower charges might have to be built into subsequent decisions involving facilities and equipment. This means the minimum-cost solution cannot be attained. Moreover, dischargers cannot be expected to react favorably to a regulatory program that changes frequently.

EFFLUENT CHARGES AND SUBSIDIES IN EUROPE

Effluent charges have been used for many years in the water quality management programs of France,

TABLE 10.3 Nonuniform Effluent Charges to Attain Ambient Standard at Minimum Cost

Trial Value of p_1 (and p_2) ($/1000 lb)	**$p_3 = 1.5\,p_1$ ($/1000 lb)**	**$X_1 = (1.81\,p_1)^{1.11}$ (%)**	**$X_2 = (2.87\,p_2)^5$ (%)**	**$X_3 = (2.29\,p_3)^2$ (%)**	**Decrease in DO at α (mg/l)**
1.0	1.5	1.93	100	11.8	1.52
2.0	3.0	4.17	100	47.2	0.98
1.977	2.966	4.12	100	46.13	1.00

Germany, and the Netherlands.[25] None of these countries set charges based on the theoretical principles economists employ to rationalize the use of charges. Moreover, in each country, polluters were motivated to reduce their wastes by factors other than effluent charges. Typically, charges were integrated into a command and control scheme including discharge standards and government *subsidies* to offset partially the costs of pollution abatement.

What do experiences from France, Germany, and the Netherlands illustrate about how effluent charges have been used in practice? Results from France demonstrate that charges have supported new water management institutions created to reinvigorate what was once an ineffective discharge permit system. The Dutch experience shows that, if charges are set high enough, they significantly influence the behavior of polluters. Finally, data from Germany demonstrate that even when effluent charges are well below average cleanup costs, charges can still cause firms to search for innovative waste reduction schemes.

Charges Support River Basin Agencies in France

The effluent charge system in France generates revenues that support six French river basin agencies created in the 1960s.[26] These agencies carry out planning and research and provide loans and grants for water and wastewater management projects, but they do not construct facilities or issue regulations.

Effluent charges are set through a complex negotiation process involving a river basin agency's staff, the agency's "basin committee," and a host of government officials. The basin committee includes representatives from various French ministries and from municipalities and other water users. Negotiations in setting charges account for how much money will be needed to reach different cleanup targets, since a portion of the charges collected are used by the river basin agency to subsidize wastewater treatment by firms and municipalities.[27] Although the diversity of interests represented in establishing charge levels sometimes slows the decision-making process, it increases the chance of developing a program of action that is fair, implementable, and not challenged in court.

The French effluent charge system includes the following parameters: oxidizable material as measured by a combination of chemical oxygen demand (COD) and biochemical oxygen demand (BOD), suspended solids (SS), phosphates, and toxic materials.[28] For any one year, a base effluent charge for a river is set to yield the total revenues needed by the river basin agency to meet water quality goals for the year. Suppose, for example, the base effluent charge for suspended solids for one year is the equivalent of about $10 per year per kg/day of SS discharged. This charge is multiplied by the pollutant discharge (in units of kg/day of SS) to compute a polluter's annual fee for suspended solids. The pollutant discharge used in calculating the annual fee is based either on periodic measurements (for large sources) or standard tabulated values per unit of output (for example, 10 gms of SS per kg of finished kraft paper). When tabulated values are used, it is necessary to know the quantity of finished product produced per unit time.

Effluent charges can be raised or lowered, depending on changes in the scope of the action program for a particular river basin. In addition, charges can vary from one zone to another. Thus waste sources in zones with major water quality problems and costly cleanup programs may have to pay relatively high charges. However, those waste dischargers may be eligible for subsidies, grants, and loans that the river basin agency provides using revenues from charges it collects.

---

[25] For introductions to the charge schemes used in these and other European countries, see Johnson and Brown (1976) and Andersen (1994).

[26] This discussion of the effluent charge program in France draws heavily from Barré and Bower (1981).

[27] Because water users and wastewater dischargers are represented on basin committees, they influence how much river basin cleanup will be attempted and how high the effluent charges will be. Many organizations are represented on basin committees, and committee members form alliances to block effluent charges viewed as being too high or too low.

[28] Chemical oxygen demand and biochemical oxygen demand are standard measures of the organic matter present in wastewater. Masters (1991) gives details on how BOD and COD are calculated. In the French effluent charge system, toxics are measured in terms of "equitox," a unit based on effects of effluents on the survival of daphnids; see Barré and Bower (1981) for details.

During the 1970s, the effluent charges used by the basin agencies were much lower than required to motivate dischargers to treat their wastewaters. Why, then, did so many polluters implement high levels of treatment during that period? The answer is found by considering a complementary part of French water quality management strategy: effluent standards. Since 1917, the prefects of French "departments" have issued permits controlling the discharge of wastewaters.[29] Penalties for permit violations include fines and court-imposed sanctions on dischargers. Before the basin agencies were formed in the 1960s, the discharge permit system was not implemented effectively. Even though basin agencies do not issue permits, their programs of charges and subsidies have motivated polluters to reduce their wastes in compliance with permit conditions. The use of a two-part strategy—economic incentives administered by basin agencies, and permit requirements administered by prefects—has led to substantial water quality improvements.

Dutch Experience with How Charges Influence Firms

As in France, new institutions for managing water quality were created in the Netherlands in the 1960s. During that period, a law requiring adoption of a nationwide system of effluent charges was adopted.[30] In the Dutch system, effluent charges are levied on oxidizable material (as measured, for example, using chemical oxygen demand) and heavy metals, such as mercury and zinc. The Dutch effluent charge scheme was initiated primarily to generate government revenues to subsidize waste management efforts. However, the charge itself has proven to be effective in motivating firms to abate pollution. One reason the Dutch charge scheme has affected behavior is that effluent charges are comparable to the incremental cost of waste reduction for some dischargers.[31]

Soon after the introduction of effluent charges, wastewater releases decreased dramatically in the Netherlands. For example, between 1970 and 1980, the discharge of oxygen-consuming wastes by industry dropped by more than half, even though industrial output expanded significantly over the same period. Could the falloff in pollution be attributed to the use of effluent charges? Or were there alternative explanations for the cutbacks?

Several research studies were conducted to determine the influence on polluters of the Dutch effluent charge system. Some studies used statistical methods to identify factors correlated with the sharp decrease in water pollution in Holland during the 1970s. Other investigations relied on survey questionnaires and interviews with government officials and corporate decision makers. While the research studies yielded some differences in details, they were consistent in pointing to effluent charges as a key factor affecting the behavior of polluters. For example, statistical analyses and survey-based research showed effluent charges as being much more significant than the Dutch permit requirements in causing industry to reduce organic pollution. Moreover, subsidies for pollution control, the original motivation for developing the charge system, seemed less important than either discharge permits or effluent charges in encouraging firms to abate pollution.

Results of interviews with environmental officials indicate that the Dutch charge system has provided regulators with an influential addition to the command and control scheme that had been in place. The response by one government official characterizes how effluent charges complement other regulatory instruments in the Netherlands: "When I'm going to have a talk with a company about the abatement of their pollution, I always take my pocket calculator along. I calculate their potential savings on charges and invariably get an interesting conversation going."[32]

Lessons from the German Experience with Charges

The earliest applications of effluent charges were by river basin agencies in the Ruhr Valley of Germany

---

[29] France has over 90 departments, and each performs as a local unit of the national government. Each department is headed by a prefect (officially called a Commissionaire of the Republic) representing the national government. For an introduction to the traditional French administrative system and how it relates to the six river basin agencies, see Andersen (1994, chap. 5).

[30] This section is based on Bressers (1988) and Andersen (1994, chap. 7).

[31] Interestingly, the Dutch charges are several times higher than charges used in France.

[32] This quotation is from Bressers (1988, p. 513).

in the 1920s.[33] During the 1970s, the application of effluent charges was extended to all of what was then the Federal Republic of Germany (FRG or West Germany). An effluent charge law passed in 1976 required the Länder (which corresponded to states in the FRG) to levy charges on effluents released into public waterways. Uniform charge rates were set for the nation as a whole, but implementation was carried out by individual Länder.

Administration of the charge scheme was tied to a discharge permit system. The Länder issued permits to sources of wastewater discharge. Based on the permit system, a polluter is given a right to release specified quantities of wastewater, but the concentration of pollutants must be below those specified by uniform national discharge standards, or by local discharge standards. (The latter may be set more rigorously than national standards.) A second component of a permit details the data needed to calculate a polluter's waste discharge bill. Using a rate schedule set at the federal level, effluent charges are levied on the following parameters: settleable solids, chemical oxygen demand, cadmium, mercury, and toxicity for fish. Details of the computation of total charges are complex, because the basic charge is in units of deutsche marks per "damage unit." The effluent charge law spells out how to convert from quantities of pollutant to damage units (for example, 45.45 kg of chemical oxygen demand corresponds to one damage unit). The same charge per damage unit is applicable to all polluters in all regions of the country.

Based on their analysis of the FRG effluent charge system, Brown and Johnson (1984, pp. 962–63) argue that to be politically viable and administratively attractive, an effluent charge system should have the following characteristics:

(1) It covers a small number of pollutants;
(2) It is combined with permit systems;
(3) The charges begin at some specified level and escalate during a transition period;
(4) The charge levels result from a process involving the participation of interested parties including those benefitted and harmed by waste discharge;
(5) Measures and levels of volumes and pollution concentrations are simplified;
(6) Effluent charge revenues are made available for abatement related expenditures . . . ;
(7) Hardship clauses are provided to protect dischargers or industrial sectors under exceptional circumstances; [and]
(8) Care is taken to demonstrate how the effluent charge program actually can be implemented.

The charge system used in the FRG had features encouraging firms to meet effluent standards. Under the FRG scheme, if applicable wastewater discharge standards were met, a polluter's effluent charge bill would be cut in half. If standards were violated, a polluter lost the opportunity to save 50% of total charges *and* faced fines and other penalties for violating standards. In addition, revenues from charges were used to subsidize waste reduction activities by both firms and municipalities, and subsidies further encouraged polluters to meet discharge standards. Firms could even apply for subsidies to offset the cost of waste-reducing changes in production processes. As a result of actions at least partially motivated by the effluent charge program, more than half the waste dischargers in the FRG met effluent standards in 1981, and in one Länder the figure was 90%.[34]

In appraising the influence of the charge scheme on pollution abatement by firms, Brown and Johnson (1984) cited the experience of BASF, a large chemical company in the FRG. The firm made numerous innovative efforts to reduce wastes, even though effluent charges in the FRG were widely acknowledged as being much lower than required to induce firms to achieve the nation's water quality goals. Of special note was BASF's development of an *intra-firm* effluent charge scheme for reducing chemical oxygen demand. The firm computed an internal effluent charge bill for each of its branches by introducing an accounting price per unit of waste and multiplying it by the total effluent generated by the branch. Brown and Johnson (1984, p. 944) summarized the results from applying the internal charge scheme over a seven-year period:

> The response to the introduction of an internal liability system has been a 20 percent decrease in

[33] The discussion of the effluent charge system in Germany is based on Brown and Johnson (1984).

[34] These figures are from Brown and Johnson (1984, p. 945).

> discharge. Rather than mandate physical decreases the intra-firm charge elicited a "voluntary" decrease in effluent discharge achieved through process change, recycling of solvents, improved pretreatment facilities and replacement of old facilities.

Brown and Johnson went on to argue that "Even if the [effluent] charge is modest, it induces cost savings."

DISCHARGE FEES AND SUBSIDIES IN CHINA

Although it may not be surprising that effluent charges have been embraced by countries with market economies, charge schemes have also been adopted by nations with centrally planned economies.[35] As an example, consider the use of pollutant discharge fees in China during the early 1980s, a time when China started modifying its economic system to rely more on markets.

The impetus for introducing fees on pollutant discharges came from Chinese environmental management experts familiar with effluent charges in Europe.[36] During the late 1970s, these experts argued that China could benefit from a discharge fee program because the fees would enhance productive efficiency and give enterprises incentives to abate pollution. China's 1979 Environmental Protection Law included a system of pollutant discharge fees. Subsequent policy guidance and additional legislation provided details that local environmental protection bureaus (EPBs) needed to implement the system.[37] Although the guidance is for a national charge scheme, EPBs can modify the national system to accommodate local conditions, provided the local system is at least as demanding as the national scheme.

Fees and Their Influence on Enterprises

In a typical application of the national pollutant discharge fee system, an enterprise pays fees *only if* applicable discharge standards are violated. The amount owed by an enterprise violating standards is based on the extent of violation.[38] For example, consider the fee schedule for chemical oxygen demand used in Guangdong Province in southern China. If a firm's wastewater discharge has a COD concentration between one and two times the applicable COD effluent standard, 110 mg/l, the enterprise pays 0.04 yuan (RMB) per cubic meter (¥/m^3) of discharge.[39] However, if the enterprise's COD exceeds the standard by a factor of ten, the applicable fee is 0.06 ¥/m^3. Guangdong Province uses formulas to calculate unit fees (in ¥/m^3) based on the extent of violation of the standard.

The influence of the Chinese discharge fee system on the behavior of polluters differs depending on whether the discharges consist of nonhazardous organic wastes or hazardous materials such as heavy metals. First, consider fees on organic wastes (measured using COD). During the 1980s, pollutant discharge fees on organic wastes in China were generally too low to affect polluters. Although many factories cut back their waste discharges, they did so because of pressures unrelated to the fee system.[40] Fees were not influential because operation and maintenance costs for treating organic wastes were often much

---

[35] This section benefited from the contributions of three of my colleagues: Xiaoying Ma, Barbara Sinkule, and Kimberly Warren.

[36] This observation is from Ross's (1988) description of the early stages of the Chinese pollutant discharge fee program. Interestingly, economists in the Soviet Union were advocating use of pollutant discharge fees during the same period (National Academy of Public Administration, pp. 76–77).

[37] The role of EPBs in implementing policy was detailed in the section on Environmental Policy Implementation in China in Chapter 9.

[38] For information on how these fees are calculated for COD and other pollution parameters, see Sinkule and Ortolano (1995). As they point out, enterprises may be subject to additional charges; for example, a penalty of 0.1% per day of lateness for fees not paid on time.

[39] In China, wastewater flowrates are commonly exxpressed in units of m^3/day or metric tons/day; one cubic meter of water equals one metric ton. "RMB" is short for renminbi, which is Chinese for "the people's money," and yuan is the basic unit of money. In 1991, one yuan was approximately equal to 0.20 U.S. dollars.

[40] Support for this assertion is given in factory case studies reported by Sinkule and Ortolano (1995). However, for some enterprises in some Chinese cities, discharge fees (on organic wastes) are high enough to influence waste management decisions. Examples can be found in Anyang City in Henan Province. In the early 1990s, Anyang City revised its basis for calculating fees so that enterprises paid fees for discharges below applicable standards plus fees for discharges exceeding standards. The revised scheme in Anyang City yields average discharge fees that are much higher than those associated with either the national fee system or the fees in Guangdong Province.

greater than applicable fees. In such cases, even if the costs of constructing treatment plants at factories had been fully subsidized, fees would still have been ineffective as an incentive to clean up. It was much cheaper for the factories to simply pay fees.

Discharge fees for hazardous materials were often relatively high (compared to fees on COD), and they had a greater influence on polluters. For example, consider the release of cadmium, chromium, and other dangerous metals from electroplating factories. Managers of electroplating enterprises often faced intense pressure for cleanup from local EPBs and residents downstream of wastewater discharges. In this context, factory managers viewed pollutant discharge fees as one more reason to abate pollution. This outcome is consistent with some experiences in Europe: discharge fees can motivate cleanup even when fees are lower than incremental costs of waste reduction.[41]

Use of Revenues Generated by Discharge Fees

About 80% of the revenue generated by China's pollutant discharge fee system is placed in a fund to subsidize waste treatment by enterprises that pay fees.[42] Subsidy programs have varied over time. At the outset, factories simply received 80% of the fees they paid as a rebate to be used in reducing pollution. However, many factories spent their rebates on activities unrelated to pollution control.

Once EPBs learned that the 80% rebates were being misused, they reformed their subsidy programs. A common modification was to offer a subsidy as a low-interest loan instead of a nonrepayable rebate or grant. Another change involved approving an enterprise's application for a subsidy only if the enterprise could show it was able to undertake the treatment work detailed in its application. For example, an enterprise might be required to demonstrate that it had a substantial portion of the funds needed for treatment plant construction. Some EPBs used the following scheme to minimize the misuse of subsidies: If a factory built the treatment facility it detailed in its application for a subsidy, *and* if the facility passed a field test of its effectiveness, the factory would have to pay only the interest (but not the principal) on the loan.

As noted, discharge fees (for nonhazardous organic pollutants) are often lower than the costs of operating treatment plants. In such cases, even if treatment plant construction could be fully subsidized, it would not make financial sense to reduce wastes by building and operating a treatment facility.

If discharge fees and subsidies based on fees don't often motivate Chinese enterprises to reduce pollution, what accounts for the enthusiasm with which EPBs have implemented the discharge fee and subsidy program? The answer is revenues. Local environmental protection bureaus typically retain about 20% of the fees they collect.[43] In principle, revenues that EPBs retain from the discharge fee program are earmarked to purchase monitoring equipment and otherwise support EPB inspection programs. In practice, however, EPBs have used fee-generated revenues for many other purposes, including hiring new staff and covering routine operating expenses.[44] Indeed, some EPB staff are said to "eat pollutant discharge fees," a phrase used when staff positions are funded entirely from discharge fees.

Fee-generated revenues to EPBs have grown so large that some observers worry about EPBs becoming dependent on them as an income source.[45] If enterprises comply with discharge standards, fees collected only for discharges exceeding standards would

---

[41] These observations on the influence of fees on pollution abatement at electroplating factories are based on Warren (1996). Comparable European outcomes (regarding the influence of fees on polluters) were reported in the discussion of effluent charges in Holland and Germany.

[42] The 80% figure applies in Guangdong Province. Some provinces in China have negotiated different percentages with national authorities. For example, in Henan Province, only 75% of the fee is used for subsidies; the remaining 25% is divided between local EPBs (17%) and the provincial EPB (8%). Also, the 80% figure applies only to pollutant discharge fees per se. Other revenues are generated by the fee program, and they are often used to fund EPB activities. For example, enterprises pay penalty charges for late fee payments, and they pay "double fees" if they violate discharge standards *and* have treatment works that were approved under the three synchronizations policy (see Chapter 9).

[43] In addition, EPBs often retain 100% of other fee-related payments, such as those mentioned in the previous footnote.

[44] Sinkule and Ortolano (1995) report on one city in southern China where the 1990 revenues from fees accounted for 60% of the EPB's entire operating budget.

[45] This point is documented in Wang and Blomquist (1993, p. 71).

disappear. Does this give EPBs an incentive to relax their efforts in forcing enterprises to meet standards? This question has not been debated extensively because there are so many enterprises out of compliance with regulations in China. Nevertheless, it is notable that the issue has even been discussed.

Fees and Subsidies As Supplements to Command and Control

If discharge fees and subsidies are not sufficient to cause most Chinese enterprises to abate pollution, what motivates waste cutbacks? Analyses of about 25 case-study enterprises conducted by Sinkule and Ortolano (1995) and Warren (1996) highlight numerous motivating factors:

- Enterprises violating discharge standards are subject to fines and other penalties that EPBs can impose.
- For factories undergoing major expansions (and for new factories), EPBs can often arrange to have operating licenses withheld unless acceptable waste treatment systems are in place.[46]
- Some enterprises implement pollution prevention measures such as materials recycling because these measures enhance profits.
- Enterprises discharging heavy metals and other hazardous materials often face intense pressure from affected citizens who complain directly to factories or to EPBs.
- Cities sometimes conduct areawide cleanup programs in which a city mayor and relevant city bureaus provide financial and technical assistance to support pollution abatement by enterprises.

As in European countries that use effluent charges, the discharge fees and fee-based subsidies in China complement other policies and programs used to control pollution.

Lessons About Pollutant Discharge Fees from China

Except for the case of hazardous wastes, the main influence of the Chinese system of pollutant discharge fees in the 1980s appears unrelated to the behavior of polluters. Many enterprises paid fees as a cost of doing business, but did little else in response to the fee system. Arguably, the most significant influence of the Chinese system has been in fueling the growth of local environmental protection bureaus. Many EPBs in China have used revenues generated by discharge fees to hire staff and expand their data collection and inspection programs. Moreover, some EPBs have used revenues from fees to cover expenses that were different from those officially allowable by guidelines on the use of funds.

The Chinese fee program demonstrates how participants in a regulatory system can manipulate it to meet their own goals. In the Chinese experience during the 1980s, the participants taking the greatest advantage of the charge scheme were the regulators themselves. In addition, many enterprises eligible for subsidies based on fees worked the system by using subsidies earmarked for waste reduction to carry out unrelated activities.

The Chinese experience also demonstrates the difficulty of setting fees at levels required to enhance productive efficiency. Officials of the Chinese National Environmental Protection Agency (NEPA) knew the original discharge fees were too low to affect many polluters, but the agency lacked the political clout to set fees higher. NEPA could only propose discharge fees. The final decision on fees was made by the powerful Ministry of Finance, which was concerned with the economic health of industries and was not willing to set the fees as high as NEPA wanted them.[47]

In addition to influencing the level of fees, the economic health of enterprises affects whether fees are even collected. The case of a large state-owned paper mill that was not prospering during the early 1990s illustrates problems in collecting fees. At the time, the paper mill employed several hundred workers and was paying retirement benefits for many former employees.[48] Under the circumstances, the enterprise refused to pay the discharge fees it owed the

[46] The three synchronizations policy is the principal mechanism for ensuring that new (or renovated) factories have waste treatment systems (see Chapter 9).

[47] During the first 11 years of the fee program's existence, the Chinese government raised the fees only once (in 1990). Even then, the 25% rate increase did not approach the increased costs of waste treatment due to inflation in the same period.

[48] For large, old state-owned factories in China, the payment of benefits to retirees is an important and substantial obligation.

city EPB. The poor financial health of the mill and its role in employing a large workforce and paying retirement benefits to former employees convinced the EPB not to try to collect the fees. Although this case is not typical, it is not rare.

Experiences in China and Europe show that discharge fee systems cannot easily be made to yield productively efficient outcomes. In practice, government regulators have long recognized the difficulty and uncertainty in setting prices "correctly." Thus discharge fee systems are invariably used in tandem with a host of traditional regulatory instruments, including ambient standards and discharge standards. Even though practical discharge fee programs cannot yield the idealized outcomes associated with economic theory, fees can provide economic incentives to spur cleanups. Moreover, fee programs can provide revenues for chronically underbudgeted environmental protection agencies and for subsidy programs that make it easier for polluters to undertake cleanups.

Key Concepts and Terms

Effluent Charges and Productive Efficiency

Polluter pays principle
Social cost vs. private cost
Pigouvian tax
Internalizing external costs
Marginal cost of waste reduction
Marginal benefit
Productively efficient level of pollution abatement
Equalization of marginal costs

Effluent Charges to Meet Ambient Standards

Cost-effectiveness analysis
Trial-and-error computation of charges
Present value of costs
Capital recovery factor
Firms' response to fees
Transfer coefficients
Nonuniform effluent charges

Effluent Charges and Subsidies in Europe

Motives for implementing charges
French river basin agencies
Charges based on cleanup goals
Discharge permit systems
Charge per damage unit
Fee-based subsidy programs
Characteristics of viable charge schemes

Discharge Fees and Subsidies in China

Charges in relation to discharge standards
Influence of fees on enterprises
Charges as revenues for agencies
Grants and low-interest loans
Misuse of subsidies
Political opposition to high fees

Discussion Questions

10-1 Government policy analysts would like to use an effluent charge to bring about compliance with an ambient standard. They want to estimate what each firm would do in response to a charge of 10¢ per unit of waste discharged. The analysts estimate that one of the firms in question has a raw waste load of 20 units per day, and that total cost of waste reduction for the firm is as follows:

Percent of Raw Load Reduction	Present Value of Total Cost ($)
25	100
50	400
75	1000
85	1500
95	2500

For purposes of converting present value of total cost to an equivalent uniform series of annual costs, the analysts assume the firm would use a discount rate of 12% and a time horizon of 20 years.

(a) Which of the percent reductions in the table would the firm use if it were faced with a charge of 10¢ per unit waste and it wanted to minimize the sum of waste reduction costs and effluent charges?

(b) How much higher would the effluent charge have to be before the firm shifted to the next higher percent reduction?

10-2 You are asked to perform a comparative analysis of effluent charges and effluent standards as policy instruments for attaining ambient water quality standards at a minimum total cost of pollution abatement. Possible attributes to use in a comparison include incentives to firms to develop new pollution abatement technologies, and requirements for monitoring by environmental agencies. What other attributes would you include in a comparative analysis? Use your list of attributes as the framework for comparing effluent charges and effluent standards.

10-3 You are hired as a consultant to the Ministry of Environment in Colombia. The ministry is planning to design and implement an effluent charge program. How would you advise the ministry on how to proceed based on lessons learned using effluent charges in Europe and China?

10-4 In 1993, the Ministry of Environment in Colombia explored how it could use revenues generated by effluent charges to increase the rate at which firms and municipalities were abating pollution. What advice would you offer the Ministry on the following questions: Should fee-based revenues be used to subsidize waste treatment by municipalities only or by both firms and municipalities? Should subsidies be in the form of grants or low-interest loans?

10-5 American economists have frequently urged the U.S. Congress to implement a national program of effluent charges, but these efforts have failed. Provide a possible explanation for why Congress has not legislated a national effluent charge scheme.

10-6 The environmental economics literature often makes a sharp distinction between policies based on command and control, and market-based policies that depend on pollutant discharge fees. Is such a distinction useful given the way discharge fee programs are implemented? Do these policies differ in terms of the level of government involvement in policy implementation?

References

Andersen, M. S. 1994. *Governance by Green Taxes: Making Pollution Prevention Pay.* Manchester, U.K.: Manchester University Press.

Barré, R., and B. T. Bower. 1981. "Water Management in France, with Special Emphasis on Water Quality Management and Effluent Charges." In *Incentives in Water Quality Management, France and the Ruhr Area,* ed. B. T. Bower, R. Barré, J. Kühner, and C. S. Russell, 31–209. Research Paper R-24. Washington, DC: Resources for the Future.

Bressers, H. T. A. 1988. A Comparison of Incentives and Directives: the Case of Dutch Water Quality Policy. *Policy Studies Review* 7 (3): 500–18.

Brown, G., and R. W. Johnson. 1984. Pollution Control by Effluent Charges: It Works in the Federal Republic of Germany, Why Not in the U.S. *Natural Resources Journal* 24 (October): 929–66.

Grant, E. L., W. G. Ireson, and R. S. Leavenworth. 1982. *Principles of Engineering Economy,* 7th ed. New York: Wiley.

Gue, R. L., and M. E. Thomas. 1968. *Mathematical Methods in Operations Research.* London: Macmillan.

Hahn, R. W. 1989. *A Primer on Environmental Policy Design.* Chur, Switzerland: Harwood Academic Publishers.

Hudson, J. F., E. E. Lake, and D. S. Grossman. 1981. *Pollution-Pricing: Industrial Responses to Wastewater Charges.* Lexington, MA: Lexington Books.

Johnson, E. 1967. A Study in the Economics of Water Management. *Water Resources Research* 3(2): 291–305.

Johnson, R. W., and G. W. Brown, Jr. 1976. *Cleaning up Europe's Waters: Economics, Management and Policies.* New York: Praeger.

Masters, G. 1991. *Introduction to Environmental Engineering and Science.* Englewood Cliffs, NJ: Prentice–Hall.

National Academy of Public Administration (NAPA). 1994. *The Environment Goes to Market: The Implementation of Economic Incentives for Pollution Control.* Washington, DC: NAPA.

Noll, R. G. 1994. "Instrument Choice in Environmental Policy." Unpublished manuscript, Dept. of Economics, Stanford University, Stanford, CA.

Oates, W. E. 1996. *The Economics of Environmental Regulation.* Cheltenham, U.K.: Edward Elgar.

Opschoor, J. B., and H. B. Vos. 1989. *Economic Instruments for Environmental Protection.* Paris: Organization for Economic Cooperation and Development.

Pigou, A. C. 1932. *The Economics of Welfare,* 4th ed. London: Macmillan.

Ross, L. 1988. *Environmental Policy in China.* Bloomington: Indiana University Press.

Sinkule, B., and L. Ortolano. 1995. *Implementing Environmental Policy in China.* Westport, CT: Praeger.

Wang, X., and R. Blomquist. 1993. The Developing Environmental Law and Policy of The People's Republic of China: An Introduction and Appraisal. *The Georgetown International Environmental Law Review* 5: 25–75.

Warren, K. A. 1996. *Going Green in China: An Organization Theory Perspective on Pollution Prevention in Chinese Electroplating Factories.* Ph.D. diss., Dept. of Civil Engineering, Stanford University, Stanford, CA.

Chapter 11

Tradable Pollution Permits

An innovative way of charging for use of the environment's ability to assimilate waste involves creating legal *rights to pollute* and allowing the rights to be bought and sold like ordinary commodities.[1] Rights are set out in pollution permits. A system of tradable pollution permits can ensure that ambient environmental quality targets are met while allowing polluters to manage their wastes cost-effectively. The approach accommodates economic growth by requiring new sources of waste to buy pollution permits from others before the new sources are allowed to release wastes. In addition, political resistance to the tradable permits idea can be minimized by giving the initial set of permits to existing polluters.

The theory behind tradable (or marketable) permits to pollute is developed in the first half of this chapter. The theory demonstrates that under certain conditions, marketable permits to pollute can yield productively efficient abatement levels. Later sections of the chapter examine applications of marketable permits to reduce air pollution in the United States. The chapter ends by comparing tradable permits to pollute with two other policies: pollutant discharge fees, and command and control.

[1] This idea dates at least as far back as Dales (1968), whose book on the subject is widely viewed as a seminal work in this field.

BASIC THEORY: BENEFITS INDEPENDENT OF WASTE SOURCE LOCATION

To establish that a productively efficient outcome is attainable using marketable discharge permits, consider the first example presented in Chapter 10. The example involved two firms, and Figure 10.4 shows the marginal cost of waste reduction for each firm. The example assumes that incremental benefits of a unit of pollution abatement are independent of which firm carries out the abatement.[2] This assumption is satisfied, for instance, in controlling sulfur dioxide (SO_2) emissions from coal-fired power plants to reduce acid rain. Such SO_2 control programs focus on total emissions, not on which particular sources contribute to the total.

Suppose an environmental regulatory agency had perfect knowledge of the marginal costs and benefits in Figure 10.5. (Recall from Chapter 10 that the marginal abatement cost in the figure represents the minimum incremental cost of waste reduction for *both* firms to attain a given level of abatement A.)

[2] The more complex case, in which discharge location influences benefits, requires the use of models linking waste reductions to improvements in ambient quality. This case, which is taken up later in this chapter, can be handled using transfer coefficients that reflect results from transport modeling studies.

As demonstrated in the previous discussion of Figure 10.5, a productively efficient level of pollution abatement occurs where the curves of marginal cost and marginal benefit intersect. Chapter 10 showed that a regulatory agency can induce the two firms to attain this efficient level of abatement by imposing a Pigouvian tax. This is a *priced-based* approach. The analysis that follows demonstrates how this same efficient abatement level can be reached using a *quantity-based* approach: a fixed number of tradable permits to pollute is issued, and permit holders buy and sell permits until no opportunities for mutually beneficial exchanges exist.

In a tradable pollution permits scheme, a waste discharger can only release a quantity of waste corresponding to the number of pollution permits it holds. Assume this requirement is strictly enforced. Otherwise, dischargers might not have sufficient incentive to comply with the requirement to have pollution permits.

In designing a tradable permits scheme, the regulatory agency must choose the total number of pollution permits to issue. How can the agency decide? To examine this question, consider the example associated with Figure 10.5.[3] Suppose a right to discharge waste is spelled out in a tradable permit called a "pollution allowance." For the example in Figure 10.5, a firm holding a *pollution allowance* has a legal right to release one unit of waste. (Here the word "allowance" signals that a permit to pollute has a particular meaning in the context of an example.)[4] Assume that a pollution allowance has no value except as permission to release waste, and that firms do not speculate in allowances or hoard them. Assume also that firms can buy and sell only integer values of allowances; sales of fractions of allowances are not permitted. Furthermore, assume that transactions costs are zero. In this context, *transactions costs* include the costs for buyers and sellers to identify each other, negotiate over price, and finalize a sale. For the example in Figure 10.5, the agency can ensure that pollution abatement is at the productively efficient level if it issues only 15 pollution allowances. As indicated in Figure 10.5, this number is the difference between the 20 units of waste load generated and 5 units of waste discharge reduction. The latter corresponds to the efficient level of pollution abatement identified in Chapter 10.

To understand why the productively efficient outcome will result, consider the decision processes of individual firms. If firm i holds N_i allowances and has a raw load of R_i, it can do one of three things. The firm can discharge N_i units of waste, which requires it to reduce its waste load by $R_i - N_i$. Alternatively, it can sell one or more of its allowances. However, if the firm sells an allowance, it must increase its abatement level by one unit. Finally, the firm can buy some allowances and do less pollution abatement. Assume that in deciding on a course of action, a firm's goal is to minimize the cost of dealing with its waste.

In analyzing what a firm holding N_i pollution allowances would do, it is convenient to introduce the concept of a *trading equilibrium,* a distribution of allowances in which firms will not find it advantageous to make trades.[5] At a trading equilibrium, the maximum price that any one firm would offer to purchase an allowance is less than the minimum price that any other firm would accept to sell an allowance. One way to identify whether firms would trade pollution allowances is to calculate the maximum offer price and minimum acceptance price for each firm. These calculations are possible, because the value to a firm of buying an allowance is the reduction in its total abatement cost, and the cost of selling an allowance is the corresponding increase in its total abatement cost.

Computations to identify whether a trading equilibrium has been reached are demonstrated by considering an arbitrary distribution of the 15 pollution allowances issued by the regulatory agency: $N_1 = 5$ and $N_2 = 10$. Consider firm 1's options. Because the firm holds 5 allowances, it would have to cut its raw load of 8 units down by 3 units in order to comply with environmental requirements. Instead of making a 3 unit reduction, firm 1 could try to purchase another pollution allowance. In that case, the firm's required abatement level would drop from 3 to 2, and its abatement cost would go down by 5.

[3] This presentation of the theory of tradable pollution permits is adapted from an unpublished manuscript by Noll (1994).

[4] In this chapter, the words "right" and "permit" are used in a general sense, whereas "allowance" is used to refer to a permit that grants a particular right to pollute.

[5] It will be shown that a minimum total cost of cutting back waste by the required amount occurs at a trading equilibrium.

TABLE 11.1 Maximum Offer Prices and Minimum Acceptance Prices for Pollution Allowances

	Possible Trades Exist		Trading Equilibrium	
	Firm 1	**Firm 2**	**Firm 1**	**Firm 2**
Number of allowances held	5	10	6	9
Raw load	8	12	8	12
Abatement level	3	2	2	3
Maximum offer price	5	2	3	$3\frac{1}{3}$
Minimum acceptance price	7	$3\frac{1}{3}$	5	$4\frac{2}{3}$

(This reduction in abatement cost corresponds to the area under the MC_1 curve in Figure 10.4, between abatement levels 3 and 2.)[6] Thus firm 1's *maximum offer price* for an additional pollution allowance is 5. Alternatively, firm 1 could sell one of its allowances. In this case, the firm would have to abate one more unit of pollution. The resulting increase of 7 in its total abatement cost is the firm's *minimum acceptance price,* the minimum purchase offer it would accept for selling a pollution allowance.

Firm 2 has similar options, and its maximum offer price and minimum acceptance price are calculated in the same way. Results for both firms are summarized on the left side of Table 11.1. The prices in the table demonstrate that the distribution of 5 allowances for firm 1 and 10 allowances for firm 2 is *not* a trading equilibrium. A mutually advantageous trade can occur because firm 1's maximum offer price is higher than firm 2's minimum acceptance price.

Extending this reasoning, it can be shown that the trading equilibrium for the two firms is $N_1 = 6$ and $N_2 = 9$. At this point, the maximum offer price for each firm is less than the minimum acceptance price of the other (see right side of Table 11.1). No mutually advantageous trades are possible. This equilibrium corresponds to the abatement levels at which total waste reduction costs are minimized.

The result that a minimum cost outcome occurs at a trading equilibrium can be generalized beyond this example. However, in the generalization, the restriction that firms cannot buy and sell fractions of a pollution allowance is dropped. Without that restriction, a trading equilibrium can be identified by inspection of Figure 10.4. The equilibrium occurs where $MC_1 = MC_2$ *and* the total pollution reduction equals 5, which is consistent with having 15 pollution allowances in circulation.[7]

If fractions of an allowance can be traded (which is not customary in practice), the marginal cost at an arbitrary abatement level represents both a firm's maximum offer price and its minimum acceptance price at that abatement level.[8] Moreover, a mutually advantageous trade is possible for any distribution of allowances that correspond to $MC_1 \neq MC_2$. When a distribution with $MC_1 \neq MC_2$ exists, the firm with the higher marginal cost can offer a price that is above the minimum acceptance price of the other firm. Opportunities for mutually advantageous trades are exhausted only when both firms abate pollution to the point that their marginal waste reduction costs are identical. As demonstrated in Chapter 10, the condition that $MC_1 = MC_2$ is necessary to attain a productively efficient outcome (assuming incremental benefits of waste reduction are independent of waste source location). This condition on the equality of marginal costs generalizes to situations involving an

[6] In general, the area under the marginal cost curve between any two points represents the increase in total cost in moving from the lower to the higher point. This provides the basis for computing each of the maximum offer prices and minimum acceptance prices in this example.

[7] Although $MC_1 = MC_2$ for any horizontal line in Figure 10.4, only the horizontal line corresponding to a marginal cost of 4 yields the required 5 units of pollution abatement.

[8] Using calculus, the marginal cost is the first derivative of the total cost. Thus marginal cost represents the change in waste reduction cost for an arbitrarily small change in pollution abatement level.

arbitrary number of waste dischargers.[9] However, as will be shown, the requirement that marginal costs be equal does not apply when incremental benefits of pollution abatement depend on the locations of sources that cut back their waste.

MODIFYING THE BASIC THEORY TO ACCOUNT FOR SOURCE LOCATION

In many cases, the locations of waste releases affect the incremental benefits of waste reduction. For instance, the benefits of reducing wastewater discharges typically depend on source location. The example in Figure 8.1 involving three sources of waste is used to show how locational effects can be handled in a scheme based on tradable pollution permits.

Recall that the example in Figure 8.1 included three sources of biochemical oxygen demand (BOD) discharging into a hypothetical stream. Two of the sources are on opposite sides of the stream and the third is located downstream (see Figure 8.1). The influence of each waste discharge on stream quality is reflected in a set of transfer coefficients, given in units of 1000 lb of BOD/day per mg/l of stream-dissolved oxygen.

The design of a tradable pollution permits scheme requires determining how many BOD allowances to issue without violating the ambient standard, which is set at 6 mg/l of dissolved oxygen at a point (labeled α) downstream of the three sources. The holder of a BOD allowance is entitled to release a fixed amount of BOD, but the amount varies with the location of the discharge. To decide how many allowances to issue, an environmental agency must first compute the quantity of BOD that can be released without violating the ambient standard. This BOD release depends on the locations of the effluents. To compute the maximum permissible BOD discharge, assume temporarily that source 3 does not exist. Under these circumstances, the maximum allowable release multiplied by the transfer coefficient, $\phi_{1\alpha} = \phi_{2\alpha} = 0.002$, must be less than or equal to the allowable 1 mg/l decrease in DO at point α. This relationship indicates that, at most, 500 units of BOD can be released at the location of source 1 or 2.[10] A unit of BOD is expressed in 1000 lb/day.

The environmental agency could issue 500 BOD allowances, where one allowance grants a right to discharge 1000 lb/day of BOD at the location of source 1 or 2. These BOD allowances could not be used by source 3 without first adjusting for the way a BOD release from source 3 influences DO at point α. This adjustment is readily made. A ratio of transfer coefficients,

$$\frac{\phi_{1\alpha}}{\phi_{3\alpha}} = \frac{0.002}{0.003} = \frac{2}{3}$$

reflects the relatively greater influence that BOD from source 3 has on dissolved oxygen at α. Thus, for source 3, a BOD allowance can only be used to release two-thirds of a unit of BOD (667 lb/day). To demonstrate the validity of this adjustment, suppose that source 3 has all 500 allowances. In this case, there are no discharges from sources 1 and 2, and source 3 releases two-thirds of 500 units of BOD. This quantity, multiplied by $\phi_{3\alpha} = 0.003$, yields the allowable DO decrease of 1 mg/l.

Next consider the initial distribution of the 500 BOD allowances. If the transaction costs of buying and selling BOD allowances are zero, the initial allocation of allowances does not influence the cost-effectiveness of the outcome.[11] Dischargers will make mutually advantageous purchases and sales of allowances. The resulting trading equilibrium will be *cost-effective:* polluters will reduce their BOD loads to

[9] For an extension of the theory (presented here) that considers an arbitrary number of waste sources, see Noll (1994). Other theoretical treatments are widely available; see, for example, Tietenberg (1985).

[10] The 500 units results if the total BOD release is set *equal* to the maximum allowable DO decrease (1 mg/l). In this case, the maximum allowable BOD is

$$\frac{1 \text{ mg/l}}{0.002 \dfrac{\text{mg/l}}{10^3\text{lb/day}}} = 500\ (10^3\text{lb/day}).$$

[11] As shown by Krupnik, Oates, and Van De Verg (1983), a trading equilibrium corresponding to a minimum total cost of pollution reduction will result regardless of the initial allocation of tradable pollution permits if: (1) the total number of permits is limited such that the ambient standard is attained; (2) for each source, the incremental cost of pollution cutback increases with increasing removal percentages; and (3) the transaction costs associated with buying and selling pollution permits are zero.

attain the ambient DO standard at a minimum total waste reduction cost.

To demonstrate that a trading equilibrium corresponds to a cost-effective outcome, suppose the environmental agency initially allocates the 500 BOD allowances by auctioning them off one at a time to the highest bidder. Assume that only the three dischargers participate in the auction and that each knows its *own* costs of BOD reduction. Furthermore, assume that the highest sum a polluter would offer for any BOD allowance is the amount its treatment cost would be reduced by owning the allowance. The dischargers do not speculate in allowances. They purchase only what they currently need. Under these conditions, the agency's auction will result in removal percentages that minimize the *total* treatment costs for the three waste sources.

For each waste source, Table 11.2 shows the maximum and minimum treatment cost savings that would result from purchasing one additional BOD allowance. In this example, the cost of removing an additional unit of BOD increases as the removal percentage increases. The first allowance purchased is the most valuable one because it leads to the greatest decrease in treatment costs. The least valuable allowance is the one purchased when the discharger is only one short of having all the allowances it can use to reduce treatment costs.

The computations of incremental savings in treatment cost are demonstrated for source 1 when it owns no BOD allowances. If source 1 purchased one allowance, its savings would equal the cost of removing 100% of its raw BOD load minus the cost of removing 99%. Using the cost equations in Figure 8.1, this difference is

$$1000(100)^{1.9} - 1000(99)^{1.9} = \$119{,}000.$$

A slightly more complex computation is required for source 3 since a BOD allowance covers only two-thirds of a unit of discharge. For example, if source 3 owns 499 allowances, it is entitled to discharge two-thirds of 499 units of BOD. Its required BOD reduction is 167.3 units, the difference between its raw load (500) and its permissable discharge. This corresponds to $X_3 = 33.47\%$. If source 3 purchased one more allowance, its required removal would decrease to 33.33%, and its cost savings would be

$$5000(33.47)^{1.5} - 5000(33.33)^{1.5} = \$6000.$$

Table 11.2 shows that the incremental savings in BOD reduction costs for source 2 are always lower than the corresponding savings for source 3. Therefore, source 3 could outbid source 2 for each of the 500 BOD allowances being auctioned. Consequently, source 2 purchases no allowances and removes 100% of its raw load.

The competition for BOD allowances between sources 1 and 3 involves only the last 100 allowances auctioned. This occurs because the raw load at source 1 is only 100 units. Suppose source 3 has 400 allowances in hand as it begins to compete with source 1. To determine how much each source would bid for allowances, it is necessary to compute the incremental savings in treatment costs if one more allowance is purchased. For each of the first 96 allowances, the incremental savings are greater for source 1 than for source 3. Consequently, these allowances are purchased by source 1. If source 1 obtained the ninety-seventh BOD allowance, its treatment costs would be reduced by

$$1000(4)^{1.9} - 1000(3)^{1.9} = \$6000.$$

If source 3 obtained this allowance, it would have to reduce 46.67% instead of 46.53% of its raw load. The corresponding cost saving would be

$$5000(46.67)^{1.5} - 5000(46.53)^{1.5} = \$7000.$$

TABLE 11.2 Range of Incremental Savings in BOD Reduction Costs

Waste Source	Hypothesized Number of BOD Allowances Held	Cost Savings from Holding One More BOD Allowance
1	0	119,000
	99	1,000
2	0	3,000
	499	2,600
3	0	10,000
	499	6,000

Because its incremental savings would be higher, source 3 purchases the ninety-seventh BOD allowance. This outcome can also be shown to hold for each of the remaining 3 allowances. The final trading equilibrium has source 1 with 96 allowances, source 2 with no allowances, and source 3 with 404 allowances. This outcome is essentially equivalent to the minimum cost removal percentages (see Table 10.2).

In general, attainment of ambient standards at minimum cost is only one of the advantages of employing marketable pollution permits. A marketable permits program is also attractive because, in theory, it *automatically* adjusts for either growth in raw waste loads at existing sources or introduction of completely new sources. The ambient standard is met regardless of future conditions because the number of available pollution permits is fixed at a level consistent with attaining the standard. The tradable pollution permits approach also gives firms an incentive to invent new pollution abatement technologies. If a firm reduces the amount of waste it discharges, the firm can profit from selling pollution permits it no longer needs.

ROLES FOR GOVERNMENT IN A PERMIT MARKET

Defining Rights to Pollute and Facilitating Exchanges

Environmental agencies play several key roles in a tradable permits program. For example, an environmental agency must make a clear and legally enforceable statement about what the bearer of a tradable pollution permit is entitled to do. In the three-waste-source example, the holder of a BOD allowance was entitled to discharge 1 unit of BOD at the location of source 1 or 2, or two-thirds of a unit at the location of source 3. Tradable pollution permits cannot be defined this simply in practice. Moreover, for a right to discharge waste to be meaningful, the agency must ensure that those without pollution permits do not discharge. This requires the agency to keep track of market transfers of permits, monitor discharges, and enforce the tradable permit program's rules.

The environmental agency also needs to decide how many pollution permits to issue.[12] This decision is relatively easy if the locations of discharge points do not affect the quantity of waste the holder of a permit is allowed to discharge. If polluter location is a key factor, the decision on how many permits to issue is problematic. In the three-waste-source example, information on transfer coefficients provided the basis for the decision to issue 500 BOD allowances. In practice, ambient standards are established at numerous places, not just a single location such as point α in the example. Thus, to decide on the number of tradable permits, the environmental agency must know the transfer coefficients for all combinations of waste sources and locations at which standards are set. Information on transfer coefficients is often unavailable. Even when there is a basis for predicting how discharges affect ambient environmental quality, the predictions are often uncertain. Using inaccurate information to determine the total number of permits leads to uncertainties regarding whether the ambient standard will be met when rights to pollute are exercised.

After allocating pollution permits, an environmental agency may work to facilitate market transactions. As in any other market, buyers and sellers of permits must spend time and money to locate each other and learn about potential offers and selling prices. For a product as unusual as tradable pollution permits, the cost of obtaining this information may be high. An environmental agency can act as a broker by bringing buyers and sellers together. The cost of exchanging a permit could include a fee to offset the agency's expense to administer the program. Even if the brokerage function is undertaken by private parties, there is a continuing role for government in recording exchanges.

Price stability is important since firms and municipalities cannot make good decisions on how much

[12] In practice, governments sometimes phase in tradable permits to pollute by issuing permits that allow a fixed discharge during one year only and reissuing permits annually, but with a decrease in the total allowable discharge in successive years. Under these circumstances, dischargers need advance notice of how the total number of permits will change over time so they can plan their investments in waste reduction.

to invest in waste reduction if prices of pollution permits fluctuate widely and unexpectedly. An environmental agency helps stabilize prices by being consistent in how it defines rights to pollute. To maintain stable prices, the agency could purchase pollution permits if sellers have difficulties in finding buyers, or if too many permits are put up for sale at any one time. Dales (1968) describes ways an agency might rig the market to maintain stable prices for pollution permits.

Initial Allocation of Tradable Permits

One of the most contentious issues in designing a tradable pollution permits scheme is the *initial* allocation of permits. Because rights to pollute have economic value, the initial allocation of permits changes the distribution of wealth, and it can also affect whether new sources of pollution can enter a region.

Governments can distribute marketable pollution permits in at least two ways:[13] give permits away for free, or sell permits using an auction process. If permits are given away, how should the initial allocation be decided? Because substantial transfers of wealth may be involved, the initial allocation is best established through a political process. One approach involves granting discharge permits to existing polluters in proportion to current allowable waste releases. A proportional scheme is politically attractive, because it minimizes disruption to (and resistance from) existing waste dischargers.

If the initial set of transferable discharge permits is to be sold at an auction, then how should the auction be designed? A traditional auction involving sales to the highest bidders will yield revenues for government, but it will be resisted by current polluters. Existing dischargers will be forced to pay potentially large sums for rights to pollute that they once enjoyed for free.[14]

Another type of auction, the *Hahn-Noll zero revenue auction* (ZRA), aims to accomplish the initial allocation of tradable pollution permits in an economically and politically defensible way. The term "zero revenue" is used because the auction does not yield new revenue for the government. After auctioning off the permits, the government returns all auction receipts to dischargers who were legally entitled to pollute before the tradable permits were allocated.

The Hahn-Noll zero revenue auction has the following characteristics.

- A clear price signal results.
- The equilibrium price is a meaningful measure of the value of permits to permit holders.
- The auction process yields a Pareto-efficient outcome.[15]
- The resulting redistribution of wealth is not drastic, and thus the auction may be politically acceptable.

Operation of the Hahn-Noll ZRA is demonstrated using a hypothetical example involving an area with only two polluters, Alpha and Beta.[16]

1. An environmental agency decides on the total number of pollution permits that are to be allocated; in this example, five permits, called "pollution allowances," are distributed.
2. The environmental agency decides on a procedure for the pre-auction distribution of the five allowances. This allocation is *provisional* and used primarily as a basis for distributing revenue from the auction. A commonly suggested procedure for provisional allocation involves distributing pollution permits in *proportion* to either (a) amounts polluters are currently discharging, or (b) amounts polluters were legally entitled to release just before the tradable permit scheme was introduced.[17]

[13] Other approaches are possible, but the two schemes introduced here are widely discussed.

[14] For an analysis of issues to be considered in designing an auction for marketable pollution permits, see Heggelund (1991).

[15] For a discussion of Pareto-efficient resource allocations, see Chapter 5. Further details on zero revenue auctions are given by Hahn and Noll (1983).

[16] This example was developed based on personal communication with Robert Hahn of the American Enterprise Institute, Washington, DC, December 3, 1993.

[17] Using this proportional approach to allocating permits, ambient environmental quality can be improved by restricting the total allowable discharge, as defined by the number of permits issued, to be less than the smaller of the following quantities: (a) total amounts polluters are currently discharging or (b) total amounts polluters were legally entitled to release just prior to the issuance of tradable permits.

In this example, the pre-auction (or provisional) allocation of pollution allowances is

Alpha: 2 allowances,

Beta: 3 allowances.

3. Participants in the auction are requested to give the environmental agency a sealed *bid schedule* indicating how much the participant would be willing to pay for 1 allowance, 2 allowances, and so forth. Suppose the bid schedules submitted by the two participants are as shown in Figure 11.1a.
4. The environmental agency rank-orders the individual bids for 1 allowance, 2 allowances, and so forth to obtain an aggregate bid schedule that reflects aggregate willingness to pay (or demand) for rights to pollute. Results of the aggregation are shown in Figure 11.1b.
5. The environmental agency allocates the five allowances to participants by selling allowances to highest bidders as reflected in the aggregate bid schedule. As indicated in Figure 11.1, Alpha purchases 3 allowances and Beta purchases 2 allowances. The selling price is the price at the intersection of the vertical supply curve for allowances and the aggregate bid schedule; in this case, the selling price is 6.
6. The environmental agency allocates all revenues collected from the sale in proportion to the pre-auction (provisional) holdings of the participants. The total revenues equal the product of the equilibrium price (6) and the total number of allowances (5). Table 11.3 shows the redistribution of wealth resulting from the auction.

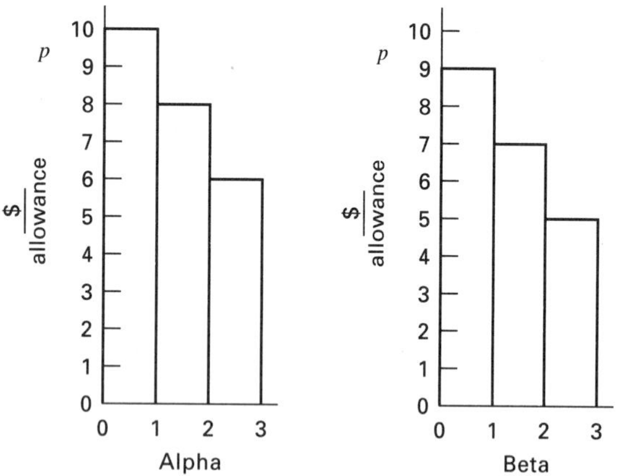

(a) Individual bid schedules

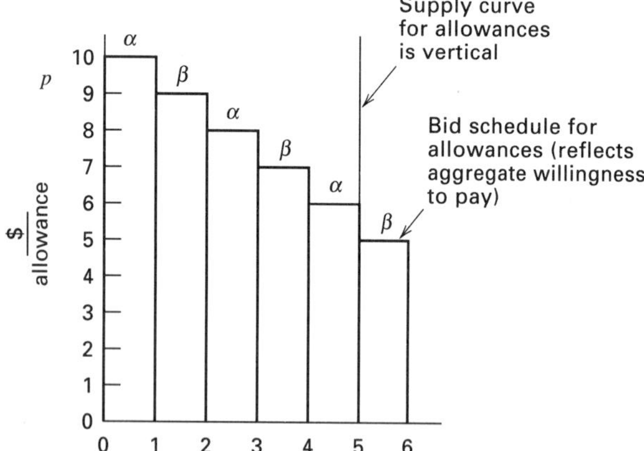

(b) Aggregate bid schedule: "α" means the bid was made by Alpha and "β" means the bid was made by Beta

Figure 11.1 Bid schedules for two firms participating in a Hahn-Noll zero revenue auction.
This figure was developed based on a personal communication with Robert Hahn of the American Enterprise Institute, Washington, DC, December 3, 1993.

Another way of looking at the outcome in Table 11.3 is that, in essence, Alpha purchased one allowance from Beta for a price of 6, and the government obtained zero revenue from the allocation process. As detailed later in this chapter, a variation of the Hahn-Noll ZRA has been implemented in connec-

TABLE 11.3 Redistribution of Wealth Using Hahn-Noll Zero Revenue Auction

Waste Source	Status	Pollution Allowances Held	Asset Value[a]
Alpha	Pre-auction	2	12
	Post-auction	3	18
Beta	Pre-auction	3	18
	Post-auction	2	12

[a] Asset value is calculated as the product of the number of pollution allowances held and the equilibrium price, which is 6 in this example.

tion with a system of tradable permits to emit SO_2 in the United States.

TRADING BASED ON EMISSION REDUCTION CREDITS

During the 1970s, the U.S. Environmental Protection Agency (EPA) created an innovative system of tradable rights to pollute based on Emission Reduction Credits (ERCs). These credits can be bought, sold, and banked for future use. The policy on ERCs evolved over a decade as EPA struggled with pressures to reach national ambient air quality standards (NAAQS) in cost-effective ways. Another factor affecting the policy on ERCs was the need for rules allowing economic growth in *nonattainment areas,* regions violating the NAAQS.

Although EPA's policies on Emission Reduction Credits evolved over a long period, they stabilized by the mid-1980s. By then, ERCs had become a basic unit of accounting in four EPA policies for air quality management (see Table 11.4).[18] The four policies applied only to pollutants for which NAAQS had been established (for example, nitrogen dioxide). An ERC is created when an existing source or proposed major new source reduces air pollution emissions *below* levels stipulated by regulations. A major source of air pollution is one with emission above a numerical cutoff value, such as 100 tons per year.

Trading in Emission Offsets

EPA's initial tradable permits policy concerned requirements for *proposed* major sources of air pollution in nonattainment areas. The Clean Air Act Amendments of 1977 required major new sources in nonattainment areas to satisfy stringent technology-based emission standards.[19] Moreover, those dischargers had to meet an additional requirement concerning pollutants for which the NAAQS were violated in the region: the new polluter's emissions had to be offset by reductions in emissions from *other sources.* Collectively, the new emissions and emission reductions (or *offsets*) had to represent "reasonable progress" toward meeting the NAAQS.

Use of emission offsets (later termed Emission Reduction Credits) and their relationship to tradable pollution permits are clarified by an example from the San Francisco Bay region. Following passage of the 1977 Clean Air Act Amendments, the Wickland Oil Company applied to the Bay Area Air Quality Management District (BAAQMD) for a permit to construct a new 40,000 barrel per day terminal at a site on San Francisco Bay in Contra Costa County.[20] Fuel oil, received via tanker ship or pipeline, was to be stored in floating-roof tanks on the site for later shipment by tank trucks or pipeline. At the time the permit application was processed, Contra Costa County was a nonattainment area for three air pollutants: sulfur dioxide, carbon monoxide (CO), and hydrocarbons (HC). Moreover, the proposed terminal was a significant new source of air pollution and thereby subject to requirements calling for emission offsets.

The base conditions used in calculating the necessary emission offsets were the technology-based emission requirements for new sources in nonattainment areas. The first row in Table 11.5 shows the discharges that would result *after* Wickland Oil implemented various measures to meet these requirements. The expected releases of CO were low enough to be disregarded. However, the emissions for SO_2 and HC shown in Table 11.5 were so high that BAAQMD required offsets.

Before the Wickland Oil Company could receive approval from the BAAQMD, it had to either (1) reduce its emissions below the levels used by the BAAQMD to designate a new source as "significant," or (2) obtain enough emission offsets to demonstrate that the net effect of its actions would be progress in attaining the NAAQS for sulfur dioxide and hydrocarbons. The first option was prohibitively expensive. Wickland Oil therefore set out to acquire the necessary offsets by the following means:

[18] One of the four policies, *netting,* is not discussed further since it is intended largely to provide firms with a means of relief from exacting regulations on major new stationary sources of air pollution (see note 23). For a history of EPA's policies on Emission Reduction Credits, and many other aspects of the programs that use ERCs, see Tietenberg (1985).

[19] The 1977 amendments included definitions clarifying which new sources would be covered by these technology-based standards.

[20] The discussion of the Wickland Oil example is based on an unpublished report by Lance M. Goto, prepared while he was a student at Stanford University, Stanford, California.

TABLE 11.4 Four Uses of Emission Reduction Credits in EPA's Air Quality Management Program

Offset Policy	Applies to firms interested in locating major new stationary sources in regions that violate national ambient air quality standards for pollutants discharged by the new sources.[a] Firms may locate in such areas only if they obtain ERCs that, when subtracted from proposed emissions of a pollutant, yield a net decrease in emissions of the pollutant in the region.
Banking	Firms creating ERCs in one time period may save them for use in a future time period. Stored ERCs can be either retained for the firm's own use or sold to other firms.
Bubble Policy	Applies to multiple existing facilities owned by the same organization. Requirements for existing sources within a designated area (or under an imaginary bubble) may be aggregated. Instead of meeting requirements for each individual source within the bubble, the organization can meet an aggregate requirement on the maximum allowable discharge leaving the bubble.
Netting	Applies to existing facilities that plan to expand. Additional emissions associated with proposed plant changes can be offset by reductions from existing sources within the facility. If the net increase is less than the threshold discharge amount that defines *major* new sources, the firm can avoid stringent requirements imposed on major new sources.

[a] A major source is one that has discharges exceeding a numerical cutoff value, such as 100 tons per year.

1. Reducing or eliminating emissions from an existing Wickland Oil facility near the proposed terminal. These are generally termed *internal offsets,* and they are created by lowering emissions below regulatory standards.
2. Utilizing *banked offsets,* emission offsets obtained by Wickland Oil in the past for bringing emissions at nearby facilities below levels required by the BAAQMD.
3. Reducing or eliminating emissions from existing facilities of *other firms.* Wickland Oil could get credit for *external offsets* by either paying other firms to decrease emissions or buying their facilities and closing them down. The BAAQMD would not accept external offsets at full value; for the Wickland Oil proposal, 1.2 tons/year of external emission reduction was required to offset 1 ton/year at the proposed terminal.
4. Purchasing offsets that had been banked by other firms.

The Wickland Oil Company arranged for a 181.7 tons/year reduction in hydrocarbon emissions at the City of Paris dry cleaners in San Francisco. Wickland

TABLE 11.5 Wickland Oil Proposal—Estimated Emissions and Proposed Offsets (Tons per Year)[a]

	SO_2	HC	CO
Total estimated emissions before offsets	24.7	83.2	1.31
City of Paris offset—new equipment		−151.4	
Virginia Chemicals, Inc., offset—plant shutdown	−7.4		
Offsets from ships and vehicles burning low-sulfur fuel	−22.2		
Total estimated emissions after offsets	−4.9	−68.2	1.31

[a] From "Notice Inviting Written Public Comment on the Authority to Construct the Wickland Terminal." Bay Area Air Quality Management District, San Francisco, CA, March 15, 1979. Reprinted with permission.

Oil agreed to purchase new dry cleaning and solvent reclamation equipment that would bring the City of Paris emissions substantially below those required by the BAAQMD. The oil company was credited with 151.4 tons/year of hydrocarbon offsets (181.7 tons/year divided by the 1.2 factor applied to external offsets). The sulfur dioxide offset requirements were met in two ways. First, Wickland Oil purchased and closed down a plant owned by Virginia Chemicals, Inc., that was on the site to be developed by Wickland Oil. After applying the 1.2 factor, this accounted for an offset of 7.4 tons/year. Second, Wickland Oil arranged for a number of oceangoing oil tankers and surface motor vehicles to switch to the use of low-sulfur fuels. This provided an additional 22.2 tons/year of offsets for SO_2. As indicated in Table 11.5, the net result of these offsets was an estimated *reduction* of 4.9 tons/year of sulfur dioxide and 68.2 tons/year of hydrocarbons.[21]

The Wickland Oil Company example demonstrates that transactions costs can be very high using a system of tradable emission offsets (or ERCs). According to a report by the U.S. General Accounting Office (1982), Wickland Oil contacted over 150 firms in its efforts to secure the needed offsets for its hydrocarbon emissions.

Transactions costs have also been high for firms in other parts of the United States that sought emission offsets to meet requirements of the 1977 Clean Air Act Amendments. During the late 1970s, regional air quality agencies such as the BAAQMD introduced "offset banks" as a mechanism for reducing transactions costs and for helping to form active markets in offsets.[22] These banks were created to provide a convenient first stop for buyers seeking sellers of offsets. The banks also played an important role in clarifying the legal status of offsets and the offset policies of regional air quality agencies. Without such clarification, firms with potential offsets would hesitate to make the investments in pollution abatement needed to create offsets.

A Bubble Policy to Reduce Costs

In 1979, the EPA extended the concept of emission offsets, as used in nonattainment areas, to a different context: multiple sources of air pollution generated at a single site.[23] This extension, known as the *bubble policy,* is illustrated in Figure 11.2. The figure depicts a firm that must control releases from smokestacks at two adjacent plants. Before the bubble policy, the firm had to comply with emission standards that allowed only 100 tons/day from each plant. The total discharge was 200 tons/day. The unit cost of emission controls for plant A was much higher than that for plant B, but the emission requirements were insensitive to these cost differences. Using the bubble policy, the firm is free to decide how to reduce waste releases at each plant. The only restriction is that the total discharge from an imaginary bubble surrounding the two plants must be no greater than 200 tons/day. In the early 1980s, the original bubble policy was extended to include plants that were not at the same location ("multiplant bubbles"). In addition, Emission Reduction Credits were broadened so they could apply to bubbles, banking, netting, and emission offset requirements.

The aim of the bubble policy is to have firms decrease the costs of achieving a particular level of residuals reduction. The policy allows plant managers to exchange decreases in emissions at easily controlled sources for increases in emissions at sources that are expensive to control. Opportunities for cost savings are illustrated by a case involving the Narragansett Electric Company in Rhode Island.[24] The utility had two power plants located one-quarter mile

[21] The Wickland Oil Company was granted a permit by the Air Pollution Control Officer (APCO) of the BAAQMD in 1979. In May 1980, the APCO's decision was overruled by the BAAQMD hearing board based on objections raised by citizens. Subsequent negotiations between the Wickland Oil Company and the objecting citizens led, eventually, to a mutually acceptable solution. Wickland Oil Company agreed to eliminate two large storage tanks from its original proposal and thereby reduce the expected emissions of hydrocarbons by an additional 10 tons/year. This new proposal was acceptable to the hearing board, and a permit to construct the oil terminal was issued.

[22] For more information on the introduction of offset banks, see Liroff (1980).

[23] The EPA also allows firms to engage in a process called *netting,* which applies only to existing firms that plan to expand. Netting allows a firm planning an expansion to reduce emissions from its existing facilities beyond what is required by regulations. Those extra reductions can then be subtracted from the emissions at the new facility, and the net result may be so low that the firm can avoid stringent emission requirements imposed on significant new sources. For more information on netting, see Hahn (1989).

[24] This description of the Narragansett Electric Company case is from the National Commission on Air Quality (1981, p. 277).

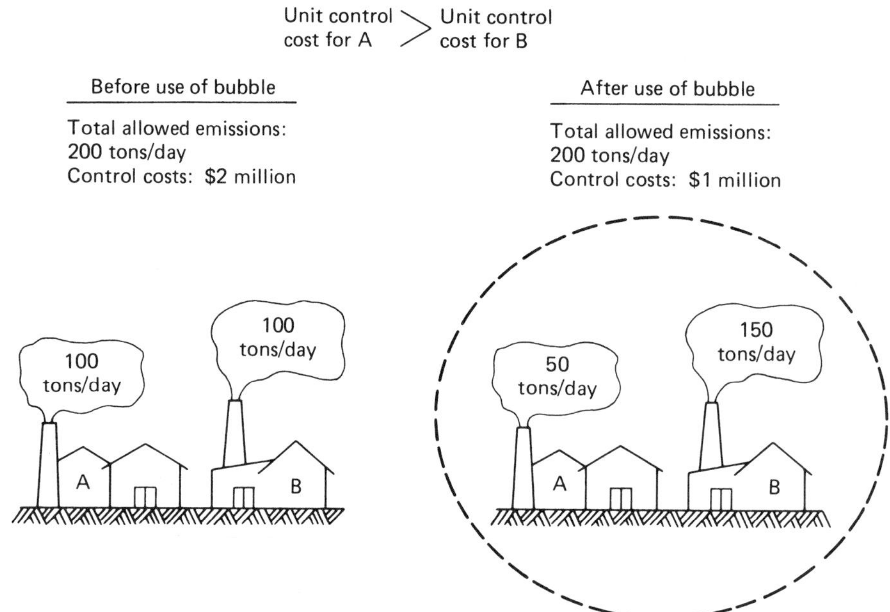

Figure 11.2 Reducing costs using the bubble policy. Source: U.S. Environmental Protection Agency (1982).

apart. Before applying the bubble policy, the sulfur dioxide releases at each plant resulted from using fuel oil with a 1% sulfur content. In applying the bubble concept, the utility planned to switch to a high-sulfur oil (2.2% sulfur) at one plant. In exchange, the SO_2 emissions at the other plant would be reduced by either burning natural gas or not operating when the high-sulfur fuel was used. According to EPA (1982), this application of the bubble policy would save Narragansett $3 million a year. It would also decrease overall SO_2 emissions from the two plants by 10%.

The federal government's Emission Reduction Credit policies involve some, but not all, of the concepts used in the theory of tradable pollution permits. In that theory, polluters are not required to meet discharge standards. In contrast, the federal ERC policy rests on a foundation of requirements: ERCs are defined as reductions in emissions that *exceed* cutbacks stipulated by environmental authorities. Another difference concerns stability in the definition of rights granted to the holder of a pollution permit. In the theory, the rights granted to a permit holder are defined just once. However, under the ERC policy, rules defining and governing use of ERCs are subject to change. For example, in the Los Angeles emission offsets market, arguably the most active one in the United States, rules defining use of ERCs changed 59 times in a 12-year period.[25] Such changes introduce uncertainty in the nature of the property right embodied in an ERC.

TRADING IN SO_2 EMISSION ALLOWANCES

During the 1980s in the United States, the question of whether and how to reduce SO_2 emissions from coal-fired electric power plants impeded efforts to

[25] The impacts of rule changes on trading activities are described by Foster and Hahn (1995). They report one case of a rule change that reduced to 20% of face value all banked ERCs that had been created by shutting down facilities emitting air pollutants. For a case study of the use of emissions trading in the Los Angeles region, see National Academy of Public Administration (1994, pp. 24–72).

control acid rain. A proposal for an innovative program of tradable SO_2 pollution permits provided a breakthrough. The proposed program was supported by conservative legislators who felt that environmental regulations needed to be cost-effective. Moreover, the program had the crucial endorsement of an influential U.S. environmental group, the Environmental Defense Fund.

The Clean Air Act Amendments (CAAA) of 1990 include a cap on the annual aggregate emissions of sulfur dioxide from electric power plants in the 48 contiguous states. The law proposed to reduce annual SO_2 emissions by 10 million tons (compared to 1980 levels) over a 10-year period ending in the year 2000. The SO_2 reduction goal was scheduled to be reached in two-phases. The first phase, a five-year period beginning January 1, 1995, affected 110 of the highest-emitting power plants in the United States. In the second phase, beginning in the year 2000, the program will include most existing electric-generating units using fossil fuels within the 48 contiguous states.

The management scheme in the CAAA of 1990 requires a regulated electric utility to have a number of SO_2 allowances sufficient to cover its SO_2 discharge in any year (starting in 1995). An *SO_2 allowance* is a right to emit one ton of sulfur dioxide during one year.[26] The allowable discharge is either for the year indicated on the allowance or a later year. Each allowance has an identification number, so both EPA and the utilities can keep track of individual allowances. Sulfur dioxide allowances can be bought and sold, and they can be saved for future use. If, at the end of a particular year (or an accounting date set for early the following year), a utility does not have enough allowances to cover its emissions for the year, it faces a penalty (which resembles an emissions fee) of $2000 per ton of SO_2 emitted. In addition, the utility must make up for its shortage in allowances by giving up an equivalent number of allowances from those it obtains in future years.

Initial Allocation of Allowances

Designers of the SO_2 allowance program faced tough questions: How many allowances should there be, and how should they be distributed? The total number of allowances was based on targets for SO_2 cutbacks: 10 million tons per year, with reductions phased in over a 10-year period.

The number of allowances received (free of charge) by each of the 110 utilities in Phase I is based primarily on each plant's *baseline fuel consumption,* defined as average annual fuel usage in millions of British thermal units (MBtu) during the 1985–87 period. A single generating plant typically has several units, and EPA obtained values for annual average fuel use for each unit. The number of SO_2 allowances earmarked for any one unit equals the fuel use (in MBtu per year) multiplied by an emission factor: 2.5 lb of SO_2 per MBtu. For Phase II of the program, the allocation procedure is similar, except the emission factor is cut back to 1.2 lb of SO_2 per MBtu. The allowances computed using this procedure are supplemented based on bonus provisions and other features included in the CAAA of 1990 to placate special interests.[27]

Auctions and Direct Sales

Designers of the trading program for SO_2 emissions were concerned about how *new* electric-generating facilities (producers that received no initial allocation of allowances) could enter the market. Barriers to free entry of new power producers would reduce competition and decrease productive efficiency. Designers of the trading scheme were also concerned about the need to provide potential buyers and sellers of allowances with opportunities to make exchanges without incurring high transactions costs, such as costs to find trading partners and negotiate trades. Transactions costs could grow substantially if a robust market for SO_2 allowances failed to develop and market prices for allowances were difficult to determine.

To keep transactions costs low and permit entry of new power producers, Congress included requirements for an annual auction and direct sales of allowances. Provisions of the CAAA of 1990 require EPA to create a "special allowance reserve" by withholding 2.8% of all SO_2 allowances that would otherwise

[26] The right is *limited* in that Congress can modify the SO_2 allowance program at any time.

[27] There are many exceptions to the basic allocation procedure. The numerous bonuses, extensions, exceptions, and special options in this program are too complex to be introduced here. Details are given in Title IV of the CAAA of 1990.

TABLE 11.6 Number of SO_2 Allowances Available in Auction and Direct Sales[a]

	Auction		Direct Sale	
Year of Purchase	**Spot Allowances[b]**	**Advance Allowances[c]**	**Spot Allowances**	**Advance Allowances**
1993 (to 1995)	50,000	100,000	–	25,000
1995 (to 1999)	150,000	100,000	–	25,000
2000 (and after)	100,000	100,000	25,000	25,000

[a] Direct sale allowances are available for sale at $1,500 per ton, adjusted using the Consumer Price Index.

[b] Allowances sold in the spot auction (or spot sale) in any year may only be used in that year (unless banked for use in a later year). An exception applies for allowances sold in spot auction (or spot sale) between 1993 and 1995; those allowances are not usable until 1995.

[c] Allowances sold in the advance auction (or advance sale) in any year may only be used in the seventh year after the year in which they are first offered for sale (unless banked for use in a later year).

Source: U.S. Clean Air Act Amendments of 1990 (Public Law 101-549), Section 416.

be distributed to utilities in a given year. Allowances in this reserve are sold at auction or in direct sales (at $1500 per ton, adjusted using the consumer price index), and the resulting revenues are returned (on a *pro rata* basis) to the utilities from which the allowances were originally withheld.

Allowances sold at auction and in direct sales are of two types: spot and advance. A *spot allowance* may be used to comply with SO_2 requirements in the year it is sold, whereas an *advance allowance* cannot be used until seven years after the year of purchase. Either type of allowance can be traded immediately after purchase. Spot allowances can assist utilities with short-term compliance problems, whereas advance allowances give utilities an opportunity to do long-term planning. Table 11.6 indicates numbers of allowances available through auctions and direct sales.

EPA's annual auction involves the following steps: Each potential buyer submits one (or more) sealed bids indicating how many allowances would be purchased at a fixed price. For example, in the first spot auction, held in March 1993, Kentucky Utilities Company bid $201 (per allowance) for 1000 allowances, and $151 for 8900 allowances. Bids are ranked from highest to lowest based on bid price. Allowances from the government's reserve are then sold to participants based on bid price, starting with the highest bid and continuing until all the government's allowances are sold. The procedure departs from an ordinary market in that successful bidders do not pay an equilibrium price; rather, they pay the price they bid.[28]

Private parties also sell allowances at EPA's annual auction, but these sales begin after the government has sold all allowances from its reserve. Sellers specify a minimum acceptance price for their allowances, and their proposals are ranked starting with the lowest price. A sale is made by matching the seller specifying the lowest minimum acceptable price with the highest remaining bid price. The seller receives the bid price. This rewards sellers who are able to cut back SO_2 emissions at low cost.

There are no restrictions on who can purchase SO_2 allowances during an auction. In EPA's first spot allowance auction, electric utilities entered just over half the spot bid quantities and accounted for about 95% of the spot purchase quantities. Other bidders included private firms (20%), brokers (8%), and public interest groups (6%). Prices of successful bids ranged from $131 to $450.[29]

Since 1993, the prices paid for SO_2 allowances at EPA's annual auction have continued to drop. In the auction held in March 1996, the price paid for an SO_2 allowance dropped to about $68 per ton.[30] Why

[28] In the terminology of economics, the buyers do not enjoy a *consumers' surplus.* All of the surplus is directed to the sellers.

[29] A complete listing of participants and bid prices in the government's first auction is given in *Public Utilities Fortnightly* (vol. 131, no. 9, May 1, 1993, p. 38).

[30] The $68/ton figure is from a news service clipping in the *New York Times,* national edition, (March 28, 1996, p. C19).

are prices for SO_2 allowances so much lower than the $750 per ton EPA expected in 1990, and the $1500 purchase price set by the CAAA of 1990 for direct sales?[31]

A key explanation for the unexpectedly low price of allowances in EPA's annual auction is the productive inefficiency of federal SO_2 emissions requirements before 1990.[32] The pre-1990 federal regulations on SO_2 emissions forced many electric utilities burning coal to make end-of-pipe emission reductions, often with costly flue gas desulfurization systems (commonly known as scrubbers).[33] Cost data from the 1980s were used by EPA to estimate the price of allowances, and these data yielded price projections that were much higher than the auction prices.

Under the CAAA of 1990, a utility that is out of compliance with regulations need not install scrubbers to meet requirements. Many options are available and some may be much cheaper than scrubbers. A utility can use one or more of the following in meeting federal SO_2 regulations:

- Install scrubbers either to meet requirements or to overcomply with requirements and thereby create excess allowances that can be sold.
- Use low sulfur coal or blends of coals with varying sulfur content.
- Switch from coal to another energy source, such as natural gas or oil.
- Purchase needed allowances from other utilities, brokers, or other external parties.
- Make intrautility transfers of allowances from one generating unit to another.
- Use relatively clean generating facilities more intensively.
- Promote energy conservation to reduce demand for electricity.
- Purchase more electricity from other utilities.

Once the CAAA of 1990 created new SO_2 regulatory compliance options for electric utilities, providers of inputs needed to implement these options began to compete with each other. For example, manufacturers of scrubbers were in competition with providers of low sulfur coal. This competition created new incentives for input providers to discover cost-cutting production methods. Indeed, technological innovations led to substantial decreases in prices of both scrubbers and low sulfur coal.

Part of the explanation for the low auction price of SO_2 allowances centers on changes in regulations governing rail transportation. At the time the CAAA of 1990 was enacted, analysts believed that bottlenecks in rail transport would make it too costly to ship low-sulfur coal to East Coast utilities from the huge Powder River Basin coal deposits in Wyoming and Montana. Forecasts of SO_2 allowance prices made in 1990 were based on prices for low-sulfur coal from Appalachia that was easily available to eastern utilities. However, the potential bottlenecks failed to materialize because deregulation of railroads in 1980 gave railroads increased flexibility in setting tariffs and higher motivation for cutting costs. Drops in transport costs made it possible for eastern utilities to take advantage of relatively inexpensive low-sulfur coal from the West. Moreover, in the mid-1990s, the price of low-sulfur Appalachian coal was much lower than estimated in 1990.

In addition to motivating input suppliers to innovate, the SO_2 emission trading program encouraged utilities themselves to investigate new compliance options. For example, before 1990, industry analysts believed that blending coals with different sulfur content would adversely affect equipment, such as fuel firing systems. The flexibility in compliance options provided by the CAAA of 1990 motivated experiments with fuel blending. Results showed that the adverse effects of using blended coals to generate electricity were less than originally supposed.

Other explanations for the low price of SO_2 allowances are institutional in nature. For example, some analysts believe utilities have been unwilling to make high bids on allowances because they are uncertain about future government actions. Because Congress can amend the allowance program at any time, the right to discharge SO_2 is less secure than an ordinary property right. Moreover, many state public utilities commissions (the bodies that regulate electric utilities) have not ruled on how profits on sales of allowances should be divided between ratepayers (via lower electricity prices) and stockholders. Questions also remain regarding how gains from

---

[31] The $750/ton estimate by EPA is as reported by Burtraw (1996, p. 84).

[32] The following explanation for the surprisingly low auction price for SO_2 allowances relies heavily on Burtraw (1996).

[33] Burtraw (1996, p. 86) reports that EPA's rules implementing the Clean Air Act Amendments of 1977 "imposed a minimum SO_2 reduction standard of 90 percent on high sulfur coal and 70 percent on low sulfur coal, effectively requiring the use of scrubbers on all coal and eliminating the incentive for use of low sulfur coal."

transactions involving allowances will be taxed. Still another explanation of low prices for allowances centers on the incentives that the auction process gives bidders to understate the true value of SO_2 allowances. Because sales of allowances take place at bid prices, potential buyers attempt to submit the lowest bid they think will be accepted.[34]

Other Forms of Trading

The purchase and sale of SO_2 allowances in the EPA's annual auction is only one of many opportunities for trading. The Chicago Board of Trade (CBOT), which administers the government's annual auction, provides other outlets for potential traders. In particular, CBOT conducts periodic private auctions of SO_2 allowances to complement the government's auction and direct sales activities. During the early 1990s, CBOT announced the creation of options and futures markets.[35] Ordinary investors can now trade in SO_2 allowances the same way they trade in other commodities at CBOT. All of these activities by CBOT help keep transactions costs down and encourage an active market in allowances.

In addition to the opportunities for trading created by CBOT, organizations can negotiate private trades. The first two announced trades took place in 1992. Both involved sales by Wisconsin Power & Light Company (WP&L), a utility that had to meet stringent SO_2 reduction requirements (similar to those in the CAAA of 1990) by 1993 in order to satisfy Wisconsin's environmental requirements.[36] Because of its early compliance with SO_2 reduction requirements, WP&L was able to sell part of its allotment of allowances.

The motivations of the two buyers of WP&L's allowances demonstrate the increased flexibility utilities gained under the CAAA of 1990. One of WP&L's trading partners, the Tennessee Valley Authority (TVA), purchased 10,000 allowances as insurance against possible delays in scrubber installation at one of its plants. If the scrubbers went in on schedule, TVA would be in a position to resell the allowances it purchased from WP&L. The second of WP&L's trading partners, Duquesne Light Company, wanted the 15,000 to 25,000 allowances it purchased to reduce its dependence on low-sulfur coal. In both sales, the price of allowances varied between \$250 and \$400 per ton of SO_2 (per year).

The effects to date of the SO_2 allowance program have been paradoxical: utilities are not relying heavily on trades to comply with CAAA of 1990, but their compliance costs are nonetheless lower than under the pre-1990 regulatory approach. Evidence of the relatively low reliance of utilities on SO_2 allowances is abundant. Three different studies reported by Burtraw (1996, p. 90) show that well over 50 percent of the electricity generating units regulated under Phase I of the SO_2 allowance trading program comply with SO_2 regulations by fuel switching or fuel blending. In contrast, less than 15 percent of the regulated units purchased allowances to meet SO_2 requirements.

Documentation of major cost savings in meeting SO_2 reductions goals is also substantial. For example, a 1994 study by the General Accounting Office estimated the cost of meeting targeted reductions with the SO_2 allowance scheme at one-half to one-third of SO_2 reduction costs using a command and control approach.[37]

The cost reductions under the SO_2 allowance trading program are not due to extensive trading in allowances. However, as argued by Burtraw (1996), productive efficiency provides a more appropriate basis for evaluating tradable permit programs than whether or not trading is extensive. Based on productive efficiency, the SO_2 allowance program is an apparent success because of the low-cost options it gives utilities for meeting SO_2 reduction goals. Even though the SO_2 allowance program does not satisfy

[34] Consider an example from the 1993 spot auction. If American Electric Power Service had offered the highest recorded bid price (\$450) instead of its actual bid (\$131), it would have paid about \$156,000 more for the 488 allowances it bought at the auction. Since electric utilities must justify their actions to public utility commissions, they are wary of paying relatively high prices at auction.

[35] In a futures market, investors pay a small fraction of the purchase of a commodity for the right to purchase the commodity at a fixed price at a specified future date.

[36] Facts on this sale are as reported by Greenberger (1992).

[37] This fact is from Burtraw (1996, p. 92). He reports projected annual costs of compliance from a 1994 study by the U.S. General Accounting Office (GAO). Under a command and control approach, the estimated annual cost of meeting SO_2 cutbacks mandated by the CAAA of 1990 is \$4.3 billion. The GAO estimates the annual cost to meet SO_2 cutbacks from \$1.4 to \$2.5 billion using the SO_2 allowance program.

TABLE 11.7 Practical Considerations in the Design of Tradable Permit Systems[a]

- Number of permits to pollute must be limited.
- Number of permits should not change unexpectedly.
- Rights granted to permit holders should be unambiguous and stable.
- Government expropriation of permits to pollute should be rare.
- Trading of permits should be on a one-for-one basis if possible.
- Impediments to free trade should be minimal.
- Permits to pollute should be usable in future years.
- Costs of transactions should be low.
- Government enforcement of permit trading rules must be credible.

[a] This table extends a presentation by Hahn and Noll (1990). Reprinted with permission.

the many assumptions made in the underlying theory of tradable pollution permits, the program's contributions to productive efficiency have been impressive.

LESSONS FROM EXPERIENCE WITH TRADABLE PERMITS

During the past few decades, many agencies have tried to implement tradable pollution permit systems that meet ambient standards in cost-effective ways. The experience in implementing those schemes suggests several conditions that can facilitate system development.[38] The conditions are listed in Table 11.7 and highlighted in this section.[39]

Several practical issues surround both the definition of permits to pollute and the number of permits issued. The total number of permits to pollute must be limited in order to meet ambient environmental quality targets. If and when the number of permits in circulation changes, it is important that those participating in the market not be caught by surprise. The system of SO_2 allowances under the CAAA of 1990 demonstrates that it is possible to gradually decrease the total number of pollution permits without disrupting the workings of a tradable permits system. The key is that everyone knows, in advance, when and how the number of permits will change. The SO_2 allowance system also includes an unambiguous definition of what the holder of an allowance is entitled to do. Contrast this with the air quality management scheme involving Emission Reduction Credits. There are many cases where rule changes made by environmental agencies had the effect of reducing the value of ERCs. In at least one case, the face value of ERCs was discounted.[40] Such activity by government reduces the attractiveness of participating in markets for tradable pollution permits.

Selection of the pollutants used in a tradable permit program influences the ease of implementation. When pollutants are such that the location of sources affects the terms of trade, the system will often involve high transaction costs. In the case of SO_2 allowances under the CAAA of 1990, trades are made on a one-for-one basis, which simplifies the program. Compare this with a marketable permit program for releasing BOD to streams; only BOD sources close to each other could make one-for-one trades. For sources that are far apart, it would be necessary to conduct modeling studies to establish transfer coefficients, and these would affect the terms of trade between sources.

Restrictions imposed on traders also influence the functioning of a pollution rights market. This is illustrated by the system of tradable BOD discharge permits on the Fox River in Wisconsin. Eligible traders included municipalities and pulp and paper mills discharging to the Fox River. Waste sources were grouped in two clusters, each with six or seven dischargers. Trading was allowed only among sources within a cluster. This limitation, together with a host of complex restrictions governing trades, greatly discouraged the formation of a market. Over a six-year period only one trade occurred, and the anticipated

[38] There are many reviews of experiences with marketable permits, such as those by Hahn (1989), Opschoor and Vos (1989), and Heggelund (1991).

[39] The following discussion uses the presentation by Hahn and Noll (1990) as a point of departure.

[40] For numerous examples of the extraordinary uncertainty introduced by government rule changes in the ERC program, see Foster and Hahn (1995).

cost savings in meeting ambient water quality standards did not materialize.[41]

Market formation is also impeded when discharge permits cannot be saved for use in future time periods. This was demonstrated in the ERC program, where concerns over useability of ERCs in future years led to creation of ERC banks. The existence of these banks gave firms incentives to abate pollution beyond required levels whenever it was timely. A firm's banked ERCs could be used in later years to support plant expansions, or they could be sold to other firms. In most regions, markets for ERCs have not been very active; firms with ERCs often save them to ensure that they will have ERCs available when they build new facilities.

The use of banks as a mechanism for expanding market activity relates to the more general issue of transactions costs. Experience with ERCs demonstrates that transactions costs can be high when participants in a market for pollution rights have trouble finding trading partners. Foster and Hahn (1995), in their analysis of emissions trading in the Los Angeles region, reported that finding a suitable trading partner took firms anywhere from one day to a year and a half. Once a partner was found, approval of a proposed trade by the regional air quality management agency usually took between 5 and 12 months. Moreover, many proposed trades were not approved because of problems in certifying claimed emission reductions, and disagreements on the benchmark used for defining ERCs. High transactions costs reduce both the potential gains to traders and the likelihood that potential cost-effectiveness gains will be realized.

The SO_2 allowance scheme under the CAAA of 1990, with its allowance auctions and direct sales, illustrates ways that transactions costs can be contained. The auction and direct sales also provide utilities with assurance that they can find sources of allowances when they need them. In addition, under the SO_2 allowance program, the role of the government in certifying trades is minimal because the basis for defining allowances is clear-cut.

For a scheme based on tradable rights, sanctions must be imposed on firms that don't abide by the rules. Otherwise firms might not have adequate incentives to own and use permits. Hahn and Noll (1990, p. 353) argue that for a marketable permit "system to operate effectively, the marginal cost of a violation must exceed the marginal cost of participating in the trading system." As they point out, because permits are durable assets, noncompliance penalties must account for future changes in technology and demands for environmental quality. The problem of defining penalty structures is only part of the enforcement problem; environmental authorities must also be able to monitor discharges.

The potential cost reductions associated with marketable pollution permits are a feature that is likely to foster their more widespread use. During the mid-1990s, substantial attention was being given to creating international pollution permit markets for CO_2 and other gases associated with global climate change.[42] In addition, many regions in the United States were actively experimenting with innovative tradable permit schemes.

COMPARATIVE ANALYSIS: STANDARDS, FEES, AND TRADABLE PERMITS

Discharge fees and marketable pollution permits are sometimes grouped under the label of *incentive regulation* and then compared to the principal policy tools of *command and control regulation:* discharge standards.[43] The term *policy instrument* refers here to each of the following three schemes: discharge standards, discharge fees, and tradable permits.

This section analyzes the *instrument choice* problem: Which policy instrument should be used to attain an ambient standard? In practice, this problem is not solved by picking one instrument or the other. Sophisticated regulatory schemes are hybrids that use combinations of policy instruments, with incentive regulation viewed as a complement (not a substitute) for discharge standards.

A basis for comparing alternative policy instruments is provided by Table 11.8, which lists attributes often used in arguing that discharge standards have

[41] A summary of the experience with the BOD discharge permit scheme on the Fox River is given by Hahn (1989).

[42] For an analysis of the feasibility of using marketable permits in this context, see Heggelund (1991).

[43] This section benefited greatly from Noll's (1994) unpublished paper, "Instrument Choice in Environmental Policy."

TABLE 11.8 Attributes Used in Comparing Environmental Policy Instruments

- *Cost-effectiveness.* Can an ambient standard be met at a minimum waste reduction cost under conditions of perfect information?
- *Uncertainty in discharges.* Assuming (realistically) that pollution abatement costs are imperfectly estimated, can waste discharges be predicted?
- *Uncertainty in costs.* Assuming imperfect information on abatement costs, can an upper bound be placed on costs to the regulated community?
- *Delayed compliance.* Are there inherent incentives for polluters to delay implementing abatement measures?
- *Distribution of costs.* Are there differences in the cost burdens shouldered by members of the regulated community?
- *Accommodation of growth.* What program changes are needed to continue to meet an ambient standard if new sources of waste enter a region?
- *Technological innovation.* Are there strong incentives for research and development to improve monitoring or waste reduction capabilities, or both?
- *Rights to pollute.* Are accepted views on who has a legal right to pollute accommodated?

advantages over incentive regulation or vice-versa. Table 11.9 summarizes how discharge standards compare with discharge fees and marketable permits using attributes in Table 11.8 as a basis for comparison. The following discussion elaborates on the table entries.

Consider the first attribute: *cost-effectiveness.* Imagine a hypothetical world in which transactions costs are zero and an environmental agency has perfect knowledge of cleanup costs and transfer coefficients. In such a world, discharge standards, fees, or marketable permits could each be used to meet an ambient standard at minimum cost. In real settings, the cost-effectiveness of outcomes depends on how policies are designed. For example, discharge standards based on *best available technology* will not be cost-effective if they take no account of a discharger's location. In addition, standards set uniformly with all sources removing the same percent of their raw loads will be inefficient because they ignore differences in marginal costs of waste reduction among polluters.

Incentive regulation is often advocated because of cost-effectiveness, but cost-effectiveness is not ensured in practice. For example, discharge fees are often set too low to influence the behavior of polluters. In contrast, marketable pollution permits seem to have practical advantages in reducing costs of meeting environmental goals, even when information available to government authorities is incomplete. This is demonstrated by the impressive cost reductions documented for EPA's policies based on Emission Reduction Credits.[44]

The second and third items in Table 11.8 concern *uncertainties in outcomes* under different conditions. Consider a circumstance in which cleanup costs are uncertain, but transfer coefficients are well known. In this case, assuming program implementation is not delayed, discharge fees are inferior to discharge standards because the response of polluters to fees is unknowable a priori. Once fees are set, an environmental agency must wait and see how polluters respond. Because of the uncertanties regarding what it costs polluters to reduce discharges, the initial fees might be so low that ambient standards are violated.

For tradable pollution rights schemes in which the location of waste release does not affect incremental benefits (for example, the case of SO_2 control to reduce acid rain), discharge uncertainty is not an issue. If polluters comply with requirements, the total waste discharge can be determined because it is constrained by the number of pollution permits issued by the environmental agency. However, if incremental benefits are sensitive to waste source location, then discharge uncertainty comes into play because it is impossible to know a priori which sources will purchase pollution rights and have relatively high dis-

[44] For details on cost savings using ERCs, see Hahn (1989).

TABLE 11.9 A Comparison of Discharge Standards, Discharge Fees, and Tradable Permits

Attribute	Discharge Standards	Discharge Fees	Tradable Permits
Cost-effectiveness	Attainable only with perfect information	Attainable only with perfect information	Attainable only with perfect information; cost savings likely with incomplete information
Uncertainty in discharges	Discharges known if polluters comply	Discharges uncertain; ambient standard may be violated	Aggregate discharge known if polluters comply; location of releases uncertain
Uncertainty in costs	Often no accurate upper bound on costs	Upper bound on costs can be computed easily	Market forces affect costs of compliance; transactions costs difficult to predict
Delayed compliance	Financial incentives to delay compliance	Reduces advantages of delayed compliance	Flexibility in compliance options reduces incentives to stall compliance
Distribution of costs	Dischargers pay only for abatement	Dischargers pay fees plus abatement costs; environmental agency may obtain revenues	Allocation of costs depends on how permits are distributed initially; acceptability can be enhanced by distributing initial permits to existing sources for free
Accommodation of growth	Some discharge standards made more stringent	Fees must increase	Built-in mechanism to accommodate new loads
Technological innovation	Distorts research programs; new technology may become new standard	Incentives to improve technologies for monitoring and waste minimization	Incentives to improve technologies for monitoring and waste minimization
Rights to pollute	Rights are implicit and seem unobjectionable	Rights to pollute are explicit and consistent with polluter pays principle	Rights to pollute are explicit and consistent with polluter pays principle

charges. This is the basis for the *hotspot problem,* which occurs with tradable permits if changes in the distribution of abatement activities make ambient quality in some locations worse instead of better. The hotspot problem can be avoided by imposing discharge standards in addition to requiring waste sources to hold a number of pollution permits consistent with their releases. However, the result of adding discharge standards is a reduction in the cost savings that a tradable pollution rights scheme might otherwise deliver.

While discharge standards have an apparent advantage regarding uncertainties in releases of pollutants, the opposite holds regarding uncertainties in costs. In setting discharge standards to meet ambient standards, an environmental agency's knowledge of potential compliance costs is limited since the best information comes from regulatees who may not provide adequate or accurate cost information. An environmental agency is often unable to estimate accurately the maximum cost to meet discharge standards, and actual compliance costs for the regulated community can be extraordinary. In contrast, a discharge fee scheme imposes an upper bound on costs: total fees paid when all polluters engage in no waste reduction.

With a tradable permit scheme, there is uncertainty in the cost of compliance, but the environmental agencies a priori estimates of cost are likely to be too high. The increased flexibility afforded to waste dischargers, and the influence of competition among those promoting new ways of complying with regulations, may reduce compliance costs below initial estimates. This was the initial outcome in the U.S. program based on SO_2 emission allowances. The possibility of using scrubbers to make SO_2 emission allowances available for sale influenced competition between scrubber manufacturers and providers of low-sulfur coal, and the cost of compliance with environmental requirements was several hundred dollars per ton of SO_2 below EPA's original estimates. The effects of market forces, together with uncertainties associated with how SO_2 allowances would be treated by public utility commissions and tax agencies, made it difficult for EPA to estimate the cost of compliance accurately. Another important element of compliance, transactions costs, are also difficult to predict with a marketable permits scheme. If markets are "thin," in the sense that few trades take place, transactions costs can be high.

Policy instruments also differ in the degree to which they encourage (unintentionally) *delays in compliance.* While discharge standards can be designed to meet an ambient standard, difficulties often arise in meeting those standards because of the many opportunities for polluters to delay compliance. For example, polluters may claim to be ignorant of requirements or threaten to close down (with attendant losses in local employment and tax revenues), or they may agree to meet compliance deadlines without intending to. Lawsuits (or threats thereof) may also be used to delay compliance. Polluters generally have opportunities to challenge discharge standards or engage in protracted negotiations with environmental agency staff over precisely how standards should be set. Incentives to delay are strong. The longer a polluter can stall, the longer funds required for compliance can be put to other uses.

With a discharge fee scheme, polluters have no inherent motivation to delay compliance. If they fail to abate pollution, they pay higher total discharge fees. And with a marketable permits scheme, the incentives for polluters to delay are reduced because they have enhanced flexibility in deciding how to comply. Notwithstanding these apparent advantages for discharge fees and marketable permits, implementation of either of those schemes may, as in the case of discharge standards, be stymied by protracted legal battles during rulemaking, and by shortcomings in monitoring and enforcement.

Next consider how cost burdens are distributed. In comparison to discharge standards, the *distribution of costs* using discharge fees falls more heavily on polluters. To illustrate, consider a situation where polluters reduce their releases to levels consistent with a particular set of discharge standards. If discharge standards are used, polluters use some of the environment's waste receptor services for free. In contrast, with discharge fees, polluters pay for waste reduction *and* for any releases they make. This difference in the way costs are distributed often leads polluters to argue against the introduction of discharge fees. In contrast, environmental agencies sometimes support discharge fee systems because they can provide a new source of revenue to support agency operations and fund waste reduction subsidy programs.

Who bears the costs of a tradable permits scheme depends on how permits are distributed at the outset. Because the cost-effectiveness of a trading equilibrium does not depend (in theory at least) on the initial allocation, any politically attractive initial distribution can be used. When the initial allocation favors existing polluters (as it does in the SO_2 emissions allowance scheme), they receive a windfall gain that can enhance the political acceptability of a marketable permits scheme.

The distribution of compliance costs changes when new waste sources enter a region or existing polluters expand their operations. Neither discharge standards nor fees automatically accommodates *growth in sources of waste.* To meet an ambient standard when growth occurs, discharge standards may need to be tightened. In practice, new waste sources often face requirements that are more stringent than those imposed on existing sources. This is generally inefficient inasmuch as marginal waste reduction costs for existing sources are often relatively low. An existing source facing weak abatement requirements may be able to reduce its wastes at a lower cost per unit than a new source that faces stringent waste reduction requirements because it is new to the region. Pollution discharge fee programs also require modifications to deal with growth. To continue to meet an ambient standard in the face of regional growth that increases raw waste loads, discharge fees must be raised. And if *all* fees increase, existing polluters incur increased costs to accommodate regional growth. If it is assumed that an environmental agency has perfect information and is not forced to discriminate in favor of existing sources, then cost-effective outcomes can be obtained in response to regional growth with either discharge standards or fees. In real settings, neither of these assumed conditions is likely to hold.

In contrast to both discharge standards and fees, tradable permit schemes have a built-in mechanism for accommodating growth. Since the number of permits is fixed by the environmental agency, new waste sources can only enter a region by purchasing the permits they need. The price of permits may increase as a result of the increased demand for permits, and a rise in price provides potential gains to existing polluters. The accommodation of new growth can be impeded, however, if existing permit holders choose not to sell permits as a way of blocking potential competitors from entering a region. This can occur, for example, if one polluter in a region holds a large fraction of available permits and chooses to exercise its power over the permit market. The annual auction and direct sales of SO_2 emissions allowances by EPA illustrates one way of reducing problems of this sort.

Policy instruments have different effects on *technological innovation.* Consider, for example, differences between discharge standards and fees in motivating improvements in pollutant-monitoring devices. Using discharge fees, both environmental agencies and polluters have an economic interest in ensuring accurate monitoring. Polluters don't want to pay more charges than they must, and environmental regulators don't want to lose potential revenues from charges. If existing monitoring devices are believed to overestimate measurements of pollution, dischargers have an incentive to invent new ones. If underestimation of values occurs, the environmental agency would be motivated to improve monitoring.

There is also an incentive to improve monitoring under a tradable permit scheme because the scheme creates an economic asset (a tradable right to pollute) that has a value that depends on accurate monitoring. As in the case of discharge fees, parties with an incentive to improve monitoring may differ from case to case, depending on whether existing monitoring devices are believed to underestimate or overestimate actual discharges.

Policy instruments also differ in the extent to which they motivate research on new, cheaper ways to reduce wastes. Incentive regulations give individuals greater freedom to decide on levels and methods of waste reduction. Under either marketable permits or discharge fees, polluters gain financially as they discover and implement more-efficient techniques for abating pollution. By comparison, discharge standards constrain the decision-making ability of polluters and can distort research on waste reduction methods. Distortions in research agendas are especially problematic when technology-based standards are used, since creation of new pollution abatement methods can be used as a basis for more stringent technology-based discharge requirements in the future.

Another issue relevant in a comparison of discharge standards and incentive regulation concerns the ethical basis for *rights to pollute.* Some people object to discharge fee schemes because they grant an explicit legal right to "pollute if you pay." The same can be said of marketable permit schemes. These arguments are not convincing, however, since discharge standards grant an *implicit* right to release wastes.[45]

Some people advocate the use of discharge fees and marketable permits because they embrace the *polluter pays principle.* This principle assumes that no one has an inherent right to pollute. If wastes are released, then those responsible should pay the costs their pollution imposes on others. In this sense, the polluter pays principle can be interpreted as an argument supporting both productive efficiency and eq-

[45] One version of the argument against letting those who pay pollute is as follows: An effluent charge system sends the wrong message to industry. If an overall goal is to promote a healthy environment, it is inconsistent with that goal to permit some companies to discharge massive amounts of pollution as long as they are willing to pay for it. Setting up a charge system runs counter to the idea that it's the responsibility of every person and every company to work toward a healthy environment.

uity. Discharge fees and marketable permits force polluters to internalize the external costs they impose, and it is fair that polluters account for those costs.

This comparative analysis helps explain why practical applications of incentive regulation are added as complements to (not substitutes for) traditional command and control regulations based on discharge standards. As incentive regulation becomes more widely known, and as government decision makers become more familiar with advantages of implementing them, they will undoubtedly be used more. Hybrid regulatory schemes involving combinations of discharge standards, fees, marketable permits, and subsidies may become increasingly common.

Key Concepts and Terms

Basic Theory: Benefits Independent of Waste Source Location
Marginal cost of abatement
Marginal benefit of abatement
Rights to pollute
Benefit of holding additional rights
Mutually advantageous trades
Maximum offer price
Minimum acceptance price
Trading equilibrium

Modifying the Basic Theory to Account for Source Location
Linking discharge allowances to locations
Transfer coefficients
Cost-effectiveness

Roles for Government in a Permit Market
Definition of rights for permit holders
Number of permits issued
Initial distribution of permits
Tracking market transfers of permits
Monitoring and enforcement
Brokerage functions
Instability in price of permits
Zero revenue auction
Aggregate bid schedule

Trading Based on Emission Reduction Credits
Baseline for defining ERCs
Nonattainment regions
Offset policy
Bubble policy
Banking
Netting
Internal vs. external offsets

Trading in SO_2 Emission Allowances
Definition of SO_2 emission allowance
Initial distribution of allowances
Auctions and direct sales
Chicago Board of Trade
Spot allowance vs. advance allowance
Options markets and futures markets
Compliance options for electric utilities
Scrubbers
Low sulfur coal and coal blends
Deregulation of railroads
Competition in markets for compliance options

Lessons from Experience with Tradable Permits
Legal enforceability of rights
Limits on number of permits
Transactions costs
Impediments to free trade
Expropriation of permits
Future use of permits
Enforcement of trading rules

Comparative Analysis: Standards, Fees, and Tradable Permits
Instrument choice problem
Incentive regulation
Cost-effectiveness
Uncertainty in costs and discharges
Hotspot problem
Delayed compliance
Incidence of costs
Accommodation of growth in waste loads
Incentives for technological innovation
Ethical basis for rights to pollute
Polluter pays principle

Discussion Questions

11-1 Suppose the initial allocation of BOD allowances in the three-waste-sources example is as follows: source 1 has 100 allowances, source 2 has 400 allowances, and source 3 has no allowances. Demonstrate that this is not a trading equilibrium. Describe a mutually advantageous purchase and sale of BOD allowances that would cause a change from this initial allocation.

11-2 Emission offsets (measured using Emission Reduction Credits) provide a mechanism for allowing new economic growth in areas not meeting ambient air quality standards, while still making progress toward meeting ambient standards. What other strategies could be used to allow new growth in an area that has no additional waste receptor services to offer?

11-3 In the United States, pollution discharge fees have been seriously considered as a supplement to discharge standards since the 1960s. However, discharge fees have not been adopted. In contrast, the federal government has made a substantial commitment to the use of marketable permits to pollute. How would you explain why marketable permits, which are a relatively new invention, have been adopted in the United States, while pollution discharge fees have been consistently rejected.

11-4 You are hired as a consultant to the European Environment Agency. Your task is to present an argument for or against the use of tradable permits to pollute as a policy instrument for controlling SO_2 to reduce problems with acid rain in Europe. What position would you adopt?

11-5 The problem of limiting global climate change associated with the greenhouse effect centers on controlling total, worldwide emissions of carbon dioxide (CO_2) and other *greenhouse gases.* Assume that reducing sources of CO_2 at any location contributes to reducing climate change. An international system of tradable permits to emit CO_2 is one strategy for reducing total CO_2. What makes tradable permits an attractive strategy for limiting total CO_2 emissions? How might the theory presented in the first section of this chapter be modified to accommodate the contributions of photosynthetic activity of forests in offsetting CO_2 emissions? Photosynthetic activity must be considered, because options for reducing global climate change include programs for afforestation and for maintenance of existing forests.

11-6 In July 1996, representatives of about 150 nations that are signatories to an international treaty on global climate change met in Geneva to begin negotiations on air pollution reduction targets to deal with projected climate change problems. The American position in the negotiation was that each country should have the flexibility to decide on its own strategy for meeting the new targets. Timothy Wirth, Undersecretary of State for Global Affairs in the Clinton Administration, "signaled that the Administration intends to use market-based approaches, like the trading of pollution permits, to help its industries [meet pollution cutback targets] in whatever fashion they deem most efficient."[46] What evidence is there to support the Clinton Administration's claims about the efficiency advantages of tradable permits? Would you advocate use of tradable permits for dealing with climate change problems?

11-7 Tradable permits have a prominent place in *Reinventing Environmental Regulation,* a statement of priorities issued (in March 1995) by President Bill Clinton and Vice-President Al Gore. One of 25 "high priority actions" highlighted in the statement singles out trading in wastewater discharges:

> Effluent trading in watersheds. EPA will place top priority on promoting use of effluent trading to achieve water quality standards. . . . Trading can be used to achieve higher levels of water quality in watersheds at lower cost than inflexible discharge requirements for individual sources.

Reinventing Environmental Regulation encourages EPA to issue "policy guidance" to facilitate development of effluent trading programs. Based on U.S. experience with emission

[46] This quotation is from the *New York Times,* July 17, 1996, Cushman, J. W., Jr., "U.S. Will Seek Part on Global Warming" (national edition, p. A6).

trading, what issues should EPA address in the policy guidance?

References

Burtraw, D. 1996. The SO_2 Emissions Trading Program: Cost Savings Without Allowance Trades, *Contemporary Economic Policy.* XIV: 79–94.

Dales, J. H. 1968. *Pollution, Property and Prices.* Toronto: University of Toronto Press.

EPA (Environmental Protection Agency). 1982. *The Bubble and Its Use with Emission Reductions Banking.* Washington, DC: Office of Policy Resource Management.

Foster, V., and R. W. Hahn. 1995. Designing More Efficient Markets: Lessons from Los Angeles Smog Control. *Journal of Law and Economics* XXXVIII: 19–48.

Greenberger, L. S. 1992. The Birth of an Allowance Trading Market. *Public Utilities Fortnightly* 129(12): 31–32.

Hahn, R. W. 1989. *A Primer on Environmental Policy Design.* Chur, Switzerland: Harwood Academic Publishers.

Hahn, R. W., and R. Noll. 1983. Barriers to Implementing Tradable Air Pollution Permits: Problems of Regulatory Interactions. *Yale Journal on Regulation* 1: 63–92.

———. 1990. Environmental Markets in the Year 2000. *Journal of Risk and Uncertainty* 3: 351–67.

Heggelund, M. 1991. *Emissions Permit Trading: A Policy Tool to Reduce the Atmospheric Concentration of Greenhouse Gases.* Calgary: Canadian Energy Research Institute.

Krupnik, A. J., W. E. Oates, and E. Van De Verg. 1983. On Marketable Air Pollution Permits: The Case for a System of Pollution Offsets. *Journal of Environmental Economics and Management* 10(3): 233–47.

Liroff, R. A. 1980. *Air Pollution Offsets: Trading, Selling and Banking.* Washington, DC: The Conservation Foundation.

National Academy of Public Administration (NAPA). 1994. *The Environment Goes to Market: The Implementation of Economic Incentives for Pollution Control.* Washington, DC: NAPA.

National Commission on Air Quality. 1981. *To Breathe Clean Air.* Washington, DC: U.S. Government Printing Office.

Noll, R. G. 1994. "Instrument Choice in Environmental Policy." Unpublished manuscript, Dept. of Economics, Stanford University, Stanford, CA.

Opschoor, J. B., and H. B. Vos. 1989. *Economic Instruments for Environmental Protection.* Paris: Organization for Economic Cooperation and Development.

Tietenberg, T. H. 1985. *Emissions Trading: An Exercise in Policy Reform.* Washington, DC: Resources for the Future.

U.S. General Accounting Office. 1982. *A Market Approach to Air Pollution Control Could Reduce Compliance Costs Without Jeopardizing Clean Air Goals.* Summary No. PAD-82-15A (March 23). Washington, DC.

Part Three

Environmental Management Programs in the United States

Part Three demonstrates concepts and theories from previous chapters by examining several environmental management programs in the United States. The emphasis is on what can be learned from past experiences with U.S. federal programs rather than what the latest environmental regulations call for.

Chapter 12 investigates federal efforts to control water pollution, and Chapter 13 provides a parallel examination of air pollution control. Each chapter adopts a historical perspective and demonstrates that U.S. pollution control policies have been revised frequently in response to new information and shortcomings in performance. Federal air and water quality programs provide numerous examples of the command and control approach to regulation. To a lesser extent these programs also demonstrate regulations based on economic incentives: subsidies to control water pollution and marketable permits for emissions of sulfur dioxide.

Problems associated with hazardous waste are increasingly significant in many countries, and Chapter 14 examines the U.S. response to these problems. The chapter details key provisions of a statute regulating generators and handlers of hazardous waste. It also analyzes the Superfund program, which relies on a liability scheme to finance the cleanup of hazardous waste storage and disposal sites that have contaminated the environment. The chapter also examines why many firms favor pollution prevention over waste treatment as a way of meeting hazardous waste requirements.

Chapter 15, the final chapter in Part Three, analyzes federal requirements to assess the environmental impacts of projects. In addition to evaluating how these requirements have been implemented, the chapter compares federal requirements with analogous state-level impact assessment programs. U.S. environmental impact assessment procedures are particularly notable because of their influence on impact assessment activities in other countries.

Chapter 12

Water Quality Management

Analysis of how the U.S. government has tried to control water pollution demonstrates the difficulty in designing effective environmental programs. The history of federal water pollution control efforts is filled with policy experiments and revisions. What started, in the 1940s, as a reluctant federal intervention to assist states wound up, in the 1970s, as a dominant federal presence in pollution control. Federal influence receded somewhat during the Reagan administration, and it shows signs of being further diminished as Congress, in the mid-1990s, makes efforts to shift more authority and responsibility for pollution control back to the states.

This chapter's chronological account of federal water quality management does not dwell on details. Because federal environmental regulations are revised so frequently, a presentation emphasizing the latest fine points would be obsolete in a short time.[1]

[1] Up-to-date discussions of U.S. water quality management programs are given in recent issues of *Environmental Science and Technology* and *Water Environment and Technology*. Environmental law case books that are updated frequently also provide timely information on water quality requirements; see, for example, Grad (1995), and Novic, Stever, and Mellon (1995). In addition, the annual reports of the Council on Environmental Quality provide yearly summaries of environmental management activities in the United States. Finally, the Internet provides quick access to the latest requlatory information and data; for details on how to access environmental management information on the Internet, see Katz and Thornton (1996).

WATER POLLUTION CONTROL BEFORE THE 1980s

Early Dominance of State and Local Programs

Before 1948, state agencies played the central roles in water pollution control, and activity by the federal government was minimal.[2] State institutions to control pollution began to appear in the last half of the nineteenth century, as scientists began clarifying links between contaminated water and the spread of disease. In that period, water pollution control programs were administered by state boards of health. The Massachusetts Board of Health, formed in 1869, was the first state agency with responsibilities for water pollution control. A major concern of early state programs was control of waterborne infectious diseases such as typhoid fever and cholera.

Although an effective technology for treating municipal wastewater existed by the late 1870s, only 4% of the nation's population had its wastes treated as of 1910.[3] By 1939, about half of the nation's urban

[2] The federal role in water pollution control began with the Public Health Service Act of 1912, a law that established the Streams Investigation Station at Cincinnati to carry out water pollution research. The Oil Pollution Act was passed in 1924 to prevent oily discharges on coastal waters.

[3] Facts in this paragraph are from Hey and Waggy (1979, p. 128).

TABLE 12.1 Key Federal Laws Controlling Water Pollution

Year	Title	Selected *New* Elements of Federal Strategy[a]
1948	Water Pollution Control Act	Funds for state water pollution control agencies Technical assistance to states Limited provisions for legal action against polluters
1956	Federal Water Pollution Control Act (FWPCA)	Funds for water pollution research and training Construction grants to municipalities Three-stage enforcement process
1965	Water Quality Act	States set water quality standards States prepare implementation plans
1972	FWPCA Amendments	Zero discharge of pollutants goal BPT and BAT effluent limitations NPDES permits Enforcement based on permit violations Permits for dredge and fill activities
1977	Clean Water Act (CWA)	BAT requirements for toxic substances BCT requirements for conventional pollutants
1981	Municipal Waste Treatment Construction Grants Amendments	Reduced federal share in construction grants program
1987	Water Quality Act	State Nonpoint Source Management Program National Estuary Program NPDES permits for stormwater discharges Toxic hotspots program State Revolving Funds

[a] The table entries include only the *new* policies and programs established by each of the laws. Often these provisions were carried forward in modified form as elements of subsequent legislation.

population still discharged its waste untreated. According to Hey and Waggy (1979), two factors impeded use of wastewater treatment technology: (1) the ability to protect drinking-water supplies economically using chlorine for disinfection, a practice initiated in 1911; and (2) the widely shared attitude that natural waterways were appropriate receptors for wastes.

During the 1930s and 1940s, Congress debated whether the federal government should take an influential role in controlling water pollution. These deliberations led to the limited expansion of federal powers expressed in the Water Pollution Control Act of 1948. Table 12.1 highlights the key features of this act and other important federal water quality laws.

The 1948 act, administered by the U.S. Public Health Service, provided funds and technical services to states to strengthen their water quality control programs.[4] Although the 1948 act allowed the federal government to move against polluters, federal action could not be taken without the "consent of the state in which pollution was alleged to originate." The 1948 act formally recognized the primacy of the states in water quality management. Its preamble declared a national policy "to recognize, preserve and protect the primary responsibilities and rights of the states in preventing and controlling water pollution." Despite

[4] Information in this paragraph is from Cleary's (1967, pp. 251–52) summary of early federal water pollution control laws.

this declaration, the decades that followed saw many of the traditional rights and responsibilities of the states overtaken by an increasingly dominant set of federal interventions to restrain water pollution.

Early Federal Strategy: Enforcement Conferences and Construction Grants

The Federal Water Pollution Control Act (FWPCA) of 1956 was the cornerstone of early federal efforts to reduce pollution. (Acronyms in this discussion are listed in Table 12.2.) The 1956 law expanded the 1948 act's basis for federal action against polluters, and it introduced a new program of subsidies for municipal treatment plant construction. Increased funding for state water pollution control efforts and new support for research and training activities were also provided. Each of these programs—federal action against polluters, subsidies for treatment plant construction, and funds for research and training—was continued in the several amendments to the Federal Water Pollution Control Act in the 1960s and 1970s.

Expansion in the federal government's ability to act against polluters took the form of a *three-stage enforcement process.* The FWPCA of 1956 allowed the surgeon general of the Public Health Service to move against firms and municipalities whose wastewaters were interfering with the health or welfare of water users in a downstream state. The first stage of the enforcement process involved a *conference* of federal and state water pollution control officials. During the conference, the Public Health Service presented evidence documenting the interstate water pollution problem. Agreements were then reached with relevant state agencies regarding appropriate remedial measures and a timetable for their imple-

TABLE 12.2 Acronyms Used in Federal Water Quality Management

BADT	Best available demonstrated control technology
BAT	Best available technology economically achievable
BCT	Best conventional pollution control technology
BPT	Best practicable control technology currently available
CSO	Combined sewer overflows
CWA	Clean Water Act
EPA	Environmental Protection Agency
FSA	Food Security Act
FWPCA	Federal Water Pollution Control Act
HEW	U.S. Department of Health, Education, and Welfare
NEP	National Estuary Program
NPDES	National Pollutant Discharge Elimination System
NPS	Nonpoint source
NSPS	New source performance standards
POTW	Publicly owned treatment works
SRF	State Revolving Fund
USDA	U.S. Department of Agriculture

mentation. If state activities failed to yield the agreed-upon actions, the surgeon general could implement the second stage of the enforcement process by holding a *hearing* on the matter. The third stage, entered only if the hearing failed to yield a satisfactory outcome, involved *lawsuits* by the federal government. The second and third stages of the enforcement process were not used frequently.[5] Often, when the deadlines set at a conference were not met, new conference sessions were convened and deadlines were renegotiated. Although dischargers were encouraged to clean up, the resulting cutbacks in effluents were often much less than expected.

Cities felt that the federal government should help pay for the treatment plants required by federal enforcement activities. The 1956 act responded to this pressure from cities by creating a program of *construction grants,* subsidies to help build municipal treatment facilities. Grants made under the 1956 act could subsidize as much as one-third of the construction cost for a municipal plant, but the maximum grant for a single project was limited to $600,000. A total of $50 million per year was authorized for the program.

The effectiveness of the construction grants program was challenged frequently. Critics argued that the total budget authorized for grants was woefully inadequate, and that the maximum funds available for any one project were too small to influence the decisions of large cities. The numerous modifications to the FWPCA of 1956 throughout the 1960s and 1970s responded to these concerns by substantially increasing both the total funding and the maximum subsidy for any one project.

Another criticism of the construction grants program was that municipalities sometimes used the lack of available federal funds as an excuse for not constructing facilities quickly. Because so many cities needed to construct treatment plants, there was often a long wait to receive the limited grant monies. In addition, delays were caused by complex administrative procedures for processing grant applications and obligating funds.[6]

The following additional points were raised in assessments of the construction grants program during the 1960s and 1970s.

1. The program was biased toward end-of-pipe treatment since mainly treatment plant construction was subsidized. Other options, such as changing the timing or location of discharges, might have yielded the same ambient water quality at a much lower cost.
2. Many federally subsidized treatment plants were operated inefficiently, because cities failed to maintain equipment and hire adequately trained plant operators.[7]
3. The grants provided an indirect subsidy to firms discharging to municipal sewers, and this weakened the incentives firms had to seek minimum-cost waste management schemes. (This indirect subsidy was substantially reduced by amendments to the FWPCA, which required firms to pay charges to municipalities reflecting the additional costs of waste treatment. In addition, industries discharging to publicly owned treatment works (POTWs) were required to *pretreat* their wastes prior to discharge to ensure that industrial wastewaters would not interfere with proper operation of POTWs.[8])
4. Since municipal treatment plants are traditionally designed with capacities much higher than their initial wastewater loadings, the construction grants program subsidized future growth that could lead to further deterioration in environmental quality.[9]

As detailed later in this chapter, the federal government eventually dismantled its construction grants program.

A Shift in Strategy to Ambient Standards

The Water Quality Act of 1965 carried forward many provisions of the FWPCA of 1956, generally with increased levels of funding. The 1965 act also intro-

[5] Freeman, Haveman, and Kneese (1973, p. 116) report that between 1956 and 1965 only one federal enforcement case reached the third stage.

[6] Whipple (1977, pp. 4-5) cites an analysis by the National Utility Contractors Association (1975) indicating that treatment plant construction was sometimes delayed a few years because of the grant application process.

[7] Freeman, Haveman, and Kneese (1973, p. 119) give statistics documenting this inefficiency in plant operations.

[8] To implement this requirement, EPA issued both general *pretreatment standards* and specific pretreatment standards for various industrial categories. For a summary of EPA's minimum pretreatment requirements, see Vincoli (1993, pp. 100–101) and Gallagher (1995, pp. 157–158).

[9] Binkley et al. (1975) examine this subject in detail.

duced important *new* requirements for states to establish ambient water quality standards, and detailed plans indicating how the standards would be met. The act also shifted responsibilities for administering the federal water quality program from the Public Health Service to a separate agency, the Federal Water Pollution Control Administration, within the U.S. Department of Health, Education, and Welfare (HEW). This was a short-lived change. In 1970, a presidential reorganization order placed water pollution control activities and several other federal environmental programs in a newly created U.S. Environmental Protection Agency (EPA).

States had established *ambient water quality standards* for many watercourses long before the Water Quality Act of 1965. In general, ambient standards were set by first determining the water uses to be protected on a given stretch of river. Professional judgment was employed to establish *water quality criteria,* bases for assessing the suitability of water for different activities. Knowing the uses to be accommodated on a particular river section, and the criteria associated with protecting those uses, it was possible to set ambient standards.

The Water Quality Act of 1965 required states to establish standards for all interstate waters by June 1967. If states failed to meet the deadline, the secretary of HEW was required to promulgate standards. Many states had trouble meeting the June 1967 requirement because the time allotted for setting standards was short and existing water quality data were inadequate. In addition to establishing ambient standards, states had to prepare *implementation plans.* For each waste source, these plans specified the steps required to reduce wastewater discharges so that ambient standards would be met in a timely manner.

The three-stage federal enforcement process was modified by the 1965 act to reflect the new importance of ambient standards. Federal lawsuits could be brought against dischargers violating ambient standards on interstate waters. Even though the federal government could initiate litigation 180 days after notifying violators, the government's ability to sue polluters was not widely used. By the end of 1971, only 27 notifications of impending court action had been issued. A reason the courts were not used more frequently is the difficulty in proving that a particular waste source caused a violation of downstream ambient standards. Establishing such causal relationships was problematic when there were many nearby sources of waste, and when the behavior of contaminants released into natural waters was poorly understood. In these circumstances, alleged violators of ambient standards could easily argue that a different waste source caused the problem, or that there was an inadequate scientific basis for determining how the contaminants were transported.

Major Changes in Strategy During the 1970s

In amending the Federal Water Pollution Control Act in 1972, Congress introduced five major program elements: (1) national water quality goals, (2) technology-based effluent limitations, (3) a national discharge permit system, (4) federal authority to sue polluters violating permit conditions, and (5) controls on dredge and fill activities affecting navigable waters.

The 1972 amendments aimed to restore and maintain "the chemical, physical and biological integrity of the nation's waters." The amendments specified, as a *national goal,* that the "discharge of pollutants into navigable waters be eliminated by 1985." They also included an "interim goal" of "fishable and swimmable waters":

> [W]herever attainable, an interim goal of water quality which provides for the protection and propagation of fish, shellfish and wildlife and provides for recreation in and on the water [should] be achieved by July 1, 1983.

The goal of eliminating the discharge of pollutants to waterways was criticized because it did not account for the ways in which water pollutants might be shifted to air and land. For example, requiring substantial reductions of solids in municipal wastewaters led to increases in municipal sludge disposal problems. Another criticism centered on the high cost of "eliminating" the discharge of pollutants. Abatement costs often increase sharply when the percentage of waste removal goes beyond about 90%.

The 1972 amendments gave enormous power to the EPA administrator, who was made responsible for defining "pollutants" in the context of the national goal, and for making operational the various effluent restrictions that the amendments described in general terms. Effluent limitations guidelines were specified by the administrator on an industry-by-industry basis. Because they applied nationwide, the effluent limitations did not depend on the particular circumstances in which a discharge occurred.

The EPA administrator was required to set effluent restrictions that met the following general requirements of the 1972 amendments: by 1977, all dischargers were to achieve the *best practicable control technology currently available* (BPT); and by 1983, all dischargers were to have the *best available technology economically achievable* (BAT). After delays caused by numerous legal challenges to the EPA administrator's effluent limitations guidelines, the BPT provisions were implemented.[10]

The principal criticism of the original BAT effluent limitations was that the cost of reducing pollutants by the very high required percentages would be much greater than the benefits. In defining BAT, costs were considered, but only in the general context of affordability by industry. Computations of the social benefits of stringent effluent controls were not a central factor. Congress presumed that the benefits of eliminating water pollutants would be substantial. Congress's insistence on very strict effluent limitations can also be interpreted as an effort to guarantee the rights of Americans to high-quality waters.[11]

The best available technology requirements were so heavily disputed that Congress modified them in the Clean Water Act (CWA) of 1977. Congress responded to critics of BAT by requiring it only for toxic substances. A different requirement was introduced for *conventional pollutants,* which included well-understood parameters such as biochemical oxygen demand, suspended solids, fecal coliform bacteria, and pH. The effluent limitations guidelines for these pollutants were to be based on the "best conventional pollutant control technology" (BCT).

Table 12.3 indicates the main types of effluent limitations the EPA administrator set under the FWPCA amendments as modified by the Clean Water Act of 1977. As indicated in the table, the BPT/BCT/BAT system of requirements apply only to existing discharges. Waste sources that satisfy EPA's definition of a *new* source are required to meet *new source performance standards* (NSPS) based on *best available demonstrated control technology* (BADT). Application of BADT could include the complete elimination of waterborne pollutant discharges. Although BADT limitations are often similar to BAT requirements, they can be more stringent. "In addition, because EPA must consider alternative production processes and operating methods, NSPS can also effectively dictate the choice of production processes used by a new source" Gallagher (1995, p. 150).[12]

The Clean Water Act of 1977 strongly endorsed the view that waterborne toxic substances must be controlled. The act included a list of 65 substances and classes of substances that was used as the basis for defining toxics. This list resulted from a 1976 settlement of a legal action in which several environmental organizations sued the EPA administrator for failing to issue toxic pollutant standards under the FWPCA amendments of 1972.[13] The list was subsequently refined by EPA to include 129 specific materials, referred to as *priority pollutants.*

Effluent limitations required by the FWPCA amendments of 1972 (and later the Clean Water Act of 1977) formed the basis for issuing *National Pollutant Discharge Elimination System* (NPDES) permits. As explained by Quarles (1976), the permit system idea stemmed from actions taken by the Department of Justice in the late 1960s. With the support of a favorable interpretation by the U.S. Supreme Court, Justice's attorneys relied on the 1899 Refuse Act to prosecute industrial sources of water pollution. The 1899 act, which was originally interpreted to prohibit deposits of refuse in navigable waters to keep them clear for boat traffic, was interpreted in the 1960s as applying to liquid wastes as well. In December 1970, the EPA administrator issued an order calling for a water quality management program using permits and penalties based on the Refuse Act of 1899. Although this program was delayed by court challenges in 1971, Congress made it a central part of the federal strategy embodied in the FWPCA amendments of 1972.

NPDES permits prescribe treatment requirements, construction schedules, and the timing of ef-

[10] Illustrations of effluent restriction based on BPT are given in Table 8.3.

[11] Velasquez (1982, p. 191) examines Congress's strict effluent limitations in the context of moral rights to a high-quality environment.

[12] Gallagher (1995, p. 150) provides an example of how production processes can be dictated: "the NSPS for the steam electric category prohibits the discharge of fly ash transport water, effectively requiring 'new source' plants in this category to use a dry ash handling system."

[13] This litigation is discussed by Zener (1981, pp. 99–100).

TABLE 12.3 Technology-Based Effluent Limitations: Examples from the Clean Water Act of 1977

Publicly Owned Treatment Works[a]

Requirements for 85% BOD removal, with possible case-by-case variances that allow lower removal percentages for marine discharges.

Industrial Discharges—Existing Sources

Toxic pollutants—Best available technology.

Conventional pollutants—Best conventional technology. In determining required control technology, EPA is directed to consider "the reasonableness of the relationship between the costs of attaining a reduction in effluents and the effluent reduction benefits derived."

Nonconventional pollutants (pollutants that are not classified as either conventional or toxic)—Best available technology. However, possible case-by-case variances allow lower degrees of treatment.

Industrial Discharges—New Sources

New source performance standards based on best available demonstrated control technology.

[a] Industries that discharge into publicly owned treatment works are regulated by pretreatment standards that form the basis for pretreatment permits. For an introduction to the pretreatment program, see Gallagher (1995, pp. 157–158).

fluent discharges.[14] The permits are for pollutant discharges to surface waters, but discharges from agricultural lands, such as flows running off of irrigated fields, are not covered by the program. Recognizing the responsibilities of the states in water pollution control, the 1972 amendments included provision for transferring the administration of the NPDES permit program from EPA to the states. For a state to assume responsibility for issuing permits, it had to have adequate staff and resources and the statutory authority needed for program implementation. By 1980, more than half the states issued NPDES permits directly.

The FWPCA amendments of 1972 gave EPA substantial power to enforce the NPDES permit program. EPA was authorized to seek court injunctions against permit violators. In addition, EPA could issue administrative orders requiring compliance with permit conditions. Failure to comply with these orders exposed the discharger to *criminal sanctions* and *civil penalties.* Criminal fines of $25,000/day ($50,000/day for second offenses) could be imposed on any person who "willfully or negligently" violated a permit condition or discharged without a permit. In addition, civil penalties of up to $10,000/day could be assessed. The 1972 amendments also included provisions for *citizens suits* in federal district courts by persons "having an interest which is or may be adversely affected" by violations of discharge requirements.

The permit-based enforcement strategy used after 1972 had an advantage over the earlier enforcement process based on ambient water quality standards. Using ambient standards, the government had to show that a particular discharge caused a violation in a downstream standard. As mentioned previously, it is often difficult to establish these causal relations. The permit-based approach was easier to implement, because only effluent monitoring was needed to establish that a particular source was discharging more than its NPDES permit allowed.[15]

At the same time it created the NPDES in 1972, Congress created another permitting program to regulate *dredged and fill material.* Dredged material includes muds, slurries, and other substances excavated from a water body. Fill material has been defined broadly to include any substance whose primary purpose is to replace an aquatic area with dry land, or to change the bottom elevation of a water body. Permits are issued by the U.S. Army Corps of Engineers in cooperation with EPA. The permit program covers

[14] For details on the process of issuing NPDES permits, see Gallagher (1995, pp. 140–159) and Hunter and Waterman (1996, pp. 34–39).

[15] For analyses of EPA's implementation of the NPDES permit program, see Hunter and Waterman (1996) and Mintz (1995, pp. 23–27).

not only disposal of dredged and fill material, but also their placement for development purposes. The reach of the permit program extends to all navigable waters of the United States, and it includes wetlands.[16]

ISSUES IN WATER QUALITY MANAGEMENT AFTER 1980

During the first several years of the 1980s, Congress was unable to reach a consensus on how to amend the Clean Water Act of 1977.[17] Moreover, cuts in EPA funding and other administrative actions during President Reagan's first term slowed progress in meeting national water quality goals. Finally, in 1987, the CWA was revised. In the 1987 amendments to the Clean Water Act, which were collectively known as the Water Quality Act of 1987, toxic waste and pollution from nonpoint sources (NPS) received more attention than they had previously, and stormwater discharges from urban areas and industrial facilities were included in the NPDES permit program. The act also altered the construction grants program for publicly owned wastewater treatment works.[18] For simplicity, this section is organized by issue areas and a strict chronological ordering is not followed.

Controlling Nonpoint Sources

While federal water pollution control laws prior to 1987 made strides towards controlling wastewater releases from discrete or point sources such as municipal wastewater treatment plants, significant diffuse or *nonpoint* sources often remained unregulated. Nonpoint sources of pollution include stormwater flows from urban and industrial areas, as well as runoff from farms, mining operations, and commercial logging areas. Flows from nonpoint sources can contain pesticides, fertilizers, sediments, metals, and oils that eventually contaminate water bodies. The 1972 amendments to the Federal Water Pollution Control Act had created a planning and regulatory program to reduce pollution from nonpoint sources, but it was widely viewed as ineffective, and Congress stopped funding it in 1980.[19]

In amending the Clean Water Act of 1977, Congress displayed renewed interest in nonpoint source pollution and launched several programs to control it. The most prominent of these was the *State Nonpoint Source Management Program* set up by the 1987 amendments. Under this program, states were to identify waterways that would not be expected to meet ambient standards because of nonpoint sources. For these areas, states were to develop NPS programs that identified *best management practices* for the categories of nonpoint sources involved. States had to propose plans to carry out those management practices, provide implementation schedules, and show they had funds to execute their plans. Because EPA had little or no authority to regulate many nonpoint sources, it relied upon state and local governments to implement controls.

In 1992, EPA's Office of Policy, Planning, and Evaluation reviewed the State Nonpoint Source Management Program and found it to be plagued by inadequacies in funding, monitoring, and enforcement.[20] While the program fostered communication among state agencies responsible for managing nonpoint source pollution, states had different conceptions of what an NPS management program should accomplish, and many states had no authority for administering nonpoint source control plans. The Office of Policy, Planning, and Evaluation study suggested that EPA clarify the program's goals and strategies and move toward a *watershed protection approach:* one that tailors NPS pollution control

[16] The Clean Water Act of 1977 also authorized states to create permit programs for dredge and fill activities in non-navigable waters. For further information on the dredge and fill permit program, see Gallagher (1995) and Novick, Stever, and Mellon (1995).

[17] Sambhav Sankar assisted in researching and writing early drafts of this section and the one that follows. Mr. Sankar was a graduate student in the Department of Civil Engineering at Stanford University at the time the work was carried out.

[18] In addition, the Water Quality Act of 1987 included stiffer penalties than the CWA of 1977. For example, the 1987 law included new penalties for *knowing endangerment:* "A person who knowingly violates a permit or other requirement of the Act and knows at the time that he thereby places another person in imminent danger of death or serious bodily injury is subject to a fine of up to $250,000 ($1,000,000 for an organization) and imprisonment for up to 15 years. Penalties are doubled for second offenses. An action for such 'knowing endangerment' may, for example, be brought if a person knowingly contaminates a water supply or deliberately dumps hazardous materials into sewers or waterways (Gallagher, 1995, p. 167)."

[19] A summary of early efforts to manage nonpoint sources is given by Percival et al. (1992, pp. 945–946).

[20] For more on the Office of Policy, Planning, and Evaluation study, see Adler, Landman, and Cameron (1993, p. 189).

strategies to fit conditions in particular watersheds and gives state and local governments flexibility. In addition to engineering controls on nonpoint source pollution, management strategies might include outreach and education programs[21] as well as government action to change land use patterns and activities that contribute to NPS pollution. The watershed protection approach attempts to improve coordination by bringing together agencies whose jurisdictions include significant nonpoint sources, as well as environmental groups and industries within a watershed.

The 1987 amendments also created the *National Estuary Program* (NEP), which provides a basis for managing NPS pollution. Even though the program was largely voluntary, it had potential to foster coordinated action by state and local agencies. Any state could declare an estuary to be of "national significance," and then apply for funds to convene a management conference that brings together representatives from EPA, other federal agencies, state and local government, and industry to develop estuary protection plans. As with the State Nonpoint Source Management Program, the NEP relied on existing statutory authority at the state and municipal level to accomplish program goals. Success of the NEP hinges on the motivation and authority of local agencies. As of the early 1990s, the NEP had achieved success in some locations. For example, management measures developed in the Chesapeake Bay region promised to cut nutrient discharges by 40%.[22]

A third element of the congressional response to nonpoint source pollution is contained in the Coastal Zone Management Act. Under the Act's 1990 amendments, coastal and Great Lakes states must submit coastal nonpoint source management plans to EPA and the U.S. National Oceanographic and Atmospheric Administration, the agencies that administer the act jointly. The NPS management plans required by the Coastal Zone Management Act must identify land uses contributing to degradation of impaired coastal waters and adjacent areas. The plans must also specify abatement measures that are "economically achievable" and yet "reflect the greatest degree of pollutant reduction achievable through the application of the best available nonpoint pollution control practices, technologies, processes, siting criteria, operating methods, or other alternatives."[23]

An additional response to nonpoint source pollution in the 1980s was the Food Security Act (FSA) of 1985. In this act, Congress laid the groundwork for the U.S. Department of Agriculture (USDA) to tackle the problem of NPS pollution from agriculture. Under the FSA, farmers who cultivated highly erodible lands were to implement conservation plans that used best management practices to reduce runoff. However, USDA interpreted the act as an erosion control measure, rather than as a broad mandate to control runoff-related pollution. According to some critics, instead of using the Food Security Act to curb excess use of pesticides and fertilizers, USDA approved management plans without close scrutiny and allowed farmers to omit runoff control measures that the farmers found overly expensive.[24]

Regulating Stormwater Discharges

Although stormwater runoff from municipal and industrial facilities is a nonpoint source of pollution, it becomes a point source if it is collected and discharged through sewers. In 1987, Congress mandated NPDES permits for releases from many municipal and industrial storm sewers. The 1987 amendments to the Clear Water Act of 1977 required NPDES permits for municipalities with populations over 100,000, discharges associated with industrial activities, and any other stormwater discharge that significantly pollutes a water body.[25] The *permit program for stormwater* was implemented in phases: first, large urban areas and industry were included, then mid-sized municipalities, and finally smaller cities. Stormwater releases from agricultural areas were not covered. Rules implementing the permit program were unpopular with industries and local governments because they called for significant data-gathering, and

---

[21] For example, outreach and education programs might be used to encourage land developers to account for stormwater runoff problems when planning new construction, and to demonstrate pesticide reduction measures and sound grazing practices to farmers.

[22] For more on the NEP, see Patton, Boggs, and Blow and Roy F. Weston (1990, p. 168), and Adler, Landman, and Cameron (1993, p. 218).

[23] Coastal Zone Management Act, as quoted by Percival et al. (1992, p. 948).

[24] For an analysis of USDA's implementation of the water quality protection measures in the Food Security Act, see Rosenthal (1988).

[25] See Gallagher (1995, pp. 155–156) and Adler, Landler, and Cameron (1993, p. 153) for more on requirements applicable to stormwater discharges.

some feared that costly treatment for stormwater discharges might be required to meet effluent limitations.[26]

NPDES permits for municipal stormwater discharges require controls "to the maximum extent practicable," using management practices and engineering methods.[27] Municipalities must also eliminate all non-stormwater inflows to the sewers. This is important in some older cities, where illicit wastewater discharges were connected to storm sewers before wastewater treatment was practiced.

For industrial activities, EPA's definition of runoff was broad, covering industrial plant yards, manufacturing buildings, access roads, and more.[28] Nearly any operation in which materials could be exposed to rain or snow fell under the NPDES regulations. Some facilities were exempted from NPDES permit requirements, but most were forced to file for coverage under a general permit, a group permit, or an individual permit.[29] A general permit application imposed the lightest administrative burdens on firms. Using this scheme, EPA and the states wrote rules for broad groups of facilities. Firms then applied for coverage under the general permit's guidelines.

If a number of similar firms could not obtain coverage under an existing general permit, they could file together to create a group permit. Under the group application provision, industrial facilities with similar processes could file for a single permit, and only a portion (usually 10%) of the group members would need to provide quantitative data on their discharges. If a facility could not be covered under either a group or general permit, it had to file for an individual permit. In this case, the facility was required to undertake detailed studies to characterize the amount and quality of stormwater released, describe the topography and processes at the site, and enumerate the plant's materials handling and waste management practices.

As of the early 1990s, many municipalities had applied for their initial stormwater discharge permits and were beginning to develop stormwater management programs. Critics argued that the stormwater permit program did not specify "substantive performance targets for the permits. . . ."[30] Additionally, while many large urban centers could reasonably obtain permits under the program, some smaller towns expected to have difficulty meeting the exhaustive stormwater characterization and monitoring requirements. Based on a survey of 59 large cities conducted by Gebhart and Lindsey (1993), the cost of completing the permit *application* process averaged almost one million dollars.[31]

Preserving and Restoring Wetlands

In the 1980s, Congress began to recognize both the pivotal role of wetlands in many ecosystems and the staggering rate at which wetlands were being filled for development. Earlier (in 1972) the government had established a wetlands protection program: the dredge and fill permit program created by the Federal Water Pollution Control Act amendments and administered by the U.S. Army Corps of Engineers (the Corps) in conjunction with EPA. However, the potential of dredge and fill permits as a tool for wetlands protection was not fully realized until the permit program was interpreted by the courts to cover all navigable waters of the United States, and interpreted by the Corps to cover both disposal *and* emplacement of dredged material. Permits are granted only if applicants take all reasonable measures to mitigate adverse environmental effects.[32]

In 1989, the Bush administration adopted a *no net loss of wetlands policy,* and implemented it by directing the U.S. Army Corps of Engineers to avoid adverse impacts to wetlands whenever possible in its dredge and fill permitting decisions.[33] The Corps was to en-

[26] Information on the cost of compliance with NPDES stormwater permitting rules is available in Gebhardt and Lindsey (1993).

[27] The quoted phrase is from section 402 of the 1987 amendments to the CWA.

[28] This explanation of NPDES stormwater permit application procedures was provided by Eugene Bromley, U.S. EPA Region 9, San Francisco, in a telephone interview conducted by Sambhav Sankar, August 22, 1994.

[29] Examples of exempted facilities are food processors and printers.

[30] This quote is from a critique of the program by Adler, Landman, and Cameron (1993, p. 196).

[31] In 1995 EPA formed an Urban Wet Weather Flows Advisory Committee which is trying to deal with some of the concerns cities have raised about the high costs and complexity of the stormwater discharge permit process. For more on the committee's activities, see 60 Federal Register 21189 (1995).

[32] From permit issuance guidelines, developed jointly by the Corps and EPA pursuant to section 404, as paraphrased by Percival et al. (1992, p. 967).

[33] For more on the no net loss of wetlands policy and other aspects of wetlands protection during the Bush administration, see Adler, Landman, and Cameron (1993, pp. 205–216).

courage permitees to avoid or minimize impacts rather than compensate for avoidable impacts by using "off-site mitigation." Some critics of the Bush administration's approach charged that it protected areas of dubious ecological value at great expense, while others noted that less than 20% of all wetlands losses are due to activities subject to the permit program.[34]

The amount of land protected by the dredge and fill permit program depends on how wetlands are defined. The importance of the definition is demonstrated by provisions in a bill to amend the permit program that was debated in the U.S. House of Representatives in May 1995. The House bill required that, to be protected as wetlands, land had to be saturated by water at the surface for 21 consecutive days during the agricultural growing season. According to the chair of a National Academy of Sciences committee of prominent wetlands experts, the definition in the House bill would reduce "the total acreage of wetlands [protected] in the United States by 50 percent or more. . . ."[35] In contrast to the House bill, the experts on the National Academy of Sciences committee felt a more accurate definition of wetlands is "one that would look at periodic saturation for shorter periods, in the root zone of plants rather than at the surface, and not necessarily in the months between the spring thaw and the autumn frost."

In addition to the dredge and fill permit program, the *swamp buster* provision of the Food Security Act was intended to protect wetlands. This provision denies valuable U.S. Department of Agriculture benefits (such as disaster relief payments and price supports) to farmers who plant on wetlands that were converted to cropland after 1985. The Soil Conservation Service, which administers the swamp buster program of the FSA, has been charged with lax enforcement of the provision's penalties. According to Rosenthal (1988), of 500 investigations of swamp buster violations, only five farmers lost agricultural benefits during the first five years of the program.

Legislating Against Toxic Water Pollution

Framers of the Water Quality Act of 1987 were dissatisfied with the nation's progress in reducing discharges of toxic chemicals. Under the Clean Water Act that had been passed 10 years earlier, dischargers of toxic substances in wastewater had to install the best available technology for pollution control, and releases were supposed to be further restricted in areas where conditions (for example, low streamflow) made BAT insufficient to meet ambient water quality standards. However, EPA was slow to produce guidelines to fully implement the toxic waste provisions of the 1977 Clean Water Act.[36] Congress responded by including requirements in the Water Quality Act of 1987 to regulate toxic releases more carefully.

In what is sometimes referred to as the *toxic hotspots* provision of the 1987 Water Quality Act, Congress directed states to develop ambient water quality standards based on federal criteria for toxic pollution. States were to list waters where ambient standards for toxics were violated and identify sources of the violations. If nonattainment was due "entirely or substantially"[37] to point source discharges of a pollutant designated as toxic by EPA, states were required to identify the responsible polluters and propose plans for each that would lead to attainment of the ambient standards. Requirements for implementing the new plans were to be written into NPDES permits. In response to the toxic hotspots program, over 17,000 waters were listed as impaired by toxic releases from either point or nonpoint sources, or both. However, EPA found that only 595 of these waters were impaired by toxic releases from point sources. A significant number of those point sources were municipal storm sewers and publicly owned wastewater treatment plants.[38]

The toxic hotspots program has been subject to the same problems as many other regulatory programs that set effluent limits based on ambient water quality standards. Dischargers often dispute their alleged contributions to ambient quality degradation and challenge models used to predict how discharge cutbacks would affect ambient quality. Monitoring data is unavailable for many water bodies, and states

[34] These criticisms are as reported by Percival et al. (1992, p. 967).

[35] Quoted material in this paragraph is from an article in the *New York Times* (May 10, 1995, sec. D, p. 17, by J. H. Cushman, Jr.). For more information on the definition of wetlands, see Adler, Landman, and Cameron (1993, pp. 208–9).

[36] Additional information on EPA's implementation of BAT requirements is given by Patton, Boggs, and Blow and Roy F. Weston (1990, p. 62).

[37] The interpretation by EPA of what is meant by "due entirely or substantially to discharges from point sources" is discussed in detail by Patton, Boggs, and Blow and Roy F. Weston (1990, p. 67).

[38] These figures are derived from Patton, Boggs, and Blow and Roy F. Weston (1990, p. 72).

have had difficulty in determining and adopting the numerical criteria called for by the Act. Moreover, criteria often varied widely from state to state. For example, a hypothetical discharger of mercury in New York was allowed to release about 145 kilograms per year (kg/yr) of mercury, while the same entity in Michigan was allowed only 1 kg/yr of discharge.[39] Additionally, a study by the General Accounting Office showed that in many cases the toxic hotspots program had not led to more restrictive conditions in NPDES permits, and that some firms diverted their wastes to municipal sewers to evade compliance with the program.[40]

Reformulating Federal Subsidies

During the 1980s, the federal program for subsidizing construction of wastewater treatment plants came under scrutiny. Shortcomings of the program, such as the bias it provided for end-of-pipe treatment, are noted in the previous section of this chapter. Because of these deficiencies and the program's high price tag (as much as $4 billion per year during some years in the 1970s), the construction grants program was targeted by budget cutters. In 1981, Congress reversed its earlier pattern of increased funding by reducing the portion of treatment plant costs eligible for subsidies. It also gave local officials a greater role in establishing priorities for federal grants.

In 1987, Congress decided to phase out the construction grants program entirely. Legislators replaced the grants to municipalities with a series of grants to state governments to establish *state revolving funds* (SRFs). Public entities could borrow money from SRFs at low interest rates to cover costs of treatment plant construction. Although the loans were targeted for the construction of publicly owned wastewater treatment works, they could also be used to help with nonpoint source controls and estuary management plans. Some states have used this funding source inventively. For example, South Dakota used its SRF to fund a *solid*-waste management program, reasoning that because municipal landfills are nonpoint pollution sources and are regulated under the State Nonpoint Source Management Program, landfills should be eligible for SRF loans.[41] The federal grants to support SRFs began in 1989 and had been scheduled to continue through the 1994 fiscal year. Instead of discontinuing the flow of funds while it debated proposed amendments to the Clean Water Act, Congress (as of mid-1996) was appropriating funds for SRFs on a year-by-year basis.

As of the early 1990s, the states' response to the change in the construction grants program was mixed. Under the new program, states had to match 20% of the federal contribution to SRFs, and they were forced to make tough loan approval decisions. Many municipalities whose treatment works were inadequate and were out of compliance with their NPDES permit requirements did not have funds to pay back the loans needed to build treatment plants. Problems were exacerbated for small communities, which had neither the staff to evaluate wastewater treatment alternatives nor the money to hire engineering consultants for assistance. Moreover, some small communities lacked the financial expertise to establish programs to fund repayment of loans from SRFs. States often found themselves either extending SRF loans to resource-scarce communities and thereby risking loan defaults, or effectively preventing communities from complying with water quality requirements by refusing to make chancy loans. To compound the problem, the SRF program includes burdensome requirements for loan eligibility, including, for example, a requirement that municipalities pay the "prevailing wage" for construction work. Application of this requirement has led to increases in the cost of municipal treatment plant construction of up to 20%.[42]

CONTINUING CHALLENGES IN MANAGING WATER QUALITY

Despite cleanup efforts that have been made, conventional point sources of water pollution remain a sub-

[39] This illustration is from Adler, Landman, and Cameron (1993, p. 163).

[40] These observations regarding the General Accounting Office study are reported by Adler, Landman, and Cameron (1993, p. 167).

[41] Presentation by James Raysor, South Dakota Dept. of Natural Resources, at the "Challenge of Watershed Protection" conference, convened by the Terrene Institute, San Francisco, July 27, 1994.

[42] These observations are based on telephone interviews conducted by Sambhav Sankar with Jerry Bock, U.S. EPA Region 9, San Francisco (July 25, 1994), and Wayne Aerni, Administrator, State Revolving Fund Program, Arizona Department of Environmental Quality (July 25, 1994).

stantial problem. While the $128 billion spent from 1972 to 1989 on municipal wastewater treatment resulted in over half of the U.S. population being served by "secondary" wastewater treatment, many others relied on "primary" treatment.[43] In 1990, EPA estimated that meeting additional municipal wastewater treatment needs through the year 2010 would require another $110 billion (in 1990 dollars).[44] At the same time, Congress reduced federal funding for municipal wastewater treatment plants. It is not clear where funds needed for legislatively mandated wastewater treatment will come from.

An additional concern over municipal discharges occurs in cities where wastewater and stormwater flow in the same sewer network. During a heavy rainfall, flow in these "combined sewers" can overload the capacity of the system's wastewater treatment plant. Excess wastewater is discharged without treatment through what are known as *combined sewer overflows* (CSOs) and can result in significant pollution. About 1100 communities, comprising 43 million Americans, have combined sewers, which discharge about 1200 billion gallons of untreated sewage annually.[45] While estimates of the cost of controlling combined sewer overflows have been as high as $160 billion, EPA released a policy in 1994 that projected the cost of CSO control at only $41 billion. Reductions in the estimated cost were primarily a result of tailoring control requirements to the financial needs of individual municipalities and the amount of damage caused by their discharges. The principal means of enforcing the combined sewer overflow policy is the NPDES permit system; overflows from combined sewer are point sources that require permits.

Industrial dischargers have also spent vast sums on wastewater treatment. From 1973 to 1986, industry spent over $57 billion for water pollution controls and cut back releases of "priority" toxic organic compounds by 99%. Overall, industry has reduced emissions of toxic chemicals in wastewater by over 1 billion pounds per year.[46] Some analysts wonder, however, whether these reductions were achieved by shifting the destination of pollutants to land and air. If industries and municipalities discharge less pollutants to waterways by disposing of them in landfills or incinerating them, total damage may not necessarily be reduced. In any case, the toxic water pollution problem still remains. According to EPA's 1990 *Toxic Release Inventory,* industries discharged 200 million pounds of toxics that year, including 2.5 million pounds of carcinogens.[47]

Although implementation of the NPDES permit program has reduced pollution from point sources, cutbacks in wastewater discharges have *not* consistently yielded corresponding improvements in ambient water quality.[48] One of the main reasons is the growing significance of nonpoint source pollution. An enormous number of areas are believed to be threatened or degraded by NPS pollution: 40,000 river miles, 2 million acres of lakes, 52,000 acres of wetlands, 1.2 million acres of coastal waters, and 5,000 square miles of estuaries.[49] Much nonpoint source pollution is from agricultural sources,[50] which may require costly controls that are strongly opposed by powerful agricultural interests.

Nonpoint source pollution is difficult to regulate even when all parties agree to attempt solutions. Instead of simply applying waste reduction technology and monitoring discrete flows, NPS pollution control often requires changes in land use and management, such as reducing fertilizer use and adding buffer

---

[43] *Primary treatment* most often consists of physical processes such as screening and settling, which remove suspended pollutants from wastewater before discharge. *Secondary treatment* generally refers to the use of chemical and biological processes to remove nutrients and biochemical oygen demand after primary treatment. For more details on various wastewater treatment methods, see Tchobanoglous and Schroeder (1985, pp. 443–670).

[44] All cost estimates in this paragraph are from Adler, Landman, and Cameron (1993, p. 14).

[45] Facts in this paragraph are from an article in the *Boston Globe* (April 12, 1994, Metrol/Region, p. 21, by S. Allen), and EPA's 1994 CSO Control Policy (see 59 *Federal Register* 18688 [1994]).

[46] All facts in this paragraph are from Adler, Landman, and Cameron (1993, p. 16). The term "priority toxics" refers to substances listed as toxics under a consent decree between the Natural Resources Defense Council and EPA during the 1970s.

[47] As explained in Chapter 14, EPA's *Toxic Release Inventory* is developed under provisions of the 1986 Emergency Planning and Community Right to Know Act.

[48] For example, from 1978 to 1987, four times as many streams deteriorated as improved with respect to nitrogen and dissolved solids. Other parameters, such as phosphorous and metal content, improved over the same period. This long-term water quality data is from the U.S. Geological Survey's National Stream Quality Accounting Network, as described by Adler, Landman, and Cameron (1993, p. 19).

[49] These estimates are based in EPA data as reported by Adler, Landman, and Cameron (1993, pp. 173–74).

[50] For more on NPS pollution from agriculture, see Rosenthal (1988, p. 15).

zones between crops and waterways. Land use controls are also necessary if the pollution results from logging, grazing activities, or urban development. The necessary land use changes often involve the administrative jurisdictions of agencies such as the U.S. Department of Agriculture, whose mission and primary goals may be at odds with placing controls on NPS pollution. Frequently, government agencies must work together to implement schemes since the problems typically cross jurisdictional boundaries. The complexity of NPS program development and administration combined with the potential cost to powerful industrial and agricultural interest groups make it likely that efforts to control NPS pollution will proceed slowly, if at all.

Congress's goal of eliminating the discharge of pollutants into navigable waters has been contentious since it was introduced in the FWPCA amendments of 1972, and it will continue to be debated. Critics question whether the zero discharge goal is appropriate, because surface waters can accommodate some wastewater without causing great damage, and restricting *all* pollutant discharges into water bodies transfers releases to land and air. The Clean Water Act of 1977 recognizes some criticisms of the zero discharge goal. Discharges into deep ocean waters are subject to less stringent standards than are river discharges, and conventional pollutants such as biochemical oxygen demand and suspended solids are not regulated with the best available technology approach used for toxics. It is likely that Congress and the EPA will continue to deliberate the feasibility of meeting the zero discharge goal and the amount of money the nation should spend in meeting clean water targets.

The interconnections among "media" receiving waste—air, land, and water—present special challenges to regulators concerned with water quality. Some significant sources of water pollution are not regulated by water quality laws at all. For example, lake damage caused by acid rain is a water quality problem, but the damage is caused by emissions of sulfur and nitrogen oxides into the atmosphere. Acid rain is regulated by air pollution legislation, not water quality laws. Similarly, groundwater contamination by toxic chemicals is another issue that is not adequately dealt with by water quality legislation. Instead, laws such as the Resource Conservation and Recovery Act; the Comprehensive Environmental Response, Compensation, and Liability Act; and the Toxic Substances Control Act govern the use, storage, and disposal of potentially hazardous groundwater contaminants. Environmental agencies generally implement programs regulating pollution to air, land, and water by having separate administrative units for each, and this makes it difficult to account fully for *cross-media effects.*

As this chapter has demonstrated, many technical, financial, and political hurdles continue to impede efforts to improve water quality in the United States. The significance of these hurdles is reflected in the recent, acrimonious debates in Congress as that body tried (unsuccessfully as of mid-1996) to amend the Water Quality Act of 1987.

Key Concepts and Terms

Water Pollution Control Before the 1980s

State water pollution control agencies
Three-stage enforcement process
Construction grants for municipal facilities
Water quality criteria
Ambient water quality standards
Fishable and swimmable waters goal
Effluent limitations guidelines
Technology-based effluent standards
Best practicable technology (BPT)
Best available technology (BAT)
Best conventional technology (BCT)
New source performance standards (NSPS)
Best available demonstrated control technology (BADT)
Pretreatment standards

Issues in Water Quality Management After 1980

Nonpoint sources of pollution
Best management practices
Watershed protection approach
Stormwater discharges and NPDES
Wetlands preservation and restoration
No net loss of wetlands policy
Toxic hotspots program
State Revolving Fund (SRF)

Continuing Challenges in Managing Water Quality

Groundwater contamination
Combined sewer overflows

Zero discharge of pollutants goal
Cross-media effects
Acid rain

Discussion Questions

12-1 Water quality management in the United States involves strong roles for both federal and state governments. The stringent requirements set out by Congress and detailed by the U.S. Environmental Protection Agency are often implemented by state agencies. During the past half century, the relative power of federal and state agencies involved in water pollution control has had a tendency to ebb and flow. How would you characterize these shifts in power? Do you think the federal government currently has too much or too little authority for managing water quality? What balance of power would be appropriate in the control of water pollution?

12-2 Consider the three-stage enforcement process introduced as part of federal water quality programs in the 1950s and 1960s. In what sense could the introduction of these programs be viewed as a bold step forward? Why was the three-stage process *not* used as the central thrust of the enforcement programs employed by EPA in the 1970s?

12-3 In the mid-1960s, federal water quality management strategies included the following elements: (1) establishment of ambient water quality standards by states, and (2) development of implementation plans in which states indicated measures to be taken to meet the ambient standards. This approach was modified fundamentally in the 1970s by the establishment of technology-based effluent standards and their use in tandem with ambient water quality standards. What were the reasons for this shift in strategy?

12-4 The following pattern is common in the United States: Congress sets out a general water quality management policy, an overall policy implementation strategy, and selected numerical targets and timetables. The EPA administrator is left to work out implementation details by engaging in rulemaking (discussed in Chapter 3) and publishing regulations that have the force of law. What are the problems with this approach to environmental policy design and implementation? Is this approach unavoidable?

12-5 The effluent limitations in Table 12.3 indicate that industrial wastewater sources are treated differently depending on whether the sources are existing or new and whether or not they discharge into sewer lines connected to publicly-owned treatment works. Explain the differences in the way the aforementioned types of industrial sources are treated. Are the differences in regulatory treatment fair? Do they contribute to cost effectiveness?

12-6 What strategies has Congress adopted for dealing with nonpoint sources of water pollution? How successful have they been? List factors that contribute to difficulties in managing NPS. Give particular attention to runoff from cities, industrial facilities, and agricultural areas. Does the "watershed protection approach" advocated by EPA resolve the difficulties you have identified?

12-7 Congress's goal for eliminating the discharge of pollutants to navigabale waters has been debated since it was promulgated in 1972. What are the central issues in this debate? What is your personal position regarding the suitability of the goal of zero discharge of pollutants?

12-8 The Ministry of Environment in Colombia is responsible for implementing an effluent charge program. The ministry will collect a substantial amount of money and wants to direct a large fraction of the money to municipalities as subsidies for wastewater management. You are called in as a consultant to advise the government on how it should set up its subsidy program. You may advise the ministry on any points you want, but the ministry expects to learn about

a. the types of wastewater management activities that should be subsidized;
b. what fraction of construction (or operation) costs should be funded and whether any repayment should be required; and
c. how priorities should be set for distributing funds among cities.

What advice would you offer the ministry based on your knowledge of the construction grants program administered under U.S. federal clean water laws?

References

Adler, R. W., J. C. Landman, and D. M. Cameron. 1993. *The Clean Water Act 20 Years Later.* Washington, DC: Island Press for Natural Resources Defense Council.

Binkley, C., B. Collins, L. Kanter, M. Alford, M. Shapiro, and R. Tabors. 1975. *Interceptor Sewers and Urban Sprawl.* Lexington, MA: Heath.

Cleary, E. J. 1967. *The ORSANCO Story—Water Quality Management in the Ohio Valley under an Interstate Compact.* Baltimore, MD: Johns Hopkins University Press for Resources for the Future.

Freeman, A. M., III, R. H. Havemen, and A. V. Kneese. 1973. *The Economics of Environmental Policy.* New York: Wiley.

Gallagher, L. M. 1995. "Clean Water Act." In Sullivan, T. F. P., ed. *Environmental Law Handbook,* 11th ed. Rockville, MD: Government Institutes, Inc. pp. 135–168.

Gebhardt, A. M., and G. Lindsey. 1993. NPDES Requirements for Municipal Separate Stormwater Systems: Costs and Concerns. *Public Works* 124(1): 40–42.

Grad, F. G. 1995. *Treatise on Environmental Law,* vol. 1. New York: Mathew Bender & Co.

Hey, D. L., and N. H. Waggy. 1979. Planning for Water Quality: 1776–1976. *Proceedings of the American Society of Civil Engineers, Journal of the Water Resources Planning and Management Division* 105, no. WR1 (March): 121–131.

Hunter, S., and R. W. Waterman. 1996. *Enforcing the Law: the Case of the Clean Water Acts,* Armonk, NY: M. E. Sharpe.

Katz, M., and D. Thornton. 1996. *Environmental Management Tools on the Internet.* Delray Beach, FL: St. Lucie Press.

Mintz, J. A. 1995. *Enforcement at the EPA: High Stakes and Hard Choices.* Austin: University of Texas Press.

National Utility Contractors Association. 1975. *Fulfilling a Promise—An Analysis of the 1972 Clean Water Act's Construction Grants Program.* Washington, DC: National Utility Contractors Association.

Novick, S. M., D. W. Stever, and M. G. Mellon, eds. 1995. *Law of Environmental Protection,* vol. 2. Deerfield, IL: Clark, Boardman, Callaghan.

Patton, Boggs, and Blow and Roy F. Weston Inc. 1990. *Clean Water Handbook.* Rockville, MD: Government Institutes, Inc.

Percival, R. V., A. S. Miller, C. H. Schroeder, and J. P. Leape. 1992. *Environmental Regulation: Law, Science, and Policy.* Boston: Little, Brown.

Quarles, J. 1976. *Cleaning up America: An Insider's View of the Environmental Protection Agency.* Boston: Houghton–Mifflin.

Rosenthal, A. 1988. Going with the Flow: USDA's Dubious Commitment to Water Quality. *Environmental Forum* 5: 15–17.

Sullivan, T. F. P. (ed.), 1995. *Environmental Law Handbook* (13th ed.), Rockville, MD: Government Institutes, Inc.

Tchobanoglous, G., and E. D. Schroeder. 1985. *Water Quality.* Reading, MA: Addison–Wesley.

Velasquez, M. G. 1982. *Business Ethics, Concepts and Cases.* Englewood Cliffs, NJ: Prentice–Hall.

Vincoli, J. W. 1993. *Basic Guide to Environmental Compliance.* New York: Van Nostrand Reinhold.

Whipple, W., Jr. 1977. *Planning of Water Quality Systems.* Lexington, MA: Lexington Books–Heath.

Zener, R. V. 1981. *Guide to Federal Environmental Law.* New York: Practicing Law Institute.

Chapter 13

Air Pollution Control

As in the case of water quality, the U.S. government moved into air quality management slowly and reluctantly. However, by the 1970s a strong federal presence in air pollution control had been established. Since then, key policy debates have centered on finding an appropriate balance between federal and state air pollution control initiatives. The history of air quality management differs from that of water quality in that most state agencies had practically no air pollution control programs before the first federal interventions during the 1950s. This chapter adopts a historical perspective and highlights key policies and shifts in federal strategies to improve air quality.[1]

AIR QUALITY MANAGEMENT PRIOR TO 1990

Similarities exist in the ways that air and water quality management programs have evolved in the United States. In each case, federal activity before the 1950s was minimal. Between 1955 and 1970, federal air quality management strategies were often modeled on water pollution control policies that had been developed a few years earlier. This occurred because the same congressional committees formulated air and water quality laws.

Early Control Strategies: Municipal Ordinances and Nuisance Law

The earliest programs to manage air quality used municipal ordinances to regulate emissions from smokestacks. The first anti-smoke ordinance in the United States was issued by Chicago in 1881. It prohibited the emission of "dense smoke" from stacks and provided fines of up to $50 per day for offenders.[2]

A notable air pollution control effort was initiated by Pittsburgh around the turn of the century. In 1906, when it was still widely known as "Smoky City," Pittsburgh set up a smoke-control program based on education campaigns to improve combustion equipment and techniques used by industries. This approach was not fully effective. City ordinances passed during the 1940s established emission standards to limit discharges of smoke, fumes, soot, and cinders. A special-purpose municipal agency admin-

[1] As in Chapter 12, this chapter does not dwell on details or the latest fine points in regulations. Up-to-date discussions of U.S. air quality management programs are given in recent issues of *Environmental Science and Technology* and the *Journal of Air and Waste Management.* In addition, environmental law case books are updated frequently and provide timely information on air quality requirements; see, for example, Grad (1995), and Novick, Stever, and Mellon (1995). For yearly summaries of federal air quality management activities, see the annual reports of the Council on Environmental Quality. Recent regulatory information is also available on the Internet (Katz and Thornton, 1996).

[2] For a description of these early municipal programs, see Bibbero and Young (1974, pp. 1–6).

istered the ordinances and monitored industrial emissions. Programs like the one in Pittsburgh have since been established in many urban areas.

Early efforts to decrease air pollution also relied on private litigation based on the law of public nuisance. Under common-law doctrine, a *public nuisance* is generally defined as "un unreasonable interference with a right common to the public."[3] Public nuisance law has been used by both individual citizens and government officials to reduce emissions of air pollutants and force dischargers to provide financial compensation for alleged damages. Typically, a judge handling a nuisance lawsuit tries to balance the damages to injured parties against the social value of the activities causing the nuisance. Under these circumstances, judges make case-by-case determinations of what emission standards should be and what pollution control devices must be installed. Thus judges are forced to make technical decisions with little guidance to assure that similar air pollution cases are treated in similar ways. This is one reason nuisance law is not an effective basis for an air pollution control program.

During the 1950s, there was a shift away from nuisance law and municipal ordinances as bases for air quality management.[4] Two factors stimulated the development of additional air pollution control strategies. One was a tragic episode at Donora, Pennsylvania. In 1948, a combination of poor atmospheric ventilation and concentrated emissions from steel mills, smelters, and other plants in and around Donora caused death to 20 people and illness to several thousand. The second factor was the growing recognition of the linkage between automobile exhausts and photochemical smog. In the early 1950s, researchers at the California Institute of Technology identified a photochemical reaction through which hydrocarbon vapors and nitrogen oxides from motor vehicle exhausts combined to yield ozone and a number of other highly reactive organic oxidants. Photochemical smog, which requires sunny days, atmospheric stability, and the type of emissions produced by internal combustion engines, is a major problem in Los Angeles and many other cities.[5]

[3] This definition is used by Stewart and Krier (1978, p. 210). They describe how public nuisance law has been used in the United States to control environmental quality.

[4] This paragraph is based on Kneese (1978); statistics describing the Donora air pollution tragedy are from Turk (1980, p. 261).

Early Federal Air Quality Laws: 1955–1967

The reliance on nuisance law and local agencies to manage air quality began to change with passage of the Federal Air Pollution Control Act of 1955 (see Table 13.1). This act established a program of federally funded research grants administered by the U.S. Public Health Service. Although the 1955 act represented an expansion of the federal role, it was a very limited one. The legislative history of the act indicates Congress's intent to limit federal involvement and to respect the rights and responsibilities of states, counties, and cities in controlling air pollution.[6]

The federal role was further extended by the Clean Air Act of 1963, which allowed direct federal intervention to reduce interstate pollution. The form of intervention followed the three-step enforcement process in the Federal Water Pollution Control Act of 1956. This process involved (1) a conference of interested parties to gain agreements regarding necessary pollution controls, (2) a public hearing to be convened if the measures agreed upon at the conference were not implemented, and (3) federally initiated litigation as a last resort. Like its counterpart in the water pollution control program, this enforcement process had an undistinguished record. Fewer than a dozen conferences were called under the 1963 act. Only one case reached the court action stage, and it required three years to get there. Even though the Clean Air Act of 1963 permitted federal intervention, the act followed earlier legislation by emphasizing the role of state and local governments in air quality management. The 1963 law also continued federally funded research, and it set up a program of grants to strengthen state and local air pollution control activities.

[5] Engine exhaust emissions are not the only vehicle-based contributions to photochemical smog. Another contributing factor is the emission of hydrocarbons due to evaporation of vehicle fuels. Evaporative fuel emissions occur while a vehicle is moving, standing, and refueling.

[6] This discussion of the history of federal efforts to control air quality relies on Bibbero and Young (1974, pp. 140–63), and Stewart and Krier (1978, pp. 333–37).

TABLE 13.1 Key Federal Laws Controlling Air Pollution

Year	Title	Selected *New* Elements of Federal Strategy[a]
1955	Air Pollution Control Act	Funds for air pollution research
1960	Motor Vehicle Exhaust Study Act	Funds for research on vehicle emissions
1963	Clean Air Act	Three-stage enforcement process Funds for state and local air pollution control agencies
1965	Motor Vehicle Air Pollution Control Act	Emission regulations for cars beginning with 1968 models
1967	Air Quality Act	Federally issued criteria documents Federally issued control technique documents Air quality control regions (AQCRs) Requirements for states to set ambient standards for AQCRs Requirements for state implementation plans
1970	Clean Air Act Amendments	National ambient air quality standards New source performance standards Technology-forcing auto emission standards Transportation control plans National emission standards for hazardous air pollutants
1977	Clean Air Act Amendments	Relaxation of previous auto emission requirements Vehicle inspection and maintenance programs Prevention of significant deterioration areas Emission offsets for nonattainment areas Lowest achievable emission rate for major sources in nonattainment areas
1980	Acid Precipitation Act	Development of a long-term research plan
1990	Clean Air Act Amendments	Requirements for nonattainment areas set by categories of areas Strengthening of reasonably available control technology requirements Alternative fuels program Requirements for low emission vehicles New controls for toxic air pollutants Tradable SO_2 emission allowances National permit program

[a] The table entries include only the *new* policies and programs established by each of the laws. Often these provisions were carried forward in modified form as elements of subsequent legislation.

The first federal restrictions on motor vehicle emissions came with the Motor Vehicle Air Pollution Control Act of 1965.[7] Based on earlier vehicle emission control efforts in California, the 1965 law gave the Secretary of the Department of Health, Education, and Welfare (HEW) authority to establish permissible emission levels for *new* motor vehicles beginning with the 1968 model year. The control of emissions from older vehicles was left to individual states. (Acronyms used in this chapter are listed in Table 13.2)

Regional Ambient Standards and State Implementation Plans

In 1967, still another element was added to the federal air quality management program. The Air Quality Act of 1967 continued earlier federal grants supporting research and state and local programs, and it also continued the three-stage federal enforcement process. In addition, the 1967 law borrowed concepts from the Water Quality Control Act of 1965 by requiring states to develop *ambient air quality standards* and *state implementation plans* (SIPs) to achieve the standards. Implementation plans were to include emission requirements for controlling air pollution, and a timetable for meeting those requirements. Deadlines were set for submitting ambient standards, which were to be established on a regionwide basis. If satisfactory standards were not issued on time, they would be developed by the federal government.

As a precursor to state activity in setting standards, the secretary of HEW was required to do the following:

1. Designate *air quality control regions* (AQCRs) centered around metropolitan areas.
2. Publish *air quality criteria documents* for particulates, sulfur oxides (SO_x), carbon monoxide (CO), hydrocarbons, nitrogen oxides (NO_x), and photochemical oxidants. The documents were to include the latest scientific knowledge about how these pollutants influenced human health and welfare.
3. Publish *control technique documents* for various substances indicating the feasibility and cost-effectiveness of different emission reduction methods.

For each of the AQCRs designated by the secretary of HEW, *states* had to adopt ambient air quality standards using a process involving extensive public involvement and the information in the criteria and control technique documents. In addition to setting standards for these AQCRs, states were required to indicate how the standards would be met. This information was to be part of the state implementation plans.

Although this process for setting standards and developing plans appears straightforward, it required much more time to implement than most people anticipated. By April 1970, only 28 AQCRs had been designated by the secretary of HEW, and some of the required criteria documents and control technique documents had not been issued. Because of these delays, states did not make great strides in setting standards and developing SIPs. By early 1970, not one state had adopted a complete set of ambient air quality standards and implementation plans. This slow progress, together with the high level of public concern for environmental quality, led Congress in 1970 to establish uniform national standards and redirect the federal air pollution control effort.

National Ambient Standards and Federally Imposed Emission Standards

Although the Clean Air Act Amendments of 1970 continued many of the research and state aid programs established by prior legislation, several aspects of the amendments reflected major changes in strategy. Initiatives included (1) a requirement that the administrator of the U.S. Environmental Protection Agency (EPA) set *national ambient air quality standards* (NAAQS) and emission standards for selected categories of new industrial facilities, and (2) the explicit delineation (by Congress) of *technology-forcing* auto emission standards.

In 1970, Congress abandoned the process of setting standards on a regional basis. Instead, it required the EPA administrator to prescribe national primary and secondary ambient air quality standards for each of six *criteria pollutants,* the substances for which air quality criteria documents had been required under the 1967 act. *Primary* standards were limits that, in the

[7] The Motor Vehicle Exhaust Study Act of 1960 provided funds for research on vehicle emissions, but it did not regulate vehicle emissions.

TABLE 13.2 Acronyms Used in Federal Air Quality Management

AQCR	Air quality control region
BACT	Best available control technology
CFC	Chlorofluorocarbon
CMAQ	Congestion Management and Air Quality Improvement Program
CO	Carbon monoxide
CO_2	Carbon dioxide
EPA	U.S. Environmental Protection Agency
GACT	Generally available control technology
HEW	Department of Health, Education, and Welfare
ISTEA	Intermodal Surface Transportation Efficiency Act
LAER	Lowest achievable emission rate
LEV	Low emission vehicle
MACT	Maximum achievable control technology
NAAQS	National ambient air quality standards
NSPS	New source performance standards
NO_x	Oxides of nitrogen
PSD	Prevention of significant deterioration
RACT	Reasonably available control technology
SIP	State implementation plan
SO_2	Sulfur dioxide
SO_x	Oxides of sulfur
TCP	Transportation control plan
VOC	Volatile organic compound
ZEV	Zero emission vehicle

judgment of the EPA administrator, were required to safeguard the public health. *Secondary* standards, which were to be at least as stringent as the primary standards, were intended to "protect the public welfare from any known or anticipated adverse effects." For example, the primary standards for sulfur oxides were to safeguard the population from various respiratory diseases, whereas secondary SO_x standards were to protect against damage to vegetation.

The primary NAAQS are summarized in Table 8.2. As shown there, the current version of the standards reflects revisions made since the early 1970s. For example, an ambient standard for lead was added in 1978. Two aspects of the NAAQS are particularly notable.[8]

1. The NAAQS are widely viewed as *health-based standards;* in mandating NAAQS, Congress made no mention of either cost or technological feasibility.

[8] For details on how NAAQS are established, see Novick, Stever, and Mellon (1995, pp. 11-8 to 11-14).

2. The NAAQS are established to safeguard the health of sensitive subgroups within the U.S. population. For example, ambient standards for lead were set to protect the health of children.

The use of federally determined ambient standards represented a major departure from earlier reliance on state and local agencies to set standards. States retained the authority to decide on how to achieve the NAAQS, and these decisions were to be reflected in state implementation plans. However, states were not entirely free to make decisions because the SIPs were subject to federal approval. If a state did not comply with the 1970 Clean Air Act Amendments, it risked the loss of federal funds for air pollution control programs, and the prospect of having the federal government establish its state implementation plan.

Another manifestation of the expanded federal role in air quality management is the 1970 amendments' requirement that the EPA administrator issue *new source performance standards* (NSPS), emission limits on new stationary sources categorized by the administrator as contributing significantly to air pollution. Examples of the types of facilities subjected to NSPS include portland cement plants, nitric acid plants, and municipal incinerators. In setting performance standards for each category of facilities, the EPA administrator was required to determine the "best system of emission reduction which (taking into account the cost of achieving such reduction)" had been "adequately demonstrated." These regulations were very controversial. The EPA administrator was often sued by industries who argued that the proposed NSPS were not technologically feasible or that they were too costly in light of benefits from improved air quality. Sometimes opponents charged that a proposed NSPS would cause environmental problems; for example, the large quantities of solid waste generated when stiff requirements were placed on removal of airborne particulates.

The 1970 amendments also changed the federal approach to controlling automobile emissions. In frustration over the auto industry's lack of rapid progress in controlling emissions, Congress mandated a 90% decrease in hydrocarbon and carbon monoxide emissions (from 1970 levels) by 1975, and a 90% decrease in emissions of nitrogen oxides (from 1971 levels) by 1976. These reductions in auto emissions were technology forcing in that they went beyond what car manufacturers viewed as feasible at the time the amendments were passed. Whether or not this congressional action forced the auto companies to do more effective research on controlling emissions is disputable. The original 1975–76 standards were not attained on schedule. Moreover, the Clean Air Act Amendments of 1977 relaxed the emission requirements somewhat and extended the compliance deadlines into the early 1980s. Although the modified requirements were eventually met, the early 1980s was a period of continuing debate concerning the need for additional emission cutbacks.

Transportation Control Plans and Vehicle Inspections

Because restraints on stationary sources and motor vehicle emissions might not be enough to attain the national ambient air quality standards, the Clean Air Act Amendments of 1970 required that SIPs include land use and transportation controls "where necessary to achieve and maintain air quality standards." The EPA administrator responded cautiously to this portion of the 1970 amendments since land use controls in the United States are traditionally the province of local governments. In fact, the EPA administrator acted on this matter only after he was sued by environmental groups who argued that EPA was required to issue *transportation control plans* (TCPs) if the national ambient air quality standards would not be met using other means.[9]

One of the first transportation control plans announced by the EPA administrator was for the Los Angeles region. In 1970, the NAAQS for photochemical oxidants were exceeded in the Los Angeles area on 250 days, and there was little hope of attaining the standard by regulating only stationary sources and motor vehicle emissions. The California SIP should have included a transportation control plan for the Los Angeles area, but it did not. In response to the previously mentioned lawsuit, EPA issued a plan that included inspections of vehicle emission controls, retrofitting of old vehicles with pollution control devices, and an 82% reduction in *vehicle miles traveled* by means of gasoline rationing, car pooling,

[9] A brief account of this lawsuit is given in Chapter 3. For details, see Stewart and Krier (1978, pp. 442–43).

and vehicle-free zones. This and other transportation control plans issued by EPA under the 1970 amendments were never implemented. The negative reaction to the federally imposed TCPs was so strong that Congress revised the TCP requirements in the Clean Air Act Amendments of 1977.

Under the 1977 amendments, any state that could not meet the primary NAAQS by 1982 could obtain a five-year extension if it (1) implemented a program for the inspection and maintenance of vehicle emission control devices, and (2) revised its implementation plan to include appropriate TCPs. These transportation control plans were to be designed by the people who would be responsible for implementing them: local transportation planners and elected officials.

The importance of vehicle inspection and maintenance was highlighted by "in use" emissions data in a report by the National Commission on Air Quality (1981). These data indicated that although vehicles meet applicable emission standards when they are new, they often fail to meet those standards after a year or two on the road. Factors contributing to these failures include lack of proper maintenance, deliberate tampering with emission control systems, and fouling of catalytic converters by use of leaded gasoline.[10] Under the 1977 amendments, states that did not implement required vehicle inspection programs could lose federal subsidies for highway and wastewater treatment plant construction. Despite this, many states had serious problems in setting up inspection programs. States were concerned about imposing additional regulations and the costs of inspections and follow-up repairs.

In order to gain extensions in deadlines for meeting NAAQS, many states developed transportation control plans, which typically included the following approaches to reducing the number of vehicle miles traveled:

1. *Pooling.* Providing incentives for car and van pooling.
2. *Flex time.* Staggering work hours to reduce peak-hour traffic.
3. *Priority lanes.* Designating highway lanes for use only by buses and car and van pools.
4. *Parking.* Making it more expensive to park; for example, by eliminating parking subsidies to employees.
5. *Public transit.* Developing and improving rail and bus systems.

Many regional transportation agencies estimated that the *maximum* pollution reductions obtainable by their plans would be 5 to 8%.[11] Recently, the Environmental Protection Agency and the Department of Transportation (1993, p. 9) estimated the mobile source emissions were reduced by only 1 to 2% using traditional measures of the type noted.

New Sources and the Attainment of Standards

Soon after national ambient air quality standards were promulgated, two important growth-related issues began to be debated. One concerned areas where the ambient air quality was *better* than required by the national standards. Should new sources of emissions be permitted to degrade the air quality in such areas down to NAAQS levels? A second issue involved *nonattainment areas,* regions that had not yet attained the national ambient standards. Since the standards were not met in nonattainment areas, should those areas be allowed to accept *any* new air pollution sources? These questions were debated extensively during the 1970s. The positions Congress took on these matters in 1977 are summarized here.

Prevention of Significant Deterioration

Congress was unwilling to accept the prospect of allowing areas with air quality already above the national ambient standards to degrade to the lower limits set by the standards. At the same time, Congress would not exclude the possibility of having some air quality degradation in those areas. A compromise position was based on the concept of *prevention of significant deterioration* (PSD).

The Clean Air Act Amendments of 1977 required that an area meeting the national ambient standards for a given air pollutant be declared a "PSD

[10] *Catalytic converters* are afterburning devices in the exhaust systems of motor vehicles. Converters use transition metals (for example, platinum) as catalysts to achieve significant reductions in emissions of CO and hydrocarbons. (CO and hydrocarbons are oxidized into carbon dioxide and water vapor.) Different catalysts are used in reducing nitrogen oxides. The role of lead in fouling converters was a major factor leading to removal of lead from gasolines in the United States (Grad, 1995, p. 2–578).

[11] These figures are from Seltz-Petrash (1979).

area" for that pollutant. The amendments also defined three classes of PSD areas. For each class, numerical limits indicated the maximum permissible increment of air quality degradation from all new (or modified) stationary sources of pollution in an area. Class I, which was mandatory for areas containing important national resources such as the Grand Canyon, had the lowest allowable increments of degradation. For example, the allowable increase for total suspended particulates was 5 mg/m^3 measured on an annual average basis. The highest allowable changes were for Class III areas. In these cases, the total suspended particulates increment was 37 mg/m^3 measured on an annual average basis. Except for special places such as national parks, each state could classify its own PSD areas.

The 1977 amendments required that state implementation plans include provisions for preconstruction review of "significant" new stationary sources in PSD areas. A source was considered significant if it would emit more than 100 tons per year. For a significant new source to be allowed into a PSD area it had to meet stringent emission reduction levels specified by the EPA administrator. These specifications required use of the *best available control technology* (BACT) for each pollutant regulated under the Clean Air Act.[12] In addition, the increment of air quality degradation caused by the source had to be within the numerical limits of allowable degradation available in the region. Additional requirements were imposed on new sources in PSD areas where reduction in *visibility* was an important issue. These added restrictions were intended to protect spectacular vistas that were threatened by air pollution in the western states.[13]

Offsets in Nonattainment Areas

The *Clean Air Act Amendments* of 1977 also indicated that significant new sources could locate in areas that did *not* meet the NAAQS, but only if certain conditions were satisfied. The amendments required that a significant new source locating in a nonattainment area had to meet very strict standards based on *lowest achievable emission rate* (LAER) requirements developed by the EPA administrator.[14] In addition, discharges from the new source had to be more than offset by reductions in emissions from other sources in the region. After the *emission offsets* were applied, the net effect had to be reasonable progress toward meeting the NAAQS in the region. The Wickland Oil Company example in Chapter 11 illustrates the use of emission offsets (see Table 11.5).[15]

During the 1980s, it became apparent that further changes were needed in the Clean Air Act. The law was difficult to administer, and the high cost of meeting its requirements provided pressure to devise more economical strategies.[16] In addition, problems with toxic air pollutants and acid rain were worsening. Despite these pressures for change, Congress was unable to act.

CLEAN AIR ACT AMENDMENTS OF 1990

A stumbling block to passage of new clean air legislation in the 1980s was the Reagan administration's anti-regulation stance, particularly its position on acid rain.[17] The response of the Reagan administration to the acidification of lakes and forests in much of eastern North America was a call for additional research. The administration was joined in supporting more research by senators from states whose power plants relied on high-sulfur coal. Sena-

[12] The Clean Air Act of 1977 defined BACT as the "maximum degree of [emission] reduction . . . achievable," taking into account economic, energy, and environmental factors.

[13] For a summary of visibility protection requirements, see Brownell (1995, pp. 120–122).

[14] LAER is a particularly rigorous standard. For any waste source, LAER is determined by considering the more demanding of two conditions: (1) the most stringent emission limitation contained in the SIP for any state for such a source (unless those limits can be shown to be inapplicable), and (2) the most stringent limitation achieved in practice (Brownell, 1995, p. 112). Existing major polluters in nonattainment areas are treated more leniently. They are required to install *reasonably available control technology,* which "in most cases is less stringent than new source performance standards" (Percival et al., 1992, p. 1296).

[15] The discussion in Chapter 11 also highlights three related EPA programs: bubble policy, banking, and netting. These programs are discussed in the context of Table 11.4.

[16] The Council on Environmental Quality (1980) estimated that $22 billion was spent meeting federal clean air requirements in 1979.

[17] Thanks to Sambhav Sankar for his assistance in researching and writing early drafts of this section and the one that follows. Mr. Sankar was a graduate student in the Department of Civil Engineering at Stanford University at the time the work was carried out.

tors from midwestern states were particularly concerned about the cost to electric utilities of proposed cutbacks in sulfur dioxide emissions to reduce acid rain. The result was that the only federal air quality legislation during the Reagan years was the Acid Precipitation Act of 1980, which set up an interagency task force to establish a research program. Because most proenvironment legislators would not consider a new Clean Air Act complete without provisions to control acid rain, Congress remained deadlocked during the Reagan administration. The election of George Bush as president in 1988 and changes in Senate leadership opened the door to new air pollution control legislation—the Clean Air Act Amendments of 1990.[18]

The 1990 amendments, totaling over 700 pages of text, included precise emission limitations for numerous chemicals, and compliance schedules with strict deadlines. The law's specificity was an attempt by Congress to reduce the discretion of EPA, and the deadlines were accompanied by provisions that gave EPA incentives to issue new regulations on time.[19] Among the act's most significant features were compliance requirements for urban areas that did not meet the national ambient air quality standards.

Promoting NAAQS Attainment

The Clean Air Act Amendments of 1977 had provided a 1987 deadline for attainment of the NAAQS everywhere in the nation. However, between 1987 and 1989, 96 areas still violated ozone standards, 41 failed to satisfy carbon monoxide standards, and 58 did not meet standards for airborne particulate matter.[20] In enacting the 1990 amendments, Congress sought to speed progress in attaining the NAAQS by dividing nonattainment areas into categories and using the categories as a framework for detailing requirements.

The Clean Air Act Amendments of 1990 used severity of pollution as a basis for sorting nonattainment areas into categories.[21] The law defined six classifications for ozone nonattainment areas, from "marginal" to "extreme," as well as two categories each for areas that violated standards for carbon monoxide and particulate matter. Compliance deadlines were tied to the classification scheme for nonattainment areas. For example, regions classified as having marginal ozone pollution had to satisfy the NAAQS for ozone by 1993, while extreme regions had until 2010. (As of 1994, only Los Angeles had been classified as an extreme area for ozone pollution.) Similarly, deadlines for attainment of CO and particulate standards ranged from 1994 to 2001, depending on an area's classification.

New requirements for nonattainment areas were also based on the classification scheme. For example, ozone nonattainment areas classified as marginal had to ensure that vehicle inspection programs were functioning properly, submit inventories of emissions sources every three years, conduct stringent reviews of new pollution sources, and revise their definitions of *reasonably available control technology* (RACT). (The Clean Air Act Amendments of 1977 had required RACT for *existing* major air pollution sources in nonattainment areas, and the 1990 amendments tightened the RACT requirements.) For areas classified as marginal, the threshold for determining new sources that were significant and thus subject to emission offset requirements remained at 100 tons per year, the same cutoff used in the 1977 amendments.

At the other end of the regulatory spectrum, ozone attainment areas classified as extreme had to comply with a host of new provisions, of which only a few are listed here. The threshold for defining significant new sources dropped tenfold to only 10 tons per year. Instead of being able to offset the

---

[18] See Cohen (1992) for an account of the political struggles waged to gain passage of the 1990 Clean Air Act amendments.

[19] An example is contained in the alternative fuels program. Congress directed EPA to develop regulations for clean-burning gasolines to reduce auto emissions in areas where air quality was poor. These regulations were to go into effect on January 1, 1995, at which time no gasoline could be sold in those areas unless it met EPA's regulations. The motivation for EPA to issue rules on a timely basis was that if no regulations were produced by January 1995, no gasoline could be sold in areas with poor air quality. By including this provision, Congress provided a powerful incentive for the petroleum industry to ensure that EPA produced the needed regulations on time. This is an illustration of a *hammer provision.* Other examples are detailed in Chapter 14 in the context of the Hazardous and Solid Waste Act of 1984, the first law where Congress used hammer provisions to prod EPA to act.

[20] These figures are from Percival et al. (1992, p. 810).

[21] Materials in this paragraph and the two that follow were drawn from Percival et al. (1992, pp. 811–13).

pollution from proposed new facilities with equal reductions in emissions elsewhere, industries were forced to offset 1.5 times the amount of emissions that significant new sources would produce. (The corresponding figure was 1.1 for marginal areas.) If one portion of an area classified as extreme was located in an urban zone, the entire metropolitan area became subject to new-source review requirements under the nonattainment program. This provision, together with the new threshold for defining significant sources, increased the number of sources brought under stringent regulatory control.

In addition to meeting new requirements for stationary sources, areas with severe carbon monoxide and ozone air pollution problems might also be forced to impose severe transportation control measures. For urban areas, mobile sources comprise up to 50% of total NO_x emissions, 80 to 90% of CO discharge, and 40 to 50% of the hydrocarbons that combine with NO_x to form ozone.[22] An example of the new demands for transportation control measures in the 1990 amendments concerns areas classified as severe or worse for ozone. For these areas, the legislation required employers to create programs reducing vehicle traffic caused by employee commutes. In addition, these areas had to have transportation control measures that would offset emissions caused by growth in vehicle miles traveled.

Cleaning Up Mobile Sources

Along with requirements for transportation control plans in areas with poor air quality, the 1990 amendments augmented the motor vehicle emission control requirements contained in earlier legislation. Technology-forcing provisions for automobile tailpipe emissions within the Clean Air Act Amendments of 1970 had been effective. Between 1940 and 1970, auto emissions of volatile organic compounds (VOCs) had more than doubled and carbon monoxide emissions had nearly tripled. However, the 1970 amendments' mandates reversed this trend.[23] Even though the number of vehicle miles traveled in the United States had increased by 81% from 1970 to 1988, emissions of VOCs decreased by 46%, CO emissions dropped by 48%, and nitrogen oxide emissions gradually declined.

Vehicle manufacturers and environmentalists each interpreted these pollution reduction figures to their advantage. Industry claimed that it had already borne the brunt of efforts to reduce air pollution and should not be forced to go further, whereas environmentalists argued that additional tailpipe emission controls were a cost-effective way to protect urban air quality. After debating these conflicting positions, Congress eventually included in the 1990 amendments extensive new federal requirements for vehicle manufacturers. For example, in a program for cars and light trucks, vehicle manufacturers were required to improve emission control equipment on 100% of their 1996 model-year product line: nitrogen oxide emissions had to be reduced to 0.4 grams per mile (gpm), down from a 1990 value of 1.0 gpm; and hydrocarbon emissions had to drop to 0.25 gpm (from a 1990 value of 0.41 gpm). Also, the manufacturers' warranties on emission control equipment had to be extended to 10 years or 100,000 miles. The new amendments gave states the option to choose between federal vehicle emissions standards and the more stringent standards that had been set earlier in California.

The 1990 Clean Air Act Amendments also mandated an *alternative fuels program* in areas where motor vehicle emissions were particularly troublesome. Congress was originally hesitant to include an alternative fuels measure, because the petroleum industry had lobbied vigorously against it and EPA had not championed its inclusion. Impetus for the program came from an unlikely source: petroleum giant ARCO announced in 1989 that it had developed a reformulated gasoline to lower emissions from older cars that were not equipped with catalytic converters.[24] ARCO's announcement spurred Congress

[22] Information in this paragraph was drawn from Environmental Protection Agency and Department of Transportation (1993). A list of possible transportation control plans appears in the Clean Air Act, section 108(f)(1).

[23] These figures and those in the following paragraph are from Bryner (1993, p. 132). For an analysis of long-term trends in U.S. highway emissions, see Flachsbart (1995).

[24] For more information on the impact of ARCO's announcement, see Bryner (1993, p. 134).

to include provisions requiring EPA to promulgate new standards for gasoline in nonattainment areas where air pollution problems were due largely to motor vehicles.[25] The 1990 amendments required use of *reformulated gasolines* in the nation's nine worst areas of ozone pollution, starting in 1995 and phasing in over five years. In all ozone nonattainment areas, *oxygenated fuels* were to be sold during peak pollution months. Ozone nonattainment areas classified as serious or worse (as well as some carbon monoxide nonattainment areas) had to ensure that a gradually increasing portion of vehicle fleets met standards for *low emission vehicles* (LEVs) that had been set previously by California.[26]

Provisions of the Clean Air Act Amendments of 1990 regulating mobile-source emissions seemed cautious in comparison to initiatives adopted by the California Air Resources Board (CARB). California has some of the nation's most polluted air, and much of its pollution is due to vehicular traffic. Controls that the 1990 amendments specified for cars in the 1994 model year had been required in California by 1993. In addition, CARB adopted low-emissions and clean fuels regulations in 1990 that would force large automobile manufacturers to make both low emission vehicles and *zero emission vehicles* (ZEVs) available for sale.

[25] The differences between "alternative fuel," "reformulated gasoline," and "oxygenated fuel" are easily confused. Reformulated gasoline, as defined by the 1990 amendments, refers to gasoline that has been altered to reduce emissions. Reformulated gasoline must have lower benzene content and higher oxygen content than normal gasoline, and it must result in improved carbon monoxide and hydrocarbon emissions. Reformulated gasoline can be produced by adding ethanol or other additives to normal gasoline. So-called oxygenated fuels must have an even higher oxygen content than reformulated gasoline. Both of these types of fuels are usable in normal automobiles without engine modifications. In contrast, alternative fuels often require anything from slight vehicle modifications to radical redesign. Alternative fuels include compressed natural gas, hydrogen, methanol, and ethanol. While electricity, in itself, is not a fuel, it is often considered an alternative fuel for mobile-source emissions purposes. The Clean Air Act Amendments of 1990 give detailed performance specifications for each of these types of fuels. This footnote is based on Percival et al. (1992, pp. 839–40), and Bryner (1993, pp. 135–44).

[26] A fleet is defined as a group of more than 10 vehicles that are owned by a single operator and centrally fueled; for further details on requirements for fleets, see Percival et al. (1992, p. 840), and Bryner (1993, p. 125).

California's 1990 regulations compelled automakers to produce combinations of LEVs and ZEVs to achieve specified average fleetwide emission requirements, but the ZEV mandate went further. By 1998, 2% of each large carmaker's light-duty fleet offered for sale in California had to be composed of ZEVs. This figure increased to 5% in 2001 and 10% in 2003.[27] During the early 1990s, the only vehicles that qualified as ZEVs were those powered by electricity. Critics of the ZEV requirement observed that electric cars are not pollution free since they shift emissions from vehicle tailpipes to power plant smokestacks. And the batteries used are dangerous to dispose of and must be replaced regularly. Automakers contended that ZEVs would cost more than gasoline-powered vehicles and would have lower performance. They also argued that ZEVs might not be ready for large scale production in time.

In March 1996, the California Air Resources Board voted to water down its requirements.[28] CARB's original 2% policy would have translated into about 22,000 electric vehicles sold per year in California beginning with the 1998 model year. With its March 1996 vote, CARB agreed to a requirement for 3750 electric cars annually by 2003. The 2003 model-year cars would have a range of 125 miles between battery charges. In contrast, General Motors Company had already announced plans to sell in 1996 a two-seat electric car with a range of 70 to 90 miles between charges.

Controlling Acid Rain

One of the most inventive elements in the 1990 Clean Air Act Amendments was a sulfur dioxide (SO_2) emissions trading scheme to control acid rain. Although acids from the atmosphere can also be transferred to water and land by snowfall and by dry deposition, the term *acid rain* is commonly used to refer to all forms of wet and dry deposition of acidic substances.[29] By the late 1980s, damage caused by acid

[27] These figures are from State of California Air Resources Board (1994, p. 3).

[28] Facts in this paragraph are from M. W. Wald, "California Eases Requirements on Number of Electric Cars," *New York Times,* national edition, March 30, 1996, p. A6.

[29] Technical aspects of the acid rain problem are introduced in Chapter 22.

rain had become one of the main sources of tension between the United States and Canada, and the new Bush administration was determined to reduce that tension by committing the United States to making a 50% reduction in sulfur dioxide emissions by the year 2000 compared to 1980 levels. Half of the SO_2 reduction was to occur by the end of 1995.

The innovative aspect of the Bush administration's SO_2 reduction proposal concerned its provisions encouraging cost-effective goal attainment. As detailed in Chapter 11, coal-fired power plants included in the acid rain control program were allowed to buy and sell *SO_2 emission allowances* so that cost-effective means of SO_2 reduction could be employed. Utilities that burned coal at their power plants could meet Clean Air Act requirements in several ways: cut back demand for power; reduce production and make up the difference by purchasing power from other utilities; install additional pollution abatement technologies; and purchase SO_2 emission allowances. In addition, power plants that burned high-sulfur coal could meet SO_2 reduction requirements by switching to low-sulfur coal, blended coals, or other fuels. Under the acid rain provisions of the Clean Air Act Amendments of 1990, electric utilities had strong incentives to discover cost-effective ways to meet emission reduction targets.

Controlling Toxic Air Pollution

Another air pollution problem demanding attention in the 1980s was toxic pollutant releases. By 1989, EPA had regulated only seven toxic air pollutants: arsenic, asbestos, benzene, beryllium, mercury, radionuclides, and vinyl chloride. These were controlled under the *National Emission Standards for Hazardous Air Pollutants* program established by the Clean Air Act Amendments of 1970. Though pressure for progress in regulating toxic releases had been building for over 20 years, a turning point came in 1988, when new data were released indicating that toxic air emissions were much more significant than industry had previously claimed. These data, collected in the process of developing EPA's *Toxic Release Inventory,* estimated that over 2.7 billion pounds of toxic chemicals were released into the air in 1987.[30]

[30] The first Toxic Release Inventory was published in 1989; the estimate of 2.7 billion pounds of toxics in 1987 is from Percival et al. (1992, p. 433).

The toxic air pollutants catalogued in the annual Toxic Release Inventory included known human carcinogens and mutagens. EPA estimated that exposure to this subset of toxic air pollutants probably resulted in between 1000 and 3000 deaths annually during the late 1980s.[31] Congress responded to these health risks by including in the 1990 amendments requirements for EPA to promulgate emission limits for major toxic air pollution sources. (Major sources were defined as those that emit either more than 10 tons per year of any one toxic air pollutant, or 25 tons per year of any combination of toxins.) EPA was also given discretion to regulate *area sources,* defined as any source not classified as a major source (except for vehicles). This area source provision was aimed at sources that emitted toxic air pollutants in small quantities, but were so numerous as to warrant attention in some contexts. Examples of area sources include dry cleaning facilities, chromium electroplating factories, and wood-burning stoves.

The 1990 amendments included an initial list of 189 hazardous chemicals to be regulated. EPA was instructed to identify categories of processes that produce those 189 chemicals and issue standards to govern emissions from each such category. The standards, which required the *maximum achievable control technology* (MACT) to reduce emissions, were to "provide an ample margin of safety to protect public health," taking into account costs, energy, and safety.[32] To hasten the reduction of toxic air pollutant discharges, Congress allowed companies who voluntarily reduced their emissions by 90% to escape MACT requirements for six years. For area sources, the 1990 amendments give EPA authority to require use of "generally available control technologies [GACT] or management practices."[33]

Illustrations of emission standards for hazardous air pollutants are given by EPA's regulations governing perchloroethylene (PCE) emissions from dry-cleaning operations.[34] For new major sources of PCE

[31] This estimate is reported by Percival et al. (1992, p. 861).

[32] This material from the Clean Air Act Amendments of 1990 is as presented by Bryner (1993, p. 153).

[33] This reliance on management practices to control area sources is analogous to the use of best management practices to regulate nonpoint sources of water pollution. And the MACT standard is analogous to the federal requirements for best available technology to control toxics in point sources of wastewater.

[34] The regulations for dry cleaners are published in 40 *Federal Register* 49354 (1993).

consisting of dry-cleaning facilities using "dry-to-dry machines," it is necessary to install both refrigerated condensers and carbon adsorbers to control PCE releases. These are MACT requirements. In contrast, GACT requirements for new *area* sources using dry-to-dry machines need only use refrigerated condensers.

Improving Enforcement

The extraordinary complexity of federal air pollution control requirements often resulted in confusion for both polluters and regulators. In an attempt to reduce confusion, the Clean Air Act Amendments of 1990 established a federal *permit program* for major air pollution sources. For any particular discharger, a permit would include all requirements that the Clean Air Act and its amendments set out for the source, including emission limitations, monitoring requirements, and even maintenance procedures. By clarifying requirements for each emission source, the permits would also allow regulators to clearly identify who was in violation.

The permit program is only one of several enforcement initiatives contained in the 1990 amendments. The amendments went further than previous laws by including a *bounty hunter provision:* EPA was authorized to pay up to $10,000 to individuals providing information leading to convictions in cases alleging Clean Air Act violations.[35] The amendments also allowed any person to bring suit against dischargers for failing to comply with their emission permits or any other portion of the act, and to seek civil penalties for noncompliance. Additionally, citizens could sue state or federal officials who failed to take actions prescribed by the Clean Air Act, and obtain court orders that compel officials to perform their duties. Finally, criminal enforcement provisions of the 1990 amendments include higher fines and longer prison sentences than previous versions of the Clean Air Act.[36]

[35] Facts in this paragraph are from Bryner (1993, pp. 157–58) and Vincoli (1993, pp. 88–89).

[36] For example, "Knowingly causing a release of air toxics will subject an individual to fines of up to $250,000 per day and 15 years imprisonment; businesses can be fined up to $1 million. . . ." (Vincoli, 1993, p. 89).

UNRESOLVED AIR POLLUTION PROBLEMS

By the mid-1990s, there was growing concern that controls on motor vehicle emissions might be insufficient to meet the NAAQS in many urban areas. Although tailpipe emissions of carbon monoxide and other pollutants declined during the 1980s, vehicle miles traveled increased concurrently to offset air quality gains. In addition, the growing popularity of sport utility vehicles as replacements for traditional automobiles does not bode well for the environment.[37] These utility vehicles, which sometimes provide less than 14 miles per gallon, have been termed the "new gas guzzlers." And because the federal government classifies them as light trucks, utility vehicles are permitted to emit more pollutants than most of the cars they replace. Urban air pollution due to vehicular emissions is likely to remain problematic as long as two variables remain unchanged: the low price of gasoline (which supports the popularity of gas guzzlers) and the reliance of individuals on passenger vehicles as opposed to mass transit systems.

As of the mid-1990s, efforts to reduce motor vehicle use were so politically charged that use of costly technological solutions—such as zero emissions vehicles—appeared to be the strategy of choice. However, some new attempts were being made to reduce vehicle miles traveled. Under the Intermodal Surface Transportation Efficiency Act (ISTEA) of 1991, for example, an urban area's federal transportation funding was tied to its air quality attainment status. Areas with poor air quality could obtain federal funds for transportation projects under ISTEA's *Congestion Management and Air Quality Improvement Program* (CMAQ). Funds for this program ($6 billion over six years) were aimed at reducing CO and ozone pollution. Transportation control measures detailed in the Clean Air Act and listed in a state implementation plan for reducing airborne pollutants were given highest priority for funding under CMAQ. The overall goal was to "support travel demand management and services which provide al-

[37] For more on environmental problems associated with sport utility vehicles, see K. Bradsher, "What Not to Drive to the Recycling Center," *The New York Times,* national edition, July 28, 1996, Sec. 4, p. 2.

ternatives to single-occupant vehicle use in a central city or suburban activity center."[38]

Although new transit system projects may help cut back vehicle miles traveled, some analysts feel that needed reductions will not be attained without more innovative measures. The existence of massive subsidies for auto use, such as free parking and government funds for highway construction, make it difficult for alternative modes of transportation to compete with the private auto on equal terms. Some analysts believe that the cost of private auto use should be increased to reflect the social cost it imposes in the form of air pollution, traffic congestion, and other external effects. Accordingly, increased attention has been given to a host of economic measures, including *congestion fees* for auto travel during peak commute hours, higher fees on parking in urban centers, and increased gasoline taxes. In addition, *smog taxes* on either vehicle emissions or miles traveled have been suggested. In the most heavily polluted areas, attainment of the NAAQS may require gas rationing and restrictions on driving that would be so severe that they would effectively "remake life" in some places.[39]

Another unresolved air quality problem is *indoor air pollution* and the public health risks it imposes. Common indoor air pollutants include asbestos, carbon monoxide, formaldehyde from building materials, radon, volatile organic compounds from solvents and paints, and numerous biological organisms.[40] Even though the concentrations of these pollutants may be low, the long exposure time that residents and workers endure can create significant health risks.

Part of the indoor air pollution problem relates to tobacco use. At least 50 known or suspected carcinogens are contained in tobacco smoke,[41] and a fierce debate raged in the early 1990s on how to control health risks from so-called "secondhand smoke." In some microenvironments, such as bars, restaurants, automobiles, and airplanes, the concentration of respirable particulate matter that originates from cigarettes can reach from 700 to 1000 $\mu g/m^3$, far higher than 150 $\mu g/m^3$, the 24-hour average particulate maximum specified in the NAAQS.[42] Numerous studies have linked involuntary inhalation of tobacco smoke with the increased chance of contracting lung cancer.[43] By the mid-1990s, many states and cities had taken steps to restrict smoking in restaurants, workplaces, and a variety of public spaces.

Unresolved air quality problems also exist on a global scale.[44] For example, chlorofluorocarbons (CFCs), often used as refrigerants, solvents, and aerosol propellants, can destroy the earth's protective *stratospheric ozone layer.* Because CFCs are chemically stable, they can persist and do damage for long periods. In the 1990 Clean Air Act Amendments, Congress committed the United States to exceed the requirements of the international agreement to reduce CFCs as set out in the Montreal Protocol[45] by banning nonessential CFC uses and all production by the year 2000. The 1990 amendments also instituted programs to recycle and reuse CFCs in appliances, and gradually phase out the use of even the less-offensive hydrochlorofluorocarbons by 2030. Although new requirements are in place to restrain CFC emissions, it is not clear how long previous emissions of CFCs will continue to cause ozone depletion in the stratosphere. Moreover, problems exist in enforcing existing bans on the use of CFCs. A notable example is the problem of preventing illegal imports of CFCs into the United States.[46]

As of the mid-1990s, air pollution problems related to *global climate change* had not yet been addressed successfully. Many scientists believed that anthropogenic additions of *greenhouse gases,* such as carbon dioxide (CO_2) and methane, to the earth's

[38] This quotation is from Federal Transit Administration (1994, p. 21).

[39] EPA, Approval and Promulgation of Implementation Plans; California—South Coast Air Basin; Ozone and Carbon Monoxide Plans (1988) as described by Percival et al. (1992, p. 807).

[40] For details, see Samet and Spengler (1991).

[41] Samet and Spengler (1991, p. 41) cite a U.S. Department of Health and Human Services report that indicates that more than "4,500 compounds have been identified from burning tobacco, and 50 of these are known or suspected carcinogens."

[42] Data on particulate levels is from Samet and Spengler (1991, p. 134).

[43] Samet and Spengler (1991, p. 147) list a sample of the many case-control studies that have "demonstrated statistically significant associations between involuntary smoking and lung cancer."

[44] Technical dimensions of the global air pollution problems discussed here—stratospheric ozone depletion and global climate change—are analyzed in Chapter 22.

[45] The Montreal Protocol and other international agreements to limit CFCs were introduced in Chapter 3.

[46] For more about the illegal import of CFCs, see Chin (1996).

atmosphere were causing a gradual increase in average temperatures that would result in numerous adverse effects. The pollutants of concern are airborne, and many, like CO_2, were not regulated under any program in the United States.[47]

The United States joined many other nations in signing the United Nations Framework Convention on Climate Change. This international agreement, which was ratified in 1994, requires signatory nations to gather inventory data on sources and sinks of greenhouse gases. The framework convention also calls for development of regional and national plans to prevent or mitigate climate change and its adverse effects.[48] The Clinton administration's "Climate Change Action Plan" responds to the requirements of the convention. The plan promises to reduce greenhouse gas emissions to 1990 levels by the year 2000. The Clinton administration's proposed cutbacks reflect a voluntary commitment, not one required by the framework convention.

Many initiatives exist for cutting back carbon dioxide emissions. For example, a number of U.S. electric utilities agreed voluntarily to reduce CO_2 emissions. An illustration of the effect of this voluntary program is a complex trade entered into by two utilities, Arizona Public Service Company (APS) of Phoenix, Arizona, and Niagara Mohawk Power Corporation of Syracuse, New York. Under conditions of the agreement, APS gave up 25,000 SO_2 emission allowances to Niagara Mohawk and received 1.75 million tons of CO_2 reductions in exchange.[49] The CO_2 reductions were recorded with the Department of Energy as a credit toward APS's voluntary commitment to cut back on greenhouse gas emissions. In this complex deal, Niagara Mohawk decided not to use the SO_2 allowances it received. Instead, it donated the allowances to environmental nongovernmental organizations and anticipated a $1 million tax benefit from the donation. Niagara Mohawk had plans to invest the tax savings in both domestic and international projects to further reduce CO_2.

Large differences in costs of air pollution abatement exist across and within countries. Analyses of the global warming problem often highlight differences across nations in costs of making improvements. It is commonly noted that developing countries have relatively low-cost opportunities to reduce CO_2 and other greenhouse gases and to create CO_2 sinks, for example, by planting or preserving forests. However, developing countries are often not prepared to undertake such projects because of other demands on their resources, and because many of them view global warming as a problem caused by high energy use in industrialized nations.

The differences across countries in costs of projects to reduce CO_2 releases (or create CO_2 sinks) have led to international discussions of *joint implementation:* parties in one country pay for a project in a second country that reduces CO_2 emissions or creates CO_2 sinks.[50] In exchange, the party providing funds for the project gets credit for the CO_2 reduction in the context of the CO_2 regulations it faces in its own country. Attempts at joint implementation are likely to increase as major emitters of CO_2 in industrialized countries seek ways of reducing releases without making significant cutbacks in output.

Key Concepts and Terms

Air Quality Management Prior to 1990

Public nuisance law
Local ordinances controlling air quality
Health effects of air pollution
Auto emissions and photochemical smog
Air quality criteria
National ambient air quality standards (NAAQS)
Technology-forcing auto emission standards
New source performance standards (NSPS)
Transportation control plans
Vehicle inspection and maintenance
Prevention of significant deterioration (PSD)
Best available control technology (BACT)

[47] The Clean Air Act Amendments of 1990 include relatively minor provisions relating to global climate change. In particular, all air pollution sources subject to the amendments' acid rain controls must report their CO_2 emissions to EPA.

[48] For details on the Framework Convention on Climate Change and the process of negotiating the agreement, see Pulvenis (1994).

[49] Facts in this paragraph are from an article ("Innovative Utility Accord Expected to Reduce Greenhouse Gases and Eliminate Sulfur Dioxide Emissions") in *Environmental Manager,* vol. 1, pp. 33–34.

[50] The concept of joint implementation was endorsed by the United Nations Framework Convention on Climate Change. For more on joint implementation, see Bodansky (1994).

Visibility in PSD areas
Nonattainment areas
Lowest achievable emission rate (LAER)
Emission offsets
Emission reduction credit (ERC)

Clean Air Act Amendments of 1990

Categories of nonattainment areas
Reasonably available control technology (RACT)
Required transportation control measures
Alternative fuels program
Low emission vehicle (LEV)
Zero emission vehicle (ZEV)
Acid rain
SO_2 emission trading
Air toxics
Maximum achievable control technology (MACT)
National air pollution permits
Bounty hunter provisions

Unresolved Air Pollution Problems

Vehicle miles traveled
Congestion fees
Indoor air pollution
Secondhand smoke
Stratospheric ozone layer
Chlorofluorocarbons
Greenhouse gases
Global climate change
Joint implementation

Discussion Questions

13-1 As of 1950, the public and private resources devoted to managing wastewater in the United States were greater than those devoted to managing air quality. Moreover, the resources for controlling conventional air pollutants were much greater than those for controlling hazardous substances. Why do you think this pattern of resource allocation existed at that time?

13-2 Air quality management in the United States involves strong roles for federal, state, and local governments. Often the stringent requirements set out by Congress and detailed by the U.S. Environmental Protection Agency are implemented by state and regional agencies. During the past half century, the relative power of federal, state, and local agencies involved in air pollution control has had a tendency to ebb and flow. How would you characterize these shifts in power? What arguments can be made for giving the federal government more (or less) power to control air pollution than it currently has? What balance of power would be appropriate in the control of air pollution?

13-3 In the United States, standards on the quality of ambient air are set for the nation as a whole, but standards for water quality vary by state and, more typically, by individual water body (or section thereof). How would you explain these differences in approaches to setting ambient standards for air and water?

13-4 Federal clean air regulations require much more abatement from proposed new sources of air pollution than from existing sources. Is this approach cost-effective? Is it fair? Why is it used?

13-5 Explain the difference between a "technology-based" standard and a "technology-forcing" standard. In what context have federal technology-forcing emission standards been used in the United States? Have they worked in the past? Are they likely to be effective in the context of California's requirements for zero emission vehicles?

13-6 The nonattainment provisions of the Clean Air Act Amendments of 1977 made the basis for standards for new waste sources different from those for existing sources: significant new sources had to meet standards based on *lowest achievable emission rates* while major existing sources were asked to retrofit with *reasonably available control technology.* Why are the two types of sources treated differently? How does the difference in standards for new and existing sources influence cost effectiveness?

13-7 It is often argued that cost-effective reduction in emissions of volatile organic compounds (VOCs) can be attained by getting heavy-polluting, older motor vehicles off the roads.

Sketch out a program that would contribute to retiring older vehicles in the United States. How does your proposed program affect drivers with low incomes?

13-8 Some analysts believe that cost-effective solutions to some of the nation's air quality problems could be achieved if programs existed for trading VOC emission reduction credits for both mobile and stationary sources. What would stand in the way of implementing such programs? Would these programs affect the system you proposed in response to question 13-7?

13-9 Indoor air pollution is recognized as a significant problem, yet it remains unregulated. How would you explain this? Is regulation the only approach to improving the quality of indoor air? In responding, consider possible roles for building codes and professional norms governing the design of heating and ventilating systems.

13-10 In analyzing alternative regulatory programs to reduce emissions of greenhouse gases, particularly CO_2, a link between industrialized and developing countries is generally recognized. Many relatively low-cost options for CO_2 management (including reforestation or maintenance of existing forests) involve actions that may be efficiently undertaken in developing countries. This has led to suggestions for CO_2 control programs based on the concept of "joint implementation." What are some of the advantages and disadvantages of this approach?

References

Bibbero, R. J., and I. G. Young. 1974. *Systems Approach to Air Pollution Control.* New York: Wiley.

Bodansky, D. 1994. "The United Nations Framework Convention on Climate Change." In *Greening International Law,* ed. P. Sands. New York: The New Press.

Brownell, F. W. 1995. "Clean Air Act." In *Environmental Law Handbook* 13th edition, ed. T. F. P. Sullivan, 101–34. Rockville, MD: Government Institutes.

Bryner, G. C. 1993. *Blue Skies, Green Politics.* Washington, DC: Congressional Quarterly.

Chin, K. 1996. Customs, EPA Crack Down on CRC Crooks. *Chemical Engineering* 103 (4): 43 and 51.

Cohen, R. E. 1992. *Washington at Work: Back Rooms and Clean Air.* New York: Macmillan Publishing Co.

Council on Environmental Quality. 1980. *The Eleventh Annual Report of the Council on Environmental Quality.* Washington, DC: Council on Environmental Quality.

Environmental Protection Agency and U.S. Department of Transportation. 1993. *Clean Air Through Transportation.* Washington, DC: EPA, U.S. DOT.

Federal Transit Administration. 1994. *Intermodal Surface Transportation Efficiency Act: Flexible Funding Opportunities for Transit FY'94.* Washington, DC: U.S. Department of Transportation.

Flachsbart, P. G. 1995. Long-Term Trends in United States Highway Emissions, Ambient Concentrations, and In-Vehicles Exposure to Carbon Monoxide in Traffic. *Journal of Exposure Analysis and Environmental Epidemiology* 5 (4): 473–95.

Grad, F. G. 1995. *Treatise on Environmental Law,* vol. 1. New York: Mathew Bender & Co.

Katz, M., and D. Thornton. 1996. *Environmental Management Tools on the Internet.* Delray Beach, FL: St. Lucie Press.

Kneese, A. V. 1978. "A Commentary on Needed Changes in the 1970 Air Quality Act Amendments." In *Approaches to Controlling Air Pollution,* ed. A. F. Freidlaender, 433–47. Cambridge, MA: MIT Press.

National Commission on Air Quality. 1981. *To Breathe Clean Air.* Washington, DC: U.S. Government Printing Office.

Novick, S. M., D. W. Stever, and M. G. Mellon, eds. 1995. *Law of Environmental Protection,* vol. 2. Deerfield, IL: Clark, Boardman, Callaghan.

Percival, R. V., A. S. Miller, C. H. Schroeder, and J. P. Leape. 1992. *Environmental Regulation.* Boston: Little, Brown.

Pulvenis, J.-F. 1994. "The Framework Convention on Climate Change." In *The Environment After Rio: International Law and Economics,* ed. L. Campiglio, L. Pineschi, D. Siniscalo, and T. Treves, 71–110. London: Graham & Trotman.

Samet, J. M., and J. D. Spengler, eds. 1991. *Indoor Air Pollution: A Health Perspective.* Baltimore, MD: Johns Hopkins University Press.

Seltz-Petrash, A. 1979. Transportation Planners Join Battle for Cleaner Air. *Civil Engineering* 49 (11): 84–88.

State of California Air Resources Board. 1994. *Staff Report: 1994 Low Emission Vehicle and Zero Emission Vehicle Program Review.* Prepared by Mobile Source Division, CARB, Sacramento, CA.

Stewart, R. B., and J. E. Krier. 1978. *Environmental Law and Policy.* Indianapolis: Bobbs–Merrill.

Turk, J. 1980. *Introduction to Environmental Studies.* Philadelphia: Saunders.

Vincoli, J. W. 1993. *Basic Guide to Environmental Compliance.* New York: Van Nostrand Reinhold.

Chapter 14

Hazardous Waste Management

Before the mid-1970s, hazardous wastes were not highly regulated and often handled in a haphazard way. Frequently, unlabeled metal drums filled with dangerous liquids were sent to municipal dumps and private land, where they eventually corroded and released their contents to both land and groundwater. In addition, toxic industrial wastewaters were often diverted to on-site ponds or lagoons, where they would gradually seep into the ground and sometimes contaminate aquifers.[1]

As this chapter demonstrates, hazardous waste disposal rules have changed dramatically during the past few decades. The chapter begins by examining the system the U.S. Congress established to regulate ongoing activities involved in the transport, storage, treatment, and disposal of hazardous waste. This is followed by an analysis of the Superfund program, which deals with contamination resulting from past hazardous waste management practices. The chapter ends by describing what has become, for many, the hazardous waste management strategy of choice: pollution prevention via waste source reduction and materials recycling and reuse.

TRACKING WASTE AND REGULATING WASTE MANAGEMENT FACILITIES

During the 1970s, Congress recognized that, while the types and quantities of hazardous waste were increasing, there was little control on where the waste was being sent and how it was being managed.[2] This absence of control led Congress to include Subtitle C in the Resource Conservation and Recovery Act (RCRA) of 1976, which established a major federal program for the management of hazardous solid waste. The statute defines *solid* unconventionally to include liquids and other nonsolids. The regulatory framework established in Subtitle C was expanded in 1984 when Congress passed the Hazardous and Solid Waste Amendments. Administration of the federal hazardous waste management program was left to the U.S. Environmental Protection Agency

[1] Federal regulations on toxics in wastewater don't apply unless the wastewater is discharged to a stream or lake or marine waters. On-site disposal of wastewater using ponds or lagoons was not covered by federal restrictions.

[2] Michael Hingerty of the U.S. Environmental Protection Agency's Region 9 office in San Francisco, California, offered many suggestions for improving this section.

(EPA), but the law allows states that satisfy certain conditions to implement regulations under RCRA, and nearly all states do so.

Definition of Hazardous Waste

In implementing RCRA, a key question concerns what constitutes a waste. The statute defines *solid waste* using the phrase ". . . discarded materials, including solid, liquid semisolid, or contained gaseous materials . . . ." In its regulations, EPA expanded upon the statute by defining discarded material to include material that is abandoned or "inherently wastelike;" dioxin wastes exemplify substances deemed to be inherently wastelike by EPA.[3]

One aspect of EPA's definition of solid waste that has caused confusion concerns materials that are recyclable. While EPA's definition clearly excludes substances recycled in closed-loop production processes, it is ambiguous in treating materials intended for recycling at some future time. Complexities in interpretation of EPA's definition arise, in part, because of practices related to the speculative accumulation of materials with a *potential* to be recycled.[4]

In addition to defining solid waste, EPA had to respond to the question: What makes a solid waste hazardous? This deceptively simple question took four years to answer. In 1980, EPA finally issued regulations indicating that a solid waste would be considered hazardous if it either exhibited a hazardous characteristic or appeared on a list of hazardous substances and waste sources.

The term *characteristic waste* refers to a waste that is declared hazardous because it is ignitable, corrosive, reactive, or toxic. Each of these characteristics is defined by a standard laboratory test detailed by EPA.[5]

In contrast, the term *listed waste* refers to a waste considered hazardous because it appears on one of several lists issued by EPA.[6] Some of EPA's lists include "nonspecific sources" such as spent halogenated solvents (for example, carbon tetrachloride) which are used in degreasing, and wastewater treatment sludges from electroplating operations. Other lists mention specific sources, such as oven residue from the production of chrome green pigments.[7] EPA's lists also include a category of substances deemed "acutely hazardous," and those substances are subject to some requirements that are more stringent than rules summarized in the following paragraphs.

Those responsible for managing hazardous waste may be able to avoid RCRA's requirements by transforming waste on-site using a process such as dilution. This raises a key question: Is the transformed waste hazardous or not? EPA has issued the "mixture rule" and the "derived-from rule" to answer this question.[8] In broad terms, the *mixture rule* says that if a listed waste is mixed with another substance, the mixture will be considered a hazardous waste.[9] However, if a characteristic waste is mixed with another substance, the mixture may or may not be hazardous. The result depends on whether the mixture itself is ignitable, corrosive, reactive, or toxic.

The *derived-from rule* clarifies whether material that remains after a hazardous waste is treated is itself hazardous.[10] Here, also, the distinctions between characteristic waste and listed waste are crucial. If a listed waste is treated, the remaining material

---

[3] EPA's definition of solid hazardous waste is too complex to be summarized here. For a comprehensive definition, see Fortuna and Lennett (1987, pp. 61–109).

[4] EPA is developing a proposal to redefine solid waste in ways that would clarify the status of recycleable materials.

[5] An overview of the test procedures is given by Vincoli (1993, pp. 112–17).

[6] For details on these lists and where they are located in the *Code of Federal Regulations,* see Percival et al. (1992, pp. 240–44), and, more generally, Fortuna and Lennett (1987).

[7] These examples of listed items are from EPA rules summarized by Fortuna and Lennett (1987, pp. 31–33).

[8] The mixture rule and derived-from rule have been controversial. After a court challenge, EPA repromulgated these rules on an emergency basis pending resolution of ambiguities in the definition of hazardous waste. In 1995, EPA proposed instituting the rules on a permanent basis in the context of the proposed Hazardous Waste Identification Rule mentioned in the next footnote.

[9] EPA has procedures that can be used to have particular substances removed from lists of hazardous materials (a process known as "delisting"). Thus a generator may attempt to have a mixed waste delisted, but the generator has the burden of proving that the mixed waste is not hazardous. Moreover, EPA has proposed the Hazardous Waste Identification Rule, 60 *Federal Register* 66344 (Dec. 21, 1995), which would simplify procedures for dealing with "low-risk" solid wastes designated as hazardous.

[10] The derived-from rule applies to waste storage and waste disposal as well as treatment. However, its common applications concern treatment. As with the mixture rule, generators can try to gain exemptions from results of applying the derived-from rule to listed wastes. The exemptions require use of EPA's procedures for delisting.

is still considered to be hazardous. However, if a characteristic waste is treated, the material that remains may or may not be hazardous. As with the mixture rule, the result depends on whether the material that remains after treatment is ignitable, reactive, corrosive, or toxic. As an example, if a listed waste is incinerated, the residual ash is a hazardous waste. But if a characteristic waste is incinerated and the ash is not ignitable, corrosive, reactive, or toxic, the ash is not hazardous.

Because of the financial and record-keeping burdens in disposing of hazardous waste, entities that create hazardous waste (referred to as *generators*) often go to great lengths to render them nonhazardous; for example, by diluting or treating waste on-site. However, the key to effective dilution or treatment is to make sure the transformation process does not cause the generator to become a *treatment, storage, or disposal* (TSD) facility under RCRA. As illustrated in the following section, RCRA's requirements for TSD facilities are generally much harsher than those for generators.[11]

Using the previously mentioned definition of hazardous solid waste, the majority of the wastes regulated under RCRA are liquids. And a very large fraction of the regulated wastes are associated with chemical manufacturing, petroleum refining, and metal fabrication and processing. Notably, some waste streams that may be hazardous are exempted from RCRA. These include irrigation return flows,[12] wastes from households, and wastes from facilities generating less than 100 kilograms per month of hazardous waste.[13]

[11] There are ingenious ways for a generator to transform hazardous wastes into nonhazardous wastes without becoming TSD facilities, and several common methods are detailed in Percival et al. (1992, pp. 252–57). An illustration is where chemicals are added to strong acids stored properly in barrels for a short time. The neutralized acid might qualify as nonhazardous provided it was not a listed waste.

[12] Selenium in water draining from irrigation projects provides an example. Such drainage has been linked to significant adverse effects on waterfowl at the Kesterson reservoir, which receives irrigation return flows from farms in the Central Valley in California.

[13] Strictly speaking, facilities generating less than 100 kg of hazardous waste per month are "conditionally exempt." Such facilities face less stringent regulations provided they comply with certain conditions. For a more complete list of wastes exempted from RCRA, see Vincoli (1993, pp. 119–21).

In broad terms, the hazardous waste management program established by RCRA (as amended in 1984) is based on two sets of requirements. One set concerns the *manifest system* for "cradle-to-grave" tracking of hazardous wastes. In this context, the point at which the waste is generated is the cradle and the point of disposal is the grave. The second set of requirements are for the design, operation, and performance of facilities used for the treatment, storage, and disposal of hazardous waste. The two sets of requirements are linked: all hazardous waste subject to RCRA (and therefore tracked by the manifest system) can only be treated, stored, or disposed of at facilities that meet RCRA's standards for design, operation, and performance.

Manifest System for Tracking Hazardous Waste

The cradle-to-grave waste tracking system gets its name from the word manifest, which refers to multiple-copy forms used in the shipping industry to detail the contents, origin, and destination of shipments. In the tracking system set up by RCRA, multiple-copy forms accompany a waste shipment from when it leaves a generator through all intermediate stages of transport, storage, and treatment to the point of final disposal. Under this scheme, the generator is responsible for determining whether any of its wastes are *hazardous* (using EPA's definitions in the context of RCRA). If some wastes are hazardous, the generator is responsible for preparing a Uniform Hazardous Waste Manifest, the standard form that accompanies a generator's hazardous waste until it reaches a RCRA-licensed treatment, storage, or disposal facility (see Figure 14.1). The number of copies of the manifest must be large enough to allow all parties who handle the waste to retain a copy, and to allow the treatment, storage or disposal facility that accepts the waste to send a copy back to the generator to indicate that the waste has been received at an approved RCRA facility.[14]

Participants in the manifest system include generators and transporters, as well as owners (or operators) of facilities for treatment, storage, and disposal. In each case, the waste handler must

[14] For more on the Uniform Hazardous Waste Manifest and how it is completed, see Karnofsky (1992).

State of California—Environmental Protection Agency
Form Approved OMB No. 2050–0039 (Expires 9-30-94)
Please print or type. *Form designed for use on elite (12-pitch) typewriter.*

Department of Toxic Substances Control
Sacramento, California

UNIFORM HAZARDOUS WASTE MANIFEST

1. Generator's US EPA ID No.: CA9722218511720 — Manifest Document No.: 00001
2. Page 1 of

Information in the shaded areas is not required by Federal law.

3. Generator's Name and Mailing Address
AJAX METAL PROCESSING
27 Main Street, Big Town, California

A. State Manifest Document Number

B. State Generator s ID

4. Generator's Phone ( )

5. Transporter 1 Company Name
MAC S HAULING

6. US EPA ID Number
CA1240001111

C. State Transporter s ID

D. Transporter s Phone

7. Transporter 2 Company Name

8. US EPA ID Number

E. State Transporter s ID

F. Transporter s Phone

9. Designated Facility Name and Site Address
SAFE TREATMENT, INC.
Wildcat Road, No Name, Ohio

10. US EPA ID Number
OH2222411111122

G. State Facility s ID

H. Facility s Phone

11. US DOT Description (including Proper Shipping Name, Hazard Class, and ID Number)	12. Containers No.	12. Containers Type	13. Total Quantity	14. Unit Wt/Vol	I. Waste Number
a. WASTE CYANIDE SOLUTION, NOS UN1935	030	DM	00055	G	State / EPA/Other
b.					State / EPA/Other
c.					State / EPA/Other
d.					State / EPA/Other

J. Additional Descriptions for Materials Listed Above

K. Handling Codes for Wastes Listed Above
a. b.
c. d.

15. Special Handling Instructions and Additional Information

16. **GENERATOR'S CERTIFICATION:** I hereby declare that the contents of this consignment are fully and accurately described above by proper shipping name and are classified, packed, marked, and labeled, and are in all respects in proper condition for transport by highway according to applicable international and national government regulations.

If I am a large quantity generator, I certify that I have a program in place to reduce the volume and toxicity of waste generated to the degree I have determined to be economically practicable and that I have selected the practicable method of treatment, storage, or disposal currently available to me which minimizes the present and future threat to human health and the environment; OR, if I am a small quantity generator, I have made a good faith effort to minimize my waste generation and select the best waste management method that is available to me and that I can afford.

Printed/Typed Name	Signature	Month	Day	Year
SANDRA SMITH				

GENERATOR

17. Transporter 1 Acknowledgement of Receipt of Materials

Printed/Typed Name	Signature	Month	Day	Year

18. Transporter 2 Acknowledgement of Receipt of Materials

Printed/Typed Name	Signature	Month	Day	Year

TRANSPORTER

19. Discrepancy Indication Space

20. Facility Owner or Operator Certification of receipt of hazardous materials covered by this manifest except as noted in Item 19.

Printed/Typed Name	Signature	Month	Day	Year

FACILITY

Figure 14.1 EPA's Uniform Hazardous Waste Manifest. Source: U.S. Environmental Protection Agency Form 8700-22.

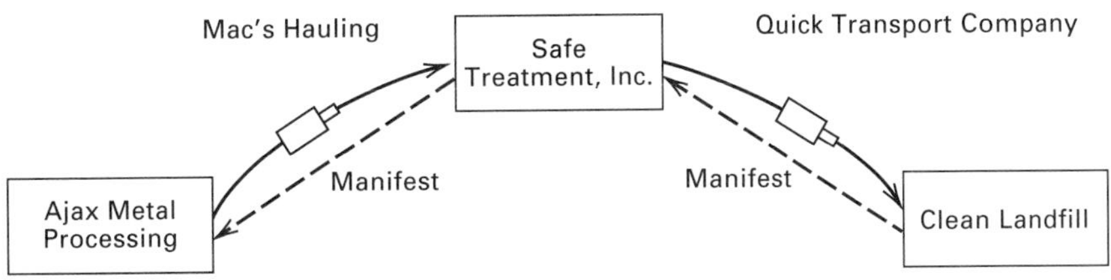

Figure 14.2 Application of the manifest system.

- Obtain an identification number from EPA or the state agency implementing the federal program.
- Train its staff in procedures for hazardous waste management and emergency response.
- Ensure that the hazardous waste shipment is handled in a manner consistent with applicable RCRA standards for facility design, operation and performance.
- Satisfy record-keeping and reporting requirements.

As will be detailed, RCRA's standards are used as a basis for issuing permits to TSD facilities. Only facilities with RCRA permits are allowed to treat, store, or dispose of hazardous wastes.

Operation of the manifest system is clarified by Figure 14.2, which diagrams a case involving treatment prior to disposal. The generator, Ajax Metal Processing, initiates the waste-tracking process by completing portions of the manifest. As illustrated by Figure 14.1, the manifest includes information on the generator, the transporter, and the TSD facility that is to receive the waste.[15] Each of the waste handlers has an EPA ID number and those numbers are listed on the manifest. The manifest also details the contents of the shipment based on codes for different wastes set out by the U.S. Department of Transportation.[16]

The generator keeps one copy of the manifest, and the hauler takes the remaining copies. When the waste arrives at a TSD facility—Safe Treatment, Inc., in this example—one copy of the final manifest is sent back to the generator. This provides the generator with proof that it has met its obligations under RCRA's manifest system. If the generator does not receive the copy from the TSD facility within a time specified by EPA, the generator is obligated to track down the shipment and send an *exception report* to EPA (or the appropriate state agency), indicating that a departure from the required procedure has occurred.

In the example in Figure 14.2, the waste arrives at a TSD facility that is not the point of ultimate disposal. In this illustration, the treatment facility, Safe Treatment, Inc., will treat the waste received from Ajax Metal Processing and many other generators. Assuming the material derived from this treatment is itself hazardous, Safe Treatment, Inc. must prepare a manifest to transport the derived waste to another TSD facility. The definition of a generator used by EPA covers owners and operators of TSD facilities when they ship hazardous wastes from their facilities.[17] In this example, Safe Treatment, Inc., completes a manifest and ships the derived waste to Clean Landfill, a RCRA disposal facility. As with any other generator, Safe Treatment, Inc., must receive a copy of the final manifest from the recipient of its waste (Clean Landfill). If the final manifest does not arrive on time, Safe Treatment, Inc., must file an exception report with EPA or the appropriate state agency.

In addition to exception reports, EPA (or the state) may also receive reports at two other points in the process. If hazardous wastes are released unintentionally during transport, the hauler is obliged to

[15] RCRA also covers the transport of hazardous wastes outside the United States. For a summary of requirements for exported hazardous wastes, see Fortuna and Lennett (1987, pp. 125–26).

[16] The Department of Transportation (DOT) has several responsibilities in implementing RCRA. These responsibilities are established under the Hazardous Materials Transportation Act, and they are summarized by Kokoszka and Flood (1989, pp. 295–316). Requirements set out by DOT relate to proper packaging, labeling, and placarding of hazardous waste shipments. DOT has detailed requirements for vehicles that carry particular types of waste.

[17] For details on the applicable regulations supporting this interpretation, see Cooke (1995, pp. 3–11).

report details of the incident. In addition, if the waste received by a transporter or a TSD facility is notably different from the waste detailed on the manifest, the waste handler must file a report describing the discrepancy.

Operation of the manifest system imposes substantial information processing requirements on all participants, including the responsible environmental agency. Day-to-day operation of the system requires only that the environmental agency be alerted when something is clearly out of order (as when waste is spilled in a transport accident). In addition, every two years (each year in some states) the responsible environmental agency receives a report from each hazardous waste generator that has participated in the manifest system. These reports summarize information on wastes that were shipped by the generator and the various transporters, and TSD facility operators that handled the generator's shipments.[18] Details on the contents shipped are also summarized. Generators are required to maintain copies of their manifests, biennial reports, and exception reports for at least three years, and many think it prudent to retain this information for longer periods. The information explosion that has accompanied implementation of the manifest system has taxed more than one environmental agency.

Requirements for TSD Facilities

As of the late 1980s, there were several thousand TSD facilities receiving hazardous wastes in the United States, and they were subjected to what many view as RCRA's most burdensome requirements. These facilities include treatment units, which accomplish such things as incineration, dewatering, and waste solidification. Disposal facilities include landfills, surface impoundments, and underground injection wells.

All TSD facilities that receive solid hazardous waste (as that term is used in implementing RCRA) must obtain a TSD facility hazardous waste permit. Requirements that must be satisfied to obtain a permit include the following:[19]

- *Records and Reports.* Requirements to keep records and notify authorities of releases of hazardous wastes.
- *Security.* Provisions to ensure that unauthorized persons (or livestock) do not enter the facility.
- *Design, Operation, and Performance.* Technical specifications of many types, including, for example, a requirement that landfills include liners to prevent hazardous materials from entering groundwater.
- *Monitoring.* Activities to detect releases of hazardous materials, as illustrated by requirements that incinerator operators measure emissions of gases, and landfills include groundwater monitoring wells to detect leaks of hazardous substances.
- *Corrective Action.* Responsibility to clean up releases from a facility that poses risks to human health and the environment, such as leaks from a landfill that may contaminate a drinking water supply.
- *Closure and Post-Closure Plans.* Detailed plans for closing the facility and maintaining and monitoring it for a specified period (typically, 30 years) after closure.
- *Financial Responsibility.* Assurance, by means of bonds, insurance, and other financial devices, that facility owners have funds sufficient to pay for damages during operations and implementation of closure and post-closure plans.

In the amendments to RCRA in 1984, Congress imposed new requirements on the disposal of hazardous substances on land. These requirements have led to a complex set of additional rules: land disposal restrictions.[20] Although the term *land ban* is commonly used to describe these restrictions, Congress did not prohibit the disposal of hazardous material on land. Instead, it instructed EPA to issue requirements on the treatment required *before* hazardous materials could be placed safely on land units such as landfills, surface impoundments, injection wells,

[18] Every two years, EPA summarizes information received from all states in a National Biennial Hazardous Waste Report. The information from the biennial reports of individual generators is entered into computers, and the national biennial report is available from EPA on the Internet.

[19] Details on requirements that must be met by TSD facilities are given in Cooke (1995, chap. 5).

[20] For details on EPA's land disposal regulations, see Cooke (1995, pp. 5-302.30 to 5-302.68).

and so forth.[21] EPA responded to this mandate by issuing regulations on *best demonstrated available technology* (BDAT). Hazardous materials may be placed on land if they undergo treatment using BDAT, or if they are treated to yield results (in terms of maximum concentrations of contaminants) equivalent to what could be attained using BDAT. However, if EPA specifies that a particular technology must be used, there is no choice involved.

EPA's land disposal regulations also impose requirements on generators of hazardous waste. Before sending waste to a TSD facility, a generator is obligated to determine if the waste is subject to pretreatment requirements under EPA's land disposal regulations. The generator is also required to provide information to the TSD facility regarding whether the waste satisfies conditions imposed on wastes destined for land disposal. If the waste does not meet these requirements, the generator is obligated to include a notification of the treatment needed before the waste can be disposed of on land. If the waste is destined for disposal on land, the facility that provides the requisite treatment must certify that treatment was carried out, and a notice of this certification must accompany the waste to the place of disposal.

ENVIRONMENTAL MANAGEMENT BASED ON LIABILITY RULES

RCRA is intended to prevent hazardous waste from endangering human health and the environment.[22] But what about problems at hazardous waste sites that released contaminants *before* RCRA was passed? Many of those sites have been abandoned for years. Congress responded to calls for cleanups at contaminated hazardous waste sites by enacting the Comprehensive Environmental Response, Compensation and Liability Act (CERCLA), which is often referred to as *Superfund* because the act created the Hazardous Substances Response Trust Fund. In 1980, when the law was enacted, Congress allocated $1.6 billion for the fund. Six years later, the Superfund Amendments and Reauthorization Act (SARA) authorized an appropriation of $8.5 billion over a five-year period. In 1990, Congress reauthorized the Superfund program again, this time at a funding level of $5.1 billion. The trust fund is fed by an assortment of sources, but primarily by taxes on crude petroleum and chemical feedstocks used by firms, and by a corporate environmental income tax.[23]

Despite the substantial size of the Superfund, Congress knew that the billions of dollars in the fund would not be large enough to clean up all the troublesome hazardous waste sites in the United States. Major funding for the cleanup was not intended to come from the trust fund; rather, it was to be provided by parties that could be held legally liable for having created the problem sites. Thus, to understand how the Superfund program works, it is necessary to examine CERCLA's principal liability provisions.

CERCLA's Core: Liability

Congress passed CERCLA to deal with releases of hazardous substances into the environment from facilities at which hazardous wastes were disposed of, treated, or stored. The definition of *hazardous* under CERCLA is much broader than the corresponding definition in the Resource Conservation and Recovery Act. It includes substances defined as hazardous under RCRA, as well as substances defined as hazardous under numerous other federal air, water, and toxic substance control laws.

Under CERCLA, when a facility is subject to cleanup (either an emergency removal of materials or a long-term *remedial action*), all *potentially responsible parties* (PRPs) may be held liable for either conducting the cleanup or reimbursing others for the cost of cleanup. In this context, the term party can include a person, a corporation, a lending institution,

[21] Congressional concern regarding land disposal of hazardous wastes also resulted in new requirements for municipal solid waste landfills that receive wastes not legally defined as hazardous. Congress recognized that municipal landfills receive hazardous materials that are exempt from RCRA, such as hazardous wastes from households and from firms that generate wastes in quantities too small to make them subject to RCRA. For these reasons, Congress instructed EPA to impose requirements on both the design of municipal solid waste landfills and the treatment required before hazardous wastes can be disposed of in those landfills.

[22] This section benefited from conversations concerning CERCLA with Patti Collins and Michael Hingerty of EPA's Region 9 office in San Francisco.

[23] The three principal taxes used to provide income for the trust fund are detailed by Probst et al. (1995, pp. 54–58).

an insurance company, or an agency of federal, state, or local government. For a particular hazardous waste facility, PRPs may include the following:[24]

- *Generators.* Parties who arranged for treatment, storage, or disposal of hazardous substances at the hazardous waste facility.
- *Transporters.* Parties who accepted hazardous substances for transport to the facility *and* participated in the selection of the treatment, storage, or disposal site.
- *Owners and Operators.* Current owners and operators of the facility, and, under some conditions, former owners and operators who were involved with the facility during times when hazardous substances were stored, treated, or disposed there. Former owners and operators are liable only if there was a release of contaminants at the time of their ownership or operation.

These categories of potentially responsible parties were refined under SARA. For example, state and local governments could be excluded as PRPs if they acquired a facility undergoing cleanup under CERCLA as a result of bankruptcy and had no other connection to the facility.

The U.S. federal courts have interpreted liability under CERCLA as being *retroactive, strict,* and, typically, *joint and several.*[25] Liability established by a statute is retroactive if it applies to activities that took place before a statute was enacted. Under CERCLA, liability extends to activities causing contamination before the law was passed in 1980. In the U.S. legal system, strict liability means that liability is established whether or not the party involved was meeting all applicable laws and regulations or was negligent or intended to cause harm. Thus, for example, an allegedly liable party cannot claim, as a defense, that standard industry practice for waste disposal had been used, or that all government regulations and discharge permit conditions in force at the time were satisfied.

The doctrine of joint and several liability often applies when damages are caused by the actions of multiple parties, and it would be a practical impossibility to identify individual contributions to the total problem. When this doctrine is invoked, as it generally is in the context of CERCLA, any one party can be held liable for the entire cleanup, even if many parties contributed to the waste at a site. Although CERCLA does not explicitly call for joint and several liability, the legislative history and subsequent court interpretations make it clear that joint and several liability will apply in most cases. Otherwise, the burden on EPA to identify how much each particular waste source contributed to a problem would be overwhelming.

Under Section 107 of CERCLA, a "responsible party" is liable, among other things, for all costs of a hazardous waste site cleanup undertaken by EPA. This liability provision provides a basis for what has been termed EPA's *fund start strategy.* (An alternative, more colorful description is "shovels first, lawyers later.") Using the fund start strategy, EPA employs the trust fund to pay for cleanup of a site. Following the cleanup, EPA can try to recover costs using Section 107, which gives EPA authority to sue responsible parties in order to recover the costs of cleanup, attorneys' fees, and associated administrative costs.

Section 106 of CERCLA provides a basis for another of EPA's principal implementation strategies: *enforcement start* (or "lawyers first, shovels later"). Under Section 106, EPA is authorized to issue administrative orders requiring responsible parties to either conduct studies and develop cleanup plans or implement the cleanup plans EPA set out in the orders. Failure to comply exposes the party receiving an order to substantial penalties: fines can be as high as $25,000 per day of violation of the order, and punitive damages may be imposed at up to three times the costs incurred by EPA as a result of the party's failure to carry out the order.

Although EPA itself can issue administrative orders under Section 106, it must use the federal courts to enforce its orders. As is common for federal agencies, EPA must refer its case for legal action to attorneys in the Department of Justice (DOJ), and the latter sues the alleged responsible party. This extra

---

[24] The question of who may be liable under CERCLA is extraordinarily complex and well beyond the scope of this test. For a summary discussion of the subject, see Gaba (1994, pp. 176–78) and Lee (1993, p. 290).

[25] Although CERCLA itself does not spell out the liability scheme as retroactive, strict, and joint and several, that is how liability under CERCLA has generally been interpreted by the courts. The CERCLA liability scheme is sometimes characterized as being based on the *polluter pays principle:* companies are responsible for cleaning up the pollution they caused.

step further increases the financial and administrative burdens of bringing suit. It also adds time, since the priorities of DOJ may not coincide with those of EPA. In addition, for any particular case, attorneys at DOJ and EPA must spend time to coordinate their objectives and approaches. For example, coordination is necessary because attorneys from the two agencies often have differing attitudes about what constitutes on acceptable settlement in a case.

Parties receiving Section 106 administrative orders cannot formally challenge the contents of an order unless there is a court action. This may occur if a party under orders fails to comply *and* EPA and DOJ go to court to enforce the order. A second mechanism for contesting an order is for a party to comply with the order and then sue EPA to recover some or all of its costs of compliance.

Court actions based on Sections 106 and 107 do not commonly proceed through a full trial. Settlement is the norm. Both EPA and potentially responsible parties are often motivated to settle rather than endure a protracted and costly trial. In addition, settling gives each of the parties in dispute a chance to influence the outcome.

A settlement with EPA is typically set out either as an administrative order on consent (*consent order*), or a consent decree. In cases detailing a remedial action, a consent decree is filed with a federal court. In signing a *consent decree,* a judge indicates the court's acknowledgment that a legally valid agreement was reached.[26] A consent decree has the advantage that a neutral party—the judge—can be called upon later to resolve any disputes that arise as conditions of the decree are implemented. A consent order does not require judicial approval and is not used to spell out EPA's final selection of a remedial action.[27]

Joint and Several Liability, and Equity Issues

In discussing CERCLA, one thing PRPs complain about bitterly is the apparent inequity of the statute's liability provisions.[28] The potential for unfair outcomes in applying joint and several liability has been an issue since the congressional deliberations leading to passage of CERCLA. In testimony before Congress, a representative of the Chemical Manufacturers Association presented the following scenario in arguing against joint and several liability:

> Let's envision, for a moment, that an orphan dump site has been discovered. It is known that at least 50 disposers have used the site. But the Justice Department decides that just one of those companies—likely to be one with sales exceeding several billion dollars per year—should be singled out for attack. . . . [It] would be entirely possible that one company might have to bear the entire cost of the cleanup, while at the same time it has been making fee payments into the cleanup fund. This is what is meant by joint and several liability.[29]

For reasons that will be explained, the scenario described by the representative of the Chemical Manufacturers Association is a most unlikely one.

Although EPA can, in principle, bring suit against a single, well-endowed, potentially responsible party when there are tens or even hundreds of parties involved, the potential for gross inequities is reduced by the explicit provision (in the modifications made to CERCLA in 1986) of a "right of contribution." This provision grants the right to any PRP held liable for implementing a cleanup (or paying the cost of a cleanup) to sue other PRPs at the same site for a "contribution" equal to an appropriate portion of the total cleanup burden. In practice, these suits, referred to as *contribution actions,* are not heard until after a court resolves the government's original case against the one or more PRPs that constitute the "primary defendants."

The language of CERCLA (as amended in 1986) directs courts to apply "equitable factors" in allocating costs among jointly and severally liable parties in a contribution action. According to Lee (1993, p. 311), the following factors appear to have influenced the courts in most cases:

---

[26] In a typical case, the Department of Justice as plaintiff files with the court both the complaint and the consent decree. A consent decree does not bind the court if the dispute is later brought back to court by a party mentioned in the decree.

[27] Some actions detailed in a consent order may contribute to the design or performance of a remedial action. However, the remedial action itself is placed on record in a consent decree, not a consent order.

[28] For a comprehensive analysis of equity issues surrounding CERCLA, see Probst et al. (1995). Their analysis also examines economic efficiency and the influence of CERCLA on the insurance industry.

[29] Testimony of Louis Fernandez on behalf of the Chemical Manufacturers Association as cited by Barnett (1994, pp. 65–66).

1. The volume of hazardous substances contributed by each party.
2. The relative degree of toxicity of each party's wastes.
3. The extent to which each party was involved in the generation, transportation, treatment, storage, or disposal of the substances involved.
4. The degree of care exercised in handling the hazardous substances.
5. The degree of cooperation by the parties with government officials in order to prevent any harm to public health or the environment.

Another mechanism for enhancing the possibility for equitable outcomes is the *de minimis* settlement. Under provision of the 1986 amendments to CERCLA, EPA can settle with small contributors (for example, those contributing less than 1%, in terms of amount and toxicity of the hazardous substances at a site) by requesting relatively modest sums, say several thousand dollars. In return, *de minimis* settlers are released from all liability at the site, including contribution actions brought by other PRPs; the latter advantage is termed *contribution protection.* Similar provisions apply for *de micromus* settlements. These typically involve payments of only a few thousand dollars, and they are used by EPA to settle with very small contributors; for example, those contributing less than 0.1% of the hazardous waste at a site.

Economic Inefficiency of CERCLA Implementation

In addition to being labeled as inequitable, CERCLA's liability provisions have been criticized for leading to economic inefficiencies in hazardous waste management. However, the opposite claim has also been made: in the long term, CERCLA will improve productive efficiency. These conflicting perspectives can be made consistent by considering the time frame of each argument.

Internalization of External Costs

From a long-term perspective, CERCLA's liability provisions provide a basis for improving productive efficiency in the future. Many generators and handlers whose hazardous waste created problem sites were able to use the environment's waste receptor services at much less than the social cost of waste disposal. In the language of economics, many generators of hazardous waste and those involved in its storage, transport, and disposal were not forced to *internalize* the *external costs* imposed by the waste. CERCLA has changed that.

The liability provisions of CERCLA are what forces the internalization of external costs. Philip Cummings, chief counsel of the Senate Environment Committee when CERCLA was drafted, highlighted the statute's intended influence on future actions:

> The main purpose of CERCLA is to make spills or dumping of hazardous substances less likely through liability, enlisting business and commercial instincts for the bottom line in place of traditional regulation. It was a conscious intention of the law's authors to draw lenders and insurers into this new army of quasi-regulators, along with corporate risk managers and boards of directors.[30]

Although it is difficult to quantify the extent to which external costs have been internalized as a result of CERCLA, there is qualitative evidence that internalization of costs has occurred. Many firms have developed programs to minimize the generation of hazardous wastes, in part because they fear being held liable for future site remediation.[31] In addition, the insurance industry has played a significant role in forcing internalization of external costs. This takes place indirectly via insurance rates. Dover (1990, pp. 183–84) describes the process:

> If insurance markets are operating properly, high-risk generators of hazardous wastes would pay higher premiums for liability coverage than low-risk generators or disposers. An economic incentive is thereby created to minimize the sum of disposal and insurance costs. As a result, a firm would select a disposal method and a level of liabil-

[30] Philip Cummings as cited by Percival et al. (1992, p. 290). Not everyone agrees with this position. Acton and Dixon (1992, pp. 5 and 6) note that some of CERCLA's critics believe "a liability exposure that is too sweeping may encourage greater amounts of deception and illegal practices, so that parties cannot be traced for their activities in disposing of hazardous waste in the future. Thus, rather than improving disposal practices, the sweeping liability may encourage less safe practices."

[31] As explained later in this chapter, fear of being held liable for future cleanups is not the only factor motivating firms to adopt waste minimization programs.

ity coverage that would begin to balance marginal costs and benefits.

In examining the productive efficiency aspects of CERCLA, most commentators do not focus on long-term improvements. Instead, they point to the resources being wasted in ongoing efforts to clean up problematic hazardous waste facilities using CERCLA. There are several potential sources of inefficiency. Congress's call for permanent solutions to problems at hazardous waste disposal sites has required substantial expenditures, as has its call for use of rigorous cleanup standards. Critics frequently charge that the costs of meeting cleanup requirements established by CERCLA are less than the benefits in reduced risk to human health and the environment.

Transactions Costs

Arguably the most virulent charges of inefficiency have targeted the enormous transactions costs involved in some site remediations. In the context of CERCLA, *transactions costs* include "expenses that are directly related to determining issues of liability and in forcing private parties to live up to their obligations" under the statute.[32] The following are examples of transactions costs:

- EPA's cost to identify PRPs and its efforts to identify the relative share of wastes contributed by different parties.
- The costs to PRPs, EPA, and other government regulators in negotiating cleanup standards and remedial actions.
- The costs to the government in preparing administrative orders and the costs to PRPs in responding.
- The costs to the government, PRPs, and insurance companies in litigating disputes.

A report by Acton and Dixon (1992) of the Rand Institute for Civil Justice documented transactions costs for inactive hazardous waste sites subject to action under CERCLA and state programs to clean up abandoned sites. The data for the study came from separate surveys of insurance companies and PRPs; the main results are summarized as follows.

[32] This definition of transactions costs is from Church and Nakamura (1993, p. 37); the examples of transactions costs cited are from the same source.

Consider, first, transactions costs for insurance companies. Acton and Dixon examined several years of cost outlays of 4 of the top 25 insurance companies providing general liability coverage in the United States. They divided costs into four categories:

- *Indemnification.* "[S]ums paid in connection with the obligation to indemnify a policyholder or otherwise resolve a claim."[33] Typically these funds were used either to clean up a site or compensate parties allegedly injured by the insured party.
- *Coverage Disputes.* Payments to the insurer's external legal counsel in disputes with policyholders. Insurance companies and insured PRPs often disagreed over whether a particular insurance policy applied to hazardous waste site cleanups, or what cleanup costs or legal fees were covered.
- *Defense of Policy Holder.* Payments to external legal counsel for costs incurred in representing the policyholder against the liability claims of government, other PRPs, or other private parties.
- *Internal Staff Costs.* Costs for time spent by the insurer's internal staff to process claims, coordinate lawsuits related to claims, or supervise the resolution of claims.

Acton and Dixon argue that, *except* for indemnity payments, all of the above outlays are transactions costs. Using this definition, they estimated that 80 to 96% of insurance company outlays on hazardous waste cases (between 1986 and 1989) were transactions costs.[34] Acton and Dixon (1992, p. xi) provide a context for interpreting their findings: "If the sampled firms are representative of the insurance industry as a whole, insurers spent 410 million [out of a total of 470 million] on transactions costs in 1989."

The Rand study also analyzed transactions costs incurred by five PRPs, each of which was a firm on

[33] This quotation is from Acton and Dixon (1992, p. 21). See the same source for a full description of the four categories of costs.

[34] On average, "88 percent of expenditures were transactional, with 37 percent for defense of the policy holder and 42 percent for coverage disputes." (Acton and Dixon, 1992, p. 24). Action under CERCLA took place at 26% of the sites analyzed by Acton and Dixon and involved 40% of the total cost outlays. However, the breakdown of expenditures into categories for the sites subject to action under CERCLA did not differ substantially from the breakdown for all sites in the Acton and Dixon study.

the *Fortune 100* list of U.S. companies.[35] The database assembled by Acton and Dixon included 144 inactive hazardous waste sites where the five firms had been contacted by a federal or state environmental agency.[36] For these sites, costs directed toward site cleanup, designated as "investigative and remediation" costs, were distinguished from transactions costs. The latter, which do not contribute directly to a cleanup, typically involved legal expenses or costs incurred in assigning financial liability (for example, searching for other PRPs, and duplicating technical studies to check or contest results found by others). Acton and Dixon (1992, p. xiii) summarized their findings as follows:

- Between 1984 and 1989, transactions costs averaged 21% of the total outlays of each firm.
- The split between transactional and remedial expenditures was reasonably consistent across these firms, with individual firm shares ranging from 15 to 31%.
- The bulk of the transactions costs were for legal representation.

Acton and Dixon (1992, p. xiii) went on to characterize the transactions costs at the 73 sites where the five PRPs had spent more than $100,000:

> Eighty-five percent of their total outlays occurred at these sites. Seventeen percent of the total outlays at these sites was transaction costs. . . . Sites with the largest expenditures generally have the lowest transaction-cost shares. Transaction-cost shares varied considerably across sites.

The Brownfields Problem

CERCLA also affects productive efficiency via its unintended influence on the marketability of urban real estate. This side effect of CERCLA is manifested in the large inventory of *brownfields,* abandoned and underutilized industrial properties thought to be contaminated by hazardous materials. What EPA calls the *brownfields problem* concerns how actual or perceived liability for cleanup of brownfields inhibits the redevelopment and reuse of these sites. The regulatory risk for investors and developers interested in brownfields can be extraordinary, since purchase and development of a brownfield site opens the door to liability under CERCLA and analogous state laws.

In addition to impeding economic development in urban areas, brownfields contribute to urban sprawl. For both lenders and developers, CERCLA liability risks can be eliminated by developing *greenfields,* previously undeveloped lands in suburban and rural areas. Two sets of problems result when developers select greenfields over brownfields to avoid potential cleanup liability. First, abandoning brownfield sites in urban areas contributes to unemployment, decreased municipal tax revenues, and lower property values. Second, developing greenfield sites may lengthen job commutes, destroy valued open space, and contribute to urban sprawl.

In some states, the significance of the brownfields problem has caused legislatures to act. In California, for example, several laws have been passed since 1990 to help reduce impediments faced by developers and financial institutions interested in brownfield sites. The California laws are intended to speed contaminant cleanup and make cleanup standards consistent with planned future uses of brownfield sites. In addition, lenders and developers are able to obtain limited immunity from liability in exchange for undertaking site cleanup activities approved by state agencies. While the California statutes help reduce liability under state laws, they offer no protection from liability under CERCLA.

Given the significance of the brownfields problem, what has EPA done? In an effort to create new jobs in cities, EPA under the Clinton administration developed a *brownfields initiative.* This program includes dozens of EPA demonstration grants (of up to $200,000 each) intended to "encourage community groups, investors, lenders, developers and other affected parties to work together to clean up contaminated sites and return them to productive use."[37] EPA's brownfields initiative also provides technical assistance to state and local governments and improves coordination among federal programs supporting job training and local economic development. The next reauthorization of CERCLA will provide

[35] These firms consisted of a mixture of petroleum, chemical, and manufacturing firms.

[36] Of the 144 sites, about 50% involved total cost outlays (for a PRP) of less than $1000, which typically included costs of responding negatively to inquiries about whether the company had sent any waste to a particular site.

[37] This quotation is from Wise (1995, p. 151). Wise summarizes several elements of EPA's brownfields initiative.

Congress a chance to guide EPA's efforts to reduce the brownfields problem.

Implementation Strategies

The productive efficiency dimensions of CERCLA have not played a major role in shaping EPA's Superfund strategy. Indeed, it is difficult to identify a single EPA approach. Based on case studies of six hazardous waste site cleanups, Church and Nakamura (1993) identified three distinct EPA implementation strategies: prosecution (or "enforcement start"), public works (or "fund start"), and accommodation.

According to Church and Nakamura, the *prosecution* approach was well illustrated by an EPA regional office that had a reputation summarized by the phrase "cut no deals, take no prisoners." As described by Church and Nakamura (1993, p. 24), the "central tenet of the prosecution approach is that the selected PRPs can and must be made fully responsible for all cleanup costs." Implementation of this approach is characterized by strict adherence to rigid procedures and schedules, and an aversion to developing informal relationships with PRPs.

The *public works* strategy was followed by an EPA regional office in which "the mandate [was] to clean up the site now and let the lawyers and accountants worry about money and legalities later" (Church and Nakamura 1993, p. 95). This approach leaves little room for negotiating agreements with PRPs on cleanup standards and remediation measures. The public works strategy relies on EPA's authority (under CERCLA) to initiate *removal actions* at sites posing an "immediate and substantial danger to public health or the environment." Removal actions must be relatively modest in scope: under the 1986 amendments to CERCLA, a removal action should normally cost less than $2 million and be completed within a year. Exceptions are possible, however, and EPA has been able to use them to continue some removal actions beyond the $2 million/one year constraints.[38]

The third implementation strategy, *accommodation,* was heavily used in a third EPA regional office studied by Church and Nakamura. This office favored extensive negotiations with PRPs to settle disputes related to site cleanups. Compared to the other two EPA regional offices, the office emphasizing accommodation was reputed to be "more likely to share technical information with PRPs, and to take their input seriously in decisions embodied in the remedial investigation and feasibility study . . . or in the choice of a remedy . . ." (Church and Nakamura 1993, p. 68).

The accommodation strategy can be effective because PRPs often find it advantageous to cooperate and discharge their legal obligations without being forced to by administrative orders or court actions. In voluntarily settling with EPA on terms of a site remediation, a PRP may influence both what it must do and when it must do it. In addition, a settlement can offer protection from lawsuits by either the government or other PRPs. However, there may be good reasons not to settle. For example, a PRP may wish to retain the right to challenge EPA's site remediation scheme in court, and this option is foreclosed by settling.

The analysis by Church and Nakamura examines a key question: What EPA implementation approach works best in a given circumstance? Based on their research, Church and Nakamura include the following factors among those that should influence EPA's implementation strategy in a particular circumstance:[39]

- Size, technological difficulty, and projected cost of the cleanup.
- The feasibility of various remedial alternatives.
- The number, financial resources, and cooperativeness of PRPs associated with the site.
- The strength of the evidence against the responsible parties.
- The stance taken by state officials.
- The need to expedite cleanup because of dangers posed by the site.

In addition to being site-specific, an EPA regional office's strategy may also change over time. Policies issued by EPA headquarters are updated frequently, and those policies guide actions at regional offices.

[38] Several years ago, EPA initiated the Superfund Accelerated Cleanup Model, which emphasized increased use of removal actions and other approaches for immediate risk reduction at large sites.

[39] Except for the last item, this listing is a quote from Church and Nakamura (1993, p. 140). The last item is paraphrased from the same source. The quoted material has been reformatted.

CLEANUP OF A SUPERFUND SITE

The remediation of hazardous waste sites under CERCLA often involves multiple parties, complex technological issues, and a plethora of regulations.[40] The cleanup process can be costly and lengthy, and negotiation and litigation play key roles in determining outcomes. The step-by-step process used to clean up a contaminated site in Mountain View, California, provides a basis for demonstrating how EPA has implemented CERCLA.

History of Contamination at the MEW Site

Discovery of the Site

Cleanup efforts at the "Middlefield-Ellis-Whisman" (MEW) site, named for the streets surrounding the site (see Figure 14.3), began shortly after CERCLA was enacted. While the U.S. Environmental Protection Agency was struggling through the beginning stages of implementing the ambitious Superfund cleanup program, the MEW site remained within the jurisdiction of the local office of a state agency, the San Francisco Bay Regional Water Quality Control Board (RWQCB). The MEW site was one of a number of sites with contaminated groundwater that had been discovered during the early 1980s on the peninsula just south of the city of San Francisco.

The MEW site included properties of more than 20 companies, including Raytheon Company, Intel Corporation, Fairchild Semiconductor Corporation, NEC Electronics, and Siltec Corporation. The firms were involved in semiconductor manufacturing, metal finishing, parts cleaning, aircraft maintenance, and other activities requiring a variety of chemicals. Believing that their operations were relatively innocuous, companies routinely stored raw and waste solvents in tanks beneath the ground and utilized acid-neutralization sumps. These practices were consistent with regulations existing at the time. In the early 1980s, after many of the companies had been operating at the site for about 20 years, people began to realize that chemicals were seeping into the soil and groundwater, and they worried about the safety of local drinking water sources. Contamination was first discovered at the site in late 1981, but at that time, the RWQCB considered the MEW site a low priority and concentrated its efforts on more troublesome toxic waste sites.

With the discovery of additional contamination at various locations in the San Francisco Bay Area, citizens and environmental groups demanded further action. Additional testing was conducted at the MEW site and extensive contamination was found. About 70 different chemicals were eventually discovered in the groundwater and soil at the site; the primary contaminant was trichloroethene (TCE), a volatile organic compound commonly used as a degreaser. Several of the contaminants at the MEW site are suspected carcinogens and are believed to cause adverse health effects including depression, nausea, rashes, and possibly birth defects.

During the early 1980s, contamination at the MEW site appeared limited. Five companies identified as "dischargers" drilled groundwater wells.[41] Most of the wells were *monitoring wells,* which pump out small samples of water for testing to measure contaminant concentrations. Others were *extraction wells,* designed to pump large quantities of water to the surface for treatment by methods such as aeration, air-stripping, or carbon filtration. Extraction wells are also used to control the hydraulic movement of *plumes* of contaminated groundwater. Each company focused on its own property, and there was no coordinated effort to characterize the contaminant plume and conduct a comprehensive cleanup.

The subsurface environment at the MEW site is extremely complex, with numerous *aquifers,* bodies of underground water, separated by impermeable clay layers called *aquitards.* During the early 1980s, the configurations of these aquifers and aquitards were unknown. Thus, while TCE and other organic compounds had been found in some shallow areas, no one knew whether this contamination would affect deeper aquifers used for drinking water. Moreover, even if the drinking water aquifers were hydraulically

[40] This section is based on research conducted by Jean Buo-Lin Chen and Cecilia Takayama while they were students at Stanford University. Ms. Chen prepared the original case study analysis, and Ms. Takayama helped rework the materials in the form of a first draft for this section. Patti Collins and Michael Hingerty of EPA's Region 9 office in San Francisco offered many comments to help improve this section.

[41] Installation of a well involves boring holes through the soil into the groundwater, inserting a pipe surrounded by a well casing into each hole, and then backfilling the spaces with gravel.

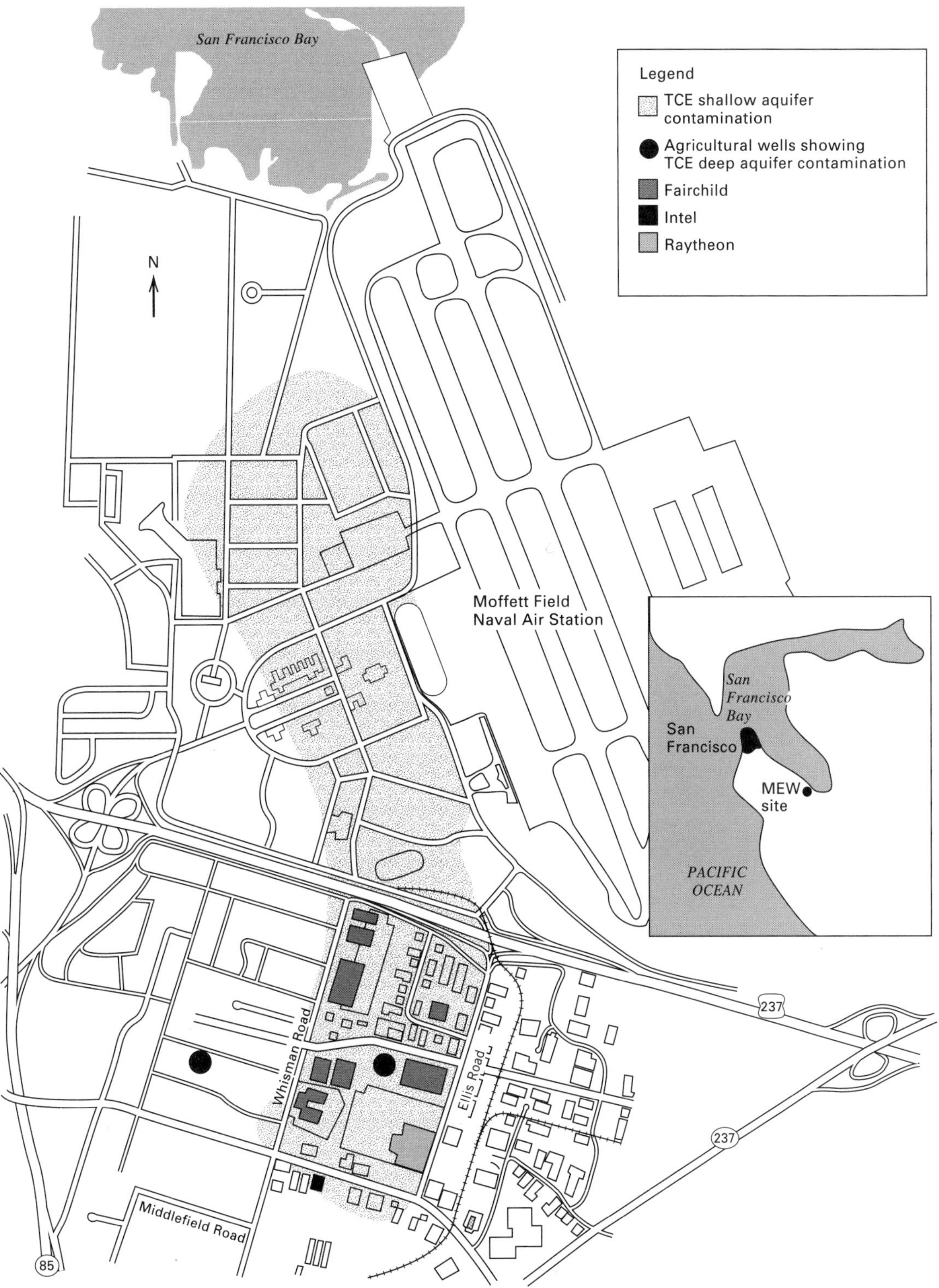

Figure 14.3 MEW Plume and Study Area, Mountain View, CA. Source: Fact sheet for MEW site prepared by the U.S. Environmental Protection Agency, Region 9, San Francisco, CA.

separated from the contaminated shallow aquifers, openings in the aquitards might exist. Such conduits could eventually serve as pathways for contaminated waters to mingle with sources of drinking water. Furthermore, the extent of contamination at different locations and depths was not known, so the RWQCB could not adequately assess the risks to health and the environment posed by the contamination.

By 1982, it was clear that the problem was more serious than the companies or the RWQCB had previously thought. Fairchild's monitoring efforts showed that TCE concentrations in several non-drinking water wells exceeded both the federal *Maximum Contaminant Level* (*MCL*) and the *State Action Level* for TCE of 5 parts per billion (ppb). Federal MCLs are enforceable drinking water standards promulgated under the Safe Drinking Water Act. California State Action Levels are ambient water quality goals set by the State Department of Health Services to protect public health.

Tests performed in 1984 showed unexpectedly high concentrations of contaminants (especially TCE) spread over an extensive area. However, the cleanup effort was stalled as companies, the RWQCB, and the community argued about what should be done and who should pay the costs. By 1984, the situation had reached an impasse because the regulators' and companies' perspectives differed, and because the companies could not agree among themselves.

In early 1985, the RWQCB evaluated three possible changes in its strategy. The first was to issue *waste discharge requirements,* which the RWQCB was authorized to do under California laws for managing water quality. Waste discharge requirements would include definitive standards and time schedules that the state attorney general could enforce. A second option was to take direct enforcement action under state water quality laws by issuing a "cease and desist order," which would include a strict time schedule for cleanup. The third option considered by the RWQCB was to refer the case to EPA for cleanup under CERCLA. Each of these actions provided for enforcement through the courts.[42]

The RWQCB proceeded by asking five companies to sign a single consent order requiring them to undertake further investigation of groundwater quality, and to prepare and implement plans for interim containment and cleanup of the entire area. The companies refused to sign. Consequently, the RWQCB issued the companies separate waste discharge requirements, each of which applied only to the individual company's facilities. At the same time, having failed to organize a comprehensive site cleanup, the RWQCB referred the MEW site to EPA for cleanup under CERCLA. At this point, the RWQCB knew that the concentration of contaminants was high and the plume was spread over an extensive area. Referral to EPA appeared to provide the best hope for a timely cleanup.

Site Placed on the National Priorities List

Figure 14.4 provides an overview of the process followed by EPA in implementing remedial action under CERCLA. In the case of the MEW site, the *site discovery* step was completed when the RWQCB referred the site to EPA. Based on results from *preliminary assessment* and *site inspection* activities, EPA placed three of the properties at the MEW site on the National Priorities List (NPL), a list of the most serious hazardous waste sites in the nation. The NPL was developed under CERCLA to direct federal attention to hazardous waste sites ranked according to their need for cleanup. Only sites listed on the NPL can use Superfund money for construction, operation, and maintenance of a remedial action.[43]

In order to determine which sites should be placed on the National Priority List, EPA uses its *hazard ranking system* (HRS), a procedure to evaluate contamination at a site in terms of four contaminant migration pathways: groundwater, surface water, air, and soil. Application of the HRS to a particular site yields a score of between 1 and 100 points. Sites that accumulate more than 28.5 points are considered serious candidates for the NPL. By 1986, about 1200 sites had been placed on the NPL because of the score they were assigned under HRS.[44] While sites on the NPL often become Superfund sites

[42] This paragraph is based on an internal RWQCB memorandom by Thomas J. Berkins, a staff engineer at the RWQCB (re: Fairchild Semiconductor Corporation; Intel Inc.; NEC Electronics, Inc.; Siltec Corporation; Mountain View, Santa Clara County, dated 22 April 1985).

[43] EPA can, however, use Superfund money for many activities at sites not listed on the NPL. For example, the Superfund can support enforcement activities and studies at *unlisted* sites.

[44] This estimate is from Mazmanian and Morell (1992, p. 31).

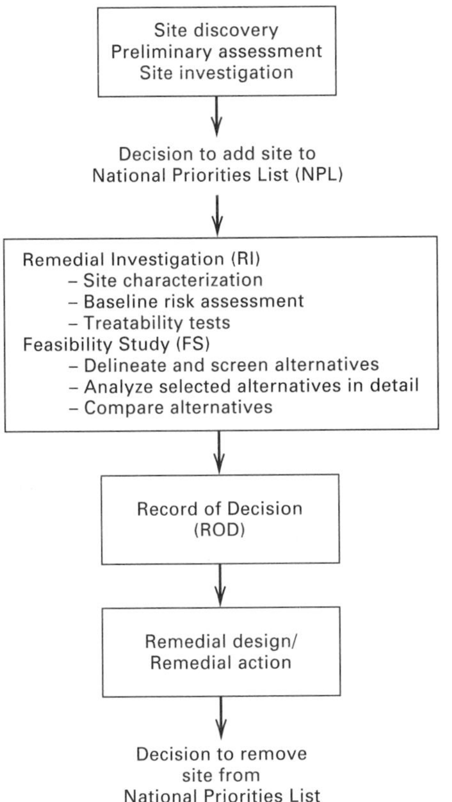

Figure 14.4 Process for implementing CERCLA at a Superfund site.

(and thus capable of being cleaned up using the trust fund), they do not automatically attain this status.

Implementing the Superfund Process at the MEW Site

Using authority granted by the Superfund legislation, EPA appointed a project manager for the entire MEW site and required companies to conduct a *remedial investigation/feasibility study* (RI/FS). Notwithstanding EPA's broad authority to organize action, the existence of numerous potentially responsible parties and deficiencies in information on the extent of contamination made it impossible to effect a rapid cleanup. A massive remedial investigation and feasibility study had to be completed, and remedial actions had to be decided by the EPA before a final cleanup could be designed and implemented. During the RI/FS there was substantial duplication of effort (for example, companies hiring engineering consultants to verify results obtained by others), and the large number of regulators and affected parties complicated negotiations.

Remedial Investigation and Feasibility Study

The RI/FS, which is a standard part of EPA's site remediation process, is used to produce a recommendation for a final cleanup plan. The complexity of the RI/FS process was apparent at the MEW site, where the study lasted from 1985 to 1988. As a result of negotiations, three of the five companies identified as potentially responsible parties (in 1985) at the MEW site entered into an administrative order on consent with EPA to fund and conduct an RI/FS according to a time schedule established by EPA. The parties—Intel, Raytheon, and Fairchild—also agreed to perform an operable-unit feasibility study to evaluate *interim actions,* measures taken to prevent immediate threats to health or the environment while an RI/FS is being conducted. The two other PRPs—NEC and Siltec—did not enter into an agreement with EPA.

Remedial investigations involve three phases (see Figure 14.4). During the first phase, *site characterization,* hundreds of monitoring wells are typically drilled, and soil borings and water samples are collected at close spatial intervals. Laboratory studies, statistical analyses, and computer models are used to characterize the soil stratigraphy and groundwater plume and to define the extent of contamination.

The second phase of a remedial investigation, *baseline risk assessment,* assesses the toxicity of the contaminants present and the extent to which people and the environment will be exposed to them. At the MEW site, the baseline risk assessment concluded that contaminated groundwater used to supply drinking water posed the greatest threat to public health.

Finally, during the third phase of the remedial investigation, engineers conduct *treatability tests,* which involve pilot tests of potentially applicable technologies to see if they would be effective in remediating contaminants at a specific site.

A remedial investigation is accompanied by a feasibility study, which is performed to develop, evaluate, and select remediation alternatives.[45] Technical

[45] Parts of the remedial investigation and feasibility study are carried out simultaneously, and the activities are closely linked.

specialists identify relevant pollution containment or disposal requirements and then identify potential treatment technologies. Some of the alternative technologies are subjected to more detailed studies, and alternatives are compared to each other. At the MEW site, several methods were considered for treating soils, and still other methods were examined for treating groundwater.[46] In general, a number of alternative plans (one of which is "No Action") are presented in the feasibility study report, along with analyses of each and a recommended plan of action. EPA selects its proposed cleanup plan from among these alternatives.

CERCLA requires EPA to submit its proposed remedial action to public comment for a minimum of 30 days. An informal public meeting is also conducted to allow interested parties to express their opinions. In reviewing and responding to public comments, EPA may decide to modify the action that was orginally proposed. Public acceptance of a remedial action is one of several criteria used by EPA in selecting a final plan.

EPA's Record of Decision

EPA announces its final plan in a *Record of Decision* (ROD), which documents the decision of what, if anything, is to be done at a contaminated site. If action is required, EPA's record of decision indicates the technologies to be used and the cleanup standards to be met. In a case involving groundwater contamination, setting cleanup standards generally involves answering questions about current and projected uses of the groundwater, and the feasibility of providing an alternative water supply.[47] Standards and other aspects of a remedial action must be consistent with all "applicable or relevant and appropriate regulations" that govern potential risks to health and the environment posed by contamination at a site.

EPA's Record of Decision for the MEW site, issued in June of 1989, outlined the remediation methods to be used. The ROD included separate cleanup plans for contaminated soils and groundwater. The soil cleanup program relied on the use of *vapor extraction wells,* which pump air containing volatile organic compounds from the soil. Contaminants in the extracted air may (if necessary) be treated with carbon filters, and the extracted air is discharged through a tower. The plan for soil cleanup also included excavating contaminated soil and exposing it to the air.

The groundwater cleanup plan involved a "pump and treat" scheme. Contaminated groundwater is pumped to the surface and then passed through *stripping towers,* which causes contaminants to evaporate. The plan calls for reusing the treated groundwater to the extent feasible and discharging the water that is not reused to local streams.[48] The entire remediation effort was estimated to cost between $49 million and $56 million.[49]

After issuing the ROD for the MEW site, EPA began negotiating with PRPs to determine which parties would implement the various remedial measures contained in the ROD. Negotiations continued for about a year. Although there were more than 20 PRPs at the MEW site, EPA focused its efforts on the same five companies that had been targeted by the RWQCB *and* six others. Of the total of eleven PRPs, nine chose not to settle, and they received Section 106 administrative orders from EPA (see Table 14.1). The two PRPs that settled with EPA—Raytheon and Intel—formalized their agreement by signing a consent decree stipulating actions they would take to clean up the site (see Table 14.2).[50]

The decision as to whether a PRP should enter into an agreement with EPA to clean up a site is complex. Advantages and disadvantages of settling (or not settling) are summarized in general terms in Table 14.3. For example, PRPs that settle with EPA (such as Raytheon and Intel at the MEW site) may

[46] While the RI/FS was being conducted for the MEW site, EPA engaged in further efforts to identify PRPs. Although EPA identified additional PRPs, the process was complicated because a number of companies in the area had released toxic chemicals, and their contaminants had entered the groundwater below their sites to form a *merged plume.* It was difficult to distinguish the sources of pollution in the plume or the amounts contributed by each company.

[47] For more information on the process of setting cleanup standards under CERCLA, see Barth, Hanson, and Shaw (1988).

[48] A permit under the National Pollution Discharge Elimination System would be required for the direct discharges to surface water.

[49] Details of the remediation effort at the MEW site are given in a "fact sheet" for the MEW site issued (in 1989) by the EPA Region 9 Office.

[50] As in the case of a ROD, the process of issuing a consent decree also involves a period of public comment. Public comments and EPA's response to those comments become part of the administrative record documenting EPA's decision making at a Superfund site.

TABLE 14.1 Selected Elements of the Administrative Order Issued by EPA to Nine PRPs at the MEW Site[a]

- Assume responsibility for the long-term operation and maintenance of the areawide cleanup system until the MEW Study Area is cleaned up to levels required in the ROD.
- Investigate the current configuration of the contamination in the groundwater (known as the plumes).
- Install a system of controls to prevent further migration of the contamination.
- Control air emissions to comply with the Bay Area Air Quality Management District and EPA standards.
- Locate any conduits, such as old wells and storm drains, that could allow the further migration of the contamination and close each conduit; or, if closing is not possible, prevent any further contaminant migration through the conduit.
- Conduct further investigations to determine the chemistry of the clean groundwater surrounding the MEW contamination.
- Find and develop options for the reuse of all the water that has been treated by the treatment system.
- In addition, EPA may assess penalties against the companies for failure to comply with the terms of the order.

Source: MEW Study Area Fact Sheet #3, U.S. Environmental Protection Agency, Region 9, San Francisco, CA (May 1991).

[a] The following PRPs received the order: Fairchild Semiconductor Corporation; Schlumberger Technology Corporation; National Semiconductor Corporation; NEC Electronics, Inc.; Siltec Corporation; Sobrato Development Companies; General Instrument Corporation (GIC); Union Carbide Chemicals & Plastics; and Tracor X-Ray, Inc.

TABLE 14.2 Selected Elements of the Consent Decree for Implementation of the ROD at the MEW Site

Terms of the consent decree require Raytheon Company and Intel Corporation to:

- Pay for design and construction of an areawide groundwater cleanup system.
- Pay for 35% of the cost of operating and maintaining the areawide groundwater cleanup system from the end of an "interim period" until cleanup is complete.[a]
- Install a system to control further migration of the contaminated groundwater while the cleanup system is being designed and constructed.
- Control air emissions from air strippers to comply with the Bay Area Air Quality Management District and EPA standards.
- Clean up soil contamination at their respective facilities.
- Install further monitoring systems in the Silva Well area, a deep water well that is now closed.
- Reimburse EPA for past oversight costs.
- Reimburse EPA for the past costs of closing two deep water wells.
- Pay future oversight costs.
- In addition, EPA may assess penalties against the companies for failure to comply with the terms of the agreement.

Source: MEW Study Area Fact Sheet #3, U.S. Environmental Protection Agency, Region 9, San Francisco, CA (May 1991).

[a] The end of the "interim period" is defined as the time at which the costs paid by Raytheon and Intel have balanced the costs paid by PRPs under administrative orders (see Table 14.1) such that "PRPs under the Consent Decree have paid 35%, and the PRPs under the Order have paid 65%." After the interim period, costs are split 35%–65%.

sue PRPs that did not settle (in a *contribution action*) to recover costs the non-settlers avoided by not participating in the cleanup. However, settlers cannot themselves be sued by other PRPs; they are said to have "contribution protection." Consider, also, some advantages for non-settlers. In the MEW case, the nine PRPs that did not settle and were placed under administrative orders retain the right to challenge EPA's remedy if they enter into litigation with EPA. Those PRPs that did not settle and were *not* placed under administrative orders may avoid paying any part of the cleanup. However, that outcome is highly uncertain. Those non-settlers may eventually be placed under administrative orders or be subjected to a contribution action brought by PRPs that participated in the cleanup.

TABLE 14.3 Factors Influencing PRP Decisions on Whether to Settle with EPA in Implementing Elements of a ROD[a]

The Option to Settle

Advantages—Settlors:

- Participate in decisions on details of the work to be performed (for example, scheduling) and whether costs will be divided between PRPs and EPA.[b]
- Obtain *contribution protection,* which prevents suits by non-settlors.
- Obtain a *covenant not to sue* from the government, which, for example, precludes suits for additional work on all matters covered by the settlement.[c]

Disadvantages—Settlors:

- Agree not to challenge the selected remedy in court, and to continue remediation activities until agreed upon cleanup standards are met.
- Agree (usually) to pay some or all of EPA's past and future oversight costs.
- Incur the burden of filing a *contribution action* against non-settlors to recover costs other PRPs have avoided by not settling.

The Option Not to Settle

Advantages—Non-settlors:

- Enjoy the possibility that the government will neither issue administrative orders under Section 106, nor sue to recover costs under Section 107.
- Have the option, if placed under Section 106 administrative orders, of challenging in court the details of EPA's remedy, including its cleanup standards.

Disadvantages—Non-settlors:

- Risk being named in a Section 106 administrative order, which may include substantial penalties for noncompliance with the order.
- Risk being sued by the government in a Section 107 action to recover EPA's past costs.[d]
- Risk being sued by settlors who bring contribution actions to recover their costs.
- Lose the ability to influence implementation details associated with the remedial action, such as scheduling when work is to be performed.

[a] Based on unpublished notes prepared by Patti Collins, U.S. Environmental Protection Agency, Region 9, San Francisco, CA (Spring 1994), for Stanford University, Civil Engineering 266, *Environmental Policy Design and Implementation.*

[b] Costs are divided among PRPs and EPA under a provision of CERCLA (as amended) that permits "mixed funding." Congress allows Superfund monies to be used to cover costs attributable to parties that are unknown or insolvent. In some circumstances, EPA may also use the fund to cover costs of known, solvent PRPs who refuse to settle; EPA has the option of bringing suit under Section 107 to recover costs from those parties.

[c] In general, when EPA grants a covenant not to sue, the covenant is accompanied by a provision (termed a "reopener") that allows EPA to sue the settlor if information gathered in the future demonstrates that the remedy in the settlement no longer protects public health or the environment. Both the covenant not to sue and contribution protection relate only to remediation issues covered in a settlement document.

[d] In a typical circumstance where this occurs, EPA chooses not to issue a Section 106 order requiring implementation of its selected remedial action. Instead, EPA preforms the work itself and sues under Section 107 to recover past costs. Non-settlors run the risk that these costs may be relatively high, since EPA would have less incentive to control costs than would a PRP performing the work itself.

Complexity of the Superfund Process

During the five years that the MEW site was under RWQCB enforcement, the cleanup effort moved slowly. However, even with the remediation process guided by EPA's regulations, considerable time passed before EPA was able to issue its Record of Decision specifying the terms of the selected remedy. Why was the process so lengthy? What factors contribute to delays at complex Superfund sites such as the MEW site?

Multiple Parties Involved in Superfund Cleanups

One reason cleanup of toxic waste at places such as the MEW site is so costly and lengthy is the large number of parties involved. Differences in motivation among participants in the cleanup process contribute to difficulties in determining liability and negotiating details of the work to be performed. The main groups of actors in a Superfund cleanup as complex as the MEW site include the following:

- Regulatory agencies, including EPA project managers, and attorneys for both EPA and the Department of Justice.
- PRPs, including project managers, in-house attorneys, and in-house technical staff.
- Local residents, local governments, and environmental advocacy groups.
- Technical consultants for the PRPs and EPA.
- Outside attorneys for the PRPs.

Relationships between PRPs and regulators in a Superfund remediation are often adversarial. Regulatory agencies tend to view sites from a broad perspective and seek to protect human health and restore the quality of the environment. In contrast, PRPs try to protect their own interests by attempting to reduce their liability to pay for what they often view as needlessly costly cleanups.

Even within a regulatory agency or a PRP that is an organization, technical consultants, attorneys, and project managers tend to view differently what is in the organization's best interests. Different attitudes and interests are often held by each group.[51]

[51] The following observations on attitudes and interests of technical consultants, attorneys, and project managers are based on an interview with Patti Collins, one of EPA's project managers for the MEW site, February 5, 1992.

Technical consultants perform investigative work, prepare technical documents, and act as liaisons between PRPs and regulatory agencies. EPA hires its own technical consultants. Most companies at the MEW site hired technical consultants to manage investigations on their properties. Although technical consultants to PRPs work in their clients' interests to keep costs down, manage an efficient cleanup, and appease regulators, the consultants themselves also have their own perspectives. For example, environmental engineering consultants have differing views on the utility of certain analysis procedures and treatment technologies. In the case of the MEW site, differences among technical consultants were apparent in disputes (described below) over the value of mathematical models used to describe how contaminants move in groundwater.

Attorneys representing PRPs are responsible for minimizing the legal liability of their clients, and they are sometimes less concerned with efficient cleanup than with making sure that proper procedures are followed. In the context of the MEW site, company attorneys (both in-house and external) prepared and reviewed documents, encouraged careful documentation of all actions, and advised company project managers about what they should or shouldn't say in communications with regulatory agencies.

Because PRPs have many concerns besides the legal and technical aspects of remediating contamination on their properties, company managers in charge of cleanup projects do not always agree with their attorneys and technical consultants. Project managers must consider a variety of factors, including the potential costs of cleanup, the company's financial status, other projects in progress, the company's reputation, and the nature of the surrounding community. At the MEW site (and at many other Superfund sites), differences in the roles, incentives, and personalities of project managers, technical consultants, and attorneys added an element of tension to the inherently adversarial relationship between EPA and the PRPs.

Technical and Regulatory Complexity

Another factor contributing to the complexity of Superfund site cleanups is the technical and regulatory complexity associated with site remediation and the selection of cleanup goals. Typically, there is a shortage of data describing a site, the subsurface environment is complex, it is difficult to predict contaminant behavior, and there are technical disagreements over which remedial actions are best.

Consider, for example, controversies that surrounded the use of models to predict contaminant flow at the MEW site. Mathematical models were used to determine the present and historical groundwater flows in the area, and to describe the behavior of private wells as open conduits between aquifers. Models were also used to "back-cast" contaminant sources. In back-casting, data describing a contaminant plume is analyzed to determine which sources contributed to the plume and what amounts were contributed. Companies objected to modeling study outcomes based on claims that (1) the data were incomplete, (2) the parameters in the mathematical models were estimated incorrectly, and (3) modeling was inappropriate for a set of aquifers as complex as those at the MEW site. The companies viewed the modeling effort as a costly procedure of dubious value that was delaying remedial action; "no technical consensus" was reached regarding the use of mathematical modeling.[52]

The use of slurry walls as an interim remedial action at the MEW site further illustrates how technical disagreements complicate the cleanup process. Raytheon and Fairchild Semiconductor wanted to construct slurry walls to constrain the movement of contaminants below their property (see Figure 14.5). NEC Electronics objected to Raytheon's proposal, arguing that slurry walls might divert greater contamination beneath NEC's property and make it harder to determine who was responsible for the contamination. Similarly, Siltec opposed Fairchild's proposal to construct slurry walls, fearing that contaminated groundwater would be diverted toward Siltec's property. Altering the flow of groundwater had serious implications for the PRPs, because at the time of slurry wall construction, the sources of groundwater contamination had not been fully determined. Any change in existing conditions, including the use of monitoring and extraction wells, could make it more difficult to pinpoint contaminant sources. In this instance, EPA dismissed objections and approved construction of the slurry walls.

[52] Jim McClure, principal engineer for the MEW site, Harding Lawson Associates, Novato, California, telephone interview conducted by Jean Chen of Stanford University, February 11, 1992.

The process of selecting cleanup standards adds an additional layer of difficulty. Most complex hazardous waste sites are subject to overlapping regulations at federal, state, and local levels. Differences among these regulations can lead to disputes over cleanup standards. Remedial actions at Superfund sites must comply with all *applicable or relevant and appropriate requirements* (ARARs) and consider other pertinent federal criteria, advisories, and guidelines, and state standards.[53] Considerable effort is often required to determine which regulations are relevant and appropriate ones.

At the MEW site, ARARs were a source of confusion for the companies and complicated the EPA's determination of "how clean is clean?" For example, the RWQCB disagreed with the companies' applications of California laws in the RI/FS. The RWQCB cited problems with interpretations of (1) the Basin Plan, which governed potential uses of water; (2) the state's procedure to identify water quality objectives for the cleanup of polluted groundwater; and (3) the definition of "best available technology" requirements. EPA acted as a liaison between state regulators and the companies. Disagreements over the language of the ROD added new complications: while the original ROD specified cleanup "goals," EPA later gave these "goals" the status of "final cleanup standards." This difference in terminology had a bearing on whether an ARAR waiver could be invoked, and it also was a factor leading to a lawsuit against EPA by Fairchild Semiconductor, one of the PRPs.[54]

As demonstrated at the MEW site, the large number of parties involved combined with the regulatory and technological complexity of the remediation process can confound what is inherently a difficult problem. Unless relationships among the parties are improved, cleanup standards are lowered, hazardous

[53] Under limited circumstances, noncompliance with an ARAR can be justified. ARAR waivers are described in Sheridan (1987, p. 23).

[54] Fairchild Semiconductor believed that in changing the "goals" to "standards," EPA nullified the previous position that final cleanup levels would depend on technical practicability. Although Fairchild was one of the three PRPs that participated in preparing the RI/FS, the changes EPA made in amending the ROD led Fairchild to refuse to sign the consent decree to implement the remedial action. Both EPA and Fairchild incurred substantial litigation costs, and, in the end, the federal courts refused to review EPA's decision on the matter.

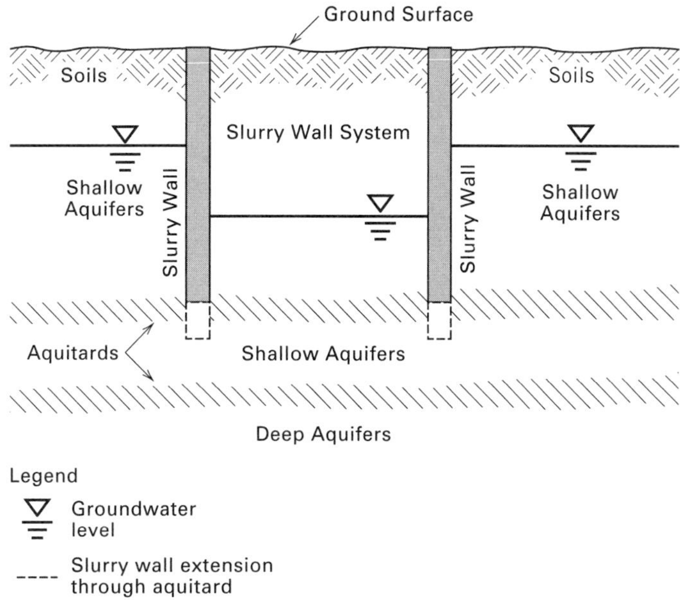

Figure 14.5 Slurry wall system at the MEW site. Source: Fact sheet for MEW site prepared by U.S. Environmental Protection Agency, Region 9, San Francisco, CA.

waste regulations are simplified, or remediation technologies become less expensive, the remediation process at Superfund sites is unlikely to become easier or faster.

Is a Lengthy Cleanup Process Unavoidable?

The high cost and long delays in cleaning up Superfund sites are a continuing concern. Are there prospects for improving the remediation process? One possible direction involves changing the liability rules. Many question whether the CERCLA liability system provides the most effective way of funding cleanups. Others have advocated increased use of *alternative dispute resolution* methods such as mediation and arbitration.[55] Still others have suggested that cleanup standards should be relaxed because current standards sometimes force actions that are at the limit of what technology can accomplish, and involve costs that far exceed estimated monetary benefits of reduced risk to health and the environment.

Notwithstanding continuing efforts to reduce costs and decrease the time for cleanups, there are technological limitations to speeding up remediation of contaminated soils and groundwater. Even under an efficiently managed cleanup process, groundwater behavior is difficult to predict, and its movement is slow. Cleanups can take decades, and the results of pumping and treating contaminated groundwater and soils are often uncertain. Experts in hazardous waste site remediation argue that the public has unrealistic expectations about what can be accomplished by a cleanup effort. For example, McCarty (1990, p. 51) argues that "regulators, policy makers and the public" need to understand the "technical difficulties, uncertainties, required time and costs for effective remediation" and thereby develop "more realistic remediation goals and expectations."

In addition to constraints imposed by technical factors, the decision process itself places limits on the efficiency and speed of cleanups. The RI/FS procedure that EPA designed is responsive to the congressional mandate set out in CERCLA. But that process is complex and its implementation can take years, especially when many PRPs are involved. Experience at the MEW site is illustrative. EPA stepped in during 1985 and it took until 1988 to complete the remedial

[55] Alternative dispute resolution is discussed in Chapter 18.

investigation and issue a draft of the feasibility study report. It was another year before EPA's regional administrator signed the ROD (in June 1989) outlining the required cleanup methods. And it took another few years to issue the Section 106 administrative orders (which became effective in December 1990), and the consent decree with Raytheon and Intel (which was made available for public review in May 1991).

The time involved to complete studies and reach agreements at the MEW site are not out of the ordinary for a site with numerous PRPs. Although estimates vary, a typical time line for remediation of a Superfund site involves four years before EPA's remedial action begins to be implemented.[56]

Can the remediation process be made shorter? The CERCLA liability scheme has been at the center of debate on how to improve the Superfund site remediation process. But is changing CERCLA's liability system the key to getting timely results? Consider the following hypothetical circumstance: EPA and other parties contributing funds to the MEW cleanup each believes its cost burden is fair. Determining who pays for cleanup is not at issue. Even in this hypothetical setting, many problems remain concerning how scientific and engineering information is gathered and evaluated, what cleanup standards are appropriate, and which technical remedies are feasible. Arguably, the length and contentiousness of the Superfund remediation process is due more to technical matters than to deficiencies in CERCLA's liability scheme.

POLLUTION PREVENTION AS A WASTE MANAGEMENT STRATEGY

Rigorous hazardous waste laws, such as RCRA and CERCLA, along with demanding requirements in the control of toxic wastewater and air discharges, have led many companies to adopt *pollution prevention* as a keystone in their efforts to manage wastes.[57] Although there is no single, uniformly adopted definition of pollution prevention, Table 14.4 illustrates common usage of the term.[58] The table presents a hierarchy of measures with *waste source reduction* at the top level. Many firms aim to minimize the quantity and strength of waste as a way to reduce costs of meeting waste discharge requirements. The next two levels down in the hierarchy include *materials recycling and reuse,* with the preferred position given to recycling that takes place at the site where the waste is generated. The preferred form of on-site recycling involves a "closed loop," where materials are recovered from intermediate waste streams and returned for reuse in the same production process. Off-site recycling has a lower position because of the need to transport the materials, and uncertainties in markets for recycled materials.[59] In addition, off-site

TABLE 14.4 Hierarchy of Pollution Prevention Measures[a]

- *Waste source reduction.* Minimize amount and harmfulness of wastes by
  - improving operating procedures
  - changing inputs
  - redesigning production processes
  - reformulating products
- *Onsite recycling and reuse.* Reduce final waste discharges by processing intermediate wastes to recover materials valuable for use as process inputs or for sale to others.
- *Off-site recycling and reuse.* Transport wastes off-site for processing to recover materials valuable for use as process inputs or for sale to others.
- *End-of-pipe treatment.* Use chemical, physical, and biological processes and incineration to transform wastes into less harmful substances.

[a] Terminology is not yet standardized, and some analysts view only source reduction and on-site recycling and reuse as pollution prevention; see, for example, Hirshhorn and Oldenburg (1991, p. 40).

[56] This estimate is from Helman and Hawkins (1988, p. 101).

[57] Scott R. Leopold contributed to the initial draft of this section while he was a student at Stanford University.

[58] The term *eco-efficiency* is sometimes used as a synonym for pollution prevention.

[59] For an analysis of why off-site recycling is low in the hierarchy of pollution prevention measures, see Hirschhorn and Oldenburg (1991, pp. 47–51).

recycling facilities can become major sources of wastes themselves. Traditional *end-of-pipe treatment* is at the bottom level of the hierarchy.

Illustrative Pollution Prevention Measures

The literature on pollution prevention is vast, and many examples exist to illustrate each of the items in Table 14.4.[60] Because waste source reduction is arguably the least familiar of the main categories in the table, a few examples are given. Wastes can be reduced using *improved operating procedures,* including efforts to conserve water and energy, to maintain equipment (e.g., by fixing leaks from valves and seals), and to reduce inventories of input chemicals that have limited shelf life. Chemicals that expire because they remain in inventory too long become wastes that must be disposed of. Other examples of improved operating procedures are materials handling methods that minimize spills of hazardous chemicals.

The potential for reducing waste by making simple changes in operations is illustrated by a plumbing equipment manufacturer that modified its product quality control routine.[61] The company's original practice was to inspect metal parts after they had been electroplated to improve corrosion resistance and appearance. This meant that all chemicals used to plate defective parts were wasted and generated unnecessary toxic discharges. By inspecting parts *before* electroplating rather than afterward, the company made a substantial reduction in the hazardous waste it generated.

Waste source reduction via *input change* includes efforts to use inputs that pose fewer problems when they enter waste streams. An example involves a paper company that was required by U.S. Environmental Protection Agency regulations to diminish the concentration of polychlorinated biphenyls (PCBs) in its wastewater discharge. The company examined two treatment schemes and rejected both. Instead of relying on treatment, the company got rid of the PCBs by changing the mix of recycled office paper used as inputs to its production process. By eliminating the recycled paper containing PCBs (especially carbonless copy paper), the firm cut its discharge of PCBs to zero. It was cheaper for the firm to change its inputs than to use either of the treatment schemes[62]

Experience at an electroplating facility illustrates the *redesign of production processes* to cut back the amount of waste generated.[63] In electroplating, metal parts on racks (or in barrels) are submerged into plating tanks where they are coated with other metals, such as zinc and chromium. When a rack is lifted out of a plating tank it typically "drags out" some of the plating solution with it. The electroplating company redesigned its operations to include a drag-out tank where the rack of parts would be set very briefly before being placed in a rinse tank. The new drag-out tank provided two advantages. First, chemicals dragged out from the plating solution could be collected in the drag-out tank and returned to the plating tank. Second, chemicals used in the rinse tank would last longer because there was less contamination by impurities carried in from the plating tank. The overall effect was that the company generated less hazardous waste and saved on the amount of chemicals used.

The final category of waste source reduction measures in Table 14.4, *product reformulation,* is illustrated by a paper company's decision to manufacture the same mix of bond paper, paper towels, and other products, but with a reduced level of brightness.[64] Producing white paper of high brightness requires more chemicals, water, and energy than producing an unbleached paper that is otherwise similar in quality. By making paper products with lower brightness, the discharge of some wastes can be diminished substantially.

[60] For an introduction to the literature, see Fischer and Schot (1993), and Hirschhorn and Oldenburg (1991).

[61] This example is from Hirschhorn and Oldenburg (1991, p. 77).

[62] The paper company example is from Bartos (1979). Another instance of changing inputs as an alternative to end-of-pipe treatment involves electric power production using coal-fired power plants. By shifting from high-sulfur coal to low-sulfur coal, electric utilities have been able to generate the same quantities of electricity with lower discharges of sulfur oxides. Other cases in which inputs were modified as a pollution reduction strategy are given by Kneese and Bower (1979); more generally, see the numerous case examples given by Hirshhorn and Oldenburg (1991).

[63] The electroplating example is from Massachusetts Institute of Technology (1991).

[64] The example involving product reformulation is from Kneese and Bower (1979).

Regulations Motivate Pollution Prevention

A linkage exists between government regulations and corporate decisions to adopt pollution prevention. Some companies see pollution prevention as a strategy for complying with regulations in a cost-effective manner. Other companies are pro-active and view pollution prevention as a means to cut costs and risks by reducing pollution below levels required by regulations. Some companies have even gone as far as adopting the goal of eliminating all waste discharges.

Table 14.5 lists some regulatory programs and statutes that are linked to the surge in corporate interest in pollution prevention within the United States. Two items on the list have been discussed in previous sections: liability for future cleanup costs under CERCLA, and cost of compliance with regulations issued under RCRA. The complex regulatory maze created by RCRA is not the only one faced by U.S. industries. As detailed in previous chapters, firms also face stringent cleanup requirements under federal air and water pollution control laws. And penalties for violating environmental laws include both substantial fines and the possibility of imprisonment for corporate staff. Collectively, these regulations and penalties for noncompliance have motivated many companies to embrace waste source reduction as a way of avoiding the paperwork, time, and expense to satisfy waste management regulations. In addition, waste source reduction provides a way for companies to buffer themselves against the uncertainties associated with satisfying regulations that are subject to continual modification and reinterpretation by environmental agencies and the courts.

TABLE 14.5 Examples of Federal Laws and Programs Encouraging Source Reduction

- CERCLA's liability provisions
- Federal laws concerning air and water pollution control and hazardous waste management
- RCRA's requirement for certifying that waste reduction programs are in place
- EPCRA's Toxic Release Inventory
- Pollution Prevention Act of 1990

Table 14.5 also lists three measures passed by Congress that have fostered pollution prevention.[65] One measure is a requirement under RCRA for generators of hazardous waste to certify (on shipping manifests) that a program is in place "to reduce the volume and toxicity of waste generated" where economically practicable (see Figure 14.1).

A second regulatory measure that has led firms to prevent pollution is the *Toxic Release Inventory* (TRI), which EPA issues annually under the Emergency Planning and Community Right to Know Act (EPCRA) of 1986.[66] Each year, firms releasing more than minimal quantities of one or more of hundreds of toxic chemicals listed by EPA must report those releases to EPA. The annual Toxic Release Inventory is made available as a written report and in digital form (on a CD-ROM), and this has provided governments, interested citizens, and environmental groups with details of toxic discharges by individual companies.

Soon after the first Toxic Release Inventory was issued in 1988, environmental groups publicized information on which firms had the largest releases of toxics. The resulting adverse publicity led many companies to improve their ranking in the TRI by reducing discharges of toxics.[67] Public pressure, which is often heightened by information on toxic releases, has been a factor leading some companies to take a pro-active stance regarding pollution prevention. The TRI can work in favor of companies, since improvements in rankings among companies listed in the TRI are a source of positive publicity and allow firms to gain the public's confidence.

---

[65] The list of government programs in Table 14.5 is illustrative, not comprehensive. There have been numerous other government initiatives that encourage pollution prevention. For example, EPA's Green Lights Program provides technical assistance to firms that voluntarily choose to modify lighting as a means of reducing use of electricity.

[66] Although EPCRA was created as Title III of the Superfund Amendments and Reauthorization Act, it is a separate statute that is independent of the Superfund program. In addition to mandating the Toxic Release Inventory, EPCRA requires firms to prepare emergency response plans, and to notify state and local authorities in the event of an emergency release of regulated substances.

[67] For example, in 1988, the Monsanto Corporation pledged a 90% reduction of releases of toxic pollutants from its facilities following publication of the TRI (Novick, Stever, and Mellon, 1995, pp. 13–209).

A third federal mandate fostering pollution prevention is the Pollution Prevention Act of 1990. The act directs EPA to establish a pollution prevention office, and to compile and disseminate information on techniques for waste source reduction. In addition, the act requires companies to report annually to EPA, providing details of their waste minimization and recycling plans when they report their toxic wastes under EPCRA.[68] Several states have also enacted laws that focus on pollution prevention. For example, Oregon's Toxics Use Reduction and Hazardous Waste Reduction Act requires companies to develop and implement pollution reduction plans.[69]

Cleaner Production Can Increase Profits

The items listed in Table 14.5 emphasize regulation as a factor motivating pollution prevention by firms. However, industry has economic incentives to pursue pollution prevention that extend beyond reducing the costs of compliance with regulations. Many companies have found that pollution prevention makes financial sense even in the absence of regulations. Firms that are proactive about preventing pollution often reexamine their production operations and discover better ways to carry out their work. One of the best-known programs, called Pollution Prevention Pays, is run on a worldwide company basis by the Minnesota Mining and Manufacturing Company (3M). The company has reported annual cost savings in excess of $400 million (worldwide) as a result of savings in manufacturing costs as well as savings in the costs of complying with regulations.[70]

Recently firms have discovered that pollution prevention is only one way to enhance profits while responding to environmental concerns. Many firms throughout the world are discovering the potential profit from *cleaner production:* "the continuous application of an integrated preventive environmental strategy to processes and products to reduce risks to humans and the environment."[71] Pollution prevention focuses on the *process* of production. The cleaner production concept is broader, because it targets clean *products* as well as clean production processes.

The potential to profit by manufacturing clean products is tied to increased consumer demand for environmentally-friendly products. However, the formation of *green markets* (that is, markets for environmentally-sound products) has been stymied because of misleading advertising and its consequence: consumer confusion. Although some countries (for example, Germany) have reduced consumer confusion by implementing nation-wide, voluntary standards for environmental labeling, a similar labeling standard has not been established fully in the United States.[72] With the expansion of international trade, interest in international standards for environmental labeling has increased. The European Union has responded by developing an "eco-labeling" program. However, no universally accepted environmental labeling standards have yet been established.[73]

Industry's interest in cleaner production is further manifested by the emergence, in 1996, of a series of voluntary environmental management standards issued by the International Standards Organization (ISO). The ISO is "a worldwide federation founded in 1946 to promote the development of global standards for manufacturing, trade and communications. . . . While ISO standards are voluntary, widespread adoption and trade imperatives have a way of making them virtually mandatory" (Bird 1995, p. 95).

The new series of environmental standards (ISO 14000) has a centerpiece: ISO 14001, the specifications used to determine if an organization

[68] Under the Pollution Prevention Act of 1990, EPA must publish a biennial report to Congress that contains an analysis of reductions in TRI wastes achieved by pollution prevention measures taken by industry.

[69] For more on Oregon's program, see Hansen (1989).

[70] For more on the 3M program, see Buchholz (1993, pp. 375–79). He also describes a similar program by Dow Chemical U.S.A. entitled "Waste Reduction Always Pays."

[71] This definition is from the introduction to a special "Cleaner Production" issue of *Industry and Environment* (Vol. 17, No. 4, p. 4), a publication of the United Nations Environment Programme (UNEP). The cleaner production concept blossomed in the early 1990s. It has been heavily promoted by UNEP, which launched its Cleaner Production Programme in 1990.

[72] According to Perrone (1996, p. 420) more than 25 countries had environmental labeling programs as of early 1996. For information on attempts to develop environmental labels in the United States, see Bick and Howett (1992). Details on environmental labels used in Europe are given by the Organization for Economic Cooperation and Development (1991).

[73] Work of the International Standards Organization (described below) may lead to an international standard for environmental labels.

has an acceptable *environmental management system* (EMS). As of 1996, ISO 14001 was the only standard in the ISO 14000 series that contained requirements that may be objectively audited for certification by an external body.[74] The ISO 14001 specifications concern a wide range of subjects, including an organization's environmental policies, its environmental objectives and targets, and its capabilities to monitor its own environmental performance. Organizations that can demonstrate to external auditors that they meet the specifications in ISO 14001 receive a certificate and are said to be "certified to ISO 14001."

The focus of ISO 14001 is on "ensuring that company systems and actions are in *conformance* with established company environmental policies and objectives" Cascio (1996, p. 31). When a firm receives ISO 14001 certification, enhanced environmental quality is not guaranteed. However, environmental quality may improve as companies generate and digest the information they produce to comply with ISO 14001. For example, if a firm's internal audits reveal that failure to satisfy environmental requirements exposes the firm to heavy penalties, the company may choose to improve its environmental performance.

A number of additional standards in the ISO 14000 series, termed "guidance standards," are intended to help organizations develop their environmental management systems. Some ISO 14000 guidance standards concern the EMS per se, while others are to help organizations evaluate the environmental characteristics of their products. An example of a guidance standard concerning a company's EMS is the standard for *environmental management system audits,* which are audits companies undertake to assess their compliance with their own environmental objectives and targets and other aspects of an EMS. Environmental management system audits are different from *environmental compliance audits;* the latter are internal audits that companies perform to determine their compliance with environmental laws and regulations.

An example of an ISO 14000 guidance standard dealing with product characteristics is the standard for *life cycle assessments.* In a *life cycle assessment,* designers and engineers consider the environmental impacts associated with a product. The assessment considers all stages of production, beginning with the extraction of raw materials and the manufacture of inputs. Life cycle assessments also consider production, distribution, consumption, maintenance and reuse of the product, as well as pollution prevention, waste treatment, and facility decommissioning. Another example of a product-related ISO 14000 standard is the guidance standard for environmental labels. Eventually, ISO's labeling standard may lead to a harmonization of environmental labeling programs in different countries.[75]

Several factors motivate firms to rely on ISO 14000 standards and obtain ISO 14001 certification.[76]

- Opportunities to discover cost savings associated with pollution prevention;
- Chances to reduce exposure to penalties for noncompliance with environmental regulations;
- Ability to sell products in countries that have environmental labeling programs based on ISO standards;
- Potential profits from selling products and services to firms and governments that require suppliers to be certified to ISO 14001; and
- Public relations values, since ISO 14001 certification provides a means for companies to demonstrate their commitment to environmental protection.

Although the impact of the ISO 14000 standards will be impossible to gauge until companies have a chance to react, previous experience with ISO quality standards suggests that effects on both industry and the environment may be substantial.[77]

A number of factors—stringent regulations, opportunities for cost savings, emerging green markets, the new ISO environmental management standards, and public demands for environmental improvement—have combined to make pollution prevention an integral part of many corporations. Leaders of many corporations have encouraged their staffs to

[74] This discussion of the ISO standards is based on Cascio (1996).

[75] According to Cascio (1996, p. 18), the first of several ISO *guidance documents* related to labeling is scheduled to be finalized in 1997.

[76] For information on additional benefits of ISO 14001 certification, see Cascio (1996, pp. 10–11).

[77] For more on the process of developing the ISO 14000 standards and the influence of earlier ISO quality standards, see Crognale (1995, p. 13)

"design for environment," and it is anticipated that enormous numbers of firms will adopt the voluntary standards in the ISO 14000 series.[78] The challenge to any company committed to improving its environmental performance is to transform corporate environmental goals and policies into improvements in day-to-day operations by staff at all levels of the organization.

Key Concepts and Terms

Tracking Waste and Regulating Waste Management Facilities
- Generators regulated by RCRA
- Characteristic vs. listed waste
- Mixture and derived-from rules
- Manifest system
- Transporters regulated by RCRA
- Exception reports
- TSD facilities
- Land disposal restrictions
- Best demonstrated available technology (BDAT)

Environmental Management Based on Liability Rules
- CERCLA and SARA
- Hazardous Substances Response Trust Fund
- Potentially responsible parties (PRPs)
- Removal vs. remediation
- Retroactive liability
- Strict liability
- Joint and several liability
- Fund start vs. enforcement start
- Consent order vs. consent decree
- Contribution action
- Equitable factors
- *de minimis* settlements
- Equity vs. efficiency in implementing CERCLA
- Internalizing external costs
- Transactions costs
- Brownfields vs. greenfields
- EPA's brownfields initiative
- EPA's CERCLA implementation strategies

Cleanup of a Superfund Site
- Plumes of contaminants
- Monitoring vs. extraction wells
- Aquifers and aquitards
- National Priorities List (NPL)
- Hazard ranking system
- Remedial investigation/feasibility study (RI/FS)
- Interim actions
- Remedial action
- Record of Decision (ROD)
- Contribution action
- Contribution protection
- Parties involved in Superfund cleanups
- Technical and regulatory complexity
- ARARs
- Negotiating settlements at Superfund sites
- Section 106 administrative orders
- Section 107 cost recovery suits

Pollution Prevention as a Waste Management Strategy
- Motivation for pollution prevention
- Source reduction strategies
- Redesign of products and processes
- Recycling and reuse
- On-site vs. off-site recycling
- Toxic Release Inventory (TRI)
- EPCRA
- Cleaner production
- Green markets
- Environmental labeling
- ISO 14000 standards
- ISO 14001 certification
- Life cycle assessments
- Environmental management system audits
- Environmental compliance audits

Discussion Questions

14-1 In Figure 14.2 there are two manifests: one covers the transport from Ajax Metal Processing to Safe Treatment, Inc., and a second covers the shipment from Safe Treatment, Inc., to Clean Landfill. The waste in the second transport step is derived from treating the original waste from Ajax Metal Processing and similar wastes from other generators. In this example, Mac's Hauling chose the

[78] For more on design for environment, including its link to what has been termed "industrial ecology," see Allenby and Richards (1994).

treatment facility and Quick Transport Company selected the landfill.

Suppose Clean Landfill became a site on the National Priorities List. Under the circumstances, which "generator(s)" and other waste handlers in Figure 14.2 might be liable (under CERCLA) for remediation efforts at Clean Landfill? How can the manifest system enable the responsible environmental agency establish liability for cleanup based on joint and several liability.

14-2 Imagine that you are advising a developing country that wants to implement a cradle-to-grave hazardous waste tracking system. Suppose also that computers are not widely used by the environmental agency that would implement the proposed program. Would you recommend the manifest system used in the United States? What changes in the manifest system would you suggest to keep the environmental agency from being overwhelmed by the information processing burdens imposed by the proposed tracking system?

14-3 Describe the intent of the land disposal regulations under RCRA. Who is responsible for implementing these regulations? Can you think of a simpler way to satisfy the intent of Congress to require treatment of hazardous waste prior to disposal on land?

14-4 CERCLA enables EPA to carry out its obligations by conducting cleanups and then suing PRPs later to recover costs. What practical difficulties prevent EPA from using this approach in *all* cases?

14-5 In designing policies for the Superfund program, EPA headquarters has given serious consideration to at least two broad approaches: fund start and enforcement start. One can imagine "mixed strategies" that involve combinations of the two. Propose a Superfund implementation strategy to EPA headquarters. Your proposal can be one of the two strategies mentioned, or an approach of your own invention. In responding, point out how your proposal eliminates or reduces disadvantages of alternative strategies.

14-6 You are asked to advise an EPA regional office investigating a Superfund site with the following characteristics: high projected cleanup costs; large, profitable PRPs, all of whom are uncooperative; and no special need for a speedy remediation. Which of Church and Nakamura's three implementation strategies—accommodation, public works, or prosecution—would you recommend to EPA in this circumstance?

14-7 Suppose you represent a chemical industry group responsible for lobbying Congress to change the Superfund program. What aspects of the program would you want Congress to modify and why? How would your answer change if you lobbied on behalf of the insurance industry? How would it change if you represented an environmental nongovernmental organization?

14-8 Imagine that you are hired to advise the president of a large, economically healthy company on general steps that could be taken to enhance the company's environmental performance. Prepare lists of possible actions and criteria you would use to evaluate those actions. Based on the analysis used in preparing your lists, how would you advise the company president?

14-9 Bick and Howett (1992, p. 312) report that "[i]n one survey, 47 percent of American consumers dismissed all environmental claims [on product labels] as 'mere gimmickry'." If you were a consultant to a major corporation intent on penetrating green markets, how would you advise the company to respond to consumer distrust and confusion over green labeling? Would you recommend that corporate officials lobby state legislators to set up a state-level program for environmental product labeling? Would you advocate uniform, national government regulations on labeling? What would you say about private efforts to develop seals of approval, such as Green Cross and Green Seal in the United States?

14-10 Environmental markets include markets for goods and services related to waste management, air and water pollution control, soil decontamination, noise control, and so on. Expenditures in U.S. environmental markets during the period from 1970 to 1990 were increasing at a faster rate than the GNP during the same period. Markets for pollution control and waste management systems have also increased in Europe and in the newly industrialized countries of Asia.

Suppose you are asked to testify before a congressional committee on steps the government might take to improve the competitive position of U.S. environmental technology companies in growing overseas markets. Do you think there is any role for government in this arena? What points would you emphasize in your testimony?

14-11 The 10 *Valdez Principles* are used by investors to gauge the degree to which major corporations are behaving responsibly toward the environment. The principles call for reduction and disposal of waste, the wise use of energy, and the marketing of safe products, among other things. Companies are invited to agree to some or all of the principles on a voluntary basis. The last of the Valdez principles concerns environmental audits:

> We will conduct and make public an annual self-evaluation of our progress in implementing these [Valdez] principles and in complying with all applicable laws and regulations throughout our worldwide operations. We will work toward the timely creation of independent environmental audit procedures which we will complete annually and make available to the public.[79]

As of 1992, no major corporation had signed off on the complete set of Valdez Principles, and, according to White (1992, p. 50), Principle 10 "causes trepidation." Why would corporations be reluctant to adopt this principle on environmental audits?

14-12 Explain the following statement: Pollution prevention is a component of cleaner production; the two terms are not synonyms.

14-13 What does it mean for a company to be certified to ISO 14001? What are the advantages of certification to large companies that compete in international markets? How might small- and medium-sized companies benefit from being certified to ISO 14001?

[79] From the Valdez Principles as reproduced in White (1992, p. 50).

References

Acton, J. P., and L. S. Dixon. 1992. *Superfund and Transaction Costs: The Experiences of Insurers and Very Large Industrial Firms.* Report R-4132-ICJ. Santa Monica, CA: Rand—The Institute for Civil Justice.

Allenby, B. R. and D. J. Richards, eds. *The Greening of Industrial Ecosystems.* Washington, D.C.: National Academy Press.

Barnett, H. C. 1994. *Toxic Debts and the Superfund Dilemma.* Chapel Hill: University of North Carolina Press.

Barth, E. F., III, W. Hanson, and E. A. Shaw. 1988. "Establishing and meeting groundwater protection goals in the Superfund Program." In *Hazardous waste site management: water quality issues: report on a colloquium sponsored by the Water Science and Technology Board. Feb. 19–20, 1987.* Washington, DC: National Academy Press.

Bartos, M. J., Jr. 1979. EPA Goes to BAT Against Toxic Industrial Wastewater. *Civil Engineering* 51(9): 87–89.

Bick T., and C. Howett. 1992. "The Green Labeling Phenomenon: Issues and Trends in the Regulation of 'Environmentally Friendly' Product Claims." In *The Greening of American Business: Making Bottom-Line Sense Out of Environmental Responsibility,* ed. T. F. P. Sullivan, 309–45. Rockville, MD: Government Institutes, Inc.

Bird, W. A. 1995. ISO 14000: With an Anticipated Rollout Date of January 1996, It's Time to Put This Program Decision in the Hopper. *Chemical Engineering* (September): 94–96.

Buchholz, R. A. 1993. *Principles of Environmental Management: The Greening of Business.* Englewood Cliffs, NJ: Prentice–Hall.

Cascio, J., 1996, "Background and Development of ISO 14000 Series." In J. Cascio, ed. *The ISO 14000 Handbook.* Fairfax, VA: CEEM Information Services, pp. 4–24.

Church, T. W., and R. T. Nakamura. 1993. *Cleaning Up the Mess: Implementation Strategies in Superfund.* Washington, DC: The Brookings Institution.

Cooke, S. M. 1995. *The Law of Hazardous Waste Management: Management, Cleaning Liability and Litigation,* vol. 1. New York: Mathew Bender.

Crognale, G. G. 1995. Environmental Management: What ISO 14000 Brings to the Table. *Total Quality Environmental Management* (Summer): 5–15.

Dover, R. C. 1990. "Hazardous Wastes." In *Public Policies for the Environment,* ed. P. R. Portney, 151–94. Washington, DC: Resources for the Future.

Fischer, K., and J. Schot, eds. 1993. *Environmental Strategies for Industry: International Perspectives on Research Needs and Policy Implications.* Washington, DC: Island Press.

Fortuna, R. C., and D. J. Lennett. 1987. *Hazardous Waste Regulation: The New Era.* New York: McGraw–Hill.

Gaba, J. M. 1994. *Environmental Law.* St. Paul, MN: West Publishing Co.

Hansen, F. 1989. Pollution Prevention Planning: A New Mandate for Oregon's Environment. *The Environmental Forum* (Sept./Oct.): 30–34.

Hellman, T. M., and D. A. Hawkins. 1988. "How clean is clean? The need for action." In *Hazardous waste site management: water quality issues. Report on a colloquium sponsored by the Water Science and Technology Board, Feb. 19–20, 1987.* Washington, DC: National Academy Press.

Hirschhorn, J. S., and K. V. Oldenburg. 1991. *Prosperity Without Pollution: The Prevention Strategy for Industry and Consumers.* New York: Van Nostrand Reinhold.

Karnofsky, B., ed. 1992. *Hazardous Waste Management Compliance Handbook.* New York: Van Nostrand Reinhold.

Kneese, A. V., and B. T. Bower. 1979. *Environmental Quality and Residuals Management.* Baltimore: John Hopkins University Press for Resources for the Future.

Kokoszka, L. C., and J. W. Flood. 1989. *Environmental Management Handbook.* New York: Marcel Dekker.

Lee, R. T. 1993. "Comprehensive Environmental Responses, Compensation and Liability Act." In *Environmental Law Handbook,* 12th ed., J. G. Arbuckle et al. Rockville, MD: Government Institute Inc.

Massachusetts Institute of Technology. 1991. *Preventing Pollution: Focus on Organization and Management.* Report of the Technology, Business, and Environment Program, Cambridge, MA.

Mazmanian, D., and D. Morell. 1992. *Beyond Superfailure: America's Toxics Policy for the 1990s.* Boulder, CO: Westview Press.

McCarty, P. L. 1990. "Ground water and soil contamination remediation: toward compatible science, policy, and public perception." In *Report on a colloquium sponsored by the Water Science and Technology Board.* Washington, DC: National Academy Press.

Novick, S. M., D. W. Stever, and M. G. Mellon, eds. 1995. *Law of Environmental Protection.* Deerfield, IL: Clark, Boardmand, and Callaghan.

Organization for Economic Cooperation and Development. 1991. *Environmental Labelling in OECD Countries.* Paris: OECD.

Percival, R. V., A. S. Miller, C. S. Schroeder, and J. P. Leape. 1992. *Environmental Regulation: Law, Science, and Policy.* Boston: Little Brown.

Perrone, M. A. 1996. "Trade Barriers, GATT, and ISO 14000." In J. Cascio, ed. *The ISO 14000 Handbook.* Fairfax, VA: CEEM Information Services, pp. 411–424.

Probst, K. N., et al. 1995. *Footing the Bill for Superfund Cleanups: Who Pays and How?* Washington, DC: The Brookings Institution and Resources for the Future.

Sheridan, E. 1987. How Clean is Clean: Standards for Remedial Actions at Hazardous Waste Sites Under CERCLA. *Stanford Environmental Law Journal,* vol. 6. Stanford, CA: Stanford University School of Law.

Vincoli, J. W. 1993. *Basic Guide to Environmental Compliance.* New York: Van Nostrand Reinhold.

White, M. A. 1992. "Effect of the Green Movement on Investors." In *The Greening of American Business: Making Bottom-Line Sense of Environmental Responsibility,* ed. T. F. P. Sullivan. Rockville, MD: Government Institutes, Inc.

Wise, J. C. 1995. EPA's Brownfields Economic Development Initiative. *Land Use & Environment Forum* 4(3): 151–53.

Chapter 15

Environmental Impact Statements and Government Decision Making

During the late 1960s, many people felt that government projects were significantly degrading the quality of the environment.[1] Among the objectionable works were highways that destroyed scenic landscapes, power plants that belched pollutants from ugly smokestacks, and water projects that drained ecologically valuable wetlands. Critics argued that environmentally destructive projects were being built because government agencies considered only economic and technical factors in making their decisions. Environmental factors were being ignored.

In response to such criticism, many nations established policies requiring government agencies to assess environmental impacts when designing their own projects and when granting approval for private projects. This chapter concentrates on the first of these government environmental impact assessment requirements: the program established by the U.S. National Environmental Policy Act of 1969 (NEPA).[2] Following the enactment of NEPA, many states and countries introduced their own environmental impact assessment programs, some of which are introduced later in this chapter.

NEPA'S OBJECTIVES AND PRINCIPAL PARTS

The fundamental aim of NEPA was to force all agencies of the federal government to integrate environmental concerns into their planning and decision making.[3] As shown in Table 15.1, the act contains provisions that created the following: a national environmental policy; "action-forcing" requirements for federal agencies; and a new organization, the Council on Environmental Quality (CEQ).

[1] Some sections of this chapter are adapted from material that appeared originally in Ortolano and Shepherd (1995). I am grateful to Professor Anne Shepherd of the Georgia Institute of Technology for her contributions.

[2] Although NEPA was passed by Congress in 1969, it did not become law until it was signed by the president in 1970. The act retains "1969" in its title because of conventions used by Congress in naming legislation.

[3] For a more extensive introduction to NEPA and its implementation, see Clark and Canter (forthcoming).

TABLE 15.1 Principal Parts of the National Environmental Policy Act

Declaration of national environmental policy:
- Broad policy statement—Section 101(a).
- Responsibilities of the federal government—Section 101(b).

Establishment of action-forcing requirements—all agencies of the federal government shall
- Utilize an interdisciplinary approach to planning—Section 102(2)(A).
- Develop procedures to give environmental factors "appropriate consideration" in decision making—Section 102(2)(B).
- Prepare environmental impact statements—Section 102(2)(C).

Creation of the Council on Environmental Quality.

The first portion of NEPA contains a declaration of national policy. In Section 101(a), Congress declares

> [that] it is the continuing responsibility of the Federal Government . . . to use all practical means and measures . . . to create and maintain conditions under which man and nature can exist in productive harmony and fulfill the social, economic and other requirements of present and future generations of Americans.

Section 101(b) lists six ends to be served by this "continuing responsibility" of the federal government. These concern such things as assuring "safe, healthful, productive, and aesthetically and culturally pleasing surroundings" and maintaining, "wherever possible, an environment which supports diversity, and a variety of choices." Table 15.2, which contains the six goals in Section 101(b), makes it clear that Congress used the term environment in a broad sense to include social and cultural factors as well as physical and biological components.

Section 102(1) of NEPA is a model of efficiency:

> The Congress authorizes and directs that, to the fullest extent possible: (1) the policies, regulations and public laws of the United States shall be interpreted and administered in accordance with the policies set forth in this Act. . . .

TABLE 15.2 Section 101(b) of the U.S. National Environmental Policy Act

In order to carry out the policy set forth in this Act, it is the continuing responsibility of the Federal Government to use all practicable means, consistent with other essential considerations of national policy, to improve and coordinate Federal plans, functions, programs, and resources to the end that the Nation may

1. fulfill the responsibilities of each generation as trustee of the environment for succeeding generations;
2. assure for all Americans safe, healthful, productive, and esthetically and culturally pleasing surroundings;
3. attain the widest range of beneficial uses of the environment without degradation, risk to health or safety, or other undesirable and unintended consequences;
4. preserve important historic, cultural, and natural aspects of our national heritage, and maintain, wherever possible, an environment which supports diversity, and variety of individual choice;
5. achieve a balance between population and resource use which will permit high standards of living and a wide sharing of life's amenities; and
6. enhance the quality of renewable resources and approach the maximum attainable recycling of depletable resources.

With that one sentence, Congress made the consideration of environmental impacts an integral part of all other federal laws, policies, and programs.

The first three items in Section 102(2) are among the principal action-forcing provisions of NEPA. Section 102(2)(A) requires federal agencies to utilize a systematic interdisciplinary approach that will insure use of the natural and social sciences and the environmental design arts in planning and decision making that may impact the human environment.

Section 102(2)(B) of NEPA indicates that Congress did not intend to prohibit federal agencies from making decisions that had adverse environmental consequences. Instead, Congress required a balancing of environmental values and other considerations. Section 102(2)(B) instructs federal agencies to

> identify and develop methods and procedures . . . which will insure that presently unquantified environmental amenities and values may be given *appropriate consideration* in decision making along with economic and technical considerations. [Emphasis added]

Section 102(2)(C) requires agencies to prepare a "detailed statement" of environmental impacts for "major Federal actions significantly affecting the quality of the human environment." This provision, which is the one most commonly associated with NEPA, increased the amount and type of information an agency must consider and disseminate before making decisions. The preparation and external review of these *environmental impact statements* (EISs) "form the cornerstone of NEPA's system of environmental protection" (Bass and Herson, 1993, p. 3).

Finally, the last part of NEPA created the Council on Environmental Quality, which consists of three members (appointed by the president) to advise the president on environmental matters. The Council has a small staff to carry out duties specified in NEPA, which include assisting the president in preparing an annual environmental quality report, and appraising federal agency performance in implementing the action-forcing provisions of Section 102(2).[4]

ENVIRONMENTAL IMPACT STATEMENT PREPARATION AND USE

The decision-making process used by federal agencies was modified fundamentally by the EIS requirements of Section 102(2)(C). These requirements apply to virtually all federal agencies.

NEPA applies not only to federal agency projects, but also to federal decisions on programs, permits, grants, and loans. An example is the Nuclear Regulatory Commission's decision to give an electric utility a license to construct a nuclear power plant. In this case, the utility would need permits from several federal agencies. The agencies themselves decide which one is to serve as the *lead agency* in preparing an environmental impact statement.[5] Another example of an action covered by NEPA is a decision by the Department of Housing and Urban Development to provide a loan for construction of low-income housing. NEPA does not apply to a project by a private party or a nonfederal agency unless the project involves a decision by a federal agency.

In a typical year, the majority of EISs are prepared for specific projects such as federal reservoirs and federally funded highways or housing developments. Less frequently, an EIS might involve not an individual action but a group of actions that are viewed collectively as an agency program. Examples of actions that have been the subject of *programmatic EISs* include the Forest Service's vegetation control activities in Arizona and New Mexico, and the Energy Research and Development Agency's liquid metal fast breeder reactor program.

Steps in the NEPA Process

Procedures to be followed in preparing an EIS are set out in regulations issued by the Council on Environmental Quality (CEQ, 1986). Figures 15.1 and 15.2 provide an overview of these procedures, which, collectively, have come to be called the *NEPA process*.

The CEQ regulations instruct agencies to issue individualized, agency-level NEPA procedures and to identify categories of actions that are minor and unlikely to have significant environmental impacts. An agency can declare such actions as being categorically excluded from all Section 102(2)(C) requirements of NEPA. Examples of actions listed as *categorical exclusions* by the U.S. Army Corps of Engineers include minor dredging projects to maintain navigation channels, and real-estate grants for government-owned housing.[6] Use of categorical ex-

[4] The Council on Environmental Quality, which is part of the Executive Office of the President, is completely distinct from the U.S. Environmental Protection Agency. The council is much smaller, and the only environmental regulations it issues are those detailing obligations imposed on federal agencies by NEPA.

[5] In this context, the other agencies are called *cooperating agencies.*

[6] These examples of categorical exclusions apply to "civil works projects" of the U.S. Army Corps of Engineers and are from a listing compiled by Canter (1996, p. 13). For the dredging example, there is an added condition: dredged materials must be deposited at an existing disposal site.

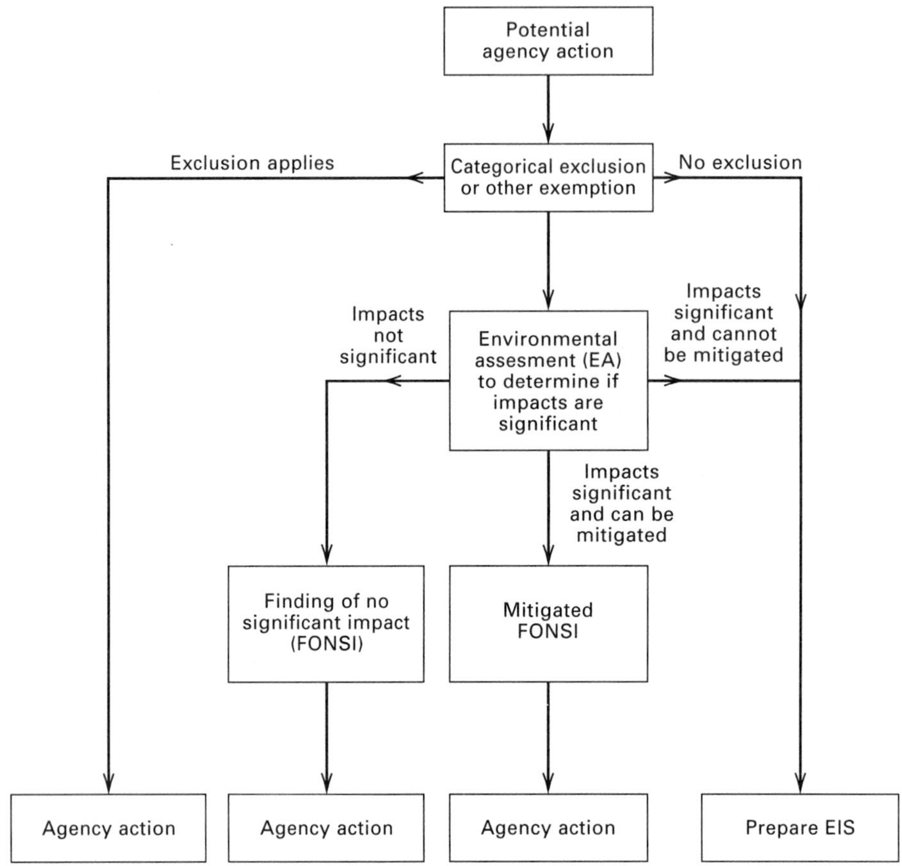

Figure 15.1 Decision to prepare an EIS using the NEPA process.

clusions allows agencies to avoid using resources to assess projects with trivial environmental impacts.

For all federal actions that are not categorically excluded, a preliminary analysis may be used to determine whether anticipated impacts will be significant enough to require an EIS. This analysis is documented in an *environmental assessment* (EA). In principle, an EA is a concise public report containing sufficient information to determine whether the proposed federal action will significantly affect the quality of the environment. If a proposed action has the potential to cause significant adverse impacts, an EIS must be prepared. However, the agency may determine that impacts are minor and use the EA as a basis for issuing a *finding of no significant impact* (FONSI). In this case, an EIS is not prepared. As will be explained, citizens can use the federal courts to challenge an agency's decision to issue a FONSI.

In practice, agencies often view the EA as a document that can be used to justify (and defend) issuance of a FONSI, and some EAs have the length and appearance of a full environmental impact statement. An environmental assessment may indicate that potentially significant impacts can be reduced to less-than-significant levels if mitigation measures are taken; in this case, an agency may issue a *mitigated FONSI.*

Although the term *mitigation* is used widely in the context of environmental impact assessment programs, it is confusing and requires elaboration. The CEQ regulations include several types of mitigation, one of which involves eliminating environmentally damaging elements of a proposed plan. In this sense, "to mitigate" is to avoid having the damage take

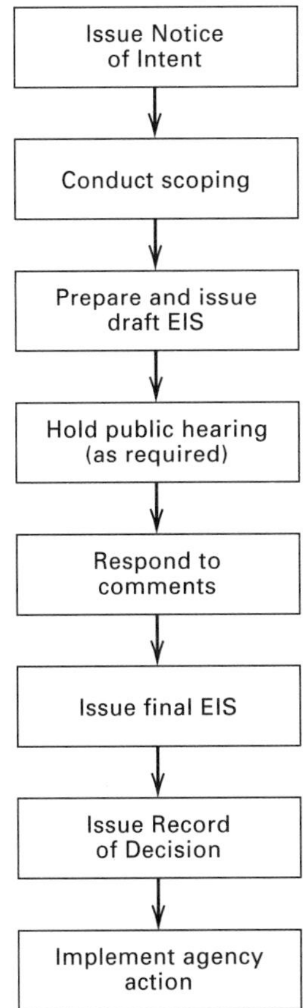

Figure 15.2 Steps in preparing an EIS using the NEPA process.

place. Other forms of mitigation mentioned by the CEQ regulations include the following:[7]

- Minimizing adverse effects by limiting the degree or magnitude of an action (e.g., reducing habitat loss by lowering the height of a proposed dam).
- Repairing, rehabilitating, or restoring the affected environment (e.g., replanting with native vegetation areas that have been cleared for installation of a pipeline).
- Reducing impacts over time by preservation and maintenance operations (e.g., adding fish ladders to a dam to allow anadromous fish to reach upstream spawning areas).
- Creating or acquiring environments similar to those adversely affected by an action (e.g., donating wetlands to a public trust to compensate for wetlands destroyed by a project).

If an environmental assessment makes clear that neither a FONSI nor a mitigated FONSI can be issued, then the lead agency must complete an environmental impact statement. The steps involved are shown in Figure 15.2 and summarized here.

- Issue a public *Notice of Intent* (NOI) that an environmental impact statement will be prepared. The notice describes the proposed action and possible alternative actions and is issued to individual citizens, interest groups, and other agencies likely to be concerned with the proposal. The NOI also gives interested parties details on how to get more information and whether, when, and where any "scoping meeting" will be held.
- Conduct *scoping* to identify the range of actions, alternatives, and impacts to be considered in the EIS. The scoping procedure allows interested parties to make known their concerns, and to contribute their views on what the EIS should contain. Although public meetings are often held as part of a scoping exercise, they are not required. Sometimes scoping is based largely on written comments received in response to an NOI. The scoping process has been characterized as one that helps ensure "that real problems are identified early and properly studied; that issues that are of no concern do not consume time and effort; that the draft [EIS] when first made public is balanced and thorough; and that delays occasioned by redoing an inadequate draft are avoided."[8]
- Prepare a *draft EIS* in accordance with parameters established during the scoping process and consistent with requirements detailed in both the Council on Environmental Quality's regulations for implementing NEPA and the lead agency's guidelines on the preparation of an EIS. Part or all of the draft EIS may be prepared by applicants for agency grants and permits, and by consultants working under contract for the lead agency. Regardless of who conducts the environmental im-

[7] Items in this list are paraphrased from the CEQ regulations; the examples are not a part of those regulations.

[8] This quotation is from "scoping guidance" issued by CEQ (1981).

pact study, the lead agency is legally responsible for the EIS.

- Circulate the draft EIS for *review and comment* by interested parties, including citizens, nongovernmental organizations (NGOs), and various federal, state, and local units of government. The draft EIS must be provided to agencies with jurisdiction over the proposed action, and to anyone requesting a copy. A minimum of 45 days is allowed for review, and the review period starts when the U.S. Environmental Protection Agency (EPA) publishes a notice in the *Federal Register* listing its receipt of the draft EIS.
- Conduct *public hearings* on the draft EIS if required by the lead agency's own EIS guidelines or if one of the following conditions holds: the proposed action is controversial, there is substantial interest in conducting a hearing, or another agency with jurisdiction over an action requests a hearing.
- Prepare the *final EIS* to include responses to all substantive comments received during the review of the draft EIS.
- Circulate the final EIS for review. The final EIS must be provided to a variety of interested agencies, to all citizens and NGOs who submitted substantive comments on the draft EIS, or to anyone requesting a copy of the final EIS. There is a 30-day time period for public review of a final EIS, and the time period starts when EPA publishes a notice in the *Federal Register* indicating that it has received the final EIS. The lead agency's final decision on the proposed action cannot be issued until after this 30-day period. During this period, opponents of the proposed action may sue the lead agency based on inadequacies in the agency's implementation of NEPA.[9]
- Issue a *Record of Decision* (ROD) and begin implementation. The ROD consists of a public statement that explains the decision made by the lead agency. It spells out how environmental factors were balanced against economic and technical factors in making the final decision. The ROD also indicates actions the lead agency will take to mitigate adverse environmental effects.

[9] As explained later in this chapter, citizens and nongovernmental organizations can use the federal courts to challenge agency decisions using the Administrative Procedure Act.

Elements of EIS Preparation

Suppose a federal agency decides that an action it proposes requires an EIS. In this instance, the agency typically forms a team of specialists to carry out the analyses identified during the scoping process. The appropriate composition of the team depends on the type of project involved. For complex projects, an ideal team might include natural and social scientists, engineers, and design professionals. For a team to function effectively as a unit, communications among members must be good. Communications can be fostered by a skillful team leader. Sometimes a team does not work as a unit, and members contribute their skills to the environmental impact study by working individually with the team leader.

The starting point for determining the content of an EIS is Section 102(2)(C) of NEPA, which requires that an EIS include information on the following:

> (i) environmental impact of the proposed action,
> (ii) any adverse environmental effects which cannot be avoided should the proposal be implemented,
> (iii) alternatives to the proposed action,
> (iv) the relationship between local short-term uses of man's environment and the maintenance and enhancement of long-term productivity, and
> (v) any irreversible and irretrievable commitments of resources which would be involved in the proposed action should it be implemented.

In the context of NEPA, the term *environment* refers to more than just the biophysical environment. Thus an EIS is required to consider impacts on individuals, groups, and communities as well as effects on historic, cultural, and scenic resources. The term *social impacts* is often used to refer to impacts other than those affecting the natural environment.[10]

Judicial interpretations of NEPA have influenced the required content of an EIS. During the 1970s there were over 1000 cases in federal courts claiming that agencies failed to comply with NEPA. In many of these cases, plaintiffs argued that an agency's EIS was incomplete. Federal courts were thus put in the position of deciding what constituted an adequate EIS. Their judgments influenced the reg-

[10] Social impact assessment is a field of professional activity and there is a vast literature on the subject. For an introduction to the field, see Burdge (1994) and Taylor, Bryan and Goodrich (1995).

ulations of the Council on Environmental Quality (CEQ, 1986), which specify generally what an EIS should contain. The courts also influenced the more specific EIS content requirements elaborated in the NEPA guidelines of individual federal agencies.

Preparation of an EIS requires identifying actions that can be considered as *alternatives* to the action proposed by the agency? According to the CEQ regulations, the part of an EIS that treats alternative actions constitutes "the heart of the environmental impact statement." The regulations call for more than a mere listing of actions the agency considered. They require a comparative analysis of the environmental consequences of the alternatives. The comparison must include the *no-action* alternative, which is defined by the related activities likely to occur if the federal agency does not take positive action. Suppose, for example, that the proposed action in an EIS is an agency decision granting a license to a railroad. In this case, the no-action alternative is made up of the predictable public and private actions expected to take place if the license is not granted. This includes projected increases in freight transport by competing transport modes such as trucks, barges, and airplanes.

Determining the environmental impacts of the proposed action (and alternatives to the proposed action) requires making *forecasts,* a subject treated in the next chapter. For any particular action, an *impact* is defined as the difference between the future state of the environment if the action took place and the state if no action occurred.[11] Impacts may be direct or indirect. *Direct impacts* are those that follow immediately and obviously from an action or project. *Indirect impacts* of an action may be far removed in both time and distance. For example, consider a proposed action that includes construction of a dam on a coastal stream. Direct impacts include replacement of portions of a free-flowing stream with a reservoir. A possible indirect effect of the dam is the destruction of ocean beaches. This may occur over many years as sediments that once flowed to the ocean to replenish the beaches settle out in the reservoir behind the dam.

Forecasts of environmental impacts provide a basis for two activities: identification of mitigation measures and evaluation of alternative actions. This evaluation includes environmental, economic, and engineering considerations. As explained in Chapter 17, the process of evaluation requires making value judgments about how much environmental quality is to be "traded off" to gain increases in economic benefits or other dimensions of human well-being. The NEPA process involves many such judgments. The outcome is a recommendation by the lead agency regarding the action it wishes to pursue ("the proposed action" in an EIS).

COMPLIANCE WITH NEPA'S REQUIREMENTS

In the months following NEPA's passage, many federal agencies complied with the statute's action-forcing provisions in a perfunctory way. Many early environmental impact statements were brief, uninformative documents whose only apparent purpose was to jump the administrative hurdle of preparing an EIS. In the first year after NEPA's passage, token compliance with EIS requirements was the norm. This performance is not surprising in retrospect, considering that agencies are generally enthusiastic about promoting their own projects. The NEPA process was intended to force agencies to analyze and disclose adverse effects. The information disclosed could (and did) threaten agencies' abilities to carry out their plans without interference by those who opposed their actions.

Because NEPA included no provisions to enforce agency compliance with its action-forcing provisions, the initial, perfunctory compliance by many federal agencies might have continued. However, lackluster performance of the agencies did not persist. This section introduces three factors that led federal agencies to comply thoroughly with the procedural requirements of NEPA: public reviews of a draft EIS, the CEQ referral process, and citizen-initiated court challenges to agency performance.

Public and Agency Reviews of an EIS

Section 102(2)(C) of NEPA obligates an agency preparing an impact statement to "consult with and obtain the comments of any Federal agency which has jurisdiction by law or special expertise with respect

[11] This widely used definition of an environmental impact is from Munn (1979, p. xvii). He clarifies much of the specialized terminology used in the environmental impact assessment literature.

to any environmental impact involved,'' and to make the EIS and agency comments available to interested parties. These requirements of NEPA have yielded an elaborate process involving circulation of the EIS in draft form, preparation of review comments by recipients of the draft EIS, revision of the draft by the issuing agency, and distribution of a final EIS (see Figure 15.3). The review process provides opportunities for a full scrutiny and critique of the lead agency's environmental analysis methods and its rationale for selecting a proposed action. It also provides a forum for opposing views and can be a source of new information.

Comments resulting from the review of a draft EIS cover a wide range. Some reviewers may laud the agency's proposed action, whereas others may be disapproving. Critics might question the accuracy of facts in the draft EIS or the procedures used in forecasting impacts. They might also accuse the agency of failing to consider important environmental effects. Some commentators might request that the agency either undertake studies of actions that were not con-

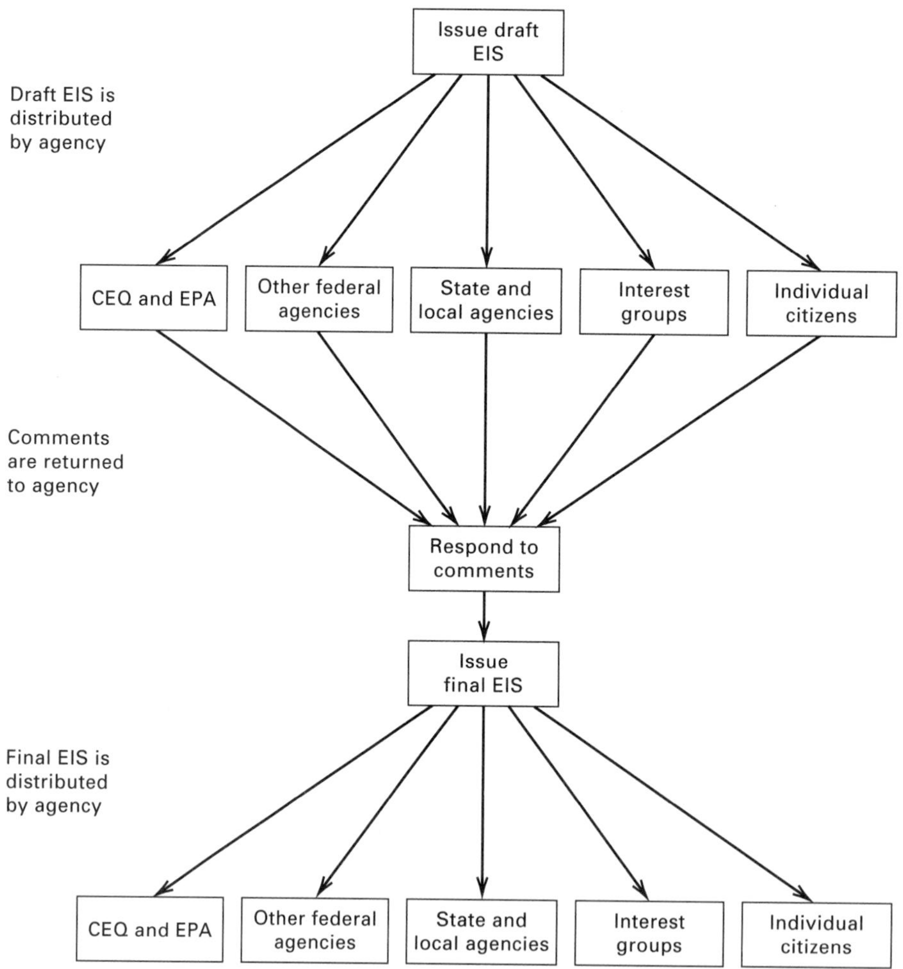

Figure 15.3 Process of review and comment on a draft EIS. Adapted from Jain, Urban, and Stacey, *Environmental Impact Analysis, A New Dimension in Decision Making,* 2d ed., Van Nostrand Reinhold Co., Copyright © 1981 by Litton Educational Publishing, Inc.

sidered or abandon its proposal in favor of an alternative.

The agency issuing the EIS must respond to each significant point raised during the review of the draft. If the agency feels that a comment warrants no action, it must give a reason for its position. More substantive responses take many forms. The agency might add to its EIS by examining new alternatives or by analyzing environmental impacts not considered previously. If reviewers point out significant impacts that were omitted from the draft EIS, the agency might respond by proposing mitigation measures. (As previously noted, these are actions that either reduce an adverse impact or compensate for it in some way.) In some cases, the agency might substantially modify its proposal or even abandon it entirely.

As indicated in Figure 15.3, a final EIS is issued after the agency has responded to comments on the draft. The agency cannot implement the action proposed in the final EIS until interested parties have had a chance to see how the agency responded to review comments. Those who are dissatisfied with the final outcome and wish to stop or modify the agency's action must, at this point, resort to traditional mechanisms for influencing administrative processes: political pressure, and litigation. Pressure can be applied, for example, by exerting influence through CEQ or others in the executive branch. Private parties may attempt to influence an agency's actions by bringing a lawsuit after a final EIS has been distributed (or after a FONSI has been issued).

Because the review of an EIS is so extensive, agencies often try to issue FONSIs instead of EIS's. If a FONSI is issued, the degree of public comment is generally modest if the environmental assessment that forms the basis for the FONSI is not circulated for review. Although the CEQ regulations call for public availability of environmental assessments, CEQ reports "that agencies consistently involve the public in fewer than half of their EA preparations."[12] Agencies often provide only official public notice that a FONSI has been issued and indicate how further information can be obtained. If an EA is not circulated for review, public involvement in agency decision making is not fostered. In some cases, agencies issue *mitigated FONSIs* because they feel they can avoid the public involvement requirements associated with preparation of an EIS.[13]

The CEQ Referral Process

Private parties who disapprove of a lead agency's action can go to court to oppose it, but that option is unavailable to other federal agencies. Although federal agencies attempt to resolve differences among themselves, sometimes interagency conflicts remain after a final EIS is issued. Under these circumstances, a federal agency opposed to a lead agency's proposal may ask CEQ to help resolve the dispute. This referral must occur no later than 25 days after the final EIS is made available.[14] Although CEQ can mediate interagency disputes, it cannot impose a settlement.

The *CEQ referral process* has its origins in Section 309 of the Clean Air Act of 1970. In that section, Congress gave EPA a nondiscretionary obligation to comment in writing on every EIS and "to refer to CEQ any matter determined to be unsatisfactory from the standpoint of health, welfare or environmental quality" (CEQ, 1984, p. 524). In response, EPA developed a system for rating both proposed actions and draft EIS's submitted to it for review. If EPA gives a proposal a rating of "environmentally unsatisfactory," the proposal is "referred to CEQ unless the unsatisfactory impacts are mitigated."[15] If EPA finds that the draft EIS has significant deficiencies in either impact analysis or the range of alternatives considered, it may also refer the matter to CEQ.

The possibility of referring proposals to CEQ was expanded in 1978 to apply to all federal agencies, not just EPA. The CEQ regulations specified criteria for making referrals and emphasized the need to restrict use of the referral process to matters of national importance.

[12] This quotation is from Blaug (1993, p. 56), based on results from a CEQ survey of agency practice in implementing NEPA.

[13] Blaug's (1993, p. 57) analysis of data from a CEQ survey of federal agencies supports this point. She reports that "CEQ believes that a major reason for agencies [to use mitigated FONSIs to avoid] EIS preparation is the erroneous view that preparation of EAs does not require public involvement."

[14] Recall that an agency cannot issue its record of decision until 30 days after issuing its final EIS.

[15] This quote and a description of EPA's rating system are in Bass and Herson (1993, p. 57).

Although the CEQ referral process has been effective at resolving interagency disputes, it has not been used often. As of 1992, there were only 24 cases referred, or about one proposal a year on average. For most of these cases, CEQ mediated the interagency conflict in ways that reduced the adverse effects of proposed projects. Despite its lack of extensive use, the CEQ referral process, together with EPA's rating scheme, puts pressure on federal agencies to comply with NEPA. However, that pressure pales in comparison to effects of actual or threatened lawsuits.

Role of Courts in Enforcing Procedural Compliance

During the early 1970s, legal scholars viewed the courts as the principal enforcers of NEPA. This is a curious outcome since NEPA contains no explicit provision for judicial oversight.[16]

Under the circumstances, how did the courts get involved? The legal mechanism used in nearly all NEPA-based court actions was the Administrative Procedure Act (APA) of 1946, a law that governs activities of federal agencies. The APA allows citizens to sue agencies that violate federal laws. Citizens and nongovernmental organizations were able to use the APA to sue agencies for not preparing an environmental impact statement as required by NEPA.[17] Plaintiffs were able to establish their legal *standing to sue* by arguing that an agency's actions—for example, not preparing an EIS as required by NEPA—caused environmental or aesthetic harms.[18]

In hearing NEPA cases, the courts made it clear that NEPA granted court-enforceable procedural obligations. The most common issues involved in more than 2000 NEPA-related court actions between 1970 and 1989 were failure of an agency to prepare an EIS when required to do so by NEPA, and failure of an agency to prepare an EIS that was adequate in a legal sense.[19] The position of the federal judiciary was made clear in a series of Supreme Court opinions that describe the EIS as serving two purposes: "(1) ensuring the agency will have environmental information in reaching its decision, and (2) . . . making the information available to a 'larger audience,' which may also play a role in the process of making and implementing the decisions."[20]

The significance of the courts' enforcement of NEPA has diminished as a result of decisions made by the U.S. Supreme Court during the Reagan and Bush administrations. As of the early 1990s, the Supreme Court held the view that while NEPA grants procedural rights, it does not prevent an agency from deciding "that other values outweigh the environmental costs."[21] The Supreme Court's position has been interpreted to mean that NEPA imposes only procedural (not substantive) obligations on federal agencies. Some feel that the Court misconstrued NEPA in reaching this position. For example, Yost (1990) argues that Section 102(1) of NEPA requires federal agencies to act "to the fullest extent possible" to meet the six policy goals in Section 101(b). (See Table 15.2.) In contrast, the Supreme Court's position allows an agency to ignore the substance of Section 101(b) as long as the agency follows the procedural requirements of the NEPA process introduced in Section 102(2) and detailed in the CEQ regulations.

A particularly significant NEPA decision rendered by the Supreme Court is *Robertson v. Methow Valley Citizens Council.* The Court held that although NEPA's procedures "are almost certain to affect the agency's substantive decision, it is now well settled that NEPA itself does not mandate particular results, but simply prescribes the necessary process. . . ."[22]

[16] For more on this view of the courts, see Anderson's (1973, p. 16) analysis of the role of the courts in enforcing NEPA.

[17] Although individual citizens and NGOs are the most common plaintiffs in NEPA-related cases, NEPA suits have also been brought by state and local governments, Native American tribes, and business groups.

[18] Hoban and Brooks (1996, p. 189) define standing to sue as "the right of an individual to seek relief in court. The very first requirement faced by a plaintiff seeking legal remedies is that he or she must be seeking relief from something recognized by the court as an injury to the plaintiff's interests." Changes in requirements to establish standing to sue during the 1960s paved the way for the NEPA cases. During that decade, federal courts loosened earlier requirements in which plaintiffs needed to show *specific* injury. For an expanded discussion of standing to sue in the context of environmental law, see Hoban and Brooks (1996, Chapter 13).

[19] See annual reports of the Council on Environmental Quality.

[20] This quote is from Yost (1990, p. 545).

[21] Supreme Court Justice Stevens as quoted by Yost (1990, p. 546).

[22] This quotation and the one that follows are from *Robertson V. Methow Valley Citizens Council* as quoted by Ferester (1992, p. 222).

In addition to characterizing NEPA only in terms of procedural requirements, the Court further eroded NEPA's significance by its response to a central question in *Methow Valley:* Was the Forest Service required to develop and assess specific measures to offset adverse impacts before issuing a special use permit for a ski resort in a national forest? In arguing that no specific mitigation was required, the Court reasoned as follows:

> There is a fundamental distinction . . . between a requirement that mitigation be discussed in sufficient detail to ensure that environmental consequences have been fairly evaluated, on the one hand, and a substantive requirement that a complete mitigation plan be actually formulated and adopted, on the other. . . . [*I*]*t would be inconsistent with NEPA's reliance on procedural mechanisms—as opposed to substantive, result-based standards*—to demand the presence of a fully developed plan that will mitigate environmental harm before an agency can act. [Emphasis added.]

The Supreme Court's views on the limited responsibility of the Forest Service completed what Ferester (1992, p. 221) has termed "the mechanization of judicial review" under NEPA.

Notwithstanding the Supreme Court's restricted view of NEPA's intent, it is possible for citizens to challenge the correctness of an agency's decision using the Administrative Procedure Act. However, the conditions under which a court will review an agency's decision (as opposed to its procedures) are limited. Plaintiffs must show that the agency action was "arbitrary, capricious, an abuse of discretion, or otherwise not in accordance with the law."[23] In *Overton Park v. Volpe,* the U.S. Supreme Court interpreted this provision of the APA to mean that an agency must provide evidence that it considered all relevant facts and factors, and that its decision was free from clear errors of judgment.

The ability of citizens and NGOs to sue agencies under the Administrative Procedure Act has led to a high rate of compliance with NEPA's procedural requirements. Federal agencies anticipate that if they fail to satisfy requirements to prepare environmental assessments and EISs, they may be sued by a project's opponents. Even the threat of a suit is often sufficient to cause agencies to use appropriate environmental impact analysis methods and examine a complete range of impacts. However, satisfying procedural requirements provides no assurance that NEPA's policy goals will be met.

PERENNIAL PROBLEMS WITH NEPA'S IMPLEMENTATION

This section details problems that have been associated with NEPA since its passage. Some of the problems will persist because many project proponents view the NEPA process primarily as an administrative hurdle to be cleared along the way to project implementation. This hurdle imposes risks because project proponents must make a public disclosure of impacts, and the information on impacts can strengthen the hand of a project's critics. When opposition to a project is likely, agencies often view the NEPA process in terms of staying out of court. If litigation appears unavoidable, the process often centers on "the production of huge stacks of paper as evidence that environmental issues have been given appropriate consideration."[24]

NEPA Process Often Has Limited Influence on Decisions

During more than 25 years since NEPA's enactment, the law has been critiqued frequently for establishing "little more than a bureaucratic exercise that requires federal agencies to complete paperwork they subsequently file and ignore" (Fogelman, 1993, p. 79). The argument is frequently the same: environmental impact assessment is not well integrated into decision making, it is often done after planners and decision makers begin advocating a particular proposal, and it serves largely to mitigate adverse effects of a project already selected. As Ensminger and McLean (1993, pp. 48–49) point out, "major decisions, including the action to be carried out and its location, are often made before the EIS is prepared and . . . the EIS is then drawn up to support those decisions."

[23] This quote is from Section 706(2)(A) of the Administrative Procedure Act.

[24] This quotation is from Ensminger and McLean (1993, p. 50). Their observation is based on a review of the literature and an analysis of survey responses from 56 "NEPA practitioners."

This use of the EIS as an *ex post* rationalization for decisions persists because, in many contexts, a project proponent will not prepare an EIS until after a project is well defined, and there is a high likelihood that the project will be funded and approved.[25] Many project planners feel it would be irrational to use resources to prepare an EIS is a proposed project is not likely to go forward. In addition, many project developers don't give the same weight to environmental objectives as they give to economic performance measures such as the benefit-cost ratio. If project proponents gave environmental impacts the attention they typically give to economic decision criteria, environmental assessments would be initiated along with financial studies at the earliest stages of project conceptualization.

A common explanation for the emphasis that federal agencies place on meeting formal procedural requirements (as opposed to what some term "the spirit of NEPA") is that federal courts require little more. In some cases, however, agencies have reorganized their planning procedures to improve their consideration of environmental factors; illustrations are given in the next section.

EISs Are Often Not Done for Programs or Policies

The influence of the NEPA process could be greater if it were applied more frequently at the level of programs, which are collections of individual projects, such as a coordinated series of dams or an integrated set of research investigations. Some have even argued that EISs should also be done for proposed policies and legislation.[26]

An EIS for a program or policy (termed a *programmatic EIS*) would provide opportunities to mitigate or abandon environmentally unsound concepts before they were turned into projects.[27] In addition, programmatic EISs can enhance interagency coordination and yield efficiencies. If an EIS were done for a program, then any future action consistent with the program could proceed more efficiently by means of *tiering:* the ability, when preparing an EIS for an action within a program, to summarize (and include by reference) impact assessments that appear in the EIS for the program.[28] For example, suppose a program consisted of a set of land development proposals and that the impact on traffic congestion of these development proposals was carefully analyzed in a programmatic EIS. In this case, the EIS for a specific land development project consistent with the program could summarize and refer to the traffic impact assessment in the programmatic EIS. A programmatic EIS does not eliminate the need for separate EISs for individual projects because projects considered in a programmatic EIS are often specified in very general terms.

Even though programmatic EISs have advantages, they are not prepared frequently.[29] One reason is that program and policy decisions often evolve over time, making it difficult to identify what constitutes "the program." The scope of a program may be difficult to define, both spatially and temporally, and this makes assessing impacts even more uncertain than usual. Even when spatial boundaries can be delineated, the land areas involved may be huge and may involve numerous agencies with decision-making authorities. In addition, agencies or private developers that are trying to promote an entire program may be wary of giving potential opponents a complete perspective on program impacts. Other difficulties associated with programmatic EISs relate to data and analysis methods. These problems are taken up in the discussion of a related topic, cumulative impact assessment.

Cumulative Impacts Are Often Not Assessed

The CEQ regulations provide a widely cited definition of cumulative impacts.

[25] For a further discussion of this point, see Nelson (1993).

[26] The term "strategic environmental assessment" (SEA) is often used to mean the application of environmental impact assessment (EIA) in the design of policies, plans and programs. Terminology is not yet standardized however, and other terms, such as policy impact assessment and programmatic EIS, are often used interchangeably with SEA. Based on her review of SEA activities in Europe, North America, Australia and New Zealand, Partidário (1996) finds there is still no consensus on what constitutes SEA. Her review highlights advantages of SEA and common barriers to SEA implementation. For case studies of SEAs for land use plans and infrastructure development programs, see Therivel and Partidario (forthcoming).

[27] Decisions made at the policy or program level often foreclose some types of alternatives. For example, a program-level decision to build dams or enlarge channels to control floods rules out the consideration of flood-proofing structures or using flood plain zoning to reduce flood damages.

[28] This conception of tiering is from regulations issued by CEQ (1986).

[29] For examples of programmatic EISs, see Sigal and Webb (1989).

> "Cumulative impact" is the impact on the environment which results from the incremental impact of the action when added to other past, present and reasonably foreseeable future actions regardless of what agency . . . or person undertakes such other actions.

Cumulative impacts are of concern because a series of small projects that are decided case-by-case can have collective impacts of great significance, even if each individual project has minor effects. Consider, for example, decisions to permit filling in the edges of San Francisco Bay for real estate development. For any one project, the impact on the bay may be trivial. However, hundreds of similar, small projects, if allowed to go forward, could "nibble" away a notable fraction of the bay. Indeed, by the early 1960s, the bay had been substantially reduced in area as a result of governments granting permits for development. When individual cities or counties issued permits, they were not accounting for permits being issued by others. The creation of the San Francisco Bay Conservation and Development Commission in the late 1960s put a halt to the nibbling away of the bay.

The need for cumulative impact analysis in the context of NEPA has taken on increasing importance as more information has been gathered on the extent to which environmental assessments are used. A recent CEQ survey of federal agency practice estimated that the number of EAs was over 40,000 per year.[30] This is correlated with the extensive use of FONSIs. Because there is relatively little public debate of FONSIs, there are "numerous opportunities for significant cumulative impacts of individually insignificant actions to go unnoticed."[31]

Although programmatic EISs are not prepared frequently in the United States, they represent one approach to managing cumulative impacts.[32] Individual projects are defined in general terms, and their cumulative impacts are assessed in a programmatic EIS. This approach is feasible, however, only if there is a single government body with authority to decide on projects that, when viewed collectively, constitute a program. As individual projects are proposed in the future, they are checked for consistency with the program, and they will typically require their own environmental impact studies. However, proposals consistent with the program can be analyzed efficiently using the previously mentioned tiering process. In contrast, future proposals that are not consistent with the program require a more thorough environmental study to identify their contribution to cumulative impacts.[33]

Cumulative impacts may receive more attention by federal agencies in the future. As noted by Herson and Bogdan (1991, p. 101),

> [i]n the last few years [the late 1980s], the courts have been increasingly willing to scrutinize the analysis of the effects of agency action, combined with other relevant actions, and to reject NEPA documents because of inadequate cumulative impact analyses.

If the courts increase their scrutiny, agencies may be forced to make substantive improvements in methods for cumulative impact analysis.

Minimal Requirements to Mitigate and Monitor Adverse Effects

It is routine for an EIS to include actions to mitigate adverse impacts of a proposed project. What is far less common is to have assurances that a proposed mitigation will be implemented. Indeed, in some cases, mitigations recommended in an EIS consist of actions that the project proponent has no authority to implement. (This is exemplified by the mitigation measure that calls on residents near a proposed road to install double-glazed windows to offset increased traffic noise.) Moreover, there have been cases in which project proponents disregarded mitigations that they could have implemented.

The degree to which proposed mitigation measures are ignored is significant. For example, many of the NEPA specialists participating in a survey by Ensminger and McLean (1993) cited the "lack of

[30] Results from this survey are summarized by Blaug (1993). While there have been over 40,000 EAs annually in recent years, the comparable number of EIS's is less than 500 per year.

[31] This quotation is from Ensminger and McLean (1993, p. 53).

[32] Another approach to managing cumulative impacts involves the concepts of *carrying capacity* used by urban and regional planners. Carrying capacity represents a limit to the development that an area can accommodate while still meeting community-based goals for environmental quality. For an example of how an analysis of carrying capacity can be used to manage cumulative impacts, see Dickert and Tuttle (1985).

[33] Examples of programmatic EISs and their relationship to cumulative impacts are detailed by Sigal and Webb (1989).

guidelines and action-forcing mechanisms" to ensure effective development of mitigation measures as an important deficiency of the NEPA process. Moreover, congressional proposals to amend NEPA in the early 1990s (which were not enacted) would have required that "environmental mitigation and monitoring measures [discussed in the context of an agency's NEPA process for a project] . . . shall be implemented by the appropriate agency."[34]

In recent years, some federal agencies have developed programs to ensure that mitigations specified in an EA or EIS are actually undertaken. An example is the Department of Energy's (DOE) "mitigation action plan": if an EA or EIS indicates that a mitigation will be implemented, the DOE will track the project to ensure that the mitigation is carried out.[35] Although experience with the DOE mitigation action plan has been limited, it may provide precedents for other agencies.

Lack of follow-up on whether mitigation measures are implemented is part of a broader problem: few investigations are conducted to monitor the impacts of projects after they are implemented. In commenting on the state of post-EIS project monitoring, Culhane (1993, p. 67) characterized the "baseline" approach:

> [Agency managers seem to] (1) fight their way through the thickets of project planning and environmental review to obtain a favorable decision, (2) implement the decision by, for example, seeing the project through its construction phase, then (3) move on to planning and implementing their next major project, (4) leaving behind no official with any particular stake in monitoring the effectiveness or dysfunctions of the implemented project.

According to Culhane (1993, p. 69), "Relatively few post-EIS audits have been conducted by anyone."

There have been many calls for post-project impact monitoring since the 1970s. One argument in support of monitoring is that evaluating impacts after a project is built provides opportunities to reduce adverse effects. If monitoring reveals unacceptable impacts, measures to mitigate them can be undertaken. Continued monitoring can also indicate whether the mitigation measures were effective. This strategy was advocated during the 1970s by Holling (1978) and his colleagues under the rubric of "adaptive environmental management." Holling's contentions are no less compelling today.

Monitoring can also assist in advancing the practice of impact forecasting. In principle, the process of conducting an environmental impact study can be treated as part of a scientific experiment in which a predicted impact constitutes a hypothesis about environmental changes caused by the project.[36] The hypothesis can be tested by gathering data on impacts that occur after the proposed project is built. Monitoring such impacts provides a basis for assessing the accuracy of predictions and may lead to improvements in methods of forecasting.

Although post-project monitoring is not commonly undertaken, its value has been demonstrated. For example, the Tennessee Valley Authority (TVA) monitors river flow and several water quality variables affected by its reservoirs on the Tennessee River. TVA uses these data to "improve water quality and aquatic habitat by increasing minimum flow rates and aerating releases from TVA dams to raise dissolved oxygen levels.[37] The data are also used to investigate how changes in the operations of TVA's reservoirs influence dissolved oxygen in downstream waters.

Insufficient Attention to Risk Assessment

Although the field of environmental risk assessment has advanced considerably in the past few decades, those advances are not reflected in many EISs or EAs.[38] Risk assessment yields estimates of the probability and magnitude of adverse effects, and it is especially appropriate in appraisals of projects that can have disastrous consequences if low-probability events take place. Examples of such projects include

[34] This material is from Diana Bear, as cited by Smith (1993, p. 83).

[35] The DOE program is summarized by Culhane (1993, p. 73). As of 1993, four such post-project follow-ups had been conducted.

[36] This view is elaborated by Culhane (1993, p. 69) and, more generally, by Caldwell (1982). In practice, many predictions of environmental impacts are so vague that the notion of treating them as scientific hypotheses is unworkable.

[37] This quotation is from Canter's (1993a, p. 80) summary of the TVA monitoring program.

[38] The same statement can also be made about social impact assessment. For an analysis of factors contributing to the underassessment of social impacts, see Rickson et al. (1990).

nuclear power plants and liquefied natural gas terminals.

The steps in performing an environmental risk assessment have their roots in two older fields: reliability engineering, which includes method for analyzing the probability of accidents and system failures at industrial facilities; and *toxicology,* "the study of the adverse effects of chemicals on health and of the conditions under which those effects occur. . . ." (Rodricks, 1992, p. 38). (Environmental risk assessment is treated in Chapter 17.) Risk assessment results can point to measures that either decrease the risk of accidents or reduce damages if an accident or natural disaster occurs. For example, rigorous equipment maintenance programs can cut risks of industrial accidents, and emergency response procedures can be used to reduce damages.

Although risks to human health and the environment have often been given only superficial treatment in the NEPA process, there are EISs that demonstrate the contributions risk assessment can make. Canter (1993b) provides examples based on human health risk assessments in EISs for a nuclear power station, and for plans to manage vegetation in national forests.

As environmental risk assessment has advanced as a discipline, environmental specialists have urged the Council on Environmental Quality to take a leadership role in promoting risk assessment as part of the NEPA process. At the same time, researchers have attempted to streamline risk assessment procedures so they can be conducted more quickly and at lower expense.[39]

INFLUENCE OF NEPA ON PROJECTS AND ORGANIZATIONS

Has all the effort in implementing the NEPA process led to positive results? Beneficial outcomes take several forms: effects on individual projects, changes in administrative procedures and interagency relations, and modifications within development-oriented agencies.

Perhaps the most common examples of project-level effects are the many cases in which the NEPA process has resulted in mitigation measures to offset adverse environmental effects. In some cases, projects with particularly negative impacts have been dropped. What is more difficult to document is the influence of the NEPA process in keeping a project proponent from even proposing an environmentally damaging project for fear that it would never survive a review of its adverse impacts.

While the NEPA process often yields suggestions for mitigation measures, the process does not commonly influence fundamental decisions concerning the size or location of a development project.[40] This occurs because EISs are often prepared after important project siting and sizing decisions have been made.

NEPA as an Impetus for Administrative Change

In addition to influencing projects, NEPA has changed administrative procedures. The Act has been particularly successful in opening the decision-making processes of federal agencies to public scrutiny. Citizens and nongovernmental organizations are able to obtain much information on agency proposals, and they have opportunities to influence those proposals. Agencies know that if they fail to provide full disclosure of impacts as required by NEPA, they may be forced to do so as a result of NEPA-related court actions.

NEPA has also influenced administrative processes by enhancing interagency coordination. The NEPA process requires that environmental impact statements be reviewed by EPA and other interested agencies at all levels of government. This facilitates the exchange of information about proposed actions and their impacts.

The process of interagency review also affects power relations among agencies. Before NEPA, it was easier to apply political pressure to squelch one agency's critique of another agency's proposed actions. Indeed, the requirement that the Environmental Protection Agency comment on *every* EIS originated in response to charges that the Nixon administration had suppressed an EPA critique of the supersonic transport proposal.[41] Using the NEPA

[39] For examples of ways to streamline risk assessments, see Canter (1993b).

[40] This issue is taken up in detail by Hill and Ortolano (1978).

[41] For details on the origins of EPA's responsibilities to comment on EISs, see Taylor (1984).

process, all agency comments on an EIS are made public.

Although the review and comment on a draft EIS provides new opportunities for agencies to bargain among themselves, much of NEPA's influence on relations among agencies involves behind-the-scenes activities. Personnel at CEQ and EPA often have informal working relationships with NEPA implementation units within federal agencies. These relationships enable the exchange of information on agency proposals before the process of preparing EISs is initiated. These informal relationships also allow staff at CEQ and EPA to assist environmental specialists in development-oriented agencies to enhance the quality of agency projects.[42]

The influence of CEQ as a political force within the executive branch has also been notable at times, depending on the interest of the president in environmental matters. For example, the Council on Environmental Quality had high status during the Carter administration, but it suffered a substantial diminution of power during the 1980s when the Reagan budget reduction led to massive cuts in staff.

Effects of NEPA on Project Proponents

The NEPA process also affects organizations that propose projects. However, because an agency may experience different forms of pressure to improve its environmental performance, NEPA's influence may be difficult to isolate. This is illustrated by organizational changes in the U.S. Army Corps of Engineers. During the early 1970s, the Corps was coping with NEPA's requirements, pressure from the public to improve its environmental record, and new requirements to include enhancement of environmental quality as an objective of federal water resource projects. In addition, many Corps projects were halted or delayed by NEPA-related lawsuits initiated by NGOs.

Several noteworthy changes in the Corps of Engineers took place in the 1970s.[43] Each Corps district and division office was augmented by creating a new environmental unit to meet NEPA and related requirements. Several hundred environmental specialists were hired. For the first time, specialists with disciplinary training in environmental science and environmental engineering were integrated (at some level) into the Corps' project planning process. While many of the new environmental specialists were hired to produce EISs, some learned how to influence engineers responsible for project design. Environmental specialists often promoted innovative solutions; for example, flood-proofing structures (instead of building dams) to reduce flood damages.

The Corps' new environmental specialists often had to balance their loyalty to the Corps with their commitment to enhancing environmental quality. According to Taylor (1984, p. 127), this balance was sometimes struck by discriminating carefully to find the right projects to change.

> [For a Corps of Engineers environmental specialist, the] secret of appearing reasonable to the managers is not "crying wolf" all the time or "too often." The analysts build up credit [with their managers], recognizing the unchangeable [projects], smoothing the way for the harmless, doing the little things that grease the procedural wheels in order to appear cooperative and "discriminating" rather than uniformly "anti-project." [When they identify a project they are concerned about, Corps environmental specialists engage in what some of them describe as] "calling in our poker chips," the chips being organizational credits gained by "reasonable" behavior in the past.

Following this strategy, some environmental specialists in the Corps learned when and how to engage in internal politicking to derail or modify environmentally damaging proposals.

In addition to creating new units and hiring environmental specialists, the Corps rewrote its planning procedures to (1) elevate the importance of environmental quality as a planning objective, (2) accommodate requirements of the NEPA process, and (3) facilitate the participation of citizens in some aspects of project planning.[44] Changes in the Corps indicate how

---

[42] The importance of informal relationships among environmental specialists at CEQ, EPA, and other agencies is documented by Taylor (1984).

[43] For details, see the extensive documentation of change within the Corps of Engineers provided by Mazmanian and Nienaber (1979), and Taylor (1984).

[44] The changes in the Corps of Engineers were extraordinary, given that its enormous bureaucracy was dominated by engineers with a tradition of building, and many of its congressional allies were interested in promoting new projects for their home districts.

NEPA's requirements, coupled with societal pressures, have affected the structure and behavior of some agencies.

STATE ENVIRONMENTAL IMPACT REPORTING REQUIREMENTS

Six months after NEPA was enacted, the Commonwealth of Puerto Rico initiated the first of what came to be called "state NEPAs." These are laws modeled on NEPA, and they direct state (and some local) agencies to consider the environmental effects of their actions. By 1979, a total of 18 states had comprehensive environmental impact reporting requirements established either by legislation or by executive or administrative orders (see Table 15.3). In addition, nine states had special-purpose environmental assessment requirements. An example of the latter is Nebraska's use of environmental impact reporting for state-funded highway projects.

For most state programs, the points covered in environmental impact documents are similar to those in Section 102(2)(C) of NEPA. A few states require consideration of additional items. For example, New York's State Environmental Quality Review Act calls for an assessment of the growth-inducing and energy impacts of proposed actions, and a description of mitigation measures that could minimize adverse impacts.

State programs often include steps similar to those in the NEPA process, but they differ from the federal program in several ways. For one thing, arrangements for administering requirements vary. In some states, the EIS program is managed by a department of natural resources. Other states rely on their environmental protection agencies. Another administrative form is an environmental council or board in the governor's office, such as the Office of Environmental Quality Control in Hawaii.[45]

TABLE 15.3 States with EIS Requirements[a]

States Using Comprehensive Statutory Requirements[b] (14)
California, Connecticut, Hawaii, Indiana, Maryland, Massachusetts, Minnesota, Montana, New York, North Carolina, South Dakota, Virginia, Washington, and Wisconsin

States Using Comprehensive Executive or Administrative Orders (4)
Michigan, New Jersey, Texas, and Utah

States with Special-Purpose EIS Requirements (9)
Arizona, Delaware, Georgia, Kentucky, Mississippi, Nebraska, Nevada, New Jersey,[c] and Rhode Island

[a] Based on information in Mandelker (1993, sec. 12.02) and Council on Environmental Quality (CEQ, 1979, pp. 595–602).

[b] The Commonwealth of Puerto Rico and the District of Columbia also have comprehensive statutory requirements.

[c] New Jersey has EIS programs established under a state executive order *and* special "procedural rules" for implementing state laws governing the use of wetlands and coastal areas.

State Environmental Assessment Requirements and Private Development

A more fundamental difference between NEPA and some state programs concerns effects on private land development. In California and Washington, for example, EIS requirements extend to actions of local governments, thereby subjecting private developers to impact assessment requirements.[46] Proposals by developers typically require building permits and other approvals from cities and counties. The actions of local governments to grant such approvals are decisions that may affect the environment. And if potential impacts are significant, environmental impact assessments are called for.

The influence of state NEPAs on private land development is illustrated by the California Environmental Quality Act (CEQA), which was passed in 1970 and has been amended periodically. The act and the guidelines used to implement it are known for their thoroughness.[47] When CEQA was first enacted, few imagined that it would eventually influence pri-

[45] Tryzna (1974) describes the many organizational arrangements used in administering state NEPAs.

[46] Other states where EIS requirements apply to local government include Hawaii, Massachusetts, Minnesota, and New York. In North Carolina, local governments may require EISs for major development projects undertaken by private parties. Further information on the programs in these states is given by Mandelker (1993).

[47] According to Mandelker (1993, p. 12–7), CEQA "elaborates more extensively than any other state environment policy act on NEPAs environmental decision-making requirements."

vate developers. The language of the original statute seemed to apply only for projects that state agencies built, such as dams constructed by the California Department of Water Resources. CEQA did not appear applicable to projects that local agencies approved. That changed in 1972, when the California Supreme Court decided the case of *Friends of Mammoth v. Board of Supervisors of Mono County.*

Friends of Mammoth v. Mono County concerned a conditional use permit that had been issued for a condominium project by the Mono County Board of Supervisors in 1971.[48] Mono County is in eastern California near Yosemite National Park and was characterized in the California Supreme Court's opinion on the case as "one of the nation's most spectacularly beautiful and comparatively unspoiled treasures." Plaintiffs argued that Mono County's action in granting the conditional use permit would lead to significant adverse environmental impacts, and that the decision should have been accompanied by an environmental impact report. (An *environmental impact report* under CEQA is similar to an *environmental impact statement* under NEPA.) In deciding in favor of the plaintiffs, the court concluded that the "legislature necessarily intended to include within the operations of the act, private activities for which a government permit or other entitlement of use is necessary." This position was taken in September 1972, and the California Environmental Quality Act was amended soon thereafter. The amendments indicated that CEQA's environmental impact requirements applied to a broad range of decisions by local governments.

Because local governments make enormous numbers of routine approvals, the workings of those governments would have ground to a halt if certain types of decisions had not been exempted from CEQA. Both the act and the guidelines used in implementing it provide bases for exemptions. For example, the statute exempts *ministerial projects,* which are decisions that involve the application of "fixed standards or objective measures."[49] No personal judgments are involved. A public official merely applies the law to the facts as presented and uses no discretion or judgment in reaching a decision.[50] An example of a ministerial project is issuance of a building permit where a local ordinance requires the permit to be issued as long as the applicant pays the necessary fee and meets local zoning and building code requirements. Decisions that are not ministerial are termed *discretionary;* a city council's decisions to annex new land or change a zoning ordinance are examples of discretionary actions.

Application of CEQA to private land development is illustrated by a case involving Stanford University. During the 1980s, Stanford requested a modification to its general land use permit from Santa Clara County. As part of its permit application, the university provided a general description of proposed facility types, sizes, and locations. To meet its responsibilities under CEQA, Santa Clara County called for an environmental impact report describing effects of the land use plan proposed in the permit application.[51]

The EIR for Stanford's land use plan identified a number of cumulative impacts, particularly effects of land development on traffic congestion. Before it approved Stanford's permit application, the County obtained assurances that the university would monitor cumulative traffic impacts of its new projects. Stanford agreed to gather data on traffic flows on an annual basis and take steps to reduce the number of its employees that commute to work in single-occupancy vehicles. Overall trip reduction targets were set, and the university agreed to expand its road network if its trip reduction plans did not allow maintenance of stipulated levels of service at several intersections.[52]

Procedure versus Substance in State NEPAs

Another difference between NEPA and analogous state programs is that some states impose more rigor-

[48] This presentation of the *Friends of Mammoth* case is based on Rodgers (1976, pp. 67–58).

[49] This quote is from Remy, Thomas, and Moose (1991, p. 35). That source also contains details on the guidelines implementing CEQA. Other state environmental policy acts also exempt ministerial actions.

[50] The reason for exempting purely ministerial acts is to eliminate the time and effort of preparing an assessment that cannot affect an agency's decision. The guidelines implementing CEQA contain lists of actions that have been exempted by either the statute or the guidelines used in implementing the act.

[51] An EIR for a land use plan is much like a programmatic EIS under NEPA.

[52] The EIR for Stanford's land use plan demonstrates the process of tiering. If individual projects Stanford proposed in the future were consistent with projects outlined in its general land use plan, the County would refer to the EIR for the general plan for an assessment of at least some impacts. Accordingly, environmental impact reports prepared for new projects could be briefer.

ous obligations on public decision makers.[53] The California program is used again for illustrative purposes. The California Environmental Quality Act requires that if an EIR identifies significant environmental impacts, the lead agency must either (1) call for changes that avoid (or substantially reduce) the adverse effects, or (2) provide evidence that it is infeasible to do so.[54]

If an agency approves a project that will cause significant adverse effects that are not "at least substantially mitigated," CEQA requires the agency to issue a *statement of overriding considerations.* In this statement, the agency must demonstrate why "the benefits of the proposed project outweigh the unavoidable adverse effects." Benefits, in this context, include reasons for approving the project such as the need to create jobs, generate tax revenue, and provide housing. If the agency approval is challenged in court, CEQA requires courts to use the *substantial evidence role* to decide the case. Under this rule, a reviewing court will defer to an agency's judgement if information on record would allow a reasonable argument to be made in support of the agency's approval of the project.[55]

Another substantial obligation imposed by CEQA is a 1989 amendment that requires a public agency to adopt a *monitoring program* if it finds that mitigation measures will be used to reduce impacts to less than significant levels.[56] This amendment was added in response to studies indicating that many agencies implementing CEQA did not know whether the mitigation measures they called for were being implemented.

As of 1993, the CEQA requirements for monitoring programs had not been spelled out. Although agencies that call for mitigations in EIRs were required to adopt a monitoring program, what constituted an acceptable monitoring program remained unclear.[57]

ENVIRONMENTAL IMPACT ASSESSMENT OUTSIDE THE UNITED STATES

The influence of NEPA has not been limited to the United States. By 1990, dozens of countries had environmental impact assessment (EIA) requirements.[58] Some, such as the EIA system in the Philippines under former president Ferdinand Marcos, included language similar to Section 102(2)(C) of NEPA. Others, such as the Chinese EIA program, were quite different and reflected deliberate efforts to tailor EIA requirements to the local political context. Some countries set up their EIA programs using laws, while others, such as Taiwan, relied on administrative orders. In addition to national-level programs, the states (or provinces) in some countries established their own EIA requirements. Not surprisingly, there are wide variations in the scope and quality of EIA among and within countries.

Requirements for EIAs are even imposed in countries that have no formal programs because development assistance organizations such as the World Bank often require EIAs on projects they support.[59] Although international aid agencies have spotty records in implementing their own EIA requirements, they have been under pressure from nongovernmental organizations to improve the way EIAs are conducted for projects they fund.

[53] California, Washington, and New York are in this category. For an analysis that compares NEPA with the state environmental impact reporting requirements in California, Washington, and New York, see Ferester (1992).

[54] In this context, the CEQA guidelines define "feasible" to mean "capable of being accomplished in a successful manner within a reasonable period of time, taking into account economic, environmental, legal, social and technological factors."

[55] Under the substantial evidence rule, the record of facts might conceivably be interpreted to favor a decision different from the agency's. However, courts will defer to the agency provided that a reasonable person might accept the evidence on record as adequate support of the agency's position.

[56] Remy, Thomas, and Moose (1991, p. 511) provide details on the monitoring program requirements and the studies that led up to the 1989 amendment of CEQA. These studies indicated that independent site visits to ensure that mitigation took place were not frequently made. Also, many agencies did not require developers to let them know that mitigation requirements had been satisfied. A monitoring program is also needed if an agency issues the CEQA equivalent of a mitgated FONSI in the NEPA process.

[57] As of 1993, detailed guidelines had not been issued, and the courts had not yet heard cases centering on the adequacy of monitoring.

[58] Outside of the United States, it is common to use the acronym "EIA" broadly, in referring to both assessment programs and assessment documents such as environmental impact statements. Robinson (1992) reports that over 40 countries have EIA requirements; Canter (1996) indicates the number is close to 80.

[59] For information on the EIA requirements of multilateral and bilateral lending institutions, see Mikesell and Williams (1992), and Kennedy (1988), respectively.

Each national, state, or local EIA program can be viewed as a policy experiment whose results can inform the EIA programs of others. Until recently, EIA specialists from different areas learned from each other primarily through articles in environmental journals, meetings of professional associations, and ad hoc communications among staffs of government agencies. Information exchange was augmented in the mid-1990s, when "The International Study of the Effectiveness of Environmental Assessment" was conducted. The study included surveys and appraisals of EIA practices in different countries.[60]

Although it is infeasible to summarize the many forms of EIA requirements outside the United States, some typical features can be highlighted. This section uses the EIA system in Rio de Janeiro State to introduce an EIA system design that is common outside of the United States: an environmental protection agency defines the scope of an EIA, evaluates its contents, and makes recommendations to decision makers. The discussion following the Rio de Janeiro example further illustrates EIA requirements outside of the United States by examining key questions in EIA program design.

EIA Program in Rio de Janeiro

In 1986, a resolution issued by the Conselho Nacional do Meio Ambiente (CONAMA, the national council for environmental improvement) required that each of Brazil's 26 states establish programs that make EIAs compulsory for major public and private projects.[61] Responses to the 1986 resolution differ from state to state. The program for the state of Rio de Janeiro is among the most sophisticated in Brazil.

The EIA program for Rio de Janeiro is administered by the state's environmental protection agency: Fundação Estadual de Engenharia do Meio Ambiente (FEEMA). The state's program works as follows: If a project proponent (either public or private) proposes a major facility, FEEMA organizes a multidisciplinary EIA team from among its staff. The team prepares the *terms of reference* for the EIA, which include the issues to be studied by those conducting the assessment. FEEMA requires the project proponent to hire an independent consultant to conduct the EIA.[62] The consultant submits the completed EIA to FEEMA where it is evaluated by the team that prepared the original terms of reference. The team may, at its discretion, call for additional information or analysis. In addition, if requested by local citizens, a public meeting may be held to provide citizens with an opportunity to register their concerns.

FEEMA's evaluation of the consultant's EIA provides a basis for its recommendation to the decision-making body: an environmental commission made up of the president of FEEMA and the heads of various state ministeries. FEEMA's recommendation can take one of three forms: acceptance or rejection of the project as proposed, or (as is common) acceptance subject to project modifications to mitigate adverse impacts. The state's environmental commission has often adopted FEEMA's recommendations.

The extent of compliance with FEEMA's EIA procedures is often high for several reasons. First, the state's regulations implementing the 1986 resolution of CONAMA require that, for major projects, a proponent cannot build or operate a facility without a license from the state's environmental commission. The commission, in turn, will not grant a license until it receives FEEMA's recommendations. Second, the Brazilian Constitution of 1988 includes sections requiring EIAs for projects that may cause significant degradation of the environment, and Law 7347 of 1985 gives private associations the right to take "public civil actions" to enforce "responsibility for damages caused to the environment" (Findley, 1991). Citizens of Brazil have taken advantage of these legal provisions, and they have used the courts to stop projects that violate EIA requirements. Third, and finally, FEEMA can sometimes increase its political

[60] This study was a joint initiative of the International Association for Impact Assessment and Canada's Federal Environmental Assessment Review Office (Sadler, 1996).

[61] The requirement for EIAs was established by Resolution No. 1 of 1986 of CONAMA, pursuant to Brazil's National Environmental Policy Act, Law 6938 of 1981. This discussion of EIA in Brazil is based, in part, on information presented by Tania Muniz Ferreira and Angela Maia of the Fundação Estadual de Engenharia do Meio Ambiente at a United States/Brazil Workshop on EIA and Risk Assessment, Federal University of Rio de Janeiro, Brazil, July 29–August 1, 1991. The remainder of this chapter is adapted from Ortolano (1993, pp. 355–58).

[62] This requirement for an independent consultant is intended to enhance objectivity. However, objectivity is not ensured since the proponent selects the consultant and controls the budget for the assessment.

clout by gaining support from the environmental movement in Brazil. The ability of FEEMA to have its recommendations taken seriously is enhanced when its positions are endorsed by politically influential nongovernmental organizations.

Notwithstanding the generally positive role that FEEMA has played in implementing its EIA procedures political forces sometimes interfere with its environmental assessment work. For example, the Linha Vermelha, a highway that connects the airport serving Rio de Janeiro with downtown Rio, was exempted from EIA requirements, in part because of political pressure to build the highway in time to serve the 1992 United Nations Conference on Environment and Development.

Another case in which political forces influenced FEEMA's work concerns a proposed overflow tunnel to control flooding in the area near Lake Rodrigo de Freitas (Brito and Moreira, 1992). In 1989, the Superintendencia Estadual de Rios e Lagoas (SERLA, the state agency for lakes and rivers) obtained a World Bank loan for an overflow tunnel that had been designed in the 1970s. After SERLA received the loan, it pressed FEEMA for a cursory and speedy EIA review because it was paying interest on the loan for the tunnel and wanted to begin construction. In this case, FEEMA withstood the pressure. Its EIA process revealed that the proposed tunnel would not solve the flooding problem, and that the environmental impacts of the project had not been adequately considered. After analyzing the situation, a World Bank mission recommended that efforts to implement the tunnel be postponed. As these Brazilian examples suggest, political pressures can play an important role in EIA implementation, and the political clout of an environmental agency is a key factor in determining EIA effectiveness.

Differences Among EIA Programs

One way to understand how EIA programs differ is to consider how the following questions are answered:

1. Who decides if an EIA is required for a particular proposal?
2. Who prepares the EIA?
3. Who decides what impacts an EIA for a proposed project must consider?
4. When is the EIA conducted relative to project location and design decisions?
5. How may citizens participate in the EIA process?
6. What resources are available to the unit evaluating the EIA?
7. What authority does the unit evaluating the EIA have to modify the project?

Decision to Require an EIA

The process for determining if a proposed action requires an environmental impact assessment is called *screening,* and it varies among EIA systems. One approach to screening uses lists of types of projects that must have an EIA. Often the lists include threshold criteria based on project size or cost. For example, a decree issued to implement an EIA program in France includes several lists.[63] One contains projects, such as mining operations, that automatically require an environmental impact study. Another list has projects that require an EIA if they exceed certain size limits. For example, an EIA is required for hydroelectric power facilities that have a capacity greater than 500 kilowatts. Still another list includes projects, such as airports, that require an EIA if project cost exceeds six million francs. The environmental studies required by the EIA program in France become part of the project application materials evaluated by the prefect of the *département* (similar to a U.S. county) responsible for making a decision.[64]

Some countries leave the decision on whether a project requires an EIA up to the discretion of government officials. In these circumstances, proposed projects with significant environmental impacts may be exempted from EIA requirements. This is illustrated by EIA programs in Australia, where a small fraction of development projects are subject to EIA procedures.[65] Critics argue that this relatively low usage of EIA is a result of the discretionary nature of the EIA programs. For example, in analyzing the Commonwealth of Australia's EIA program between 1975 and 1985, Formby (1987) found that less than 10 environmental impact statements per

[63] The national EIA program in France was established by Law No. 76-629 of July 10, 1976, regarding the protection of nature.

[64] Details on implementation of the French EIA program are given by SEPIA (1981).

[65] In addition to having an EIA system at the commonwealth level, Australia has several EIA programs at the state level.

year were called for. This constituted only 4% of the proposals considered significant. The small number of environmental assessments resulted because the decision to initiate the commonwealth's EIA process was in the hands of the minister proposing an action, not the minister responsible for environmental affairs. In addition, citizens not directly harmed by a project are unable to gain standing to sue government agencies in the courts of the commonwealth.[66]

The examples from France and Australia illustrate two approaches for screening out proposed actions requiring environmental impact assessment: (1) lists of actions (with specified size or cost thresholds) that require an EIA and (2) procedures giving government officials discretion to decide if an EIA is needed. A comparative analysis by Wood (1995, p. 15) of EIA in several countries shows that "most EIA systems adopt a hybrid approach involving lists, thresholds and the use of discretion."

Preparation of EIA

A central question in designing an EIA program is: Who should prepare the impact assessment? In Rio de Janeiro, project proponents are not permitted to prepare EIAs because of concerns related to bias. In contrast, the EIA systems in Australia allow proponents to prepare their own environmental impact statements, but the environmental units that evaluate impact statements prepared by proponents take account of possible bias. According to Porter (1985, p. 222),

> [government environment assessment units in Australia] often regard the [environmental impact statement] primarily as a marketing document that "sells" the advantages of the development, but which includes some concessions to environmental protection, and their [that is, the government unit's] assessment report as a means of emphasizing environmental impacts in an effort to win as many concessions from the proponent as possible. In this role-playing exercise, the minister is then seen as the "judge" who ultimately makes the decision on behalf of the community.

The question of who has the technical qualifications to conduct an EIA has been treated in some countries. In China, for example, national regulations list several hundred research institutes and other organizations licensed to prepare environmental impact assessments.[67] Another example is provided by Japan, which recently developed requirements for licensing "EIA experts" for certain types of projects.

Environmental Effects to be Considered

Statutes and regulations provide a framework for determining, in general terms, what impacts on EIA must examine, but how are detailed EIA content requirements set for a specific proposal? Environmental professionals invariably play a key role in making judgments about the scope of work of a particular EIA, but the process for determining the scope of a study differs among countries. The previously described scoping process used in the U.S. federal EIA system provides one approach, but interesting alternatives to the U.S. procedure exist.

Many countries convene panels of experts to decide the scope of work (or *terms of reference*) for an EIA. Consider, for example, the EIA system in the Netherlands. The system is administered by an EIA Commission, which includes a small, full-time staff and about 200 environmental experts who are available as consultants to assist the staff in defining the scope of EIAs and evaluating results of impact studies. EIAs for proposed projects are carried out by project proponents, using guidelines issued by the competent authority (that is, the public official or agency responsible for making decisions on the proposed project).

When the EIA Commission is notified of a project proponent's intention to prepare an EIA, the Commission appoints a panel of environmental experts to recommend guidelines detailing the subjects to be covered in the EIA. The EIA Commission submits the panel's results to the competent authority as recommended guidelines. Often, but not always, the competent authority adopts the recommendations of the EIA Commission. In all cases, the scope

[66] There are jurisdictions in Australia that rely on courts to enforce EIA requirements. Consider, for example, the state of New South Wales, which established an EIA system in response to its Environmental Planning and Assessment Act (EPAA) of 1979. The act enables citizens to initiate court actions based upon alleged breaches of the EPAA (Wright, 1985).

[67] In China, an organization's EIA qualifications are based on its ability to conduct impact monitoring and on the technical training and experience of its staff.

of work for the EIA is defined by the competent authority's guidelines. In the Dutch system, the public has an opportunity to comment on the guidelines before they are adopted by the competent authority.

Timing of EIA

As argued by Armour (1991, p. 29), it is a "principle" among students of EIA that "for the EIA process to be meaningful and useful, impact analysis must occur early in the planning process and prior to any decision to proceed with a project or action." Armour also observes that this principle is seldom applied in practice.

Notwithstanding the difficulties in having project proponents prepare environmental assessments in a timely fashion, EIA programs differ in terms of when the agency appraising the EIA does its work. This is illustrated by differences between the national EIA program in Japan and the program in Shiga, a Japanese prefecture. The Environment Agency of Japan evaluates EIAs produced under the national EIA program, but the evaluation occurs so late in the decision process that it is viewed as a "ceremony."[68] The appraisal by the Environment Agency of Japan is conducted as the final stage of the EIA process, at a time when it is often too late to have any influence beyond the recommendation of minor changes.[69] In contrast, the EIA systems developed by Shiga prefecture and other local governments in Japan call for independent evaluations of EIAs by "neutral expert committees" at an earlier stage in planning. Thus the locally based systems are frequently more effective than the national EIA system.

Public Participation in EIA

Many countries with EIA programs have mandated some level of public participation in the EIA process. As an example, consider the public involvement requirements of member states of the European Union (formerly the European Community). Directive 85/337/EEC required each member state to develop, by 1988, an EIA program consistent with the general framework in the directive. Although the directive mandates public involvement in EIA, the requirements are modest. At a minimum, citizens must be made aware that an environmental impact statement has been prepared and be given a chance to comment before final decisions are carried out.

The degree of public involvement in EIA varies greatly among member states of the European Union. In some countries—for example, Spain—public involvement is minimal. The national EIA program in Spain requires that citizens be made aware of the existence of environmental impact statements and be permitted to consult them.[70] As of 1993, it was not common practice to provide copies of EISs to citizens. In contrast, other member states have more elaborate requirements. For example, citizens in the Netherlands may review and comment on both guidelines for an EIA and the environmental impact statement that results from an assessment.[71]

Although requirements for public involvement in EIA exist in many countries, those requirements do not often provide citizens with a chance to be involved in key decisions. Even when citizens are permitted to participate in scoping, it is often too late to affect decisions concerning project location and scale. Frequently, by that time, project proponents are attached to a particular course of action. Public involvement may then be reduced to public relations for purposes of defending a decision that has already been made. In this context, many citizens focus on getting project proponents to mitigate adverse effects.

This description of public involvement does not reflect the full range of experiences. In countries where governments are authoritarian, there is often no public participation in EIA. At the other extreme, some project developers (for example, Hydro-Québec in Canada) occasionally involve citizens in the early stages of project conceptualization, well in advance of project design.

Resources Available to Units Evaluating EIAs

In many countries, resources available for performing EIA appraisals are inadequate. For example, the EIA

---

[68] This characterization of the Japanese EIA system is from Shimazu (1992). The following point about EIA in Shiga prefecture is also from Shimazu.

[69] For more information on EIA in Japan, see Barret and Therivel (1991).

[70] Information on the EIA process in member states of the European Union is from Kinhill Engineers (1994).

[71] In many countries with democratic political traditions, opportunities for public involvement in EIA are similar to the opportunities within the European Union. Sometimes citizens may participate in scoping, and frequently they may comment on an EIS in writing or at a public inquiry or both.

process in Rio de Janeiro (and other Brazilian states) suffers from significant limitations on data, budget, and human resources. Observers of the EIA process in Brazil point to difficulties in gathering enough data to conduct meaningful analyses. They also note the absence of sufficient numbers of environmental specialists to conduct environmental impact studies, and to critically evaluate the work of EIA consultants. These difficulties are common for EIA programs in developing countries (Kennedy, 1988).

Resource and data limitations have also constrained the effectiveness of EIA in highly industrialized nations. For example, Shimazu (1992) indicates that the EIA system in Japan suffers from a shortage of environmental specialists to prepare and evaluate the increasing number of EIAs being conducted.

Authority of the Unit Evaluating EIAs

A basic requirement of a meaningful EIA system is that final decisions on a proposal should not be made until the unit evaluating an EIA has completed its work. That unit may recommend to decision makers that the proposal be accepted or rejected. Commonly, however, a unit evaluating an EIA recommends changes that would mitigate the adverse impacts of a proposed action. Whether the recommendations of the unit evaluating an EIA are taken seriously depends on numerous factors, including the economic importance of the proposed action and the power of the environmental unit. If the environmental unit's recommendations interfere with goals of powerful development ministries and departments, those recommendations may be ignored. In some cases—for example, the national EIA system in Japan—the environmental unit may not even conduct an appraisal unless it is requested to do so by the governmental body responsible for a proposed project.

The importance of the political power and negotiating skill of environmental units is demonstrated by the Chinese experience with EIA. Within China, city and county environmental protection bureaus (EPBs) implement national EIA regulations. An EPB, which is a bureau of local government, evaluates EIAs and makes recommendations to local government leaders. If an EPB considers an impact assessment for a development project to be unacceptable, the EPB may be opposed by pro-development bureaus that are a part of the same local government. In this context, the ability of an EPB to have its recommendations followed depends on its influence with local government leaders, and its ability to negotiate compromises with pro-development bureaus.

In some countries, for example, the Netherlands, the units evaluating EIAs have considerable influence. In the Dutch system, the component authority is required (by statute) to incorporate results of the EIA in its decision making. The competent authority must explain in writing the basis for its decision, and this explanation must be made public. A procedure exists for third-party appeals against decisions made by competent authorities, and this procedure has been used frequently. Although decisions of competent authorities may be challenged in court, the number of court actions has been small. Assessments of the Dutch EIA system indicate that the "openness of decisions" has led to improvements in the way environmental factors influence decisions.[72]

As this section has demonstrated, a variety of EIA programs exist at different levels of government throughout the world. In some contexts, EIA has been reduced to an exercise in paper shuffling. However, in many countries, EIA has contributed significantly to raising the environmental awareness of private developers, government officials, and the public at large.

Key Concepts and Terms

NEPA's Objectives and Principal Parts

National environmental policy
Action-forcing provisions
Interdisciplinary planning
Appropriate consideration of environmental values
Environmental impact statement (EIS)
Council on Environmental Quality (CEQ)

Environmental Impact Statement Preparation and Use

CEQ regulations
Lead agencies
Actions requiring an EIS
Categorical exclusion

[72] This characterization of the Dutch system is based on Wood (1995, pp. 189–190).

Environmental assessment (EA)
Finding of no significant impact (FONSI)
Mitigation measures
Mitigated FONSI
Scoping process
No-action alternative
Required content of an EIS
EIS review procedures
Draft vs. final EIS
Record of decision (ROD)
Social impact assessment
Direct vs. indirect impacts
Forecasting vs. evaluation

Compliance with NEPA's Requirements

Agency response to comments on draft EIS
EPA's rating of EISs and projects
CEQ referral process
Role of courts in NEPA's implementation
Administrative Procedure Act
Standing to sue
Procedural obligations to disclose impacts
Substantive obligations to meet NEPA's goals

Perennial Problems with NEPA's Implementation

Influence of NEPA process on decisions
Programmatic EIS
Tiering
Cumulative impact assessment
Carrying capacity analysis
Obligations to mitigate adverse effects
Post-project monitoring of impacts
Adaptive environmental management
Risk assessments in EISs

Influence of NEPA on Projects and Organizations

Project-level influences
Interagency coordination
Power relations among agencies
Full disclosure of impacts
Citizen participation in agency decision making
Environmental staff within development agencies
Revisions in agency planning procedures

State Environmental Impact Reporting Requirements

NEPA as a model for states
Differences among state environmental policy acts
EISs for private development projects
California Environmental Quality Act (CEQA)
Ministerial vs. discretionary actions
Substantive obligations under CEQA

Environmental Impact Assessment Outside the United States

Alternative forms of EIA programs
Screening
Administrative discretion and decisions to perform EIAs
Project size and cost thresholds
Scope of work
Terms of reference
Licensing EIA specialists
Enforceability of EIA procedural requirements
Politics and negotiation as elements of EIA
Resource constraints on EIA implementation
Influence of EIA recommendations

Discussion Questions

15-1 Give four illustrations of actions that are both subject to NEPA requirements and do not involve construction of a facility by a federal agency. At least one example should involve a project constructed by a private party.

15-2 Suppose you are working in the office of a U.S. federal agency and your principal task is to prepare environmental impact statements for actions proposed by your office. How would you explain the most important aspects of your job to a visitor from a foreign country? Suppose, further, that the EISs you prepare are for projects your agency constructs. Does this compromise your ability to perform objective environmental impact assessments? Does the review process delineated in Figure 15.3 ensure the objectivity of EISs prepared by your agency?

15-3 In what ways do the requirements for review and comment on environmental impact statements contribute to meeting NEPA's policy goals? Would you be in favor of minimizing those requirements as a way of reducing the cost of complying with NEPA?

15-4 How would you characterize the role of the courts in the implementation of NEPA? Why were the courts called upon so frequently to litigate cases involving agency compliance with NEPA?

15-5 In 1972, the United States Court of Appeals for the Eighth Circuit considered whether the U.S. Army Corps of Engineers should be permitted to proceed with its Gillham Dam project on the Cossatot River in Arkansas. In reaching its decision, the court held that NEPA was "intended to effect substantive changes" in administrative decision making because it required more than a full disclosure of environmental impacts. In *Environmental Defense Fund v. Corps of Engineers,* the Court wrote,

> [t]he purpose [of the EIS procedure] is to "*insure that the policies enunciated in Section 101 are implemented.*" The procedures included in [Section] 102 of NEPA are not ends in themselves. They are intended to be "action forcing."[73]

Elaborate on the court's position by clarifying the linkage between Sections 101 and 102 of NEPA. To what extent did the U.S. Supreme Court agree with the Eighth Circuit in NEPA-related cases decided by the Supreme Court through the early 1990s. Do you agree with the position of the Eighth Circuit?

15-6 How might a programmatic EIS be used to consider cumulative impacts? What other approaches can be used to account for cumulative impacts?

15-7 Suppose Congress wanted to amend NEPA to impose unambiguous substantive obligations on agencies to meet the policy goals in the act. What might Congress do to clarify the meaning of the enviromental objectives spelled out in Section 101(b)? What might Congress learn from experience with state EIA programs that impose substantive obligations on government decision makers.

15-8 Suppose the Council on Environmental Quality were to amend its NEPA regulations. Which problems in NEPA implementation might the CEQ address? What regulatory changes could it make to improve implementation of NEPA?

15-9 Between 1970 and 1990, both the number of EISs prepared under NEPA and the number of NEPA-related lawsuits trended downward. For example, Ferester (1993, p. 225) reported that during the 1970s, an average of about 1460 EISs per year were prepared, whereas only 640 EISs were prepared annually during the 1980s. He also observed that from 1985 to 1989, an average of about 73 NEPA-related lawsuits per year were filed, which is 47% lower than the average number of cases filed from 1975 to 1979. How might these trends be explained? Consider the following in preparing your response: the U.S. Supreme Court's positions on substantive vs. procedural obligation under NEPA, the use of "mitigated FONSIs," and the internal changes that NEPA may have prompted within agencies (for example, hiring large numbers of environmental scientists and environmental lawyers).

15-10 To what extent do state environmental policy acts differ from each other and from NEPA? In what sense is the California Environmental Quality Act more demanding than NEPA?

15-11 Suppose you were called upon to advise the ministry of environment in a developing country on whether it should advocate adoption of a law like NEPA. What advice would you offer based on what you know about experiences with NEPA and state NEPAs in the United States?

15-12 At a session of the 1994 annual meeting of the International Association for Impact Assessment, the subject of how federal agencies in the United States attempt to avoid compliance with NEPA was discussed. The Bush administration's refusal to prepare an EIS for the North American Free Trade Agreement was used as an example. During the session, a participant from a developing country argued that attempts by U.S. decision makers to avoid the NEPA process was hypocritical. He asked, "How can the United States urge developing countries to undertake EIAs for their projects, when it does not carry out EIA

[73] The court's opinion is as quoted by Ferester (1993, p. 215). The material in italics is from a U.S. Senate Report from 1969. For further details and citations, see Ferester.

for its own projects?"[74] How would you respond to this question?

References

Anderson, F. R. 1973. *NEPA in the Courts.* Baltimore: Johns Hopkins University Press.

Armour, A. 1991. Impact Assessment and the Planning Process: A Status Report. *Impact Assessment Bulletin* 9(4): 27–33.

Barrett, B. F. D., and R. Therivel. 1991. *Environmental Policy and Impact Assessment in Japan.* London: Routledge.

Bass, R., and A. I. Herson. 1993. *Mastering NEPA: A Step-by-Step Approach.* Point Arena, CA: Solano Press Books.

Blaug, E. A. 1993. Use of the Environmental Assessment by Federal Agencies in NEPA Implementation. *The Environmental Professional* 15(1): 57–65.

Brito, E. J. N., and I. V. D. Moreira. 1992. "The World Bank-Supported Drainage Network of Urban Watersheds in Rio de Janeiro." Presented at the Twelfth Annual Meeting of the International Association for Impact Assessment, August 19–22, Washington, DC.

Burdge, R. J. 1994. *A Community Guide to Social Impact Assessment.* Middleton, WI: Social Ecology Press.

Caldwell, L. 1982. *Science and the National Environment Policy Act.* University: University of Alabama Press.

Canter, L. 1993a. The Role of Environmental Monitoring in Responsible Project Management. *The Environmental Professional* 15 15(1): 76–87.

———. 1993b. Pragmatic Suggestions for Incorporating Risk Assessment Principles in EIA Studies. *The Environmental Professional* 15(1): 125–138.

———. 1996. *Environmental Impact Assessment,* 2d ed. New York: McGraw–Hill.

CEQ (Council on Environmental Quality). 1979. *The Tenth Annual Report of the Council on Environmental Quality.* Washington, DC: U.S. Government Printing Office.

[74] The United States has promoted EIA for developing countries by its influential role in pressuring the World Bank to have EIAs done for projects that the World Bank funds. In addition, the U.S. Agency for International Development has required EIAs for its projects for many years.

———. 1981. "Memorandum: Scoping Guidance." April 30, 1981. Washington, DC: CEQ.

———. 1984. *The Fifteenth Annual Report of the Council on Environmental Quality.* Washington, DC: U.S. Government Printing Office.

———. 1986. *Regulations for Implementing the Procedural Provisions of the National Environmental Policy Act.* 40 CFR, Parts 1500–1508 (as of July 1, 1986).

Clark, E. R. and L. W. Canter. Forthcoming. *Environmental Policy and NEPA: Past, Present and Future.* Delray Beach, FL: St. Lucie Press.

Culhane, P. J. 1993. Post-EIS Environmental Auditing: A First Step to Making Rational Environmental Assessment a Reality. *The Environmental Professional* 15(1): 66–75.

Dickert, T. G., and A. E. Tuttle. 1985. Cumulative Impact Assessment in Environmental Planning: A Coastal Wetlands Watershed Example. *Environmental Impact Assessment Review* 4(1): 37–64.

Ensminger, J. T., and R. B. McLean. 1993. Reasons and Strategies for More Effective NEPA Implementation. *The Environmental Professional* 15(1): 46–56.

Ferester, P. M. 1992. Revitalizing the National Environmental Policy Act; Substantive Law Adaptations from NEPA's Progeny. *Harvard Environmental Law Review* 16: 207–69.

Findley, R. W. 1991. "The Evolution of Environmental Law in Brazil." Paper presented at Hastings International and Comparative Law Review Tenth Annual Symposium (February 23), Hastings College of the Law, San Francisco, CA.

Fogleman, V. 1993. Toward a Stronger National Policy on Environment. *Forum for Applied Research and Public Policy* 8(2): 79–84.

Formby, J. 1987. The Australian Government's Experience with Environmental Impact Assessment. *Environmental Impact Assessment Review* 7(3): 207–26.

Gilpin, A. 1995. *Environmental Impact Assessment: Cutting Edge for the Twenty-first Century.* Cambridge, U.K.: Cambridge University Press.

Herson, A. I., and K. M. Bogdan. 1991. Cumulative Impact Analysis Under NEPA: Recent Legal Development. *The Environmental Professional* 13: 100–106.

Hill, W. W., and L. Ortolano. 1978. NEPA's Effect on the Consideration of Alternatives: A Crucial Test. *Natural Resources Journal* 18(2): 285–311.

Hoban, T. M., and R. O. Brooks. 1996. *Green Justice: The Environment and the Courts* 2nd ed. Boulder, CO: Westview Press.

Holling, C. S., ed., 1978. *Adaptive Environment Assessment and Management,* New York: John Wiley & Sons.

Kennedy, W. V. 1988. "Environmental Impact Assessment and Bilateral Development Aid: An Overview." In *Environmental Impact Assessment: Theory and Practice,* ed. P. Wathern, 272–85. London: Unwin Hyman.

Kinhill Engineers. 1994. "Public Participation in the Environmental Impact Assessment Process." Report prepared for the Commonwealth Environmental Protection Agency by Kinhill Engineers Pty. Ltd., Turner ACT 2601, Australia.

Mandelker, D. R. 1993. *NEPA Law and Litigation,* 2d ed. Deerfield, IL: Clark Boardman Callaghan.

Mazamanian, D., and J. Nienaber. 1979. *Can Organizations Change? Environmental Protection, Citizen Participation, and the Corps of Engineers.* Washington, DC: Brookings Institute.

Mikesell, R. F., and L. F. Williams. 1992. *International Banks and the Environment.* San Francisco: Sierra Club Books.

Munn, R. E., ed 1979. *Environmental Impact Assessment, Principles and Procedures,* 2d ed. Chichester, England: Wiley.

Nelson, R. E. 1993. "A Call for a Return to Rational Comprehensive Planning and Design." In *Environmental Analysis: The NEPA Experience,* ed. S. G. Hildebrand and J. B. Cannon, 66–70. Boca Raton, FL: Lewis Publishers.

Ortolano, L. 1993. Controls on Project Proponents and Environmental Impact Assessment Effectiveness. *The Environmental Professional* 15(4): 352–63.

Ortolano, L., and A. Shepherd. 1995. "Environmental Impact Assessment." In Vanclay, F. and D. Bronstein, *Environmental and Social Impact Assessment,* ed. F. Vanclay and D. Bronstein, chap. 1. Chichester, England: Wiley.

Partidário, M. R. 1996. Strategic Environmental Assessment: Key Issues Emerging From Recent Practice. *Environmental Impact Assessment Review* 16(1): 31–55.

Porter, C. F. 1985. *Environmental Impact Assessment: A Practical Guide.* St. Lucia Brisbane, Australia: University of Queensland Press.

Remy, M. H., T. A. Thomas, and J. G. Moose. 1991. *Guide to the California Environmental Quality Act,* 5th ed. Point Arena, CA: Solano Press Books.

Rickson, R. E., et al. 1990. Institutional Constraints to Adoption of Social Impact Assessment as a Decision-Making and Planning Tool. *Environmental Impact Assessment Review* 10(1,2): 233–43.

Robinson, N. 1992. International Trends in Environmental Impact Assessment. *Boston College Environmental Affairs Law Review* 19(3): 591–621.

Rodgers, J. L., Jr. 1976. *Environmental Impact Assessment, Growth Management and the Comprehensive Plan.* Cambridge, MA: Ballinger.

Rodricks, J. V. 1992. *Calculated Risks: the Toxicity and Human Health Risks of Chemicals in Our Environment.* Cambridge, MA: Cambridge University Press.

Sadler, B. 1996. *Environmental Assessment in a Changing World: Evaluating Practice to Improve Performance.* Final report of the International Study of the Effectiveness of Environmental Assessment, Cat. No.: EN 106-37/1996E. Canada: Minister of Supply and Services.

SEPIA (Sociétépour l'Environnement le planification et l'Ingéniérie Atkins). 1981. *Environmental Impact Assessment and Nuisance Control in France. Report prepared for Ministerie van Volksgeezondherd en Milieuhygiëne, the Netherlands. Paris: SEPIA.*

Shimazu, Y. 1992. "The Current EIA System and Operation Procedures of Japan." In *Workshops on Environmental Impact Assessment,* chap. 3. Environmental Protection Administration, Government of the Republic of China, Taipai, Taiwan (June 1992).

Sigal, L. L., and J. W. Webb. 1989. The Programmatic Environmental Impact Statement: Its Purpose and Use. *The Environmental Professional* 11: 14–24.

Smith, E. E. 1993. "Future Challenges of NEPA: A Panel Discussion." In *Environmental Analysis: The NEPA Experience,* ed. S. G. Hildebrand and J. B. Cannon, Boca Raton, FL: Lewis Publishers.

Taylor, C. N., C. H. Bryan and C. G. Goodrich. 1995. *Social Assessment: Theory, Process and Techniques,* 2nd ed. Christchurch, Australia: Taylor Baines & Associates.

Taylor, S. 1984. *Making Bureaucracies Think.* Stanford, CA: Stanford University Press.

Therivel, R. and M. R. Partidário. Forthcoming. *The Practice of Strategic Environmental Assessment.* London: Earthscan.

Trzyna, T. C. 1974. *Environmental Impact Requirements in the States: NEPA's Offspring.* Report prepared for Office of Research and Development. U.S. Environmental Protection Agency, Washington, DC.

Wood, C. 1995. *Environmental Impact Assessment: A Comparative Review.* Essex, England: Longman Scientific and Technical.

Wright, C. 1985. "Incorporating Environmental Factors at All Levels of Planning." In *Pollution in the Urban Environment,* ed. M. W. H. Chan et al. London: Elsevier Applied Science Publishers.

Yost, N. C. 1990. NEPA's Promise—Partially Fulfilled. *Environmental Law* 20(3): 681–702.

Part Four

Forecasting and Evaluation in Environmental Planning

As the previous chapters demonstrate, the process of forecasting and evaluating environmental effects plays a key role in both designing environmental regulations and appraising the impacts of proposed projects. Part Four surveys methods used in forecasting and evaluation. *Forecasting* involves making predictions of environmental changes, but no judgments are made about whether the predicted changes are good or bad. Those types of judgments are part of *evaluation,* the process of placing values on the significance of projected impacts of alternative actions, and synthesizing those values to help decision makers choose from among those alternative actions.

Chapter 16 concerns forecasting questions that play a central role in environmental decision making: What effect will a proposed regulation have on ambient quality? And how will a proposed project impact the environment? The chapter introduces categories of methods used to answer these questions. Some techniques are qualitative and rely heavily on expert judgment. Other forecasting procedures are quantitative and rest on fundamental scientific principles. Still other forecasting methods use mathematical models derived from statistical analyses of data.

When analyzing a proposed policy or project, forecasts of impacts on air quality, noise, and so forth provide valuable information. But how can this information be synthesized to assist in decision making? This question is tackled using methods of evaluation, a topic taken up in Chapter 17. The chapter surveys methods used to solve *multicriteria decision problems;* that is, problems in which two or more criteria are used as a basis for choosing among a set of alternative actions. Multicriteria decision problems arise commonly in evaluating proposed regulations and projects. Chapter 17 also introduces *risk assessment,* a process used increasingly in formulating environmental regulations and appraising proposed projects.

Chapter 18, which concludes Part Four, analyzes the role of citizens and nongovernmental organizations in planning and decision processes of public bodies. In many countries, citizens participate directly in decision-making processes affecting environmental regulations. Citizen participation can also play an important role in decisions on public and private development projects. Chapter 18 describes how citizen participation programs are designed and implemented. It also introduces processes for resolving disputes between private parties and agencies making decisions affecting the environment.

Chapter 16

Forecasting Environmental Effects of Proposed Projects and Regulatory Actions

Many of the methods used to forecast how proposed regulations affect the environment are also used to make predictions in an environmental impact assessment (EIA). When EIA programs emerged in the 1970s, environmental analysts attempted to synthesize and categorize forecasting methods from different professional fields. As the EIA field matured, analysts recognized that no single method would be appropriate for conducting environmental assessments for all projects. Indeed, even in an EIA for a simple project, it is common to use several forecasting procedures.[1] For the past few decades, manuals and textbooks on EIA have included dozens of different forecasting techniques.[2] This reflects the diversity of topics treated in environmental assessments, and the wide range of procedures available for conducting an assessment related to any one topic.[3]

[1] See Lemons and Porter (1992) for results of a survey of methods used in conducting EIAs.

[2] See, for example, Canter (1996) and Jain et al. (1993).

[3] A survey of methods used for predicting impacts on noise, traffic, water, and so forth is given by Morris and Therivel (1995); a more compact treatment of many of the same topics is given in Part Five of this book.

AIDS TO IMPACT IDENTIFICATION

Much of the early literature on techniques for forecasting in the context of EIA emphasized methods for identifying types of impacts typically associated with a particular kind of project or activity.[4] These techniques help in organizing environmental impact assessments, and in suggesting topics needing further study in an EIA.

A *checklist* is a simple aid to impact identification. Many federal agencies in the United States have produced checklists to help their staffs review draft environmental impact statements (EISs) prepared by other agencies. The EIS guidelines developed by the Region 10 Office of the U.S. Environmental Protection Agency (EPA, 1973) provide an example. These guidelines include separate checklists for highways, dredging and spoil disposal, land management, air-

[4] Early work on EIA methods occurred during the few years after passage of the U.S. National Environmental Policy Act of 1969. The early literature, which is reviewed by Jain et al. (1993), also provided innovative tables and diagrams for displaying results of an impact assessment.

ports, water resource developments, nuclear power plants, and pesticide use. Each list indicates the issues and analyses that EPA reviewers expect to see in an EIS involving a project (or activity) of a given type.

Some checklists denote only broad categories of impacts that may be associated with a particular type of project. An example is the list developed by the United Kingdom Department of the Environment (1989) to assist land developers in England and Wales who seek government consent for proposed projects likely to have significant environmental impacts. Examples of the broad categories of effects included in the checklist are "flora, fauna and geology" and "air and climate." For each category, there is a list of items that a developer should consider in preparing an EIA as part of the approval process used by local planning authorities in England and Wales.

Table 16.1 illustrates the types of questions that may appear under individual categories of a checklist. The table is from the "noise and vibration" category in a checklist prepared by Clark et al. (1981) containing 23 categories of impacts from major development proposals. The questions are to be raised (where applicable) for both the construction and operational phases of a proposed development.

An *impact matrix* provides another basis for identifying impacts. Figure 16.1 contains a matrix devised to identify the likely impacts of industrial development projects. Columns of the matrix contain characteristics of the *construction phase* of a proposed development, and rows include characteristics of the proposed site and its surroundings. The matrix was developed by Clark et al. (1981) who intended its use as follows: "When it is thought that the interaction of components on the vertical and horizontal axes is likely to result in an impact, the appropriate intersection is marked." Figure 16.1 illustrates this usage. Clark and his colleagues also developed a second matrix, one that demonstrated the likely impacts associated with the *operational phase* of industrial development projects. Answers to the types of questions on the checklist in Table 16.1 can be helpful in completing each matrix.

In addition to checklists and matrices, *flowcharts* are sometimes used to identify impacts. Flowcharts are particularly valuable in highlighting impacts that are two or more steps removed from the initial (or *direct*) effect of a project. The use of a flowchart to identify secondary (or *indirect*) impacts is illustrated in Figure 16.2, which sketches the physical and ecological impacts of activities that compact soil. As the figure shows, the indirect effects of soil compaction involve sequences of events. For example, increased compaction decreases water percolation into soils, which in turn can cause other impacts, such as an increase in flooding during rainy periods.

Information useful in identifying environmental effects is also given in *surveys of literature* on impacts for a given project type. Berns's (1977) review of

TABLE 16.1 Questions to Guide Assessment of Noise and Vibration Impact from Construction and Operational Phases of Proposed Development[a]

Noise and Vibration

(a) Will the proposed installation significantly alter background noise levels?
(b) If these levels increase, will the introduced noise levels be of a magnitude to cause complaints from residents either during day or night-time?
(c) Will the levels have any adverse effect on the functioning of schools, hospitals, and old people's homes or on informal recreation areas either during day or night-time?
(d) Are the levels likely to have a significant effect on the wildlife of a Nature Reserve, Site of Special Scientific Interest, Local Nature Reserve or high quality habitat of local significance?
(e) Will the levels exacerbate an already existing situation of "creeping" ambient noise levels?
(f) If so, is this significant in the local context, especially if other installations are likely to follow?
(g) Will vibration from blasting, pile-driving, etc., cause human discomfort and annoyance?
(h) Will vibration cause structural damage to ancient monuments and other old buildings?
(i) Will vibration cause structural damage to other buildings, especially houses, schools, etc.?

[a] This is one of 23 categories in a checklist developed by Clark and his associates. Crown copyright is reproduced with the permission of the Controller of HMSO. Clark et al. (1981, pp. 45–46).

	CHARACTERISTICS OF PROPOSED PROJECT															
Characteristics of the Existing Situation	**Immigration**	**Severance**	**Transport of raw materials**	**Transport of employees**	**Site preparation**	**Dust and particulates**	**Employment**	**Local expenditure**	**Water demand**	**Vibration**	**Noise**	**Gaseous emissions**	**Odours**	**Aqueous discharges**	**Solid waste disposal**	**Hazard**
Climate																
Land uses		X														
Water quality					X	X								X		
Landscape quality					X									X	X	
Ecological characteristics																
Population density																
Tourism																
Employment structure							X									
Unemployment							X									
Local economy								X								
Traffic			X	X												
Water supply									X							
Sewerage																
Finance	X															
Education	X										X					
Health service facilities	X															
Housing	X									X	X					
Emergency services	X															X
Community structure	X															
Culture	X															

Figure 16.1 Impact assessment matrix for construction phase of industrial development projects. Source: Clark et al. (1981, p. 14). Crown copyright is reproduced with the permission of the Controller of HMSO.

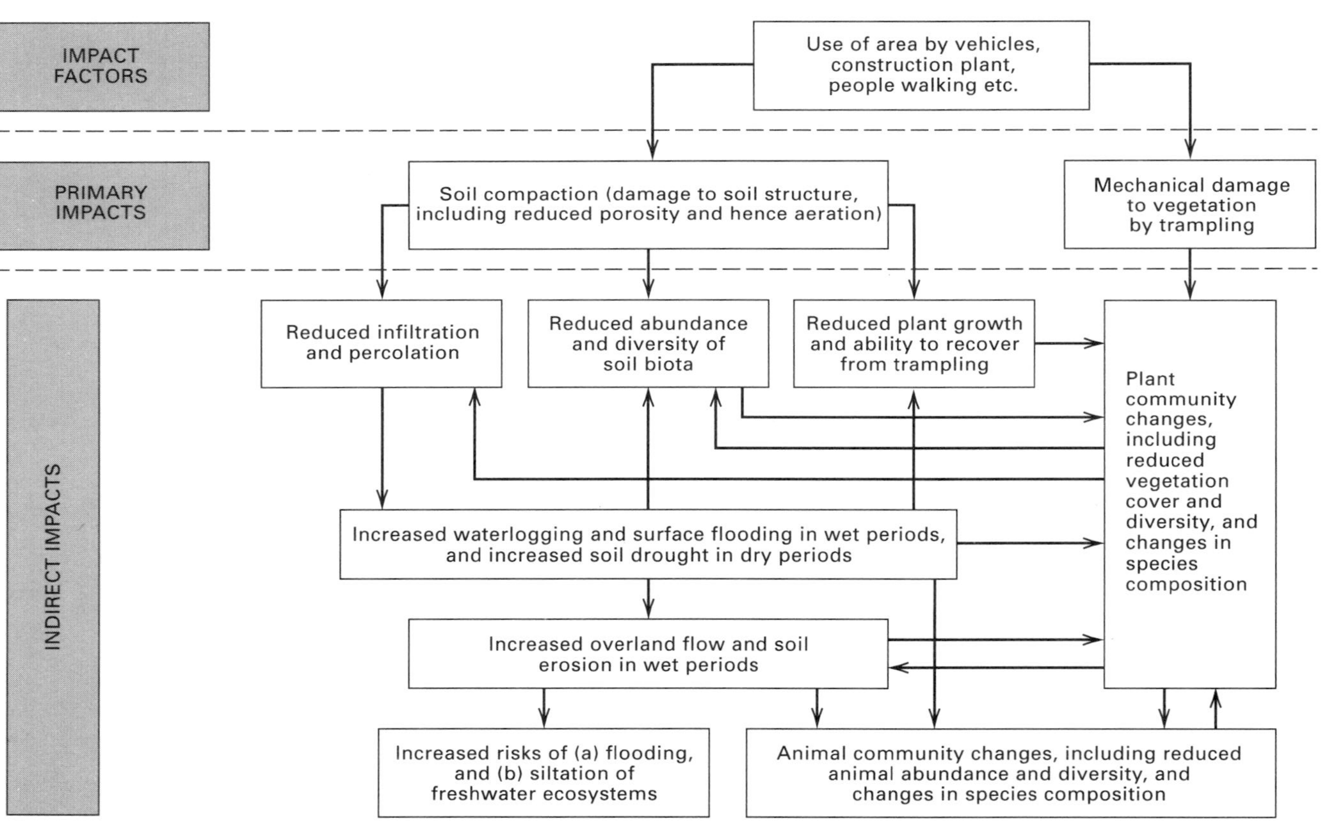

Figure 16.2 Interrelationships between ecological impacts associated with trampling and soil compaction. From Morris and Therivel, *Methods of Environmental Impact Assessment,* p. 215, © 1995 UCL Press, Ltd., London, U.K. Reprinted with permission.

impacts from land development projects provides an example.[5] Literature reviews often catalogue information on environmental changes caused by past projects, and sometimes they include case studies. For a given type of impact, reviews often refer to documents describing commonly used forecasting procedures.

Since the 1970s, computer specialists have contributed to the impact identification process by developing software to make information concerning impacts for a particular project type accessible to noncomputer specialists.[6] Some computer tools to assist in impact identification rely on sophisticated technology. For example, Fedra, Winkelbauer, and Pantulu (1991) have developed an "environmental screening" program that integrates state-of-the-art computer tools: expert systems, geographic information systems, and computer graphics.[7] In an interactive session, a user of the system responds to queries about a proposed project by selecting items on a menu. The session's output includes detailed information on the types of project impacts likely to be important. A prototype of this computer system handles several categories of development projects being considered for the lower Mekong River basin.

Although many sophisticated computer tools for impact identification have been developed, most are research prototypes and few are used routinely in practice. An example of a tool that has been used in a practical setting is SCREENER™, a software package developed by ESSA Technologies, Ltd., of Canada.[8] SCREENER™ uses information about a proposed project and its environmental setting to provide an analysis of potential impacts. The system's user receives a report indicating whether impacts are likely to be significant, and whether a full-scale environmental impact study is called for. This software has been implemented at about 40 offices of government departments in Canada, and it is being expanded to a second-generation system (called Calyx™). The new system, which is being developed by ESSA Technologies for the Asian Development Bank, contains extensive information on environmental impacts of projects in four sectors: irrigation, urban wastewater and water supply, electric power, and transportation.[9]

Impact identification can be expedited by means of *scoping,* the process of inviting agencies and interested citizens to help an organization responsible for an EIA choose which impacts to study.[10] In a basic form of scoping, parties with a potential interest in a proposed project are invited to write to the responsible organization offering suggestions for what impacts the EIA for the project should consider. Other forms of scoping are interactive and rely on public meetings designed to delineate topics to be included in an assessment.

JUDGMENTAL APPROACHES TO FORECASTING

Forecasting plays a central role in designing and implementing environmental regulations and in conducting EIAs. Indeed, an environmental impact is generally defined as a projected change in the value of one or more measures of environmental quality. Forecasts are often made for a number of alternative

[5] Many "how-to" manuals on EIA identify impacts typically associated with various types of projects. See, for example, Clark et al. (1981), and Carpenter and Maragos (1989). Also, for abstracts of the hundreds of EIA guidelines and manuals that have been prepared, see Roe, Dalal-Clayton, and Hughes (1995).

[6] An example of an early computer-based tool to assist in impact identification is one developed for the U.S. Army. According to Jain, Urban, and Stacy (1981, p. 82), the system employs a "matrix approach to identify potential environmental impacts"; it is applicable to projects in nine functional areas, including construction, research and development, training, and operation and maintenance. In addition to describing potential impacts, the computer tool also gives information on applicable mitigation measures.

[7] *Geographic information systems* are used to store, analyze, and display spatial data, and they are described in the appendix to Chapter 19. *Expert systems* are computer programs that solve problems by using, among other things, the qualitative decision rules employed by experts in the problem area. Use of expert systems in EIA is summarized by Geraghty (1993).

[8] Information on SCREENER™ and Calyx™ (mentioned below) is based on personal communication with Robert Everett, President, ESSA Technologies, Ltd., Vancouver, British Columbia, Canada, July 28, 1995. For other examples of computer-based tools for impact identification, see Julien, Fenves, and Small (1992), Antunes and Câmara (1992), and Geraghty (1993). Some of these tools integrate mathematical models for impact prediction along with checklists and other aids to impact identification.

[9] The system being developed by ESSA Technologies, Ltd., is designed to do more than identify impacts. For example, the system also issues recommendations for mitigating adverse impacts of a proposed project.

[10] As noted in the previous chapter, scoping was included in the U.S. environmental impact assessment procedures issued by the Council on Environmental Quality (CEQ, 1986).

plans, including the *no-action alternative,* which consists of related actions likely to be taken if the decision-making agency (or project proponent) pursues none of *its* alternatives. The no-action forecasts provide a base against which the environmental conditions for other alternatives can be compared.

The most frequently used approaches to forecasting environmental effects rely heavily on expert opinion. In this context, an *expert* is someone with special knowledge useful in forecasting. Thus, for example, a real estate agent might be considered an expert in predicting the land use changes induced by a proposed highway. More generally, the term expert refers to a person with academic training and practical experience relevant to the forecasting task. An expert's familiarity with impacts caused by similar projects in analogous settings commonly plays a key role in forecasting.

To illustrate how expert opinion is used in predicting environmental changes, consider a proposal to fill a marsh and construct a housing project on it. Suppose a biologist with extensive knowledge of marsh ecosystems is asked to predict the biological impacts of the project. The biologist would employ standard scientific references and field investigation methods to characterize the project area from a biological perspective. If time were available, she might gather information on observed effects of housing projects in similar settings. The forecast would consist of an opinion based on the collected information and the biologist's understanding of how marsh ecosystems function.

The judgments of specialists are also commonly used in forecasting *social impacts* of proposed projects.[11] McCoy's (1975) approach to predicting effects of a highway proposed for the Georgetown neighborhood in Lexington, Kentucky, is illustrative. The city highway department and the mayor's office were consulted for details regarding the proposed highway. Statistical descriptions of Georgetown were obtained from the U.S. census, and field visits provided further information about local residents and their activities. A questionnaire was administered to find out how residents spent their time, how they perceived their neighborhood, and how much they knew about the proposed highway and the process of relocating their residences. In addition, the literature on projects having social impacts on urban neighborhoods was reviewed. All of this information provided the basis for McCoy's projections of how the highway would affect residents of Georgetown.

Individual specialists sometimes make forecasts of environmental impacts by working in small groups. In a group, experts can build on each other's ideas and thereby provide a more useful forecast. However, there are potential difficulties in using ordinary meetings as a format for conducting group forecasting exercises. For example, a few vocal individuals may dominate the deliberations, and some specialists may feel pressured to give opinions in conformity with others in the group.

The *Delphi method* is one of the many procedures developed to increase the effectiveness of experts making forecasts as a group.[12] The method avoids the disadvantages of group meetings by using mail questionnaire surveys to obtain the opinions of experts. Throughout a Delphi exercise, the anonymity of individuals providing responses to the questionnaires is preserved. Several iterations (or "rounds") of the questionnaire survey are used to give individuals a chance to revise their previous forecasts based on the judgments of others in the group. Opinions are shared using statistical summaries of responses from preceding rounds. The summaries are mailed out with each successive round of the questionnaire. Statistical measures, such as the median and the interquartile range of the forecasts, are employed in these summaries to reduce group pressure to arrive at a consensus forecast.

The Delphi method was used by Cavalli-Sforza and Ortolano (1984) to forecast changes in land use resulting from each of three transportation projects in San Jose, California. They organized a group of 12 experts including transportation planners, engineers, economists, city officials, and representatives of citizens' groups. Each specialist received a questionnaire requesting numerical estimates of future population, housing, employment, and transit use in each of several zones that would be impacted by the proposed

[11] For an introduction to methods used in a social impact assessment, see Canter (1996, Chapter 14), Burdge and Vanclay (1995), and Jain et al. (1993, Chapter 10).

[12] Armstrong (1978) and Porter et al. (1980) discuss numerous techniques to elicit forecasts from groups of experts. The procedures are sometimes referred to collectively as "judgmental" (or subjective) forecasting methods.

transportation facilities. Results from the first-round questionnaire were summarized and mailed to each of the original respondents with a request that they revise their previous projections where appropriate. Experts whose forecasts deviated significantly from the median values were asked to rationalize their responses, and their explanations were included anonymously as part of the between-round summary information. This entire process was repeated three times. Although Cavalli-Sforza and Ortolano experienced difficulties in maintaining the enthusiasm of participants to complete the long and complex questionnaire, the final land use forecasts appeared reasonable. As in the case for most impact predictions, follow-up studies were not conducted to determine the accuracy of the projections.

PHYSICAL MODELS AND EXPERIMENTS

Physical models are scaled-down, three-dimensional representations of reality, and they have been used to make predictions for thousands of years. A centuries-old example is an architect's model of a proposed building.

In addition to depicting individual buildings, physical models can show how the appearance of an entire landscape or cityscape would change with the addition of a new project. This is demonstrated by the model of downtown San Francisco developed by the Environmental Simulation Laboratory at the University of California, Berkeley (see Figure 16.3). Using special camera equipment, films can be made showing what a pedestrian would observe when walking through various parts of the city. The model was applied in forecasting the visual effects of proposed high-rise office buildings in San Francisco. Scale models of the buildings were set within the model of the city, and films were made to portray views with the new office buildings in place. Models such as this have simulated the visual impacts of proposed projects in a variety of settings. They have also been used to estimate how proposed high-rise buildings would cast shadows and thereby decrease sunlight on city streets.

Physical models can also serve in predicting the impacts of projects on water bodies. Models are especially useful in analyzing tidal estuaries because the complex mixing of fresh and salt water that occurs in estuaries makes it difficult to apply other forecasting techniques. Experiments on physical models of estuaries have been used to forecast how navigation projects and port facilities influence water surface elevation, current velocity, salinity, and waste dispersion characteristics. Predictions have also been made of the effects of reducing freshwater inflows, such as occurs when a water supply reservoir is developed upstream of an estuary.

Estuary models generally include an array of electrical and mechanical devices to represent the effects of tides and the inflow of fresh water. Instruments are included for measuring water depth, velocity, temperature, and concentrations of salinity. Dyes can be used to simulate wastewater discharges. Figure 16.4 shows a small portion of the U.S. Army Corps of Engineer's model of Chesapeake Bay, the largest estuary in the United States. This model is built to a horizontal scale of 1 ft = 1000 ft and occupies about nine acres. It is scaled such that one year of real time can be simulated in 3.65 days.

The design of estuary models is based on scientific principles. However, these principles alone are not sufficient to guarantee an accurate representation of reality. Extensive measurements of water circulation and quality characteristics in a real estuary must be obtained to "verify" the model of that estuary. Verification of a physical model is a time-consuming process in which model features, such as the roughness of the model's surface, are changed gradually until the behavior of the model replicates that of the estuary. Models can be adjusted to provide accurate forecasts of how physical changes within an estuary influence water surface elevation and velocity. However, many modeling experts feel that these models should not be relied on for quantitative predictions of changes in water quality variables.[13]

Small-scale physical models are also used to forecast how airborne pollutants will be transported after they are discharged. Figure 16.5 shows results from an air pollutant dispersion modeling study using a wind tunnel at a U.S. Environmental Protection Agency research laboratory. Model dimensions were determined using physical principles, and the pollutant dispersion patterns in the model provided a reasonable approximation to corresponding patterns in

[13] The limitations of physical models of estuaries are elaborated by Tracor, Inc. (1971, p. 494).

Figure 16.3 Berkeley Environmental Simulation Laboratory's model of downtown San Francisco. (Photograph by Kevin Gilson. Courtesy of the Environmental Simulation Laboratory, University of California, Berkeley.)

the atmosphere. The model used fans to simulate wind conditions and tracers to represent airborne contaminants. Electronic instruments measured the velocities and concentrations of emissions downwind of the discharge point.

Models can be used to investigate how *plumes* of air pollutants vary under different conditions of stack height and building width. Figure 16.5 provides an example. The plume associated with the narrow building is almost the same as the plume with no building at all. However, the wide building causes a significant amount of *plume downwash,* the descending movement of airborne contaminants downwind from the discharge. The results shown qualitatively in the figure were confirmed by quantitative measurements of tracer concentrations at different points downwind of the smokestack. This type of wind tunnel experiment can serve to forecast concentrations of airborne contaminants emitted from smokestacks within cities.

Scientists sometimes define the term "physical model" broadly to include *laboratory tests* and *in situ experiments,* such as a study of how a small section of a salt marsh responds to increasing doses of an insecticide. Laboratory studies and in situ experiments allow scientists to investigate phenomena in controlled settings where disturbances from processes irrelevant to the objectives of a study can be minimized or eliminated. An example of scientists' broad conception of physical models is given by Suter and Barnthouse (1993, p. 34), who define them as "material representations of some object or system that is not itself subject to manipulation, or that cannot be manipulated as easily or with as much control as the models."

Although experiments used to make forecasts are commonly conducted in laboratories or on small portions of real systems, sometimes experiments are conducted for the entire system that would be influenced by a proposed project. An example is the use of tracer dyes to simulate the movement of pollutants. In a typical case, a harmless dye would be injected into an estuary to allow scientists to predict how

Figure 16.4 Portion of the physical model of Chesapeake Bay. (Courtesy of the U.S. Army Corps of Engineers, Baltimore District.)

a nonreactive contaminant would be transported within the estuary.[14]

Suter and Barnthouse (1993, p. 34) interpret the term physical model as including what are often referred to as *analog studies:* investigations of full-scale "physical systems that are exposed to similar contaminants or stress" as the system under study. They cite an impact assessment for a particular reservoir to illustrate the use of analog studies. Suter and Barnthouse argue that the most convincing evidence that the proposed reservoir would "become eutrophic" was the circumstance of a similar, nearby reservoir that was eutrophic. *Eutrophication* is a natural process in which lakes are transformed, over hundreds of years, into bogs or marshes as a result of increased inflows of plant nutrients and sediments.[15] The exam-

[14] Experiments with harmless dyes have been conducted on both real water bodies and physical models of water bodies.

[15] More information on eutrophication is given in Chapter 23.

Figure 16.5 EPA model studies of air pollutant dispersion. (a) Building width twice its height. (b) Building width one-third its height. (c) Stack without building. Reprinted with permission from *Atmospheric Environment,* vol. 10, Snyder and Lawson, Jr., "Determination of a Necessary Height for a Stack Close to a Building—A Wind Tunnel Study," with kind permission from Elsevier Science Ltd, The Boulevard, Langford Lane, OX5 1GB. (Photos supplied courtesy of the Fluid Modeling Facility, U.S. Environmental Protection Agency, Research Triangle Park, North Carolina.)

ple given by Suter and Barnthouse concerns the acceleration of this natural process that occurs when lakes and reservoirs receive high levels of plant nutrients (typically, compounds of phosphorous and nitrogen) as a result of human actions such as increased use of chemical fertilizers on lawns and agricultural fields.

FORECASTING WITH MATHEMATICAL MODELS

For decades, engineers and scientists have used mathematical models to predict environmental changes. For some environmental quality indicators, such as chloride concentration in surface waters, successful experience with mathematical models has encouraged regulatory agencies to use them. For example, models for predicting biochemical oxygen demand and dissolved oxygen are used routinely in implementing water pollution control laws. Such models are also used in conducting environmental impact assessments, but the time and expense involved in building mathematical models often constrains their use in EIA work. Later chapters detail many mathematical models that are used in implementing environmental regulations and conducting EIAs. The discussion here introduces basic modeling concepts.

Mathematical models used to predict environmental effects generally consist of one or more algebraic or differential equations. They are usually based on scientific laws, statistical analyses of data, or both. A group of mathematical models frequently used in environmental studies is termed *fate and transport models.* These models predict the transformation and distribution of contaminants after they have been introduced into the environment. Other mathematical models estimate the impacts on plants and animals of the increased concentrations of contaminants predicted with fate and transport models. Models of how environmental changes affect humans consider the nature of a person's *exposure* to pollutants, the *dose* of pollutants ingested, and the resulting adverse impacts on health and welfare.[16]

[16] Models of how environmental impacts influence people are considered in Chapters 6 and 17. Chapter 6 adopts the perspective of economists and evaluates effects of pollution in monetary terms. In contrast, Chapter 17 focuses on how calculations of exposure and dose are used in gauging human health risks.

Three examples are introduced here to demonstrate aspects of mathematical modeling. In the first example, a fundamental scientific principle, the law of conservation of mass is used to model changes in concentration of bacteria in a stream. The second example uses statistical analysis methods to develop an equation for predicting how increased highway traffic would change ambient concentrations of carbon monoxide.[17] The third example introduces an approach to modeling the transport of chemical pollutants from soil to air and water.

Predicting Bacterial Concentrations Using Scientific Principles

Application of the law of conservation of mass (that is, *mass balance analysis*) is the foundation for most models used in forecasting changes in water and air quality. In conducting a mass balance analysis, a modeler must first choose a *control volume.*[18] Experienced modelers are often able to select control volumes that simplify the mass balance for a substance. In conducting a mass balance, modelers take account of *transport* of the substance across the boundaries of the control volume, and all processes that produce or reduce the substance within the control volume. Processes of production and reduction are termed *sources* and *sinks,* respectively.

[17] Environmental modeling specialists sometimes use the term *mechanistic model* for models that employ fundamental scientific principles "to describe in quantitative terms the relationship between some phenomenon and its underlying causes." (Quotations in this note are from Suter and Barnthouse, 1993, p. 37). Scientific laws commonly used in building fate and transport models include, in addition to the law of conservation of mass, Newton's laws of motion, and the first and second laws of thermodynamics. Mechanistic models are frequently contrasted with statistical models. In the former, model coefficients "have real operational definitions and are (at least in principle) subject to independent measurement." In contrast, the coefficients in statistical models have "no intrinsic meaning."

[18] In their introduction to mass balance analysis, Hemond and Fechner (1994, p. 6) describe the process of choosing a control volume: "While the choice of a good control volume is somewhat of an art and depends on both the chemicals and environmental locations that are of interest, the control volume's boundaries are almost always chosen to simplify the problem of determining chemical transport rates into and out of the control volume."

For any time interval, the mass balance expression for a control volume is written as follows:

Change in storage = Inflow − Outflow
+ Mass produced by sources
− Mass eliminated by sinks.

In the simplest cases, there are only inflow and outflow terms and they balance. This is demonstrated by an equation for calculating the concentration of a water pollutant at a point immediately downstream from where the pollutant enters a stream. In this case, which is illustrated in Figure 16.6, the following quantities are known:

Q_w = rate of discharge of wastewater
Q_u = rate of streamflow upstream of the wastewater discharge
N_w = concentration of pollutant in the wastewater
N_u = concentration of pollutant upstream of the wastewater discharge

Wastewater discharge and streamflow are *volume flowrates* represented in units of volume per time—for example, cubic feet per second (cfs)—and pollutant concentration is typically in units of mass per volume, such as milligrams of pollutant per liter of water (mg/l). The analysis below is for *steady state conditions,* which means that all flowrates and concentrations are constants; they do not vary over time.

The simplest mass balance analysis to calculate pollutant concentration immediately downstream of the discharge point assumes that the wastewater discharge and the upstream flow mix completely and instantaneously when they meet. With complete mixing, the pollutant concentration is a constant immediately below the discharge point. As shown in Figure 16.6, the inflows and outflows of water into a control volume at the point of discharge must balance, and therefore,

$$Q_d = Q_u + Q_w \qquad \textbf{(16-1)}$$

where Q_d is the volume flowrate just downstream of the point of discharge.

A second mass balance equation can be written. The *mass flowrate* of pollutant (in units of mass per time) is the product of the volume flowrate (of water or wastewater) and the corresponding pollutant concentration.[19] A mass balance equation for the pollutant entering and leaving a control volume at the point of wastewater discharge is

$$Q_d N_d = Q_u N_u + Q_w N_w \qquad \textbf{(16-2)}$$

where N_d is the concentration of pollutant immediately downstream of the point of wastewater discharge. Solving equation 16-2 for N_d and using equation 16-1 to find Q_d, the concentration of pollutant downstream is

$$N_d = \frac{Q_u N_u + Q_w N_w}{Q_u + Q_w}. \qquad \textbf{(16-3)}$$

A more complex mass balance analysis is needed when there is production or decay of the pollutant, or a change in storage (for example, as a result of pollutants settling to the bottom of a stream). This is illustrated by a water pollution analysis that accounts for the decay of coliform bacteria, a commonly employed indicator of contamination by feces. Although coliform bacteria are usually not harmful in themselves, they signal the possible presence of microorganisms capable of causing diseases such as typhoid fever and cholera.

A model to forecast the coliform concentration (in number of organisms per volume of water) downstream of a wastewater discharge is developed by writing a mass balance equation for bacteria entering and leaving a control volume downstream of the discharge (see Figure 16.7). For simplicity, suppose both streamflow and wastewater discharge are constant. Under these steady state conditions, there is no change in storage of bacteria within the control volume. The only other component needed to build the

[19] The units involved are clarified by noting: mass flowrate (mass/time) = volume flowrate (volume/time) × concentration (mass/volume).

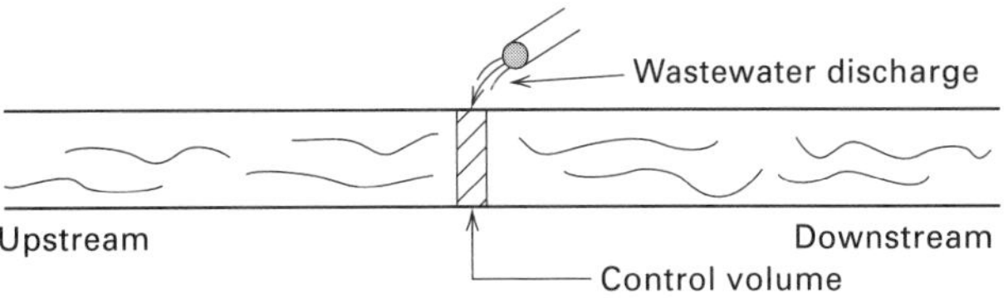

(*a*) Wastewater discharge to stream

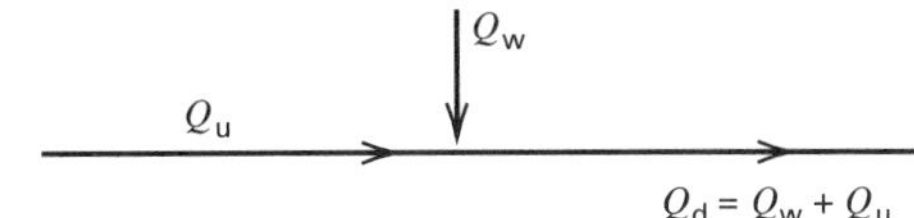

(*b*) Volume flowrates

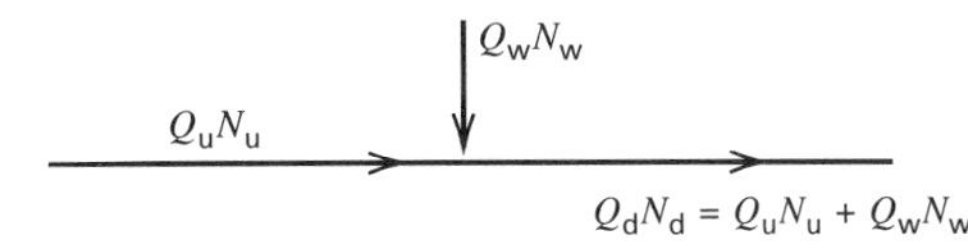

(*c*) Mass flowrates

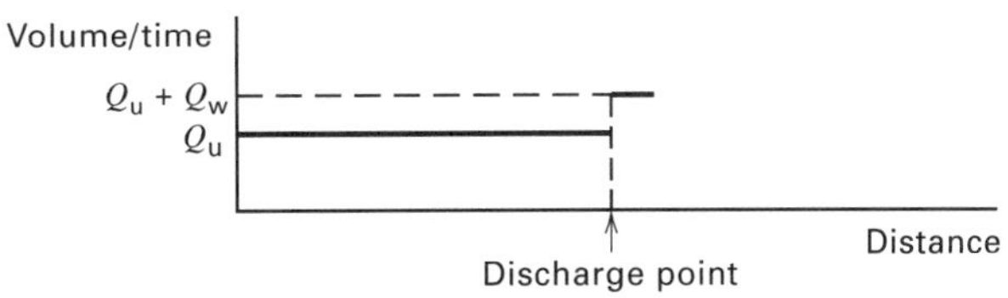

(*d*) Volume flowrate vs. distance

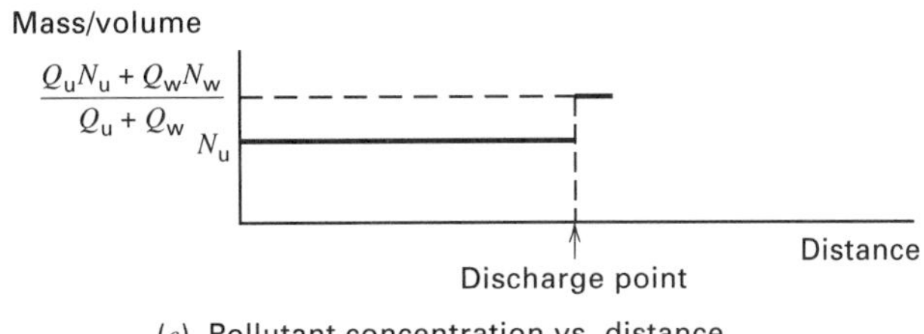

(*e*) Pollutant concentration vs. distance

Figure 16.6 Mass balance analysis to determine concentration immediately downstream of discharge point.

forecasting model is a statement about how bacteria are transformed within the stream. Field studies indicate that changes in the number of bacteria often follow a regular pattern that can be described mathematically. Many representations can be used, depending on how accurately the real bacterial transformation process is to be simulated. A simple formulation, and one that is often considered adequate for making forecasts, assumes that the number of bacteria will decrease according to a first-order reaction as the water flows downstream. The phrase *first-order reaction* is a shorthand way of saying that the *rate* of bacterial die-off is proportional to the number of bacteria present.[20]

Figure 16.7 summarizes components of the model building process. A typical volume unit located x length units downstream of the point of discharge is selected. Since the velocity of streamflow is a constant (u), this volume unit can also be described as having traveled for t time units downstream of the discharge. *Travel time* is related to distance and velocity using

$$t = x/u. \quad \textbf{(16-4)}$$

A mass balance equation is written for the bacteria entering and leaving the control volume. If the in-stream bacterial transformation process is represented as a first-order decay reaction, the solution to the differential equation representing the mass balance is[21]

$$N(t) = N_o e^{-kt} \quad \textbf{(16-5)}$$

where

t = time of travel downstream of the discharge point
$N(t)$ = concentration of coliform at time t
N_o = concentration of coliform just below the point of discharge ($t = 0$)
k = bacterial die-off coefficient

Figure 16.8a, which is a plot of equation 16-5 using Cartesian coordinates, shows how concentration falls off exponentially with increasing values of travel time. If the vertical axis is taken as the logarithm of N_t/N_o, the resulting plot is linear, and this is illustrated in Figure 16.8b.[22] The slope of the line in Figure 18.8b is proportional to the value of the bacterial die-off coefficient.

Equation 16-5 provides a way to predict how downstream bacterial concentrations will change in response to either a new source of wastewater or a change in the discharge from an existing source. However, before the equation can be used to make forecasts, the bacterial die-off coefficient (k) must be determined. Empirical studies of bacterial decay in streams provide a basis for estimating k. In a typical field investigation, samples are retrieved from several locations downstream of a waste source under conditions when both the streamflow and the wastewater discharge are approximately constant. Laboratory measurements of the coliform concentration are made for each sample. The laboratory data provides N_t for several values of t including $t = 0$. The observed values of log (N_t/N_o) versus travel time are plotted, and a straight line is passed through the data points. The die-off coefficient is estimated based on the slope of the line. Figure 16.9 is a plot of coliform concentration data for the Tennessee River below Chattanooga. The figure demonstrates the use of a formula for computing k from the slope of the line passing through data points for the first several days of travel time. Because the points trace out a nonlinear curve beyond a travel time of about 5 days, the assumption of a first-order bacterial decay process is not valid for t greater than 5 in this case. The entire die-off process may be represented with two straight-line segments. In this case, a separate value for the die-off coefficient is calculated for the period beyond the fifth day.

A typical application of equation 16-5 is to predict the effect of a proposed environmental requirement. For example, what would be the effect on downstream coliform levels of a requirement that forced a municipality to chlorinate its wastewater? The following steps could be taken to estimate downstream coliform concentrations, assuming the munici-

---

[20] The term *first-order* is used because the bacterial decay process is described using a differential equation that indicates the rate of reaction (dN/dt) is proportional to the concentration (N) raised to the power of one. If the reaction rate is assumed to be proportional to N^2, the reaction is termed *second-order.* For a more general definition of the *order of reactions,* see Davis and Cornwell (1991, pp. 137–39).

[21] The parameter e in equation 16-5 is the number whose natural logarithm equals 1. The value of e, to three decimal places, is 2.718. The analysis yielding equation 16-5 presumes that the physical transport of bacteria is accomplished only by the average motion of the stream. There is no transport of bacteria due to diffusion or dispersion in the direction of flow. The differential equation for this case is

$$u[dN/dx] + kN = 0$$

subject to the initial condition $N = N_o$ at $x = 0$. For simplicity, equation 16-5 shows the solution to this equation with time of travel substituted for (x/u). The associated form of the differential equation is $dN/dt = kN = 0$.

[22] The linearity of a plot of the logarithm of (N_t/N_o) versus travel time can be established by taking the logarithm of equation 16-5 and using standard formulas for manipulating terms involving logarithms. If natural logarithms are used instead of logarithms to the base 10, then the absolute value of the slope equals the bacterial die-off coefficient.

The Physical Conditions

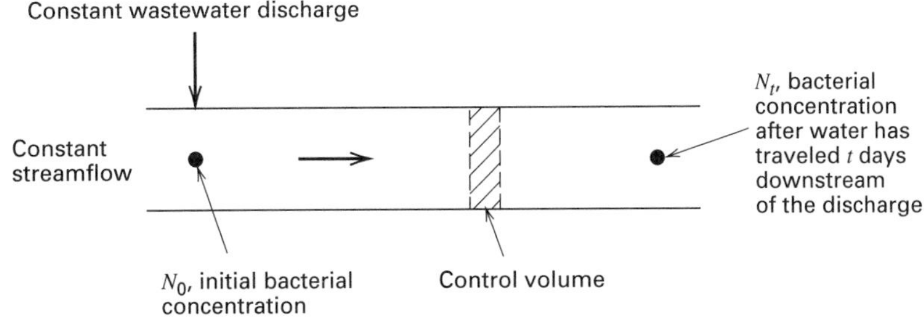

Relationship between travel time and distance

With a constant stream velocity (u), travel time (t) and distance downstream from the discharge (x) are related by

$$t = \frac{x}{u}$$

Mass balance for control volume

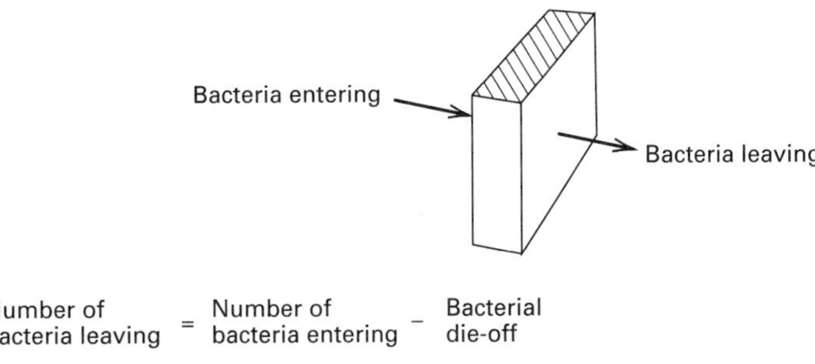

Representation of bacterial die-off process

First- order reaction: die-off *rate* is proportional to number of bacteria present, where k is the die-off coefficient.

Resulting forecasting model

$$N_t = N_0 e^{-kt}$$

Figure 16.7 Modeling bacterial concentration in a stream.

pality complied with the chlorination requirement. The value of k is estimated using procedures discussed in connection with Figure 16.9, or using results from studies determining bacterial die-off coefficients on similar streams. The existing coliform concentration in the wastewater discharge and in the stream immediately above the point of discharge are obtained by direct measurement. The expected percentage reduction of bacteria due to chlorination together with equation 16-3 provide a basis for computing N_0, the bacterial concentration just below the discharge point *after* chlorination. Once k and N_0 are determined, equation 16-5 is used to calculate values of bacterial concentration at selected downstream locations.[23]

[23] For more information on how coliform models are used in practice, see Thomann (1972), and Thomann and Mueller (1987).

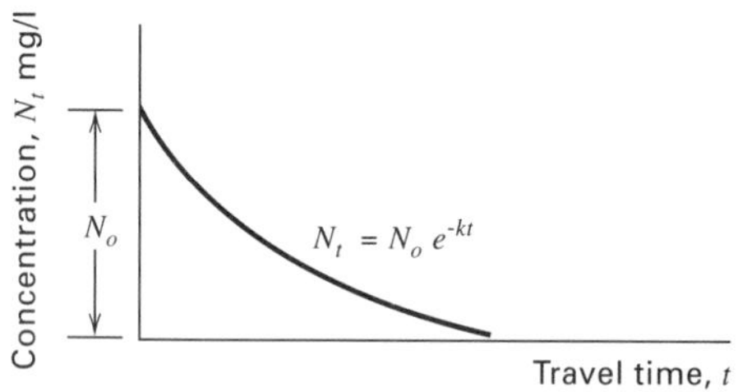

(*a*) Exponential decay downstream of discharge point

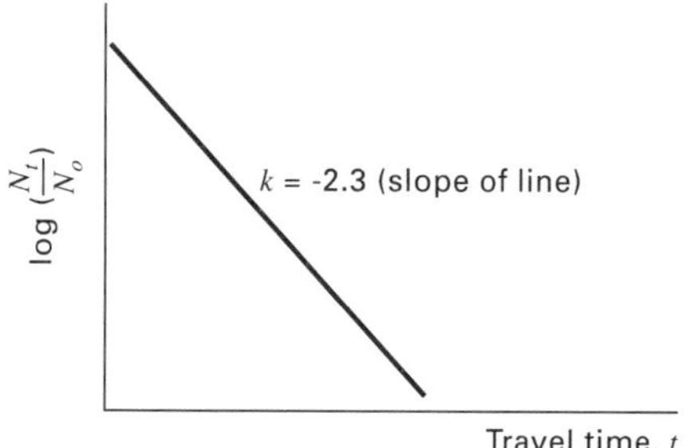

(*b*) Linear form resulting from semi-log plot

Figure 16.8 Concentration downstream of discharge point based on first-order decay.

This model for the transport and decay of coliform bacteria assumes that as the bacteria move downstream, the concentration in any cross section of the stream is uniform. The model presupposes complete mixing of coliform bacteria from top to bottom and side to side, but *no* mixing in the direction of flow; that is, the longitudinal direction. (The term *advection* is used to describe the transport of a contaminant by streamflow when there is no longitudinal mixing.)[24] In many circumstances, equation 16-5 provides useful forecasts of contaminant concentration even though real rivers and streams don't satisfy the aforementioned assumptions related to mixing. A contrasting situation is where coliform bacteria enter an estuary. In this case, tidal motion and nonuniformities in velocity cause significant contaminant mixing in the direction of flow. The result is a *longitudinal dispersion* of bacteria that must be accounted for along with advection and biological decay.[25] In some water bodies, lateral and vertical gradients in velocity and concentration are significant enough to require models that account for contaminant transport in three directions (in contrast to the one-dimensional analysis used in developing equation 16-5).

[24] The term advection has a more general meaning in models of the fate and transport of contaminants. MacKay and Patterson (1993, p. 133) define the term in circumstances, treated later in this section, where a contaminant migrates from one environmental medium, such as water, to a second, such as air. They describe *advective processes* as those "in which a chemical is conveyed from one medium to another by virtue of 'piggy backing' on 'carrier' material which is moving between media for reasons unrelated to the presence of the chemical." Examples include deposition of sediments in a reservoir. Chemicals from the water that are absorbed onto particles suspended in the water are transferred to bottom sediments as the particles settle.

A Statistical Model for Predicting Carbon Monoxide Levels

Statistical forecasting models are often used when there is insufficient scientific knowledge to model the processes affecting pollutant transport. In developing a statistical model, intuition, previous studies, and scientific theories are used to postulate a relationship among what are considered to be the key variables. The assumed relationship contains constants (*model parameters*) that are estimated based on measured values of the variables.

The statistical approach to forecasting is illustrated by Tiao and Hillmer's (1978) model for predicting carbon monoxide (CO) concentrations at a particular monitoring site 25 feet from the San Diego Freeway in southern California. Their modeling effort built on previous studies that identified variables affecting CO in the ambient air. Some of these variables concerned traffic flow, especially traffic speed and volume, and others described local meteorological and topographic conditions.

Based on an analysis of factors likely to influence CO at their site, Tiao and Hillmer identified two key variables, traffic density and wind speed. They postulated that carbon monoxide concentrations were related to these variables using

[25] Experts in mass balance modeling sometimes describe mixing processes using the term *diffusion* instead of dispersion. Distinctions are often ambiguous. Environmental modeling specialists frequently use *dispersion* as a general term that refers to spreading of a pollutant due to many phenomena, including diffusion.

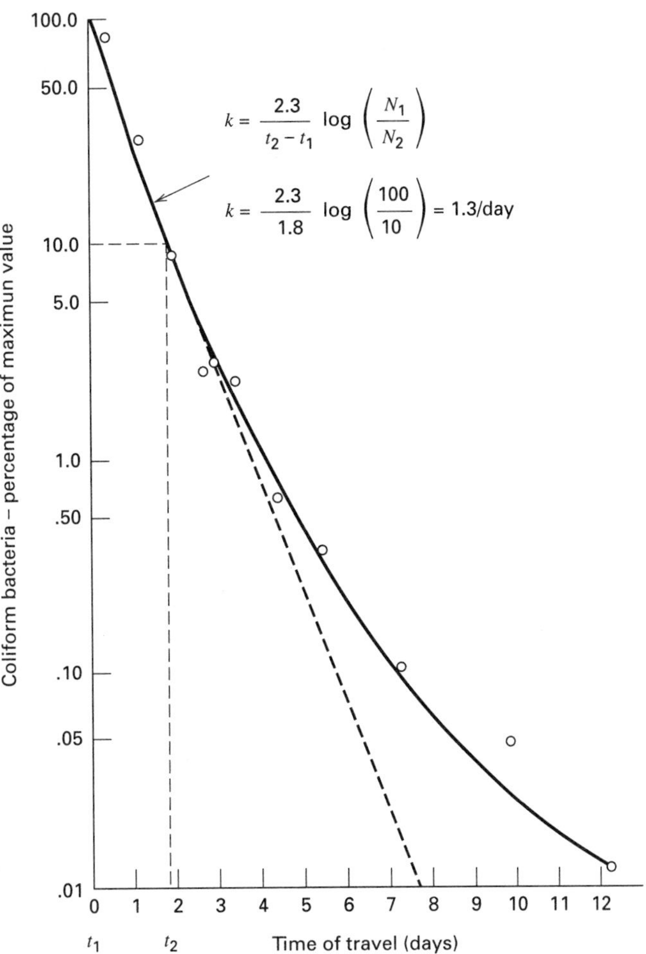

Figure 16.9 Estimation of bacterial die-off coefficient: Tennessee River below Chattanooga. From Thomann (1972). Reprinted with permission.

$$C_t = a + MD_t \exp\left[-b(W_t - W_o)^2\right] \quad \textbf{(16-6)}$$

where

C_t = CO concentration at the monitoring site for hour t (parts per million)

D_t = traffic density on the adjacent freeway for hour t (vehicles per hour)

W_t = wind speed in the direction perpendicular to the freeway for hour t (miles per hour)

The constants, a, b, M, and W_o, were determined using past measurements of the model variables, C_t, D_t, and W_t. The parameter a represents the background CO concentration that exists independent of the freeway traffic. The exponential term in the equation accounts for the transport of carbon monoxide caused by atmospheric turbulence. Carbon monoxide emissions from motor vehicles on the freeway are reflected in M, a proportionality constant that is multiplied by traffic density.

Once the form of the equation relating the variables had been assumed, the next step involved estimating the parameters, *a, b, M,* and W_0. This was done by obtaining hourly measurements of C_t, D_t, and W_t, for a two-and-one-half year period beginning in June 1974. A preliminary analysis of the data indicated that hourly CO concentrations at the monitoring site were significantly influenced by both the season and the day of the week. Therefore, the measurements of C_t, D_t, and W_t were divided into groups corresponding to different seasons and days of the week, and these groupings were maintained in all subsequent stages of model building.

To see how the model parameters were estimated, consider the set of data for Sundays during "summer" (June through October). Hourly measurements of C_t, W_t, and D_t during the five-month period beginning in June 1975 were used to determine the parameters for the version of equation 16-6 applying to summer Sundays. A standard statistical procedure, the method of least squares, was used to compute *a, b, M,* and W_0. The procedure is called *least squares* because it yields model parameters that give the smallest value to the sum of the squares of the errors. An *error* is defined as the difference between estimated and observed values of C_t for a specific hour on a given Sunday within the five-month measurement period. Consider, for example, the hour from noon to 1 P.M. on the first Sunday in June. The *estimated* value of C_t is computed by substituting into equation 16-6 values of wind speed and traffic density measured at that particular hour and day. The *observed* value of C_t is the carbon monoxide concentration recorded at the monitoring site for that precise hour. The differences in estimated and observed CO concentrations are computed for each summer Sunday hour for which there are measurements in the five-month period. They are then squared and added together. Values of *a, b, M,* and W_0 are selected such that the sum of the squared errors is a minimum.

Tiao and Hillmer applied the method of least squares to the data for summer Sundays in 1975 and calculated the following parameter values: $a = 1.79$, $b = 0.013$, $M = 0.019$, and $W_0 = 2.54$. These values were substituted into equation 16-6 to provide a basis for estimating CO concentrations at the monitoring site during summer Sundays. Figure 16.10 provides an indication of how well this model represents the hourly variations in carbon monoxide. The line labeled "observed values" delineates the average hourly CO measured at the site during Sundays in the summer of 1975. The circles depict CO concentrations estimated with equation 16-6. The values of D_t and W_t used in estimating CO with equation 16-6 were also average hourly values based on all of the data for Sundays in the five-month summer period. As indicated in Figure 16.10, the statistical model closely replicates the observed data.

Once Tiao and Hillmer had estimated the model parameters, they could have used equation 16-6 to predict how changes in traffic density would influence carbon monoxide at the monitoring site. However, this equation applies only to summer Sundays at a particular site. For other time periods and physical locations, the four model parameters must be reestimated using measurements appropriate to the new site and time period. Because the form of equation 16-6 is not based on physical laws, it may not provide a suitable representation of how C_t, D_t, and W_t are related in settings different

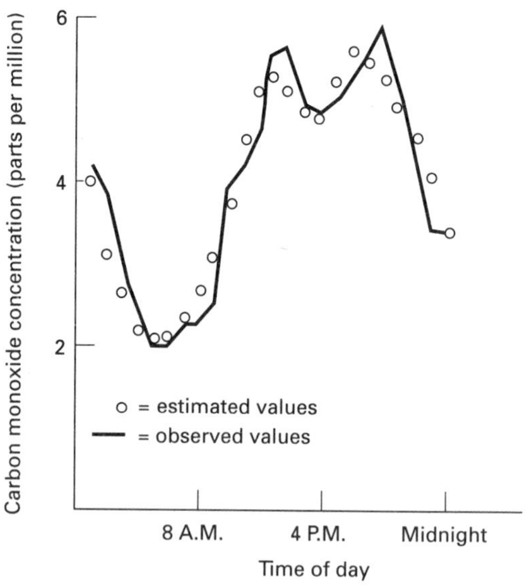

Figure 16.10 Estimated versus observed values of carbon monoxide for summer Sundays in 1975 at the San Diego Freeway monitoring site. Reprinted with permission from Tiao and Hillmer, *Environmental Science and Technology,* vol. 12, pp. 820–28. Copyright © 1978, American Chemical Society.

from the San Diego Freeway site. The lack of transferability from one location to another is a principal shortcoming of statistical models.[26] An important advantage is that a forecasting equation applicable to particular conditions can often be obtained, even though scientific theories relating the variables are incomplete.

Models of Contaminant Transport Across Media

The two models just considered are *single medium models:* the first concerned water as an "environmental medium," and the second involved air. In the past few decades, environmental modeling specialists have attempted to account for *intermedia transport,* particularly as new research demonstrated the significance of such transport. The history of studies of DDT is illustrative. At one time, scientists believed that DDT applied to a soil would remain in the soil indefinitely. As MacKay and Patterson (1993, p. 132) point out, scientists later learned that "application of DDT to soil will result in appreciable concentrations in the atmosphere which is exposed to that soil, and that contaminated air will result in the contamination of distant areas. . . ."

Environmental scientists use *partition coefficient* to analyze the way a chemical discharged into one medium "migrates" to other media. If two media are represented by the symbols i and j, the partition coefficient (K_{ij}) for a particular chemical is defined as

$$K_{\mathrm{ij}} = \frac{C_{\mathrm{i}}}{C_{\mathrm{j}}}$$

where C_{i} = concentration of the chemical in medium i, and C_{j} = concentration of the chemical in medium j. Each of the concentrations is a long-term equilibrium value that reflects steady state conditions. Before these values are reached, a disequilibrium exists, and the rate of transfer of the chemical between the two media is proportional to $(\tilde{C}_{\mathrm{j}}K_{\mathrm{ij}} - \tilde{C}_{\mathrm{i}})$, where the

[26] As in the case of other fate and transport models, the model developed by Tiao and Hillmer is limited in that it does not characterize *exposure* to pollutants. As explained in Chapter 17, exposure calculations play a key role determining human health risks. For carbon monoxide, models have been developed to predict exposures inside a motor vehicle in traffic; see Flachsbart (forthcoming) for examples.

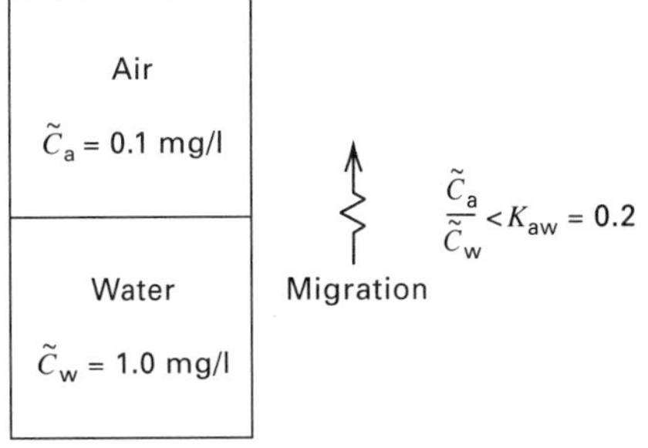

(*a*) Disequilibrium—non-steady state conditions

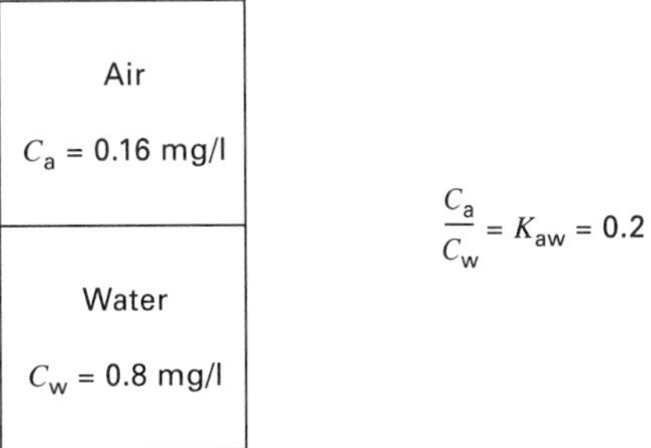

(*b*) Equilibrium—steady-state conditions

Figure 16.11 Migration of benzene from water to air. Adapted from MacKay and Patterson. Mathematical Models of Transport and Fate, pp 129–52, in *Ecological Risk Assessment,* Suter, II, (ed), Lewis Publishers, an imprint of CRC Press, Boca Raton, FL, © 1993. With permission.

symbol "~" is used to indicate that the concentrations are different from the equilibrium values C_{i} and C_{j}.

Consider, for example, the case of benzene as it migrates from water to air. The air-water partition coefficient for benzene, K_{aw}, is equal to 0.2. In Figure 16.11a, the concentrations of benzene in air and water are 0.1 mg/l and 1.0 mg/l, respectively, and the ratio of concentrations is less than the partition coefficient, 0.2. In that setting, the processes controlling movement of benzene from water to air will operate to yield a net transfer of benzene. Figure 16.11b shows an outcome in which the ratio of concentrations of benzene in air and water equals the partition coefficient; there is no net transfer. The concentrations in Figure 16.11b reflect long-term, steady state conditions.[27]

Figure 16.12 illustrates results from applying partition coefficients and mass balance analyses to build

[27] For further information on how values of partition coefficients are determined, see MacKay (1991, pp. 74–83).

a *compartment model* that determines equilibrium concentrations of a chemical in different media. In this example, the chemical is trichloroethylene and there are six compartments (or media): air, soil, water, suspended sediments, bottom sediments, and fish. The figure shows steady state mass flowrates of trichloroethylene entering directly into air, soil, and water. Trichloroethylene moves from air, soil, and water into compartments that do not receive direct discharges. Partition coefficients are used to characterize this migration. In addition, other transport phenomena, such as runoff and evaporation, must be accounted for. The equilibrium percentage distributions in the figure show that, in this example, almost all of the direct discharge to soil winds up in other compartments, primarily air.

Forecasting Models Based on "Soft Information"

The previous examples involved models to predict concentrations of bacteria and chemicals in the environment. Because concentrations can be measured, it is possible to determine whether the models' forecasts are accurate. Although it is customary to develop forecasting models for variables that can be determined solely from quantitative observations, some modeling experts find this too restrictive. They argue that the intuitive feelings that people have about ambiguous qualitative concepts such as "quality of life" should be considered when formulating government plans and policies. They also believe that mathematical procedures can be used to convert personal judgments into useful forecasts.

An application of a mathematical model called *KSIM* is used to illustrate how quantitative forecasts can be made using *soft information,* subjective opinions, and judgments about relationships among poorly defined variables. *KSIM* was developed by Kane, Vertinsky, and Thomson (1973) to help people refine their intuitions and opinions and make them more useful in a policy-making context.

The KSIM model was used in a workshop convened to assess impacts associated with a possible U.S. deep-water port system for imported oil.[28] Workshop participants had diverse backgrounds and were knowledgeable about the subject. They formulated a workshop problem statement by spelling out basic assumptions and articulating a number of "what if" questions that they would try to answer. A typical question was, "What if crude oil imports were cut off?" Participants identified what they felt were the key variables and put them in the form required by the KSIM approach: each variable was scaled to range from a minimum of 0 to a maximum of 1. *Initial conditions,* values of the variables at the beginning of the forecasting exercise, were also set. The five variables agreed upon by the participants were energy use, domestic energy supply, domestic investment in new energy sources, reliance on traditional sources of energy, and public confidence in the government. For each variable, the meaning of the values of 0 and 1 were explained in qualitative terms.

The next step involved development of a *cross-impact matrix,* a table with the five variables arrayed down the first column and across the top row.[29] For each pair of variables in the matrix, workshop members decided whether a relationship existed, and if it did, what its "strength" was on a scale from −3 to +3. For example, participants felt that the linkage between domestic investment in new energy sources and domestic energy use was positive with a strength of +2. In other words, investments in new supplies of energy would lead to an increase in energy utilization. Each number in the cross-impact matrix represented the *perceptions* of the workshop participants regarding how two variables in the matrix related to each other.

After the numbers in the cross-impact matrix had been specified, KSIM was used to make predictions. KSIM consists of a particular set of differential equations that is *assumed* to govern the interactions among the variables.[30] Details of the model are unim-

[28] This discussion of the application of KSIM to analyze policies for deep-water port development is from Mitchell et al. (1975, pp. 145–54.)

[29] The cross-impact matrix in KSIM is different from those in the "cross-impact matrix method" reported by Gordon and Hayward (1968). Their matrices have elements that represent the extent to which the probability of one "event" is enhanced (or inhibited) if a second event occurs. These probabilities are employed to provide information about which sequences of future events are the most likely to occur.

[30] The reasoning used in picking the particular set of equations used in KSIM is given by Kane, Vertinsky, and Thomson (1973).

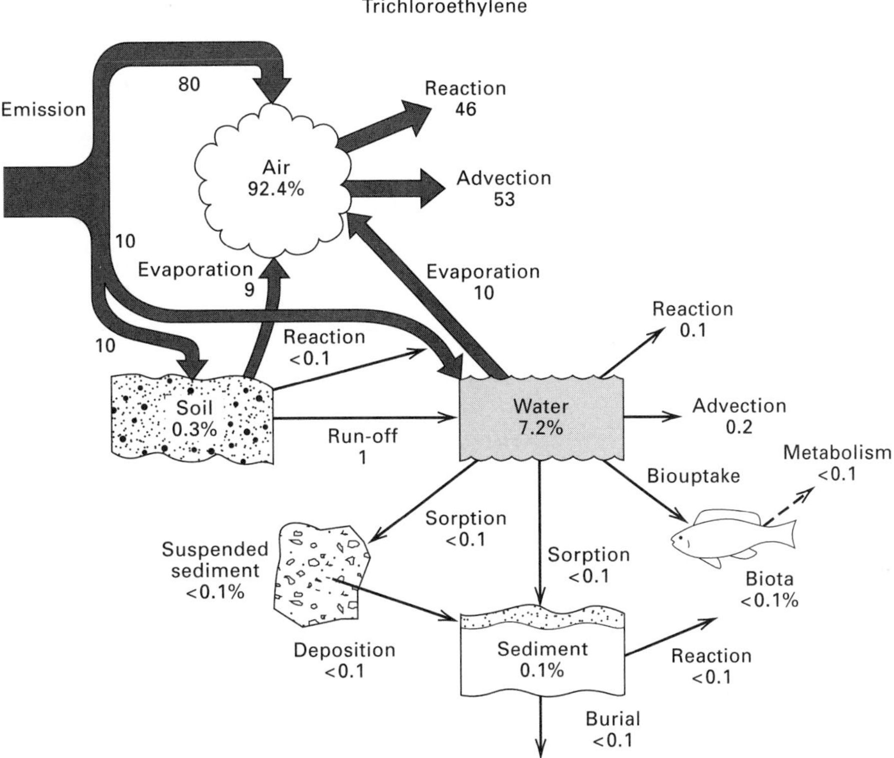

Figure 16.12 Steady state behavior of trichloroethylene in response to direct inputs to air, soil, and water. Reprinted with permission from MacKay, *Canadian Water Resources Journal,* vol. 12, p. 19, © 1987.

portant here. The key point is that the same general equations are employed in every application of KSIM. In any particular case, parameters in the equations are estimated using the cross-impact matrix and the information on initial conditions.

Solutions to the differential equations that make up KSIM consist of values of the variables over time. An illustrative result from the deep-water port workshop is shown in Figure 16.13. By examining the solutions, workshop participants were able to refine their initial impressions about cross-impacts. Participants in a typical KSIM exercise often continue to revise the cross-impact matrix until the solutions they obtain are plausible and compatible with their intuitions.

At this point in the workshop, the cross-impact matrix was enlarged to accommodate "policy interventions," and the KSIM model was solved again. Information on the impacts of interventions took the form of curves like those in Figure 16.13. The workshop exercise was *not* used exclusively as a means of producing forecasts. As Mitchell et al. (1975, p. 142) put it, "in KSIM sessions it is often the process (exchange of views, surfacing of issues, identification of concerns, and so on) that is more important than the product."

KSIM is clearly not a forecasting model in the usual sense. Its virtue is in showing the *logical consequences* (over time) of the *model users' intuitions and assumptions* about interactions among the variables. In KSIM, the forms of the equations are set. They are not modified to fit a specific problem. In any particular context, the initial conditions and numerical entries in a cross-impact matrix are changed in

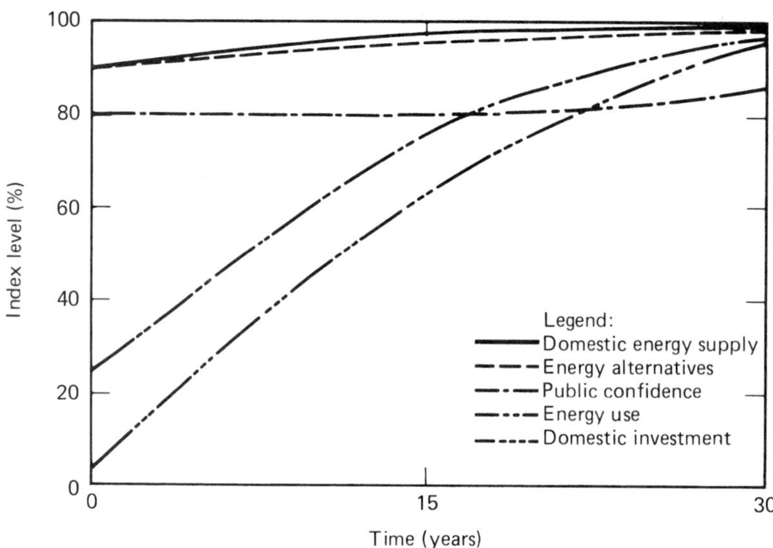

Consider: Initial projected variable interactions of the national deep water port system

Implications: Domestic investment increases steadily to near maximum.
Energy use increases steadily to near maximum.
Public confidence is stable.
Domestic energy supply increases to maximum.
Energy alternatives increase to maximum.

Figure 16.13 Sample results from the KSIM Deep-Water Port Workshop. Reprinted with permission from Mitchell et al., *Handbook of Forecasting Techniques,* IWR Contract Report 75-7.

successive applications of the model to help KSIM users refine their intuitions. Since the variables cannot be measured, the accuracy of forecasts cannot be tested by making comparisons with observed values.

Mathematical models using soft information have been disputed, especially by those who expect forecasting models to be based on empirical data and scientific theories. Critics argue that relationships among variables must be established scientifically and tailored to each problem setting. In addition, comparisons between predicted and actual outcomes are necessary to gauge a model's worth.

Those defending models based on soft information feel that a quantitative comparison with observed data is not the only way to establish a model's validity. They believe that a forecasting model should also be judged in relation to other possible means of making predictions. For example, an alternative to the KSIM projections in the deep-water port study might consist of verbal descriptions of future scenarios. Value judgments and personal opinions would be employed in generating the scenarios, but chances are that these judgments and opinions would not be made explicit. A defense of KSIM is that it requires a clear statement of judgments and assumptions. Models developed from soft information are also justified by their ability to employ whatever information exists about relationships, even if the information is scanty and of marginal quality. Traditional modeling approaches may not accommodate some important information because it is too incomplete to be acceptable in a rigorous modeling exercise.[31]

[31] Justifications for mathematical models based on soft information are elaborated by Forrester (1968, Chapter 3). For a more complete discussion of Forrester's use of such models and the controversy that his work has generated, see Pugh (1977).

Models using soft information are different in kind from those developed from physical laws and statistical analyses. They are not meant to provide verifiable, quantitative predictions of the type given by equations 16-5 and 16-6. Models such as KSIM are more appropriately viewed as tools for sharpening the qualitative discussions of the possible impacts of government actions.

Model Calibration and Validation

Before concluding this introduction to mathematical models, consider the distinctions between model validity and model calibration. A model is *calibrated* by using empirical data to estimate its parameters. Calculation of the bacterial decay coefficient in Figure 16.9 provides an example of model calibration. The process of determining the four parameters in equation 16-6 to predict carbon monoxide provides another illustration. Once a model has been calibrated, it can be used to make forecasts of the variable in question under future conditions. However, little can be said about the accuracy of those forecasts unless the model has been validated.

What does it mean to say a model is valid? In science and engineering, a model is *validated* by conducting tests to see if its forecasts are likely to be close to the outcomes eventually observed. To test a model, it is necessary to compile a set of observations of the model variables that is different from the data used in calibration. An independent data set is needed to avoid the logical inconsistency of both developing *and* testing a model using the same measurements.

A valid forecasting model is one that consistently yields predictions that are "reasonably close" to what occurs in reality. However, it is often impracticable to test a forecasting model by comparing its predictions with observed outcomes. One reason is that the expense of gathering the necessary data is frequently prohibitive. Often, what little empirical data can be gathered must be used to estimate the model parameters. Predictive accuracy is particularly difficult to judge when long-range forecasts are involved. There is typically little enthusiasm for determining the accuracy of forecasts 10 or 20 years after a model was used. Sometimes it is not even meaningful to compare forecasts with actual outcomes. This occurs, for instance, when the invention of a new technology leads to events that could not possibly have been anticipated when making the original forecasts.

In addition to comparing predicted and observed effects, there are several other ways of gauging a model's worth. Some environmental specialists consider a forecasting model valid if it "behaves" in reasonable ways.[32] For example, an air pollution model can be tested by examining predicted air quality levels under various assumptions about government pollution control regulations. A measure of the model's validity is whether these predictions are plausible and consistent with the modeler's intuitions.

Still another way of exploring a model's validity is to consider whether it produces forecasts accurate enough to aid decision makers. A model's validity is then judged in relation to its intended use. For example, in some decision-making contexts, a model that gives forecasts that are twice as high (or twice as low) as actual outcomes may be considered adequate.

Given that there are alternative conceptions of model validity, which should be used in evaluating models that forecast environmental changes? In exploring this question, Suter and Barnthouse (1993, p. 40) argue that applying predictive accuracy as a standard for gauging a model's validity is "intellectually satisfying but irrelevant to actual assessment practice." In their view, "validation may be more usefully viewed as a process of selecting among alternative approximations. . . ."[33] In other words, the issue is not whether a model satisfies a scientific criterion for being valid, but whether it gives results that are useful when compared to alternative ways of predicting the impact under investigation.

---

[32] Modeling specialists sometimes use the term *verification,* in contrast to validation, when discussing this idea. For example, Jørgensen (1994, p. 20) adopts this usage and defines verification as "a test of the *internal logic* of the model. Typical verification questions are: Does the model react as expected? Is the model long term stable?" In other words, do model variables exhibit the kinds of long-term stability associated with corresponding variables in the natural system being modeled? Jørgensen distinguishes verification from validation, which he defines as "an objective test on how well the model outputs fit [observed] data," when the latter are different from the data used to estimate model parameters.

[33] Suter and Barnthouse describe two other alternatives to using predictive accuracy to validate models: (1) publication of a model in a peer-reviewed journal, and (2) adoption of a model by environmental regulatory agencies. They claim the latter is "probably superior," arguing that the "[m]odels used to support regulatory decisions often receive much more careful scrutiny than does any journal manuscript!" (Suter and Barnthouse, 1993, p. 41).

Although this chapter emphasizes models that yield quantitative predictions, government decision makers are often forced to rely on qualitative forecasts. This confounds questions related to the validity of a forecasting procedure. For example, in an analysis of over 1100 individual forecasts in 29 environmental impact statements (EIS's) produced in response to the U.S. National Environmental Policy Act, Culhane, Friesma, and Beecher (1987, p. 97) found that less than a quarter of the forecasts were quantitative. The great majority were verbal descriptions indicating that individual impacts would (or might) be "high" or "moderate" or "insignificant." In trying to assess the accuracy of the 1100-plus forecasts, the researchers found that a large fraction of the forecasts were "accurate solely by virtue of the vagueness of the forecast." (Culhane, Friesma, and Beecher, 1987, p. 253).

Another aspects of mathematical models used to make predictions of environmental change concerns their comprehensibility. Models used in implementing environmental regulations are often presented as "black boxes," and the underlying assumptions and bases for predictions are not made clear.[34] When this occurs, those responsible for using modeling results to make decisions have no basis for determining whether the models used are valid and whether the quantitative predictions are meaningful.

Key Concepts and Terms

Aids to Impact Identification
Checklists
Impact matrices
Flow diagrams
Literature reviews by project type
Computer-based impact identification
Scoping
Impacts of project phases
Direct vs. indirect impacts

Judgmental Approaches to Forecasting
Expert opinion
Impacts of analogues past projects
Social impacts
Delphi method

Physical Models and Experiments
Modeling visual impacts
Estuary models
Verifying a physical model
Wind tunnels
In situ experiments
Analog studies

Forecasting with Mathematical Models
Fate and transport models
Mass balance equation
Steady state conditions
Mass flowrate
Volume flowrate
First-order reaction
Complete and instantaneous mixing
Advection
Longitudinal dispersion
Statistical models
Model parameters
Method of least squares
Sum of squared errors
Intermedia transport
Partition coefficients
Soft information
KSIM and cross-impacts
Model calibration
Model validity

Discussion Questions

16-1 Give an example of a physical model that can be used to forecast environmental impacts. Try to select an illustration different from those mentioned in Chapter 16. Indicate the types of variables that your illustrative model would forecast. Also indicate the types of projects or actions that could be examined with the model.

16-2 Suppose you were asked to use the Delphi method to predict the land use changes that would accompany alternative reservoir projects. What kinds of perspectives would you try to have represented among the experts that would perform the forecasts?

[34] For example, environmental impact statements produced in the United States often contain so little information about models and assumptions that "errors that are inherent in this [modeling] approach are not readily traceable, and the results are not subject to scrutiny" (Leon, 1993, p. 657).

16-3 Phosphorous and nitrogen are nutrients that support the growth of algae in streams. The distribution of these nutrients is sometimes modeled mathematically by assuming that they decay over time according to a first-order reaction. Suppose the original concentration of total nitrogen just below a wastewater discharge is 8 mg/l.

(a) Draw a graph showing total nitrogen versus time of travel if the decay coefficient is assumed to be 0.5/day.

(b) Forecast the effects of providing wastewater treatment that reduces the concentration of total nitrogen to 6 mg/l at a point just below the discharge.

16-4 Consider a situation in which a municipal wastewater treatment plant operating without disinfection discharges sewage to a stream. The plant is 5 miles upstream from a beach, and the relevant flows and concentrations are as shown in the diagram. In this exercise, *MPN* stands for "most probable number" of coliform and is estimated using standard laboratory procedures, and the symbols in the diagram are defined as follows:

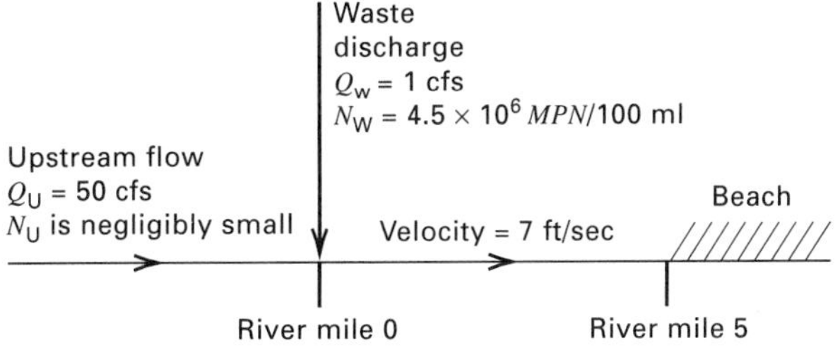

Q_u = rate of streamflow immediately upstream of discharge point, in cubic feet per second (cfs)

Q_w = rate of wastewater flow (cfs)

N_u = coliform concentration immediately upstream of discharge point (*MPN*/100 ml)

N_w = coliform concentration in wastewater (*MPN*/100 ml).

Stream survey data indicating how coliform bacteria from the discharge decays with distance (or "travel time") are as follows:

River Mile	8	16	24	32
N(*MPN*/100 ml)	46,500	16,700	9,000	2,800

(a) Calculate the decay coefficient for coliform bacteria (assuming "first-order reaction kinetics") in units of per day. This constitutes model calibration where it is assumed that the first-order decay model is an appropriate representation of the bacterial die-off process.

(b) Find the coliform concentration at the beach (5 miles downstream of the discharge), assuming that complete mixing takes place at an insignificantly short distance downstream of the point of discharge.

(c) The ambient coliform standard at the beach is 2000 *MPN*/100 ml. What percent reduction of coliform at the discharge is required to meet this standard?

16-5 The following data are a subset of data collected by Ledbetter and Gloyna for two rivers in Texas and Oklahoma.[35]

Total Dissolved Solids mg/l	Conductivity micromhos
11,800	18,000
9,000	14,500
6,200	10,200
5,000	8,000
2,200	4,000
800	2,000

Solutions containing ions are able to conduct electric current. Specific conductance or "conductivity" is defined as the reciprocal of the specific resistance of a solution in ohms. (Units of conductivity are referred to as "mhos," a backward spelling of ohms.) For a dilute solution, conductivity is approximately proportional to the concentration of ions present. This is noteworthy, because conductivity can be measured quickly in the field.

[35] The data in this exercise were taken from a summary of the original data collected by Ledbetter and Gloyna as presented by Krenkel and Novotny (1980, p. 406).

Use the data in the table to estimate the parameters (a, b) in the following postulated model of the relationship between y, total dissolved solids (TDS in mg/l), and x, conductivity (in micromhos):

$$y = a + bx.$$

The parameters a and b can be estimated from a plot of y vs. x. Alternatively, you can use the following equations derived as part of the method of least squares.[36]

$$\sum_{i=1}^{n} y_i = an + b\sum_{i=1}^{n} x_i$$

$$\sum_{i=1}^{n} x_i y_i = a\sum_{i=1}^{n} x_i + b\sum_{i=1}^{n} x_i^2$$

In this case, (x_i, y_i) corresponds to the ith data point, and the table includes a total of $n = 6$ such data points. If you choose the least squares approach, calculate the indicated sums and solve the resulting simultaneous equations for a and b.

16-6 The use of KSIM and other mathematical models with soft variables can be attacked easily by those accustomed to more traditional approaches to developing models. What criticisms would you expect to be advanced against KSIM? What kind of a defense could be mounted?

16-7 Indicate a circumstance in which you might use a mathematical model with soft variables to forecast environmental impacts.

References

Antunes, M. P., and A. Câmara. 1992, Hyper AIA—an Integrated System for Environmental Impact Assessment. *Journal of Environmental Management* 35: 93–111.

Armstrong, J. S. 1978. *Long-Range Forecasting: From Crystal Ball to Computer.* New York: Wiley.

Berns, T. D. 1977. "The Assessment of Land Use Impacts." In *Handbook for Environmental Planning,* ed. J. McEvoy III, and T. Dietz, 109–61. *The Social Consequences of Environmental Change.* New York: Wiley.

Burdge, R. J., and F. Vanclay. 1995. "Social Impact Assessment." In Vanclay, F. and D. A. Bronstein, *Environmental and Social Impact Assessment,* Chichester, U.K.: John Wiley & Sons.

Canter, L. W., 1996. *Environmental Impact Assessment,* 2nd ed. New York: McGraw-Hill.

Carpenter, R. A., and J. E. Maragos. 1989. *How to Assess Environmental Impacts on Tropical Islands and Coastal Areas.* Honolulu, HI: Environment and Policy Institute, East-West Center.

Cavalli-Sforza V., and L. Ortolano. 1984. Delphi Forecasts of Land Use-Transportation Interactions. *Journal of Transportation Engineering* 110(3): 220–37.

Clark, B. C., K. Chapman, R. Bisset, P. Wathern, and M. Barrett. 1981. *A Manual for the Assessment of Major Development Proposals.* London: Her Majesty's Stationery Office.

CEQ (Council on Environmental Quality). 1986. *Regulations for Implementing the Procedural Provisions of the National Environmental Policy Act,* 40 CFR, Parts 1500-1508 (as of July 1, 1986).

Culhane, P. J., H. P. Friesema, and J. A. Beecher. 1987. *Forecasts and Environmental Decisionmaking: The Content and Predictive Accuracy of Environmental Impact Statements.* Boulder, CO: Westview Press.

Davis, M. L., and D. A. Cornwell. 1991. *Introduction to Environmental Engineering,* 2d ed. New York: McGraw-Hill.

EPA (U.S. Environmental Protection Agency). 1973. *Environmental Impact Statement Guidelines,* rev. ed. Seattle: EPA, Region 10 Office.

Fedra, K., L. Winkelbauer, and V. R. Pantulu. 1991. *Expert Systems for Environmental Screening: An Application in the Lower Mekong Basin.* Report No. RR-91-19, International Institute for Applied Systems Analysis, Laxenburg, Austria.

Flachsbart, P. Forthcoming. "Human Exposure to Exhaust and Evaporative Emissions from Motor Vehicles," In *Motor Vehicle Air Pollution: Public Health Impact and Control Measures,* 2d ed., Schwela, D., F. Lapensee, and O. Zali, eds. Geneva,

[36] For the derivation of these equations and additional theory on the "curve fitting" reflected in this problem, see a beginning textbook in statistics, such as Freund and Simon (1992, pp. 430–43).

Switzerland: World Health Organization and Republic and Canton of Geneva.

Forrester, J. W. 1968. *Principles of Systems, Text and Workbook,* 2d preliminary ed. Cambridge, MA: Wright–Allen.

Freund, J. E., and G. A. Simon. 1992. *Modern Elementary Statistics.* Englewood Cliffs, NJ: Prentice–Hall.

Geraghty, P. J. 1993. Environmental Assessment and Applications of Expert Systems: An Overview. *Journal of Environmental Management* 39: 27–38.

Gordon, T. J., and H. Hayward. 1968. Initial Experiments with the Cross-Impact Matrix Method of Forecasting. *Futures* 1(2): 100–116.

Hemond, H. F., and E. J. Fechner. 1994. *Chemical Fate and Transport in the Environment.* San Diego: Academic Press.

Jain, R. K., L. V. Urban, and G. S. Stacey. 1981. *Environmental Impact Analysis: A New Dimension in Decision Making,* 2d ed. New York: Van Nostrand Reinhold.

Jain, R. K., L. V. Urban, G. S. Stacey, and H. E. Balbach. 1993. *Environmental Assessment.* New York: McGraw–Hill.

Jørgensen, S. E. 1994. *Fundamentals of Ecological Modeling,* 2d ed. Amsterdam: Elsevier.

Julien, B., S. J. Fenves, and M. J. Small. 1992. An Environmental Impact Identification System. *Journal of Environmental Management* 36: 167–84.

Kane, J., I. Vertinsky, and W. Thomson. 1973. KSIM: A Methodology for Interactive Resource Policy Simulation. *Water Resources Research* 9(1): 65–79.

Krenkel, P. A., and Novotny. 1980. *Water Quality Management.* New York: Academic Press.

Lemons, K. E., and A. L. Porter. 1992. A Comparative Study of Impact Assessment in Developed and Developing Countries. *Impact Assessment Bulletin* 10(3): 57–65.

Leon, B. F. 1993. "Survey of Analyses in Environmental Impact Statements." In *Environmental Analysis: The NEPA Experience,* ed. S. G. Hildebrand and J. B. Cannon, 653–59. Boca Raton, FL: Lewis Publishers.

MacKay, D. 1987. The Holistic Assessment of Toxic Chemicals in Canadian Waters. *Canadian Water Resources Journal* 12: 14–20.

———. *Multimedia Environmental Models: The Fugacity Approach.* Chelsea, MI: Lewis Publishers.

MacKay, D., and S. Patterson. 1993. "Mathematical Models of Transport and Fate." In *Ecological Risk Assessment,* ed. G. W. Suter II, 129–52. Chelsea, MI: Lewis Publishers.

McCoy, C. B. 1975. The Impact of an Impact Study, Contribution of Sociology to Decision-Making in Government. *Environment and Behavior* 7(3): 358–72.

Mitchell, A., B. H. Dodge, P. G. Kruzic, D. C. Miller, P. Schwartz, and B. E. Suta. 1975. *Handbook of Forecasting Techniques.* IWR Report 75-7, U.S. Army Engineers Institute for Water Resources, Ft. Belvoir, VA.

Morris, P., and R. Therivel. 1995. *Methods of Environmental Impact Assessment.* London: UCL Press, Ltd.

Porter, A. L., F. A. Rossini, S. R. Carpenter, and A. T. Roper. 1980. *A Guidebook for Technology Assessment and Impact Analysis.* New York: Elsevier/North Holland.

Pugh, R. E. 1977. *Evaluation of Policy Simulation Models: A Conceptual Approach and Case Study.* Washington, DC: Information Resources Press.

Roe, D., Dalal-Clayton, B., and R. Hughes. 1995. *A Directory of Impact Assessment Guidelines.* London: International Institute for Environment and Development.

Suter, G. W., II, and L. Barnthouse. 1993. "Assessment Concepts." In *Ecological Risk Assessment,* ed. G. Suter II, 21–47. Chelsea, MI: Lewis Publishers.

Thomann, R. V. 1972. *Systems Analysis and Water Quality Management.* New York: Environmental Research and Applications, Inc. Reprint. New York: McGraw–Hill.

Thomann, R. V., and J. A. Mueller. 1987. *Principles of Surface Water Quality Modeling and Control.* New York: Harper & Row.

Tiao, G. C., and S. C. Hillmer. 1978. Statistical Models for Ambient Concentrations of Carbon Monoxide, Lead and Sulfate Based on LACS Data. *Environmental Science and Technology* 12(7): 820–28.

Tracor, Inc. 1971. *Estuarine Modeling: An Assessment.* Report prepared for the U.S. Environmental Protection Agency, Washington, DC.

United Kingdom Department of the Environment, Welsh Office. 1989. *Environmental Assessment: A Guide to the Procedures.* London: Her Majesty's Stationery Office.

Chapter 17

Methods of Evaluation: Development Projects, Regulatory Programs, and Environmental Risks

The evalution (or rank ordering) of alternative development proposals and regulatory measures involves much more than environmental issues. Political, technical, and economic factors must be considered along with environmental impacts when making evaluations. Although this chapter touches upon these other concerns, it does so only to put the role of environmental issues in context. The emphasis throughout is on planning and decision making in the public sector.

The term *evaluation* is sufficiently vague to warrant elaboration. It has been defined by Lichfield, Kettle, and Whitbread (1975, p. 4) as "the process of analyzing a number of plans or projects with a view to searching out their comparative advantages and disadvantages and the act of setting down the findings of such analyses in a logical framework." Evaluation is *not* decision making. Instead, it assists decision making by highlighting differences between alternatives and providing information for subsequent deliberation. This chapter focuses on *methods of evaluation,* the analysis procedures used to articulate the positive and negative aspects of alternative courses of action.

Evaluation is based on values, and more than one set of values is relevant in deciding among alternative public sector proposals. Within democratic nations, it is commonly accepted that the values of *all* individuals who may be affected by public decisions should be considered. Adopting this perspective, evaluation also includes the process of identifying different segments of the public and ascertaining their feelings and opinions about alternative plans. There are many techniques for determining how individuals and groups value alternative public actions, and they are presented in the next chapter.

Although literally hundreds of evaluation methods exist, there is little agreement among experts about which are best. This chapter introduces procedures that are widely discussed in the public sector evaluation literature, commonly used in practice, or both. Evaluation procedures that ignore *uncertain-*

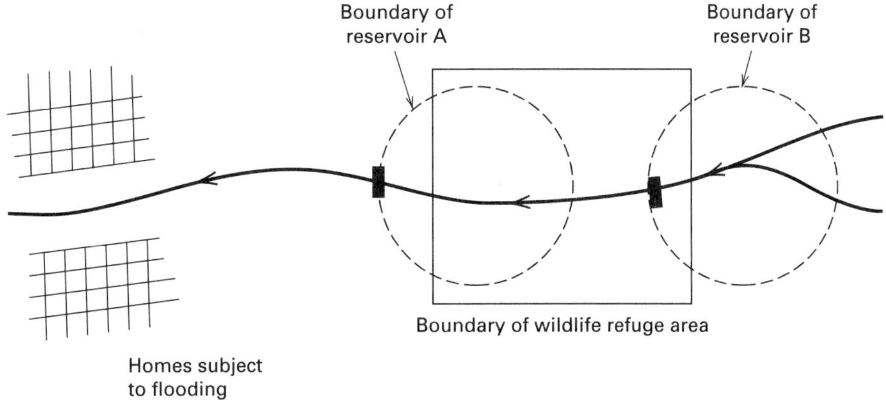

Figure 17.1 Two alternative reservoir projects.

ties associated with the prediction of environmental effects are presented first. After introducing those methods, the chapter considers *environmental risk assessment,* a collection of activities used to account for uncertainty in choosing among alternative policies and projects. The chapter shows how environmental risk assessment relates to the evaluation of both regulatory measures and development projects.

ISSUES IN MULTICRITERIA EVALUATION

The simplest circumstance for ranking alternative proposals occurs when there is only one decision criterion and all impacts are measured in the same units. Suppose, for example, the rule for choosing among several projects is maximization of net economic benefits. In this case, the difference between total benefits and costs (in monetary units such as dollars) provides an index of a project's merit and a basis for project selection. This type of single-criterion evaluation problem was analyzed in Chapter 7, and thus it is not emphasized here.

When environmental factors play a role in ranking alternatives, those factors are generally not the only criteria for decision making. Numerous criteria are often relevent, and they may be measured in units as different as dollars and number of jobs created. Such measures are incommensurable. They are not readily compared or easily combined into a single index of a proposal's overall worth.

An Example of Involving Alternative Reservoir Sites

Consider a hypothetical situation with two alternative water resource projects and two evaluation criteria. Plan A is to develop a reservoir that yields high economic benefits by reducing the likelihood of flooding to homes downstream of the dam (see Figure 17.1). However, this reservoir also inundates an important wildlife refuge area. Plan B locates the reservoir farther upstream. It destroys a smaller portion of the wildlife refuge, but it leaves the downstream homes subject to higher potential flood damages. This example is much simpler than real situations. Typically, there are many more alternative projects and numerous other pertinent impacts.

Table 17.1 summarizes the hypothetical reservoir selection problem. It contains information on the two objectives to be used in ranking the alternative plans: (1) maximization of the present value of net economic benefits, measured in dollars;[1] and (2) minimization of acreage of the wildlife refuge area destroyed by inundation. Even assuming these are the only relevant objective, there is still no obvious way to decide which project to implement. The criteria for evaluating alternative plans are expressed in dissimilar units (dollars and acres), and there is no agreed-upon technique for manipulating the evaluative information to yield a final decision.

[1] As discussed in Chapter 7, the concept of *present value* is used when costs and benefits occur at different future times.

TABLE 17.1 Effects of Alternative Reservoir Proposals

	Plan A (Downstream Site)	Plan B (Upstream Site)
Present value of net economic benefits ($ millions)	97	85
Area of wildlife refuge inundated (acres)	5000	1000

Evaluative Factors and Weights

It is convenient to view the ranking of alternative public actions in terms of *evaluative factors.* These are the goals, objectives, concerns, and constraints that various decision makers and segments of the public consider important in ranking alternatives. In the reservoir example, the evaluative factors are net economic benefits and area of wildlife refuge disturbed. Because plan ranking is strongly influenced by the choice of evaluative factors, controversy frequently arises over who should define them. In public sector planning, there are three principal sources of evaluative factors: institutions, community interactions, and technical and scientific judgments. The discussion that follows considers ways that planners delineate evaluative factors based on each of these sources.

Often, individuals who may be affected by a government proposal cannot participate directly in the evaluation process. As a matter of convenience, these people are referred to as the *nonlocal public.* Their goals and concerns are expressed *institutionally* at the national, state, and local levels in various laws, policies, and programs. Their feelings may also be reflected in the positions taken by elected officials and in policy statements of various groups such as a chamber of commerce or the Sierra Club. For example, suppose in the reservoir selection problem that a state law prohibits destruction of significant portions of wildlife refuge areas. This law allows citizens far from the project site to have an influence on the choice of a reservoir. In fact, such a law might be sufficient to force selection of the upstream reservoir, because it has much less effect on the refuge area. Planners identify institutional sources of evaluative factors by communicating with government officials and interest group representatives, and by examining relevant laws, policy statements, and regulations.

Planners can interact directly with members of the *local public* to help define problems from a local perspective and identify issues that individuals consider important in choosing among alternative proposals. To accomplish this, planners may provide information describing the problems and possible actions, as they understand them, and the impacts associated with these actions. Members of the local public can then provide feedback to planners about their own perceptions of the problems and what they consider important in ranking alternatives.

Evaluative factors are also based on *technical or scientific judgments,* which members of the public may neither appreciate nor recognize. For example, planners may deem it important to maintain the habitats of certain species in the interest of a project area's long-term ecological stability.

Sometimes planners must translate evaluative factors into technical terms that can be used to guide the formulation of alternatives, impact analysis, and evaluation. For example, the desire of citizens to maintain trout fishing in a local stream may be converted by planners into numerical constraints on stream temperatures and concentrations of dissolved oxygen.

The relative importance of different evaluative factors is often discussed in terms of *weights.* Sometimes weights are indicated explicitly, as in the assertion "maintaining the stream as a trout fishery is twice as important as the need to preserve 10 acres of open space." More frequently, proposals are selected using a consensus-building process in which weights are not stated. The concept of weights is still helpful in these cases, even though the weighting is implicit. Suppose, in the reservoir example, that the upstream project is selected in order to save 4000 acres of wildlife refuge. The cost in terms of net economic benefits foregone is $12 million (see Table 17.1). By selecting the upstream reservoir, an unstated weighting took place. The decision makers implicitly valued the refuge area at a minimum of $3000 per acre (the total net benefits foregone divided by the acres saved).

In thinking about evaluation, it is helpful to ask: Whose evaluative factors and weights are relevant, and how should they be determined and used in ranking alternatives? The next two sections of this chapter examine various evaluation techniques in terms of both evaluative factors and weights.

EXTENSIONS OF BENEFIT-COST ANALYSIS

Benefit-cost analysis has a long history of use as a method for evaluating development projects.[2] And since the 1980s, BCA has been used frequently to evaluate proposed environmental policies and regulations. In the United States, for example, Executive Order 12291 (discussed in Chapter 3) has motivated extensive use of BCA as a tool for evaluating proposed federal regulations.

Although BCA has played an important role in public sector planning, the results are *not* generally used as the exclusive basis for decision making. More commonly, BCA results are treated as one element in a broader political process aimed at yielding widely acceptable decisions. Notwithstanding that BCA has many supporters, particularly among economists, the technique also has its shortcomings. The following discussion summarizes limitations of BCA to indicate why some analysts have tried to broaden BCA to include multiple decision criteria.

Limitations of Traditional BCA

The economic benefits of a public action (for example, a development project or a regulation) are measured by the willingness of individuals to pay for the resulting outputs. The costs of undertaking the action are the gains that must be sacrificed by using the inputs (resources such as land and labor) necessary to implement the project. Economists view costs as opportunities foregone, potential gains given up by not using resources in other ways. If inputs and outputs associated with a public action are traded in competitive markets, the market prices are used to compute the costs and benefits of the action. Frequently, competitive markets do not exist. For example, in the reservoir problem of Figure 17.1, flood protection is not sold in competitive markets. When suitable market prices are unavailable, costs and benefits are estimated using procedures that rely heavily on the professional judgments of individuals performing the BCA. (For examples, see Chapter 6.) For a given situation, different analysts may choose different methods to estimate costs and benefits. Critiques of the resulting decisions are common. Government planners have been accused of making judgments that favor the proposed actions of their agencies by overestimating benefits and underestimating costs.

The use of BCA to rank alternative actions presupposes that economic benefits and costs adequately represent all of the significant effects. There are no weights involved. All effects are measured in monetary units such as dollars, and they are simply added together. (Another way of saying this is that the weights are zero for evaluative factors different from economic benefits and costs.) When maximization of net economic benefits is used exclusively to rank alternatives, a narrow, government-oriented perspective for defining the public interest is adopted. This is because the procedures for cost and benefit estimation are carried out largely by the staff of the agency proposing the action. Costs and benefits can be calculated without involving the public directly. People affected by agency proposals often find the conception of benefits and costs in a BCA incomplete.

BCA is particularly limited as a method for informing decisions that result in significant environmental impacts. An important reason is that BCA does not systematically consider impacts that cannot be described appropriately in monetary terms. This is illustrated by the reservoir example in Figure 17.1. BCA clarifies the economic gains that must be foregone to minimize the area of the wildlife refuge that is inundated. However, BCA does not provide a basis for ranking the two plans. A ranking might be possible if acres of wildlife refuge could be valued in monetary terms. Significant methodological problems arise in making such estimates. Moreover, many people are philosophically opposed to evaluating biologically important areas in terms of money.

Another shortcoming of BCA is its failure to account for equity in the distribution of benefits and costs. The emphasis is on aggregate economic effects, not on which groups gain and lose if a proposal is im-

[2] Eckstein (1958) explains the origins of BCA and how its practical application was fostered by federal water resource legislation in the 1930s. For a general introduction to BCA, see Chapter 7.

plemented. This is demonstrated by supposing that plan B in Figure 17.1 provides more flood protection to low-income families than plan A. In this instance, traditional BCA indicates that plan A is preferred because it has the higher net economic benefits. If the interests of low-income groups were to be favored, plan B might be a better choice. This income redistribution issue is not treated when only net economic benefits are examined. The limitations of BCA have stimulated efforts to modify the approach and make it applicable to a wider range of evaluation problems.

Extending BCA to Consider Multiple Objectives

Important early work in broadening BCA was carried out by Marglin (1962) and Maass (1966). They argued that it was often inappropriate to use a single objective—maximization of net economic benefits—because government programs were frequently intended to serve "multiple objectives." To evaluate them systematically, *classes of benefits* should be defined to correspond to each of the objectives.[3]

The illustrative cases discussed by Marglin and Maass often involve two objectives: maximization of net income to a nation as a whole, and maximization of net income to a particular region or group of citizens. The *national income net benefits* are defined in the same way as the net benefits included in a traditional BCA. Analysis of the second objective requires the introduction of *income redistribution net benefits,* the net income that flows to the particular region or group singled out for special treatment. For each proposed action under consideration, both national income and income redistribution net benefits are computed. Proposals may then be ranked using a weighted sum of the contributions to each objective. A variation of this approach is to maximize contributions to one objective, subject to the requirement that a minimum contribution to the second objective be provided.

Using the weighted sum of objectives technique, the evaluative factors considered in ranking the alternatives are the objectives themselves. Although Marglin and Maass often included only the national income and income redistribution objectives in demonstrating their ideas, they recognized that other objectives might need to be considered. In any particular situation, they felt that precise objectives should be specified by appropriate public decision makers.

Determining the weights to be used in computing the weighted sum of objectives has been a subject of continuing debate. Much of the large amount written on the subject of weights is "theoretical and assertive" (Steiner 1969, p. 48). The empirical work that has been done often identifies implicit weights reflected in past decisions affecting both the magnitude and distribution of national income. Past choices are analyzed to infer the weights that appear to have governed prior decisions. However, many analysts question the suitability of using weights implied by previous choices to decide what the weights should be.

Some social scientists have suggested that policymakers should articulate the relative importance of different objectives whenever a new government program is established. Agency analysts could then translate this information into weights that could be employed in evaluating alternative proposals.[4] This approach has not been widely used. Few legislators seem inclined or able to articulate specific weights on objectives such as income redistribution and environmental quality.

Extended BCA in Practice: Nam Choan Dam

Efforts to extend benefit-cost analysis to account for environmental objectives often wind up placing monetary values on as many impacts as possible, and describing the remaining impacts in non-monetary terms. This is demonstrated by an evaluation of a hydroelectric power project proposed for the Upper Quae Yai, a river in the western part of Thailand (see Figure 17.2). The centerpiece of the proposal was the Nam Choan Dam, which would create a reservoir within the Thung Yai Naresuan and Huay Kha Khaeng wildlife sanctuaries. Although these sanctuaries have two names, they constitute one contiguous zone. Their combined area of 1240 square

[3] Since the late 1960s, there has been substantial activity in a part of mathematical optimization theory that treats the multiobjective evaluation issues examined by Marglin and Maass. This branch of theory, known as *multiobjective programming,* concerns the limited set of multicriteria evaluation problems that can be described fully by mathematical equations and inequalities. For an introduction to multiobjective programming, see Cohon (1978), and Goicoechea, Hansen and Duckstein (1982).

[4] This is a procedure advocated by Freeman (1970) for programs administered by federal agencies in the United States.

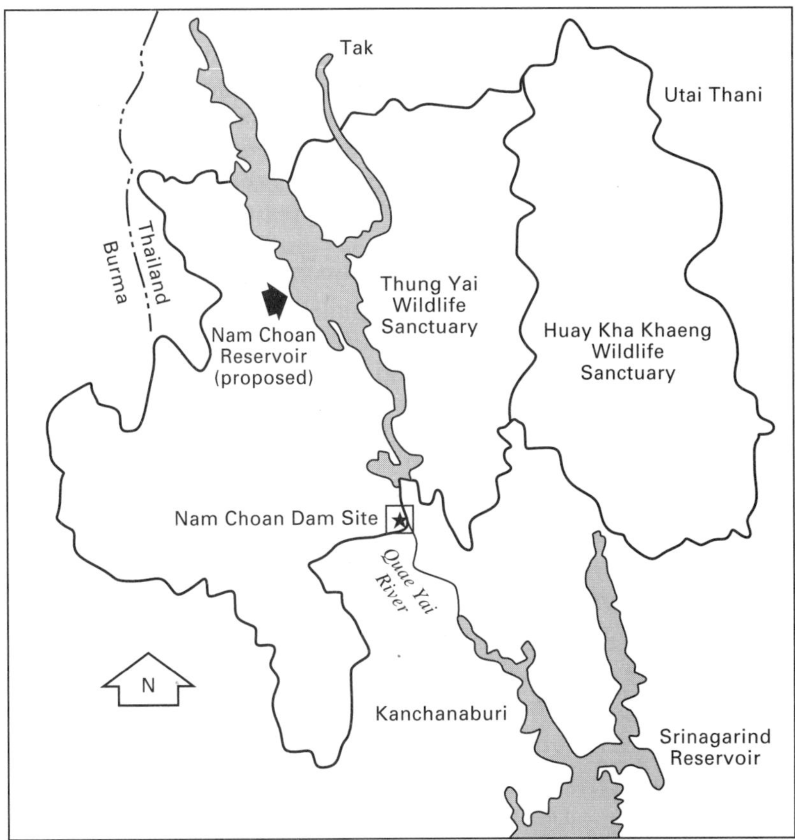

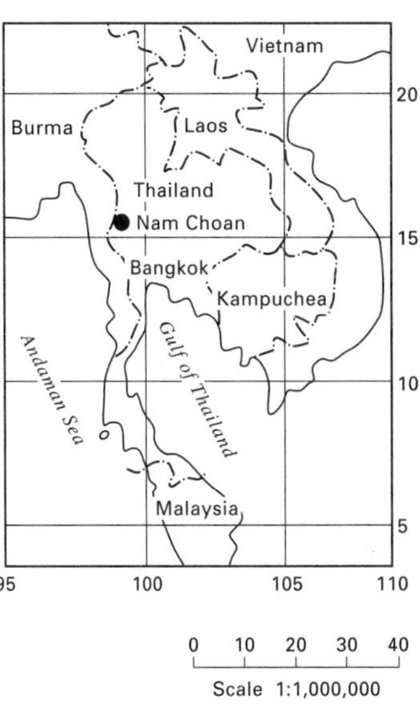

Figure 17.2 Approximate area to be inundated by Nam Choan Reservoir. Reprinted by permission of the publisher from Phantumvanit and Nandhabiwat, "The Nam Choan Controversy: An EIA in Practice," *Environmental Impact Assessment Review,* vol. 9, no. 2, pp. 137–38. Copyright 1989 by Elsevier Science Inc.

miles forms the largest wildlife sanctuary in Thailand, and the proposal to fragment the sanctuary by creating a reservoir within it met with much opposition during the late 1980s.

Controversy over the environmental impacts of the proposed Nam Choan project led the Thai government to appoint an ad hoc committee to evaluate the proposal. As part of its work, the committee undertook an analysis in which the present value of net monetary benefits was traded off against the damage to the wildlife sanctuaries, as well as other environmental effects that could not be valued in monetary terms.

Several procedures were used to estimate monetary costs and benefits, and results are summarized in Table 17.2. For example, the monetary benefits of electricity generated by the Nam Choan project were calculated by assuming that if the project were *not* built, the electricity that the project could generate would be produced by other projects. The monetary benefits were taken as the cost of the least expensive alternative means of producing the power. Other procedures for monetizing benefits involved valuation methods similar to those introduced in Chapter 6. For example, the forests cleared to make way for the reservoir would provide marketable timber products, and those products were valued using market prices. When the monetary benefits and costs in Table 17.2 were discounted to present value, the result was a net economic benefit of about U.S. $140 million per year; the benefit-cost ratio was 1.85.[5]

[5] Present value computations were made using a time horizon of 50 years and a discount rate of 12% (Phantumvanit and Nandhabiwat, 1989).

TABLE 17.2 Monetary Benefits and Costs of the Nam Choan Project[a]

Annual Benefits (U.S. $ millions per year)	
• Electric power generation.	48
• Timber cleared from reservoir and dam site.	107
• Fisheries production at reservoir.	<1
Annual Costs (U.S. $ millions per year)	
• Operation and maintenance.	18
• Compensation for environmental damages plus monitoring.	19
• Opportunity cost of loss of sustained yield forest products.	3
Construction costs (U.S. $ millions)	**253**

[a] Based on cost and benefit estimates given by Phantumvanit and Nandhabiwat (1989, pp. 143–44).

The committee performing the evaluation augmented the monetary figures by tabulating *nonmonetizable* costs and benefits of the Nam Choan project (see Table 17.3). They divided these costs and benefits into two categories, one of which involved effects that could be described quantitatively, even though reliable data were lacking. Examples of items in this category included the benefits of increased construction jobs (estimated to number about 5000), and the cost associated with the anticipated decrease in the number of tourists who would come to the wildlife sanctuary area to appreciate nature.

The second category of nonmonetized costs and benefits was termed *nonquantifiables.* Of the several nonquantifiable effects in Table 17.3, two were decisive in the final outcome: increasing risk of earthquakes, and degradation of the wildlife sanctuary area. An earthquake would threaten the integrity of the dam. There are geological faults beneath the border of Thailand and Burma, and some people believed that the reservoir would increase the likelihood of earthquakes. While there are data supporting a causal link between seismic activity and the creation of large reservoirs, there was much uncertainty and controversy about whether a causal link existed at the Nam Choan project site.

Effects on the wildlife sanctuary were also uncertain and controversial. Although the project would obstruct the movement of large animals and destroy a forest area that provided riverine habitats for rare species such as the Javan rhinoceros and the wild water buffalo, its overall effects on rare plants and animals could not be gauged because the extent to which the two sanctuaries contain rare species was not fully known. Moreover, there was disagreement over the effects of new access roads associated with the project. Some felt the roads would lead to increases in poaching, illegal logging, and land clearing for agriculture, whereas others claimed the roads would allow forest patrols to better protect the sanctuary.

In the final analysis, the Thai cabinet decided in 1988 to postpone the Nam Choan project indefinitely. Phantumvanit and Nandhabiwat (1989, p. 146) explain the outcome:

> To go ahead with the plan would defy the public sentiment. Given the prosperous state of the economy, the expected financial gain of nearly U.S. $140 million . . . could not outweigh the potential political loss.

The Nam Choan project evaluation demonstrates the usefulness of monetary valuations. The Thai officials who decided on the fate of the project used the net benefit figure as a benchmark. The decision to block the proposal was based on the judgment that the net gains of U.S. $140 million were not worth the costs that could not be valued in monetary terms.

TABULAR DISPLAYS AND WEIGHTED SUMS OF FACTOR SCORES

Many evaluation procedures rely on tabular displays of information. Table rows generally correspond to evaluative factors in one form or another. Columns correspond to the alternative proposals under consideration. A few types of table entries are commonly used. In some cases, an entry consists of a brief description of how a particular proposed action is likely to influence a given evaluative factor. In other cases, an entry is a numerical score characterizing the effects of a proposed action on a factor.

Some idea of the extent to which tables are used is given by Canter's (1979) analysis of 28 environmental impact statements (EISs) for wastewater manage-

TABLE 17.3 Nonmonetizable Benefits and Costs of the Nam Choan Project[a]

Quantifiable Factors (with incomplete and unreliable values)

Benefits	Costs
• Recreational tourism at proposed reservoir.	• Reduced visits by "tourists who appreciated nature."
• Improved access of area residents to medical facilities.	• Increased incidence of waterborne diseases.
• Increased employment for construction workers.	
• Increased water available for irrigation and other purposes.	

Nonquantifiable Factors

Benefits	Costs
• Improved access and opportunities to discover new mineral deposits and archeological sites.	• Possible loss of mineral deposits and archeological sites due to inundation.
• Hydropower as a nonpolluting source of energy.	• Degradation of wildlife sanctuary area.
	• Increased risk of earthquakes.

[a] Summary of key factors presented by Phantumvanit and Nandhabiwat. Reprinted with permission of the publisher from Phantumvanit and Nandhabiwat, "The Nam Choan Controvery: An EIA in Practice," *Environmental Impact Assessment Review,* Vol. 9, no. 2, pp. 144–45.

ment proposals. Twenty of these EISs included tables of the previously-described type. Most of these 20 impact statements used numerical scores to characterize how evaluative factors would be affected by alternative projects.

Table Entries Based on Ordinal Scales

Numerical table entries often result from rankings of how alternatives would influence an evaluative factor. Suppose, for example, that five alternative highway routes are ranked in terms of the noise they would cause in a nearby residential area. The route that would cause the least noise is assigned a 1, the route that would cause the second to the least amount of noise is assigned a 2, and so on. Another approach to making comparative observations involves categories; for example, "positive effect, no impact, and negative effect." Sometimes these categories are assigned arbitrary numerical values, such as +1, 0, and −1.

A typical display to aid in ranking alternatives is shown in Table 17.4. It was used in evaluating several wastewater management plans (labeled arbitrarily as R, S, T, U, and V) for North Monterey County, in California. Rank ordering was used to score the alternatives in terms of costs and energy consumption. For most evaluative factors, however, the alternatives were rated using four categories: adverse, beneficial, problematic (unknown or open to question), and none.

The entries in Table 17.4 are based on *ordinal scales.*[6] Only information reflecting a qualitative comparison or ordering of alternatives is given. Nothing is implied about the magnitude of the

[6] Three other measurement scales are commonly used in evaluation: nominal, interval, and ratio. A *nominal* scale provides information about categories. An example is a system for classifying soil as clay, silt, gravel, and so on. An *interval* scale has meaningful units of measurement, but a nonarbitrary origin or zero point for the scale doesn't exist. It is illustrated by altitude, which has no absolute zero. A zero altitude is assigned using a reference point such as mean sea level. A *ratio* scale has quantitative units of measurement and a nonarbitrary zero point. An example is mass. Because ordinal and nominal scales do not have quantitatively meaningful intervals, they cannot be manipulated using arithmetic operations. In contrast, all ordinary mathematical operations can be applied to variables measured using ratio or interval scales.

TABLE 17.4 Summary Evaluation of Alternative Treatment and Disposal Plans—North Monterey County, California, EIS[a]

Potential Impacts	Alternatives[b]					
	R	S	T	U	V	No Action
Physical/Biological Impacts						
Archaeological resources	P	P	P	P	P	N
Air quality	A	A	A	A	A	A
Soils and crops	N	P	P	P	P	N
Agricultural practices	N	P	P	P	P	N
Seismic risks	A	A	A	A	A	A
Groundwater quality	N	B	B	B	P	A
Surface water quality	B	B	B	B	B	A
Monterey Bay water quality	B	B	B	B	B	A
Water supply and reuse	N	B	B	B	B	N
Public health—water contamination	B	B	B	B	P	N
Public health—land contamination	N	A	A	A	A	N
Energy consumption in treatment and disposal of wastewater (rank)[c]	2	6	4	4	3	1
Aesthetics	B	B	B	B	B	A
Land use changes	N	A	P	A	A	N
Salinas River biota	A	P	A	A	A	N
Salinas River lagoon biota	B	B	B	B	B	N
Marine biota	B	B	B	B	B	N
General construction impacts	A	A	A	A	A	N
Economic Impacts						
Construction cost (rank)	3	5	4	6	2	1
Operating costs (rank)	2	6	3	5	3	1
Local cost (rank)	3	5	4	6	2	1
Overall cost (rank)	3	5	4	6	2	1
Social Impacts						
Growth inducement-accommodation	A	A	A	A	A	N
Local acceptance	A	P	P	P	P	A

[a] Adapted from Canter (1979). Reprinted by permission of Herner & Company, copyright © 1979. The original version of this table is from the "Final Environmental Impact Statement and Environmental Impact Report, North Monterey County Facilities Plan," Vol. I, issued by the U.S. Environmental Protection Agency and Monterey Peninsula Water Pollution Control Agency, San Francisco, August 1977.

[b] Key: B, beneficial; A, adverse; P, problematic (unknown or open to question); N, none.

[c] Comparative ranking from most acceptable (1) to least acceptable (6) alternative.

differences among alternatives. For example, asserting that one proposal would have a beneficial impact on air quality and another would have an adverse impact does not say anything about *how much* of a difference exists between the expected impacts of the two proposals.

Sum of Weighted Factor Scores

A common approach to summarizing evaluative information uses the sum of weighted factor scores for each alternative. First, each proposal is assigned a score for each evaluative factor. All scores must be within the same numerical limits. For example, scores might vary from 1 to 10, with 10 representing the best alternative. Weights are assigned to indicate the relative importance of each factor, and they are used in computing a weighted sum. This approach is illustrated by a case involving land use planning for the city of Palo Alto, California.

In the early 1970s, Palo Alto engaged Livingston and Blayney, a firm of planning consultants, to examine the suitability of alternative types of land development for the foothills at the edge of the city. Livingston and Blayney's (1971) procedure for evaluating alternative land use plans centered around nine evaluative factors. In their view, these factors encompassed all of the considerations (aside from "political factors") significant in ranking the alternatives.

The consultants used professional judgment to assign each factor a weight indicating its importance relative to other factors. The nine factors, along with the weights (in parentheses) are as follows: 10-year cost (5); 20-year cost (5); social impact (8); transportation needs (3); ecological impact (5); fire hazard (2); visual impact (5); geologic impact (3); and hydrologic impact (2). For each factor, the consultants assigned an ordinal score reflecting the influence of each land use plan. Scores ranged from 1 (worst) to 5 (best). For each plan, a weighted sum was computed by multiplying the score for each factor times the associated factor weight and adding up the products. The results were used in making recommendations regarding alternative land use proposals.

Evaluation procedures using a sum of weighted factor scores have one clear advantage over many other methods: they are not difficult to implement. In a typical application, problems in determining which impacts are significant and how they are valued are settled by the collective judgments of the planning specialists doing the study. Once the scores and weights are assigned by the planners, only simple arithmetic is needed to compute the sum of weighted scores for each plan.

An extensive reliance on the value judgments of planners is sometimes cited as a weakness of the sum of weighted factors approach. Planners often exercise great control over the selection of factors, the scoring of alternatives for each factor, and the assignment of weights. Critics argue that the choices of factors and weights made by planners may be inconsistent with the views of persons affected by proposed plans. A response to this criticism is to involve the public in determining factors and weights, and this is sometimes done in practice.

The validity of processes typically used to select weights has also been challenged. As shown by Hobbs (1980), the common practice of assigning weights based on an "importance scale" from 1 to 10 does not accurately reflect the preferences of those choosing the weights. Theoretically rigorous procedures for determining weights are described by Hobbs, but they are often hard to implement.

Another shortcoming is that the sum of weighted factor scores does not adequately inform decision makers. Factor scores and weights are somewhat arbitrary. In addition, when these numbers are aggregated into a single index, much useful information about impacts is buried.[7] Trade-offs among alternative proposals are not illuminated, and the value judgments made by the planning analysts are not revealed. Critics of the weighted factors approach suggest that simple prose descriptions of the main impacts of alternatives would more clearly highlight trade-offs and require planning analysts to make fewer value judgments. A problem with this suggestion is that the amount of information involved in describing the numerous impacts typically associated with several alternatives can be overwhelming.

[7] The process of computing weighted sums has also been criticized because no basis exists for adding or multiplying parameters based on ordinal scales of measurement. Elliott (1981) discusses the way alternative measurement scales are used in creating indexes for ranking alternatives. Hobbs (1980) takes up this point in connection with scales used for assigning weights.

TABLE 17.5 Hill's Goals-Achievement Matrix: An Example[a]

	Goal 1-Accessibility			Goal 2-Community Disruption		
Community weights on goals	2			1		
	Group Weights (Goal 1)	**Plan A**	**Plan B**	**Group Weights (Goal 2)**	**Plan A**	**Plan B**
Uptown group	3	+1	−1	3	−1	0
Downtown group	1	−1	+1	2	0	−1
Extent of goals achievement		+2	−2		−3	−2

[a] This table's format is adapted from Hill (1968).

The Goals-Achievement Matrix

Hill's "goals-achievement matrix" extends the logic behind the previously mentioned tabular displays and sums of weighted factor scores. His approach to scoring the effect of an alternative action on an evaluative factor uses goals and objectives as the basis for defining benefits and costs. According to Hill, benefits indicate progress toward desired community objectives, whereas costs are retrogressions from these objectives.[8]

The goals-achievement approach explicitly considers the incidence of benefits and costs. It requires identification of the various groups of individuals or establishments that may be affected by a particular proposal. Weights are assigned to indicate how much each goal is valued by the groups. Use of weights that reflect how different groups view the *same* goal distinguishes Hill's approach from the weighted factor scores technique. The goals-achievement method relies on a second set of weights to indicate the overall importance of one goal relative to another. These "community weights" are similar to the weights used in computing a sum of weighted factor scores.

Hill proposes two distinct ways of organizing data to assist decision makers in ranking proposals.[9] One involves presenting only information on scores and weights without attempting to compute an overall index of a plan's worth. This is illustrated in Table 17.5, which compares two transportation plans (A and B) in terms of "accessibility" (ease of travel between two points) and community disruption. Both an "uptown group" and a "downtown group" will be affected by the proposals. Plan A increases accessibility for the uptown group, while decreasing it for the downtown group. Plan B has the opposite effect. Also, plan A disrupts the uptown neighborhood, but it has no impact downtown. In contrast, plan B has no influence on the uptown neighborhood, but it causes disruption downtown. These impacts are translated into scores on an ordinal scale: +1 = positive effect, 0 = no effect, and −1 = negative effect. A goals-achievement matrix (Table 17.5) summarizes the information.

Hill's second approach to presenting information on scores and weights uses indexes to show how well the goals are achieved by each plan. The indexes are computed in the same way as sums of weighted factor scores, except that they also take account of how the different groups weigh each goal. The computations are explained in two parts. One involves the "extent of goals achievement" by a given plan for a particular goal. For each goal, this is calculated as a sum of products of group weights multiplied by the ordinal scores representing the effects of the plan. Consider the extent to which plan A achieves the accessibility goal. For the uptown group, the weight (3) is multiplied by the score representing plan A's effect on accessibility (+1) to yield +3. The product of the downtown

[8] The discussion here, which is based on Hill (1967, 1968), uses the terms *goals* and *objectives* interchangeably in the interests of simplicity. Hill (1967, p. 22) makes the following distinctions: a goal is "an end to which a planned course of action is directed." In contrast, an objective denotes a goal that "is believed to lead to another valued goal rather than having intrinsic value in itself."

[9] Hill's approach has been reviewed critically by McCallister (1980) and Lichtfield, Kettle, and Whitbread (1975). It has been applied widely in land use planning exercises in England.

group's weight (1) and score (−1) is −1. Summing the products for both groups yields +2, a measure of how well plan A meets the accessibility goal. For all other combinations of goals and plans, the extent of goals achievement is computed in the same way. Results are at the bottom of Table 17.5.

The second part of the calculation determines the "weighted index of goals achievement." This is done by multiplying the previously computed extent of goals achievement values and the community weights shown at the top of Table 17.5. For plan A, the weighted index of goals achievement is the sum of products of the plan A entries in the last row of Table 17.5 multiplied by the appropriate community weights,

$$(2)(+2) + (1)(-3) = +1.$$

A similar computation for plan B yields a weighted index of −6.

In considering the incidence of effects, the goals-achievement approach requires that weights be obtained for both the community as a whole and for individual groups within the community. Although Hill does not recommend a specific procedure for determining weights, he does suggest the following possibilities:[10]

1. The decision makers may be asked to weigh objectives and their relative importance for particular activities, locations, or groups in the urban area.
2. A general referendum may be employed to elicit community valuation of objectives.
3. A sample of persons in affected groups may be interviewed concerning their relative valuation of objectives.
4. The community power structure may be identified, and its views on the weighting of objectives and their incidence can be elicited.
5. Well-publicized public hearings devoted to community goal formulation and valuation can be held.
6. The pattern of previous allocations of public investments may be analyzed in order to determine the goal priorities implicit in previous decisions on the allocation of resources.

Many of Hill's suggestions involve the public directly. The next chapter elaborates on the numerous ways of identifying the goals of different groups.

[10] The following list is quoted from Hill (1967, p. 25).

Weighted sums of factor scores and the goals-achievement matrix are only two of the dozens of multicriteria decision-making procedures that have been developed to structure evaluation activities. The two methods are featured here because they demonstrate the types of methods commonly used in practical settings. Many of the more sophisticated multicriteria decision-making techniques that appear in the literature are not used in practice because they involve mathematical manipulations that appear arcane to nonspecialists.[11]

ENVIRONMENTAL RISK ASSESSMENT

Up to this point in the chapter, evaluative information has been presented as if future outcomes were certain to occur. But predictions are inherently uncertain. For centuries, mathematicians and scientists have developed theories of probability and statistics to analyze the uncertainty in predicted outcomes. With some exceptions, however, public agencies do not rely heavily on probability and statistics in evaluating environmental policies and proposed development projects. Exceptions occur in studies of projects, regulations and policies involving (1) long-term human exposure to low concentrations of toxic substances, or (2) a potential for serious accidents (such as explosions at industrial facilities) or damaging engineering system failures.

During the 1970s, risk analysis methods that had been developed in a number of fields, particularly toxicology and reliability engineering, began to be used to assess environmental risks. Increasingly, environmental specialists argued that economic development could impose enormous costs if risks associated with accidents and poor system operations were not controlled. The significance of those risks was demonstrated by widely reported accidents, including the following:

1976 An explosion at a chemical manufacturing plant near Seveso, Italy, caused the relase of dioxins, which resulted in more than 100 cases of a skin disorder, the evacuation of hundreds

[11] A qualitative introduction to multicriteria decision-making methods and a survey of their use in public sector decision making is given by Massam (1988). For a quantitative introduction to the subject, see Goicoechea, Hansen and Duckstein (1982).

of residents, the slaughter of many farm animals, and an expensive cleanup effort.

1984 A massive leak of methyl isocyanate at a chemical plant in Bhopal, India, caused at least 1700 deaths and severe problems for as many as 200,000 people.

1986 An explosion at a nuclear reactor of the Chernobyl nuclear power station in Ukraine led to several hundred deaths, the resettlement of tens of thousands of people, and psychological and physical health problems for many of those affected by the blast.

1989 A supertanker, *Exxon Valdez,* ran aground in Prince William Sound off the coast of Alaska, spilling 11 million gallons of crude oil which spread over nearly 600 miles and caused numerous adverse effects.

In addition to being alarmed by catastrophic industrial accidents, citizens and policymakers became increasingly concerned over health effects of long-term exposure to low concentrations of contaminants. In the United States, for example, effects of hazardous chemicals entering groundwater from poorly managed landfills became a major public issue.

By the early 1990s, use of risk assessment to evaluate proposed development projects and environmental policies had become institutionalized in some countries. And calls by governments for assessments of risk were accompanied by increased efforts to use analytic methods developed earlier in toxicology and other fields to create a paradigm for *environmental risk assessment.*

Although terminology is not yet standardized, there is wide agreement that an environmental risk assessment should answer the following types of questions.

- Will a decision to approve a proposed project lead to unplanned outcomes that are hazardous?
- What reduction in hazardous outcomes will accompany a decision to adopt a proposed environmental regulation?
- How can the magnitude or severity of hazardous outcomes be described?
- What is the likelihood (or probability) that potential hazardous outcomes linked to a proposed project will occur?

Questions involving likelihood of occurrence are often unanswered using the language of probability theory, as in "the probability of a major explosion at the proposed chemical plant is one in a million." Typically, numerical values of probabilities are themselves highly uncertain and sometimes (because of gaps in data or knowledge) difficult to interpret. In classical versions of probability theory, the term *probability* is interpreted as *relative frequency* of occurrence of a repeatable event.[12] For example, saying that the "probability of obtaining a head as the outcome of coin toss is 0.5" is based on the empirical fact that in a very large number of coin tosses, a head would come up half the time. In risk assessments, events such as major industrial accidents and system failures occur only rarely, and thus it is impossible to use data from large numbers of repeated trials to estimate probabilities. In these cases, experts frequently rely on *subjective probabilities,* which are estimates of the likelihood of occurrence based on expert judgment. A meterorologist's prediction that "the likelihood of rain tomorrow is 90%" illustrates the notion of subjective probability.[13]

Basics of Risk Assessment

For any particular decision context, a risk assessment consists of a process involving four phases (see Figure 17.3):[14]

- *Hazard Identification.* The central questions are: Does exposure to a particular chemical agent pose a danger to human health and the environment? How can a proposed project fail in ways that lead to hazardous outcomes? Contaminants of concern are identified along with the spatial and temporal boundaries of analysis.
- *Exposure Assessment.* Which populations or ecosystems are (or will be) exposed to the hazardous outcomes? For how long? And with what likeli-

[12] Another early definition of probability focused on games of chance in which outcomes are equally likely. For example, consider an urn filled with 15 balls, which are identical in all ways except for color. Suppose 5 of the balls are red and 10 are white. The probability of selecting a red ball at random is the ratio of possible ways of selecting a ball that is red (5 in this case) divided by the total number of outcomes (15).

[13] An introduction to the history of probability theory that highlights alternative definitions of probability is given by Raiffa (1968).

[14] While terminology in the field of risk assessment is not yet standardized, the terms in Figure 17.3 are widely used.

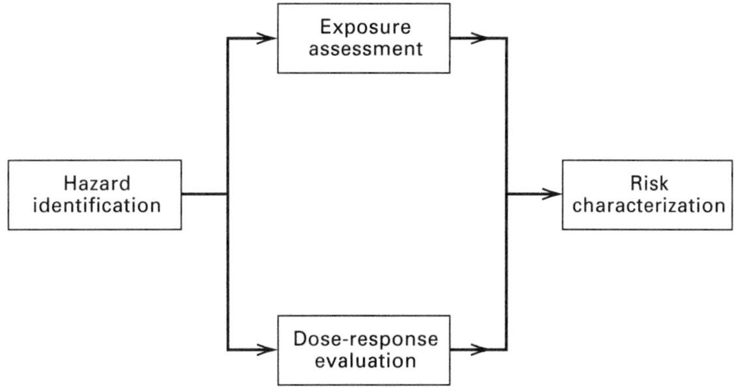

Figure 17.3 Principal activities in an environmental risk assessment.

hood? In exposure assessment, analysts use models of the fate and transport of contaminants to estimate concentrations of contaminants at different locations. In addition, calculations are made of the dose received by individuals exposed to the contaminants.

- *Dose-response Assessment.* What information exists to predict the response of humans, plants, and animals to different doses of contaminants? In *human health risk assessments,* the basic information comes from epidemiological surveys, research using laboratory animals, and studies of workers exposed to contaminants over long periods. In *ecological risk assessments,* key information comes from investigations of how selected plants and animals respond to chemical agents. Comprehensive ecological assessments also consider interdependencies among species and other relationships relevant to ecosystem structure and function.
- *Risk characterization.* Results from analyses of exposure and and dose-response assessments are synthesized to yield information on hazardous outcomes. Effects are reported along with information describing levels of uncertainty (for example, the probability of occurrence of an industrial accident). Risk characterization also includes a description of assumptions made, and sources of uncertainty associated with numerical computations of outcomes.

Activities in an environmental risk assessment are carried out in two different contexts. One involves situations where the risk is from long-term exposure to low concentrations of a dangerous material. The second concerns industrial accidents and engineering system failures, where the exposure to hazard occurs in a short time frame, but with potentially disastrous outcomes. Assessments for each of these two circumstances are described here.

Chronic Low-Level Exposure to Contaminants

Figure 17.4 provides a context for analyzing risks when exposure to contaminants occurs over long periods at low concentrations. The figure illustrates the *exposure pathways* followed by substances migrating away from an improperly managed landfill used for disposal of hazardous materials. As shown in the figure, one exposure route occurs when hazardous wastes are volatilized and transported by prevailing winds. Another occurs when substances are leached into groundwater.

Hazard Identification

In the *hazard identification* phase of a risk assessment, sampling is done to identify contaminants that may pose risks to human health and the environment. For simplicity, the example based on Figure 17.4 supposes that only a single contaminant—benzene—is identified. Risk analysts would use their knowledge of benzene's properties together with information on potential uses of the contaminated groundwater to determine, in a preliminary way, if the danger of contamination is large enough to warrant a full risk assessment.

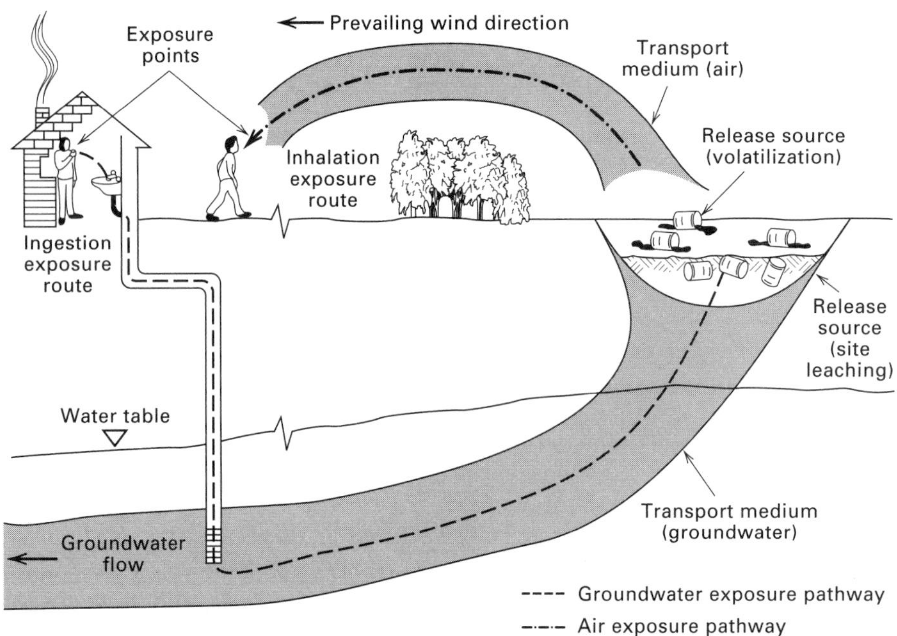

Figure 17.4 Illustration of exposure pathways. Source: U.S. Environmental Protection Agency (1986).

More generally, in assessments involving long-term exposure, hazard identification focuses on identifying (1) a list of contaminants that pose a threat, and (2) the pathways that contaminants might follow in reaching potentially affected individuals and ecosystems. Analysts rely on toxicological data to gauge risks to human health. In cases involving risk to ecosystems, expert judgment is used to identify the particular organisms or attributes of the environment thought to be at risk.

Exposure Assessment

Exposure assessment is the process of translating information on releases of contaminants at their source into estimates of doses of contaminants received by humans or other organisms. In the example involving benzene released from a landfill, fate and transport models of the type described in the previous chapter could be used to estimate how benzene would move and what its concentration would be at various locations.

Figure 17.5 puts results of fate and transport modeling in context by showing relationships between chemical releases, concentrations, exposures, and doses. Information on concentration, together with assumptions about human exposure, provide a basis for calculating dose. The latter is a key variable in estimating health risks.

Dose is computed by multiplying concentration by intake rate. The process of calculating doses of a substance received by exposed populations involves construction of *exposure scenarios.* For demonstration purposes, consider only the groundwater exposure pathway in Figure 17.4. If the benzene reached a well used as a community water supply, development of an exposure scenario would require making assumptions to answer two questions about persons consuming the water: How much contaminated water will be consumed in a typical day? Over what length of time will contaminated water be consumed? Examples of assumptions are those used by the U.S. Environmental Protection Agency (EPA) in the early 1900s to calculate dose: the average person consumes 2 liters of water per day and lives 70 years. Using EPAs guidelines, the dose of an individual exposed to 10^{-2} mg/l of benzene in water is

$$10^{-2}\,\frac{\text{mg}}{l} \times 2\frac{l}{\text{day}} = 2 \times 10^{-2}\,\frac{\text{mg}}{\text{day}}.$$

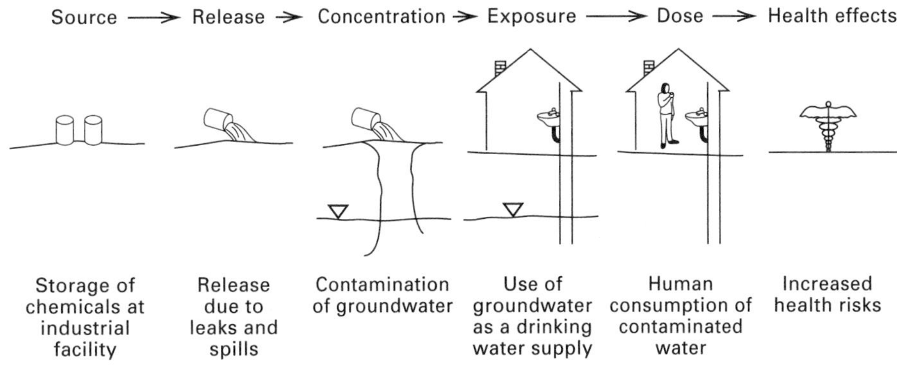

Figure 17.5 Linkages between release, concentration, exposure, dose, and health effects. Adapted with permission from Asian Development Bank (1990, p. 39).

Over a 70-year lifetime, an individual using the contaminated water would receive a total dose of 511 mg.[15]

An exposure assessment includes more than identification of exposure pathways and calculation of concentrations and doses. It also involves a description of the exposed population. For instance, in the benzene example, an exposure assessment would identify the number of individuals exposed and the existence of groups of individuals (such as infants) who might be particularly sensitive to contaminated drinking water. For ecological risk assessments, assumptions must be made about which species and ecosystems might be affected.

Dose-response Assessment

Dose-response assessment uses information on how the exposed population will be affected by the contaminants. In the case of humans exposed to contaminants in water and other environmental media, analysts can rely on databases that summarize results of toxicity studies. For example, EPA periodically publishes *Health Effects Assessment Summary Tables* that contain dose-response information for numerous chemicals.[16] For any particular substance, the EPA tables contain a *reference dose* (RfD), which is a threshold value for gauging chronic *noncarcinogenic* effects. An RfD represents a conservative estimate of the average level of exposure (in mg/kg of body weight per day) below which the health risk is minimal. Reference dose is calculated in two steps: (1) based on empirical studies, identify doses at which there have been no observable effects; and (2) divide those doses by a factor, typically between 10 and 1000, to account for uncertainty. One source of uncertainty is the use of toxicological results from laboratory animal research to calculate RfD. Uncertainty is also introduced by the common procedure of extrapolating data based on high doses (for example, those used in laboratory studies) to conditions involving low doses over long periods of exposure.

In addition to containing reference doses, EPA's *Health Effects Assessments Summary Tables* also contain *potency factors,* which are used in calculating increases in cancer cases. EPA's procedure for estimating cancer risk associated with lifetime exposure to a carcinogenic substance is based on two assumptions: no exposure to carcinogens is completely safe, and adverse effects are proportional to (or vary linearly with) dose.[17] Using these assumptions, an increase in cancer risk is calculated as the product of

[15] The total lifetime dose is calculated as

$$2 \times 10^{-2}\,\frac{\text{mg}}{\text{day}} \times 365\,\frac{\text{days}}{\text{yr}} \times 70\text{ yr} = 511\text{ mg}.$$

[16] EPA also maintains a computer database of toxicological information: IRIS, an acronym for *Integrated Risk Information System.*

[17] The use of a linear model with a zero threshold reflects policy decisions made by EPA. Other assumptions are defensible. In "Canada, for example, it is assumed there is a threshold dose for carcinogens below which adverse effects are not expected to occur" (von Stackelberg and Burmaster, 1994, p. 392). For a historical account of EPA's policy on cancer risk, see Landy, Roberts, and Thomas (1994, pp. 172–203).

potency factor and lifetime average daily intake. In other words, the "dose-response function" is a linear equation that intercepts the origin.

Application of a linear dose-response function is demonstrated using the example in Figure 17.4 involving benzene in a water supply. The potency factor reported by EPA for benzene (by either inhalation or ingestion) is 2.9×10^{-2} (mg/kg/day)$^{-1}$. The cancer risk for a person consuming 2 l/day of water contaminated with 10^{-2} mg/l of benzene is found by first calculating the lifetime average daily dose (assuming the average person weighs 70 kg):

$$10^{-2}\,\frac{\text{mg}}{l} \times 2\,\frac{l}{\text{day}} \times \frac{1}{70\text{ kg}} = 2.86 \times 10^{-4}\text{ mg/kg/day}.$$

The increase in cancer risk is then

$$2.9 \times 10^{-2}\,(\text{mg/kg/day})^{-1} \times 2.86 \times 10^{-4}\,(\text{mg/kg/day}) = 8.29 \times 10^{-6}.$$

This result suggests that over a 70-year period, there would be eight additional cancers per million people consuming water from the contaminated source.

Dose-response data is highly uncertain, and EPA's reliance on a linear model with a zero threshold (that is, no safe exposure) is deliberately conservative. According to EPA (1989), the reported value of a potency factor reflects a "95 percent confidence interval." In other words, if EPA's process for calculating potency factors is repeated 100 times, then 95 of the calculated values will be lower than the one reported by EPA. In other countries, different assumptions and policy guidance are employed in developing dose-response information used in assessing cancer risks.

Risk Characterization

Risk characterization involves synthesizing results from previous parts of the assessment and presenting them in ways that can inform decision makers. In addition to providing numerical estimates of risk to various populations, a risk characterization includes explanations of assumptions made in conducting the assessment, and gaps in information that cause risk estimates to be uncertain.

Risk characterization is intended to give decision makers a basis for interpreting numerical risk estimates. For example, the risk characterization process details assumptions and uncertainties associated with the following:

- Prediction of contaminant concentrations using fate and transport models.
- Creation of exposure scenarios for calculating doses.
- Development of dose-response curves for low dose exposure using empirical studies where subjects were exposed to high doses.
- Extrapolation of dose-response curves from laboratory animal studies to humans.

Risk analysts disagree about how the uncertainties in risk estimates should be presented. Environmental protection agencies make numerous "worst-case" assumptions in calculating health risk, but they often report only a single risk value. Critics argue that reporting only one number is misleading, particularly when it is based on a collection of conservative assumptions that, when taken together, make the reported risk extremely unlikely.[18]

This discussion has emphasized carcinogenic effects on humans, which reflects the high priority given to carcinogenic effects in risk assessments by government agencies.[19] In comparison to human health effects, agencies have placed less emphasis on risks to nonhuman species. Scientists are developing new methods of ecological risk assessment, which may lead to more effort in assessing risks for nonhuman species and ecosystems.[20]

Industrial Accidents and System Failures

Industrial accidents and engineering system failures are assessed using what is sometimes termed the "en-

[18] For an analysis of difficulties in interpreting human health risks based on a collection of conservative assumptions, see Maxim (1989).

[19] Government risk assessments have also included noncarcinogenic effects on humans. An example is provided by EPA's (1991) analysis of noncarcinogenic effects on humans of carbon monoxide in ambient air.

[20] For more on methods of ecological impact assessment, see Suter (1993), and Calabrese and Baldwin (1993).

gineering approach" to risk assessment. It differs from the method used to assess risks from long-term exposure to contaminants primarily in how hazards are identified and how exposure assessments are conducted.

The hazard identification phase of risk assessments for projects such as coastal oil terminals and nuclear power plants focuses on a basic question: What might go wrong? This may be answered by considering the hazardous substances on-site, particularly materials that are toxic, explosive, or flammable. Once dangerous materials are identified, the central question concerns how they may be released to the environment. Unplanned releases may occur because of failures in equipment, human error, or both.

The process of identifying how unplanned hazards might occur is part of exposure assessment. Techniques for constructing exposure scenarios range from exercises in brainstorming to more structured approaches used by reliability engineers.[21] Two reliability engineering methods are introduced here: event tree analysis and fault tree analysis.[22]

Event Tree Analysis

Event tree analysis is explained using an example involving risks associated with release of a toxic gas from an industrial plant. The analysis begins by considering an *initiating event,* in this case the failure of a container holding a toxic gas. The analysis begins by asking: What if the container fails and toxic gas is released? The industrial plant relies on an emergency alarm system to trigger evacuation and other hazard control procedures in case of a release. The alarm system is activated by a gas-monitoring device. If the container leaks, the monitoring device transmits a signal that sets off the alarm.

[21] In a *brainstorming* session, members of a group generate ideas spontaneously, and the ideas are neither criticized nor evaluated.

[22] Another common approach to constructing exposure scenarios involves use of Hazard and Operability (HazOps) Studies. Following this approach, multidisciplinary teams of experts in a particular facility type use "guide words" as an aid in creating scenarios. For example, in considering a particular chemical manufacturing process, guide words aid experts in considering how the original intent in design of the process might be confounded by inadvertent valve openings, flow blockages, use of incorrect input materials, and so on. The team identifies credible accident scenarios based on design information, and past operating experience with relevant engineering components and systems.

The event tree in Figure 17.6 contains all possible sequences that can yield a hazardous condition. The sequence in which the monitoring system fails and the alarm system operates has a probability of occurrence of zero because the alarm system cannot start without receiving a signal from the monitor. For each of the other sequences, probabilities of occurrence for each individual event could be estimated using statistical analysis methods. Probability theory could then be used to calculate the likelihood of a hazardous condition occurring via each combination of events represented by a path through the tree.

Fault Tree Analysis

Fault tree analysis is a second common method for identifying scenarios that lead to accidents or system failures. This technique begins by postulating a hazardous outcome and then works backward to identify combinations of events that might lead to the hazard. Figure 17.7 shows results from a fault tree analysis of the previously considered problem: hazard due to a release of toxic gas at an industrial plant. The undesired outcome—in this case the hazard associated with the release—is at the top of the fault tree. The analysis proceeds downward by answering the question: What could cause the event at a particular level in the tree? For the event labeled "hazard due to toxic gas release" to occur, two other events would have to take place: a crack in the container holding the gas *and* failure in the emergency response system at the plant. Moving one level down in the tree, consider how the emergency system might fail. This requires either a breakdown in the gas-monitoring de-

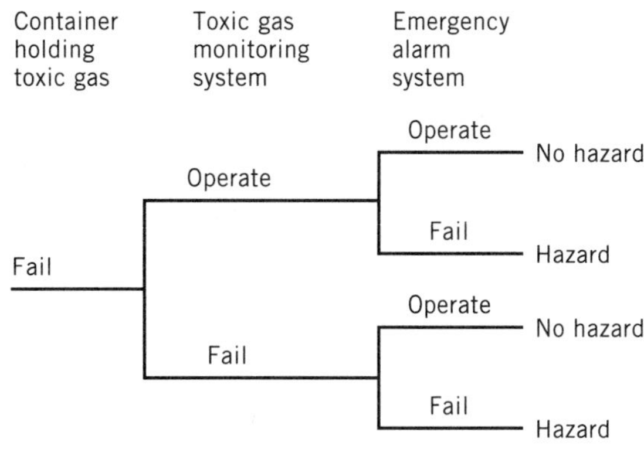

Figure 17.6 Event tree for hazard due to toxic gas release.

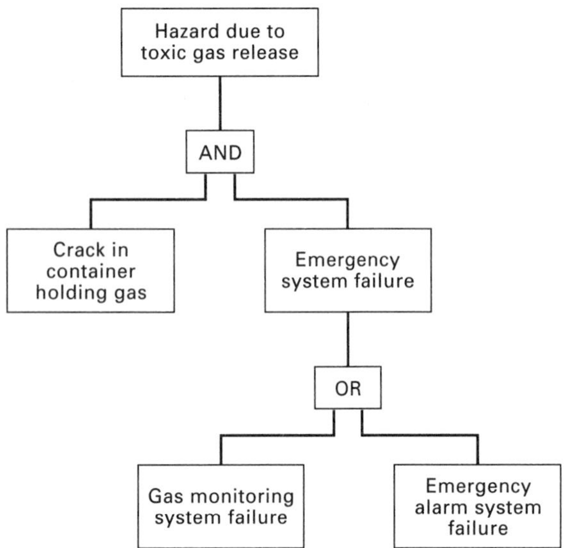

Figure 17.7 Fault tree for analyzing gas release hazard.

vice *or* a failure in the alarm itself. As in event tree analysis, probabilities can be estimated for individual failures in a fault tree, and they can be combined using probability theory to calculate the overall probability of a hazard due to a release of toxic gas.[23]

Additional Risk Assessment Steps

In addition to identifying events that could lead to hazardous conditions, a risk assessment for engineering system failures requires development of exposure scenarios. These scenarios are often generated using the same steps as in risk assessments for long-term exposure to low concentrations of contaminants. For example, in the case of a release of toxic gas from an industrial facility, exposure pathways would be delineated. Then fate and transport models and exposure calculations would be used to describe how the toxic gas might move through the environment and reach potentially affected persons and ecosystems.

The remaining steps in assessing risks from accidents and system failures involves a dose-response analysis and risk characterization. Reasoning used in these phases of assessment is analogous to that employed in corresponding phases of risk assessments for chronic exposure to low concentrations of contaminants. However, the details are different. In the case of industrial accidents and engineering system failures, the hazards often involve catastrophic losses that have low probabilities of occurrence. Quantitative investigations of risk would include probabilities of occurrence for different exposure scenarios and hazardous outcomes. Results of the assessment provide a basis for gauging the acceptability of the risks and the need for measures to mitigate risks that are judged too high to accept.

USING RISK ASSESSMENTS TO EVALUATE OPTIONS

Specialists in risk assessment commonly distinguish risk assessment from *risk management.* Risk assessment is intended to be as scientific and objective as circumstances allow, whereas risk management introduces value judgments and makes trade-offs between costs and benefits of risk reduction.[24] Some analysts have suggested that risk assessment results be included in an augmented benefit-cost analysis, an evaluative framework termed *cost-risk-benefit analysis.*

The cost-risk-benefit framework is illustrated using steps presented by Cox and Ricci (1989) for evaluating options to control chronic health risk. They describe as "common practice" the following evaluation procedure:[25]

> Step 1. Evaluate the dollar value per statistical life saved that those at risk (e.g., workers in a hazardous occupation) place on their own lives. . . . A plausible value for the 1980s might be $500,000 per statistical life saved.
>
> Step 2. Evaluate the dollar costs of alternative control options, including the option of banning the activity if that is a possibility. Include the dollar

[23] For details on the computational procedures, see Hauptmanns's (1989) application of fault tree analysis for a process used in producing hexogen, an explosive.

[24] This distinction between the "value-free," scientific risk assessment and subjective risk management does not hold up in practice. It would be difficult to claim that the numerous assumptions made in assessing risks do not involve value judgements and nonscientific policy decisions. Consider, for example, EPA's policy to use a linear dose-response curve with a zero threshold in assessing cancer risks.

[25] The steps outlined are quoted from Cox and Ricci (1989, p. 1038). For some steps, Cox and Ricci included a few supplementary remarks that are not included here.

value of benefits foregone (i.e., opportunity costs) along with the direct costs of control in calculating the total economic cost to society of each option.

Step 3. Implement those control options whose benefits (expected lives saved times imputed dollar value per statistical life saved, from step 1) exceed their costs.

Step 4. A risk is said to be acceptable, within this paradigm, if the costs of further controlling it exceed the benefits from further control. . . .

As Cox and Ricci point out, numerous arguments can be raised against use of this evaluative framework. For one thing, many applications in which risk assessments aid in decision making do not include explicit benefit comparisons. Many other objections are similar to criticisms of conventional benefit-cost analysis. For example, the cost-risk-benefit approach ignores equity considerations; it also ignores rights that individuals may have to personal safety in various contexts. As in traditional benefit-cost analysis, the cost-risk-benefit framework contains valuable elements despite its overall shortcomings. Its value is in informing, but not dominating, political decision processes that are frequently used to make social choices about acceptable levels of risk.

The following presentation describes how environmental risk assessments are used in evaluating proposed environmental policies and development projects. It distinguishes between chronic exposure to low concentrations of contaminants, and hazards caused by industrial accidents and engineering system failures.

Chronic Exposure to Contaminants

Risk assessments for chronic low-level exposure to environmental contaminants can be placed in the six categories listed in Table 17.6. Decisions in each category are typically made by balancing risk assessment results against a host of other factors, such as cost, technical feasibility, and requirements imposed by environmental statutes.

The first category in Table 17.6 concerns a threshold question: Should a particular substance be regulated? This question arises, for example, in implementing laws establishing programs to control hazardous waste. Environmental agencies must decide which substances are hazardous (and thus subject to statutory requirements), and they sometimes perform a risk assessment to help make this decision.

Another example involves regulation of carbon dioxide and other greenhouse gases. Assessments of the risk of global warming have been performed by the Intergovernmental Panel on Global Climate Change, a United Nations group of about 2500 scien-

TABLE 17.6 Decisions Influenced by Risk Assessments for Chronic Low-Level Exposure to Contaminants

- **Threshold decision to regulate**
  Should a currently unregulated substance be regulated?
- **Cleanup measures at contaminated sites**
  For a particular contaminated site, what remediation option should be implemented?
- **Ambient standards**
  For a particular substance, what concentrations in air and water pose unacceptable risks?
- **Pollutant discharge limits**
  What levels of air and water pollution discharged from a particular point source pose unacceptable risks?
- **Priority setting in environmental problem solving**
  How can information on *relative* (or *comparative*) *risk* be used in allocating resources for environmental programs?
- **Approval or denial of proposed projects**
  What risks to human health and the environment are associated with a proposed development project, and how can those risks be weighed against other project impacts?

tists that advise parties involved in negotiating international agreements to regulate greenhouse gas emissions. As of the early 1990s, the level of uncertainty associated with the panel's risk assessments was such that most governments chose to take only small steps to constrain greenhouse gas emissions.

A second category of risk assessment applications concerns the selection of measures to remediate hazardous waste sites. For example, the Superfund program for remediating contaminated hazardous waste facilities in the United States requires a risk assessment as part of the "remedial investigation/feasibility study" phase of a cleanup. Results from these assessments are used to evaluate and select cleanup measures at Superfund sites.[26]

As indicated in Table 17.6, risk assessments are used in setting both ambient standards and discharge limits. Applications in the Netherlands are illustrative. The Dutch National Environmental Policy Plan indicates that for any substance, the maximum acceptable risk limit should not be exceeded by the year 2000 (see Figure 17.8). In the Dutch setting, both ecological and human health risks are taken into account. Maximum-risk guidelines are used, for example, in regulating emissions of dioxins from waste incinerators. Risk assessment also plays a key role in decisions related to ambient standards.[27] For instance, both risk assessments and benefit-cost studies are used to evaluate alternative measures for meeting ambient standards for ozone.[28]

The next category in Table 17.6 concerns use of *comparative risk assessment* to set priorities for environmental problem solving. Support for using risk assessment in this way came from a U.S. Environmental Protection Agency (EPA, 1987) study entitled *Unfinished Business.* This report, together with various follow-on activities carried out by EPA, demonstrated that comparative (or relative) risk assessments could help environmental agencies and policymakers set priorities. Since the 1987 report, a number of comparative risk projects have been undertaken to help establish environmental priorities at national, state, and local levels of government.[29] Increased use of comparative risk assessment reflects a recognition that scarce resources for environmental protection are sometimes targeted at solving low-risk problems while more dangerous situations receive little attention. Advocates of the comparative risk approach point to its value in moving beyond reactive, incremental policy making and toward policies based on *rational risk reduction.* For example, by comparing risk reduction per dollar spent in different governmental risk management programs, resources can be allocated *cost-effectively:* a selected level of risk reduction can be attained at minimum cost.

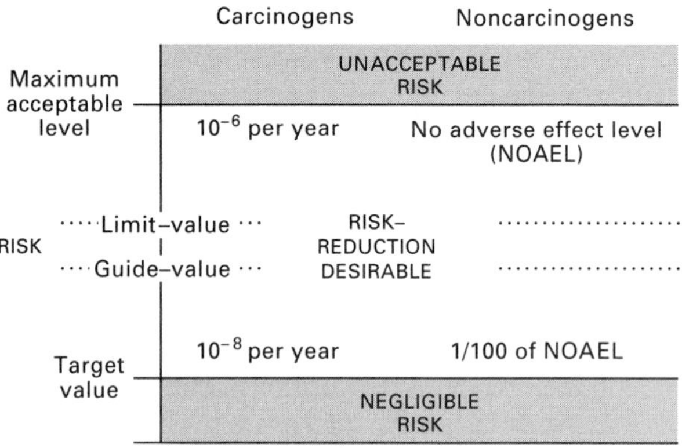

Figure 17.8 The Netherlands risk policy: upper and lower limits for exposure to chemical substances. Reprinted by permission of the publisher from *Environmental Impact Assessment Review,* vol. 15, no. 5/6, p. 428. Copyright 1994 by Elsevier Science Inc.

Rational risk reduction has its detractors. Questions raised by its opponents include the following:

- How should the many uncertainties in computing quantitative estimates of risk be accounted for?
- Shouldn't risks that are assumed voluntarily (such as eating raw oysters) be treated differently from involuntary risks; for example, the risks from breathing contaminated air?

[26] A review of EPA's experience in using risk assessments for these purposes is given by Zamuda (1989).

[27] For a case study describing how risk assessment results were used in establishing national ambient air quality standards for ozone in the United States, see Landy, Roberts, and Thomas (1994, pp. 49–88).

[28] For more on the use of risk assessment in the Netherlands, see van Kuijen (1989) and Krijgsheld (1994).

[29] For information on applications of comparative risk in the United States and elsewhere, see EPA (1993). Information on comparative risk studies within regions of the United States is given by Fiorino (1995).

- Shouldn't risks disproportionately imposed on the poor and ethnic minorities be reduced because they are unjust?

Policy debates on the suitability of rational risk reduction have revolved around such questions.[30]

The final category of decisions in Table 17.6 are those pertaining to requests for government approval of proposed development projects. In this context, risk assessment is often viewed as an element of environmental impact assessment (EIA). Use of risk assessment as a part of EIA applies both for chronic low-level exposure to contaminants and for industrial accidents and engineering system failures.

Industrial Accidents and Engineering System Failures

Assessments of accident risks and system failures for proposed development projects are frequently integrated into environmental impact assessments. As previously mentioned, risk assessments for projects that pose risks from long-term exposure to low concentrations of contaminants are also integrated into EIAs. Examples include risk assessments used in preparing EIAs for herbicide programs to manage vegetation, and for proposals to construct incinerators to burn municipal waste.[31]

Increasingly, EIA requirements are being modified to call for risk assessments in cases where development projects are hazardous because of either the potential for accidents and system failures or the exposure of populations to low concentrations of contaminants over long time periods. Because risk assessments can be costly to conduct, EIA programs typically mandate risk assessments for a limited set of project types.

Requirements of the Asian Development Bank (ADB) illustrate one approach to deciding when risk assessments should be included in EIAs. Figure 17.9 shows the two categories of ADB-funded projects that require environmental impact assessments. Category C projects are those with significant adverse impacts that require a detailed EIA, and category D projects are those directed toward solving environmental problems. Risk assessments are required for only a fraction of the projects in categories C and D, and Figure 17.9 shows the criteria used in determining if a risk assessment is needed. Risk assessments are called for in two cases. One involves projects such as petroleum refineries and offshore oil drilling platforms where the magnitude of hazard is significant, but there is a low probability that hazardous conditions will occur. The second case involves risks of chronic exposure to contaminats, such as occurs when hazardous wastes are disposed of improperly. The ADB has detailed checklists that provide a basis for deciding on whether a risk assessment is needed for any proposed project. It also has lists of projects, such as oil and gas storage facilities, that always require an environmental risk assessment.[32]

For a development project with risks high enough to justify a risk assessment, how are assessment results typically used in decision making? At the most basic level, authorities responsible for approving projects use assessment results to decide whether a project should go forward. Often, however, results are not used to say yes or no to a request for approval. Instead, risk assessment outcomes are used to suggest or require measures to reduce risks to acceptable levels.

There are many steps that can be taken to reduce risks from development projects. In the case of industrial development projects, for example, *risk mitigation measures* include the following:

- Project design changes, such as the addition of a backup container to capture toxic gas if the original containment system fails.
- Quality control measures to ensure that applicable building codes are met.
- Emergency response plans, such as procedures for evacuating staff from a plant in the event of hazardous conditions.
- Training of staff on safe materials-handling procedures and other hazard reduction measures.
- Zoning changes to increase the distance between hazardous activities and sensitive areas, such as schools and residential areas.

In the final analysis, results from risk assessments included in EIAs for proposed projects leave decision

---

[30] For an analysis of advantages and shortcomings of rational risk reduction, see Garetz (1993), and Finkel and Golding (1994).

[31] These and other examples are given by Canter (1993).

[32] Details on these checklists and other guidelines for risk assessment are given by the Asian Development Bank (1990).

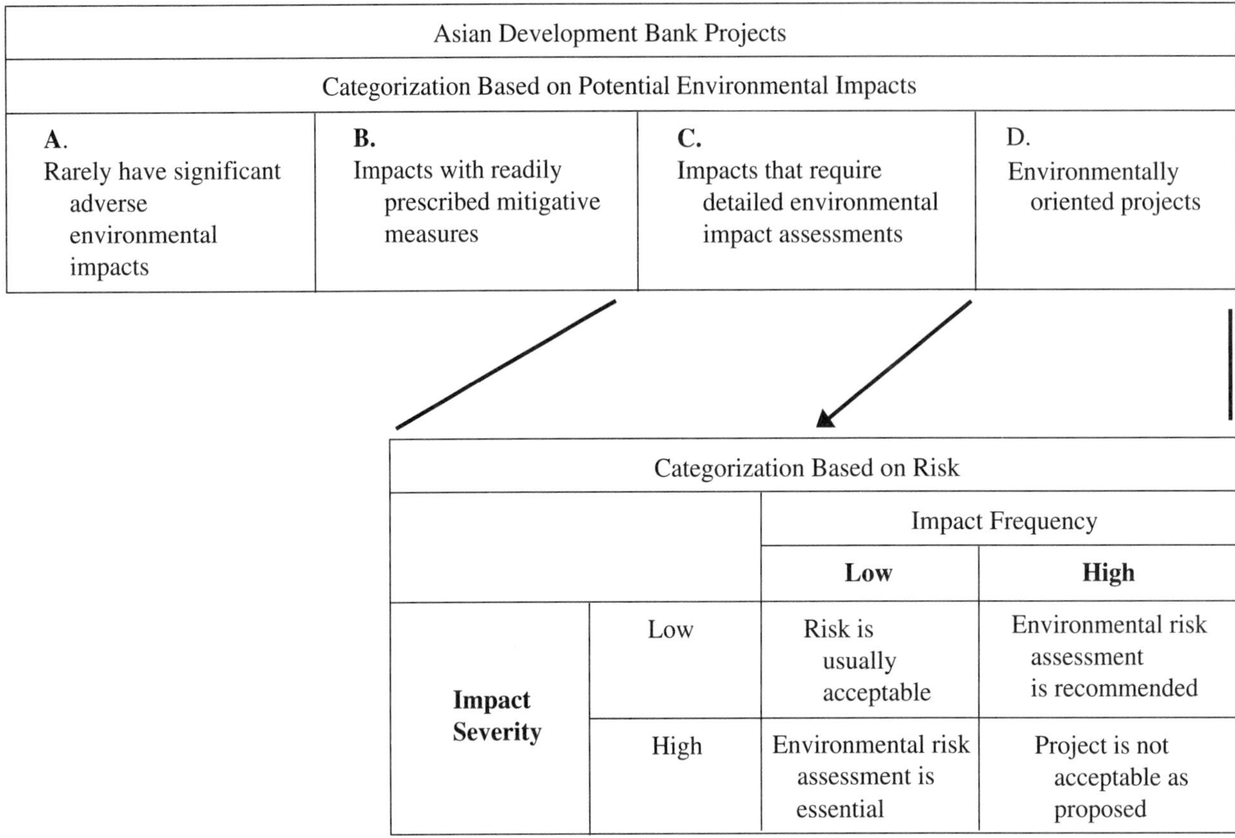

Figure 17.9 Asian Development Bank scheme for identifying proposals needing risk assessments. Reprinted with permission from Asian Development Bank (1990, p. 11).

makers with the question of deciding on whether risks are acceptable. While there are no universally accepted norms to define acceptability, governments have attempted to provide guidance for decision makers. An example is given in Figure 17.10, which shows risk acceptability criteria developed by the Dutch government during the 1980s. The figure applies to "low probability-high consequence" events, including accidents and other hazardous events at industrial facilities.[33]

Since the 1980s, risk assessments have been required for many proposed development projects. The following mandates are illustrative.[34]

1982 The European Economic Community adopted the *Seveso Directive,* which requires assessments of potential industrial hazards.

1984 Following the disaster at Bhopal, India, the World Bank issued guidelines and an accompanying manual to help control major accidents at projects it funds.

1992 More than 50 banks signed an agreement to recommend environmental risk assessment as part of their credit risk procedures in evaluating new loans.

As of the mid-1990s, there was still a significant gap between requirements for environmental risk assessments in project evaluations and the implementation of those requirements. Nevertheless, mandates of the type noted here point toward increased use of risk assessment.

[33] For examples of other guidelines used in determining risk acceptability, see Asian Development Bank (1990), Canter (1993), and Carpenter (1995).

[34] These examples are from Carpenter (1995, pp. 194–95).

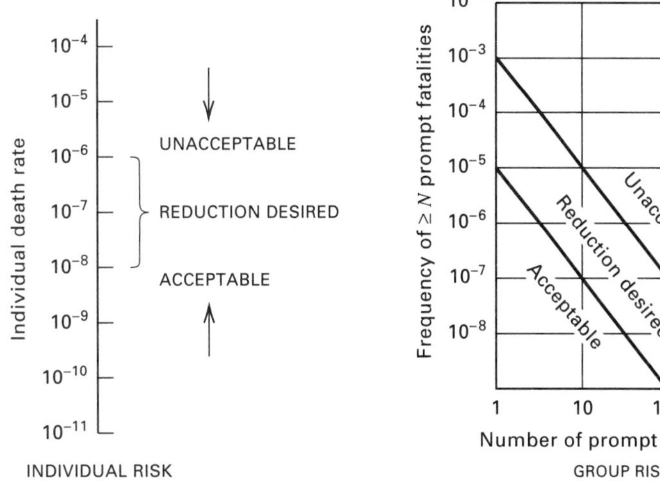

Figure 17.10 Risk criteria for the policy on external safety in the Netherlands. Reproduced from van Kuijen in Maltezou, Biswas, and Sutter (eds.), *Hazardous Waste Management,* 1989 by permission of Mansell Publishing (a Cassell imprint), London.

Key Concepts and Terms

Issues in Multicriteria Evaluation
- Incommensurate effects
- Sources of evaluative factors
- Weights reflecting trade-offs

Extensions of Benefit-Cost Analysis
- Limitations of BCA
- Benefits defined in terms of objectives
- Multiple objectives
- National income net benefits
- Income redistribution net benefits
- Weighted sum of objectives
- Bases for estimating weights

Tabular Displays and Weighted Sums of Factor Scores
- Ordinal factor scores
- Bases for scoring
- Goals-achievement matrix

Environmental Risk Assessment
- Relative frequency of occurrence
- Subjective probability
- Hazard identification
- Exposure pathways and scenarios
- Exposure assessment
- Concentration vs. dose
- Dose-response assessment
- Reference dose
- Potency factor
- Risk characterization
- Uncertainty in risk estimates
- Human health vs. ecological risks
- Event tree vs. fault tree analysis

Using Risk Assessments to Evaluate Options
- Risk management
- Cost-risk-benefit analysis
- Risk guidelines and standard setting
- Comparative risk assessment
- Rational risk reduction
- Risk assessment and EIA
- Risk mitigation measures
- Acceptable risk

Discussion Questions

17-1 A suburban community is considering several alternative land development schemes for a parcel of undeveloped land within its boundaries. The following hypothetical information represents results from a study by a consultant to the city's planning department. The ratings shown vary from 5 (best) to 1 (worst).

Factors	Weights	Ratings for Alternative Proposals		
		Plan 1	Plan 2	Plan 3
Net contribution to city tax base	10	5	3	1
Increased traffic congestion	5	1	3	5
Adverse visual effects	2	1	2	5
Attainment of low-cost housing goals	5	3	5	1
Adverse effects on school crowding	8	3	5	1

Compute the sum of weighted factors scores for each alternative and note the plan with the highest sum. What happens to the rank ordering of the three proposals if the column of weights is changed from (10, 5, 2, 5, 8) to (5, 10, 5, 5, 8)? Criticize this approach to evaluation. What defense could be offered in response to such criticism? Would you consider using this method if you were a consultant to a city planning department? Justify your position and present a different way to rank alternatives if you find the preceding procedure unsatisfactory.

17-2 The city of Fulopville has a problem regarding future use of its land adjacent to San Girolamo Bay ("the Baylands"). A creek that flows through the Baylands to the bay carries storm runoff from a residential section. This area floods periodically when high tides in the bay retard the drainage of creek flows. The county flood control engineers propose creating retention basins in the heart of the Baylands to temporarily store creek drainage during flood periods. Based on a benefit-cost analysis of five flood control plans, the engineers believe they have the optimal solution.

The engineers' proposal has elicited strong objections from two interest groups. One group has called for a complete renovation of the Baylands to create a park with tennis courts, barbecue pits, parking lots, and marinas. This group recognizes that flood control is important but feels that flood control measures should not limit recreational use of the Baylands. The second group feels that flood control should be accomplished, but without altering the Baylands. These individuals also oppose the park concept because waterfowl in the Baylands would be driven off by large numbers of people "recreating." They argue that the Baylands, in its natural form, is important to the ecological integrity of the bay.

Assist the city's decision makers by providing a critique of the evaluative criteria used by the engineers. What actions might the decision makers take to resolve the dispute?

17-3 The EPA often reports data that allow for a calculation of cancer risk in two equivalent forms: *potency factor* and *unit risk*. Given one of these, it is possible to compute the other if certain assumption are made about intake rate, exposure duration, and body weight. For this exercise, assume the following:

Intake rate = 20 m^3/day
Exposure duration = 24 hr/day for a lifetime
Body weight = 70 kg

The potency factor is the slope of a linear dose-response curve. Thus,

$$\text{risk} = \text{potency factor (mg/kg/day)}^{-1} \times \text{ADI (mg/kg/day)}$$

where ADI is average daily intake.

The unit risk for airborne materials is a number that, when multiplied by concentration (in units of $\mu g/m^3$), gives the equivalent risk of cancer. The potency factor for beryllium in the air is 8.4 $(mg/kg/day)^{-1}$. Using these assumptions, calculate the unit risk.

17-4 A risk assessment was performed for the Smuggler Mountain (Superfund) site near

Aspen, Colorado.[35] Wastes from mines in the area are the principal source of contamination. Cadmium and arsenic were found at elevated concentrations in soil at the site. Of all the exposure paths examined, only two were significant: ingestion of contaminated soil, and inhalation of contaminated airborne particles. The analysts made the following assumptions regarding exposure via ingestion:

(a) Weight of average exposed person = 70 kg.
(b) Exposure is for half of each year for a 70-year lifetime. (The site is covered by snow for six months per year.)
(c) In the "average case," the soil ingestion rate averaged over a lifetime is 30 mg/day. (The rate is 100 mg/day for children between one and six.)
(d) In the "maximum case," the soil ingestion rate averaged over a lifetime is 140 mg/day.

Observed Concentrations of Cadmium and Arsenic in Soil at Smuggler Mountain Site[a]

	Average Value (mg/kg)	Maximum Value (mg/kg)
Cadmium	26	53
Arsenic	82	303

[a] For cadmium, 13 samples were taken, and for arsenic, 6 samples were taken. Detectable values of the chemical were found in each sample.

Cadmium, when ingested, is not carcinogenic. The reference dose (RfD) for oral exposure to cadmium is 5×10^{-4} mg/kg/day. In contrast, arsenic, when ingested, is carcinogenic, and the potency factor is 15 $(\text{mg/kg/day})^{-1}$

Given these data and assumptions, determine whether the average daily intake for cadmium exceeds the reference dose, and find the cancer risk for arsenic. For each substance, perform average daily intake calculations for two scenarios: average exposures and maximum exposures. To what extent do your results indicate a significant risk to the people exposed to the arsenic and cadmium in the area?

17-5 In September 1995, *The New York Times* described draft sections of a report by the Intergovernmental Panel on Climate Change.[36] The Times reported as follows:

> Since its last full assessment in 1990, the panel has cut its estimate of expected average sea-level rise between now and 2100 from a "best estimate" of 26 inches to a little less than 20 inches, with a possible range of 10 to 31 inches. A further rise is expected after 2100. The best estimate for 2100, the draft report says, would by then put at risk tens of millions of people in low-lying areas and on oceanic islands. Many low areas, like parts of the Maldives, Egypt and Bangladesh, would be inundated, and many of their inhabitants would be cast on the world's mercies as environmental refugees.
>
> At present, the draft report says, an estimated 46 million people experience flooding because of storm surges each year. Under the best estimate for 2100, 92 million to 118 million would be so affected. Rich countries might be able to adapt, but at the cost of spending $521 million a year on sea walls and other protection, the panelists estimate. Even then, many coastal wetlands and sandy beaches would have to be sacrificed. Poorer countries would find it more difficult to protect themselves, the report says.

If you were a member of the U.S. Congress, what position would you take regarding the need for regulations requiring U.S. electric utilities to cut back on emissions of carbon dioxide to reduce the impact of global warming? What position would you take on the

[35] This question relies on data from Lagoy, Nisbet and Schulz (1989).

[36] From W. K. Stevens, "Scientists Say Earth's Warming Could Set Off Wide Disruptions," *The New York Times*, National Edition, September 18, 1995, pp. A-1 and A-5. At the time of this newspaper account, the panel's report was still in draft form.

question if you were a legislator from Bangladesh? In responding, consider the degree of uncertainty associated with the panel's predictions.

References

Asian Development Bank. 1990. *Environmental Risk Assessment: Dealing with Uncertainty in Environmental Risk Assessment.* ADB Environmental Paper No. 7. Manila: ADB.

Calabrese, E. J., and L. A. Baldwin, 1993. *Performing Ecological Risk Assessments.* Boca Raton, FL: Lewis Publishers.

Canter, L. W. 1979. *Environmental Impact Statements on Municipal Wastewater Programs.* Washington, DC: Information Resources Press.

———. 1993. Pragmatic Suggestions for Incorporating Risk Assessment Principles in EIA Studies. *The Environmental Professional* 15(1): 125–38.

Carpenter, R. A. 1995. "Risk Assessment." In *Environmental and Social Impact Assessment,* eds. F. Vanclay and D. A. Bronstein, 193–219. Chichester, England: Wiley.

Cohen, J. L., 1978, *Multiobjective Programming and Planning.* New York: Academic Press.

Cox, L. A. Jr., and P. F. Ricci. 1989. "Legal and Philosophical Aspects of Risk Analysis." In *The Risk Assessment of Environmental and Human Health Hazards: A Textbook of Case Studies,* ed. D. J. Paustenback, 1017–46. New York: Wiley-Interscience.

Eckstein, O. 1958. *Water Resource Development, the Economics of Project Evaluation.* Cambridge, MA: Harvard University Press.

Elliott, M. L. 1981. Pulling the Pieces Together, Amalgamation in Environmental Impact Assessment. *Environmental Impact Assessment Review* 2(1): 11–37.

EPA (U.S. Environmental Protection Agency). 1986. *Superfund Public Health Evaluation Manual.* Office of Emergency and Remedial Response. Washington, DC: EPA.

———. 1987. *Unfinished Business: A Comparative Assessment of Environmental Problems.* Office of Policy Analysis. Washington, DC: EPA.

———. 1989. *Health Effects Assessment Summary Tables,* First Quarter, FY 1989. Report OSWER (OS-230)/ORD (RD-689). Washington, DC: EPA.

———. 1991. *Air Quality Criteria for Carbon Monoxide.* Report EPA/600/8-90/045F, Office of Research and Development. Washington, DC: EPA.

———. 1993. *A Guidebook to Comparing and Setting Environmental Priorities.* EPA 230-8-93-003, Office of Policy, Planning and Evaluation. Washington, DC: EPA.

Finkel, A. M., and D. Golding, eds. 1994. *Worst Things First? The Debate Over Risk-Based National Environmental Priorities.* Washington, DC: Resources for the Future.

Fiorino, D. J. 1995. *Making Environmental Policy.* Berkeley: University of California Press.

Freeman, A. M., III. 1970. "Project Design and Evaluation with Multiple Objectives." In *Public Expenditures and Policy Analysis,* ed. R. H. Haverman and J. Margolis, 347–63. Chicago: Markham.

Garetz, W. V. 1993. "Current Concerns Regarding the Implementation of Risk-Based Management: How Real Are They?" In *Comparative Environmental Risk Assessment,* ed. C. R. Cothern, 11–31. Boca Raton, FL: Lewis Publishers.

Goicoechea, A., D. R. Hansen, and L. Duckstein, 1982. *Multiobjective Decision Analysis with Engineering and Business Applications.* New York: Wiley.

Hauptmanns, U. 1989. "Fault Tree Analysis and Its Application to an Exothermal Reaction." In *Hazardous Waste Management,* ed. S. P. Maltezou, A. K. Biswas, and H. Sutter, 228–41. London: Tycooly.

Hill, M. 1967. A Method for the Evaluation of Transportation Plans. *Highway Research Record* 180: 21–34.

———. 1968. A Goals-Achievement Matrix for Evaluating Alternative Plans. *Journal of the American Institute of Planners* 34: 19–217.

Hobbs, B. F. 1980. A Comparison of Weighting Methods in Power Plant Siting. *Decision Sciences* 11(4): 725–37.

Krijgsheld, K. R. 1994. Decision-Making Based on Health Impact Assessment: Strategies and Experiences in the Netherlands. *Environmental Impact Assessment Review* 14(5/6): 425–38.

Landy, M. K., M. J. Roberts, and S. R. Thomas. 1994. *The Environmental Protection Agency: Asking the Wrong Questions from Nixon to Clinton,* expanded ed. New York: Oxford University Press.

Lagoy, P. K., I. C. T. Nisbet and C. O. Shulz, 1989. "The Endangerment Assessment for the Smuggler Mountain Site, Pitkin County, Colorado: A Case Study." In *The Risk Assessment of Environmental and Human Health Hazards: A Textbook of Case Studies,* ed. D. J. Paustenback. New York: Wiley-Interscience.

Lichfield, N., P. Kettle, and M. Whitbread. 1975. *Evaluation in the Planning Process.* Oxford: Pergamon.

Livingston and Blayney, Inc. 1971. *The Foothills Environmental Design Study: Open Space vs. Development.* Final Report to the City of Palo Alto prepared by Livingston and Blayney, City and Regional Planners, San Francisco, CA. Unpublished.

Maass, A. 1966. "Benefit-Cost Analysis: Its Relevance to Public Investment Decisions." In A. V. Kneese, and S. C. Smith, 311–327. Baltimore: Johns Hopkins University Press for Resources for the Future, Inc.

Marglin, S. A. 1962. "Objectives of Water Resource Development: A General Statement." In *Design of Water-Resource Systems,* ed. A. Maass et al., 17–87. Cambridge, MA: Harvard University Press.

Massam, B. H. 1988. Multi-Criteria Decision Making (MCDM). *Techniques on Planning, Progress in Planning,* vol. 30, part 2. Oxford: Pergamon Press.

Maxim, L. D. 1989. "Problems Associated with the Use of Conservative Assumptions in Exposure and Risk Analysis." In *The Risk Assessment of Environmental and Human Health Hazards: A Textbook of Case Studies,* ed. D. J. Paustenback, 526–60. New York: Wiley-Interscience.

McAllister, C. M. 1980. *Evaluation in Environmental Planning.* Cambridge, MA: MIT Press.

Phantumvanit, D., and W. Nandhabiwat. 1989. The Nam Choan Controversy: An EIA in Practice. *Environmental Impact Assessment Review* 9(2): 135–47.

Raiffa, H. 1968. *Decision Analysis: Introductory Lectures on Choices Under Uncertainty.* Reading, MA: Addison–Wesley.

Steiner, P. O. 1969. *Public Expenditure Budgeting.* Washington, DC: The Brookings Institution.

Suter, G. W., III, ed. 1993. *Ecological Risk Assessment,* Boca Raton, FL: Lewis Publishers.

van Kuijen, C. J. 1989. "Risk Management in the Netherlands: A Quantitative Approach." In *Hazardous Waste Management,* ed. S. P. Maltezou, A. K. Biswas, and H. Sutter, 200–212. London: Tycooly.

von Stackelberg, K., and D. E. Burmaster. 1994. A Discussion of Probabilistic Risk Assessment in Human Health Impact Assessment. *Environmental Impact Assessment Review* (14) 4, 5: 385–401.

Zamuda, C. 1989. "Superfund Risk Assessments: The Process and Past Experience at Uncontrolled Hazardous Waste Sites." In *The Risk Assessment of Environmental and Human Health Hazards: A Textbook of Case Studies,* ed. D. J. Paustenback. New York: Wiley-Interscience.

Chapter 18

Public Participation and Environmental Dispute Resolution

The degree to which citizens participate in the planning and decision processes of government agencies varies over both time and place.[1] During the late 1960s, citizens in many countries demanded increased participation in agency decisions. Since then, many governments have established procedures that allow citizens to express their views about agency policies and projects before decisions are made.

Programs to engage the public in agency planning often have multiple purposes, and no simple formula exists for a successful public involvement program. In each case, a citizen participation program must be designed to fit the particular combination of project, agency, and citizenry.

The first part of this chapter clarifies the objectives of public involvement programs and looks at methods for identifying "the public"—citizens and groups that may have an interest in a proposed project or regulatory decision. It also reviews the strengths and weaknesses of commonly used public involvement techniques. An example involving a U.S. Army Corps of Engineers flood control study demonstrates how public involvement activities can be integrated into traditional planning activities.

Later sections of the chapter consider the following question: What can be done if a government agency and citizens remain in conflict, even after public involvement methods have been applied? Those sections examine the role of mediation processes in resolving environmental disputes between citizens and agencies. A case study centering on fungicide registration requirements in Canada illustrates the steps involved in mediating an environmental conflict.

OBJECTIVES OF A PUBLIC INVOLVEMENT PROGRAM

Public involvement programs derive legitimacy from the democratic ideal of allowing all citizens to be represented in public decision making. Although representation by elected officials is the norm in democracies, citizens often seize opportunities to represent

[1] The contribution of Alnoor Ebrahim and Monique van der Marck are gratefully acknowledged. While they were students at Stanford University, each assisted in preparing this chapter. Ms. van der Marck helped with the first three sections on public involvement. Mr. Ebrahim worked on early versions of the entire chapter.

TABLE 18.1 Multiple Goals of Public Involvement[a]

- Improve decisions that are likely to impact communities and the environment.
- Give citizens a chance to express themselves and to be heard.
- Provide citizens with opportunities to influence outcomes.
- Assess public acceptability of a project and add mitigation measures.
- Defuse potential citizen opposition to agency plans.
- Establish legitimacy of agency and its decision process.
- Meet legal requirements to involve citizens.
- Develop two-way communication between agency staff and citizens:
  - —Identify public concerns and values.
  - —Inform citizens of agency plans.
  - —Inform agency about alternatives and impact.

[a] Based on Ketcham (1992), FEARO (1988, vol. I, pp. 7–8), and Parenteau (1988, p. 5).

themselves and participate directly in agency planning. As the complexity of issues and number of constituencies increase, citizens often become eager to be heard directly as individuals or groups rather than through officials.

Agency Objectives vs. Citizens' Objectives

Citizens and agencies do not always approach the process of public involvement in agency decision making with the same objectives. Table 18.1 includes typical goals of public involvement. Agencies and citizens generally have some objectives in common. For example, both an agency and citizens may be interested in a mutual exchange of information. However, some objectives of agencies may be unrelated to those of citizens and vice versa. For instance, an agency may view its public involvement activities as an exercise to satisfy legal requirements,[2] whereas citizens who participate may do so because they want a voice in the agency's decision process. Citizens often view a public involvement program as an opportunity to assert a right to be heard and to share their concerns with the agency. The influence of public participation will vary, depending on whether the agency is truly interested in citizen opinions or whether it wishes to create only the appearance of public involvement.

Agencies and citizens may also enter a public involvement process with different ideas about what constitutes a satisfactory outcome. An agency seeking a mandate for a particular project may attempt to use a public involvement program to build a consensus among citizens and to harmonize potentially conflicting interests. In this way, the agency might avoid the time and expense of costly legal battles that might be waged by opponents to its project. In contrast, individuals or groups keen on protecting special interests may work to find solutions that meet their particular needs, rather than develop a consensus among all parties.[3]

In his analysis of citizen participation in planning, Parenteau (1988, p. 4) suggests that ". . . it is an illusion to think of participation as a neutral social operator, perfectly receptive to all audiences. . . . It must be understood as a special instrument in the sociopolitical arena, an instrument suitable for the exercise of certain types of political influence for the benefit of certain segments of society and designed for this purpose." Participation can thus be viewed as a tool that both citizens and agencies can use for their own purposes.

Levels of Public Participation

In a widely cited critique of agency public involvement programs, Arnstein (1969) represents the levels of citizen participation as rungs of a ladder. She groups the rungs into three categories: nonparticipation, tokenism, and citizen power (see Figure 18.1). These levels form a continuum, and they focus attention on the "difference between going through the empty ritual of participation and having

[2] Under the National Environmental Policy Act of 1969, for example, public agencies may be legally obliged to circulate a draft environmental impact statement (EIS) to interested parties and respond to their comments in preparing the final EIS (see Chapter 15).

[3] Parenteau (1988, pp. 5–6) elaborates on the different objectives of planners, political authorities, and participants in the public involvement process.

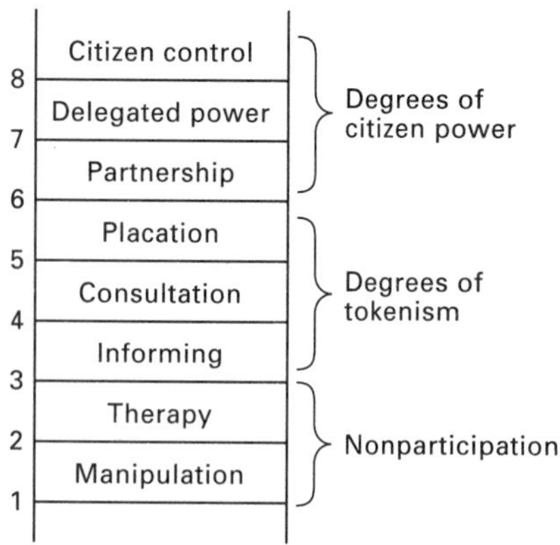

Figure 18.1 Eight rungs on a ladder of citizen participation. Source: Arnstein (1969, p. 217)

the real power needed to affect the outcome of the process."[4]

The first category, *nonparticipation,* occurs when an agency tries to coerce, manipulate, or change the minds of the public. Arnstein refers to these tactics as a substitute for genuine participation and writes, "Their real objective is not to enable people to participate in planning or conducting programs, but to enable powerholders to 'educate' or 'cure' the participants." An example of nonparticipation is provided in Iacofano's (1990, p. 76) analysis of land use planning on the Coconino National Forest in the United States. He concludes:

> The Forest Service did not seem to have any strategy for using public involvement in decision-making, even in cases where they encouraged it. If they received "good" input, they used it. If they received "bad" input they told critics they were "misinformed" and generally tried to discredit the public.

Arnstein's second category, *tokenism,* occurs when the public is allowed to participate in agency meetings of various types, but their participation has little or no effect on agency decisions. Rungs termed "informing" and "consultation" are included in this group. Arnstein argues that when informing and consultation "are proffered by powerholders as the total extent of participation, citizens may indeed hear and be heard. But under these conditions they lack the power to insure that their views will be *heeded* by the powerful."

The final category, which Arnstein calls *citizen power,* concerns the development of citizen-agency partnerships and programs that involve control by citizens. For these higher levels of Arnstein's ladder, citizen power can range from the ability to negotiate decisions to the authority to veto decisions. Examples of citizen power include the creation of neighborhood corporations to manage public projects.[5] However, these high levels of citizen participation are often unattainable, since agencies are generally not authorized to give up administrative control.[6]

IDENTIFYING THE PUBLIC

The first step in implementing a public involvement program is to identify the public. The public is not a unitary body, but a collection of numerous, continually shifting interests and alliances. Hence, there are many "publics," each forming in response to a context in which citizens' interests are affected.

Following are some of the ways citizens may be affected by public projects and regulatory decisions.[7]

- *Proximity.* People living near a proposed project may be concerned about factors such as increased pollution, decreased property values, or potential benefits to the local community.
- *Economics.* Some groups, such as land developers, may have strong economic interests in an agency regulation.

[4] This quotation is from Arnstein (1969, p. 216). Quotations from Arnstein in the next two paragraphs are from the same source on p. 217. The existence of a continuum of degrees of public involvement is widely recognized. For example, ideas similar to Arnstein's appear in a public involvement manual issued by Canada's Federal Environmental Assessment Review Office (FEARO, 1988, p. 11). The manual uses the following terminology to describe levels (beginning with the lowest level): persuasion, education, information feedback, consultation, joint planning, delegated authority, and self-determination.

[5] Other examples of citizen power are given by Parenteau (1988, p. 8).

[6] In some jurisdictions, however, these higher levels of public involvement do take place. Roberts (1995), a Canadian consultant specializing in public involvement, observes that an increasing number of government organizations have experimented with joint planning, in which members of the public are given selected voting privileges and decision-making authorities.

[7] The listing is adapted from Creighton (1992, pp. 107–8).

- *Use.* Users of existing facilities, such as hikers or hunters, may feel threatened by a new project or regulatory decision.
- *Social and Environmental Issues.* Citizens may be concerned about a proposed project's effect on social equity and cultural diversity, or risks to people and the environment.
- *Values.* Groups with strongly held beliefs (such as advocates of rights for nonhuman species) may have an interest in proposed projects and regulations.

Citizen participation specialists distinguish three ways to identify segments of the public: self identification, staff identification, and third-party identification.[8] In *self identification,* individuals and groups come forward and make their interests known. An agency proposing a project or regulation can facilitate this by holding an initial, well-advertised public hearing, or by publicizing a phone number or address of an agency contact person (for example, by placing announcements in local newspapers). For small projects, an agency may facilitate the self-identification process by leaving fliers or posters with pre-addressed, stamped response cards in places where interested parties are likely to see them, such as local supermarkets or public transit stations.

Staff identification occurs when agency personnel actively identify and contact potentially interested parties. Agency staff who have worked in an area for some time can often identify potentially interested individuals or groups. Several other staff-identification techniques are listed in Table 18.2. These rely on the use of existing mailing lists and the analysis of maps and census data. Official records of property owners and reverse telephone directories are also helpful in identifying citizens who may have an interest in an agency's project or regulatory action.[9]

The final category of techniques involves *third-party identification.* Groups or individuals may approach the agency to suggest other groups or individuals that should be involved. In addition, agency staff can interview local officials to identify persons who may want to be involved in the agency's planning process. The agency may extend this procedure using what has been termed the "snowball" approach.[10] With this method, interested parties are interviewed, and they suggest other individuals who are then contacted. The process continues until such time as few new names are mentioned. This approach is relatively expensive and time consuming, and it may only duplicate information already gathered rather than identify new individuals or groups. However, the method can be useful when an agency has little prior information about citizens or groups that may have an interest in its proposal.

Identification of the public is complicated because individuals and groups having an interest in an agency's proposed action may change as the proposal unfolds. Some people may want to be involved throughout an agency's process, whereas others may want to participate only at particular stages. Moreover, groups that are often underrepresented in government decision making, such as immigrant communities in cities, may not be accustomed to working with agencies. Special efforts may be required to include them in citizen participation programs.[11]

Even in cases where considerable energy and resources are expended to identify potentially affected individuals and groups, only a small portion of the public ever attends participation programs.[12] But the presence of a nonparticipative majority does not mean that only a minority of citizens care about a proposed project or regulation or that the "silent majority" holds a single opinion. Willeke (1976, p. 46), in commenting on the relatively low levels of citizen participation commonly observed, suggests that organized, vocal groups can act as surrogates for the general public:

---

[8] This three-part distinction is made by Willeke (1976, pp. 55–60), whose work provides the basis for this discussion of methods to identify the public.

[9] In a reverse telephone directory, entries are grouped by street location instead of by name.

[10] Willeke (1976, pp. 58–9) includes the field interview technique under staff identification. It is discussed here under third-party identification because a third party, rather than the agency, identifies interested parties.

[11] For example, in trying to reach immigrant communities affected by a project, it may be appropriate to issue materials describing an agency's plans in more than one language.

[12] Hendee and associates suggest that "[i]n our complex pluralistic society, citizens are likely to remain passive on well over 90 percent of their opportunities for public involvement" (Hendee et al., 1976, p. 142).

TABLE 18.2 Staff Identification of Parties Potentially Affected by an Agency's Proposal[a]

- Maps and reverse telephone directories
  Maps can be used to determine who will be directly affected by an agency's proposed action. For instance, a topographic map and a street map could be used together to identify residents who would be influenced by a proposed flood control project. Reverse telephone directories can be used to obtain names and addresses.
- Census data
  Citizens who have certain characteristics, such as being within a certain age bracket, can be identified using census records.
- Records of property owners
  Local records on property ownership can be used to locate homeowners likely to be affected by an agency proposal.
- Mailing lists
  Mailing lists used by the agency for planning previous actions are valuable in identifying citizens and groups who may be interested in future proposals. Mailing lists of agencies doing work in related fields can also be useful.
- Lists of local organizations
  If lists of community groups or other special interest organizations exist, they can provide a shortcut to finding citizens who may be interested in an agency's plans.
- User records
  Where an agency plans to modify areas used heavily for recreation, records such as user registration forms or permit applications can help identify interested parties.
- Newspaper stories
  An analysis of local news coverage, both recent and past, can help pinpoint potentially interested citizens and groups. Letters to the editor are another source of information.
- Staff intuition and experience
  Agency staff who have worked in an area for some time can often identify individuals and groups likely to be interested in a proposed action.

[a] Adapted from Willeke (1976, pp. 55–60), and Creighton (1980, pp. 44–45). Reprinted with permission.

When actions have low level and/or invisible impacts on a population segment, a surrogate may be the only reasonable course of action because individual citizens cannot individually bear the costs of full participation. An organized surrogate group can, on the other hand, do the necessary research, present the case to the responsible decision-makers, and muster the necessary political support.

The identification of interested citizens and groups is only one step in the design of a citizen involvement program. Once that step is completed, at least in a preliminary way, one or more techniques can be employed to engage interested parties in agency decision making.

PUBLIC INVOLVEMENT TECHNIQUES

An agency designing a public involvement program can select from a wide range of methods.[13] Typically, the following factors influence an agency's choice of public involvement techniques: the agency's objectives, time and resource constraints, the range of issues and opinions, and the geographic distribution of interested parties.

Sometimes, environmental statutes specify techniques for involving interested parties in agency decision making. This often occurs when agencies de-

[13] The discussion of techniques is based on Hendee et al. (1976), and Creighton (1980 and 1992). For a comprehensive manual on public involvement techniques, see FEARO (1988).

velop environmental regulations. For example, the U.S. Environmental Protection Agency (EPA) is obliged to employ particular techniques for involving interested parties in its rulemaking. These citizen involvement techniques, which include provisions for soliciting and responding to public comments on proposed regulations, were described in the discussion of EPA's rulemaking procedures in Chapter 3.

The discussion of public involvement techniques that follows concerns the more general case where an agency's public involvement program is not prescribed by statute. Frequently, public involvement takes place in the context of an agency's planning for a proposed project. Sometimes, private developers design programs to involve citizens in their own project planning. What techniques can a (public or private) project proponent use to involve individual citizens and groups in its decision-making process?

Involvement Techniques Based on Meetings

Project proponents often rely on one or more types of meetings to exchange information with interested parties (see Table 18.3). Agencies proposing projects often hold public hearings, which is the most rigid of the meeting types. A hearing officer generally governs the proceedings, and a stenographer makes a verbatim transcript. Presentations are formal and there is little interaction among participants. Large group meetings can be much less formal than hearings, but it is difficult for citizens, other than those most vocal, to participate directly. Nonetheless, public meetings and hearings can facilitate the presentation of large amounts of information by the agency while still enabling two-way communication through question-and-answer sessions or panel discussions.

Workshops generally focus on a specific planning task, and they are more interactive than hearings or large group meetings. In situations where opposing viewpoints differ significantly, workshops can provide a forum for conflicting parties to establish a dialogue. Workshops, however, can generally accommodate only a limited number of citizens and are more demanding on the time and resources of agencies.

Agencies sometimes rely on advisory groups to obtain citizens' perspectives when agency planning takes place over long time periods. Members are usually selected by the agency to represent a variety of interests. Powers granted to advisory committees range from making recommendations to exercising leverage over final decisions. Task forces or ad hoc committees are a type of short-term advisory group usually set up to complete a specific task and then dissolve. When advisory groups are representative of the community affected by an agency's action, they can help ensure that public interests are served, and they can enhance communication between agencies and citizens. An advisory group may involve a wide spectrum of interests, and members sometimes negotiate among themselves to arrive at recommendations for government agencies.[14]

TABLE 18.3 Meeting Types Commonly Used to Include Citizens in Agency Planning

- Public hearings
- Large public meetings
  - Official presentation followed by questions
  - Panel format
  - Informal "town meeting" structure
  - Plenary sessions and small group discussions part of time
- Public workshops
- Focus groups
- Informal small group meetings
- Advisory groups (for example, task forces and citizens' committees)

Techniques Other Than Meetings

Public involvement programs often include techniques that do not rely on meetings, and many such methods are listed in Table 18.4. The table shows several techniques for getting information to the public. These methods can be particularly useful in presenting information that allows citizens to determine if they should take advantage of other opportunities to participate in planning—for example, by attend-

[14] This use of advisory groups is demonstrated by the activities of "community resource boards" in British Columbia, Canada. These boards have provided a forum for allowing aboriginal groups to participate in land use and resource management decisions that will affect them. For details, see the Commission on Resources and Environment (1995, pp. 65–91).

TABLE 18.4 Public Involvement Techniques Not Based on Meetings

- Providing information to the public
  - Mail (direct or electronic)
  - Field trips
  - Mass media coverage (e.g. print, radio, TV, and documentary film)
  - Public notices, displays, and exhibits
  - Reports, brochures, and information bulletins
  - Pages on the World Wide Web
- Obtaining information from the public
  - Agency requests for written comment
  - Editorials and letters to the editor
  - Public opinion polls
  - Response cards in information bulletins
  - Surveys and questionnaires
- Establishing two-way communications
  - Informal contacts
  - Call-in radio/television shows
  - Interviews
  - Telephone hotlines
  - "Chat rooms" on the Internet

ing meetings. Moreover, the information allows citizens who choose to participate to do so in an informed way.

There are many opportunities for bias and confusion in the course of trying to inform the public. In the era of 30-second sound bites and MTV, it is a challenge for public agencies to convey information that is detailed enough to allow citizens to discover how they might be affected by a proposed action. Too much information can be as ineffective as too little, since people are inundated with unsolicited information and cannot be expected to sort through details in order to discover how their interests might be affected by an agency's actions. In addition, information that has the appearance of a "public relations" piece may cause citizens to question the validity of the information and the sincerity of the agency in undertaking its public involvement program.

Table 18.4 also lists several techniques for obtaining information from the public. From an agency's perspective, letters from citizens in response to information provided by the agency provide an efficient means of gathering information.[15] Surveys that include questionnaires with return envelopes encourage a large number of responses, but questionnaire and survey design require specialized skills. Amateur efforts can bias results significantly.

There are many ways, other than holding meetings, to establish two-way communications between agencies proposing actions and citizens likely to be affected by those actions. Of the two-way communication methods listed in Table 18.4, the least well-explored are those tied to recent breakthroughs in computer technology. The use of "chat rooms" on the Internet is an example.[16] While these new means of communication are likely to be efficient for many, they cannot be used to reach people without access to the requisite computer hardware and software.

The discussion of public involvement methods in this chapter emphasizes techniques that agencies design and implement. However, citizens often take the initiative to involve themselves in project decision making. The analysis in Chapter 4 of the New Melones dam in California provides examples. Opponents of the dam used a statewide ballot proposition and a court action to try and block the dam. They also lobbied state agencies, the state legislature, and the governor to gain support for their cause.

INTEGRATING PUBLIC INVOLVEMENT INTO AGENCY PLANNING: A CASE STUDY

A planning study carried out by the U.S. Army Corps of Engineers, San Francisco District Office (*the district*), demonstrates how a public involvement program can be organized to assist both project planners and citizens in dealing with important problems. The study, which was carried out in the 1970s, concerned flooding on San Pedro Creek in Pacifica, California, a small coastside community south of San Francisco.[17]

[15] In a study of public participation techniques used by the U.S. Forest Service, Hendee et al. (1976, pp. 136–37) found that soliciting letters from citizens and groups was one of the more efficient procedures used by the Forest Service to obtain information.

[16] For information on how the Internet can be used to assist in agency planning, see Zinn and Hinojosa (1994).

[17] Wagner and Ortolano (1976) provide a detailed account of the planning process used in the San Pedro Creek flood control study.

The district was committed to involving the public in each of four planning tasks:

1. Identifying the water-related problems and needs of those in the San Pedro Creek area
2. Formulating alternative plans to deal with flooding and other water problems
3. Forecasting the impacts of the various proposals
4. Evaluating the alternatives.

The district felt that citizens should be given opportunities to express their opinions throughout all stages of the planning investigation.

At the outset, the district staff identified numerous offices and agencies for inclusion in their public involvement program. Among these were the Pacifica city council, the city manager, and the state and federal fish and wildlife agencies. Local residents living either along the creek or in the floodplain were also to be involved in the study. Individuals who often played an important role in Pacifica's community affairs were interviewed to determine which citizens and groups might be interested in the district's study. Those questioned were identified initially by a review of back issues of local newspapers. The initial interviews generated the names of other people who should be contacted.

Having determined which individuals, groups, and agencies would be involved in the planning investigation, the district delineated objectives for its public involvement program. A primary goal was to keep the public informed on all aspects of the San Pedro Creek study. This required that citizens be given details on the district's perception of the water-related problems in the San Pedro Creek area. The public also needed information about possible plans to deal with those problems and the impacts of the alternative plans. Another of the district's objectives was to have two-way communications with the public. This required that citizens have opportunities to react to the district's ideas and proposals.

The San Pedro Creek study was to be carried out over a two-year period. To meet its public involvement program objectives over such a long time period, the district had to use several techniques. Not everyone with an interest in the San Pedro Creek study would either need or want to be involved on a continual basis over a two-year interval. Many individuals and groups would be content if they were consulted only when the district was about to make a key decision.

The district formed a citizens' advisory committee to maintain regular communications with at least one public entity. The committee consisted of five Pacifica residents selected by the city council. Collectively, they represented the people likely to have the greatest interest in the outcome of the San Pedro Creek investigation. These included local homeowners, merchants in a shopping center within the floodplain, and local environmental groups. The citizens' advisory committee provided information throughout the study. It also helped design other elements of the district's public involvement program.

To facilitate a two-way information flow between the district and various segments of the public, a "citizen information bulletin" was prepared a few months after the study began. A questionnaire to be returned to the district was inserted in the bulletin. Both the bulletin and questionnaire were mailed to about 1200 citizens and officials. The bulletin described the district's preliminary ideas about the San Pedro Creek flooding problem, possible alternative actions, and the likely impacts of those actions. The questionnaire considered the same topics and provided a convenient opportunity for citizens to comment on and supplement the district's preliminary concepts.

A public workshop on San Pedro Creek flood problems was held a few weeks after the bulletins and questionnaires were distributed. It was run informally by the citizens' advisory committee using a three-part format. First, participants met as a whole to hear general remarks about the planning study and the purpose of the workshop. After that, citizens were divided into small groups for discussions led by committee members. Finally, participants were reassembled for an exchange of information about what occurred in the small groups.

The workshop gave people a chance to react to the district's preliminary ideas, and to suggest additional factors that should be considered in formulating and evaluating alternative flood control plans. During the year following the public workshop, the district completed preliminary engineering, economic, and environmental studies for several proposals. Although it had met monthly with the citizens' advisory committee during this period, the district felt a need for additional communication with the public. It wanted feedback on whether all important evaluative factors had been considered in its economic and environmental impact studies. The district also wanted to know how different individuals and groups weighed the evaluative factors, and how they would rank the alternatives which the district had examined.

To provide a second opportunity to communicate with all segments of the public, another citizen information bulletin and questionnaire were prepared. Because the second bulletin summarized results from studies that had been completed since the public workshop, it was more detailed and elaborate than the first. Distribution of the second bulletin and questionnaire was coordinated with a meeting of the Pacifica city council that focused on the San Pedro Creek flooding problems. Based on information in the bulletin and presentations by the district, the city council developed its own ranking of the district's proposals. The city council's evaluation was later used by the district in judging which action should be recommended for implementation.

During the San Pedro Creek study, the public provided the district with much useful information. The citizens' comments offered insights into which factors local residents considered important in evaluating alternative plans. For example, after learning of some preliminary flood control proposals, many Pacifica residents expressed concern over creekside vegetation that would be destroyed. The district responded by formulating a plan that would reduce the flood problems without removing the valued vegetation.

The district's public involvement program helped yield a flood control plan that pleased both Pacifica and the Corps of Engineers. Even though there was no dispute over the final proposal, the plan was not implemented. This unsettling outcome resulted because the city of Pacifica was unable to generate its share of the total project costs.

A good public involvement plan seeks to establish two-way communication between citizens and a public agency in order to resolve conflicts, and to yield an agency decision that is satisfactory to as many parties as possible. When there are strong conflicting interests, conventional public involvement programs may be inadequate. In such cases, dispute resolution techniques may assist in settling conflicts over environmental resources.

RESOLVING DISPUTES OVER ENVIRONMENTAL RESOURCES

In the United States, traditional approaches for resolving serious disputes rely on litigation. However, litigation can be costly and slow. Are there alternatives? The past few decades have witnessed the emergence of *alternative dispute resolution* (ADR) methods based on consensus building: parties meet face to face to work voluntarily toward mutually acceptable outcomes.[18] Alternative dispute resolution techniques are used widely to settle many types of disputes, not just environmental disputes.[19] Although the range of ADR techniques is broad, two related ADR methods are used widely in the context of environmental problem solving: negotiation and mediation. Negotiation, of course, is a central component of traditional dispute resolution, but in ADR, negotiation is aimed at building consensus. Some authors use the term *principled negotiation* (or *consensus-based negotiation*) to distinguish the ADR approach from other forms of negotiation. Details on what makes principled negotiation different from other approaches are given later in this chapter.

[18] The term *alternative* in ADR is generally interpreted to mean alternative to litigation. However, ADR methods are often used to settle disputes that are being litigated in courts. Even though the word alternative in ADR is not particularly descriptive, the phrase *alternative dispute resolution* continues to be widely used. Terminology is not standardized in this field and some analysts prefer different phrases to describe the consensus-building dispute resolution methods described below; see, for example, Crowfoot and Wondolleck (1990).

[19] Tannis (1989) describes the many types of disputes settled using ADR techniques, and the wide range of methods used by attorneys who specialize in ADR. Methods include mini-trials and arbitration as well as the two techniques emphasized in this section: negotiation and mediation. *Arbitration* takes place when an impartial third party (an arbitrator) offers a binding settlement to a dispute. Mini-trials are private proceedings that have no fixed forms. However, *mini-trials* typically involve case presentations by disputants (or their representatives) to a neutral advisor chosen by the parties. After the case presentations, the principals meet to try and settle the dispute.

In *mediation,* disputants get together with a neutral third party (a mediator) to work out their differences. The process involves bargaining, information sharing, consensus building and, ultimately, compromising on original positions to achieve a solution acceptable to all parties.[20] When disputes involve environmental resources, the process is called *environmental mediation.*

In discussions of alternative dispute resolution, a distinction is sometimes made between facilitation and mediation. In *facilitation,* a neutral third party focuses on process and logistical concerns, sometimes acting as a moderator. The facilitator stays restricted to procedural questions and attempts to keep negotiations moving forward. In mediation, however, the third party is involved in the substance of the dispute and may have confidential interactions with disputants. Mediators are useful in situations where negotiating parties are unable or unwilling to communicate or devise options for resolving conflicts.

Beginnings of Environmental Mediation

Although mediation has long been used in a variety of contexts (for example, collective bargaining), its formal use in settling environmental disputes is relatively new. Environmental mediation originated in 1973 in Washington State where a conflict existed over a proposed flood control dam on the Snoqualmie River.[21] Construction of the dam was supported by homeowners, businessmen, and farmers, but opposed by a coalition of environmental and citizens' groups. The mediators, Gerald Cormick and Jane McCarthy, had secured funds from foundations to explore possibilities for using mediation to resolve environmental disputes. The Snoqualmie River conflict provided Cormick and McCarthy with a test case. After seven months of negotiations, the conflicting parties signed an agreement to construct a smaller dam at a new location, and to form a river basin council to coordinate planning for the entire area. Although the dam was never built, many of the land use elements in the agreement were implemented. Cormick and McCarthy demonstrated that environmental mediation could be effective. Since then, many environmental conflicts have been mediated, and they have involved both site-specific and policy-level issues related to a variety of subjects.[22]

[20] This description of mediation is based on Jacobs and Rubino (1987, p. 1). They provide an annotated bibliography on environmental mediation.

[21] This summary of the mediation of the Snoqualmie River conflict is based on Amy (1987, p. 5).

Cornerstones of Principled Negotiation

Because alternative dispute resolution rests heavily on negotiation, ADR specialists have focused on improving the ability of parties in conflict to negotiate constructively. The Harvard Negotiation Project has played an important role in devising innovative approaches to negotiation. The project's work, which is relevant to many types of disputes, is embodied in a process referred to as *principled negotiation* or *negotiation on the merits.*[23] Principled negotiation seeks to build consensus through a rational assessment of issues. It is often contrasted with *positional bargaining,* in which one or both parties take extreme positions, misrepresenting their true views for strategic reasons. In positional bargaining, small concessions are made as necessary to allow the negotiations to continue. This process can become a contest of wills that often strains or even destroys relations between parties:[24] Chances of reaching an agreement through positional bargaining diminish as the number of parties involved increases.

In contrast to positional bargaining, principled negotiation encourages cooperation. Principled negotiation can be characterized using four points.[25]

1. *Separate the people from the problem.* Because emotions often become snarled with the merits of a position, it is productive to disentangle emotional feelings from the substantive elements of

[22] For additional information on the many environmental disputes that have been mediated, see Bingham (1986).

[23] Much of the material in this section is based on Fisher and Ury (1981, chapters 2–5).

[24] On a related point, Goodpaster (1992, pp. 224–25) describes "strategic bargaining theory" in the context of litigation. This "theory asserts that litigants strategically misrepresent their willingness to settle, maneuver for advantage, and adopt bargaining strategies to help increase their settlement gain." Because of the intentional misrepresentation of positions, parties might not "realize there was a zone of agreement."

[25] This four-part characterization is from Fisher and Ury (1981).

a dispute. Experts at principled negotiation see themselves as helping parties attack the problem together, as opposed to attacking each other. Emphasis is placed on communicating clearly and trying to understand others' interests.[26]

2. *Focus on interests, not positions.* In a negotiation, positions are often different from interests. This is illustrated by a hypothetical conflict between an environmental group and a real estate development company over the use of a parcel of land. The environmental group's interest is to protect the land because of its value as wildlife habitat, but its stated position is to oppose the real estate developer's plans. The developer's interest is to attract tourism to the region, but its interest is articulated as a position: build a luxury resort. The object of a negotiation is to satisfy interests, not positions. In the land use dispute, it might be possible to satisfy both sets of interests by promoting a tourism project that does not destroy the habitat; for example, by developing an ecotourism center. For any interest, there may be several positions.
3. *Invent options for mutual gain.* Brainstorming and other techniques for generating ideas in a group setting are used to identify solutions that advance shared interests while reconciling differences.[27] Using these techniques, the creative process is not constrained by prejudging options—critical evaluation is reserved for later. In a common demonstration of failure to realize mutual gains, two siblings quarrel over an orange and finally decide to divide it in half. One sibling eats the fruit and throws away the peel, while the other discards the fruit and uses the peel in baking. Each sibling had taken the position that he or she needed the entire orange. Had the siblings moved off their positions and articulated their interests, they would have discovered a solution that was mutually beneficial.
4. *Use objective criteria for evaluating the agreement.* Mutually acceptable procedures and standards for judging alternative solutions to a conflict can help move negotiations forward. For example, in negotiations over the Law of the Sea, a conflict arose over fees charged to companies mining in the deep seabed.[28] India proposed an initial fee of $60 million per site. The United States argued that there be no initial fee at all. A contest of wills developed and neither side appeared willing to back off. Eventually, the conflicting parties learned that the Massachusetts Institute of Technology (MIT) had developed a model useful in determining the impacts of alternative fee proposals on the economics of deep seabed mining. The disputants gradually accepted the MIT model as valid. The model demonstrated that neither the Indian nor the U.S. position was reasonable, and that a fee more moderate than the Indian proposal was appropriate. Neither side had to give in, because each was able to defer to mutually acceptable modeling results.

THREE PHASES OF ENVIRONMENTAL MEDIATION

Principled negotiation often takes place in the context of environmental mediation. Although the environmental mediation process varies significantly from case to case, three phases can be distinguished: pre-negotiation, negotiation, and post-negotiation (see Table 18.5). The first phase is largely preparatory. Typically, the affected parties are identified, representatives are selected, research on the issues is conducted by the mediators, and ground rules and an agenda are set. The second phase, the negotiation itself, involves joint fact-finding, breaking down the issues, and establishing a package of actions to which all parties can commit. The third phase, post-negotiation, involves implementation and monitoring, and, possibly, renegotiation.[29]

A conflict over fungicide use in the province of British Columbia (B.C.), Canada, illustrates each of

[26] *Role playing* can be used to enhance mutual understanding. For example, in a two-party conflict, disputants could (as an exercise) assume each other's identity and try to negotiate from that perspective.

[27] In *brainstorming,* the focus is on generating ideas to solve problems, not on evaluating those ideas; brainstorming is characterized by unrestrained, spontaneous participation in discussion. See Moore (1994) for an introduction to brainstorming and other group problem-solving methods.

[28] The Law of the Sea example is from Fisher and Ury (1981, pp. 87–90).

[29] The three-phase breakdown of the mediation process is adapted from two sources: Susskind and Cruikshank (1987), and a course by S. McCreary and J. Gamman, "Negotiating Effective Environmental Agreements," taught at Berkeley, California, on April 20–21, 1994.

TABLE 18.5 Steps Within Three Phases of a Meditation Process[a]

Pre-negotiation
- Select mediator(s)
- Identify disputing parties
- Determine representatives of stakeholders
- Perform stakeholder analysis
- Develop ground rules and an agenda

Negotiation
- Engage in joint fact-finding
- Invent options for mutual gains
- Develop package of agreements
- Ratify agreements

Post-negotiation
- Implement and monitor agreed-upon plan
- Renegotiate if necessary

[a] The steps listed are in a general sequence, but variations are common; see, for example, Susskind and Cruikshank (1987, pp. 142–43).

the steps in the three phases of a mediation.[30] The dispute concerned hazardous antisapstain chemicals used in the wood products industry in B.C. Antisapstain chemicals are fungicides used to control growth of a mold that leaves bluish-black stains on lumber. While the stains do not affect the structural integrity of wood, international buyers typically either refuse stained wood or accept it only if they obtain a reduced price. Because about 90% of the coastal lumber in British Columbia is produced for overseas markets and thus treated with antisapstain chemicals, a substantial workforce is regularly exposed to dangerous fungicides.

From the 1940s to the 1980s, organochlorine compounds such as tetrachlorophenate and pentachlorophenol were the preferred antisapstain chemicals in B.C. During the late 1970s, a controversy mounted in B.C. over three aspects of pentachlorophenol: occupational exposure, persistence in the environment, and dioxin by-products during manufacture.

Agriculture Canada is the federal ministry responsible for registering and regulating fungicides and pesticides in Canada. In the early 1980s, three new antisapstain chemicals were registered by Agriculture Canada despite inadequate data on their toxicity to humans and other species.[31] Based on public response to its use of pentachlorophenol, the wood products industry began to switch over voluntarily to the new antisapstain products. However, at the time of the switch, even less was known about the new chemicals than was known about pentachlorophenol. Agriculture Canada faced a dilemma: it could neither support continued use of pentachlorophenol nor recommend a new antisapstain chemical known to pose an acceptable risk. In September 1989, Agriculture Canada issued a report considering a ban on pentachlorophenol for all antisapstain purposes "by June 1991, with the registration of alternative compounds only as they became fully supported by a complete and comprehensive database. . . ."[32] Although it was not made explicit in the 1989 report, use of the three antisapstain chemicals that had been registered in the early 1980s would be permitted under the proposed ban on pentachlorophenol.

Anticipating dissatisfaction with proposals in its September 1989 report, Agriculture Canada encouraged parties concerned about antisapstain chemicals to form a consultative group consisting of representatives of the principal *stakeholders.* A stakeholder group was formed soon thereafter. (In the context of ADR, stakeholders are organizations that are "'entitled' to be consulted as a result of legal standing or public regard and that have a prior record of involvement in the issues or localities involved."[33]) Agriculture Canada felt it could not register any other antisapstain chemicals until it had more complete data on their effects *unless* a consensus could be

[30] Information for this case was extracted from British Columbia Round Table on the Environment and the Economy (1991, app. 2, pp. 7–10), and Leiss and Chociolko (1994, pp. 219–55).

[31] The three new chemicals were 2 thiocyanomethylthio benzothiazole (TCMTB), Copper 8 quinolinolate (Cu-8), and Borax (+ Sodium Carbonate).

[32] The quoted material is from Leiss and Chociolko's (1994, p. 228) summary of Agriculture Canada's "Discussion Document on Anti-Sapstain Chemicals" (dated September 1989). Very few sawmills were using pentachlorophenol at this point, so a complete ban on it did not make much practical difference. Ironically, pentachlorophenol was the only substance for which a relatively "complete and comprehensive database" existed.

[33] This definition of stakeholder is from Leiss and Chociolko (1994, p. 206).

reached among B.C. stakeholders for an alternative course of action.

Pre-negotiation Phase of Mediation

Agriculture Canada did not officially endorse or financially support the new stakeholder group. However, if an agreement could be reached, Agriculture Canada could ensure implementation. Stakeholders believed it was essential to find a neutral party who could garner the confidence of all participants so negotiations could proceed in good faith. The group engaged a skilled *mediator,* William Leiss, to initiate a consensus-based process.[34]

An important step in the pre-negotiation stage of the mediation process is to *identify the parties* (see Table 18.5). Based on initial meetings with individual stakeholders, Leiss obtained agreement from the various interests about the need for a forum where parties could meet face to face. The next step in pre-negotiation involved *establishing representation.* Eventually, 11 parties sent representatives to negotiate, and they referred to themselves as the B.C. Stakeholder Forum on Sapstain Control (*the Forum,* see Table 18.6). Although two provincial agencies were part of the forum, Agriculture Canada and other federal regulatory agencies decided not to become members. Leiss took responsibility for keeping federal agencies informed. Significantly, group members decided not to recognize chemical manufacturers and suppliers as stakeholders. Members believed a consensus could not be reached with the chemical industry at the table.[35] Although excluding the chemical industry increased the likelihood of consensus, it compromised the openness of the process.

As part of the pre-negotiation phases, Leiss performed a *stakeholder analysis* to identify the interests and positions of the parties. Pre-negotiation also involves *developing ground rules* (or *protocols*) and an *agenda,* both of which typically occur at the first meeting of stakeholders. Examples of questions addressed at this stage include: Must everyone attend every meeting or are proxies allowed? What constitutes agreement? Who must sign the agreement and how much time will be permitted for signing once the final document is prepared? What procedure will be used to modify ground rules?[36]

A key issue at the first meeting of the forum (in December 1989) concerned the definition of consensus. Having a definition was important because Agriculture Canada would only act on the forum's recommendations if there was a consensus. The group interpreted consensus to mean endorsement by at least a bare majority. In addition, the group decided that a stakeholder could "indicate disagreement with the majority, and register that disagreement in a number of ways, without directly challenging the view that

TABLE 18.6 Parties Represented at the B.C. Stakeholder Forum on Sapstain Control[a]

Provincial Government Ministries
British Columbia Ministry of Environment
British Columbia Ministry of Forests and Lands
Industry
Council of Forest Industries
Sawmill Industry of British Columbia
Wharf Operators of British Columbia
Unions
Canadian Paperworkers Union
International Longshoremen and Warehousemen's Union
International Woodworkers of America-Canada
Pulp, Paper & Woodworkers of Canada
Environmental Organizations
Earthcare (Kelowna, B.C.)
West Coast Environmental Law Association

[a] Information in this table is from Leiss and Chociolko (1994, p. 230).

Reprinted with the permission of McGill-Queen's University Press.

[34] Dr. Leiss holds the Research Chair in Environmental Policy at the School of Policy Studies, Queen's University, Kingston, Canada.

[35] This aspect of *identifying disputing parties* actually occurred in the negotiation phase of mediation and not in the pre-negotiation phase. Table 18.5 suggests a general order, but variations are common.

[36] Experienced mediators recommend that the mediator and the parties agree on ground rules for a negotiation before it begins. For more information on the process of formulating protocols for negotiation, see Cormick (1989), and Susskind and Cruikshank (1987).

a consensus had been achieved. . . ."[37] Members of the forum also agreed to try and reach a consensus rapidly. Pressure to move quickly existed because the B.C. Ministry of Environment was planning to issue, by September 1990, stringent restrictions on discharges of antisapstain chemicals from sawmills.

Negotiation Phase of Mediation

The first step of the negotiation phase is *joint fact-finding,* an attempt to involve all parties in compiling and evaluating factual information. Joint fact-finding differs from traditional methods for bringing scientific information to bear on environmental disputes. More typically, each group conducts its own studies. This often yields competing claims that are difficult to resolve. In joint fact-finding, researchers from opposing parties gather and evaluate information together.

In the antisapstain chemicals case, forum members held fact-finding meetings with representatives of chemical manufacturers and distributors, toxicologists, and other experts. One objectives of these meetings was to arrive at a common understanding of the scientific issues and risks. Information sharing is a central theme in joint fact-finding. During one of the forum's meetings, members discovered that a representative of the B.C. sawmill industry had withheld a report on a pesticide under consideration for temporary registration. Although the report was eventually released, "the fact of its unexpected disclosure almost caused the union and environmental stakeholders to walk out of the meeting and did result in reversing the generally improving relations among stakeholders that had been in evidence until that moment."[38] This incident demonstrates a potential pitfall of negotiations involving multiple stakeholders: "a single case of perceived bad faith, especially on the part of major stakeholder, can undermine an otherwise good record of co-operation and progress towards consensus."

In the next step of negotiation, *inventing options for mutual gain,* issues are divided into separable units to allow more opportunities for bargaining. The process of negotiating mutually advantageous exchanges is illustrated by how the B.C. Stakeholder Forum determined whether insufficiently tested antisapstain chemicals should be registered. The unions and environmental groups had been reluctant to support registration of *any* antisapstain compound because little data were available on adverse environmental and health effects. These organizations made up 6 of the 11 represented stakeholders; thus they wielded some influence. Three of the unions eventually shifted from their initial position as a result of a deal negotiated with industry. The industry groups agreed to support a health-monitoring process, a long-standing demand of the unions. In return, the unions agreed to back the temporary registration of a new pesticide called "NP-1." This exchange was facilitated by dividing the larger question of whether to register antisapstain chemicals without sufficient data into several discrete issues.

The final outcome of a negotiation often consists of a *package of agreements.* In the antisapstain chemicals case, six agreements were reached. The first two established the B.C. Stakeholder Forum as a monitoring organization to be funded by the provincial Ministry of Forests. This gave the forum both authority and funding to monitor activities outlined in the other agreements. The third agreement established an information-gathering program under a health-monitoring subcommittee of the forum, and the fourth required the Council of Forest Industries to provide information each year on technologies and markets for wood products. The fifth agreement, which was *not* supported unanimously, recommended that certain sapstain control products be registered conditionally subject to annual review by Agriculture Canada. The sixth and final agreement required the B.C. Ministry of Environment to work with Agriculture Canada on establishing wastewater discharge regulations for antisapstain chemicals, and to coordinate with the forum on training in the use of those chemicals.

The final step in the negotiation phase is to *ratify the agreement.* In May 1990, the B.C. Stakeholder Forum forwarded its six agreements to Agriculture Canada, but with only 8 of 11 signatures. The Pulp, Paper and Woodworkers of Canada, and both environmental organizations, Earthcare and West Coast Environmental Law Association, refused to sign. They could not accept the fifth agreement which, in

[37] This quotation is from Leiss and Chociolko (1994, p. 231).

[38] Quotations in this paragraph are from Leiss and Chociolko (1994, p. 243).

effect, allowed use of inadequately tested chemicals. Those disagreeing with the majority felt that "Canada is generally better served by a conservative pesticide registration process which requires full safety data before it registers chemicals."[39]

Post-negotiation Phase of Mediation

Once negotiations have been completed, steps must be taken to keep an agreement from unraveling. In the *implementation* phase, accords reached during negotiation are translated into formal, binding agreements. In addition, monitoring procedures are devised. Finally, needs for *renegotiation* are assessed, and, if necessary, plans are made for reconvening parties.

In the dispute involving antisapstain chemicals, the final agreement, was implemented quickly. Agriculture Canada accepted the agreement and announced that it would withdraw pentachlorophenol from use in sapstain control, reevaluate the three other antisapstain chemicals then in use, and temporarily register several new antisapstain products. Both the forum itself and its health-monitoring subcommittee continued to meet after negotiations were completed.

APPLICATIONS OF ADR IN ENVIRONMENTAL DECISION MAKING

Since the 1973 efforts of Gerald Cormick and Jane McCarthy to settle a conflict over the damming of the Snoqualmie River in Washington, mediation and other forms of ADR have been used to settle hundreds of site-specific environmental disputes in a variety of settings. Cases subject to ADR have centered on land use, natural resources (including forests and fisheries), water resources projects, energy facilities, and hazardous substances.

ADR has also been used in environmental policy making. In 1976, the National Coal Policy Project in the United States set a precedent by bringing together 105 representatives of business and environmental interests to resolve differences in coal energy policy. The project's success in engaging conflicting parties in productive dialogue opened the way to *regulatory negotiations* (or "reg-neg") over the Toxic Substances Control Act in 1977, under the auspices of the Conservation Foundation.[40] Since then, regulatory agencies have engaged in *negotiated rulemaking,* a form of reg-neg in which regulators participate in consensus-based negotiations with affected parties to develop a proposed rule.

In a typical form of negotiated rulemaking in the United States, participants may include representatives of the regulatory agency, citizens groups, affected industries, and state governments. Participants set their own ground rules and otherwise control the negotiation process. Up-front commitments may include the agency's agreement to publish the negotiated rule as its proposed rule. In addition, nonagency participants may agree to support and not to criticize the agency's proposed rule. Nonagency participants may also agree formally not to challenge in court a final rule that is consistent with the proposed rule. The negotiation process is viable only to the point at which the agency publishes the proposed rule in the *Federal Register* and invites public comment. From that point on, rulemaking proceeds in the usual way (see Chapter 3).

Experience with reg-neg to date suggests that it can be expedient in some circumstances. Reg-neg is probably not worth trying in situations "where the number of interests needing representation is unmanageably high, or where agreement would be possible only if some participants compromised on a fundamental issue of principle."[41] As noted in the next section, some legal scholars are not convinced that the advantages of reg-neg (for example, increased speed in producing rules) are worth the potential for sacrificing the public interest in bargaining over rules.

During the past several years, EPA has promoted alternative dispute resolution methods in the context of its Superfund program (see Chapter 14). In 1988, EPA began an Alternative Dispute Resolution Pilot Project to evaluate the potential for negotiating

---

[39] This quotation is from British Columbia Round Table on the Environment and the Economy (1991, app. 2, pp. 2–10).

[40] For more information on reg-neg in the context of the Toxic Substances Control Act, see Bingham (1986, pp. 17–20). Although reg-neg is a consensus-based approach, it does not always involve a neutral third party.

[41] Quoted material is from Gelhorn and Levin (1990, p. 344). The same source provides the preceding characterization of a typical regulatory negotiation. For an analysis of the advantages and disadvantages of reg-neg, see Fiorino (1991).

disputes over cleanups at Superfund sites through assistance of a neutral third party.[42] Successes in EPA's pilot project prompted interest in ADR by other federal agencies. In 1990, Congress passed the Administrative Dispute Resolution Act "to authorize and encourage Federal agencies to use mediation, conciliation, arbitration and other techniques for the prompt and informal resolution of disputes, and for other purposes."[43]

Alternative dispute resolution has also been used in siting *locally unwanted land uses* (LULUs). By 1984, five states in the United States had enacted statutes requiring applicants for solid or hazardous waste facility permits to negotiate siting agreements with local communities.[44] Most cases under these laws have involved conventional negotiation, but mediation has also been used.

ALTERNATIVE DISPUTE RESOLUTION IN PERSPECTIVE

When is ADR Appropriate?

Alternative dispute resolution is not appropriate for all forms of environmental conflict. Some conditions under which it has been useful are listed in Table 18.7. ADR has a greater chance of being successful in cases where disputing parties perceive "[a] high level of uncertainty in terms of both timing and outcome."[45] If some stakeholders believe their objectives can be achieved over a reasonable time frame without ADR, they may have little incentive to engage in joint problem solving.

ADR is also more likely to succeed under conditions of approximately equivalent power distribution among parties, where power is "the ability to prohibit an action desired by one or more adversaries." In the case involving antisapstain chemicals in British Columbia, environmental groups and unions had relatively little influence over industry and government. Demands of the environmentalists and unions for strict registration rules were impractical from the perspective of the stronger parties. The unions, however, eventually wielded some leverage. Although they initially sided with the environmental groups, some unions later agreed to the registration of a new chemical in exchange for industry concessions on monitoring adverse health effects.

TABLE 18.7 Conditions for Use of Alternative Dispute Resolution

- High uncertainty in timing and outcome
- Equivalent distribution of power among parties
- Mediation more promising than BATNA[a]
- Disputing parties at an impasse
- Litigation ineffective

[a] BATNA is short for "best alternative to a negotiated agreement."

Another factor influencing the success of alternative dispute resolution relates to the concept of a *best alternative to a negotiated agreement* (BATNA): the best outcome a group is likely to achieve by pursuing a strategy other than negotiation.[46] As long as the prospective outcome from negotiation is more attractive than the BATNA, a stakeholder has an incentive to continue negotiating. Once prospects drop below this standard, the stakeholder would be likely to reject negotiated outcomes. For example, a group might forego negotiation if it felt it could do better by going to court or lobbying decision makers. In the B.C. antisapstain chemical case, the BATNA was Agriculture Canada's proposal of September 1989, which permitted continued use of potentially dangerous substances. The negotiation led to gains beyond this BATNA, for example, by yielding a program to monitor health effects.

Alternative dispute resolution sometimes becomes attractive when an impasse is reached in a

[42] For an analysis of pilot project outcomes, see Peterson (1992).

[43] Administrative Dispute Resolution Act (ADRA), Preamble, as cited by Peterson (1992, p. 328). For more on the ADRA, see Susskind, Babbitt, and Segal (1993).

[44] Bingham (1986, pp. 51–52) provides information on the use of negotiation in siting LULUs. According to Rabe (1994, p. 36), the use of a form of ADR in siting LULUs has been successful in some contexts; however, its "transformative potential when mandated in the volatile area of hazardous waste facility siting was seriously oversold by academic and other proponents." Rabe provides case studies of successful efforts to site hazardous waste management facilities using approaches based on extensive citizen participation.

[45] This quotation and the one in the next paragraph are from McCarthy and Shorett (1984, pp. 12–13).

[46] This definition of BATNA is from Fisher and Ury (1981, p. 104).

controversy.[47] A stalemate often signals that neither party can win on its own. Negotiating seriously then becomes an attractive option.

Another condition often associated with successful ADR is that efforts at litigation have led to lengthy and costly delays. ADR often seems appealing when attempts at settling a dispute in court are proceeding poorly. Sometimes even the *threat* of costly litigation can make ADR attractive.

Is ADR Better than Litigation?

Proponents of alternative dispute resolution argue that, compared to litigation, ADR can yield cost and time savings, and decisions that are more acceptable to disputing parties.[48] However, such claims have been challenged. Based on her study of 161 negotiated disputes, Bingham (1986) found little evidence to indicate that ADR methods are either faster or less costly than litigation. Although mediators often charge less than attorneys, Bingham (1986, p. xxv) notes that "costs of preparing for negotiation . . . may be as high as or higher than the costs of preparing for some kinds of litigation, particularly for public interest groups." Moreover, ADR is sometimes either preceded by unresolved litigation or motivated by the threat of litigation. Some ADR experts believe that ADR and litigation are more productively viewed as complementary (as opposed to alternative) approaches.[49]

The claim that the outcome of an environmental mediation is likely to be acceptable to all parties also appears exaggerated. In the B.C. antisapstain chemical case, for example, 3 out of 11 groups on the forum were dissatisfied with the final agreement and refused to sign it. Moreover, the chemical manufacturers were left out of the mediation.

Critics of ADR argue that its advantages in providing a forum for participation of citizens and environmental groups are frequently overstated. For example, Amy (1987) believes ADR may give citizens and weak parties only the illusion of power. Government agencies and industry groups have advantages because they typically possess the information, resources, and political clout to ensure that their views are heard.[50] In contrast, citizens and grassroots organizations frequently lack resources, have insufficient familiarity with negotiation processes, and often are not well organized.[51]

Other criticism of ADR focuses on whether it serves the interests of society at large. Consider, for example, Funk's (1987) analysis of EPA's regulatory negotiation leading to rules governing air pollutants from residential woodstoves. He found that many of the legal deficiencies of the proposed rule were caused or fostered by the use of reg-neg. According to Funk (1987, p. 92), ". . . regulatory negotiation finds its legitimacy in the agreement between the parties, rather than the determination under law of the public interest; in other words, regulatory negotiation substitutes a private law remedy for a public law remedy." A counterargument is that the safeguards associated with traditional rulemaking apply. Interested parties have a right to comment on a proposed rule and to challenge a final rule in court.

As the preceding discussion demonstrates, alternative dispute resolution has its advantages, but it is not a panacea. The potential usefulness of ADR in any conflict situation must be evaluated on a case-by-case basis.

Key Concepts and Terms

Objectives of a Public Involvement Program

Legal requirements for citizen participation
Two-way communication
Defusing opposition
Legitimating decisions
Influencing outcomes
Agency's objectives vs. citizens' objectives
Arnstein's levels of participation
Tokenism vs. citizen power
Information and education
Public consultation
Joint planning

Identifying the Public

Segments of the public
Self identification

[47] This point is elaborated by Amy (1987, p. 80).

[48] For more on the benefits of ADR, see McCarthy and Shorett (1984, p. ix), and Bacow and Wheeler (1984, p. 359).

[49] This position is taken by Amy (1987, pp. 89–91).

[50] For more on this point, see Amy (1987, pp. 13, 94).

[51] In some cases, the need to legitimate mediation as a process in which all parties are accorded fair representation results in assistance being provided to weaker groups.

Staff identification
Third-party identification

Public Involvement Techniques
Techniques for providing information
Methods for generating feedback
Two-way communication procedures

Integrating Public Involvement Into Planning: A Case Study
Citizen's advisory committees
Information bulletins
Public workshops
Survey questionnaires
Types of information citizens can provide

Resolving Disputes Over Environmental Resources
Alternative dispute resolution (ADR)
Environmental mediation
Principled negotiations
Positional bargaining
Interests vs. positions
Criteria for evaluating agreements
Characteristics of environmental disputes

Three Phases of Environmental Mediation
Stakeholder analysis
Protocols and agendas
Joint fact-finding
Options for mutual gain
Packages of agreements
Monitoring and renegotiation

Applications of ADR in Environmental Decision Making
Regulatory negotiations
Negotiated rulemaking
Negotiating Superfund site cleanups
Locally unwanted land uses (LULUs)

Alternative Dispute Resolution in Perspective
Distribution of power
BATNA
ADR as a complement to litigation

Discussion Questions

18-1 Suppose you are to advise a highway agency on what it can do to ensure that citizen participation in the planning and evaluation of its proposals is constructive. What advice would you offer regarding (1) the timing of citizen involvement activities, and (2) techniques used to involve the public in planning?

18-2 List three reasons why planners working for an infrastructure development agency like the U.S. Army Corps of Engineers might not be eager to involve the public in their planning. What arguments could be used to counterbalance the three reasons you offered?

18-3 Some authors distinguish between *public consultation* and *public participation.* For example, Glasson, Therivel, and Chadwick (1994, p. 165) differentiate the two as follows:

> Consultation is in essence an exercise concerning a passive audience: views are solicited, but respondents have little active influence over any resulting decisions. In contrast, public participation involves an active role for the public, with some influence over any modifications to the project and over the ultimate decision.

Do you think this distinction is useful, and if so, why? Using the definitions of Glass, Therivel and Chadwick, which of the methods used in the San Pedro Creek case study involved public consultation? Which involved public participation?

18-4 Imagine that the Corps of Engineers study of flooding on San Pedro Creek is starting now and you are in charge of designing the public involvement program. What objectives would you suggest for the public involvement program? What techniques would you use? When in the planning process would you use them?

18-5 According to Goodpaster (1992, p. 221),

> Litigating a dispute is both a major alternative to negotiating it and a way to force its negotiation. . . . [P]arties settle most lawsuits rather than [have them decided by a judge or jury]. Indeed, it is the deadline that a pending trial imposes on the possibilities of a negotiated settlement, the risk of a loss at trial, and possibly the added expense of

trial that motivates many lawsuit settlements.

Use Goodpaster's perspective to argue for or against the position that environmental mediation is a complement to (rather than a substitute for) litigation.

18-6 In an analysis of the limits of alternative dispute resolution, Grad (1989, p. 183) raises questions that arise because ADR is a private process:

> A public process such as a trial or a petition for review of [an agency's decision], involves procedural safeguards to provide opportunities for intervention or review. The determination of a case in the courts binds the public at large, unlike the agreement reached by ADR which is always subject to the question of whether all parties in interest were involved, participated and agreed.

How would the points raised by Grad apply in the case study involving sapstain control chemicals in British Columbia? Describe circumstances in which carefully tailored agreements reached using ADR might fall apart because interested parties were excluded from the negotiations.

References

Amy, D. J. 1987. *The Politics of Environmental Mediation.* New York: Columbia University Press.

Arnstein, S. R. 1969. A Ladder of Citizen Participation. *American Institute of Planning Journal* 35 (4): 216–24.

Bacow, L. S., and M. Wheeler. 1984. *Environmental Dispute Resolution.* New York: Plenum Press.

Bingham, G. 1986. *Resolving Environmental Disputes: A Decade of Experience.* Washington, DC: The Conservation Foundation.

British Columbia Roundtable on the Environment and the Economy. 1991. *Reaching Agreement: Consensus Processes in British Columbia.* Victoria, British Columbia, Canada.

Commission on Resources and Environment (CORE). 1995. *The Provincial Land Use Strategy, Volume 3: Public Participation.* British Columbia, Canada.

Cormick, G. 1989. Strategic Issues in Structuring Multi-Party Public Policy Negotiations. *Negotiation Journal* 5 (2): 125–32.

Creighton, J. L. 1980. *The Public Involvement Manual: Involving the Public in Water and Power Resources Decisions.* Washington, DC: U.S. Department of the Interior, Water and Power Resources Service, Bureau of Reclamation.

———. 1992. *Involving Citizens in Community Decision Making.* Washington, DC: Program for Community Problem Solving.

Crowfoot, J. E., and J. M. Wondolleck, eds. 1990. *Environmental Disputes: Community Involvement in Conflict Resolution.* Washington, DC: Island Press.

FEARO (Federal Environmental Assessment Review Office). 1988. *Manual on Public Involvement in Environmental Assessment: Planning and Implementing Public Involvement Programs,* vol. 1–3. Prepared by Praxis, Calgary, Alberta, Canada.

Fiorino, D. J. 1991. "Dimensions of Negotiated Rule-Making: Practical Constraints and Theoretical Implications." In *Systematic Analysis in Dispute Resolution,* ed. S. S. Nagel and M. K. Mills, 127–39. New York: Quorum Books.

Fisher, R., and W. Ury. 1981. *Getting to Yes: Negotiating Agreement Without Giving In.* New York: Penguin.

Funk, W. 1987. When Smoke Gets In Your Eyes: Regulatory Negotiation and the Public Interest—EPA's Woodstoves Standards. *Environmental Law* 18 (1): 55–98.

Gelhorn, E., and R. M. Levin. 1990. *Administrative Law and Process in a Nutshell.* St. Paul, MN: West Publishing.

Glasson, J., R. Therivel, and A. Chadwick. 1994. *Introduction of Environmental Impact Assessment.* London: UCL Press.

Goodpaster, G. 1992. Lawsuits as Negotiations. *Negotiation Journal* 8 (3): 221–39.

Grad, F. P. 1989. Alternative Dispute Resolution in Environmental Law. *Columbia Journal of Governmental Law* 14 (1): 157–85.

Hendee, J. C., R. C. Lucas, R. H. Tracy, T. Staed, R. N. Clark, G. H. Stankey, and R. A. Yarnell. 1976. "Methods for Acquiring Public Input." In *Water Politics and Public Involvement,* ed. J. C.

Pierce and H. R. Doerksen, 125–44. Ann Arbor, MI: Ann Arbor Science Publishers.

Iacofano, D. 1990. *Public Involvement as an Organizational Development Process.* New York: Garland Publishing.

Jacobs, H. M., and R. Rubino. 1987. *Environmental Mediation: An Annotated Bibliography,* no. 189. Chicago: Council of Planning Librarians.

Ketcham, D. E. 1992. "The EIS Process—Public Participation." Outline of remarks given at the advanced ALI-ABA course of study. "National Environmental Policy Act, 'Little NEPAs,' and the Environmental Impact Assessment Process," held in Washington, DC, November 13–14.

Leiss, W., and C. Chociolko. 1994. *Risk and Responsibility.* Montreal and Kingston, Canada: McGill–Queen's University Press.

McCarthy, J., and A. Shorett. 1984. *Negotiating Settlements: A Guide to Environmental Mediation.* New York: American Arbitration Association.

Moore, C. M. 1994. *Group Techniques for Idea Building,* 2d ed. Applied Social Research Methods Series, vol. 9. Thousand Oaks, CA: Sage Publications.

Parenteau, R. 1988. *Public Participation in Environmental Decision-Making.* Ottawa, Canada: Minister of Supply and Services.

Peterson, L. 1992. The Promise of Mediated Settlements of Environmental Disputes: The Experience of EPA Region V. *Columbia Journal of Environmental Law* 17: 327–80.

Rabe, B. G. 1994. *Beyond NIMBY: Hazardous Waste Siting in Canada and the United States.* Washington, DC: The Brookings Institution.

Roberts, R. 1995. "Public Involvement: from Consultation to Participation." In Vanclay, F. and D. Bronstein (eds.), *Environmental and Social Impact Assessment,* ed. F. Vanclay and D. Bronstein, 221–46. Chichester, England: Wiley.

Susskind, L., and J. Cruikshank. 1987. *Breaking the Impasse: Consensual Approaches to Resolving Public Disputes.* New York: Basic Books.

Susskind, L. E., E. F. Babbitt, and P. N. Segal. 1993. When ADR Becomes the Law: A Review of Federal Practice. *Negotiation Journal* 9 (1): 59–75.

Tannis, E. G. 1989. *Alternative Dispute Resolution that Works.* North York, Ontario, Canada: Captus Press.

Wagner, T. P., and L. Ortolano. 1976. *Testing an Iterative, Open Planning Process for Water Resources Planning.* Report No. 76-2. Ft. Belvoir, VA: U.S. Army Engineer Institute for Water Resources.

Willeke, G. E. 1976. "Identification of Publics in Water Resources Planning." In *Water Politics and Public Involvement,* ed. J. C. Pierce and H. R. Doerksen, 43–62. Ann Arbor, MI: Ann Arbor Science Publishers.

Zinn, F. D., and R. C. Hinojosa. 1994. A Planner's Guide to the Internet. *Journal of the American Planning Association* 60 (3): 389–400.

Part Five

Techniques Used in Impact Assessment and Regulation

The process of designing environmental regulations or assessing impacts of proposed development projects is often guided by forecasts, and techniques for making forecasts are the main subject of the chapters in Part Five. Methods of impact evaluation are also presented. Whereas Part Four concerned broad classes of methods, Part Five provides details on how predictions and evaluations are made in practice. Accounts of techniques are supplemented with extensive case studies.

Chapter 19 examines how biologists characterize the effects of human actions on ecosystems. Methods developed by biologists are being used increasingly to design development projects and to formulate environmental programs and policies. The appendix to Chapter 19 introduces *geographic information systems,* computer software used to store, analyze, and retrieve spatial data. As the appendix demonstrates, geographic information systems are used in conducting biological impact assessments and in carrying out other types of environmental studies.

Chapter 20 focuses on a different question: How can the visual changes expected from a proposed project be predicted and evaluated? Design professionals have developed a body of research on aesthetics and landscape preference that provides a foundation for visual impact assessment. In analyzing techniques for simulating the visual appearance of proposed projects, Chapter 19 emphasizes methods using computer technology for visual image processing.

Chapters 21, 22, and 23 concern noise, air quality, and water resources, respectively. Each of these chapters begins by describing parameters used to characterize the environment. Once these parameters are introduced, the focus shifts to mathematical models used in forecasting. Computer-based models are widely used in designing environmental regulations and assessing impacts of major projects, and thus mathematical modeling is highlighted in the chapters on noise, air quality, and water resources.

Chapter 19

Biological Considerations in Planning and Policy Making

Biological knowledge is commonly employed in assessing ecological impacts of proposed development projects. In addition, results of biological studies are used to evaluate the effects of implementing environmental regulations; examples include investigations of how changes in air quality affect the production of crops. As demonstrated in Table 19.1, biological knowledge is also used in formulating environmental programs and policies.

The chapter begins by introducing ecological concepts. It then focuses on a difficult question: What criteria should be used in evaluating human-induced changes in biological systems? This question is answered by examining factors biologists consider important when gauging the impacts of projects, programs, and policies. The chapter then examines procedures developed by fish and wildlife management agencies to evaluate proposed development projects. It concludes with a review of activities typically undertaken in assessing biological impacts of proposed actions.

BASIC ECOLOGICAL CONCEPTS

This section introduces terms and concepts used in integrating biological concerns into planning and policy making. By its nature, such an introduction must be selective.[1]

An *ecosystem,* defined as a biotic community and its abiotic (nonliving) environment, is the basic unit employed in considering biological factors in the design of projects, programs, and policies. Any change in an area will affect the ecosystem of that area. Although an ecosystem may be resilient and capable of withstanding disturbances, human or natural forces can destroy so much of an ecosystem that its integrity is not maintained and it collapses completely.

A *community* is any assemblage of living populations coexisting in a particular area. Communities are often classified by a characteristic type of vegetation; for instance, one speaks of a redwood forest community or a California chaparral community. However, the term community includes all organisms associated with the vegetative type used to name it. As an illustration, the blue-gray gnatcatcher is a member of the chaparral community. Many species, such as starlings

[1] Daniel Belik prepared an early draft of the material in this section while he was a student at Stanford University. A more complete treatment is provided by the literature on the application of biological sciences in planning and management contexts. See, for example, Edington and Edington (1978), Ward (1978), Holling (1978), Westman (1985) and McHarg (1996).

TABLE 19.1 Applications of Biological Knowledge in Formulating Environmental Programs and Policies

Policy or Program Type	Example[a]
• Renewable resource management	Managing the North Pacific halibut fishery.
• Pest control	Increasing livestock production in Latin America by reducing vampire bat populations.
• Disease control	Reducing the spread of malaria by controlling mosquito populations in West Africa.
• Species and habitat conservation	Conserving the spotted owl habitat in Washington, Oregon and northern California.
• Ecosystem protection	Managing cumulative impacts of wastewater discharges to Lake Washington at Seattle.
• Ecosystem restoration	Restoring seminatural plant communities in mined areas of the United Kingdom.

[a] Each of the examples is reported as a case study chapter in U.S. National Research Council (1986).

and dandelions, are associated with numerous types of communities. The term *habitat* refers to the physical location where an organism lives. Each species needs a habitat capable of providing the space, food, cover, and other requirements for its survival. For example, a pine forest is a suitable habitat for elk, but a desert is not.

Each organism occupies a position in a community as either prey, predator, or both. The organization of feeding patterns is represented by a *food chain.* Organisms occupying the lowest level in a food chain derive nourishment from the abiotic components of an ecosystem. A simple example is given in Figure 19.1.[2] Grasses transform solar energy into chemical energy by a process known as *photosynthesis.* This process, which uses water, carbon dioxide, and the chlorophyll in green plants, yields carbohydrates and oxygen. The grasses produced photosynthetically are eaten by herbivores such as mice, which are eaten in turn by carnivores such as hawks.

The hawks are said to be at the top of the food chain. Dead hawks are a food source for organisms such as bacteria, whose activities as decomposers of organic matter replenish nutrients in the environment. Disruption of a food chain affects both the predator and the prey. If the predator is deprived of food it will starve. The prey, with its source of population control gone, will increase in numbers. If the prey population becomes so large that it consumes its entire food source, the prey will starve in turn.

Interactions among species in a food chain are important in understanding the *flow of energy* in an ecosystem. Photosynthetic plants use the sun's radiant energy for growth and respiration. At each level up a food chain, a smaller fraction of the original solar energy is transferred. For example, only about 10% of the solar energy fixed in plants by photosynthesis is available to plant-eating species; most of the energy is used for either metabolic processes or growth of nonedible portions of plants.[3] Empirical information about how energy is transferred from plants to herbivores and from herbivores to carnivores provides a basis for modeling energy flows in ecosystems.[4] These models make it clear that substantial fractions of the solar energy fixed by plants is used in metabolic processes. An ecosystem can be maintained only if it receives continual replacements of this lost energy, either from sunlight or from organic matter from another ecosystem.

[2] Food chains typically are not as simple as the one in this example. Often several chains interconnect in complex configurations referred to as *food webs.*

[3] The 10% figure is from Westman (1985, p. 332).

[4] In addition to energy transformations, important ecosystem processes include cycling of carbon, nitrogen, phosphorous, and water. For an introduction to these cyclic processes, see an introductory textbook on ecology such as Ehrlich and Roughgarden (1987).

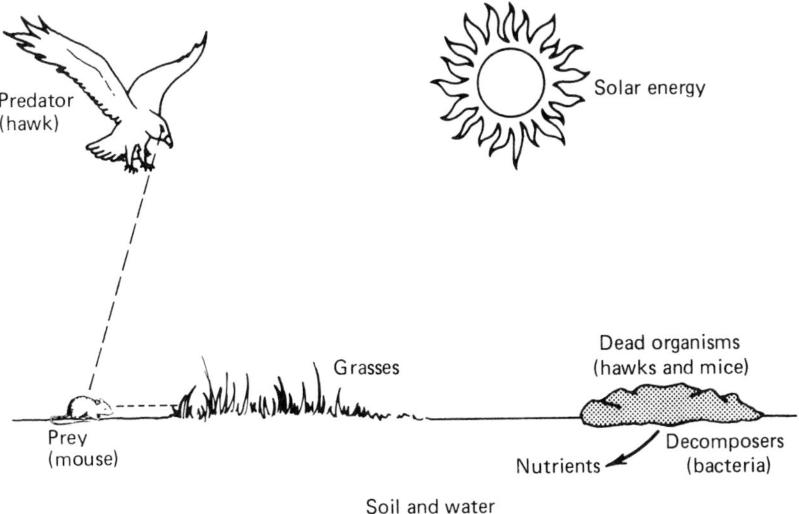

Figure 19.1 Illustrative food chain relationships.

The neglect of food chain interactions by planners and policy makers sometimes leads to unintended adverse effects. Consider an example involving field mice in Israel several decades ago. Every 10 years or so, the mice greatly increased in numbers (for just 1 year) and consumed large quantities of grain harvested in agricultural areas. During these periods, there were more eagles and falcons, birds that prey on mice. In response to a 1949–50 increase in mouse population, the Israeli Plant Protection Department of the Ministry of Agriculture implemented a program to poison the mice using thallium-coated wheat. Thallium takes several days to kill field mice and does so through slow paralysis. Partially paralyzed mice were easy prey for eagles and falcons, and before long the birds were dying from thallium poisoning. Results from the government program were not those intended: the birds of prey decreased in numbers, and the mouse population was not brought under control.[5]

Another example demonstrating the need for planners and policy makers to consider food chain interactions concerns the *biological magnification* of substances, such as DDT, that do not decompose readily. During the 1950s, many toxicological studies of individual species suggested that use of DDT for pest control would not cause significant problems. Subsequent studies revealing how DDT was passed through various food chains refuted the earlier findings. In general, much of the living matter (*biomass*) passed from lower to higher positions in a food chain is excreted or used up in *respiration,* the process of generating usable energy from the oxidative breakdown of food molecules. Only a small amount of this biomass stays in the consuming animal. DDT, however, is soluble in fats and is not easily broken down to simpler compounds. Consequently, as biomass is transferred up through the food chain, very little DDT is "lost" through each transfer. The result is that DDT concentrations often increase dramatically from one food chain level to the next (see Figure 19.2). For example, studies of marshes on Long Island, New York, showed the following magnification in DDT concentrations (measured in parts per million): 0.04 in plankton, 1 in minnows, and 75 in the tissues of carnivorous scavenging birds.[6] High DDT concentrations due to biological magnification have been a cause of mortality for some species of birds, and have interfered with reproduction for others.[7] This example shows the importance of ecosystem-level analyses of biological changes.

[5] This example is from Mendelssohn (1972).

[6] These figures, and a more complete discussion of biological magnification, are given by Woodwell (1967).

[7] Further details on the adverse effects of DDT are given by Ward (1978).

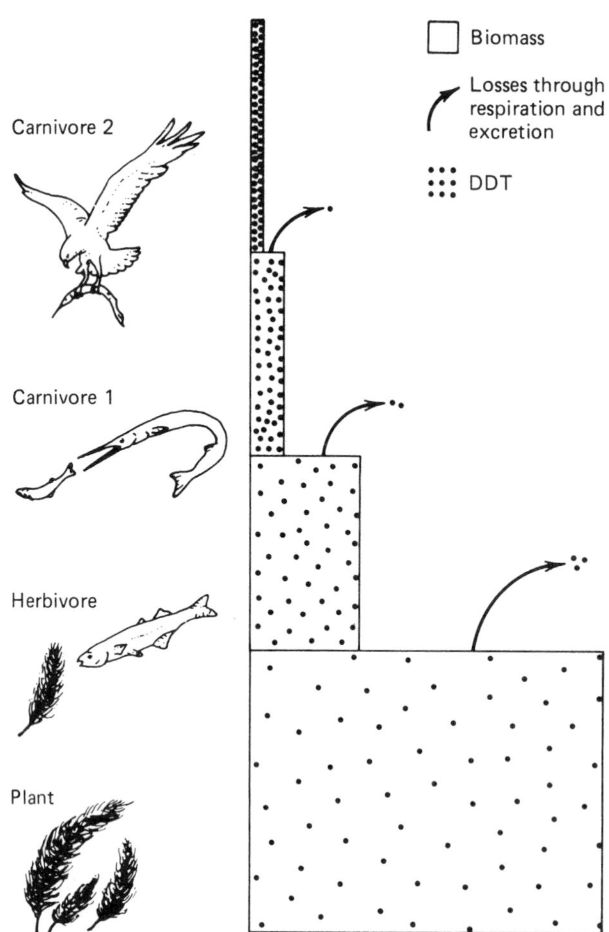

CONCENTRATION of DDT residues being passed along a simple food chain is indicated schematically in this diagram. As "biomass," or living material, is transferred from one link to another along such a chain, usually more than half of it is consumed in respiration or is excreted (*arrows*); the remainder forms new biomass. The losses of DDT residues along the chain, on the other hand, are small in proportion to the amount that is transferred from one link to the next. For this reason high concentrations occur in the carnivores.

Figure 19.2 Passage of DDT up a simple food chain. From Woodward, *Toxic Substances and Ecological Cycles* (illustration by James Egleson).

Productivity is another aspect of a biological community that merits consideration by planners and policy makers. In biology, *productivity* refers to the rate at which an ecosystem can take abiotic components and use them to produce living matter. The Arctic tundra, with sparse vegetation, is a relatively unproductive community, whereas a tropical rain forest is highly productive. Components of the abiotic environment that determine an ecosystem's productivity are water, nutrients, and energy. The energy input is generally sunlight, which is photosynthesized into chemical energy. Human actions reducing the abiotic elements used in forming organic matter can decrease an area's productivity. For example, in some parts of the western United States, groundwater is being withdrawn more rapidly than it is being replenished by natural sources. As a result, the surrounding land is becoming increasingly arid, surface water is not being held in the topsoil, and biological productivity is being decreased. The reduction in productivity affects both natural biological communities and agricultural use of the land. If enough vegetation is lost over time, the soil will not be held in place, resulting in erosion problems and dust bowls.

The change in a community's characteristics over time is termed *ecological succession.* Natural population fluctuations occur continually in a community, but succession refers to changes in the type of species and thus in the community itself. A typical example of an area's natural succession begins with the colonization of bare sand by bacteria and fungi, and then mosses and lichens. These simple organisms can break down the sand into soil that is then colonized by annual grasses and herbs. The annuals may later be crowded out by perennial grasses and herbs, which can lead to further changes in the characteristics of the soil. Changes in the soil may allow shrubs to take root. Eventually, the shrubs might be replaced by trees.[8]

Figure 19.3 shows a pattern of ecological succession observed on abandoned upland agricultural fields in the southeastern United States. Over the first 100 years, the bare field progressed through several stages to become a pine forest. After another 50 years, the community evolved into an oak-hickory forest. In this case, the oak-hickory forest represents a *climax community,* a self-perpetuating community in long-term equilibrium with its physical environment. In any situation, the particular climax community that develops is determined largely by climate and other physical factors.[9]

Ecological succession is an important concept in formulating land use plans and policies. For example, as a chaparral community matures, natural fires burn out old vegetation for replacement. Development plans for chaparral communities should consider both the risk of periodic fires and the effect that controlling fires will have on succession.

Human activities sometimes influence ecological succession in undesirable ways. Consider, for example, the case of Lake Washington at Seattle. Wastewater was released into the lake during the early 1900s. The discharges were stopped in the late 1930s, but started again in the early 1940s due to increased population around the lake. The phosphorus and nitrogen in the effluents spurred algae growth in the lake, and the lake's succession toward becoming a marshland was proceeding much faster than would be expected in an undisturbed lake. Fortunately for the residents near Lake Washington, this rapid succession has been reduced by the diversion of wastewater to another location.[10]

CRITERIA USED IN ECOLOGICAL EVALUATIONS

The application of biological concepts in designing projects, programs, and policies requires that biologists articulate their science in a form useful to planners and policy makers. For example, an analyst engaged in planning or policy making will not be enlightened by learning that the biological productivity of an area will increase as a result of a proposed action. The analyst needs to know whether higher values of biological productivity are good or bad. This places biologists in a difficult position. If a biologist insists on maintaining a purely scientific attitude and reporting only objective measures such as biological productivity, then planners and policy makers may not know how to interpret biological findings, and the biologist's contributions to decision making will be diminished.[11] However, if a biologist makes statements that planners and policy makers can comprehend (for example, decreased biological productivity is an adverse consequence from a societal perspective), then he or she is open to criticism by peers

[8] This illustrative sequence is from Weisz (1973, p. 375). He indicates that the development from sand to forest may take about 1000 years. If soil is present at the outset, the succession to forest may occur in about 200 years.

[9] The climax concept is not as simple as it appears. Biologists have developed numerous concepts to explain differences in climax communities observed in neighboring locations. For an introduction to this complex subject, see Odum (1971, pp. 264–67).

[10] The Lake Washington example is from Krebs (1972, pp. 548–51).

[11] Many people argue that ecological science itself cannot be free of value judgments; see, for example, Shrader-Frechette and McCoy (1993, chap. 4).

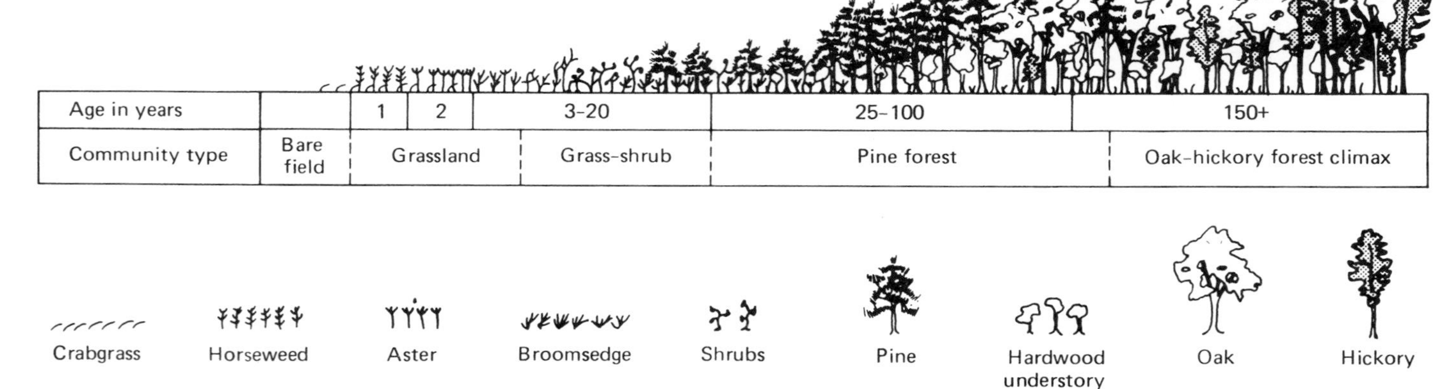

Figure 19.3 Principal plant "dominants" in natural succession on the Piedmont region of the southeastern United States. Figure adapted from *Fundamentals of Ecology*. 3d ed., by Eugene P. Odum, copyright © 1971 by Saunders College Publishing, reproduced by permission of the publisher.

who argue that such observations are value judgments that have no scientific basis.[12]

Many biologists have worked on improving the links between biological science and the formulation of development projects and environmental programs and policies. The literature on this subject recognizes that value judgments must be made to appraise the biological changes associated with proposed actions. Although there is no consensus in this literature, the following topics pertinent to evaluating changes in ecosystems caused by human actions appear frequently:

- Biodiversity and habitat fragmentation.
- Species diversity and its hypothesized linkage with ecosystem stability.
- Significant species, including keystone and indicator species.
- Ecosystem "quality" and other criteria for identifying lands to include in preservation programs.

Each of these subjects will be discussed to explore factors that biologists consider in judging the effects of projects, programs, and policies.

Biodiversity

Loss of biodiversity is a major concern in analyses of how human actions affect natural systems.[13] Biodiversity exists on four levels: genetic, species, ecosystem, and region (see Figure 19.4).[14] *Genetic diversity* refers to variations in genetic material within species and among populations.[15] Higher genetic variation makes it more probable that species can survive by evolving and adapting to changes in their environment. *Species diversity* concerns the number of different species in an area and their relative abundance. For example, Ecuador, which is estimated to contain 20,000 species of vascular plants, has a higher species diversity than Ohio, where only 2000 vascular plant species are found.[16]

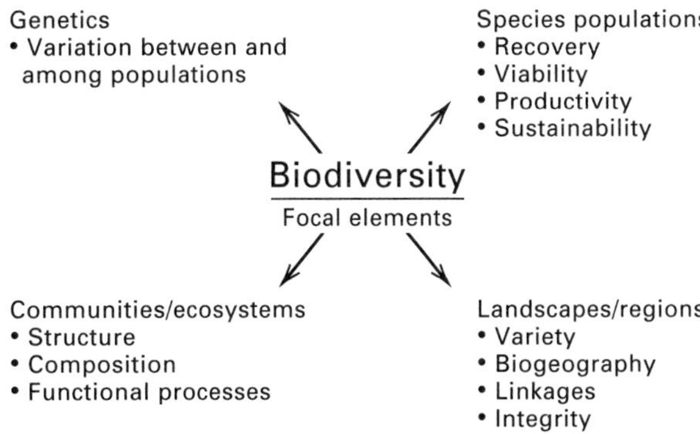

Figure 19.4 Focal elements of biological diversity might include genetic variation in selected species; viability of endangered or sensitive species; the richness, structure, and function of species and processes in biological communities; and characteristics of regional landscapes such as continuity, variety of ecosystems, and patterns of different biological communities. From Decker et al. (1991). Reprinted with permission.

At the *ecosystem (or community)* level, high diversity means high variation in the composition, structure, and function of biological communities and their nonliving environments. The final level, *regional (or landscape) diversity,* refers to variety in types of biological communities and ways in which their sizes, shapes, and interconnections allow for the movement of individual animals within a region. For some species, survival is impossible unless individuals are able to migrate over large areas.

Importance of Biodiversity

Recent losses in biodiversity are alarming. Of all the extinctions known to have taken place during the last 2000 years, more than half have occurred during the past 60 years.[17] This pattern of increasing loss is demonstrated in Figure 19.5, which shows the numbers of bird and mammal species that have become extinct since A.D. 1600. It is estimated that "between half a million and 2 million species . . . could be extin-

[12] For an expanded discussion of this point, see van der Ploeg and Vlijm (1978).

[13] An initial draft of the section on biodiversity was prepared by Melissa Geeslin while she was a student at Stanford University. Ms. Geeslin also prepared initial drafts of the following subsections: keystone and indicator species, instream flow methods, and gap analysis.

[14] Except as otherwise noted, material in this paragraph and the next are based on Decker et al. (1991, pp. 14–17).

[15] Genes are sequences of deoxyribonucleic acid (DNA) that determine the traits of every individual.

[16] This fact is from Colinvaux (1986, p. 663). Note that the area of Ohio is 41,330 square miles compared with Ecuador's 103,930 square miles. Also, the term *vascular plant* refers to plants that have tubes or vessels similar to blood vessels for transporting nutrients. Examples of nonvascular plants are mosses.

[17] This observation is from Turk (1980, p. 69).

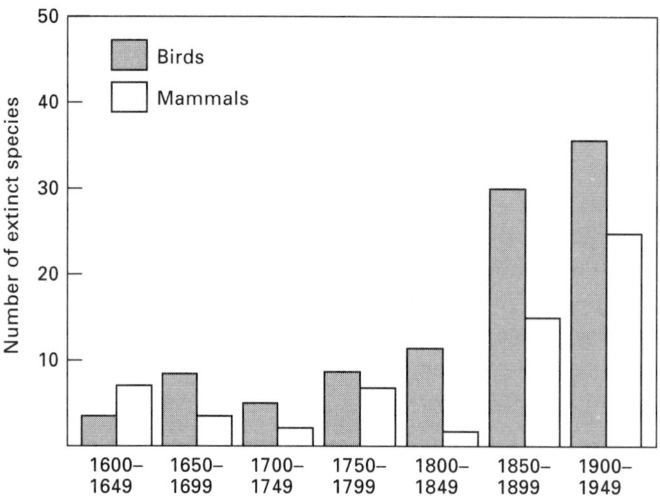

Figure 19.5 Number of species of birds and mammals known to have become extinct between A.D. 1600 and 1950. From Reid and Miller (1989, p. 34). Reprinted with permission of World Resources Institute.

guished by 2000, mainly because of loss of wild habitat, also in part because of pollution."[18]

Why is it important to maintain biodiversity? Common answers to this question reflect both anthropocentric and biocentric perspectives. Anthropocentric responses focus on the many advances in medicine, agriculture, and industry that have occurred because of high biodiversity.[19] A case involving the melon industry in the United States is illustrative. As reported by Timothy (1972), threats to the industry caused by mildew led to a worldwide search for species of melons resistant to mildew. After extensive crossbreeding, the mildew problem was considered solved. Shortly thereafter a virus attack threatened the melon industry. A worldwide search was conducted again, and a program of crossbreeding to obtain virus-resistant strains of melons was successfully implemented. The existence of many species of melon, each with a different genetic composition, made it possible to make exchanges of genetic information that maintained the industry.

In addition to being sources of commercially valuable materials, ecosystems often have economic value because they are rare. The recent growth in ecotourism provides one measure of the high value some people place on visiting ecosystems that are not ordinary. Perhaps the most evident practical reason for maintaining biodiversity concerns the value of functions (such as photosynthesis) that ecosystems perform in supporting life.

Arguments in support of conserving biological diversity can also be made from a biocentric perspective: Humans, as members of a biological community, have responsibilities to treat other members with respect. Many people feel that species and habitats should be preserved simply because they exist. As discussed in Chapter 2, the ethical view that rights should be granted to nonhuman animals is one that has a rich intellectual heritage.

Habitat Fragmentation

Many biologists consider habitat fragmentation as "one of the greatest threats to biodiversity worldwide."[20] Fragmentation occurs when a habitat is broken up into *patches.* For example, if a development project fragments a habitat, the project may form

[18] The quotation is from Barney (1980, p. 37).

[19] Ehrlich and Ehrlich (1981) provide extensive documentation of the practical value of preserving rare species and habitats.

[20] This quote is from Noss and Cooperrider (1994, p. 51), who cite numerous studies to support the contention that fragmentation is a major threat to biodiversity.

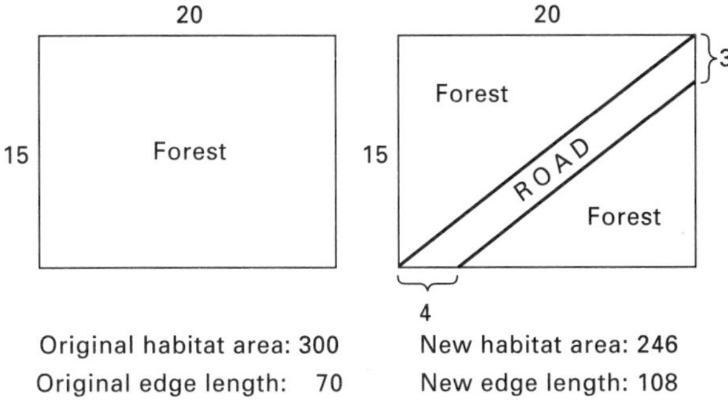

Figure 19.6 Influence of fragmentation on edge length and habitat area.

areas too small to support species that could otherwise survive there. Although creation of patches is not generally a problem for small animals, such as rodents, it can pose serious threats to large animals. Attempts have been made to develop "conservation corridors" to allow animals to migrate between two or more patches formed by a development project, but corridors are not always effective.[21]

Creation of patches and the associated reduction in habitat size are not the only negative effects of fragmentation. Other adverse effects are linked to increases in *edge area.*[22] As illustrated in Figure 19.6, fragmentation increases edge length. Adverse effects of increased edge length are illustrated by what happens when a small forest is divided up by agricultural fields. In analyzing this case, Crow (1991, p. 62) observes that species along the new forest edges are exposed to increased wind velocities and temperature extremes. In addition, plants on the new edges are more likely to be invaded by nonindigenous species from surrounding disturbed lands, and animals are more exposed to predators.[23]

[21] For an analysis of difficulties experienced in attempting to create conservation corridors, see Soulé (1991).

[22] In a review of the literature on biological consequences of ecosystem fragmentation, Saunders, Hobbs, and Margules (1991, p. 24) note that the traditional view is that edges are beneficial to wildlife. However, in the context of habitat fragmentation, Saunders and his colleagues argue that edge effects are detrimental because the edges are created by removal of vegetation, thereby placing the new edges against a completely altered surrounding environment.

[23] As an example of increased predator activity, Crow (1991, p. 62) cites the higher levels of nest predation that occur near habitat edges.

Species Diversity and Ecosystem Stability

Although most biologists are not prepared to argue that communities with high species diversity are necessarily good, species diversity has been used extensively in evaluating ecosystems.[24] The meaning of species diversity is clarified with an example. Consider three unrelated communities occupying different areas. Community A has only two species, whereas communities B and C each have four. As indicated in Table 19.2, the total number of individuals is the same for each community. The diversity of species in community A is clearly less than the diversity in either of the others.

One of the many available indexes of species diversity is defined as

$$D = \frac{N(N-1)}{n_1(n_1-1) + n_2(n_2-1) + \ldots + n_s(n_s-1)}$$

where D = Simpson's index of diversity
s = number of species
n_i = number of individuals of species i (i = 1, . . . , s), and
N = total number of individuals of all species

Simpson's measure of diversity is sensitive to both the number of species and their relative abundance. The lowest possible value, $D = 1$, occurs when there is only one species in a community. If each individual in a community belongs to a different species, D takes on an infinite value.[25]

Table 19.2 demonstrates how Simpson's index is influenced by both the number of species and their distribution. Community B has more species than community A, and this is reflected in the index values. Communities B and C have the same number of

[24] van der Ploeg and Vlijm (1978) provide numerous case examples demonstrating how diversity is used in ecosystem evaluation.

[25] Simpson's index can be interpreted by considering an experiment in which two individuals are drawn at random from a community. The value of D represents the number of randomly selected pairs that "must be drawn from the population to have at least a fifty percent chance of obtaining a pair with both individuals of the same species" (Collier et al., 1973, p. 343). An alternate form of Simpson's index represents the probability that a random selection of two individuals will be from different species. Pielou (1969, p. 223) presents this variation as well as the widely used Shannon-Wieder diversity index derived from information theory.

TABLE 19.2 Number of Individuals in Three Hypothetical Communities

	Community		
	A	**B**	**C**
Species 1	80	60	25
2	20	30	25
3	0	5	25
4	0	5	25
Number of individuals (N)	100	100	100
Simpson's diversity index (D)	1.48	2.22	4.13

Reprinted from Jørgensen, *Fundamentals of Ecological Modeling,* 1994, pp. 173–174, with kind permission from Elsevier Science—NL, Sara Burgerhartstraat, 25, 1055 KV Amsterdam, The Netherlands.

species, but the distribution of species within C is more uniform. Using Simpson's index, the diversity of community C is greater than that of B.

Although species diversity is important in itself, it has sometimes been thought of in connection with *stability,* the ability of an ecosystem to resist stress or return to equilibrium after being stressed.[26] At one time, many biologists believed ecosystem stability increased with species diversity. The explanation for this presumed correlation was that in ecosystems with high diversity, the interrelationships among species are complex and there are many alternate mechanisms for adjusting to stress. During the past two decades, however, numerous studies based on mathematical models and field observations have led biologists to abandon the stability-diversity hypothesis in its original form.[27] Notwithstanding the absence of a stability-diversity linkage, species diversity continues to receive attention in studies of how human actions affect ecosystems.[28]

Significant Species

It is impossible to delineate impacts on all species that may be affected by a proposed project. Thus, in conducting impact assessments, biologists generally employ a small number of species to characterize impacts. This raises a question: Which species should be used? The types of species described here are often singled out in assessing the biological impacts of projects and policies.

Rare Species and Habitats

"Rarity" is related to species diversity. However, rarity itself has come to occupy a singularly important position as a criterion for evaluating biological impacts. One reason to preserve rare species and habitats concerns the ability of ecosystems to respond successfully to changes in climate and other physical characteristics that inevitably occur over long periods. As previously noted, the more genetic diversity there is among and within species, the greater the likelihood that new species able to withstand the rigors of a changing environment will evolve. Another reason for preserving rare species is that scientists cannot tell which of the millions of existing plant and animal species will be of practical value in the future.

[26] Stability is often used interchangeably with *integrity.* An ecosystem is said to have integrity if after experiencing a stress, it exhibits an "organizing, self-correcting capability to recover toward an end-state that is normal and 'good' for that system" (Regier, 1993). Although stability and integrity are fundamental ecological concepts, they have been criticized for being "imprecise and vague, and therefore unable to support precise empirical laws" (Shrader-Frechette and McCoy, 1994, p. 300). Details on the numerous (and often conflicting) definitions of key ecological concepts—including stability and community—are given by Shrader-Frechette and McCoy (1993, chap. 2).

[27] For information on the studies used to repudiate the stability-diversity hypothesis, see Shrader-Frechette and McCoy (1993, chap. 2). They also describe more complex hypotheses that have replaced the original conceptions of a stability-diversity relationship.

[28] For a discussion of how the hypothesized relationships between stability and diversity are applied in planning, see Dearden (1978).

Among the less direct, but no less important, reasons for preserving rare species and habitats is the role they sometimes play in maintaining the integrity of ecosystems. Although eliminating one particular species may not have serious impacts, it is often impossible to predict what the consequences will be. The more species are eliminated, the greater the chance that the functioning of ecosystems will be impaired significantly.

Many countries have laws to protect rare species. In the United States, for example, the Endangered Species Act of 1973 established a program protecting habitats of "endangered" or "threatened" species.[29] In addition, many states have their own programs that protect individual species, preserve particular ecosystems, or both.[30]

Keystone and Indicator Species

Ecosystems sometimes change significantly when only a single species is eliminated.[31] This is illustrated in Figure 19.7, which shows differences in community composition at two groups of islands in the western Aleutian Archipelago off the coast of Alaska: one group where sea otters are present, and a second group where otters were hunted to local extinction. Sea otters feed on sea urchins, chitins, and limpets, drastically reducing their numbers and allowing seaweed to flourish in the intertidal zone.[32] On the islands where sea otters are present, nearshore fish populations that use seaweed for food and shelter are abundant, as are higher carnivores, such as harbor seals and bald eagles. Islands where the sea otters are *not* present are characterized by bare rocky intertidal zones covered by sea urchins, mussels, and octopus. Fish populations are sparse and higher carnivores are not present. In this example, the otter plays the role of a *keystone species,* a species whose removal can cause major changes in the structure of and relationships within an ecosystem.[33]

[29] Lists of such species are published periodically in the *Federal Register;* in addition, the U.S. Fish and Wildlife Service identifies specific areas known to contain "critical habitats." Much pertinent information is summarized by Golden et al. (1979).

[30] Several of these state programs are outlined by Camougis (1981, pp. 8–14).

[31] This paragraph is based on information presented by Simenstad, Estes, and Kenyon (1978), and Estes and Palmisano (1974).

[32] An *intertidal zone* is the area between the highest high tide and lowest low tide.

[33] The U.S. National Research Council (1986, p. 5) defines keystone species as "those which exert influences over other members of their ecological communities out of proportion to their abundances."

An *indicator species* is one with characteristics that make it valuable in identifying the presence of pollutants and changes in other types of environmental variables (for example, nutrient levels).[34] Some species are classified as indicators because they respond to stress relatively early. For example, mosses exhibit toxic effects from airborne pollutants before most vascular plants because mosses absorb water and nutrients directly from the air and rainwater. Other species are useful as indicators because they signal the presence of particular pollutants. Consider, for example, two common plants: bluegrass and spinach. The existence of lesions on bluegrass has been used to indicate pollution by peroxyacetylnitrate, while lesions on spinach leaves have provided evidence of ozone pollution.

Criteria Used in Ranking Areas for Preservation

The usefulness of preserving natural areas for scientific and educational purposes is widely recognized. A growing volume of literature relates the experiences of biologists from many countries in ranking natural areas in terms of their value as preserved lands. Typically, the criteria used for ranking include the size of the area involved, its availability for purchase, its utility for educational purposes, and the extent to which the area is threatened by development pressure. The *biological* characteristics used in deciding whether to preserve an area provide further insights into how biologists evaluate ecosystems.

A ranking scheme used by Wisconsin's Scientific Area Preservation Council illustrates how biological characteristics are employed in evaluating natural areas.[35] The Wisconsin approach includes biological criteria for "quality, commonness and community diversity." An area receives a numerical score for each criterion, and the sum of the scores represents a measure of its "natural area value." The sum is used with several other measures (which are *not* combined into a single index) in ranking areas for preservation purposes.

[34] Examples in this paragraph are from the U.S. National Research Council (1986, p. 83).

[35] This description of the Wisconsin ranking approach is based on Tans (1974).

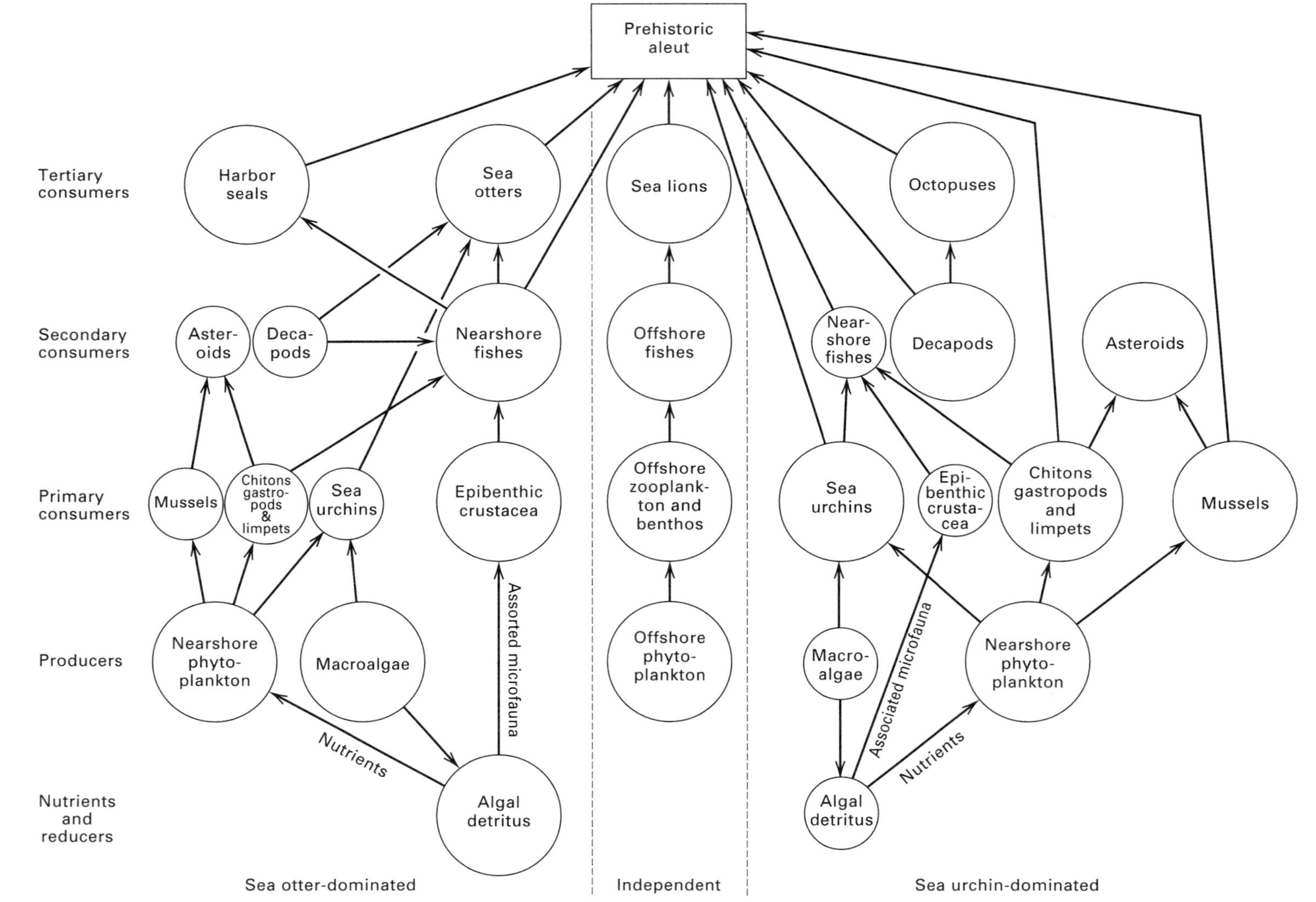

Figure 19.7 Generalized food web in the western Aleutian Islands emphasizing the effect of aboriginal Aleuts on the principal components of the nearshore community. The sizes of circles indicate relative differences in standing crop between various components of the community in the two alternate states of community organization. Arrows indicate the direction of biomass or energy flow. Reprinted with permission from Simenstad, Estes, and Kenyon, "Aleuts, Sea Otters, and Alternate Stable-Stale Communities," *Science* vol. 200, April 28, 1978, p. 404. Copyright © 1978, American Association for the Advancement of Science.

In the Wisconsin scheme, *quality* refers to the "excellence of an area's main features," and it is measured in terms of

(1) diversity of native plant or animal species present . . . ;
(2) plant community structure and integrity;
(3) the extent of significant human interference (disturbance) to the community. . . . [and]
(4) the extent to which a community corresponds with . . . [the Scientific Area Preservation Council's] concept of the identified natural community as it existed before settlement.[36]

The highest quality rating (10 on a scale from 1 to 10) is given to areas that contain no visible disturbances and satisfy the Scientific Area Preservation Council's conception of excellence of an area's features.

The second criterion, *commonness,* is rated on a scale from 1 to 6. To receive the highest score, an area must contain two or more rare or endangered species or be the only known location of a nonbotanical natural feature. An area receives the lowest score for commonness if it contains communities that occur frequently in the Wisconsin landscape.

Community diversity is the third criterion for determining natural area value. Using a 5-point scale, the highest diversity scores go to areas containing four or more different plant communities or other natural features. The lowest scores are for areas with only a single type of plant community.

There are many schemes for ranking natural areas, and they employ a variety of criteria. Rarity and species diversity are among the most widely used. Numerous indexes are available to characterize species diversity. Rarity is measured either by the existence of rare or endangered species or by the presence of a rare or unique habitat.[37]

ASSESSMENT TECHNIQUES OF FISH AND WILDLIFE MANAGEMENT AGENCIES

In many countries, fish and wildlife management agencies have responsibilities for assessing the biological effects of proposed development projects. These agencies often develop their own methods for predicting and evaluating biological changes. There are hundreds of agency procedures,[38] and the following discussion introduces three of them for illustrative purposes.

Habitat-based Evaluation Methods

Fish and wildlife conservation specialists have developed numerous habitat evaluations methods to assess impacts of proposed projects. The *Habitat Evaluation Procedure* (HEP), a method widely employed by federal agencies in the United States, is outlined here.

The U.S. Fish and Wildlife Service (1980) developed HEP to assess impacts of federal water resources projects. HEP has since been used to assess other types of projects as well.[39] The procedure assumes that the suitability of an area as habitat for a particular animal species can be determined by analyzing the area's vegetative features along with its physical and chemical characteristics. Habitat suitability is linked to the biologists' concept of *carrying capacity,* the maximum number of individuals of a species that can be sustained indefinitely by a given area.[40]

HEP evaluates how changes in habitat conditions influence an area's *potential* to support a species.[41] To apply the method, the study area is divided

[36] This quotation is from Tans (1974, p. 35).

[37] van der Ploeg and Vlijm (1978) review the several rarity and diversity measures used in 10 ecological evaluations in the Netherlands.

[38] For an introduction to the literature on impact assessment methods developed by fish and wildlife management agencies, see Hildebrand and Cannon (1993, pp. 99–220), and Morris and Therivel (1995, pp. 197–296).

[39] Canter (1996, p. 400) reports that HEP is the "most frequently used habitat-based approach in environmental impact studies in the United States. . . ." He also compares HEP with three other habitat evaluation methods developed for use in the United States.

[40] Carrying capacity has a long tradition of use in biology. An illustrative application is Odum's (1971, p. 188) analysis showing that the island of Tasmania near Australia had a carrying capacity of 1.7 million sheep during the 1800s. Recently, the carrying capacity idea has been discussed in a novel context: the computation of the "ecological footprint" of a human population. Wackernagel and Rees (1996) determine the ecological footprint of a human population "by calculating how much land and water area is required on a continuous basis to produce all the goods consumed, and to assimilate all the wastes generated, by that population." As demonstrated by Schneider, Godschalk, and Axler (1978), the carrying capacity concept is also used in city and regional planning.

[41] Providing a suitable habitat does not, in itself, guarantee that the species will develop at maximum potential density levels.

into homogeneous *cover types.* These are either terrestrial communities (such as grassland) or aquatic zones with similar chemical and physical properties. *Evaluative species* are then selected based on economic and ecological considerations. For example, a species may be chosen because it is highly valued by recreational hunters or because it plays a key role in the functioning of an ecosystem. Assessments may include several species. Consideration is given to *baseline conditions* (habitat characteristics in an area *prior* to any change in land or water use) and *future conditions* (the habitat expected following implementation of a particular development proposal). In addition to alternative development projects, HEP also considers the no-action plan.

In the remaining HEP steps, the baseline and projected habitats are described in terms of *habitat units* (HUs), calculated as a land area times a habitat suitability index (HSI). The computational procedures utilize both scientific facts and the professional judgements of biologists.

A U.S. Fish and Wildlife Service (1981) analysis demonstrates how to compute habitat suitability indexes and habitat units using the red-tailed hawk as an evaluative species. Although the agency's study included two cover types, only the grasslands portion of its analysis is considered here. Using available scientific information, the Fish and Wildlife Service concluded that three variables were important in determining the suitability of grasslands as a habitat for the red-tailed hawk:

V_1 = percentage herbaceous canopy cover
V_2 = percentage herbaceous canopy 3 to 18 in. tall
V_3 = number of trees greater than or equal to 10 in. dbh/acre[42]

For the baseline case, the agency used field data to determine values of V_1, V_2, and V_3 in the grasslands portion of the study area. Professional judgment was used to estimate these values under various future conditions.

Fish and Wildlife Service staff developed curves for computing components of habitat suitability for the red-tailed hawk. Application of these curves, which are illustrated in Figure 19.8, is demonstrated using a hypothetical example. Suppose the baseline condition in a grassland area is 25% herbaceous canopy cover. In this case, $V_1 = 25\%$, and the corresponding suitability index value is 0.39 (see Figure 19.8a). A similar process is used to calculate suitability index values for the remaining variables under baseline conditions: values of V_2 and V_3 are observed, and the curves in Figure 19.8b and 19.8c, respectively, are used to find corresponding index values.

The Fish and Wildlife Service calculates an *overall* HSI for the red-tailed hawk using mathematical formulas involving suitability indexes for V_1, V_2, and V_3, and for variables describing the second cover type. Once the overall HSI is computed, it is multiplied by the study area to yield the number of habitat units. The entire computational procedure is repeated for each scenario describing a future habitat condition.

Habitat units are used to assess alternative projects. A proposed project's impact on a particular evaluative species is adverse if the estimated HUs with the project are less than the HUs without the project. Differences in habitat units with and without a project are also used in evaluating *mitigation* features to offset negative impacts. For example, suppose a proposed Bureau of Reclamation reservoir was expected to produce a loss of 100 HUs for the red-tailed hawk. The bureau might try to compensate for this loss by managing a different land area containing a similar cover type. Its goal would be to provide 100 *new* HUs of habitat for the red-tailed hawk.[43]

Weaknesses of the Habitat Evaluation Procedure are its complexity, its narrow species orientation, and its failure to examine species diversity and ecosystem structure and function.[44] In addition, many mathematical relationships used in HEP give the impression of being more scientifically rigorous than

[42] *dbh* is defined as "diameter at breast height" (that is, 132 cm above the ground).

[43] Use of HEP to establish mitigation requirements is common, particularly in the context of the dozens of *mitigation banks* established in connection with development of wetlands in the United States. In a wetlands area where a bank has been legally established, a project proponent is able to bank "mitigation credits" (obtained, for example, by creating or restoring a wetland) to offset wetlands that may be unavoidably destroyed by a future development project. For a review of steps involved in creating and operating mitigation banks, see Canter (1996, pp. 421 and 428–31).

[44] Canter (1996, p. 415) notes also that HEP "can be both time-consuming and costly, particularly if fieldwork is required." He estimates (p. 423) that an average application takes between six months and a year and may require two to four months for data collection.

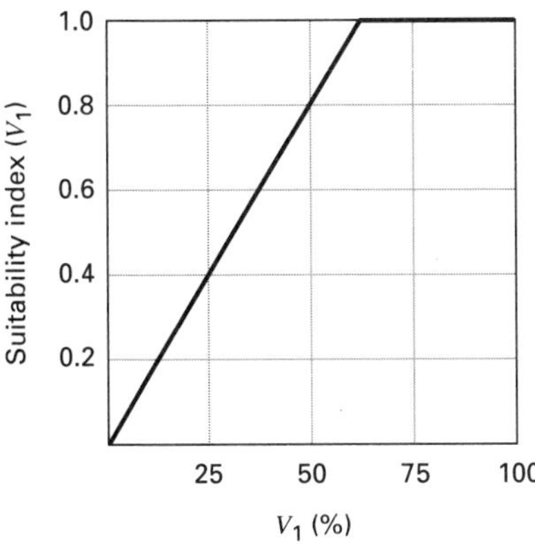

(*a*) Percentage herbaceous canopy cover.

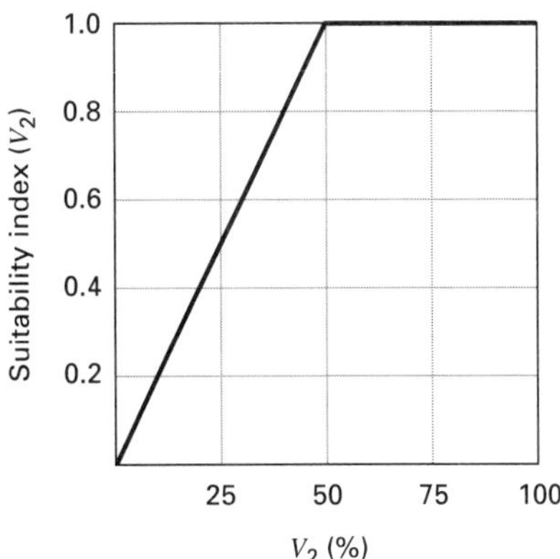

(*b*) Percentage herbaceous canopy 3 to 18 in. tall.

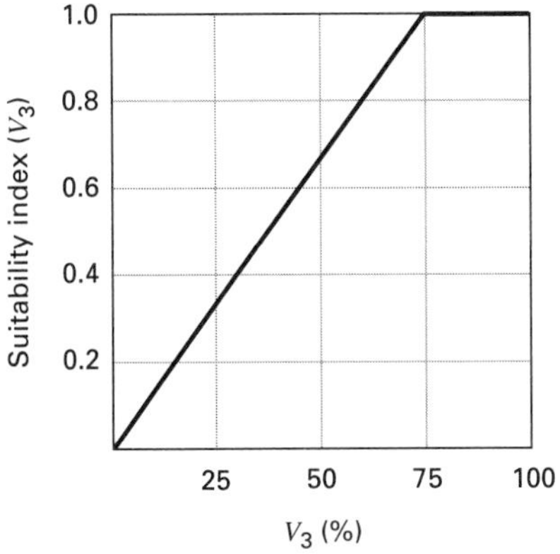

(*c*) Number of trees ≥ 10 in. dbh/acre.

Figure 19.8 Suitability index curves for red-tailed hawk example. From U.S. Fish and Wildlife Service (1981).

they are. For example, index curves such as those in Figure 19.8 suggest that habitat suitability relationships are understood precisely. In reality, suitability index curves often represent the professional judgments of fish and wildlife specialists. Despite these limitations, HEP provides a way of organizing and analyzing scientific information concerning habitat suitability.

Instream Flow Methods

Instream flow methods include procedures to predict the response of a fish habitat to changes in streamflow, stream temperature, sediment transport, or water chemistry.[45] Of the dozens of available instream flow procedures, the *Washington Method* is used here for illustrative purposes.[46] This method, which was developed by the Washington Department of Fisheries in cooperation with the U.S. Geological Survey, relies on maps showing contours of equal stream depth and velocity to determine portions of a stream preferred by a given species for spawning.

[45] As Stalnaker (1993, p. 141) explains, instream flow methods are also used to determine a minimum streamflow needed to support a fish habitat at an acceptable level.

[46] Hildebrand and Cannon (1993, pp. 99–220) describe many of these methods.

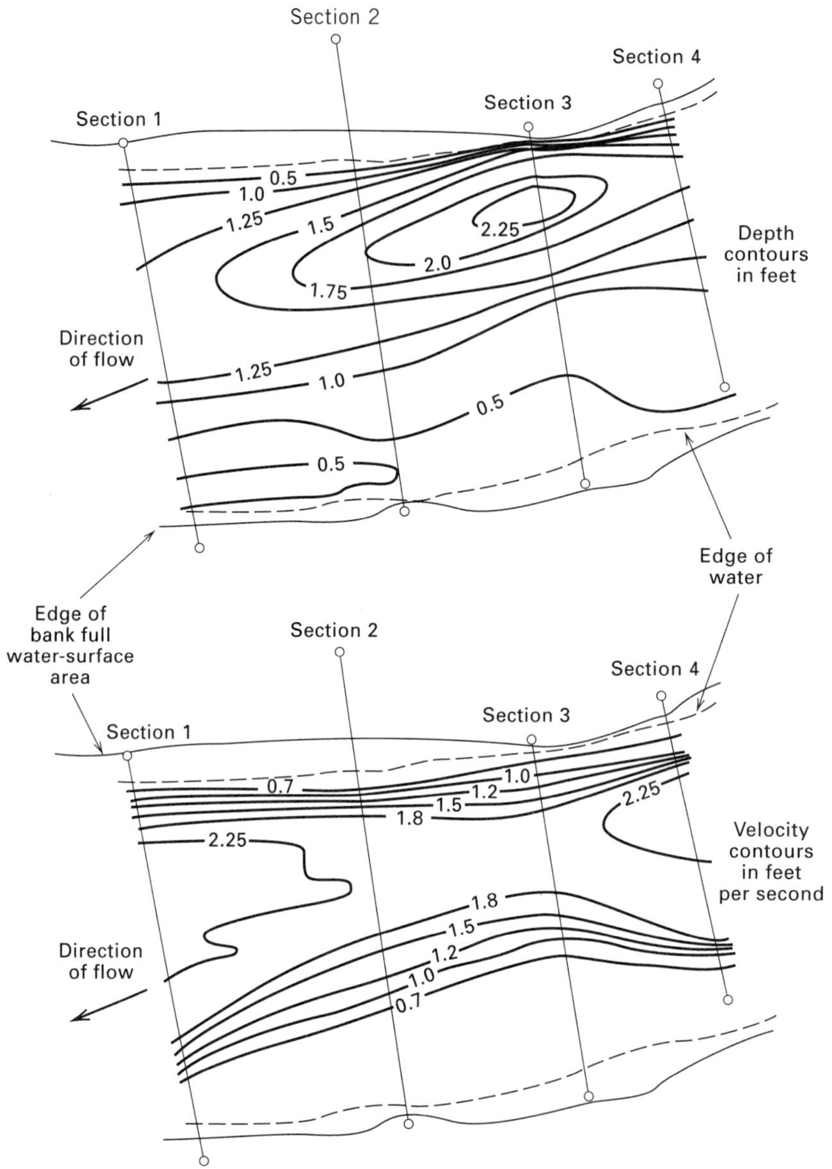

(*a*) Water depth and water velocity contours at a flowrate of 94.6 cubic feet per second.

Figure 19.9 Example of method for determining area of study reach that is preferred for spawning by fall chinook salmon at a fixed rate of river flow, North Nemah River. From Collings, "A Method for Determining Instream Flow Requirements for Fish," *Instream Flow Methodology Workshop Proceedings,* Washington State Department of Ecology, November 14, 1972.

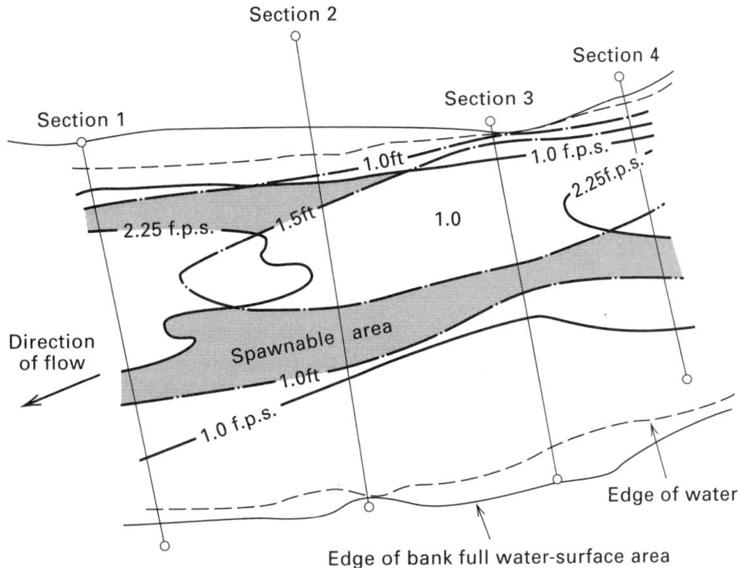

Results:
Preferred depth area; 1045 sq. ft., between 1.0–1.5 ft.
Preferred velocity area; 1706 sq. ft., between 1.0–2.25 f.p.s
Area preferred for spawning; 726 sq. ft.

(*b*) Preferred spawning area at a flowrate of 94.6 cubic feet per second.

Figure 19.9 (*Continued*)

Figure 19.9a shows examples of velocity and depth maps for a section of the North Nemah River in the state of Washington. In this case, the river had a volume flowrate of 94.6 cubic feet per second. An overlay of the two contour maps reveals portions of the stream section that satisfy criteria on both depth and velocity necessary for spawning (and overlie gravels of appropriate particle size). For example, fall chinook salmon in Washington prefer to spawn in areas that have velocities between 1 foot per second (fps) and 2.5 fps and water depths between 1 ft and 1.5 ft. Figure 19.9b shows a map delineating preferred spawning areas based on these criteria.[47]

The spawnable area in Figure 19.9b is for a single rate of flow. The analysis is repeated for several flowrates to provide points on a curve of spawnable area versus flowrate. As shown in Figure 19.10, the peak of the curve indicates the maximum spawnable area. The curve provides a basis for delineating how a proposed project that changes river flows (such as an upstream reservoir) would affect spawning of fall chinook salmon on the North Nemah River.

Numerous instream flow methods exist, and each has its own resource and data requirements and final outputs. Many agencies have developed instream flow procedures appropriate for their particular geographic settings.

Gap Analysis

Gap analysis is another assessment technique used by fish and wildlife management agencies. The principal applications of gap analysis are in identifying biologically significant areas, not in conducting impact assessments. However, gap analysis is presented in this introduction to biological impact assessment techniques for two reasons: (1) it is a map overlay procedure, and map overlays are often used in impact assessments; and (2) if a proposed project affects an

[47] This example is based on Collings (1972), and Wesche and Rechard (1980, pp. 65–72). Results in Figure 19.9 take account of the character of gravel in the streambed.

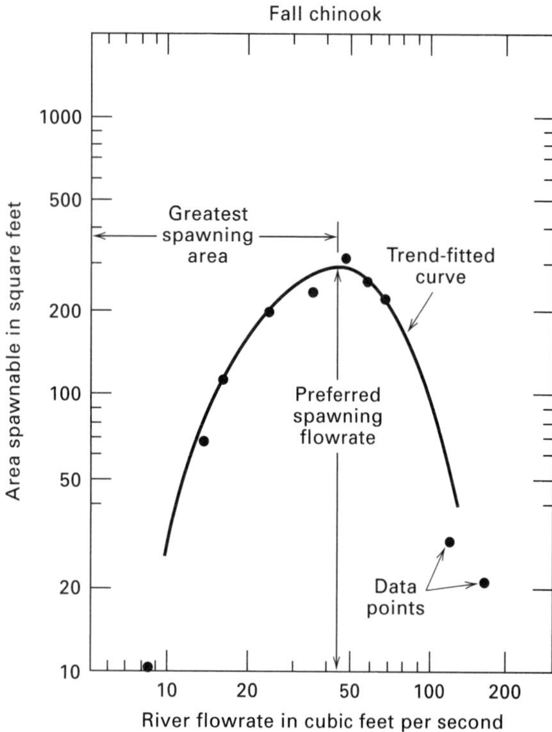

Figure 19.10 Method used to select the preferred spawning discharge, North Nemah River. From Collings, "A Method for Determining Instream Flow Requirements for Fish," *Instream Flow Methodology Workshop Proceedings,* Washington State Department of Ecology, November 14, 1972.

area identified as ecologically significant using gap analysis, the impacts are likely to be important.

Gap analysis is used to offset limitations of piecemeal efforts to identify and preserve ecologically significant areas. The approach relies on the superposition of (1) maps of land protected for conservation, and (2) maps depicting important ecosystem attributes, such as dominant vegetation types and representative species.[48] Vegetation maps are often created using satellite imagery that distinguishes among areas with different types of vegetative cover. In a gap analysis, satellite data are often presented using a grid of uniform polygons (typically hexagons).[49] Vegetative cover in each polygon is represented by a single vegetative type.

In addition to including vegetative maps, a gap analysis uses maps of distributions of animals. Species such as butterflies and large vertebrates are often selected for mapping because they represent groups of species. When applicable, maps indicating distributions of rare species are also used.[50]

In a gap analysis, information on vegetation types, species locations, and protected conservation land is entered into a *geographic information system* (GIS), a computer system that can store, analyze, and display spatial data. With the aid of a GIS, maps can be superimposed in ways that would be costly and impractical with traditional cartographic methods. Using map overlay procedures described in the appendix to this chapter, maps of vegetation, animal species, and existing conservation areas (or "preserves") are superimposed on each other. The resulting overlay map makes it possible to identify *biological hotspots,* areas presumed to have high biodiversity based on the vegetation and species distribution maps used in the analysis. The overlay map also reveals biological hotspots that have been missed by existing efforts to protect ecologically significant areas (see Figure 19.11). These areas represent "gaps" in protection (hence the term gap analysis).

In the United States, many states have conducted gap analyses to help identify ecologically significant areas that need protection.[51] Results from those analyses can be used in biological impact studies. A map of the area to be impacted by a proposed project can be superimposed on an overlay map from a gap analysis to indicate whether a proposed project is likely to affect a biological hotspot.

Gap analysis has limitations. The accuracy of species location maps depends on the quality of data

[48] This discussion of gap analysis is based on Scott et al. (1993) and Machlis et al. (1994). More generally, a gap analysis also includes maps that indicate socioeconomic factors that can suggest the vulnerability of ecologically significant areas to destruction by human activity.

[49] Vegetation types on satellite images can be identified using field surveys, aerial photography, and analyses of existing vegetation maps.

[50] Species distribution maps can be prepared using field survey data together with information from regional field guides and literature describing correlations between particular species and vegetation types.

[51] According to Machlis et al. (1994, p. 567), 32 states have gap analyses underway, and a national gap analysis program has been initiated by the National Biological Survey. Scott et al. (1993, p. 7) report that the gap analysis concept was used widely in Australia before it was applied much in the United States.

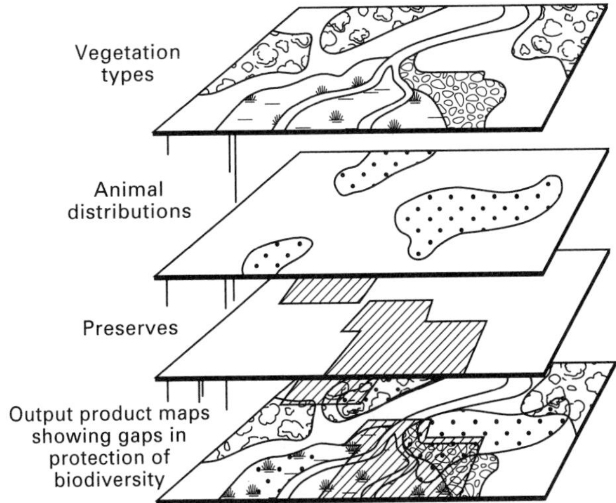

Figure 19.11 Map overlay process used in gap analysis. From Decker et al. (1991, p. 167). © Tamara Sayre.

from field surveys and other sources used in constructing them. Moreover, species location maps reveal nothing about population densities or habitat quality. The accuracy of vegetation density maps is also limited because the area of individual polygons used in preparing maps for a gap analysis is commonly larger than a few hundred acres.[52] Thus small patches of vegetation are not easily represented.[53]

ACTIVITIES UNDERTAKEN IN BIOLOGICAL IMPACT ASSESSMENTS

Biological impact assessment procedures vary greatly, depending on both the preferences of individual biologists and the context in which the work is carried out. One reason there is no standardized process is that many biological systems are not well understood. In addition, many of the available methods have been developed only during the past few decades.[54]

Although details of procedures vary, biologists often follow the same general approach to assessing impacts of proposed actions. The approach consists of seven activities that are sometimes carried out in an iterative, as opposed to sequential, fashion:

1. describe the proposed action and alternatives,
2. identify key biological concerns,
3. inventory the study area's ecosystem(s),
4. forecast biological impacts,
5. evaluate projected effects,
6. influence the decision process, and
7. conduct monitoring and mitigation activities.

Identifying Details of Proposed Actions

When conducting impact assessments, biologists sometimes have difficulties obtaining adequate project (or program) descriptions. Information important to biologists may be viewed by planners as details to be settled in the late stages of design. For example, suppose a proposed dam requires that an existing roadway be relocated. To engineers planning the dam, the key issues are the location of the dam and the storage capacity of the reservoir. The road relocation may not be finalized until late in the planning process. To a biologist, however, the impacts of the new roadway (especially if the road is routed through undeveloped land) may be as important as the effects of the dam and reservoir.

Biologists are often as concerned about ancillary facilities and construction methods as they are about major project features. For example, in constructing an electrical transmission line, the greatest biological impacts are commonly associated with access roads and not the lines themselves. In building a highway, procedures for making road cuts and conducting grading can significantly influence biological communities. The details associated with ancillary facilities and construction operations are often overlooked in proposals examined in impact assessments.

Identifying Issues of Biological Concern

Because of time and budget limits, biologists participating in an impact assessment must focus on a limited number of key issues. As explained in Chapter 17, important issues are usually identified on the basis of three sources: institutions, community interactions, and technical and scientific judgments.

Information from institutional sources is contained in laws, plans, and policy statements indicating

[52] Noss and Cooperrider (1994, p. 114) give a minimum polygon size of 247 acres.

[53] For more information on gap analysis, see Scott et al. (1993).

[54] Holling (1978) reviews widely used approaches to biological impact assessment.

the importance of specific biological resources. Examples include the lists of endangered and threatened species issued by U.S. agencies in response to the Endangered Species Act of 1973, as well as laws and regulations protecting special types of ecosystems such as wetlands.

Community interaction often includes meetings with local officials, environmental group representatives, and fish and wildlife agency personnel to identify important impact assessment issues. These meetings may also reveal the attitudes of citizens toward aspects of the local environment. For example, a particular fish species may be especially valued by local anglers.

The scientific judgments of biologists play a key role in determining issues to be examined. For instance, biologists may deem it important to assess impacts on species that are neither protected by law nor valued by the local populace. This is illustrated in a biological assessment of the influence of mosquito control chemicals on salt marsh ecosystems in New Jersey. Biologists decided to focus on two species of grasses.[55] In their judgment, adverse effects on these particular grasses would interfere significantly with the functioning of the entire salt marsh ecosystem.

Preparing Inventories to Characterize Existing Biological Systems

Biologists often rely heavily on published information to characterize the area affected by a proposed action. Time and budget constraints frequently do not permit the detailed field surveys that many biologists would like to conduct. De Santo, a biologist with considerable experience in impact assessment, provides the following "rule of caution" for inventory preparation:

> Do not give free reign to data collection. Anyone responsible for the collection and/or interpretation of data, regardless of its nature, must understand why the data is needed and what level of comprehensiveness and detail is required. In the absence of careful control over this function, great quantities of time, let alone budget, can be needlessly spent with no useful conclusion resulting from the work.[56]

[55] For more on this example, see Ward (1978).

[56] This quotation is from De Santo (1978, p. 134).

Standard manuals that list species found in different communities are typically consulted in describing baseline conditions.[57] Publications documenting past biological investigations of the study area are also useful. These include reports by local fish and game agencies, native plant societies, and school biology classes. Inventory information can also be obtained by interviewing local experts, such as bird watchers and biology teachers, who have a long-term familiarity with the area.

Many agencies routinely collect data useful for inventory preparation. For example, the National Cartographic Information Center of the U.S. Geological Survey annually publishes a series of land use and land cover maps based on photographs from orbiting satellites. In addition, the U.S. Department of Agriculture has aerial photos useful for determining vegetation, soil types, and other land features.[58]

Field trips to the site of a proposed project are another source of inventory information. For small, routine projects, a few days are typically spent in the study area observing significant habitats and natural processes. If data from literature and local experts are inadequate, a more comprehensive field investigation may be required. This might involve systematically identifying community types, endangered species, and fish and wildlife habitats, and studying structural and functional characteristics of the local ecosystem. Field surveys may also be needed to interpret data obtained using remote-sensing techniques such as infrared aerial photography.[59]

Forecasting Biological Effects

Biologists, like other environmental specialists, often rely on professional judgment in forecasting impacts of a proposed action. In these instances, projections are based on an expert's prior experience, including formal education and knowledge of the study area. An expert's judgment can also be informed by reviews of the relevant literature, including studies of impacts caused by analogous projects in similar set-

[57] For an introduction to literature useful in preparing inventories, see the references cited by Camougis (1981).

[58] Information on the many data-gathering activities of the U.S. Department of Interior is available from its Office of Library and Information Services in Washington, DC.

[59] Standard field methods used in biological impact assessment are reviewed by De Santo (1978) and Ward (1978).

tings. Sometimes forecasts are based on judgments of groups of biologists.

Ward (1978) has argued that biologists can greatly improve their forecasting abilities by going beyond intuitions and judgments and relying on systematic observations of ecosystems that have been manipulated. This position is illustrated by the alternative methods that Ward and her colleagues considered in predicting how a mosquito insecticide (temephos) would influence salt marsh ecosystems in New Jersey. The following forecasting strategies were contemplated:

- *Comparative Analysis.* Find salt marshes treated with insecticides in the past and see how they differ from untreated marshes.
- *Monitoring Approach.* Study salt marshes that the New Jersey Mosquito Control Commission had already decided to treat with insecticides, and compare them to untreated marshes.
- *Limited Experiment.* Apply experimental insecticide doses to a small section of an existing salt marsh, and contrast it with the larger untreated section of that marsh.

In this case, the limited experiment strategy was used to make predictions.

Mathematical models provide another approach to forecasting biological impacts, but their use has been limited to assessments involving major projects, programs, and policies. The limited use of models is partially explained by the substantial time and cost involved in obtaining field data needed for model calibration and validation. The complexity of ecological systems, their uniqueness, and their variability over time are further impediments to the use of modeling.[60]

Table 19.3 lists types of ecological models that have been applied to solve environmental problems. The table, which is based on a 1992 review, rates levels of modeling effort on a scale from 0 to 5.[61] A 5 represents a "very intense modeling effort," indicating that many models have been used in practice, and a 0 means there has been almost no modeling effort. Meanings of intermediate numerical values are shown in the table. Results of the survey indicate that the most intense modeling efforts have been done for agricultural lands and aquatic ecosystems. Moreover, much of the modeling on aquatic ecosystems has emphasized dissolved oxygen and contaminants such as heavy metals and pesticides. In comparison, little mathematical modeling has been conducted for some terrestrial ecosystems, such as deserts and savannahs.[62]

Environmental specialists often point to the absence of impact monitoring as an impediment to obtaining improved forecasts. Despite repeated calls by

TABLE 19.3 Biogeochemical Models of Ecosystems[a]

Ecosystem	Modeling Effort
Rivers	5
Lakes, reservoirs, ponds	5
Estuaries	5
Coastal zones	4
Open seas	3
Wetlands	4
Grasslands	4
Deserts	1
Forests	4
Agricultural lands	5
Savannas	2
Mountain lands (above timberline)	0
Arctic ecosystems	0

[a] The scale for modeling effort is: 5 = very intense; 4 = intense, 3 = "some," 2 = "a few," 1 = "one good study or a few not sufficiently well-calibrated and validated models," and 0 = "almost no modeling effort at all."

Reprinted from Jørgensen, *Fundamentals of Ecological Modeling,* 1994, pp 173–74, with kind permission from Elsevier Science-NL, Sara Burgerhartstraat, 25, 1055 KV Amsterdam, The Netherlands.

[60] Shrader-Frechette and McCoy (1994) are pessimistic about the ability to develop general models of ecosystems for making predictions. They argue that detailed case study analyses coupled with theories that apply at a local level provide a more promising direction for developing ecological impact predictions useful in solving environmental problems.

[61] For more information on the 1992 review, see Jørgensen (1994, pp. 173–74).

[62] Although mathematical models were used in ecology during the 1920s, widespread interest in ecosystem modeling began only in the 1960s. For examples of how mathematical models have been used in forecasting biological impacts, see Hall and Day (1977), and U.S. National Research Council (1986).

scientists for monitoring, follow-on studies to determine impacts of projects and programs are not often done. In a review of ecological impact assessment methods, Treweek (1995, p. 178) argues that "[t]he ability of ecologists to predict the ecological consequences of proposed actions will remain limited unless the use of operational and post-operational phase monitoring becomes more prevalent."[63]

Evaluating Biological Effects

Evaluative factors are defined in Chapter 17 as the goals, criteria, and constraints that decision makers and the public consider important in ranking alternative proposals. These factors are based on environmental laws and regulations, interactions with citizens and technical and scientific judgments of experts.

Although expert judgments are a source of evaluative factors, many biologists feel uncomfortable in making assertions about the social significance of changes in ecosystem characteristics. For example, a biologist's training may indicate that overgrazing adversely affects the long-term biological productivity of livestock pastures. However, biological science does not address whether a decrease in productivity is good or bad from a societal perspective.

Although many biologists have chosen not to argue that a change in a biological measure such as productivity is good or bad, some have taken stands on such issues. As previously noted, these positions often support the preservation of rare species, keystone species, and unique habitats, and the maintenance of high levels of biodiversity.

Influencing the Decision Process

Biological impact assessments can influence projects, programs, and policies in several ways. Results from assessments made in the early stages of planning can suggest completely new alternatives. If an assessment is conducted after much of the planning budget has been spent, the influence of the biological analyses may be limited to introducing *mitigations,* features that reduce the adverse effects of the original proposal. Consider, for example, a proposed reservoir that would ruin sport fisheries and wetlands. Project planners may try to mitigate these impacts by providing fish hatcheries and preserving wetlands similar to the ones to be destroyed. Some mitigation measures involve construction procedures, such as actions to control erosion during the building process and to revegetate impacted areas after a project is erected. Efforts to reduce adverse impacts may also be based on project operations, such as materials handling practices that reduce the likelihood of oil spills at oil terminals.

Monitoring and Mitigating Impacts

For decades, biologists have been arguing that impact assessment does not end once a project or program is implemented. For example, Holling (1978, p. 135) has suggested that traditional pre-project impact analysis activities be augmented by measuring impacts during and after construction. He and his colleagues used the term *adaptive environmental management* to emphasize that (1) impact assessment specialists should view a project or program as an experiment with uncertain outcomes, and (2) additional actions to mitigate unacceptable effects should be implemented as new impact information becomes available.

More recently, attention has been given to *ecosystem restoration,* measures to return ecosystems to states they had before being disturbed by human activity. An example is given by actions to restore stream banks destroyed by river channel modifications such as those used to reduce flood flows. Figure 19.12 shows a structure that rehabilitates eroded stream banks and creates a fish habitat. In addition to providing cover for fish, the structure narrows the stream channel, thereby increasing velocity. This stream rehabilitation structure creates downstream pools and ripples that are absent from channels subject to river engineering works.

Other examples of habitat restoration are given by Novotny and Olem (1994), and Brookes (1988). The increased interest in ecosystem restoration is

---

[63] The importance of monitoring is also highlighted in a report by a U.S. National Research Council committee of biologists who studied the application of ecological theory to environmental problem solving. The committee stressed that the success of the several case study assessments it presented depended upon "natural-history" information collected in a particular setting (U.S. National Research Council, 1986, p. 16). Elsewhere (pp. 114–15), the committee included "baseline and followup monitoring to determine the effects of the project or perturbation" as a key part of an "ecological study strategy."

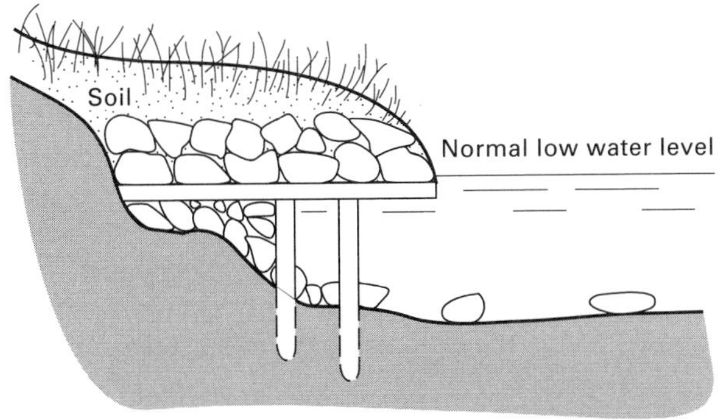

Figure 19.12 Cross-section showing deflector and bank cover device used in stream restoration. Reprinted with permission from Kumble and Shueler (eds.), Watershed Restoration Handbook, © 1991 by Metropolitan Washington Council of Governments, 777 North Capital, NE, Suite 300, Washington, DC 20002, (202) 962-3256.

reflected by the existence of specialized journals and textbooks on the subject.[64]

Biological Assessments In Practice: An Energy Facility Planning Study

The framework for impact assessment previously outlined is clarified using an energy planning example.[65] In 1973, a consortium of utilities called the Western LNG Terminal Association proposed to construct a liquefied natural gas (LNG) terminal at Point Conception along the California coast near Santa Barbara (see Figure 19.13). The utilities engaged an engineering firm, Dames and Moore, to perform biological studies and prepare an impact assessment. Dames and Moore conducted extensive field investigations, including fish and plankton sampling, and prepared a species inventory and a report analyzing the field data.

Before the LNG terminal could be constructed, the approval of the California Public Utilities Commission (PUC) had to be obtained. The PUC engaged a firm of biological consultants, Thomas Reid Associates, to provide an independent assessment. After reviewing the data assembled by Dames and Moore, the PUC's consultants concluded that the majority of significant biological impacts would involve marine organisms.

Thomas Reid Associates decided that the impacts associated with the proposed LNG project could be examined in terms of several project components, including a small boat harbor and the associated dredging to maintain the harbor, a pier and trestle for tankers, and a "kelp exclusion zone" around the LNG terminal. The latter consisted of an area where kelp harvesters would not be permitted. The consultants were also concerned about possible spills of LNG tanker fuel, and other chemicals. A seawater intake, designed to draw in 160,000 gal of water per minute, was singled out as the component likely to cause the most significant biological impacts (see Figure 19.14).

Seawater was to provide heat for vaporizing the liquefied natural gas. Officials of the California Department of Fish and Game had expressed particular concern about the possible entrainment of plankton and fish in the seawater intake. (In this context, "entrainment" refers to the process by which plankton and fish are carried into the seawater intake and transported through the intake system.) The fish that would be entrained were considered to be economically important resources.

The consultants examined both the construction and the operation of each of the project components. During the construction phase, blasting, pile driving, and kelp removal were each felt to have potentially significant effects on biological resources. Impacts related to the LNG terminal's operations centered on the seawater intake and discharge system. Other impacts associated with operations concerned the ecosystem of the kelp exclusion area. The PUC's consultants acquired detailed information on various project components, especially the seawater intake and discharge system. For example, they determined the velocity of the seawater at the ocean end of the intake pipe and the proposed method of dealing with fish inadvertently drawn into the pipe.

The next stage of the biological impact analysis was inventory preparation. Earlier studies by Dames and Moore, which included surveys of fish and plank-

[64] Among the many available resources on ecosystem restoration are Jordan, Gilpin, and Aber (1988), and the journals *Restoration Ecology* and *Restoration and Management Notes*.

[65] This discussion relies on an unpublished study by Christopher Slaboszewicz while he was a Stanford University student. The information was developed largely from Mr. Slaboszewicz's interviews of Thomas Reid and Karen Weissman of Thomas Reid Associates, Palo Alto, California.

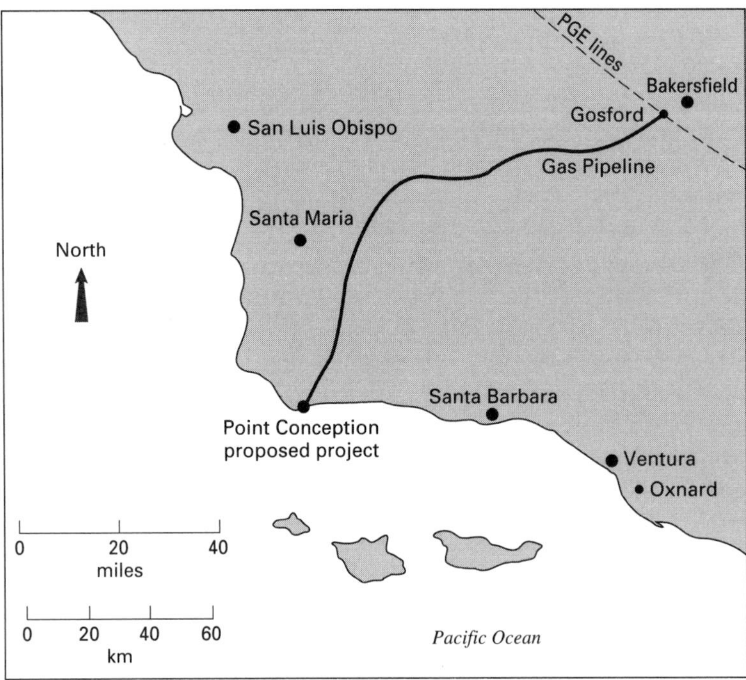

Figure 19.13 Location of proposed LNG terminal at Point Conception, California. From Thomas Reid Associates, 1978, *LNG Vaporizer Seawater System,* Technical Report No. 26 in support of the Point Conception LNG Facility DEIR, report to the California Public Utilities Commission, San Francisco, CA.

ton populations, provided much of the data needed to characterize the local ecosystem. Experts at the Scripps Institution of Oceanography and the National Marine Fisheries Service provided supplementary information regarding the plankton community. This included details on the life cycles of rare and endangered species in the area and the reproductive capacities of various planktonic species.

The impact forecasting process was guided by an analysis of biological effects of similar seawater intakes at other facilities. Much pertinent data were available because the U.S. Environmental Protection Agency had required several nearby power plants with seawater intakes to collect data on the amount and species composition of fish drawn into the intakes. Using these data, Thomas Reid Associates predicted that 40,000 lb of fish would be entrained annually. They also gave a species breakdown of the fish. The consortium of utilities had proposed the following fish return system: The fish that were sucked in through the seawater intake pipe were to be isolated in a collection tank, transported up an elevator to another pipe, and then propelled back out to sea after an 8400-ft round trip. The PUC's consultants concluded that this system would probably wind up killing many of the entrained fish.

In addition to finding the fish return system unacceptable, Thomas Reid Associates felt that the discharge of chlorine used to sterilize the seawater to prevent fouling of the vaporizers (see Figure 19.14) might have significant impacts on ocean plant and animal communities. The consultants also used a mathematical model to assess the cooling effect of the discharge water, which would be as much as 12°F colder than the ambient ocean temperature. Based on the model, and the fact that the ocean temperature fluctuated up to 9°F as a result of natural currents, no significant impact on ambient temperature was predicted. The total biomass of the plankton that would be killed by entrainment in the seawater intake

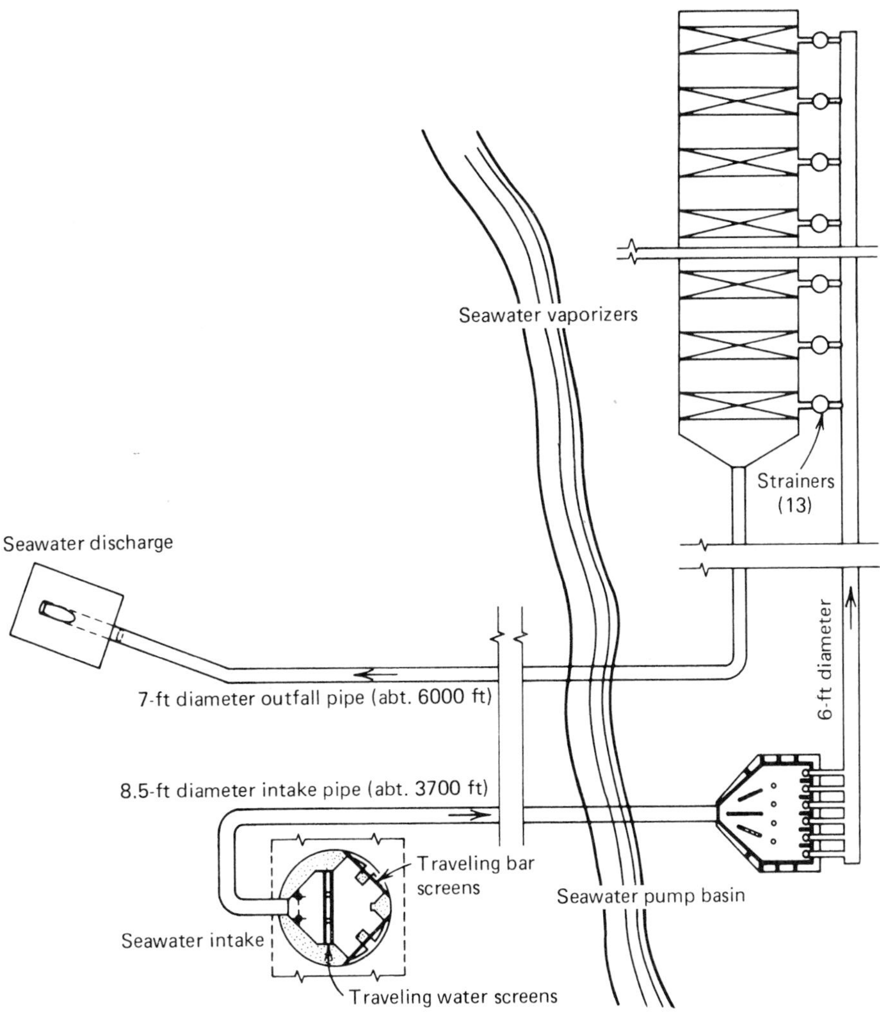

Figure 19.14 Seawater intake and discharge system for proposed LNG terminal. From Singh, D., T.S. Reid, and K.G. Weissman, 1979, *An Evaluation of Western LNG Terminal Associates' Proposed Seawater System submitted in Compliance with Condition No. 4 of the California Public Utilities Commission Decision No. 89177 for an LNG Facility at Point Conception,* report prepared for the California Public Utilities Commission LNG Task Force, San Francisco, CA.

system was also estimated. This was done using data from the California Cooperative Fish Investigation Studies performed by several universities and government agencies. The overall effect on plankton productivity was then predicted. The plankton loss was not considered significant because the amount killed would be a very small fraction of the total local plankton population and because planktonic species are able to reproduce rapidly. The PUC's consultants predicted no impact on rare and endangered species, since the rare and endangered species in the area did not have a planktonic stage in their life cycles. They also foresaw no major construction-related impacts if construction procedures were modified to use techniques other than blasting.

Another part of the impact assessment focused on measures to mitigate the negative effects of the seawater intake and discharge scheme, and the discharge of chlorine. A system of three offshore caissons, eventually modified to one, was proposed to cover the ocean end of the intake pipe. The system was designed to reduce annual fish entrainment from 40,000 to 200–300 lb. At the request of the PUC, the consultants also examined alternatives to chlorination, including the use of ozone, manual cleaning, and a newly invented antifoul rubber coating to protect the vaporizers from being damaged by impurities in the seawater. Eventually, chlorination was recommended because the alternatives were either too costly or unreliable. As a result of concerns expressed by the California Fish and Game Department regarding plankton entrainment in the seawater intake system, additional mitigation measures were examined. The PUC's consultants looked at artificial filter beds, fine-mesh screens, and other techniques to prevent plankton entrainment. They felt that the impact on saving the plankton would be only minimal, and that these techniques were neither effective nor worth the cost.

The LNG example demonstrates most of the steps mentioned in the previous discussion of approaches to biological impact assessment. In this case, much project planning had been completed before the PUC's consultants did their study. Despite this, the PUC gave its consultants a mandate to consider a wide range of alternative seawater systems. If the PUC's consultants had been brought in at the earliest stages of project planning, they would have had greater opportunities to contribute to the initial formulation of alternative plans.

Appendix to Chapter 19

Geographic Information Systems

Chapter 19 mentioned geographic information systems in the context of gap analysis, but gap analysis is just one example of how GISs are used in environmental management. The introduction to GIS in the discussion of gap analysis is extended by answering three questions: What is a GIS? How is a GIS set up? And how is a GIS used in biological impact studies?[66]

DEFINING GIS IN TERMS OF SPATIAL AND DESCRIPTIVE DATA

A standard way to define a GIS is as a "collection of computer hardware, software, geographic data, attribute data, and personnel designed to efficiently capture, store, update, manipulate, analyze, and display all forms of geographically referenced information" (Environmental Systems Research Institute, 1990, pp. 1–2). The term geographic information system can be further clarified by thinking of a GIS as a computer program that allows representations of geographic features to be connected with tabular information that describes those features. The following discussion explains how spatial and tabular data are represented in GIS, and how the two different types of data can be linked together.

Ways of Representing Spatial Data

Spatial data is information that is referenced to specific locations in two- or three-dimensional space. For example, a map contains spatial data oriented along two dimensions, a north-south axis and an east-west axis. A figure depicting concentrations of sulfur dioxide in a plume downwind of a smokestack is an example of spatial data oriented along three dimensions.

Within a GIS, the location, shape, and size of a geographic feature, such as a lake or town, can be represented using one of several *data models* (or *data structures*). A common scheme for storing spatial data is called the *vector model,* which represents geographic features as points, lines, or polygons. Each geographic feature has an "identifier" called a *record ID* or *key* that makes it unique. Consider first how a point is represented. Figure 19.15 shows four points, each of which is defined by a pair of *x,y* coordinates, such as (5,3) or (2,10). Table 19.4 associates each of these points with a record ID.

[66] Katherine Kao Cushing prepared parts of early drafts of this appendix while she was a student at Stanford University. Also, Professor David Maidment of the University of Texas provided helpful comments on an early draft.

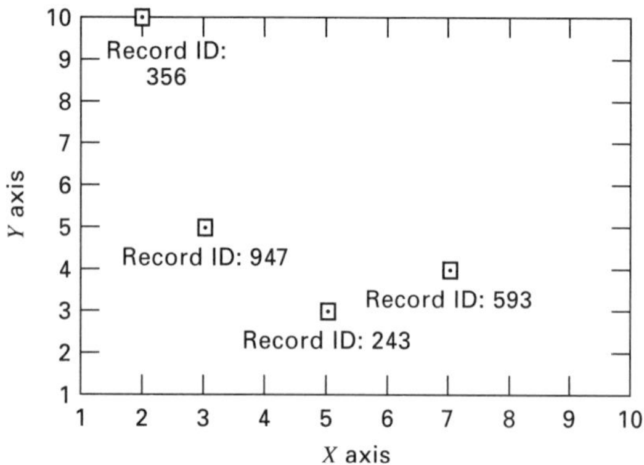

Figure 19.15 Point data in a GIS.

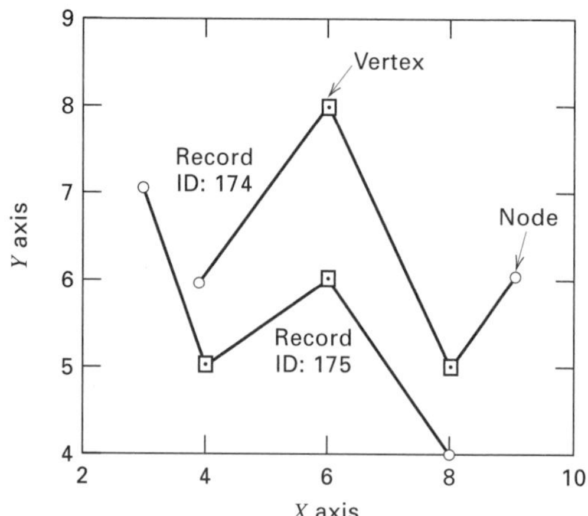

Figure 19.16 Line data in a GIS.

A line (or arc) is represented as a series of connected points. A line contains *nodes,* which are points that mark the ends of a line. Figure 19.16 shows two lines, each of which is defined by four points. Lines can be stored in a computer by linking the points that define a line to a record ID. This linkage is demonstrated in Table 19.5 using two record IDs, one for each line in Figure 19.16.

A polygon (or area) is represented by connected lines that completely surround a continuous area. Figure 19.17 contains three illustrative polygons represented by the sequence of nodes and lines shown in Table 19.6. Polygons can be used to represent a variety of objects. For example, the polygons in Figure 19.17 might depict agricultural areas within a region.

Geographic information systems that employ a vector model often include a convention for relative direction. For example, a line segment can be given direction by indicating a "from node" and a "to node." Having a way of establishing direction makes it possible to reason about relationships among spatial objects. For example, using such a convention makes it possible to identify areas to the left (or right) of a boundary line, or to route vehicles through a network of streets.

The *raster model* is another data structure commonly used to represent geographic features in a GIS. The model divides a space into rectangular cells. Each cell corresponds to an area in two-dimensional space, and the entire group of cells (which represents a particular region) is called a *grid.*

Figure 19.18 is a raster representation of soil types for a hypothetical land area that is 8 km on each side. Each of the 64 cells is one square km in area and is represented by a unique cell number from 1 to 64. In addition, each cell contains a numerical value from 1 to 3 indicating the suitability of the soil for agricultural use.

TABLE 19.4 Representing Points in a GIS

Record ID	*X* Coordinate	*Y* Coordinate
243	5	3
356	2	10
593	7	4
947	3	5

TABLE 19.5 Representing Lines in a GIS

Record ID	*X* Coordinate	*Y* Coordinate
174	4	6
	6	8
	8	5
	9	6
175	3	7
	4	5
	6	6
	8	4

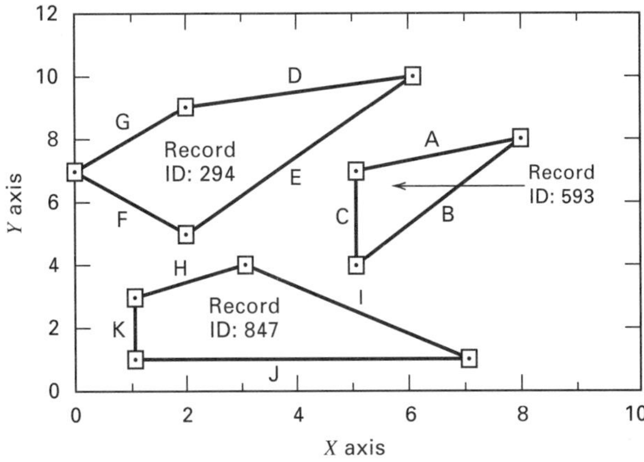

Figure 19.17 Polygon data in a GIS.

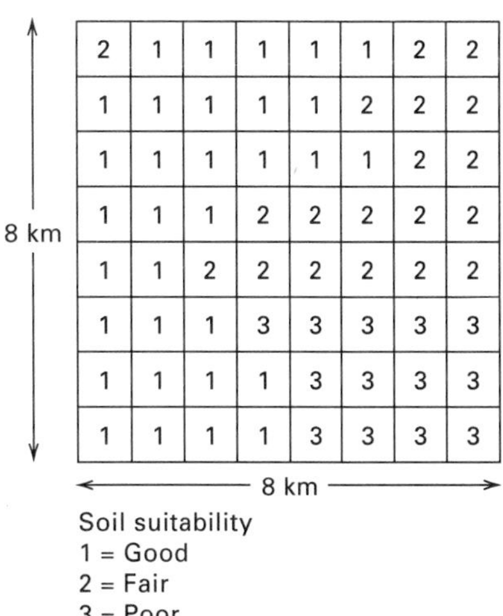

Figure 19.18 Raster representation of soil suitability for agricultural use.

A limitation of the raster system is that for any cell, an attribute has only a single value. Thus, in the example shown, each cell contains an average value of soil suitability for the entire area it represents. If a more accurate representation of the distribution of soil suitability is required, a larger number of smaller grid cells must be used.

Raster vs. Vector Data Models

Differences between raster and vector data models affect when and how they are used. One difference concerns forms of input data. Many important sources of GIS data are available primarily in raster form. An example is information from satellites. In contrast, much of the data that describes buildings, property lines, and public utilities is in vector form.

Another difference concerns the appearance of graphical outputs from the two models. (Graphical outputs include graphs, drawings, and maps.) Raster graphics are better at displaying images that have a photo-like quality. If the resolution is high enough, raster graphics can appear more realistic than vector images. Vector graphics tend to look more like line drawings.

A third difference concerns accuracy. Vector schemes, by their nature, provide precise representations, and thus they are widely used by public utility engineers and others who require high accuracy. As mentioned previously, enhancing the accuracy of data representation using a raster model requires increasing the number of grid cells, and this can be expensive. For work in forestry and land conservation, where larger areas are involved and pinpoint accuracy is not required, raster graphics are often used. Although differences between vector and raster models may influence software selection decisions, some GIS software packages give users the option of employing either raster or vector data models. Moreover, many GIS software packages are able to convert raster representations to vector and vice versa.[67]

TABLE 19.6 Representing Polygons in a GIS

Record ID	X, Y Coordinates	Lines
593	(5,7) (8,8) (5,4)	A, B, C
294	(2,9) (6,10) (2,5) (0,7)	D, E, F, G
847	(1,3) (3,4) (7,1) (1,1)	H, I, J, K

[67] Still another factor distinguishing vector and raster models is the difference in their suitability for use in mathematical modeling studies. The raster data model is easier to use for mathematical modeling because the uniform shape of the cells simplifies the process of integrating mathematical expressions involving spatial variables into a GIS. A spatial programming language called "map algebra" can be used to conduct mathematical modeling studies (Tomlin, 1990).

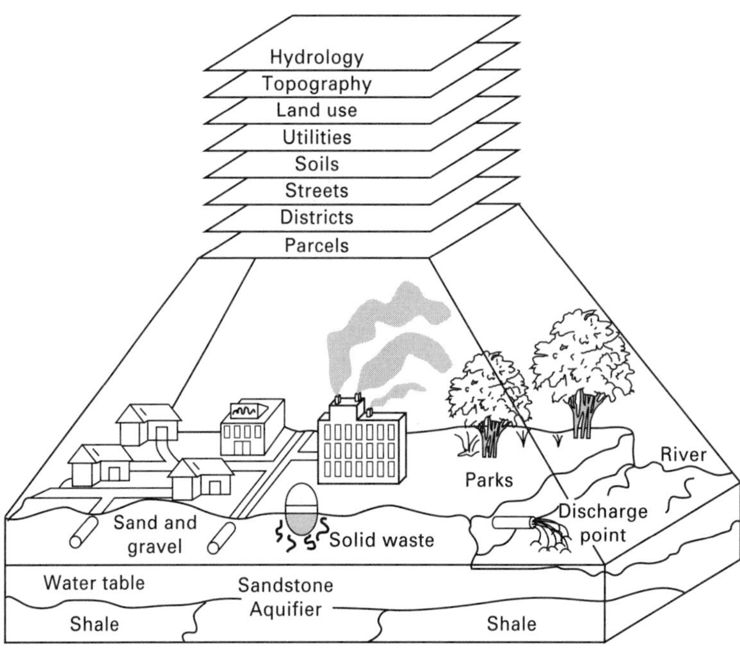

Figure 19.19 Example of layers in a GIS. GIS layers graphic provided courtesy of ESRI and is used herein with permission. Copyright © 1990 Environmental Systems Research Institute, Inc. All rights reserved.

Layers in a GIS

In a GIS, characteristics of geographic areas can be depicted as a number of related *layers* or *themes*—sets of data of a similar type. For example, a GIS operated by a water resources agency might contain separate layers for land elevation, soil type, and land use. Figure 19.19 provides additional illustration of layers in a GIS. A typical GIS database contains 200 or more layers of data, and each layer is made up of hundreds or thousands of cells or vector elements.

Figure 19.11 from the previous discussion of gap analysis illustrates how a GIS can be used to combine layers for the same land area to create a new layer. Suppose the information in Figure 19.11 were represented as separate layers in a GIS: vegetation types, animal distributions, and preserves. By using a few basic GIS analysis commands (which will be described), the three layers could be combined to yield a new layer called "gaps in protection of biodiversity." This is an application of what is sometimes called the *map overlay procedure.*

Although maps on transparent plastic could be superimposed manually, a GIS provides a more accurate representation of data and a quicker way to superimpose images.[68] The final product of analysis, such as the layer showing gaps in protection of biodiversity in Figure 19.11, can easily be sent to an output device, such as a computer screen or a plotter, or stored in a computer's memory for future use.

Linking Geographic Features and Tables of Descriptive Data

A *database* is a computer program for storing, analyzing, and retrieving information. In contrast to an ordinary database, the data in a GIS is referenced to geographic features. Consider a GIS that uses a vector model and includes a conventional database. How might the information in the database be connected to geographic features? Within a GIS using a vector model, unique record IDs are tied to each geographic feature and provide a way to link geographic features

[68] Another advantage of using a GIS to create new layers is the speed with which map overlay analyses can be redone, for example, as more accurate data becomes available.

TABLE 19.7 Example Illustrating Database Terminology

Attributes

Record ID	Crop	Annual Precipitation (in.)
593	Barley	20
294	Corn	35
847	Wheat	50

Attribute values

Record

with entries in tables of information describing those features. This linking process is illustrated with an example that also introduces the idea of a *relational database,* a type of database commonly used in GIS work.

Table 19.7 contains *records* tied via record IDs to the geographic features defined by Table 19.6 and Figure 19.17. In database systems, a record is the place where one or more attribute values are stored. In plain language, a record is a row in a table, and an *attribute* is a characteristic, such as crops or precipitation in Table 19.7. An *attribute value* consists of one unit of data; examples of attribute values in Table 19.7 are "wheat" and "50." The collection of rows is termed a *table* or *relation.* (In a relational database, the word relation refers to a table of records, not a linkage between records.)[69]

A principal characteristic of a relational database is that information in different tables can be linked together when they share a key (or record ID) that uniquely identifies each record in a particular layer. For example, the polygons in Tables 19.6 and 19.7 both contain the attribute "record ID" and have the same values for it: 593, 294, 847. Using these record IDs, the attribute data in Table 19.7 can be connected to the geographic features defined in Table 19.6. These two tables demonstrate how a GIS can combine representations of geographic features with information describing those features when a vector model is used.

[69] For details on relational databases as well as other types of databases used in geographic information systems, see Rumble and Smith (1990).

When a raster model is employed, attribute values are stored directly in grid cells. Because a separate layer is created for each attribute, much data storage capacity is needed. Storage requirements can be reduced by using relational databases with vector models. In general, GISs that use raster models are not able to store attribute information as efficiently as those using vector models.

SETTING UP A GIS

Each GIS development project, large or small, involves the same basic steps. A description of these steps and how they fit together provides an overview of how a GIS is set up.

Step 1. Define the problem. Examples of problems solved using a GIS are selecting a route for a new transmission line, and calculating a habitat suitability index. The problem-definition step includes determining the magnitude and scope of the project. Both present and potential future uses of the GIS are considered in defining a project's scope.

Step 2. Acquire data. Many sources of data for a GIS can be purchased in *computer-readable* form from government agencies and private companies. In other instances, the data needed for a GIS project may exist but may not be in a form that is immediately usable in a GIS. In this case, data must either be entered manually into a GIS or converted to computer-readable form by "scanning" or using other methods introduced in step 3. Data acquisition may also require field work.

A wide array of *descriptive data* may be included in a GIS. Data collected by government agencies are often made available in digital form. For example, the U.S. Census Bureau provides data in a system called "TIGER," which stands for Topologically Integrated Geographic Encoding and Referencing. TIGER files include demographic data as well as information from U.S. Geological Survey maps. In many countries, companies that develop GIS software sell data sets that can be used with their software.

Several forms of *spatial data* are often included in a GIS: maps, aerial and ground photographs, and data from remote sensing. *Remote sensing* is "the process of deriving information by means of systems that are not in direct contact with the objects or phenomena of interest."[70] Well-known sources of remotely sensed data are images from LANDSAT satellites, which were first launched in 1972. A relatively new form of spatial data comes from *global positioning systems* (GPS), which allow people to automatically determine their positions in terms of latitude, longitude, and elevation. A GPS relies on a device that sends and receives signals from a NAVSTAR satellite.[71]

Step 3. Preprocess and input the data. Transforming information into a form that a GIS can accept and entering it into a computer can be burdensome. This process, sometimes referred to as *data conversion,* often accounts for 80% of the total cost of developing a GIS.[72] In addition to being costly, data conversion can be labor intensive, tedious, and error-prone.

Figure 19.20 illustrates the data conversion process. The most easily managed descriptive information is data already in digital form on magnetic tapes, disks, or CD-ROMs. Other forms of descriptive data consist of printed records, such as lists of land owners' names and addresses.

As with descriptive data, spatial information is sometimes available in digital form and can be fed directly into a GIS. However, many forms of spatial data must be converted to a digital format before they can be entered.

Digitizers are tools for entering spatial information into a computer. *Digitizing* consists of moving a stylus over a map that is attached to a digitizing tablet and entering points on the map into a computer database. The stylus (also called a digitizer) contains buttons that can be pushed to signal various operations; for example, the entry of a point, or the end of an input session.[73] The tedium and error associated with using a digitizer can be avoided by using a *scanner,* a device to input information from text or images into a computer. For example, a *video scanner* functions like a television camera to create computer-readable sets of data.[74]

Step 4. Perform spatial queries and data analyses. A GIS can answer questions related to location, condition, time, spatial correlation, and modeling. Following are examples of each type of question.[75]

1. Location: What streets are within U.S. postal zip code 11218?
2. Condition: What are the addresses of all buildings in Paris whose property borders the Seine River?
3. Time: What was the total forested area in Brazil in each year between 1950 and 1995?
4. Spatial correlation: Is there a statistical correlation between the incidence of birth defects in Niagara Falls, New York, during the past 20 years and residential proximity to the hazardous waste disposal site at Love Canal?
5. Modeling: What is the rate of deforestation in Thailand likely to be during the next 20 years?

Performing analyses with a GIS is convenient because many useful operations are programmed as an integral part of GIS software. For illustrative purposes, consider some of the operations available as commands in ARC/INFO, a commonly used GIS software package. Suppose a GIS for a county in-

[70] This definition of remote sensing is from Star and Estes (1990, p. 191).

[71] Current GPS positional accuracies are approximately 5 to 10 m with standard equipment and can be as small as 1 cm with "survey grade" receivers (Goodchild and Kemp, 1990, pp. 7–8).

[72] The 80% estimate is from Goodchild and Kemp (1990, pp. 7–37).

[73] A digitizing tablet has a gridlike network of electrical wires embedded in it that transmits data from the digitizer to the GIS.

[74] Another common data conversion operation involves transforming spatial data from vector form to raster form and vice versa. This is necessary, for example, if the GIS uses a vector data model and the available input is in raster form.

[75] These question types are adapted from Environmental Research Systems Institute (1990).

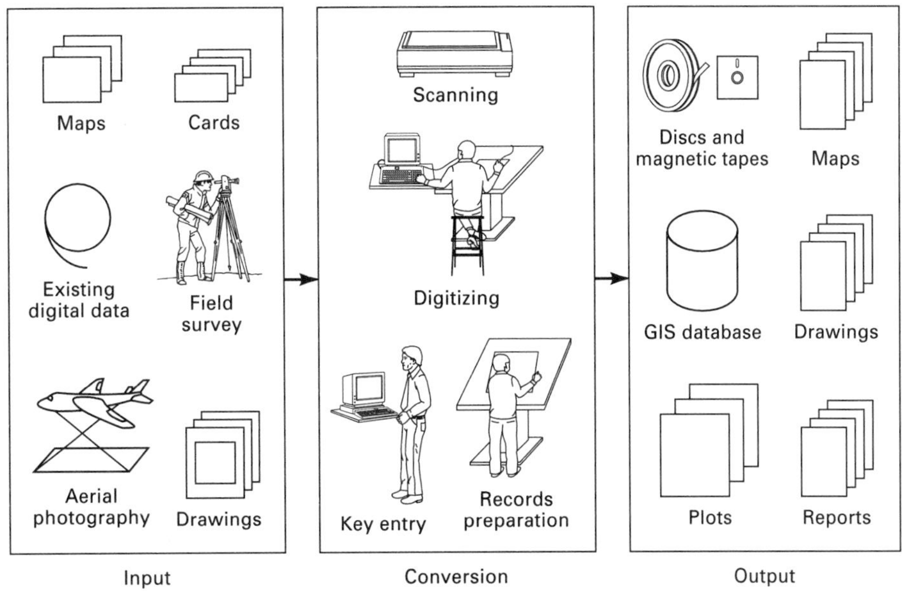

Figure 19.20 Major components of the data conversion process. Adapted from Montgomery and Schueh, (1993, p. 1). Reprinted by permission from GIS World Inc., *GIS Data Conversion Handbook,* © 1993.

cluded census data, information on land use for all parcels, and the locations of several alternative sites for a proposed incinerator. If an analyst wanted to identify all residents living within 2000 feet of alternative incinerator sites, the GIS commands called SELECT and BUFFER could be used. The analyst could employ the SELECT command to identify all the proposed sites in the county. If the analyst then selected the BUFFER command, the GIS would pose a question, What is the buffer distance? In this case, the analyst would enter 2000 ft. The buffer zone created would be a circle with a radius of 2000 ft around each potential site for the incinerator. The analyst could then employ the SELECT command again, this time to find the location of residences within each buffer zone.

A GIS can contain hundreds of different commands. Table 19.8 shows a few of the many commands that are part of the ARC/INFO system. Commands like these make it possible to answer the five previously noted types of GIS questions.

Step 5. Present the results of analysis. Although GIS outputs are often presented as maps and drawings, results from a spatial data analysis can also be provided in other forms (see Figure 19.20). A GIS generally includes programs for mapping and statistical calculations, so outputs can be generated easily. Results can be sent directly to a computer monitor, printer, or plotter to produce a paper copy. If the data are not to be displayed, but stored, then items such as CD-ROMs, magnetic tapes, or disks can be used as output devices.

USING GIS IN BIOLOGICAL STUDIES

Since their creation in the 1970s, geographic information systems have been used extensively in many contexts.[76] Applications of GIS will continue to increase as systems become more affordable, reliable, and widely known. Uses for GIS are as varied as the types of information that they may contain. This section

[76] For a summary of the vast range of applications of GIS in the public and private sectors, see Antenucci et al. (1991). A sample of GIS applications in ecological studies is given by Haines-Young, Green, and Cousins (1993). And a review of GIS in environmental impact assessment is provided by Eedy (1995).

TABLE 19.8 Commonly Used GIS Commands[a]

Command	Function
BUFFER	Delineates the zones within a given distance of a specified set of spatial features.
DISPLAY	Shows analysis results on a computer terminal screen.
CLIP	Cuts out a portion of a particular layer.
UNION	Overlays one layer over another to create a new layer that contains the features of both contributing layers.[b]
INTERSECT	Creates a new layer that keeps only those portions of two layers that fall within the spatial domain of both contributing layers.[c]
SELECT	Selects (from a layer) features that possess user-specified attributes.
ERASECOV	Removes part of the spatial coverage of a layer.

[a] These commands are among hundred of commands in the ARC/INFO GIS software package; for additional information, see Environmental Systems Research Institute, Inc. (1990).

[b] This is analogous to the operation called "union" in set theory; for example, the set created by "A union B" includes all items in set A and all items in set B.

[c] This is analogous to the operation called "intersection" in set theory; for example, the set created by "A intersect B" includes items in set A that are also in set B.

demonstrates how biologists specializing in environmental management have used GIS.

Typical GIS applications relevant to biological studies include, at the most basic level, the development of maps and inventories of various kinds. For example, it is common for fish and wildlife management agencies to use geographic information systems to create inventories of soil types, wetlands, vegetative cover, surface water features, and so forth.

Applications that build upon the analysis capabilities of a GIS include use of map overlay procedures to identify ecosystem components likely to be affected by a proposed project. Consider, for example, use of a GIS in an impact analysis of alternative reservoir projects. Using commands like those in Table 19.8, habitats that would be inundated at each of the reservoir sites could be identified. In addition, maps and other graphical displays could be created to communicate impact assessment results to decision makers.

Other applications that exploit the analysis capabilities of GIS include implementations of habitat-based evaluation procedures. Moreover, the ability to create map overlays makes it possible to identify lands suitable (or unsuitable) for certain activities. For instance, many GIS studies have been used to identify areas unsuitable for use as hazardous waste disposal sites because of their ecological sensitivity. GIS can also be used to monitor the cumulative impacts of projects, such as the cumulative effects of development projects in reducing available wetlands areas.

Three specific examples further illustrate the wide range of GIS applications in biological studies. A GIS developed for the Flathead National Forest in northern Montana was employed to delineate erosion effects of proposed timber harvesting practices. The GIS was also used to identify areas critical as elk calving habitats. This information allowed forest managers to assess the potential impacts on elk calving of alternative road closure plans (Hart, Wherry, and Bain, 1987).

In Alberta, Canada, a GIS was used to investigate the impacts of farm drainage practices. Many farmers in Alberta have drained nonpermanent wetland areas ("prairie potholes") to improve the efficiency of farm operations. Wetlands drainage has caused significant losses in waterfowl habitat and changes in runoff patterns in areas downslope of the farms where wetlands have been drained. This application took advantage of the ability to combine GIS and remote-sensing data with mathematical models to simulate hydrologic processes (MacMillan, Furley, and Healey, 1993).

A third illustration concerns habitat management. In Australia, a GIS was used to map 10,000 hectares of koala bear habitat near Brisbane using remote sensing and statistical modeling. Analyses based on these maps had a practical impact. A planned transport corridor between Brisbane and the Australian Gold Coast was rerouted to avoid disruption of core koala bear habitat (Bruce, 1992).

Future GIS applications will respond to improvements in technology. Examples of technology change include creating new devices for handling the large amounts of information supplied by remotely sensed data, and improving the accuracy and speed of data conversion processes. New GIS applications will also result from ongoing research that synthesizes GIS and environmental modeling. This synthesis exploits the capacity of a GIS to describe attributes of the environment, and to link data on environmental attributes with mathematical models of physical, chemical, and biological processes occurring within the environment.

Key Concepts and Terms

Basic Ecological Concepts

Biotic communities and habitats
Abiotic elements
Predator-prey relations
Energy flows
Food chains
Biological productivity
Biological magnification
Ecological succession
Climax communities

Criteria Used in Ecological Evaluations

Biodiversity
Habitat fragmentation
Species diversity index
Ecosystem stability
Ecosystem inertia and resilience
Species and habitat rarity
Endangered or threatened species
Keystone and indicator species

Assessment Techniques of Fish and Wildlife Management Agencies

Carrying capacity
Habitat evaluation procedure
Baseline conditions
Habitat suitability index
Instream flow method
Spawnable areas
Gap analysis
Map overlay procedure
Biological hotspots

Activities Undertaken in Biological Impact Assessments

Principal versus ancillary project features
Controlled ecosystem experiments
Ecosystem modeling
Biological inventory preparation
Impacts of analogous projects
Sources of evaluative criteria
Mitigation measures
Adaptive environmental management

Geographic Information Systems

Spatial vs. tabular data
Record ID
Vector vs. raster models
Layers or themes
Relational database
Relations and records
TIGER data
LANDSAT satellites
Digitizers and scanners
Global positioning systems
Spatial data analysis
Applications of GIS in biological studies

Discussion Questions

19-1 Pesticide chemicals have been found in nearly all kinds of foods. Explain why they are found in foods such as dairy products that are not subject to direct applications of pesticides.

19-2 In 1973, a University of Tennessee biologist surveyed the fish populations of the Little Tennessee River and discovered a previously unknown member of the perch family, the snail darter. The habitat of this fish was just above the site of the Tennessee Valley Authority's proposed Tellico Dam. The dam was about 80% complete by the time the courts began to consider whether construction should be halted to preserve the snail darter's habitat. What arguments do you think were raised by those trying to save the snail darter? After a series of court challenges the dam was eventually completed. Speculate as to why the proponents of the dam were able to implement their plans.

19-3 Suppose the U.S. Fish and Wildlife Service's Habitat Evaluation Procedure is used to assess the impacts of a proposed reservoir on deer habitat. Discuss three potentially significant biological impacts of the reservoir that would *not* be considered if HEP were the only assessment technique employed.

19-4 Ward (1978, p. 55) offers the following observations regarding the use of ecosystem modeling as a forecasting procedure:

> [M]odels often fail to predict the measured system responses. In such cases, the model is frequently useful in pointing out errors in the concepts used to develop the model. New or altered models can then be constructed. . . . A computerized mathematical model can be used to investigate the possible consequences of many options rapidly. Thus mathematical models have many advantages that should be considered regardless of the possibility of predictive failure (inaccuracy). In environmental impact analysis, precision of systems description and exploration of options, as well as predictive power, are very important.

Discuss the extent to which you agree with Ward's position. Does it encourage you to use ecosystem models in forecasting impacts?

19-5 In the United States, *mitigation banks* have been established as a way to allow the development of wetlands without reducing the total area of wetlands. Credits in a mitigation bank may be gained by converting areas into wetlands (creation); returning a disturbed wetland to a prior, less disturbed state (restoration); and modifying wetlands to provide values or functions that had not previously existed (enhancement). An example of the latter is the addition of dikes to increase the suitability of an existing wetland as a habitat for ducks.

Consider a proposed wetlands development project that would cause negative impacts that could not be mitigated. This project would be permitted as long as the developer had sufficient credits in a mitigation bank. What are the advantages and disadvantages of this mitigation banking approach? What problems would you foresee in using the habitat evaluation procedure as a basis for defining credits and debits in a wetlands mitigation banking system?

19-6 Imagine you are a planner for a state transportation agency. Suppose you are in charge of selecting a consultant to perform a biological impact assessment for a proposed highway, part of which traverses an undisturbed forest. List three questions you would ask of biological consultants who applied to carry out the assessment. What types of answers would you be looking for?

19-7 What is the difference between a relational data base and a vector data structure? How are the two related in a GIS?

19-8 The appendix to Chapter 19 describes five types of questions that can be answered using a GIS: location, condition, time, spatial correlation, and modeling. Provide an illustration of each type of question in the context of a biological impact assessment.

References

Antenucci, J. C., K. Brown, P. L. Croswell, and M. J. Kevany. 1991. *Geographic Information Systems: A Guide to the Technology.* New York: Van Nostrand Reinhold.

Barney, G. O. 1980. *The Global 2000 Report to the President of the U.S.* New York: Pergamon.

Brookes, A. 1988. *Channelized Rivers: Perspectives for Environmental Management.* Chichester, England: Wiley.

Bruce, D. 1992. GIS Helps Protect Koala Habitat. *GIS World* 5(7): 44–47.

Camougis, G. 1981. *Environmental Biology for Engineers.* New York: McGraw–Hill.

Canter, L. W. 1996. *Environmental Impact Assessment,* 2d ed. New York: McGraw–Hill.

Colinvaux, P. 1986. *Ecology.* New York: Wiley.

Collier, B. D., G. W. Cox, A. W. Johnson, and P. C. Miller. 1973. *Dynamic Ecology.* Englewood Cliffs, NJ: Prentice–Hall.

Collings, M. 1972. "A Methodology for Determining Instream Flow Requirements for Fish." In *Proceedings of Instream Flow Methodology Workshop,* 72–86. Olympia, WA: Washington Department of Ecology.

Crow, T. R. 1991. "Landscape Ecology: The Big Picture Approach to Resource Management." In *Challenges in the Conservation of Biological Resources: A Practitioner's Guide,* ed. D. J. Decker, M. E. Krasny, G. R. Goff, C. R. Smith, and D. W. Gross, 55–65. Boulder, CO: Westview Press.

De Santo, R. 1978. *Concepts of Applied Ecology.* New York: Springer-Verlag.

Dearden, P. 1978. The Ecological Component in Land Use Planning: A Conceptual Framework. *Biological Conservation* 14(3): 167–79.

Decker, D. J., M. E. Krasny, G. R. Goff, C. R. Smith, and D. W. Gross, eds. 1991. *Challenges in the Conservation of Biological Resources: A Practitioner's Guide.* Boulder, CO: Westview Press.

Edington, J. M., and M. A. Edington. 1978. *Ecology and Environmental Planning.* New York: A Halsted Book.

Eedy, W. 1995. The Use of GIS in Environmental Assessment. *Impact Assessment* 2(13): 199–206.

Ehrlich, P. R., and A. H. Ehrlich. 1981. *Extinction, the Causes and Consequences of the Disappearance of Species.* New York: Random House.

Ehrlich, P. R., and J. Roughgarden. 1987. *The Science of Ecology.* New York: MacMillan.

Environmental Systems Research Institute, Inc. 1990. *Understanding GIS: The ARC/INFO Method.* Redlands, CA: Environmental Systems Research Institute, Inc.

Estes, J. A., and J. F. Palmisano. 1974. Sea Otters: Their Role in Structuring Nearshore Communities. *Science* 185: 1058–60.

Golden, J., R. P. Ouellette, S. Saari, and P. N. Cheremisinoff. 1979. *Environmental Impact Data Book.* Ann Arbor, MI: Ann Arbor Science Publishers.

Goodchild, M., and K. Kemp. 1990. *Introduction to GIS: NCGIA Core Curriculum.* National Center for Geographic Information and Analysis. Santa Barbara: University of California.

Haines-Young, R., D. R. Green, and S. Cousins, ed. 1993. *Landscape Ecology and Geographic Information Systems.* London: Taylor & Francis.

Hall, C. A. S., and J. W. Day, Jr. 1977. *Ecosystem Modeling in Theory and Practice: An Introduction with Case Histories.* New York: Wiley.

Hart, J. A., D. B. Wherry, and S. Bain. 1987. "An Operational GIS for Flathead National Forest." In *Geographic Information Systems for Resource Management: A Compendium,* ed. W. J. Ripple. Falls Church, VA: American Society for Photogrammetry and Remote Sensing, and American Society for Surveying and Mapping.

Hildebrand, S. G., and J. B. Cannon. 1993. *Environmental Analysis: The NEPA Experience.* Ann Arbor, MI: Lewis Publishers.

Holling, C. S., ed. 1978. *Adaptive Environmental Assessment and Management.* Chichester, England: Wiley.

Jordan, W. R., III, M. E. Gilpin, and J. A. Aber, eds. 1988. *Restoration Ecology: A Synthetic Approach to Ecological Research.* New York: Cambridge University Press.

Jørgensen, S. E. 1994. *Fundamentals of Ecological Modeling.* Amsterdam: Elsevier.

Krebs, C. J. 1972. *Ecology: The Experimental Analysis of Distribution and Abundance.* New York: Harper & Row.

Machlis, G. E., J. M. Scott, D. J. Forester, and C. B. Cogan. 1994. The Application of Gap Analysis to Decision Making in the U.S. National Wildlife Refuge System. *Transactions of the 59th North American Wildlife and Natural Resources Conference,* 566–74. Wildlife Management Institute: Washington, D.C.

MacMillan, R. A., P. A. Furley, and R. G. Healy. 1993. "Using Hydrologic Models and Geographic Information Systems to Assist with the Management of Surface Water in Agricultural Landscapes." In *Landscape Ecology and Geographic Information Systems,* ed. R. Haines-Young, D. R. Green, and S. Cousins. London: Taylor & Francis.

McHarg, I. L. 1996. *A Quest for Life: An Autobiography.* New York: Wiley.

Mendelssohn, H. 1972. "Ecological Effects of Chemical Control of Rodents and Jackals in Israel." In *The Careless Technology, Ecology and International Development,* ed. M. T. Farvar and J. P. Milton, 527–44. Garden City, NY: Natural History Press.

Montgomery, G. E., and H. C. Schuch. 1993. *GIS Data Conversion Handbook.* Fort Collins, CO: GIS World Books.

Morris, P., and R. Therivel, eds. 1995. *Methods of Environmental Impact Assessment.* London: UCL Press.

Noss, R. F., and A. Y. Cooperrider. 1994. *Saving Nature's Legacy: Protecting and Restoring Biodiversity.* Washington, DC: Island Press.

Novotny, V., and H. Olem. 1994. *Water Quality Prevention, Identification, and Management of Diffuse Pollution.* New York: Van Nostrand Reinhold.

Odum, E. P. 1971. *Fundamentals of Ecology,* 3d ed. Philadelphia: Saunders.

Pielou, E. C. 1969. *An Introduction to Mathematical Ecology.* New York: Wiley.

Regier, H. A. 1993. "The Notion of Natural and Cultural Integrity." In *Ecological Integrity and the Management of Ecosystems,* ed. S. Woodley, J. Kay, and G. Francis, 3–18. Delray Beach, FL: St. Lucie Press.

Reid, W. V., and K. R. Miller. 1989. *Keeping Options Alive: The Scientific Basis for Conserving Biodiversity.* Washington, DC: World Resources Institute.

Rumble, J. R., Jr., and F. J. Smith. 1990. *Database Systems in Science and Engineering.* New York: Adam Hilger.

Saunders, D. A., R. J. Hobbs, and C. R. Margules. 1991. Biological Consequences of Ecosystem Fragmentation: A Review. *Conservation Biology* 5(1): 18–32.

Schneider, D. M., D. R. Godschalk, and N. Axler. 1978. *The Carrying Capacity Concept as a Planning Tool.* Report No. 338, Planning Advisory Service, American Planning Association, Chicago, IL.

Scott, J. M., B. Csuti, and F. Davis. 1991. "Gap Analysis: An Application of Geographic Information Systems for Wildlife Species." In *Challenges in the Conservation of Biological Resources: A Practitioner's Guide,* ed. D. J. Decker, M. E. Krasny, G. R. Goff, C. R. Smith, and D. W. Gross. Boulder, CO: Westview Press.

Scott, J. M., F. Davis, B. Csuti, R. Noss, B. Butterfield, C. Groves, H. Anderson, S. Caicco, F. D'Erchia, T. C. Edwards, Jr., J. Ulliman, and R. G. Wright. 1993. Gap Analysis: A Geographic Approach to Protection of Biological Diversity. *Wildlife Monographs* 123:1–41.

Shrader-Frechette, K. S., and E. K. McCoy. 1993. *Method in Ecology: Strategies for Conservation.* Cambridge, U.K.: Cambridge University Press.

______. What Ecology Can Do for Environmental Management. *Journal of Environmental Management* 41: 293–307.

Simenstad, C. A., J. A. Estes, and K. W. Kenyon. 1978. Aleuts, Sea Otters and Alternate Stable-State Communities. *Science* 200: April 28, 403–411.

Soulé, M. E. 1991. "Theory and Strategy." In *Landscape Linkages and Biodiversity,* ed. W. E. Hudson. Washington, DC: Defenders of Wildlife and Island Press.

Stalnaker, C. B. 1993. "Fish Habitat Evaluation Models in Environmental Assessments." In *Environmental Analysis: The NEPA Experience,* ed. S. G. Hildebrand and J. B. Cannon, 140–62. Ann Arbor, MI: Lewis Publishers.

Star, J., and J. Estes. 1990. *Geographic Information Systems: An Introduction.* Englewood Cliffs, NJ: Prentice–Hall.

Tans, W. 1974. Priority Ranking of Biotic Natural Areas. *The Michgan Botanist* 13: 31–39.

Timothy, D. H. 1972. "Plant Germ—Plasm Resources and Utilization." In M. T. Farvar and J. P. Milton (eds), *The Careless Technology, Ecology and International Development,* ed. M. T. Farvar and J. P. Milton, 631–56. Garden City, NY: Natural History Press.

Tomlin, C. D. 1990. *Geographic Information Systems and Cartographic Modeling.* Englewood Cliffs, NJ: Prentice-Hall.

Treweek, J. 1995. "Ecological Impact Assessment." In *Environmental and Social Impact Assessment,* ed. F. Vanclay and D. A. Bronstein, 171–91. Chichester, England: Wiley.

Turk, J. 1980. *Introduction to Environmental Studies.* Philadelphia: Saunders.

U.S. Fish and Wildlife Service. 1980. *Habitat Evaluation Procedure.* Report 102 ESM. Washington, DC: Department of the Interior.

______. 1981. *Standards for the Development of Habitat Suitability Index Models.* Report 103 ESM. Washington, DC: Department of the Interior.

U.S. National Research Council. 1986. *Ecological Knowledge and Environmental Problem-Solving.* National Academy Press, Washington, D.C.

van der Ploeg, S. W. F., and L. Vlijm. 1978. Ecological Evaluation, Nature Conservation and Land Use Planning with Particular Reference to Methods Used in the Netherlands. *Biological Conservation* 14: 197–221.

Wackernagel, M., and W. Rees. 1996. *Our Ecological Footprint: Reducing Human Impact on the Earth.* Gabriola Island, British Columbia, Canada: New Society Publishers.

Ward, D. V. 1978. *Biological Environmental Impact Studies, Theory and Methods.* New York: Academic Press.

Weisz, P. B. 1973. *The Science of Zoology.* New York: McGraw–Hill.

Wesche, T. A., and P. A. Rechard. 1980. A Summary of Instream Flow Methods for Fisheries and Related Research Needs. *Eisenhower Consortium Bulletin* 9.

Westman, W. E. 1985. *Ecology, Impact Assessment, and Environmental Planning.* New York: Wiley.

Woodward, G. M. 1967. Toxic Substances and Ecological Cycles. *Scientific American* 216(3): 24–31.

Chapter 20

Simulating and Evaluating Visual Qualities of the Environment

Landscape architects and other design specialists have developed techniques that allow decision makers to account for the visual impacts of proposed development projects. These techniques, known collectively as *visual impact assessment methods,* are the main subject of this chapter.[1]

A visual impact assessment for a proposed project requires judgments about visual attributes of both the existing site area (preproject conditions) and the same area after the project is constructed (postproject conditions). Although visual assessments are sometimes made after a project has been built, such assessments are not considered here. Instead, the chapter concerns visual impact assessments made before any proposal is implemented. These assessments help people decide which project, if any, to build. Even though the term *postproject* is used here, it refers to the situation that would result *if* a proposed project were carried out.

[1] Two design specialists provided many suggestions for improving early versions of this chapter: Stephen R. J. Sheppard and Patti J. Walters.

SIMULATION VERSUS EVALUATION

Visual impact studies are typically based on *visual simulations* consisting of "images of proposed projects or future conditions, shown in perspective views in the context of actual sites."[2] Simulations are typically made using one or more of the following: drawings; physical models; photographs; videos; and computer tools, such as software for computer-aided drafting.[3] Figure 20.1 shows a visual simulation based on a conventional technique: perspective drawings. The figure contains a side-by-side comparison of pre-

[2] This definition of visual simulation is from Sheppard (1989, p. 6). The term simulation is also used in describing mathematical representations of reality, such as "computer simulation models" of pollutant dispersion in the atmosphere. Because this chapter is concerned exclusively with visual impacts, the existence of multiple uses for the term *simulation* should not cause confusion.

[3] Architects and engineers frequently rely on plan, elevation, and section drawings to represent physical facilities. These drawings are more useful in providing design information than in realistically portraying the appearance of a facility, and they are not considered further.

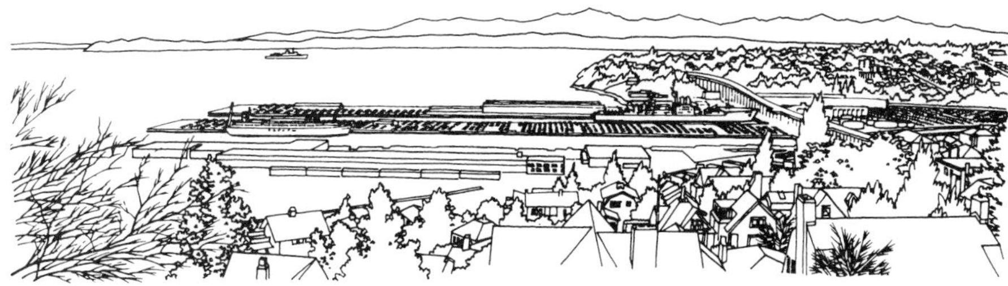

EXISTING: AUTOMOBILES / LARGE VESSEL MOORAGE / CHEMPRO

(*a*) Preproject conditions.

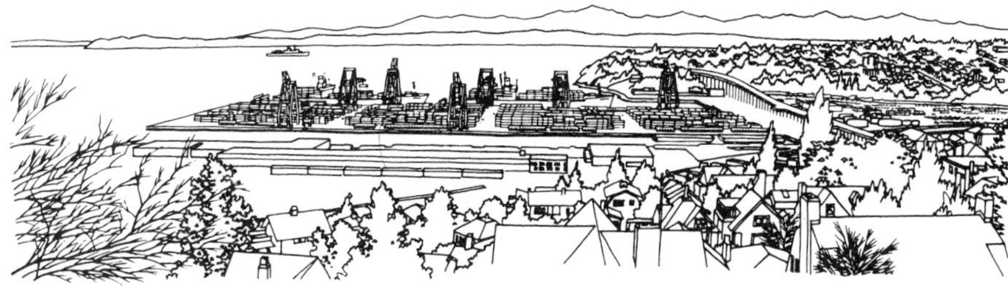

CONTAINERS / YARD CRANE

(*b*) Proposed container terminal using yard cranes.

Figure 20.1 Perspective sketches used in appraising a proposed port development in Seattle, Washington. (*a*) Reprinted with permission of Jones & Jones, Inc., Seattle, WA. (*b*) Reprinted with permission of Port of Seattle, Washington, 1981, Final Environmental Impact Statement on Alternative Uses for Terminal 91 (Pier 90/91), Seattle, WA.

and postproject conditions of a terminal at the Port of Seattle, Washington. The sketch of future conditions highlights the visual impact of yard cranes. Both drawings represent views from a location on a scenic route popular with tourists and local residents.[4] Other visual simulation techniques are described later in this chapter.

A simulation of postproject conditions is essentially a forecast, and the term *forecast* is generally defined to be free of judgments about whether the future situation is better or worse than preproject conditions. The term *evaluation* is used when value judgments are made about which circumstances are preferred. For example, imagine two perspective sketches of a flat, open field: one sketch with and one without a proposed shopping center. If the view of the undeveloped field is judged more aesthetically pleasing than the view showing the shopping center, that judgment constitutes an evaluation.

Visual impact specialists distinguish between two bases for evaluation: evaluative appraisals and preferential judgments.[5] *Evaluative appraisals* are made when the quality of a specific landscape is judged against some explicit (or implicit) standard of comparison. In contrast, *preferential judgments* reflect a personal, subjective appreciation of (or repugnance for) specific landscapes. For example, suppose a landscape architect used compatibility of the proposed project with the landscape as a criterion to judge the views of the aforementioned field with and

[4] William Blair of Jones & Jones, Inc., in Seattle helped obtain the sketches in Figure 20.1.

[5] The following definitions of evaluative appraisals and preferential judgments are from Craik (1972).

without the shopping center. That constitutes an evaluative appraisal of the project. In contrast, a local resident who feels that the landscape would be more beautiful without the shopping center is making a preferential judgment. Evaluative appraisals made by design experts and preferential judgments made by people likely to be viewing a completed project are both used in visual impact assessments.

ISSUES RAISED BY LANDSCAPE PREFERENCE RESEARCH

Psychologists and others have investigated how people perceive landscapes and cityscapes. *Landscape preference research* attempts to understand which factors influence a person's preference for one landscape (or cityscape) over another. If landscape attributes that positively or negatively influence an individual's preference could be identified, visual impact assessments could be made more objective.

Predicting Landscape Preferences

One approach to landscape preference research is illustrated by Shafer, Hamilton, and Schmidt's (1969) analysis of the "scenic perceptions" of Adirondack campers. A random sample of 250 campers participated in the study. Each was asked to rank his or her preferences for one hundred 8 × 10 in. black-and-white photographs of landscapes. The rank orderings of all 250 people were used to develop a *preference score* for each photo. Shafer and his colleagues tried to find a mathematical relationship between how much a photo was preferred, as indicated by the preference scores, and various landscape attributes. For each photo, they measured characteristics such as *perimeter of immediate vegetation,* defined as the perimeter of that part of the photo where tree barks and individual leaves could be easily distinguished; and *area of water,* defined as the area within the photo that included water. Shafer and his associates then performed a statistical analysis to discover relations between the preference score for each photo and combinations of the measured photo characteristics. The resulting statistical relationships were used to forecast preference scores for photographs of landscapes not included in the original 100 photos. The analysis by Shafer and his co-workers has been criticized for the "total lack of theoretical or even intuitive justification for the variables."[6] In other words, even though variables such as perimeter of immediate vegetation may give good predictions from a statistical perspective, little basis exists for postulating that those variables influence people's preference for one landscape over another.

Many visual assessment specialists agree that factors such as the existence of water and natural vegetation are useful predictors of landscape preference. However, there is dispute over how these variables should be measured. There is also disagreement about how preferences are affected by different combinations of factors.

Response Equivalence Between Simulations and Reality

An interesting issue examined by some researchers concerns *response equivalence,* the extent to which people's preference rankings for real landscapes are the same as their rankings for simulations of those landscapes.[7] For example, would a camper participating in Shafer's research on preferences for landscapes in the Adirondacks rank the real landscapes in the same order as the 8 × 10 in. photographs of those landscapes?

A study by Appleyard and Craik (1979) illustrates how researchers have evaluated response equivalence. Their investigation used physical modeling facilities at the Environmental Simulation Laboratory of the University of California, Berkeley.[8] Two randomly chosen groups of residents of Marin County, California, participated in the study. One group took a 9-mile tour by auto of a particular route through Marin County. A second group viewed a 16 mm color film of an equivalent tour through a three-dimensional model of the research area. The film was made by moving a remotely guided optical probe (equipped with a movie camera) through the

[6] The quoted criticism is from Kaplan (1975, p. 124). For a further discussion of the approach used by Shafer and his associates, see Brush and Shafer (1975).

[7] The term *response equivalence* is elaborated by Sheppard (1989, p. 140). In general, it applies to many different types of visual responses, not just preference rankings.

[8] The modeling facilities at the Berkeley laboratory were introduced in the discussion of Figure 16.3.

model. Both groups of participants recorded their impressions, and responses of the groups were compared. Based on this comparison, Appleyard and Craik observed that "the character of the region conveyed by the direct and simulated presentations and captured by this simulation technique is essentially identical." They also noted that, owing to the lack of human and vehicular motion, the model was characterized as being cold and barren compared to the real environment. This less positive impression was caused, in part, by limitations of the model in replicating the texture of vegetation and road and sidewalk surfaces.[9] Overall, however, individuals viewing the actual area and those viewing the color film recorded similar impressions.

According to Sheppard (1982), the high response equivalence in Appleyard and Craik's model studies of Marin County cannot be expected when less sophisticated simulation media are used. He argues that for any one simulation technique, the subject matter (or scene content), field of view, and image elements can affect responses to a simulation. *Image elements* include color, detail, and texture. Sheppard's empirical research on how image elements affect response equivalence supports his opinion that few generalizations can be made regarding this complex topic.

Preferences of Lay Persons versus Design Experts

Another issue explored by researchers is whether rankings of landscapes by professionals using aesthetic criteria would turn out the same as preference rankings of those landscapes by individuals who had no design training. This issue was investigated by Zube, Pitt, and Anderson (1975). The research involved 185 participants, each of whom viewed twenty-seven 35 mm color slides of "everyday rural landscapes" of the northeastern United States. Each person used the same procedure to describe and evaluate the 27 slides. For example, each slide was evaluated using scales to measure beauty versus ugliness, urban versus rural, and so forth. The resulting data were divided into seven groups corresponding to the occupations of the participants. The "expert group" consisted of individuals classified as "professional environmental designers." There were six groups of nonexperts corresponding to different categories of students and workers. For each group, the data were converted into scores representing groupwide summary statistics for the various descriptions and evaluations made by each participant.

Scores were analyzed to determine the extent to which groups agreed with each other. To do this, coefficients measuring the strength of correlation between the scores for each *pair* of groups were computed. (A value of 1.0 for a *correlation coefficient* indicates that the scores for two groups are perfectly correlated.) Twenty-one different pairs can be formed from seven groups. Results from this analysis are summarized by Zube, Pitt, and Anderson (1975, p. 155):

> The correlation between the seven subgroups on scenic evaluation of all 27 landscapes ranged from 0.43 (environmental design students and secretaries) to 0.91 (resource managers and teachers). The mean correlation for the 21 between-group associations was 0.77.

Based on their analyses, the researchers concluded that "overall, there was generally high agreement between the seven subgroups on both landscape evaluation and description" (Zube, Pitt, and Anderson, 1975, p. 156). These findings are for existing rural landscapes only. Whether nonexperts would agree with experts if the landscapes were modified by physical developments remains to be tested.

A subsequent, more elaborate study by Zube, Pitt, and Anderson involved 307 participants divided into 13 groups. Landscapes included in this research were from a portion of the Connecticut River basin in Massachusetts and northern Connecticut. All participants in the study were from this region. With one notable exception, congruence in the patterns of response among different groups was generally high. The exception concerned responses of a group of African-American, center-city residents from Hartford, Connecticut. Correlations between the Hartford group's responses and those of each of the other groups were consistently low. Of a total of 66 between-group correlations that were below 0.8, 58 involved the Hartford residents. These results led Zube, Pitt, and Anderson to suggest that cultural and

[9] For a discussion of this point, and other aspects of the Marin County simulation, see Appleyard and Craik (1979).

socioeconomic factors may affect landscape perception.

Other research reinforces the view that cultural and socioeconomic variables influence visual perception. Brush (1976) reviewed several investigations addressing this issue and concluded that the backgrounds and prior experiences of observers can have a substantial effect on how landscapes are evaluated. This is consistent with findings in the general literature on aesthetics and visual perception.[10]

Truth in Simulation

Landscape preference research has provided much information relevant to visual impact assessment. However, it has not thoroughly examined questions of "truth in simulation"; that is, questions of bias and misrepresentation.[11] There are several issues involved here, and they overlap partially with the previously introduced concept of response equivalence.[12]

Many decisions made in preparing a simulation can influence whether the simulation looks better or worse than the constructed project. A simulation specialist must make choices about viewing stations, simulation media (for example, drawings, photographs or models), scale, field of view, and so forth. Moreover, a simulation specialist's educational background can lead to bias inadvertently. For example, an illustrator's training may cause her to prepare aesthetically pleasing renderings, as opposed to renderings that are realistic.

Another truth-in-simulation issue concerns deliberate misrepresentations of proposed projects.[13] Examples include the addition of flashy automobiles and well-dressed people to provide a prestigious appearance, and the use of sunlight and blue skies to project visions of cleanliness and comfort.[14] Sometimes false impressions result when a simulation omits things such as traffic congestion and billboards. Perspective drawings are often criticized for being misleading, especially when building projects are involved. Appleyard (1977, p. 74) raised this objection:

> The elaborate and costly perspectives used to describe many projects are usually taken from the viewpoint that shows off the building to best advantage and minimizes its impact on the surrounding environment. Frequently, these views are not ones from which the building will be most often seen. . . . Deception, too, is relatively easy. A draftsman has only to use dramatic shadows, beautiful textures, and fine trees for most people to like any scheme he portrays.

Blatant misrepresentations are also possible by the choice of viewing location. Appleyard (1977) cites a case involving two simulations of a proposed high-rise building on the waterfront in downtown San Francisco. Project proponents showed the building from a near-overhead location several thousand feet in the air. It was hardly discernible in the cluster of downtown office buildings. Opponents of the project used an unflattering vantage point, a low "bird's-eye" view over San Francisco Bay that showed the profile of the downtown skyline. The proposed building was painted in black on an old picture of downtown San Francisco that omitted many high-rise buildings that had already been built.

Making a visual simulation appear realistic does not necessarily yield a good response equivalence. Reactions to a realistic simulation may be different from reactions to the completed project.[15] A viewer's past experience with different simulation media has an important effect on response equivalence. For example, architects can gain an accurate impression of a proposed project by viewing plan and elevation drawings that are meaningless abstractions to most people. Another example involves artist's renderings that reflect "artistic license" and show proposed projects in a favorable light. Sheppard (1982) speculates that experi-

[10] See, for example, Arnheim's (1974) work on visual perception.

[11] The phrase "truth in simulation" is from Appleyard and colleagues (1973).

[12] This discussion of truth in simulation relies on Mellander (1974), and on personal communication with Stephen R. J. Sheppard, Department of Landscape Architecture, University of California, Berkeley, October 27, 1981. Sheppard's (1982) research provides a systematic investigation of bias and misrepresentation in visual impact assessment.

[13] For accounts of the pressures on design professionals to misrepresent projects in simulations, see Bosselmann (1993). He describes steps taken by the Environmental Simulation Laboratory at the University of California, Berkeley, to alert its clients to the lab's commitment to objectivity. Bosselmann also describes defenses against those who attempt to distort or misrepresent simulations produced by the lab.

[14] These examples are from Appleyard (1977, p. 60).

[15] Sheppard (1982, p. 224) provides empirical support for this statement.

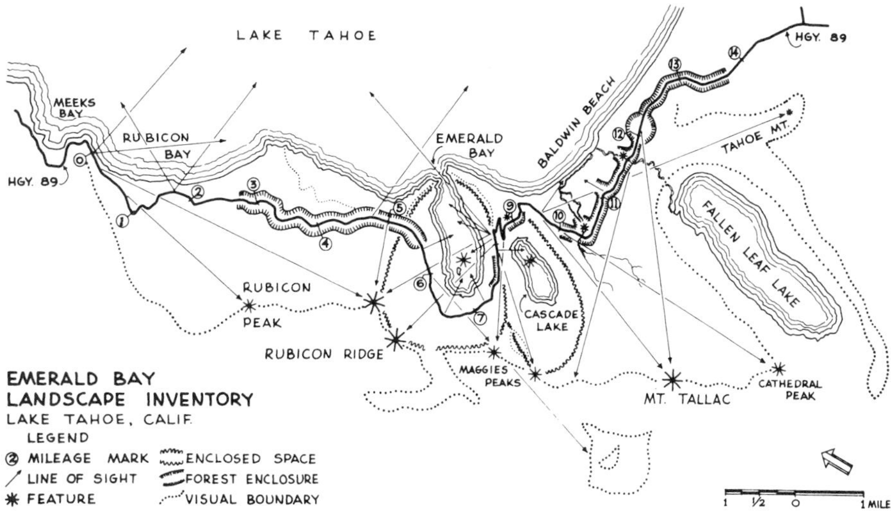

Figure 20.2 Portions of a visual inventory (route-type) of Highway 89 near Lake Tahoe. From Litton, Jr. "Natural Environments: Studies in Theoretical and Applied Analysis," in Krutilla, (ed.) *Natural Environments: Studies in Theoretical and Applied Analysis,* © 1972. Reprinted by permission of the Johns Hopkins University Press.

enced planners often view artist's renderings of projects with skepticism and may thereby avoid being misled by them. Still another complication is that a simulation that looks realistic is not necessarily unbiased. Consider an authentic-looking simulation that is inaccurate, for example, because it omits buildings adjacent to the proposed project. The simulation can be misleading precisely because it appears so credible.

PREPARATION OF VISUAL INVENTORIES

This discussion of viewer preferences, response equivalence and truth in simulation provides a foundation for examining the three main stages of a visual impact study: inventory, simulation, and evaluation. The first stage involves developing a *visual inventory,* a record of visual features likely to be affected by a proposed project. Topographic maps and aerial photos are widely used in preparing inventories. Black and white panchromatic stereo pairs of aerial photos are especially useful, because they can be viewed through stereo glasses to provide three-dimensional images. Information from topographic maps and aerial photos is often supplemented by field observations. In fact, field surveys are generally required in interpreting aerial photos. Survey data typically include sketches, notes, and ground-level photos characterizing the study area.

There are two basic inventory types: route inventories and area inventories.[16] A route inventory is appropriate when the main visual impacts involve views from a specific hiking trail, city street, highway, or waterway. An area inventory is suitable if visual effects are widespread or if views from several routes are affected by a proposal.

Figure 20.2 is part of a route inventory for a segment of Highway 89 near Lake Tahoe in the Sierra Nevadas. It includes lines of sight from selected viewing stations (or *viewpoints*) to significant visual features. The outside boundary of the inventory is defined by the limits of the landscapes visible from the route. Selection of observation points is an important part of inventory preparation. Criteria for choosing viewpoints often concern both the number of likely viewers and the quality of the scene itself. Some visual impact specialists try to include viewpoints that represent *worst-case scenarios:* locations at which large numbers

[16] These two types of visual inventories are described by Litton (1972).

TABLE 20.1 Components of a Visual Inventory[a]

- *Overall visual character.* Patterns of landform, vegetation, water, land use, and structures.
- *Scenic or visual quality.* The attractiveness or distinctiveness of the landscape, site, or built projects.
- *Viewing conditions.* The viewpoints from which the landscape, site, or project is seen; the distance and viewshed (visible area) over which it is seen; and typical lighting and visibility conditions.
- *Viewer characteristics.* Numbers of viewers, frequency and duration of viewing, and type of activity in which the viewer is taking part (for example, recreating at a particular location versus driving to or from work).
- *Viewer sensitivity.* The degree of concern that viewers have for existing visual qualities (for example, how much they care about a local mountain view or open space).
- *Visual policies.* Regulations or guidelines affecting visual resources, enacted by government agencies with responsibilities for land management and planning. Examples that can apply to a project site might include scenic highway designations, wilderness area management policies, design review guidelines, and conservation area stipulations.

[a] From Sheppard, *Visual Simulation: A User's Guide for Architects, Engineers and Planners,* 1989, p. 20, © 1989. Reprinted with permission of Van Nostrand Reinhold. References to figures and plates mentioned in the original have been omitted.

of people are likely to see very notable adverse impacts. Sketches and photos of views from important viewing locations are a part of a visual inventory.[17]

Components of a visual inventory are listed in Table 20.1. As indicated in the table, a complete visual inventory does more than describe overall visual qualities, viewing conditions, and viewer characteristics. Inventory preparation can also include the identification of *viewsheds,* areas that can be seen from a particular location. Viewsheds can be delineated in the field or by using existing topographic maps. Computer programs for determining all locations viewable from a particular spot are available.[18]

As indicated in Table 20.1, information on viewer sensitivity and visual policies also have a role in inventory preparation. Viewer sensitivity to visual resources can play a key role in identifying viewpoints. Policies and regulations protecting visual resources provide a basis for gauging whether a proposed project is compatible with applicable requirements.

[17] The choice of views to record involves more than just observer position. Other important factors include season and lighting conditions. A discussion of these and other factors that influence a view is given by Litton (1972).

[18] For a survey of procedures for plotting viewsheds using topographic maps, see Alonso, Aguilo, and Ramos (1986). They describe computer-based methods as well as traditional manual techniques.

TECHNIQUES FOR SIMULATING POSTPROJECT CONDITIONS

Postproject conditions are often represented using verbal descriptions. Sometimes the language used conveys a visual impression by evoking stereotypical images; for example, "the countryside will be filled with billboards." There are numerous opportunities for bias, as illustrated by the following sentence: "The proposed highway will obliterate the soft, rolling hills which are filled with beautiful wildflowers." Sometimes verbal descriptions employ specialized terminology that may be meaningless to persons without training in visual impact assessment. For example, consider the testimony of an expert in landscape evaluation in hearings on a proposed electrical transmission line through Riverhead, New York.[19] The expert was asked to describe the visual impact of a corridor of transmission towers along a particular route. His response included the following:

> The effect would be visually severe. The towers would overwhelm all other elements, including the tallest trees in the area. Their verticality would impinge upon the horizontal plane of the farmlands. Because the landscape is confined, a wide

[19] The expert testimony below is from Gussow (1977).

swath of such overwhelming structures would have a galvanizing impact upon the viewer.

This description can convey information to landscape design specialists, but it may have little meaning for those unfamiliar with specialized terms like "verticality" and "confined landscape."

Conventional Simulation Techniques

Perspective drawings have long been used to portray visual impacts; an example is given in Figure 20.1. Frequently, computer-aided design (CAD) software is used to create perspective drawings from different viewing angles and locations. Figure 20.3 shows computer generated perspectives depicting pre- and postproject conditions for a highway expansion proposal in Idaho. Using a computer, perspective drawings from different viewpoints can be developed rapidly. The cost of computer-generated drawings can be quite low once the computer system is in place and the terrain and project development data are in the required form.[20]

Another approach to simulation involves the manipulation of photographs. Two methods are common: *photomontage* and *photoretouching.* Effects obtainable with these procedures are demonstrated in a study of a proposed highway in Colorado. Figure 20.4a shows a preproject view from an existing road. The postproject condition is simulated in Figure 20.4b. It was produced with computer software that yielded a perspective sketch of the proposed highway from the same vantage point used in the photo of existing conditions. An artist added color and tone to the computer-generated sketch and then superimposed it on the original photo.[21]

Three-dimensional models are also used in the photomontage process. This is demonstrated by the visual analysis of pre- and postproject conditions for the Kentucky Center for the Arts in Louisville. The postproject condition was portrayed by superimposing a photo of a model of the proposed arts center on a photo of downtown Louisville. The results are shown in Figure 20.5.[22]

Sometimes postproject conditions are simulated by viewing a model directly. This is the case in Stockholm, where a three-dimensional model of the city is used to evaluate proposals for new buildings. Appropriately scaled models of proposed structures are placed in the model representing existing conditions in Stockholm. Planners and others can then assess the impacts of proposed developments.[23]

Models viewed from above provide a bird's-eye or aerial perspective, which is not representative of the way most people would see a project. Model scopes with cameras attached make it possible to take photos from an eye-level perspective within a model. Figure 20.6 shows photos made using a scope in a three-dimensional model to assess the visual impacts of a proposed bridge across a lake in Louisiana. The photographs depict pre- and postproject conditions from the same lakeside location. Compare these to Figure 20.7, photos of the real lake from the same viewing station. The postproject condition in Figure 20.7b was prepared by first photographing a scale model of the bridge. The picture of the model was then superimposed on a photo representing preproject conditions.[24]

More complex procedures are used to simulate motion using three-dimensional models. A system of structural supports is required so that movie or television cameras can be attached to the model scope, and the scope can be moved within the model to simulate walking, driving, or flying. Modeling laboratories capable of simulating motion exist in several countries (see, for example, Figure 16.3).

Digital Image Processing

During the 1980s, the increased quality and decreased cost of computer-based image editing systems made it feasible to create computerized visual

[20] Procedures for producing computer-generated sketches such as those in Figure 20.3 are given by the U.S. Department of Transportation (1978).

[21] William Blair of Jones & Jones, Inc., in Seattle helped obtain the photos in Figure 20.4. Lawrence Isaacson of the Federal Highway Administration also assisted in this effort.

[22] The photos in Figure 20.5 were provided by James J. Walters, when he was with Humana, Inc., Louisville, Kentucky.

[23] For additional information on models of Stockholm and other cities, see Appleyard (1977, pp. 77–79). Still other examples of physical models used in visual impact assessments are given by Kaplan (1993).

[24] Figures 20.6 and 20.7 were provided through the efforts of Kevin Gilson and Peter Bosselmann of the Environmental Simulation Laboratory at the University of California, Berkeley.

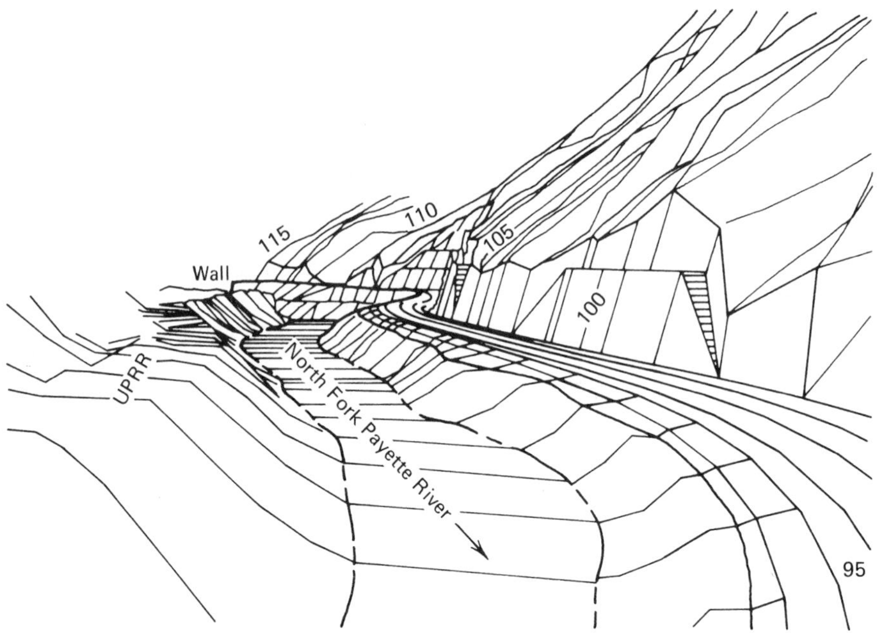

(*a*) Preproject conditions.

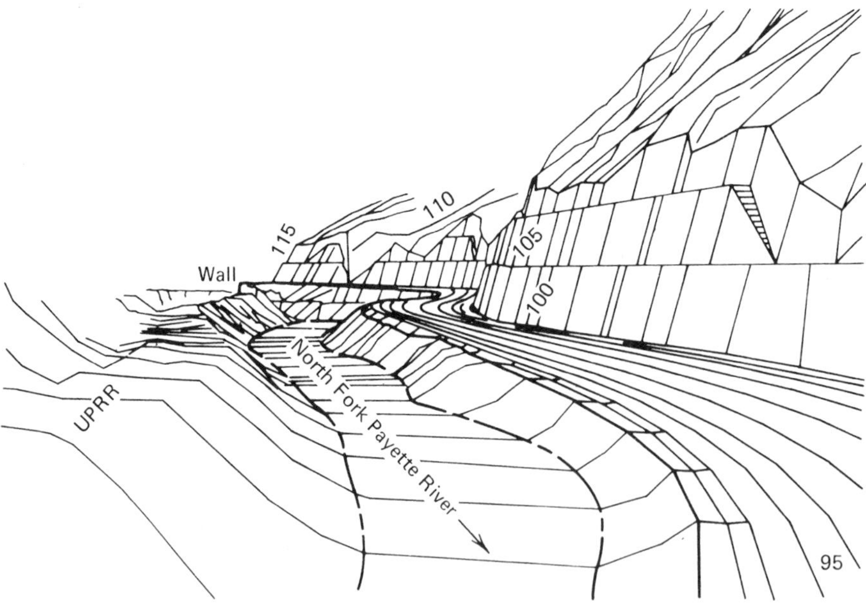

(*b*) Four-lane roadway alternative.

Figure 20.3 Computer-generated perspectives of Idaho forest highway Route 23. Source: U.S. Department of Transportation, 1978, Idaho Forest Highway Route 23, North Fork Payette River Highway, Final Environmental Impact Statement, Federal Highway Administration, Vancouver, WA.

(*a*) Preproject conditions.

(*b*) Photomontage showing postproject conditions based on a computer-generated perspective.

Figure 20.4 Photographic simulations used to assess a proposed highway in Colorado. (*a*) Courtesy of the Colorado Department of Transportation and the Federal Highway Administration. (*b*) From U.S. Department of Transportation (1981).

simulations routinely. The technique is sometimes called *digital image processing:*[25] using a computer to overlay images of proposed projects on photos (or videos) of preproject conditions, and then modifying the results with software for editing images. Digital image processing is being used increasingly in visual impact assessment because more development projects are being designed using CAD software. Another reason for the increased use of digital image processing is the widespread availability of *scanners,* devices that can convert photographic (or video) images into electric signals that a digital computer can accept.[26]

The concepts involved in creating visual simulations by digital image processing are similar to those used in preparing conventional photomon-

[25] Terminology is not standard in this field. Orland (1993) calls the process *video imaging.* Consultants who use computers to create simulations use a variety of other terms; see, for example, Tarricone (1994).

[26] For more information on scanners, see the appendix to Chapter 19.

(*a*) Preproject conditions.

Figure 20.5 Photos used to portray the Kentucky Center for the Arts in Louisville. Photos courtesy of Kentucky Center for the Arts.

tages. However, instead of manipulating a photo, the simulation specialist works with the digital representation of the photo (or video image). Because all manipulations are accomplished using a computer, changes in the appearance of both the project and the surrounding environment can be made conveniently and at relatively low cost. Sometimes decision makers are able to view changes on a computer monitor in real time. More commonly, however, visual simulations are made available for viewing in reports or as slide or video presentations.

Figure 20.8 presents the main steps in digital image processing to create visual simulations: image capture, editing, and output. *Image capture* involves entering images that describe pre- and postproject conditions into a computer. Images can include, for example, videos or 35 mm color slides of the project site from different viewing locations as well as CAD files containing the project design.[27] Simulation specialists who use digital image processing often maintain "libraries" of computer files that have images they frequently employ. Libraries might include, for

[27] Existing conditions can also be recorded using a *digital camera.* Photographic images are entered directly into a computer and no film processing is involved.

(*b*) Photomontage showing postproject conditions based on a model of the proposed arts center.

Figure 20.5 (*Continued*)

example, images of trees used to screen visually intrusive projects, and textures of different building materials. Part of the image capture phase of a simulation effort involves *digitizing:* converting photos, slides, and other images into a form that can be accepted by a computer. Digitizing is carried out using scanners and other data conversion equipment.

Once images have been entered into a computer, image editing is carried out. Using special software, perspective views of a project represented in CAD files are overlaid on views of the project site from different viewing locations. The merged images are then edited using software for coloring, shading, and so forth.

After a simulation has been set up in a computer, it can be modified easily in response to various "what if" questions. For example, the original simulation can be manipulated to show how a project's appearance would change if its size or color were modified. This ability to make changes quickly is particularly useful in planning measures to *mitigate* (or offset) unacceptable outcomes. For instance, simulations can show how well different types of vegetative screens reduce impacts. It is also possible to simulate the growth of vegetation used for screening over a multiyear period, and to evaluate the effectiveness of vegetative screens in different seasons.[28]

[28] For an introduction to numerous case-study applications of digital image processing in creating simulations, see Orland (1993) and Bosselmann (1993).

(*a*) Scale model of preproject conditions.

(*b*) Scale model including a proposed bridge.

Figure 20.6 Model simulation of a proposed bridge across a lake in Louisiana. Photos by Kevin Gilson. Courtesy of the Gilson-Berkeley Simulation Laboratory, University of California, Berkeley.

(*a*) Preproject conditions.

(*b*) Photomontage showing postproject conditions based on a model of the proposed bridge.

Figure 20.7 Photomontage used to portray a proposed bridge across a lake in Louisiana. Photos courtesy of the Gilson-Berkeley Simulation Laboratory, University of California, Berkeley.

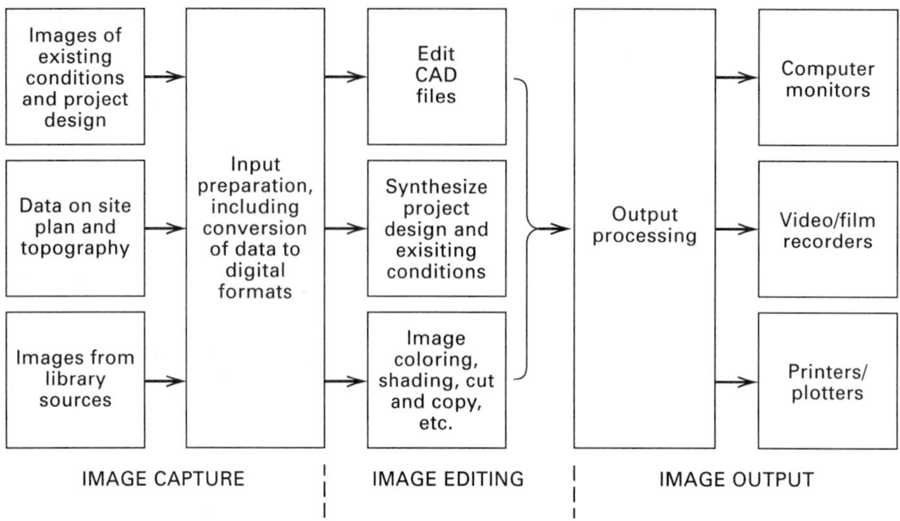

Figure 20.8 Components of a digital image processing system to create visual simulations. Source: Adapted from Orland (1993, p. 218).

The final phase of digital image processing involves creating outputs such as videos and slides. Using procedures for frame-by-frame animation, videotapes containing dynamic simulations can be created. If the expense can be justified, a videotape can include an audio component. Use of computer animation technology is illustrated by simulations conducted for an aerial, suspended light rail transit scheme proposed by the Port of Oakland, California. Tarricone (1994, p. 43) describes the video prepared to simulate the proposed transit line:

> An upside-down railcar, custom-fitted with soft rubber tires instead of metal wheels, will glide along steel trusses on its 3.5 mi route. At one point in the video, cars and passengers are bustling around outside an airport terminal, when the suspended railcars and overhead trusses suddenly appear. Even the quiet whoosh of the railcar approaching its destination is audible.

Douglas Eberhard, director of the group that produced the video, characterized the effort in these terms:[29]

> The intent [of the visual simulation] is for people to watch the tape and say "Wow, if this thing was built, this is what it would feel and sound like." It's the closest you can get to what the project is going to be like. It combines every conceivable sense other than smell.

The videotape cost $40,000 to produce. It was shown at a press conference, aired on local TV stations, and used to market the proposed project in a competition for $140 million dollars in project funding.

The same techniques used to generate the video simulations for the Port of Oakland have been used to develop *project walkthroughs,* which are simulations of what a person would see in a walk in and around a project site.[30] More advanced technologies are available to make project walkthroughs *interactive,* but they are quite expensive (at this point), and limits on computer speed and memory make it difficult to simulate smooth movement through a complex three-dimensional environment.[31] Given the extraordinary rates of change in computer technologies, the

[29] Eberhard's quotation is as it appears in Tarricone (1994, p. 44). The firm that produced the video was Parsons Brinckerhoff of Denver, Colorado.

[30] The same technology can be used to prepare *project flyovers:* simulations of views from a low-level flight over a project.

[31] An idea of what may soon be possible is given in Novitski's (1994) discussion of *virtual reality* in architecture. He uses virtual reality to mean "a collection of technologies that help participants pretend they are moving through a simulated space" (p. 122). These virtual environments include sound and "data gloves" and other paraphenalia that allow a computer to sense the participant's movement and position.

future may bring forms of audiovisual simulation that have not yet been imagined.

Although rapid advances in computer technologies have led to corresponding innovations in visual simulations, research on simulation validity and effectiveness (in terms of response equivalence) has not kept pace. Based on a review of the recent literature on landscape perception research, Zube and Simcox (1993, p. 275) conclude that

> there is little evidence that the necessary psychological research has been or is being carried out. To the contrary, the literature suggests that the development of perceptual simulation technologies is often advancing under the assumption that new technologies are de-facto meeting criteria of validity and reality, regardless of the area of application. Clearly, applications must be based on solid evidence that the simulations are valid representations of real landscapes and of the proposed landscape changes.

EVALUATING VISUAL EFFECTS AND LANDSCAPE QUALITY

As previously mentioned, visual impacts can be evaluated using both preferential judgments of potential project users and aesthetic criteria applied by design specialists. The context in which an evaluation is conducted often dictates the approach that is used. When well-defined, controversial projects are being appraised, the preferential judgments of users frequently play a dominant role. In contrast, when large landscape areas are assessed and individuals affected by a project are not easily identified, evaluations made by design specialists are often emphasized. The following discussion demonstrates some approaches applied in evaluating visual effects.[32]

Evaluative Appraisals by Design Specialists

Visual resource analysts working in private practice, universities, and land management agencies have developed many procedures for evaluating visual resources and visual impacts. These methods typically rely on professional judgments of specialists trained in the application of aesthetic criteria.

[32] Julie Lane assisted in the preparation of an early draft of this section while she was a student at Stanford University.

Use of Aesthetic Criteria

An approach developed by Litton and his colleagues at the University of California, Berkeley, has influenced several resource management agencies in the United States, and it illustrates how some design specialists carry out visual impact evaluation. The discussion below is from work by Litton and associates (1974) on the visual effects of water resources development projects.

Litton's water project evaluation technique uses a classification system composed of landscape units, setting units, and waterscape units. A *landscape unit* is regional in scale, covering areas characterized by a dominant topographic pattern or vegetative cover. An example is California's Central Valley, a relatively flat area extending for hundreds of miles down the length of California west of the Sierra Nevadas. A *setting unit* exists within a landscape unit and is a visual corridor enclosed by a group of landforms. A lake surrounded by hills on all sides provides an example; the setting unit includes both the lake and the hills. A *waterscape unit* is a topographic entity visually dominated by water, such as a lake and its immediate shore as seen by a person near the lake's edge.

In evaluating a proposed water project, two visual inventories are prepared: one for existing conditions and one for postproject conditions. Inventories for each of the three different units are based on "typologies." For example, the inventory typology for waterscape units includes the "spatial expression" of the shore, "vertical edge definition" of the shore, and so forth. The terms used in the inventory typologies are defined by Litton and colleagues (1974).

Pre- and postproject conditions are appraised by applying aesthetic criteria for three attributes—"unity, variety, and vividness"—to each element of the inventory typologies. Use of the criteria is consistent with the opinion that "any composition . . . having aesthetic merit must represent some combination of" the three attributes.[33] Litton and his associates define these criteria and provide guidelines for distinguishing between high- and low-quality unity, variety, and vividness.

Unity exists when parts of a landscape are joined into a coherent and harmonious visual entity.

[33] This quotation is from Litton et al. (1974, p. 104).

For example, a large mountain can give unity to the streams and trees that lie upon it. *Variety* is the presence of richness and diversity of objects and relationships in a landscape. However, high-quality variety is not characterized by a large number of objects per se; there "needs to be order and control over numerous and diverse parts." *Vividness* in a landscape is "that quality which gives distinction or creates a strong visual impression."[34] Such distinction is created primarily through contrast and is illustrated by an effervescent waterfall plunging into a still pool.

The three aesthetic criteria cannot be considered in isolation since they overlap to some extent. In discussing who should apply the criteria in evaluating landscapes, Litton and colleagues (1974, p. 113) observe that

> [a]esthetic criteria are not whimsical nor are they spur of the moment ideas. They are not determined by popularity contests. They represent a body of knowledge and need to be applied by those who are competent in their application.

However, an exclusive reliance on the application of aesthetic criteria by design specialists would leave no room for the preferential judgments of ordinary citizens.

Guidelines for Conducting Systematic Evaluations

Design professionals have developed numerous procedures to enhance the *reliability* of appraisals, and to make evident the aesthetic criteria used in evaluations. In this context, a reliable procedure is one that yields the same overall appraisal when applied by different evaluators to the same set of facts.

Figures 20.9 and 20.10 illustrate aspects of evaluation schemes intended to enhance objectivity and reliability. Figure 20.9 can assist analysts in evaluating *contrast.* It contains four categories—land/water, vegetation, built structures, and landscape—used in characterizing the visual contrast of the area of a proposed project. The figure provides a framework for summarizing a design professional's impressions of how a proposed project contrasts with landscape characteristics seen from a particular observation point. The rating sheet prompts the evaluator to analyze the area in terms of form, line, color, texture, and scale. For example, in observing the color of the land/water interface, the analyst is asked to consider the hue, value, and chroma of soil, rock, ice, and snow.

Figure 20.10 is an example of a numerical scheme for evaluating a project's visual impacts in terms of landscape compatibility, scale contrast, and spatial dominance. Higher numerical values indicate more pronounced impacts. For example, a project that introduced colors significantly different from those in the existing landscape would be characterized as having "severe" impacts.

Visual impact analysts have developed many schemes like the ones in Figures 20.9 and 20.10, but such schemes don't eliminate the need for professional judgment.[35] Rather, they provide a framework that makes explicit the criteria used in conducting appraisals. They also enhance reliability so that appraisals conducted by different professionals have a common bias and yield consistent outcomes.

Evaluations Based on Preferential Judgments

Preferential judgments of individuals affected by a project often play a major role in visual impact assessments, especially for projects that are controversial. These judgments can be obtained using methods for involving citizens in planning (see Chapter 18).

Projects undertaken by the Environmental Simulation Laboratory of the University of California at Berkeley illustrate how citizens can participate in evaluating visual impacts.[36] One evaluation exercise organized by the laboratory involved a waterfront development proposal in Richmond, California. The proposed project included 2000 new housing units, a commercial development, and a large marina. A three-dimensional model of the project was constructed at a scale of 1 in. = 30 ft. Then a film was produced showing aerial views of the development and views that would be seen during a drive into its center. The film had an imaginary driver get out from the car and take a stroll along a proposed shopping promenade. The driver also took a sailing trip

[34] The quotations defining variety and vividness are from Litton (1972, p. 286), and Litton et al. (1974, p. 111), respectively.

[35] See Canter (1996, chap. 13) for a review of several different numerical procedures for evaluating visual impacts.

[36] The two evaluation studies described here are from Appleyard et al. (1979).

Project Name			Date
Location	Map: ______		Scale: ______
Regional District: ______			
Strategic Planning Area: ______			
Section: ______	Range: ______		Township: ______
Longitude: ______	Latitude: ______		

Sketch Map	
	VRMA: ______ Landscape Unit: ______
	Evaluated By: ______ Checked By: ______
	Visual Resource Management Class: ______ Key Observation Point: ______

Characteristic Landscape

	Element	Descriptions	Comments
LAND/WATER	Form	Landform (3-D) water, soil pattern	
	Line	Regularity/continuity	
	Color	Soil, rock, ice, snow, hue, value, chroma	
	Texture	Clarity, grain	
	Scale	Landform/waterform mass and area	
VEGETATION	Form	Regularity, simplicity orientation	
	Line	Direction, regularity edge character	
	Color	Hue, value, chroma	
	Texture	Clarity, grain	
	Scale	Size, area surrounding objects	
STRUCTURES	Form	Regularity, simplicity orientation	
	Line	Direction, regularity continuity, simplicity	
	Color	Reflectivity, hue value, chroma	
	Texture	Clarity, grain	
	Scale	Size, height, width, surrounding areas	
LANDSCAPE	General Description	Define characteristic landscape, regional setting etc.	
	Scale	Expansive, bounded, area enclosure; visual unit	
	Spatial Composition	Focal, feature, enclosed, panoromic, canopied; weak to strong	

Figure 20.9 Visual contrast rating sheet. Adapted from Yeomans in Smardon, Palmer, and Felleman (eds.) *Foundations for Visual Project Analysis,* Copyright © 1986 John Wiley & Sons, Inc. Reprinted with permission of John Wiley & Sons, Inc. Originally appearing in Yeomans, 1983, *Visual Resource Assessment—A User's Guide.* Reprinted with permission of the Province of British Columbia, B.C. Ministry of Environment, Land and Parks.

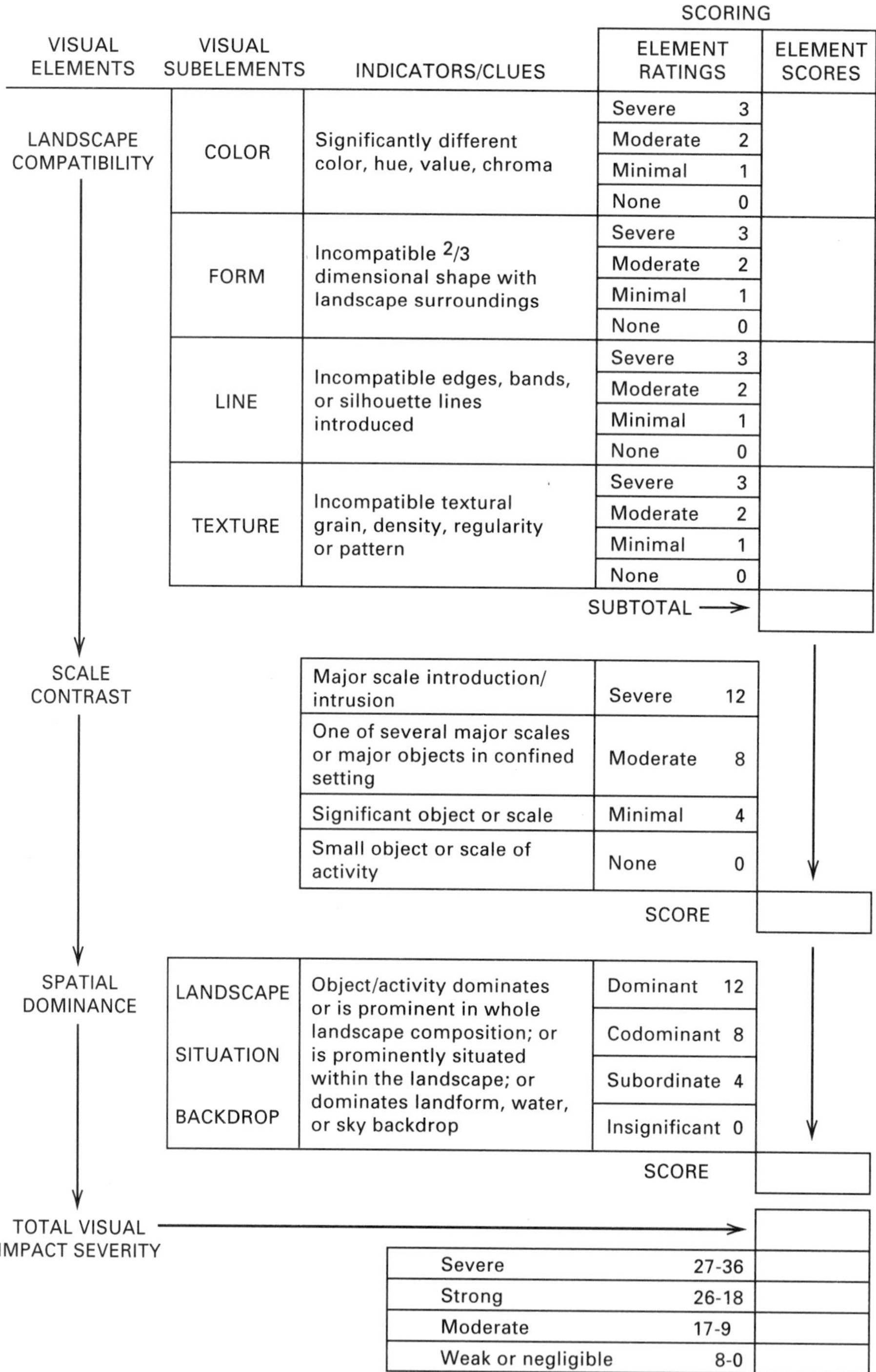

VISUAL ELEMENTS	VISUAL SUBELEMENTS	INDICATORS/CLUES	SCORING: ELEMENT RATINGS		SCORING: ELEMENT SCORES
LANDSCAPE COMPATIBILITY	COLOR	Significantly different color, hue, value, chroma	Severe	3	
			Moderate	2	
			Minimal	1	
			None	0	
	FORM	Incompatible 2/3 dimensional shape with landscape surroundings	Severe	3	
			Moderate	2	
			Minimal	1	
			None	0	
	LINE	Incompatible edges, bands, or silhouette lines introduced	Severe	3	
			Moderate	2	
			Minimal	1	
			None	0	
	TEXTURE	Incompatible textural grain, density, regularity or pattern	Severe	3	
			Moderate	2	
			Minimal	1	
			None	0	
			SUBTOTAL →		
SCALE CONTRAST		Major scale introduction/ intrusion	Severe	12	
		One of several major scales or major objects in confined setting	Moderate	8	
		Significant object or scale	Minimal	4	
		Small object or scale of activity	None	0	
			SCORE		
SPATIAL DOMINANCE	LANDSCAPE SITUATION BACKDROP	Object/activity dominates or is prominent in whole landscape composition; or is prominently situated within the landscape; or dominates landform, water, or sky backdrop	Dominant	12	
			Codominant	8	
			Subordinate	4	
			Insignificant	0	
			SCORE		
TOTAL VISUAL IMPACT SEVERITY →					
			Severe	27-36	
			Strong	26-18	
			Moderate	17-9	
			Weak or negligible	8-0	

Figure 20.10 Second revision of basic visual impact assessment form. Source: Smardon and Hunter, "Procedures and Methods for Wetland and Coastal Area Visual Impact Assessment (VIA)," in Smardon, ed., *The Future of Wetlands: Assessing Visual-Cultural Values,* Allenheld, Osmun, 1983. Reprinted with permission of the author.

through the marina to show how the project would look from a boater's perspective. Both the film-making process and the film itself led to much discussion of design details, and a debate on various aspects of the project. A citizens' committee made evaluations that led to modifications in the original proposal.

Another study undertaken at the Berkeley laboratory involved portraying views from a proposed road intended to replace San Francisco's "Great Highway" along the Pacific Ocean. One criterion for selecting the new route was that it provide "an exciting and varied sequence of views of the ocean to visitors."[37] A scale model of a prototypical 1/4-mile road segment was built to represent views from the new route. Slides taken from inside the model simulated a sequence of views from the proposed route as well as views from adjacent homes. The proposed road was controversial, and the level of public interest was high. Simulations prepared by the Berkeley laboratory allowed both citizens and design experts to evaluate the project's visual impacts.

These are just two of the many simulations that have been prepared by the Berkeley Environmental Simulation Laboratory since it was founded in the early 1970s. The laboratory has long been committed to the concept that visual simulations are to be evaluated by citizens and decision makers. Indeed, the Berkeley laboratory's primary goal is to "to provide the most accurate and comprehensible simulations possible to assist those who are evaluating plans at all stages in the planning process."[38]

In many circumstances, evaluations of visual simulations are based on both appraisals by design specialists and preferential judgments of decision makers and citizens affected by a proposed project. Use of both evaluative appraisals and preferential judgments is illustrated in the following case example.

VISUAL ASSESSMENT OF PROPOSED RAIL TRANSIT LINE

A study of a proposed rail transit line in California demonstrates how visual assessments are conducted and used in decision making.[39] In addition to clarifying the main steps in a visual assessment—inventory, simulation, and evaluation—the case highlights digital image processing.

Two Draft EIRs for Proposed Project

The Bay Area Rapid Transit (BART) provides rail transportation service in several counties of the San Francisco Bay region. The proposed project consisted of a 7.8-mile above-ground rail transit line that would run from BART's existing terminus at Fremont Station through Fremont, a city on the east side of San Francisco Bay. The line would end in Fremont's Warm Springs district (see Figure 20.11). In May 1990, BART issued a draft environmental impact report (EIR) for its proposed Warm Springs extension project. The EIR was required by the California Environmental Quality Act.

The visual analysis portion of the 1990 draft EIR relied on maps like the one in Figure 20.12. These maps, which are inventories of the local area's visual resources, do not show what the rail line would look like. The visual analysis section in the 1990 draft EIR was brief and did not mention whether visual impacts would be significant. Reviewers of the draft EIR called for an assessment of how various alternatives would change views from points along the transportation corridor. Another criticism of the 1990 draft EIR was that it did not adequately consider alternatives to the proposed rail line.

After receiving negative review comments, BART decided to revise its draft EIR. In the new draft, the proposed project would be aboveground and traverse Central Park as well as the eastern arm of Lake Elizabeth (see Figure 20.11). The new draft would also consider both underground and aerial design options. Participants involved in pre-

[37] This criterion is quoted in Appleyard et al. (1979, p. 510).

[38] The quoted material is from Bosselman's (1993, p. 296) explanation of the principles in the mission statement of the Berkeley lab.

[39] Katherine Kao Cushing conducted the research for this case study and prepared an initial draft of this section while she was a student at Stanford. The following individuals provided information for the case study: Tom Priestley of Priestley Associates, Berkeley, California; Daniel Iacofano, Carolyn Verheyen, and Nana Kirk of Moore Iacofano and Goltsman, Inc., Berkeley, California; and Martin Boyle, of the staff of the City of Fremont, California. In addition, Tom Priestley and Daniel Iacofano provided helpful comments on an early draft of the case study.

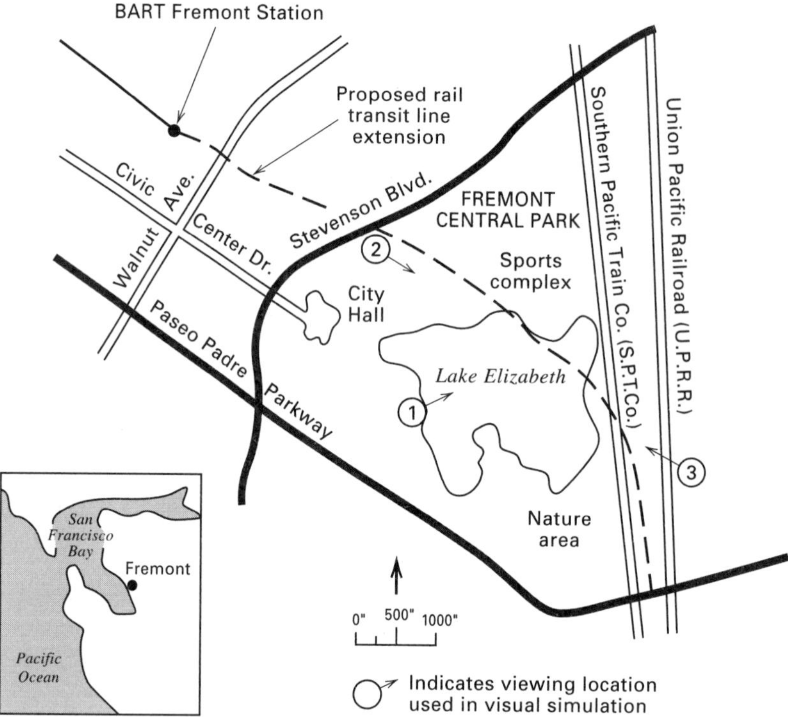

Figure 20.11 Location map for portions of study area: BART Warm Springs extension. Adapted from BART/DKS, (1991, p. 3.8–2).

paring the visual impact assessment for the new (1991) draft EIR included staff from several organizations: BART; DKS, a firm specializing in transportation planning, and the lead consultant for the project; and two consulting companies with extensive experience in doing visual assessments, Priestley Associates, and Moore Iacofano and Goltsman, Inc. (MIG).

Fremont citizens and city officials were particularly concerned about the visual impacts of the proposed rail extension on Fremont's Central Park. If the extension was built, it would cross over the eastern arm of Lake Elizabeth, a significant recreational attraction in Central Park. Many residents considered Central Park to be the showpiece of the city.

The following discussion outlines steps involved in conducting the visual assessment for the 1991 draft EIR. While the process was, for the most part, sequential, activities in various steps often overlapped.

Preparing the Visual Inventory

The visual assessment consultants initiated their work by assembling information on the proposed rail line and on the visual characteristics of the area through which it would pass.[40] After examining project drawings and surveying the area potentially affected by the proposed rail line, one of the consultants, Priestley Associates, divided the study zone into four *visual analysis areas.* Each area contained land with similar visual characteristics.

For each study area, Priestley Associates identified key viewing locations based on the degree to which views were either valued for their aesthetic qualities, seen by large numbers of people, or both. The consultants selected viewpoints based on data from local planning documents, interviews with residents and city staff, and field observations.

[40] Information was obtained from project drawings and reports, topographic and land use maps, aerial photos, local planning documents, and site visits.

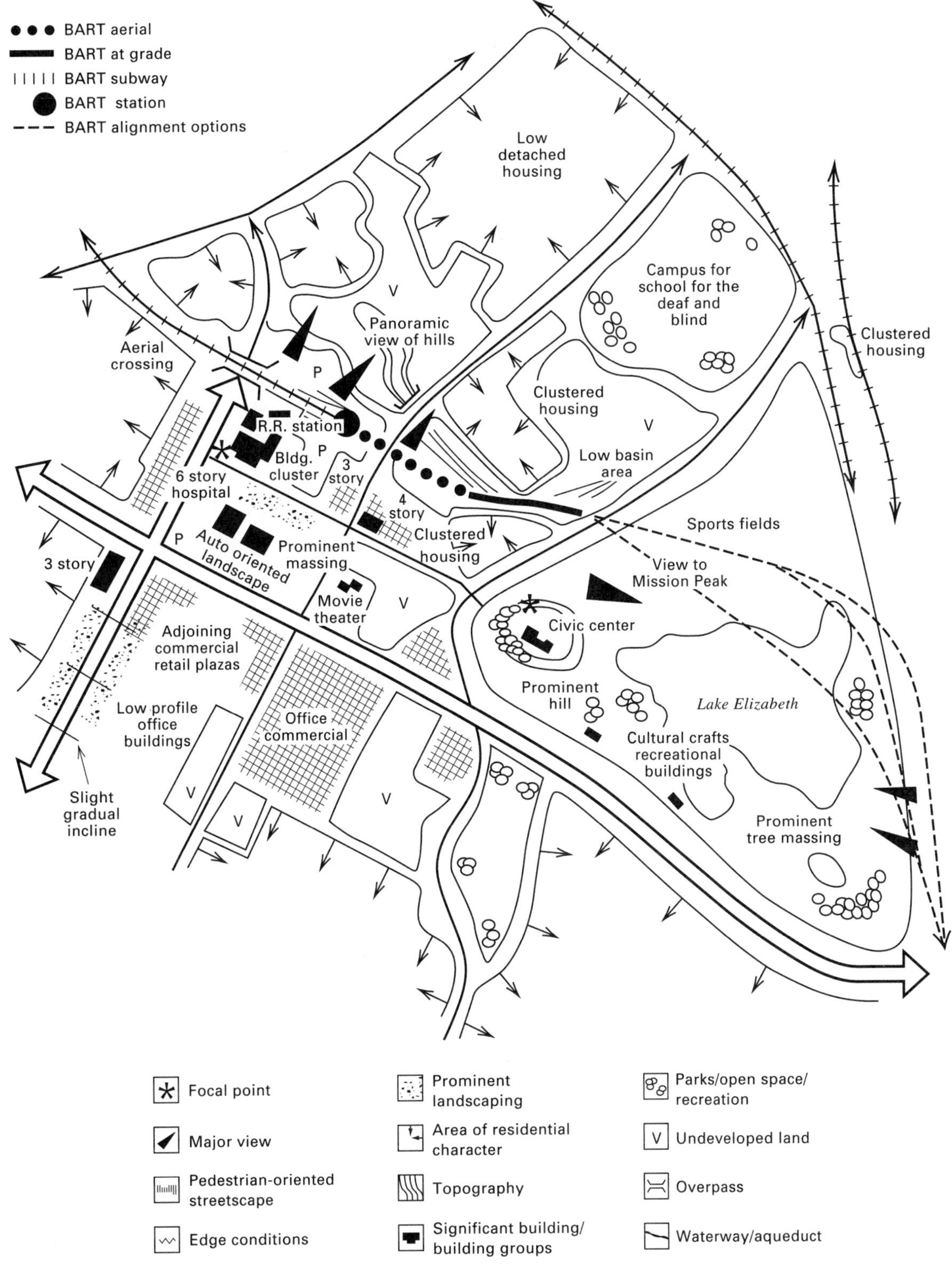

Figure 20.12 Visual inventory of Central Fremont/Central Park area. Adapted from BART/DKS, (1991, p. 3.6–2).

Three viewpoints were selected for the Central Fremont/Central Park visual analysis area (see Figure 20.11).[41]

Once viewpoints had been selected, staff at Moore Iacofano and Goltsman, Inc., created visual simulations using the digital image processing steps shown in Figure 20.8. At the outset, the MIG staff took photos from each viewpoint and used them to create 35 mm color slides of preproject conditions. Once the slides had been developed, a scanner was used to convert them into digital form.[42] At each viewpoint, staff from MIG collected information on the locations and dimensions of reference markers. In the context of visual simulation, a *reference marker* is a salient, stationary visual feature (natural or built) that is clearly visible from a viewpoint. Examples include telephone poles and trees.[43] Reference markers are used to place and scale images of proposed projects in later stages of the simulation process.

Other information used in creating simulations consisted of CAD files representing the project design. Before the visual assessment consultants started their work, BART staff had created digital images of the engineering plans and profiles for the proposed BART extension. These CAD images included drawings of the BART track and structures for the proposed project, and for key design options.[44] Using the computer files provided by BART, simulation specialists at MIG were able to put the design drawings in the form needed for creating visual simulations.

Staff at MIG then edited the images that had been captured as inputs (see Figure 20.8).[45] To prepare a simulation, staff at MIG superimposed the computer files representing project drawings onto digitized versions of photographic images taken from various viewpoints. The geographic locations of reference markers were used to place images of the proposed project (or design options) in their proper place on the digitized photos. In addition, the markers' physical dimensions were used to scale the computer representations of BART's proposed structures.

Once the superposition was accomplished, MIG staff created color slides depicting what the environment would look like if the BART structures were built. An output device called a film recorder was used to produce 35 mm color slides from computer files that contained results from the superposition exercise. Figure 20.13 is one of the 10 visual simulations that appeared in the 1991 draft EIR for the Warm Springs BART extension.

Visual Impact Evaluation

The visual simulations created by MIG were used by Priestley Associates to prepare the evaluative appraisal reported in the 1991 draft EIR. Citizens and others reading the draft EIR had opportunities to make their own judgments about visual effects. Both types of evaluation are discussed here.

Evaluative Appraisal by Design Professionals

Priestley Associates used six criteria to gauge the visual impacts of BART's proposed project (and design options): visibility, number of viewers, contrast, dominance, character, and community policies (see Table 20.2). To enhance reliability (that is, replicability), the consultants integrated the criteria into a systematic evaluation procedure.[46] In addition to reviewing the visual simulations, Priestley Associates studied BART's project designs and conducted field investigations in the study area.[47] Results from those

---

[41] Three other viewpoints were selected for other visual analysis areas. The total number of viewpoints used in the analysis was limited to six because of constraints on the time and resources available for the assessment.

[42] In the conversion process, slide film was placed in a slot inside a slide scanner and light was directed throught it. Light passing through each slide was converted into electrical signals that were interpreted as digital input data by a computer.

[43] If no reference markers are apparent, large helium-filled balloons can be attached to selected locations at a site and used instead.

[44] If the CAD files had not been available, MIG would have had to create an equivalent digital format representation of BART's engineering design in order to proceed with the digital image processing.

[45] The editing tools used by MIG included paint and draw programs. A *paint program* allows users to tint, erase, and perform numerous other graphical operations on a computer screen. In a typical paint program, a user can access the electronic counterparts of an artist's tools, such as a pencil, paintbrush, eraser, and airbrush. A *draw program* is different from a paint program in that objects created retain their identity as individual items. A typical draw program records an item's type (for example, line or rectangle), location, corner points, and other attributes, such as patterns or shading.

[46] The 1991 draft EIR referred readers to Yeoman (1986) for a description of aesthetic analysis methods used in preparing the visual assessment of the proposed BART extension. Figures 20.9 and 20.10 are examples of procedures detailed by Yeomans.

[47] For example, the consultants conducted field studies in Central Park to analyze the landscape and observe visitor use patterns.

Figure 20.13 Simulation of one of BART's design options from viewpoint 3: back yard of residence on Valdez Way. Courtesy of Moore Iacofano and Goltsman, Inc., Berkeley, CA.

TABLE 20.2 Criteria Used by Priestley Associates in Evaluating Visual Impacts of BART's Plans[a]

Criteria	Definition
• Visibility	The extent to which physical changes brought about by the project would be seen by the public.
• Number of viewers	The number of people who could potentially see the project-related changes.
• Contrast	The extent to which the form, line, color, and scale of project elements would either contrast with or be visually absorbed by the area's existing features.
• Dominance	The extent to which project elements would be dominant in views of landscapes (and cityscapes).
• Character	The extent to which the changes would be compatible with the character of the setting. Project facilities would be most compatible with industrial and commercial settings and least compatible with residential and passive recreational settings.
• Community policies	The extent to which the changes would be consistent with aesthetic policies and guidelines adopted by local governments.

[a] Adapted from BART/DKS (1991, p. 3.8–17).

evaluative activities were reported in the 1991 draft EIR.

In addition to characterizing the significance of visual impacts, the 1991 draft EIR explained how the six criteria for evaluating aesthetic effects were applied. Consider, for example, the *contrast* criterion, defined in the draft EIR as "the extent to which the form, line, color, and scale of the project's elements would either contrast with or be visually absorbed by the setting's existing features."[48] In some areas, effects on contrast were characterized as "not significant" because BART's proposed structures were similar to elements surrounding them, such as train tracks and industrial buildings. In several other areas (for example, Central Park), high contrast and other factors led the consultants to characterize the visual impacts as significant.

Preferential Judgments of Affected Parties

Fremont residents were heavily involved in planning the BART extension. Many residents received BART's Warm Springs extension newsletters, attended public meetings, participated in BART-sponsored focus groups and workshops, and submitted written comments in response to the 1991 draft EIR.

Citizens had many opportunities to learn about the visual effects of BART's proposal. In addition to presenting visual simulation results in the 1991 draft EIR, BART and its consultants showed slides of the simulations at community workshops. In the end, many citizens and business interests felt that a below-ground design was the only acceptable option for the BART extension. They frequently cited the visual and noise advantages of this option. Residents from Valdez Way who would see the proposed rail line from their homes had particularly strong objections to BART's plans (see Figure 20.13).

Most public agencies responded positively to the draft EIR and endorsed implementation of the planned BART extension. Moreover, a large number of Fremont citizens supported construction of the project, and a few residents were not concerned about whether the rail line passed over or under Lake Elizabeth.[49] However, the city of Fremont was "disappointed" that the draft EIR did not recommend use of a below-ground design option for the portion of the rail line that traversed Central Park.

[48] This quotation is from BART/DKS (1991, p. 3.8–17).

[49] Comments by citizens supporting the rail line and citizens unconcerned with visual impacts are given in BART/DKS (1991, pp. A-100 and B-33).

Fremont Acts to Block BART

As of mid-1994, long after BART had issued its final EIR for the proposed aboveground project, the city of Fremont was in the midst of a lawsuit that challenged BART's designs. The city had allowed for only a below-ground BART extension in its General Plan. Moreover, neither BART nor the city of Fremont was willing to provide the additional funding needed to put the proposed rail line underground. The lawsuit, which was based on alleged inadequacies in BART's environmental impact report, temporarily halted the project.

Notwithstanding the ambiguous outcome, the case study demonstrates how a visual assessment can provide decision makers and affected citizens with a basis for understanding the visual impacts of a proposed project. Although BART's planning process did not result in consensus on which rail line project to use (or even whether the rail line should be extended), the visual assessment was instrumental in clarifying the impacts of BART's proposal.

Key Concepts and Terms

Simulation versus Evaluation

Preproject and postproject conditions
Simulations as forecasts
Evaluative appraisals
Preferential judgments

Issues Raised Using Landscape Preference Research

Response equivalence
Expert versus nonexpert preferences
Factors affecting landscape preference
Bias due to aesthetic training
Misrepresentation in simulation
Realistic versus accurate simulations

Preparation of Visual Inventories

Topographic maps
Aerial photos
Field surveys

Route inventories
Area inventories
Viewsheds

Techniques for Simulating Postproject Conditions
Perspective sketches and sequences
Photomontage
Retouched photos
Computer-generated perspectives
Three-dimensional models
Digital image processing
Image capture, editing, and output
Digitization
Mitigation planning
Interactive project walkthrough
Simulating motion and sound

Evaluating Visual Effects and Landscape Quality
Citizen involvement in evaluation
Aesthetic criteria
Unity, variety, and vividness
Replicable evaluation methods
Landscape compatibility
Scale contrast
Spatial dominance

Visual Assessment of Proposed Rail Transit Line
Visual analysis areas
Viewing locations
Reference markers
Visual contrast

Discussion Questions

20-1 Characterize the types of judgments involved in preparing visual simulations of a proposed project. Do these judgments obscure the distinctions between simulation and evaluation?

20-2 Sheppard (1982) studied response equivalence by having one group of design specialists respond to visual portrayals of proposed projects and another group react to photographs of the same projects after they had been built. The response equivalence in the two sets of evaluations was low for three-quarters of the 30 projects included in the research. Speculate on factors that might explain this research outcome.

20-3 Identify a large-scale development project that is proposed for an area you are familiar with. How would you go about preparing an inventory of existing visual features that might be affected by the proposal? In discussing your approach, consider variables such as viewer location, time of day, and season of the year.

20-4 Alternative simulation media include photomontage, perspective drawings, and so forth. List and define five criteria you would use to characterize good media for portraying visual impacts of physical development projects.

20-5 Suppose design specialists have prepared models, sketches, and photos to portray a proposed electric power plant in a rural location. You are asked to use these simulations in designing a process for evaluating the visual impacts of the proposed facility. How would aesthetic criteria and preferential judgments fit into your evaluation scheme?

20-6 Bosselman (1993, pp. 295–96) reported on the following circumstances in which groups for and against a proposed development had their own ideas about what visual simulations should show:

> In California, a government agency entrusted with the planning of inland waterways and preservation of inland shorelines commissioned simulations from the [Environmental Simulation Laboratory] in Berkeley with the intention of strengthening its argument against a proposed development project. This development proposal was submitted by the real estate division of one of the nation's largest transportation companies. Through the simulations, this state agency hoped to influence local governments on the eastern shore of the San Francisco Bay with planning jurisdiction over the developable land. The agency wanted to convince local municipal governments to lower the amount of development potential on the transportation company's property by lowering the allowable building heights in one area from 100 feet to 40 feet. When the proponents of development learned about these simulations, they quickly commissioned their consultants to

produce their own simulations, which contradicted the previous simulations by emphasizing the small scale of 100-foot-high buildings.

What do you think would have happened if both simulations had been shown to the decision makers at the local governments? What does this example suggest about the dilemmas faced by design professionals whose clients for visual simulations have their own ideas about what the outcomes should be? What would you have done if you had represented the Berkeley lab in its interactions with the state government agency that opposed the development project?

20-7 The following four principles are from the mission statement of the Berkeley Environmental Simulation Laboratory:[50]

- The primary goal of the Berkeley Simulation Laboratory is to provide the most accurate and comprehensible simulations possible to assist those who are evaluating plans at all stages in the planning process. The secondary goal is to pursue research in improving the methods and technologies of simulations and in better understanding the role simulation can plan in a person's comprehension of and evaluation of a design or plan.
- Simulations must be open to evaluation for accuracy and confirmation. The materials and methods used in a simulation must be open to scrutiny, and independent testing.
- In all cases it should be understood that the sole purpose of the simulation laboratory is to provide information. Under no circumstances will the staff of the laboratory become involved in the process of negotiation, arbitration, or decisionmaking.
- All information produced by the laboratory is owned and protected by the Regents of the University of California. Information produced at the laboratory shall not be distorted. The University is a public institution, therefore all information produced at the laboratory is subject to the California Public Records Act and Freedom of Information Acts.

How do you think each of these principles helps ensure objectivity and "truth in simulation"? Which of these principles would be important in dealing with the kinds of pressures on simulation specialists described in the previous question?

[50] Materials from the mission statement are as given by Bosselman (1993, pp. 296–99).

References

Alonso, S. G., M. Aguilo, and A. Ramos. 1986. "Visual Impact Assessment Methodology for Industrial Development Site Review in Spain." In *Foundations for Visual Project Analyses,* ed. R. C. Smardon, J. F. Palmer, and J. P. Felleman, 277–305. New York: Wiley.

Appleyard, D. 1977. "Understanding Professional Media: Issues, Theory and a Research Agenda." In *Human Behavior and the Environment: Advances in Theory and Research,* 2, ed. I. Altman and J. F. Wohlwill, 43–88. New York: Plenum.

Appleyard, D., and K. H. Craik. 1979. "Visual Simulation in Environmental Planning and Design." Working Paper No. 314, Institute of Urban and Regional Development. Berkeley: University of California.

Appleyard, D., K. H. Craik, M. Klapp, and A. Kreimer. 1973. "The Berkeley Environmental Simulation Laboratory: Its Use in Environmental Impact Assessment." Working Paper No. 206, Institute of Urban and Regional Development. Berkeley: University of California.

Appleyard, D., P. Bosselmann, R. Klock, and A. Schmidt. 1979. Periscoping Future Scenes, How to Use an Environmental Simulation Lab. *Landscape Architecture* 69(5): 487–510.

Arnheim, R. 1974. *Art and Visual Perception.* Berkeley: University of California Press.

BART/DKS. 1991. *BART Warm Springs Extension Draft Environmental Impact Report.* Oakland: BART.

Bosselman, P. 1993. "Dynamic Simulations of Urban Environments." In *Environmental Simulation: Research and Policy Issues,* ed. R. W. Marans and D. Stokols, 280–302. New York: Plenum.

Brush, R. O. 1976. "Perceived Quality of Scenic and Recreational Environments: Some Methodological Issues." In *Perceiving Environmental Quality: Research and Applications,* ed. K. H. Craik and E. H. Zube, 47–58. New York: Plenum.

Brush, R. O., and E. L. Shafer. 1975. "Application of A Landscape Preference Model to Land Management." In *Landscape Assessment: Values, Perceptions and Resources,* ed. E. H. Zube, R. O. Brush, and J. G. Fabos, 168–82. Stroudsburg, PA: Dowden, Hutchinson and Ross.

Canter, L. W. 1996. *Environmental Impact Assessment,* 2d ed. New York: McGraw–Hill.

Craik, K. H. 1972. "Appraising the Objectivity of Landscape Dimensions." In *Natural Environments: Studies in Theoretical and Applied Analysis,* ed. J. V. Krutilla, 292–346. Baltimore: Johns Hopkins University Press for Resources for the Future, Inc.

Gussow, A. 1977. In the Matter of Scenic Beauty. *Landscape* 21(3): 26–35.

Kaplan, R. 1975. "Some Methods and Strategies in the Prediction of Preference." In *Landscape Assessment: Values, Perceptions and Resources,* ed. E. H. Zube, R. O. Brush, and J. G. Fabos, 118–29. Stroudsburg, PA: Dowden, Hutchinson and Ross.

______. 1993. "Physical Models in Decision Making for Design: Theoretical and Methodological Issues." In *Environmental Simulation: Research and Policy Issues,* ed. R. W. Marans and D. Stokol, 61–86. New York: Plenum Press.

Litton, R. B., Jr. 1972. "Aesthetic Dimensions of the Landscape." In *Natural Environments: Studies in Theoretical and Applied Analysis,* ed. J. V. Krutilla, 262–91. Baltimore: Johns Hopkins University Press for Resources for the Future, Inc.

Litton, R. B., Jr., R. J. Tetlow, J. Sorensen, and R. A. Beatty. 1974. *Water and Landscape, An Aesthetic Overview of the Role of Water in the Landscape.* Port Washington, NY: Water Information Center, Inc.

Mellander, K. 1974. *Environmental Planning: Can Scale Models Help?* American Engineering Model Society—Seminar '74 (October). 61–67. Available as Reprint No. 125, Institute of Urban and Regional Development, Berkeley: University of California.

Novitski, B. J. 1994. Virtual Reality for Architects. *Architecture* (Oct.): 121–25.

Orland, B. 1993. "Synthetic Landscapes: A Review of Video-Imaging Applications in Environmental Perception Research, Planning and Design." In *Environmental Simulation: Research and Policy Issues,* ed. R. W. Marans and D. Stokols, 213–50. New York: Plenum.

Shafer, E. L., J. F. Hamilton, Jr., and E. A. Schmidt. 1969. Natural Landscape Preferences: A Predictive Model. *Journal of Leisure Research* 1: 1–19.

Sheppard, S. R. J. 1982. *Landscape Portrayals: Their Use, Accuracy and Validity in Simulating Proposed Landscape Changes.* Ph.D. dissertation. Berkeley: University of California.

______. 1989. *Visual Simulation: A User's Guide for Architects, Engineers and Planners.* New York: Van Nostrand Reinhold.

Tarricone, P. 1994. Let's Go to The Videotape. *Civil Engineering* 64(6): 42–45.

U.S. Department of Transportation. 1978. *Highway Photomontage Manual.* Report No. FHWA-DP-40-1. Arlington, VA: Federal Highway Administration.

Yeomans, W. C. 1986. "Visual Impact Assessment: Changes in Natural and Rural Environments." In *Foundations for Visual Project Analyses,* ed. R. C. Smardon, J. F. Palmer, and J. P. Felleman, 201–22. New York: Wiley.

Zube, E. H., D. G. Pitt, and T. W. Anderson. 1975. Perception and Prediction and Scenic Resource Values of the Northeast. In *Landscape Assessment: Values, Perceptions and Resources,* ed. E. H. Zube, R. O. Brush, and J. G. Fabos, 151–67. Stroudsburg, PA: Dowden, Hutchinson and Ross.

Zube, E. H., and D. Simcox. 1993. "Landscape Simulation: Review and Potential." In *Environmental Simulation: Research and Policy Issues,* ed. R. W. Marans and D. Stokols, 253–78. New York: Plenum.

Chapter 21

Elements of Noise Impact Assessment

Hundreds of millions of people throughout the world are regularly subjected to noise that causes annoyance and interferes with activities.[1] Frequently these high noise levels are associated with construction activities and transportation facilities, especially highways and airports.

Noise is often defined as "unwanted sound."[2] *Acoustics,* the science of sound, provides a basis for defining noise indicators and making forecasts of noise. Although acoustics is a complex and specialized topic, it must be introduced to pursue even a rudimentary discussion of techniques to predict noise impacts of proposed projects. After introducing concepts from acoustics and characterizing the effects of noise on people, this chapter delineates the steps in a noise impact assessment. Noise regulations are discussed in the context of impact assessment. The concepts presented here are used in formulating noise regulations, and they are very frequently applied in analyzing noise impacts of specific projects.

[1] For example, the Organization for Economic Cooperation and Development (OECD) estimated that in the early 1980s, more than 130 million people in OECD member nations were exposed to noise levels that caused annoyance and constrained behavior patterns (OECD, 1991, p. 51).

[2] As Canter (1996, p. 305) points out, this definition of noise ignores the influence of noise on nonhuman species. Although research has been done on how noise affects animals, much of it concerns short-term impacts; for example, the effect of aircraft noise in causing large mammals to flee or hide. Long-term effects are difficult to trace. For information on the impacts of noise on animals, see Richardson et al. (1995).

SOUND AND ITS MEASUREMENT

Most people are familiar with the concentric circles of waves that result from dropping a stone into a still pond. Analogous waves propagate through the air when a metal tuning fork is struck with a hard object. The vibrations of the tuning fork push some air molecules close together and allow others to move farther apart. The resulting fluctuations in air pressure, relative to undisturbed air, are sensed by the human ear as sound.

Pure Tones and Sound Pressure Levels

Suppose that immediately after a tuning fork is struck, measurements of air pressure are made at a nearby point. A plot of the air pressure levels at this point might look like the graph in Figure 21.1. When

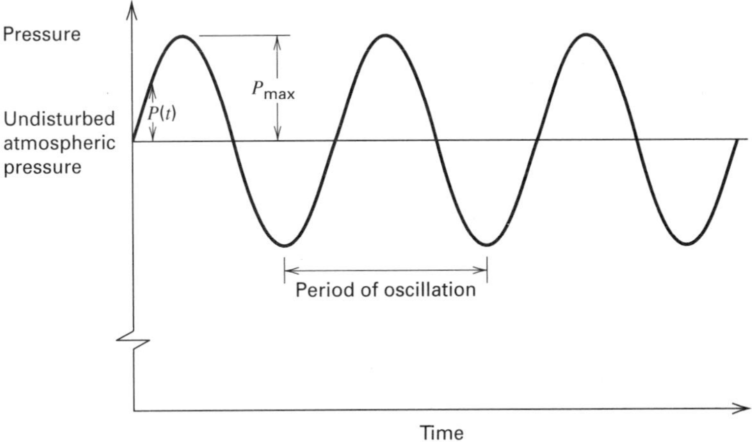

Figure 21.1 Representation of a pure tone.

air pressures associated with a sound can be described by the simple, regular shape in Figure 21.1, the sound is termed a *pure tone.* The regular pattern in the graph is called a *simple harmonic oscillation.*[3] Although most commonly heard sounds are not pure tones, they can be represented as combinations of pure tones.

Figure 21.1 introduces terms used to describe sound. The time it takes for the pressure to go through one full cycle of the pattern in the figure is called the *period of oscillation.* Its inverse is termed the *frequency of oscillation* and is generally expressed in hertz (Hz), a shorthand way of saying "cycles per second." Frequency is related to what people perceive subjectively as *pitch;* the higher the frequency, the higher the pitch.

Because sound pressures are only a minute fraction of atmospheric pressure, they are generally reported in terms of their differences from atmospheric pressure, as shown by $P(t)$ in Figure 21.1. The maximum sound pressure, P_{max}, is called the *amplitude of the oscillation.* Amplitude is associated with the subjective perception of *loudness.*[4] It is tempting to characterize the sound pressure in Figure 21.1 in terms of average pressure. However, the regular pattern in the figure makes the average pressure over a period equal to zero. Instead of a simple average, the *root-mean-square* (*rms*) *pressure* is frequently used. It is computed by first determining the square of all values $P(t)$ in a period of oscillation and then computing the average of the squared values. Finally, the square root of this average is taken; this value is rms pressure. For pure tones like the one in Figure 21.1 it can be shown that

$$P = \frac{P_{max}}{\sqrt{2}} \qquad \textbf{(21-1)}$$

where p is the rms pressure, which is generally measured in pascals (Pa).[5] Sound meters are used to record rms pressure amplitudes.

The range of audible sound pressures is enormous. At a frequency of 1000 Hz, barely audible sound has an rms pressure of 2×10^{-5} Pa, and the maximum rms pressure that will not damage the human hearing mechanism is 20 Pa. Because audible sound spans a range of 10^6 Pa, a logarithmic scale is used to describe pressure.[6] For any rms pressure, p,

[3] Harmonic oscillations can be described using trigonometric functions. Figure 21.1 has the shape of a sine curve.

[4] The links between loudness and amplitude have been determined empirically. For a discussion of these relationships, see Cunniff (1977, pp. 89–91).

[5] A Pascal is defined as the pressure exerted by a force of 1 newton over an area of 1 m^2.

[6] A logarithm to the base 10 is defined as follows: $10^a = b$ is equivalent to saying "the logarithm of b to the base 10 equals a," or $\log b = a$. Cunniff (1977, pp. 31–39) provides an introduction to how logarithms are used in noise measurements.

the *sound pressure level, L,* expressed in units of decibels (dB), is defined as[7]

$$L = 10 \log \left(\frac{p^2}{p_0^2}\right) \qquad \textbf{(21-2)}$$

where $p_0 = 2 \times 10^{-5}$ Pa, the threshold of audible sound, and the logarithm is to the base 10.

Sound pressure level is defined using the *square* of the rms pressure because the amount of power contained in sound is proportional to the pressure squared.[8] Since equation 21-2 employs the logarithm of the pressure squared, a ten-fold increase in sound pressure corresponds to a rise of 20 dB of sound pressure level.[9]

In investigating ambient noise, it is often necessary to estimate the sound pressure level resulting from several sources. As shown by Cunniff (1977), a relationship for combining the sound pressure levels of n different sources is

$$L = 10 \log \left(10^{L_1/10} + 10^{L_2/10} + \ldots + 10^{L_n/10}\right) \qquad \textbf{(21-3)}$$

where L_i is the sound pressure level of source i (i = 1, . . ., n), and L is the sound pressure level of the n sources combined. For example, the overall effect of two individual 80-dB sources ($L_1 = L_2 = 80$) is a sound pressure level of

$$L = 10 \log \left(10^{80/10} + 10^{80/10}\right) = 83 \text{ dB.}$$

Special graphs and tables are available to make it simple to calculate approximate results for the logarithmic addition in equation 21-3, and an example is given in Table 21.1.[10] Using the two-source illustra-

TABLE 21.1 Values for Combining Sound Pressure Levels Approximately Using Simple Addition

Difference in Sound Pressure Levels of 2 Sources (dB)[a]	Amount Added to the Larger Level to Estimate Sound Pressure Level for 2 Sources (dB)
0	3.0
1	2.5
2	2.1
3	1.8
4	1.5
5	1.2
6	1.0
7	0.8
8	0.6
9	0.5
10	0.4
12	0.3
14	0.2
16	0.1
>16	0

[a] If the difference in sound pressure levels is not a whole number, the amount to be added is found by interpolation. For example, if $L_1 = 63.8$ and $L_2 = 61$, the difference is 2.8, and, by interpolation, the amount added to the larger source is 1.9, which gives a total of 65.7 dB. When the table is used repeatedly to find the combined sound pressure of several sources, interpolation is often required.

Source: U.S. Department of Housing and Urban Development (1985, p. 51).

[7] Cowan (1994, p. 31) observes that the bel corresponds "to a multiple of 10 of the threshold of hearing. Because the bel (denoted B) is too large to describe the acoustic environment adequately, the unit was broken into tenths, or decibels. Therefore, 10 dB corresponds to 1 B. . . ." The bel was named after Alexander Graham Bell (1847–1922), the American inventor of the telephone.

[8] In acoustics, power is often expressed in units of watts. If a sound source has W watts of power, the sound watt level (*SWL*), or *sound power level,* in decibels, is defined as:

$$SWL = 10 \log \frac{W}{W_0}$$

where W_0, the reference power level, is often taken as 10^{-12} watts and the logarithm is to the base 10. Sound watt level is independent of location, but sound pressure level (L) is affected by distance. For this reason, sound pressure level is generally reported at a particular (unobstructed) distance from the sound source. Sound *power* level is proportional to sound *pressure* level squared; the proportionality constant depends on the impedance of the transmitting medium. For more on the relationship between sound power and sound pressure, see Sound Research Laboratories (1991, pp. 5–6).

[9] A property of logarithms is that $\log a^c = c \log a$. Therefore, equation 21-2 can be written as a $L = 20 \log (p/p_0)$. In addition, $\log 10^n = n$, and thus $\log 10 = 1$, $\log 100 = 2$, and so forth. Hence, using equation 21-2, if $p = 10\ p_0$, the decibel level is 20, whereas if $p = 100\ p_0$, the decibel level is 40.

[10] For graphs to simplify the logarithmic additions in equation 21-3, see Magreb (1975).

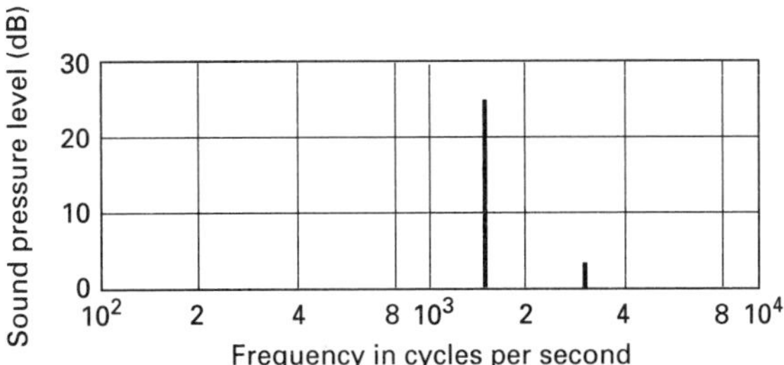

(*a*) A note from a flute (lowest frequency, termed *fundamental harmonic,* is at 1568 Hz). From *Music, Physics and engineering,* by H. G. Olson, 1967, Dover, New York.

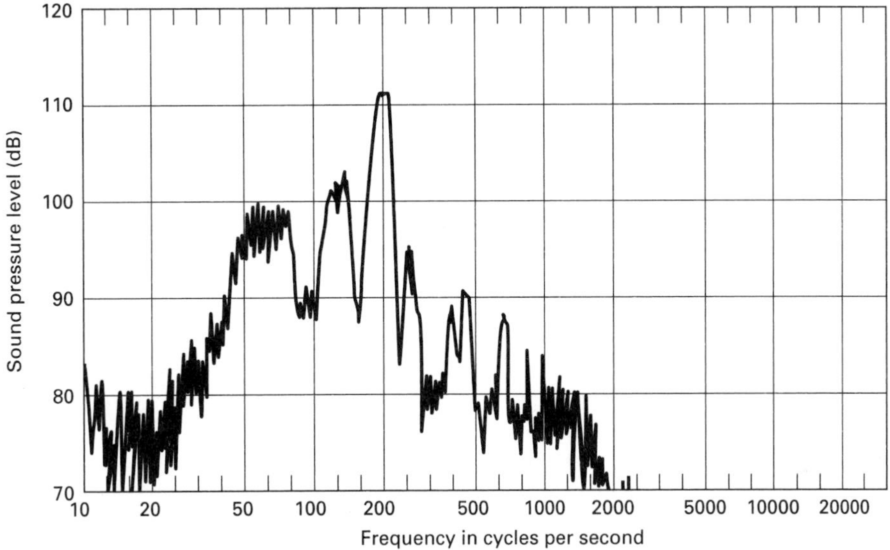

(*b*) Sound from an air compressor at a distance of 5 ft. Courtesy of C. E. Hickman, Southern Company Services, Inc., Birmingham, Ala.

Figure 21.2 Illustrations of sound spectra.

tion from the last paragraph, $L_1 = L_2 = 80$ dB, the difference in sound pressure level is zero. Table 21.1 indicates that the combined sound pressure level can be found by adding 3 dB to the larger of the two sources, which yields a combined sound pressure level of 83 dB in this case. If a third source of 80 dB were added, Table 21.1 could be used again to yield an overall level of 84.8 dB.[11]

[11] In this case, the two sound sources are the new source (80 dB) and the combined effect of the first two sources (83 dB). Using Table 21.1, the difference in sound pressure level is 3 dB; thus 1.8 is added to the larger value to yield a sound pressure level of 84.8 dB.

Sound Spectra and the A-weighted Decibel

Discussions of environmental noise usually do not focus on pure tones or even simple combinations of pure tones. Commonly heard sounds have complex frequency and pressure characteristics that can be represented on sound spectrum plots like the ones in Figure 21.2. Frequency is plotted on the abscissa, and sound pressure level is on the ordinate. Figure 21.2a, which contains two pure tones, represents a note from a flute. Figure 21.2b approximates the continuous sound spectrum at a distance of 5 feet from an air compressor.

The transmission of sound in the environment and its influence on humans depends significantly on a sound's frequency characteristics. Sound measurement devices can be used to divide any sound into frequency bands. *Octave bands* are widely used in practice. By definition, the upper frequency limit of each octave band is exactly two times the lower limit. Beginning with 11 Hz, the band limits (in Hz) are 11 to 22, 22 to 44, 44 to 88, and so forth. The sound pressure level recorded for each octave band is associated with the band's *center frequency.*[12]

The human ear is capable of hearing sounds in the range between about 20 and 20,000 Hz, but it does not perceive sounds at different frequencies in the same way. Experiments have shown that people are far more sensitive to sounds in the 500 to 5000 Hz frequency range than they are to either very low or very high frequencies. For example, the sound pressure level required to barely hear a sound at 200 Hz is substantially higher than the pressure level needed to just hear a sound at 2000 Hz. The subjective perception of a sound's loudness depends heavily on its frequency characteristics.

Sound measurement equipment has been designed to account for the sensitivity of human hearing to different frequencies. Correction factors (in dB) for adjusting actual sound pressure levels to correspond with human hearing have been determined experimentally. Several different sets of correction factors are in common use and two are shown in Table 21.2. For measuring noise in ordinary environments (*community noise*), *A-weighted correction factors* are generally employed.[13] The term *weighting* is used because sound level meters make corrections electronically using weighting networks.

Table 21.3 illustrates the computation of A-weighted sound pressure levels recorded in units of dBA (the A in dBA signifies A-weighting). The table contains measurements from a spectrum analyzer, an electronic device that separates, or *filters,* the incoming sound into frequency bands and records the sound pressure level within each band. The sound represented in Table 21.3 consists only of frequencies in the range from 44 to 355 Hz. A continuous sound has been filtered into three octave bands, and the table indicates the pressure level of the sound in each band.

Figure 21.3 plots the center frequencies and octave band pressure levels in Table 21.3 and approximates the spectrum of the incoming sound. An overall sound pressure level is obtained by substituting the octave band pressure levels (L_1 = 79 dB, L_2 = 86 dB, and L_3 = 89 dB) in equation 21-3, which gives a combined value of 91 dB. Table 21.3 also shows the results from adding A-weighting correction factors to the recorded pressure levels. When the A-weighted octave band pressure levels in the final column are entered into equation 21-3, the result is an overall level of 80.4 dBA. The A-weighted sound pressure level is much lower than the unweighted value (91 dB), because the human ear is not sensitive to the low-frequency sound in this example.

A way to learn something about the frequency characteristics of a particular sound without using

TABLE 21.2 A- and C-weighting Corrections at Octave Band Center Frequencies

Octave Band Center Frequency (Hz)	A-weighting Correction (dB)	C-weighting Correction (dB)
16	−56.7	−8.5
32	−39.4	−3.0
63	−26.2	−0.8
125	−16.1	−0.2
250	−8.6	0
500	−3.2	0
1,000	0	0
2,000	+1.2	−0.2
4,000	+1.0	−0.8
8,000	−1.1	−3.0
16,000	−6.6	−8.5

Adapted from Davis and Cornwell, *Introduction to Environmental Engineering,* 2d ed. © 1991, p. 511. Reprinted with permission of The McGraw-Hill Companies.

[12] The center frequency of an octave band is computed as $\sqrt{2}$ multiplied by the lower frequency limit of the band. As shown by Cunniff (1977, p. 39), the center frequency corresponds to the average of the logarithms of the upper and lower band limits.

[13] The literature on noise distinguishes community noise from noise in work environments (industrial noise) and from particular sources (for example, airport noise).

TABLE 21.3 Spectral Analysis and A-weighted Sound Pressure Levels

Octave Band Limits (Hz)		Center Frequency (Hz)	Octave Band Pressure Level (dB)	A-weighting Correction Factor (dB)	A-weighted Octave Band Pressure Level (dBA)
Lower Limit	Upper Limit				
44	88	63	79	−26	53
88	177	125	86	−16	70
177	355	250	89	− 9	80
Overall sound pressure level			91		80.4

a spectrum analyzer involves comparing A-weighted readings (in dBA) with C-weighted readings (in dBC) at the same location. As indicated in Table 20.2, C-weighting correction factors are notably lower than A-weighting correction factors below about 1000 Hz. Thus, if measurements using a sound level meter show that dBC is much higher than dBA, a strong low-frequency component is present in the sound's acoustic spectrum. In contrast, if the dBA and dBC readings are comparable (say within a few decibels), high frequencies dominate the spectrum.[14]

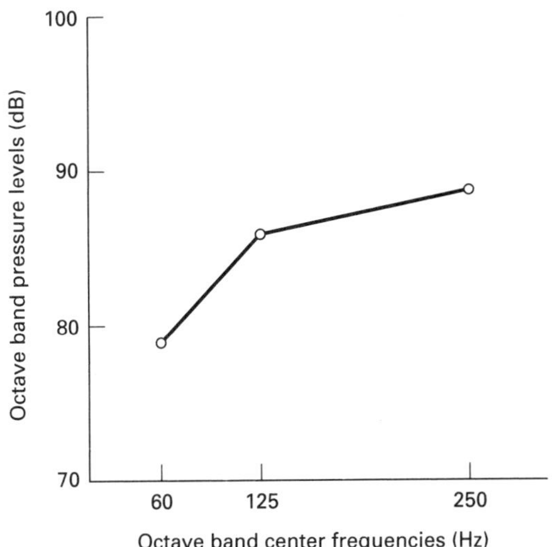

Figure 21.3 Octave band sound spectrum.

[14] This discussion of what can be learned by comparing dBA and dBC values is from Cowan (1994, p. 36).

Community noise is generally reported in A-weighted decibels. Table 21.4 indicates sound pressure levels for some familiar noise sources.

STATISTICS USED TO CHARACTERIZE COMMUNITY NOISE

Noise levels vary continually with time. For example, 10 minutes spent at a busy intersection in a residential suburb can be long enough to hear a background noise level of 50 dBA, an aircraft overhead at 65 dBA, and a sports car at 70 dBA. Parameters used to characterize community noise must account for how noise changes over time.

Histograms of Sound Pressure Levels

Variations in noise levels over time are often represented using a graph of the percentage of time the sound pressure levels falls into different dBA intervals. The resulting *histogram* is used to compute L_x, the sound pressure level that is equaled or exceeded $x\%$ of the time.[15] Values of L_{10}, L_{50}, and L_{90} are commonly interpreted as peak, median, and background noise levels, respectively.

Consider an example: Suppose a sound level meter is taken into a noisy classroom and continuous

[15] More generally, a histogram for measurements of a random variable is a bar graph having a horizontal axis showing the intervals into which the measurements can fall. The bar heights represent the relative frequencies with which the variable's measurements fall into the intervals. Figure 21.4 is a histogram in which the variable is sound pressure level.

sound pressure level readings (in dBA) are obtained for 10 minutes. Assume this continuous record is converted into 600 separate dBA readings corresponding to each second in the time period. (The use of 1 second is arbitrary; any unit of time can be employed.) A histogram for sound pressure levels during this 10-minute period is constructed by counting the number of seconds that the recorded dBA values fall within different intervals. Intervals of 5 dBA were used in this example, but they could have been longer or shorter. Table 21.5 contains the results from the counting exercise.

Figure 21.4 is a histogram based on the data in Table 21.5. By counting the percentages associated with each interval in the figure (starting from the *right*), L_{10} and L_{90} can be determined by inspection. For the two intervals on the extreme right, the percentages add up to 10. Therefore, the sound pressure level is at or above 70 dBA during 10% of the time, which means $L_{10} = 70$ dBA. Using the same reasoning, the sound pressure is at or above 50 dBA during 90% of the time, and thus $L_{90} = 50$ dBA.

Because L_{50} does not fall at an endpoint of an interval, it must be calculated by interpolation. By inspection, the area under the histogram to the right of 60 dBA accounts for 35% of the total time. Similarly, the area to the right of 55 dBA accounts for 60% of the time. Therefore, L_{50} lies between 55 and 60 dBA. Starting from the right side of the figure, the area under the histogram representing 50% of the time is

$$25y + (15 + 10 + 5 + 5)(5).$$

To determine L_{50}, this expression is set equal to one-half of the total area under the histogram, that is, $0.5 \times 5 \times 100$. Solving the resulting equation yields $y = 3$ dBA, and thus

$$L_{50} = 60 - 3 = 57 \text{ dBA}.$$

TABLE 21.4 Sound Pressure Levels of Common Noise Sources[a]

Sound Pressure Level (dBA)	Typical Source
120	Jet takeoff at 200 ft
110	Near elevated train
100	Bottling plant
90	Subway train at 20 ft
80	Pneumatic drill at 50 ft
70	Vacuum cleaner at 10 ft
60	Large store
50	Light traffic at 100 ft
40	Residential area at night
30	Soft whisper at 5 ft
20	Studio for sound pictures

[a] Based on information in Pallett et al. (1978, p. 15), and Cunniff (1977, p. 33).

TABLE 21.5 Sound Pressure Level Data Used for Histogram Construction

Interval of A-weighted Sound Pressure Levels (dBA)	Center of Interval (dBA)	Number of Seconds in Which dBA is within Interval	Percentage of Time dBA is within Interval
$45 \le L < 50$	47.5	60	10
$50 \le L < 55$	52.5	180	30
$55 \le L < 60$	57.5	150	25
$60 \le L < 65$	62.5	90	15
$65 \le L < 70$	67.5	60	10
$70 \le L < 75$	72.5	30	5
$75 \le L < 80$	77.5	30	5
Total		600	100

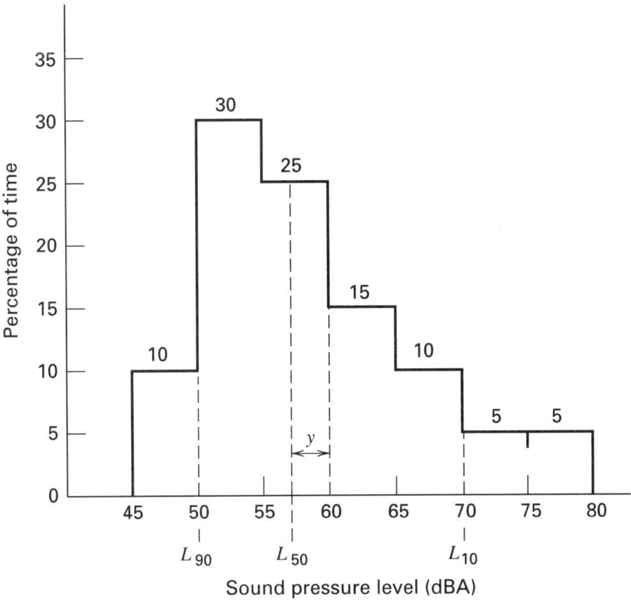

Figure 21.4 Determination of L_x from a histogram.

Another statistic characterizing the noise level during a time interval is the *energy equivalent level,* L_{eq}. For any fluctuating sound occurring during a particular time period, L_{eq} is the constant sound pressure level that has the identical acoustic energy over the period.

The basis for computing L_{eq} is that a sound's acoustic energy is directly proportional to the square of its rms pressure. Consider, for example, a 1-hour interval during which the sound has an rms pressure of p_1 for the first 30 minutes and an rms pressure of p_2 during the second 30 minutes. If the equivalent sound, which is constant for 1 hour, is to have the same acoustic energy, it follows that

$$p_{eq}^2 = \tfrac{1}{2}\, p_1^2 + \tfrac{1}{2}\, p_2^2 \qquad \textbf{(21-4)}$$

where p_{eq} is the rms pressure of the energy equivalent sound. The right side of the equation represents the average squared rms pressure of the two actual sounds. Using equation 21-4, together with the definitions of logarithms and sound pressure levels, it can be shown that

$$L_{eq} = 10 \log \left(\tfrac{1}{2}\, 10^{L_1/10} + \tfrac{1}{2}\, 10^{L_2/10}\right). \qquad \textbf{(21-5)}$$

Cunniff (1977) derives the following generalization of this relationship:

$$L_{eq} = 10 \log \left(f_1\, 10^{L_1/10} + f_2\, 10^{L_2/10} + \ldots + f_n\, 10^{L_n/10}\right) \qquad \textbf{(21-6)}$$

where f_i is the fraction of time during which the sound pressure level is constant at L_i, and $i = 1, \ldots, n$.

Equation 21-6 may be used to determine L_{eq} from noise histogram data. Consider, for example, Table 21.5. The table entries may be treated as if they resulted from seven sounds with pressure levels given by the "center intervals."[16] The appropriate time fractions are in the "percentage of time" column. Substituting these values into equation 21-6 yields $L_{eq} = 66.7$ dBA.

A variation of L_{eq} known as the *day-night sound level* (L_{dn}) is sometimes used to characterize community noise. L_{dn} is computed in the same way as L_{eq} over

[16] The noise in an interval is assumed to be adequately represented by the sound pressure level at the midpoint of that interval.

a 24-hour period, except all sound pressure levels occurring between 10:00 P.M. and 7:00 A.M. are adjusted by adding 10 dBA. The adjustment reflects the increased annoyance, particularly interference with sleep, caused by nighttime noise.

A typical computation of L_{dn} first determines the L_{eq} for each hour in a 24-hour period. Then 10 dBA is added to each hourly value between 10:00 P.M. and 7:00 A.M. Finally, the equivalent sound pressure level for this adjusted set of 24 hourly values is calculated using equation 21-6.

The indicators L_{dn}, L_{eq}, and L_x (for $x = 10, 50$, and 90) are widely used both to describe and to regulate community noise. Projected values of these statistics are employed to assess the impacts of physical development projects.

EFFECTS OF NOISE ON PEOPLE

Noise influences people adversely in several ways.[17] One negative effect is hearing loss, an important concern for some workers such as construction equipment operators and rock musicians. Hearing loss is generally reported as shifts in the *threshold of hearing,* the sound pressure level at which a tone of a given frequency is barely audible to a person with unimpaired hearing. The threshold is different at different frequencies. Aspects of hearing loss are illustrated by an experiment involving one person who was exposed to a "broad-band" sound of 103 dBA for 2 hours. Just prior to the experiment, the individual's hearing threshold was measured at several frequencies. The person's hearing thresholds were measured again at different times after the noise source was removed. For the individual being tested, there was a temporary threshold shift of 50 dBA (at 4000 Hz) soon after the noise was eliminated. In other words, compared to the pre-exposure condition, it took an additional 50 dBA before a tone at 4000 Hz was barely audible. The threshold shift was about 20 dBA after 5 hours. One day later, the person's hearing ability was almost back to normal.[18] Exposure to very high noise levels over long periods can lead to permanent threshold shifts.

In contrast to places like construction sites and rock concert halls, noise levels within the community at large do not generally pose a threat to hearing. To estimate the sound pressure levels at which community noise could impair hearing, the U.S. Environmental Protection Agency (EPA, 1974) examined data on hearing loss in work environments. It concluded that a daily average L_{eq} of less than or equal to 70 dBA would protect the public from hearing damage.

Interference with speech communication is a common effect of community noise. Research has clarified the specific noise conditions under which speech communication can be carried out. Illustrative results are in Table 21.6. It indicates maximum sound pressure levels at which conversation is feasible outdoors. For example, at 70 dBA conversation at normal voice levels is virtually impossible.

Sleep disturbance is another noise impact that has received attention by researchers. The extent to which different sound pressure levels awaken people has been studied in a variety of settings. It has been found, for example, that the probability of a person

TABLE 21.6 Conversation Levels Required to Transmit Speech Outdoors under Various Noise Conditions[a]

Distance (m)	Conversation Level		
	Relaxed[b]	Normal Voice[c]	Raised Voice
1	45	65	72
2	40	60	65
3	36	56	62

[a] Table entries are the maximum steady sound pressure levels (in dBA) at which speech is intelligible under the conditions indicated.

[b] Relaxed conversation dBA values are for 100% "sentence intelligibility"; this represents an ideal environment for speech communication.

[c] Normal and raised voice conversation figures are for 95% sentence intelligibility, which is often considered adequate for communication outdoors.

Source: Based on information in U.S. Environmental Protection Agency (1974, p. D-5).

[17] Much information is available concerning the effects of noise on people; Sáenz and Stephens (1986) provide a detailed treatment of the subject.

[18] Experiments of this type are common. The figures reported here are from a graph of experimental results in Purdom and Anderson (1980, p. 384).

being awakened by a peak sound pressure level of 40 dBA is only 5%, but it increases to 30% if the level is 70 dBA, and to about 90% for an 80-dBA noise.[19] It is, of course, common knowledge that noise can awaken people or interfere with their ability to fall asleep. It is less well known that noise can cause a shift in the stage of sleep (for example, from deep sleep to light sleep) without causing a person to wake up. The implications of these more subtle effects of noise are being investigated by sleep research specialists.

In addition to causing hearing loss and interference with speech and sleep, community noise affects people's ability to perform complex tasks. Researchers have found that noise is more likely to reduce the accuracy of work than to reduce the quantity of work performed.[20] Other areas of research center on hypothesized linkages between noise and psychological stress and physiological disorders.[21]

Instead of investigating particular effects of noise, some researchers have surveyed communities to determine the extent to which people become annoyed at different sound pressure levels. Results from different surveys are difficult to compare because they often use dissimilar measures of noise (for example, L_{50} versus L_{dn}). Moreover, rating scales defining degrees of annoyance differ from one study to the next. Schultz (1979) attempted to put the data from 11 community surveys of various forms of transportation noise on a common footing, and his results are summarized in Figure 21.5. For each survey, annoyance levels were described in terms of *percentage highly annoyed,* and sound pressure levels were put in the form of L_{dn}. The results show a clustering around the curves in Figure 21.5. Although Schultz's analysis is controversial, it represents an interesting attempt to combine disparate research results and make them more generally useful.[22]

[19] These probability estimates are from Cunniff (1977, p. 111), and Rau and Wooten (1980, p. 4–31).

[20] Other effects of noise on task performance are reported by Cunniff (1977, p. 113).

[21] Cohen et al. (1981) review the literature on how noise affects various aspects of human physiology.

[22] The extent of controversy over this analysis is reflected in the discussion published with Schultz's (1979) paper. For a discussion of recent efforts to update Schultz's work, see Cowan (1994, p. 49).

NOISE IMPACT ASSESSMENT PROCESS

A typical assessment of noise impacts from a proposed project involves six steps.

1. Establish background noise levels.
2. Determine which land uses are potentially affected by project noise.
3. Identify applicable noise regulations and criteria.
4. Forecast future noise "with and without" the proposed project.
5. Compare predicted noise with applicable criteria.
6. Modify plans, if necessary, to deal with potential noise problems.

The first step, establishing background noise levels in the vicinity of the proposed project, can sometimes be carried out by consulting governmental units such as city planning departments and transporation agencies. They may have results from recent noise surveys. Existing conditions can also be estimated by metering community noise directly. If the proposed project is near roadways or airports, background noise can be estimated using mathematical noise-forecasting models described later in this chapter. Much information about the sound pressure levels adjacent to transportation facilities is available, and it can be used with forecasting models to calculate sound pressure levels under existing conditions.

The second step in the noise assessment process is to determine land uses potentially influenced by noise from a proposed project. Existing (and expected) land uses in the vicinity of a project can usually be ascertained from local planning agencies. Special efforts are generally made to identify noise-sensitive land uses, such as schools, hospitals, and convalescent homes.

Government sources can be consulted in carrying out the third step: determining noise regulations and criteria applicable to the potentially affected land uses. Regulations vary from place to place.[23] In the United States, for example, municipal noise codes and ordinances are a source of criteria. These ordinances and codes frequently place limits on the maxi-

[23] For information on noise regulations in OECD member nations, see OECD (1991). The U.S. Department of Housing and Urban Development (1985) and Canter (1996, pp. 311–318) provide details on noise regulations and guidelines issued by federal agencies in the United States.

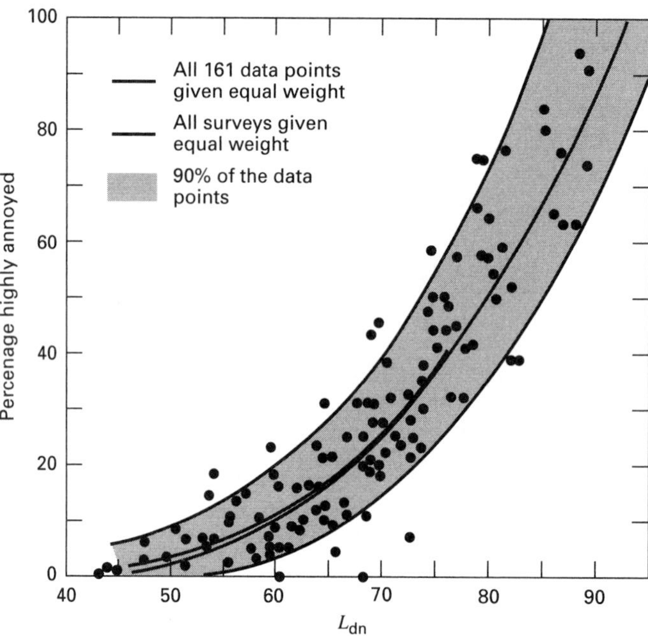

Figure 21.5 Data from 11 community surveys of annoyance with transportation noise. From Schultz in Pepper and Rodman (eds.), *Community Noise* (ASTM Special Tech. Pub. 692), 1979. Copyright ASTM. Reprinted with permission.

mum sound pressure level or the maximum allowable increase above background levels. Sometimes statewide regulations of this type exist. At the federal level, noise regulations and guidelines have been issued by several U.S. agencies, but no single agency has a comprehensive mandate to manage community noise pollution.[24]

Forecasting noise impacts, the fourth step in the assessment process, involves delineating scenarios of future conditions with and without the project. For example, if the proposal is to widen an existing highway, the scenarios would include information on average daily traffic, mix of vehicles (cars, diesel trucks, and so forth), and speed of vehicles. Conditions for both the existing highway and the proposed widened highway would be projected. For each scenario, mathematical models could be used to forecast future sound pressure levels at nearby locations.

[24] For information on federal, state, and local noise regulations and guidelines in the United States, see Cowan (1994, chap. 5).

Once noise impacts are predicted, they can be compared to applicable noise regulations and criteria, the fifth step in the assessment process. If applicable requirements are satisfied, further analysis may be unnecessary. However, if they are not met, then steps to meet the requirements are investigated. Actions to lessen noise problems can be grouped in three categories: source, path, and receptor. For example, suppose a proposed highway would violate acceptable noise levels for a nearby elementary school. The noise *source* could be reduced by decreasing the capacity of the highway. The noise *path* could be modified, for instance, by putting a 15 foot-high solid wall between the proposed highway and the school. In addition, the *receptor* could be changed: in this case, the school building could be purchased and converted to a commercial facility, and the school could be relocated.

Although modifications to both source and receptor are sometimes made, a typical solution to noise problems caused by a physical development project is to modify the sound's path. This is often done using noise barriers. As explained later in this chapter, the effectiveness of barriers is limited, and they sometimes create their own adverse impacts.

Because construction activities and transportation facilities are among the principal sources of community noise, they deserve special attention. The following sections on construction and transportation facilities illustrate aspects of the six-part assessment procedure.

NOISE FROM CONSTRUCTION AND MAINTENANCE ACTIVITIES

Virtually all physical development projects, from the extension of a driveway to the implementation of public works, involve a construction phase. One need only contemplate what it is like to converse in the vicinity of a jackhammer to imagine the possible social disruption linked to construction activities.

Maintenance activities undertaken outdoors—for example, tree cutting and shredding—can also be a significant source of noise. Table 21.7 shows sound pressure levels for typical construction and maintenance activities.

TABLE 21.7 Illustrative Sound Pressure Levels for Construction and Maintenance Equipment at about 50 Feet

Source	Sound Pressure Level (dBA)
Construction Equipment[a]	
Concrete mixers	74–86
Compressors	77–87
Jackhammers	82–96
Tractors	78–84
Maintenance Equipment[b]	
Chain saws	82
Lawn mowers	74
Leaf blowers	76
Snow blowers	84

[a] Values for construction equipment are based on bar graphs of data from the U.S. Environmental Protection Agency as presented by Canter, *Environmental Impact Assessment,* 2d ed. © 1996, p. 321. Reprinted with permission of The McGraw-Hill Companies.

[b] Values for maintenance equipment are based on U.S. Environmental Protection Agency data presented by Davis and Cornwell, *Introduction to Environmental Engineering,* 2d ed. © 1991, p. 538. Reprinted with permission of The McGraw-Hill Companies.

Forecasting Construction and Maintenance Noise

Noise from construction equipment is often characterized by a sound pressure level (L_1) at a particular distance (r_1) from the source. Forecasting construction noise typically involves using known values of L_1 and r_1 to find the sound pressure level (L_2) at a specified distance (r_2) from the source. In the simplest case, the equipment is on a flat site free of obstructions that influence sound transmission. Under these conditions, sound pressure decreases with distance primarily because of *wave divergence.* As the distance from the noise source increases, the emitted sound energy is distributed over larger and larger surface areas. The sound diminution is described using

$$L_2 = L_1 - 10 \log \left(\frac{r_2}{r_1}\right)^2. \qquad \textbf{(21-7)}$$

This relationship holds for a *point source* such as a jackhammer. A different equation is used for a *line source* such as a highway containing a steady stream of traffic. Equation 21-7, sometimes referred to as the *inverse square law,* indicates that the sound pressure level for a point source in an unobstructed, "loss-free" environment decreases by 6 dB each time the distance from the source is doubled.[25] If the noise source is obstructed by nearby walls or buildings, equation 21-7 must be modified to account for reflections of sound waves off the obstructing surfaces.

The inverse square law is demonstrated by considering an unobstructed portable air compressor on a flat construction site.[26] The compressor in this example causes a sound pressure level of 75 dBA at a

[25] To see this, use equation 21-7 with $r_2 = 2r_1$. The inverse square law does not apply when the locations involved are very close to the noise source. Magrab (1975, p. 4) gives a precise definition of "very close." Hothersall and Salter (1977, p. 106) indicate that for motor vehicles, the 6-dB reduction applies only for distances greater than about 7 meters from a vehicle.

[26] Air compressors are used to operate construction equipment such as pavement breakers and rock drills.

distance of 50 feet. Suppose it is necessary to predict the sound pressure level at a property line of the construction site, a point 200 feet from the compressor. Using equation 21-7 with $r_1 = 50$ ft, $r_2 = 200$ ft, and $L_1 = 75$ dBA yields the required estimate:

$$L_2 = 75 - 10 \log \left(\frac{200}{50}\right)^2 = 63.0 \text{ dBA}.$$

Wave divergence is not the only factor causing sound pressure levels to decrease with distance. Physical barriers, atmospheric absorption effects, and climatic influences combine to further diminish sound. As shown by Rau and Wooten (1980), the effects of this "excess attenuation" can be estimated empirically and represented as factors (in units of dB) to be subtracted from the sound pressure levels given by equation 21-7. Of the causes of excess attenuation, physical barriers receive the greatest attention in impact assessments because they are often proposed as ways of reducing noise problems.

Regulations Controlling Construction Noise

Construction-related noise has been restricted by many governments, and examples are given by federal and local regulations in the United States. The U.S. Environmental Protection Agency (EPA) limits noise from newly manufactured construction equipment under the Noise Control Act of 1972. For instance, consider the regulations for portable air compressors promulgated by EPA in 1978: compressors with rated capacities at or above 75 ft^3/min must have an average sound pressure level of 76 dBA at a point 7 meters (approximately 23 feet) from the compressor.[27] Another federal effort to control construction noise is the General Services Administration's limits on noise from equipment at federal construction projects.

At the local level, the principal mechanism for restricting noise from construction and other sources is the community noise ordinance. As an example, consider a portion of a 1971 Chicago ordinance:

> It shall be unlawful for any person to use any pile driver, shovel, hammer, derrick, hoist tractor, roller or other mechanical apparatus operated by fuel or electric power in building or construction operations between the hours of 9:30 p.m. and 8:00 a.m., except for work on public improvements and work of public service utilities, within 600 feet of any building used for residential or hospital purposes.

In addition to prohibiting selected construction activities during various periods, community noise ordinances often set a maximum value of L_{10} or L_{eq} at particular locations such as property lines.

ASSESSING AND MITIGATING HIGHWAY NOISE

For many people, motor vehicles, airplanes, and trains are the greatest contributors to outdoor noise. Thus noise assessments are often a major component of environmental impact studies for proposed transportation facilities. Transportation noise impacts can also be significant for nontransport projects such as shopping centers that generate new motor vehicle traffic. Because of the importance of road networks as a noise source, much attention has been devoted to predicting and mitigating highway noise. Both prediction and mitigation of highway noise are examined here.

Forecasting Noise from Highways

Methods exist to forecast how highway traffic affects sound pressure levels at different locations. The A-weighted sound pressure level is often employed in these procedures, since it is considered a reasonable predictor of human response to noise from motor vehicles.

Some highway-noise-forecasting models are based on physical principles such as the law of conservation of energy. The simplest of these models represents a highway as an infinitely long source emitting a constant sound intensity. It assumes the road is on level terrain with no obstructions providing excess attenuation of the sound. The model indicates that wave divergence will cause the sound pressure level to decrease by 3 dBA each time the distance from the source is doubled.[28] For example, a highway noise

[27] This compressor regulation and the following excerpt from the Chicago noise ordinance are quoted by Cunniff (1977, p. 197).

[28] This modeling result is shown by Lyons (1973).

of 70 dBA at 50 feet would diminish to 67 dBA at 100 feet, assuming wave divergence is the cause of sound diminution.

Although models based exclusively on physical laws provide important insights, they are not used extensively in practice. Mathematical models derived from *both* empirical studies and theoretical principles are more commonly employed. Numerous mathematical models have been developed for predicting highway noise, and many of them are similar to each other. To provide an illustration, the following discussion introduces a widely used forecasting approach developed for the National Cooperative Highway Research Program (NCHRP).[29]

Computing Reference Values of L_{50}

The NCHRP procedure has distinct models for passenger cars and trucks. Separate models are appropriate because the key source of automobile noise is typically tire-roadway interaction, whereas the engine is the dominant noise source for most trucks moving at moderate speeds.

The NCHRP method finds reference values of L_{50} for a given vehicle type (auto or truck). The *reference conditions* consider the roadway as an infinitely long line source at grade on flat, unobstructed terrain. Vehicles are assumed to be distributed uniformly along the road. Empirical analyses indicate that under these circumstances, L_{50} can be reasonably well predicted using the following variables:

V = volume of traffic in vehicles per hour (vph)
S = average vehicle speed in miles per hour (mph)
D = distance from centerline of roadway to sound receptor (feet).

The NCHRP studies show that, for both autos and trucks, L_{50} increases as V increases, and decreases as D increases. However, the relationship between L_{50} and S depends on vehicle type. For passenger cars, L_{50} increases approximately in proportion to the third power of speed, whereas for trucks, L_{50} typically decreases with increasing speed. Analyses of field measurement and theoretical models of traffic noise yield equations relating L_{50} (in units of dBA) to V, S, and D. An example is the following formula for autos:[30]

$$L_{50} = 10 \log V - 15 \log D + 30 \log S + 10 \log \left[\tanh\left(1.19 \times 10^{-3} \frac{VD}{S}\right)\right] + 29. \quad \textbf{(21-8)}$$

This relationship, and an analogous one for trucks, was used to prepare sets of easy-to-use curves. Figure 21.6 contains a plot of equation 21-8 with $D = 100$ ft. It applies only to autos on roadways satisfying the noted reference conditions, and it gives L_{50} for different combinations of V and S. For example, if there were 1000 autos per hour passing continually on a long, straight, flat stretch of highway at an average speed of 50 mph, the forecast of L_{50} at $D = 100$ ft would be approximately 63 dBA based on the curves in Figure 21.6.

Adjusting Reference Values of L_{50}

A noise-forecasting procedure that could only be used for highways meeting the reference conditions would be very limited. Most highways have complex vertical and horizontal configurations, and they are typically surrounded by barriers in the form of walls, buildings, and vegetation. In addition, traffic is not distributed evenly. The NCHRP method treats these departures from the reference conditions by adjusting the reference values of L_{50}.

The overall NCHRP procedure is as follows. The roadway under consideration is broken up into elements having similar slope, road surface type, and other features. Figure 21.7 shows a road that has been divided into elements. It is assumed that V and S are constant on any one segment. For each road segment, estimates are made of V and S for both autos and trucks. Reference values of L_{50} are determined using the curves in Figure 21.6 for autos and analogous curves for trucks. Empirical and theoretical relationships are used to adjust the reference L_{50}s based on the following factors:

- Roadway width and number of lanes.
- Extent of vertical elevation or depression.
- Existence of "noninfinite" roadway lengths.
- Shielding by roadside barriers, structures, and landscaping.
- Roadway slope and road surface materials.
- Interruptions in traffic flow due to stop signs and signals.

To demonstrate how adjustments are made, suppose the road element in question had an average slope

[29] This discussion is based on Hothersall and Salter's (1977) summary of the NCHRP procedure. A full description of the NCHRP approach is given by Gordon et al. (1971).

[30] The term *tanh* stands for hyperbolic tangent, a trigonometric function given in handbooks of mathematical tables and functions.

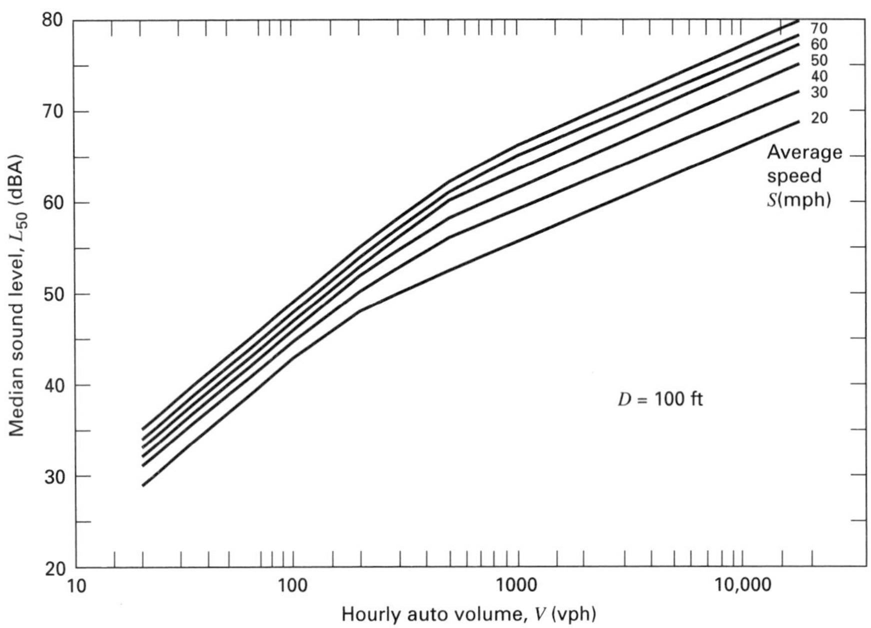

Figure 21.6 Relationship between L_{50} for automobiles and vehicle speed and volume of traffic. From Gordon et al. (1971). Reprinted with permission of Cooperative Research Programs, Transportation Research Board.

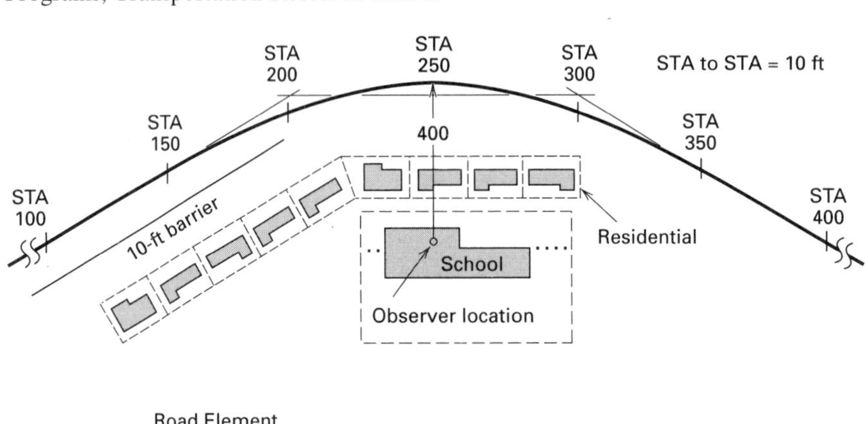

Road Element No.	Description
1	Stations 100–200 on level roadway
2	Stations 200–300 on depressed roadway
3	Stations 300–400 on depressed roadway

Ortolano/ Enviroment Figure 21.7 W-129 Wiley

Figure 21.7 Plan view of proposed highway and surroundings. Adapted from Gordon et al. (1971). Reprinted with permission of Cooperative Research Programs, Transportation Research Board.

of 5%. In this case, the NCHRP procedure adjusts the reference L_{50} for trucks by adding 3 dBA to it. No adjustment is made for the reference L_{50} for autos, because noise does not increase significantly when cars travel uphill.

Once the adjusted L_{50}s are computed for both autos and trucks, the NCHRP procedure determines L_{10}s. Again, empirical and theoretical relationships are employed for this purpose. The last step of the NCHRP approach combines the sound pressure levels obtained separately for autos and trucks. This is done using logarithmic addition as in equation 21-3.

Experience with Highway Models

The NCHRP procedure is just one of many highway noise prediction methods in use. It forecasts L_{50} and L_{10}. Other procedures, such as the model developed by Kugler and associates (1977), rely on L_{eq} and L_{dn} as descriptors of highway noise.[31]

For several highway-noise-forecasting models, comparisons have been made between predicted noise levels and corresponding observed noise levels. An example is given by Cohn and McColl (1979). They determined values of L_{10} using noise measurements adjacent to highways in the Albany, New York, metropolitan area. Fifty-two sets of short-term (15 to 20 minutes each) noise data were employed in the analysis. Cohn and McColl used the NCHRP procedure to predict L_{10}s for each of the 52 cases. The average difference between predicted and measured values of L_{10} was 2.44 dBA.[32] The analysis also indicated traffic flow conditions under which other forecasting procedures gave better predictions than the NCHRP method.

Mitigating Highway Noise Using Barriers

Many countries try to restrict noise from highways. A common form of regulation limits noise for particular sources, such as motorcycles, buses, and heavy trucks. This strategy has been used in the United States, Japan, and many European nations. Another regulatory approach involves standards on ambient noise for new highway projects. As an illustration, Table 21.8 shows a few of the design standards for new construction developed by the Federal Highway Administration in the United States.[33] Still another regulatory technique involves prohibiting use of heavy vehicles on certain routes.

What can be done if a proposed highway project will generate noise that violates applicable standards? As previously mentioned, noise mitigation measures can be grouped into actions that influence the source, path, and receptor. Mitigation actions available to highway agencies typically involve path modifications, and the measure of choice is often the *noise barrier*.[34]

Because of the way sound is redirected by a barrier, efforts to mitigate highway noise using barriers can only be partially effective. Redirected paths, which are illustrated in Figure 21.8, can be grouped into three categories: transmission, diffraction, and reflection.

The *transmission* path consists of portions of the original sound wave that go through a barrier. If a few rows of trees are used as a barrier, penetration will be high, which explains why highway agencies often rely on more substantial materials for barriers. For example, the California Department of Transportation (1992, p. 14) recommends use of materials that are durable and weigh at least 4 pounds per square foot of barrier. In these circumstances, noise from the transmission path is negligible on the receptor side of a barrier.

---

[31] Goldstein (1979) provides equations for converting from one set of noise indicators to another.

[32] The standard deviation of the differences between the observed and predicted values of L_{10} was 2.92 dBA. Observed values of L_{10} ranged from 52 to 80 dBA. Comparative assessments were also conducted for L_{50} and L_{eq} as predicted using the method in Kugler et al. (1977).

[33] According to the U.S. Environmental Protection Agency (EPA, 1978, p. D-1), the Federal Highway Administration (FHWA) "views the design noise levels [in Table 21.8] as the upper limit of acceptable traffic noise conditions, recognizing that in many cases the achievement of lower noise levels would result in even greater benefits to the community." Even if these upper limits are violated by a proposed project, FHWA may approve the project if the "adverse social, economic and environmental effects of providing abatement measures are too high."

[34] A report by Caltrans, the California Department of Transportation (1992, p. 13), makes the case for barriers as follows. Caltrans has encouraged legislation to reduce motor vehicle use (source reduction), but without success. It has also had limited success with noise source reduction via pavement redesign. Mitigation at the receptor can be accomplished via land use planning and building design (for example, using noise-insulating materials), but these are not actions that Caltrans has authority to undertake. The creation of buffer zones by purchasing additional land is another possible mitigation measure. However, it is not cost-effective. For Caltrans, the mitigation option most commonly used is the barrier.

TABLE 21.8 Examples of Design Noise Levels Developed by the U.S. Federal Highway Administration

Activity Category	Design Noise Levels[a] (dBA)		Description of Activity Category
	L_{eq}	L_{10}	
A	57 (Exterior)	60 (Exterior)	Tracts of land on which serenity and quiet are of extraordinary significance and serve an important public need, and where the preservation of those qualities is essential if the area is to continue to serve its intended purpose. Such areas could include amphitheaters, particular parks or portions of parks, open spaces, or historic districts which are dedicated or recognized by appropriate local officials for activities requiring special qualities of serenity and quiet.
B	67 (Exterior)	70 (Exterior)	Picnic areas, recreation areas, playgrounds, active sports areas, and parks which are not included in category A, and residences, motels, hotels, public meetings rooms, schools, churches, libraries, and hospitals.

[a] Either L_{eq} or L_{10} design noise levels may be used; both L_{eq} and L_{10} are hourly values in this case.

Source: Portions of Federal-Aid Highway Program Manual as reprinted by U.S. Environmental Protection Agency (1978, pp. 1–7 and D-2).

Part of the acoustic energy passing over the top of a barrier *diffracts* (or bends) downward on the receptor side. Because of diffraction, highway noise bends around the edges of barrier openings and downward from barrier tops.[35]

The third path in Figure 21.8 involves *reflection* of noise off the barrier. By selecting a barrier surface with good sound-absorptive qualities, the extent of reflection can be minimized. If sound absorption is low and two parallel barriers are used on either side of a road, waves reflected off one wall can be reflected off the other wall and onto the receptor side of the barrier.

The effectiveness of a highway noise barrier is measured by *insertion loss,* the net noise reduction at the site of the receptor. Insertion loss is the difference in sound pressure level at the receptor site with and without the barrier. In general, it is not difficult to attain insertion losses of 10 dBA, but 15 dBA is difficult, and 20 dBA is nearly impossible.[36]

While highway noise barriers have reduced noise for some people on the receptor sides of barriers, they are alleged to have increased noise problems for others. For example, in Brentwood, California, homeowners living between 0.2 and 0.4 miles from a state highway brought a lawsuit in 1987 based on claims that a newly constructed highway noise barrier caused noise to increase for them. As a result of this court action and similar complaints in other parts of the state, the California Department of Transportation conducted research to identify whether amplification of waves reflected off parallel barriers is causing noise problems as far as two miles from highways with newly installed barriers. Results to date suggest that barriers are not to blame for increasing noise, but complaints from citizens persist.

[35] To reduce effects of diffraction around openings, some highway planners suggest that barriers have a length that is at least four times the distance between the barrier and the receptor (Cowan, 1994, p. 123).

[36] In the United States, the Federal Highway Administration, as quoted by Cowan (1994, p. 123), uses the following rule of thumb for estimating insertion loss: "5 dBA when the line of sight [between source and receptor] is broken and 1 dBA more for each additional 2 ft of barrier height."

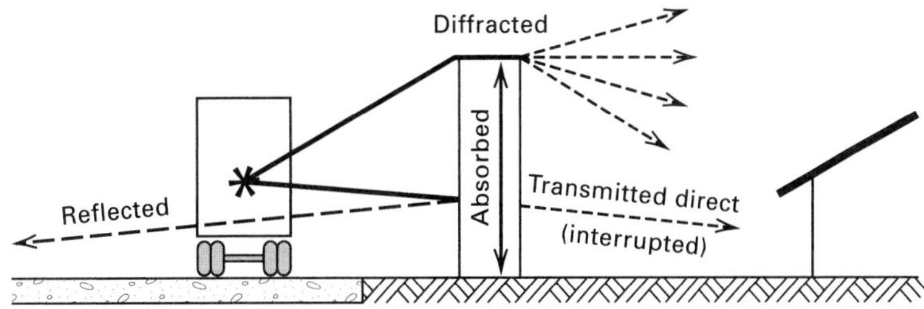

Figure 21.8 Noise paths altered by a barrier. Source: California Department of Transportation (1992, p. 13).

Increased noise is not the only problem associated with barriers. Some people are adversely affected by how barriers modify views. Environmental assessments of proposed highways sometimes use the visual simulation methods described in the previous chapter to assess visual impacts of barriers proposed as noise mitigation measures.

FORECASTING NOISE FROM AIRPORTS

The unique characteristics of aircraft noise have led to the development of special noise indicators. They fall into two related categories: (1) measures of the *perceived noisiness* of the takeoff or landing of a particular aircraft type, and (2) measures of *community annoyance* resulting from the combined effects of all aircraft using an airport in a typical time period.

Perceived Noisiness

Noise from a single aircraft flyover is often described by *perceived noise level* (PNL). The design of this indicator rests on experiments in which individuals indicated their reactions to the "noisiness" or annoyance of different sounds.[37] The experiments indicated that at higher frequencies, lower sound energies are required to produce a given level of annoyance or noisiness. In other words, high-frequency noise is judged to be noisier than equally loud noise of low frequency. Observe that noisiness is different from loudness.[38]

Perceived Noise Decibels

Perceived noise level, which is based on experimental tests measuring noisiness, is recorded in *perceived noise decibels* (PNdB). A doubling of *perceived noisiness* is equivalent to an increase of 10 PNdB.[39] Because the procedure for determining PNL is complex, it is common to approximate PNL using easy-to-measure parameters such as the D-weighted decibel. Sound level meters can automatically apply D-weighting correction factors. Sound pressure levels in different frequency bands are corrected to approximate results from experiments testing the human perception of noisiness. The "D-weighting network" involves the same kind of electronic addition of adjusted sound pressure levels used in computing the A-weighted sound pressure level in Table 21.3. An approximate relationship between D-weighted sound pressure levels (L_d) and the perceived noise level (in PNdB) is

$$\text{PNL} \approx L_d + 7 \qquad \textbf{(21-9)}$$

where L_d is the D-weighted sound pressure level recorded during an aircraft overflight.[40] The accuracy

[37] The basis for the PNL is described more fully by Goldstein (1979).

[38] Loudness is more closely related to experiments in which listeners indicate their ability to hear sounds at different frequencies. These experiments are the basis for the A-weighted decibel.

[39] This interpretation of a doubling of perceived noisiness is explained by Magrab (1975, p. 63).

[40] For more information on equation 21-9, see Cunniff (1977, p. 155).

of this approximation to PNL depends on the spectrum of the measured sound.

Effective Perceived Noise Levels

Although PNL is a useful measure of noisiness from aircraft flyovers, it does not fully account for the following two influences:

1. As the *duration* of a single-event sound increases in time, ratings by individuals of perceived noisiness increase in magnitude.
2. If a complex sound has an identifiable *pure tone,* such as the whining of a jet aircraft, it is judged to be noisier than a similar sound without such a tone.

The *effective perceived noise level* (EPNL) accounts for these two influences by "correcting" the PNL readings as follows:

$$\text{EPNL} = \text{PNL} + C_1 + C_2 \qquad \textbf{(21-10)}$$

where EPNL is measured in EPNL decibels, and

C_1 = a factor reflecting the existence of identifiable pure tones in the sound spectrum ($0 \leq C_1 \leq 3$)
C_2 = a factor accounting for the duration of a flyover[41]

Manufacturers estimate contours of EPNL while testing new aircraft under various government certification programs. These contours, referred to as *noise footprints,* represent the loci of points on the ground where EPNL is constant for a given aircraft type performing under a particular set of conditions. Figure 21.9 shows noise footprints for a takeoff and landing. Contours of EPNL depend on operating procedures followed in takeoff and landing and on conditions of temperature and relative humidity.

Community Annoyance

Indicators of the noisiness of a particular takeoff or landing are used in constructing measures of community annoyance from an airport's operations on a typical day. The *noise exposure forecast* (NEF), a widely used measure of community annoyance from airports, provides an illustration.[42] The NEF uses contours of EPNL together with "corrections" to account for the increased community annoyance caused by nighttime aircraft operations. Procedures for computing NEF generally require computers and the following information:[43]

- Airport runway configurations.
- Number of aircraft operations by aircraft type.
- Aircraft flight plans and arrival and departure data.
- Percentage use of each flight path (by aircraft type) for takeoffs and landings.

Results from an assessment of airport noise can be presented as contours of NEF (see Figure 21.10). An analysis of NEF contours indicates particular land areas likely to be affected by a proposed airport or by changes in operations at an existing airport. Sometimes L_{dn} contours are used. The two measures are related approximately by

$$L_{dn} \approx \text{NEF} + 35. \qquad \textbf{(21-11)}$$

The U.S. Environmental Protection Agency (EPA, 1974) has issued guidelines indicating the fraction of people likely to be annoyed when L_{dn} or NEF values exceed various levels.[44]

How are predictions of NEF or related noise indicators made for projects involving new airport facilities or operations? Typically, computer models based on both theory and empirical findings are employed. An example is the Integrated Noise Model (INM), which the U.S. Federal Aviation Administration developed for predicting L_{dn} contours at airports.[45] The Integrated Noise Model contains a database including noise characteristics for over 60 types

[41] The factor C_2 is computed as $C_2 = 10 \log (\Delta t/20)$, where Δt is the time interval (in seconds) during which the noise level is within 10 PNdB of the maximum instantaneous value of the perceived noise level (Cunniff, 1977, p. 157).

[42] Goldstein (1979) discusses several other indicators of community annoyance due to airport noise.

[43] For a summary of a well-documented Federal Aviation Administration computer program for predicting *NEF* and L_{dn} contours, see Cohn and McVoy (1982, pp. 178–85).

[44] Noise at U.S. airports is also controlled by means of aircraft noise standards issued by the Federal Aviation Administration. The International Civil Aviation Organization has also adopted noise certification standards for some aircraft (Ashford and Wright, 1992, p. 485).

[45] This discussion of the Integrated Noise Model is based on Ashford and Wright (1992, pp. 496–97). For information on airport noise prediction models developed by other U.S. agencies, see Canter (1996, pp. 330–32).

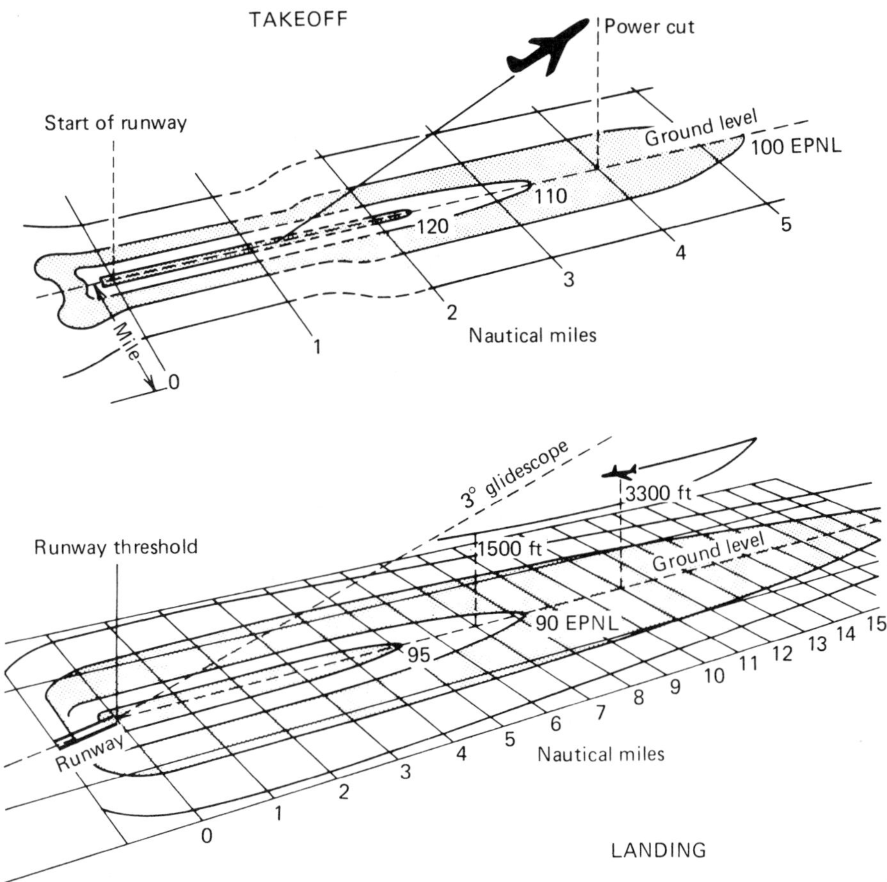

Figure 21.9 Contours of EPNL for aircraft takeoff and landing. From U.S. Department of Transportation (1972).

of aircraft. For each aircraft, the data includes points on curves of EPNL versus distance at several thrust settings. Required inputs for INM include physical characteristics of the airport (for example, runway locations and elevations) as well as details on airport operations.

In addition to predicting noise from proposed airports, models such as INM can be used to analyze how noise contours would change if mitigation measures were implemented. Actions to reduce noise at airports can involve changes in source, path, and receptor. Noise sources can be reduced by changes in type of aircraft. An example is the action taken to prohibit use of supersonic transport aircraft at most nonmilitary airports in the United States. Noise path can be changed by modifying aircraft arrival and departure procedures; for instance, by specifying takeoff speeds and turning angles. Mitigation measures that involve receptors typically involve zoning and other land use controls to keep noise-sensitive activities out of areas with high noise levels.

Although computers are often used in airport noise studies, there are methods for making simple, approximate predictions without them.[46] De-

[46] An example is the report by Pallet et al. (1978) containing worksheets and background information for making rough estimates of airport noise; the report also contains worksheets for predicting noise from highways. See, also, the prediction method presented by the U.S. Department of Housing and Urban Development (1985).

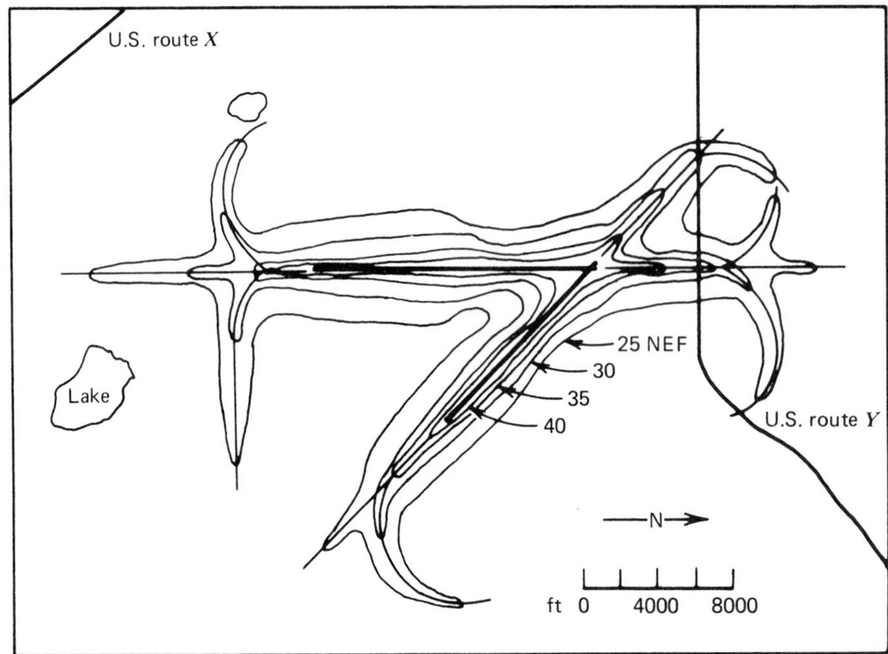

Figure 21.10 Typical NEF contours around an airport. From U.S. Department of Transportation (1972).

velopment of noise-forecasting procedures is an active field of research, and specialists in acoustics must be consulted for the most up-to-date approaches.

Key Concepts and Terms

Sound and Its Measurement
- Acoustics
- Air pressure fluctuations
- Pure tones
- Frequency of oscillation
- Cycles per second (Hz)
- Amplitude of oscillation
- Root-mean-square (rms) pressure
- Sound pressure level (dB)
- Sound spectrum
- Octave bands
- A-weighted sound pressure level (dBA)
- C-weighted sound pressure level (dBC)

Statistics Used to Characterize Community Noise
- Histogram of sound pressure levels
- Sound pressure level exceeded $x\%$ of time (L_x)
- Energy equivalent sound level (L_{eq})
- Day-night sound level (L_{dn})

Effects of Noise on People
- Shift in threshold of hearing
- Interference with speech communication
- Shifts in stages of sleep
- Community annoyance

Noise Impact Assessment Process
- Community noise survey
- Noise-sensitive land use
- Noise regulations and criteria
- Local noise ordinances
- Modifications in source, path, and receptor

Noise from Construction and Maintenance Activities
- Point source versus line source
- Inverse square law

Wave divergence
Excess attenuation

Assessing and Mitigating Highway Noise
Empirical models
Tire-roadway interaction
Engine noise
NCHRP model
Reference values of L_{50}
Noise barriers
Transmission path
Reflection and absorption
Diffraction
Insertion loss

Forecasting Noise from Airports
Noisiness versus loudness
Perceived noise level (PNL)
Perceived noise decibel (PNdB)
Effective perceived noise level (EPNL)
Noise footprints and contours
Noise exposure forecast (NEF)
Integrated Noise Model
Mitigation measures

Discussion Questions

21-1 Explain what is meant by loudness, amplitude, pitch, and frequency using musical instruments you are familiar with. Use two instruments having very different frequency ranges.

21-2 What is the purpose of filtering a sound into octave bands? Think of a circumstance where a spectrum analysis based on filtering into octave bands would be required for the design of noise mitigation measures. In responding, recognize that the effectiveness of materials used to absorb sound or block its transmission varies with sound frequency. For example, absorptive characteristics of materials are reported using an *absorption coefficient, a,* which varies from 0 to 1. A value of $a = 0$ indicates all sound energy is reflected, whereas a value of $a = 1$ indicates all sound energy is absorbed. Absorption coefficients vary with frequency.[47] Since all materials reflect and absorb some sound, a values of 0 and 1 are theoretical end points.

21-3 Data in the following table are for a metal cutter.

Center Frequency of Octave Band (Hz)	Sound Pressure Level (dB)	
	Cutter without Silencer	Cutter with Silencer
63	63	63
125	70	70
250	73	73
500	76	76
1000	92	79
2000	80	80
4000	76	76
8000	63	63

a. Use equation 21-3 to calculate the total sound pressure level (in dBA) for the cutter with and without the silencer. Correction factors needed to computed A-weighted decibels are in Table 21.2.
b. Repeat the exercise in a. using the approximate procedure for adding multiple sources of sound presented in Table 21.1. Comment on the usefulness and validity of the approximate procedure in this case.
c. Suppose the cutter will be installed in a manufacturing plant that has an existing ambient noise level of 94 dBA. Find the new ambient noise levels for two cases: (1) using the cutter without a silencer, and (2) using the cutter with a silencer.

21-4 When a compressor and a jackhammer are running simultaneously, the overall sound pressure level is 88 dBA. If the jackhammer is running separately, the overall sound pressure level is 87 dBA. Assume background noise is negligible. What is the approximate sound pressure level of the compressor?

[47] Illustrative values of a for unpainted concrete are 0.44, 0.29, and 0.25 for frequencies of 250, 1000, and 4000 Hz, respectively. Values of a for many materials used in mitigating noise impacts are tabulated by Cowan (1994, pp. 92–93).

21-5 Consider the following data for a particular street corner from 4:00 P.M. to 6:00 P.M. on a weekday.

Sound Pressure Level Intervals	Number of Minutes in which dBA is within Interval
$50 \le L < 55$	20
$55 \le L < 60$	40
$60 \le L > 65$	35
$65 \le L < 70$	13
$70 \le L < 75$	12

Compute L_{10}, L_{50}, and L_{90}.

21-6 The following data characterize the distribution of sound pressure levels at a particular point on a typical day:

Time Period	dBA
Midnight—6 A.M.	40
6 A.M.—noon	50
Noon—6 P.M.	55
6 P.M.—midnight	45

Compute L_{eq} and L_{dn}.

21-7 Calculate the equivalent sound level for the data in Table 21.5.

21-8 In the United States, community noise ordinances control maximum sound pressure levels at property lines. Cunniff (1975) reports that a typical daytime limiting value for commercial areas is 60 to 65 dBA. Explain why this regulatory approach would not be effective for controlling construction noise. What parameters (for example, L_{10} or L_{eq}) would you use to construct a more effective restraint on construction noise.

21-9 Characterize the relative difficulty in forecasting noise from construction activities, highways, and airports.

21-10 The following empirical equation has been proposed for use in predicting L_{eq} for highways:

$$L_{eq} = 42.3 + 10.2 \log (V_c + 6V_t) - 13.9 \log D + 0.13\, S$$

where

L_{eq} = energy equivalent sound level during one hour, dBA
V_c = volume of automobiles (four tires only), veh/hr
V_t = volume of trucks (six or more tires), veh/hr
D = distance from edge of pavement to receiver, m
S = average speed of traffic flow during one hour, km/hr

Consider a proposed highway in which it is estimated that 5000 vehicles per hour will travel at an average speed of 50 miles per hour. Ten percent of the vehicles will be trucks and the rest will be cars. The route of the proposed road is such that the edge of the road is 100 feet from a school. The terrain is level and there are no noise barriers. The U.S. Federal Highway Administration noise design levels indicate that new roads near schools should have exterior $L_{eq} \le 67$ dBA. Will the proposed road meet this design target? If not, what actions might be taken to improve the situation?

References

Ashford, N. and P. H. Wright. 1992. *Airport Engineering,* 3rd ed. New York: Wiley.

California Department of Transportation. 1992. *California Noise Barriers.* Report prepared by the Special Task Force on Noise Barriers, Sacramento, CA.

Canter, L. W. 1996. *Environmental Impact Assessment,* 2d ed. New York: McGraw–Hill.

Cohen, S., D. S. Krantz, G. W. Evans, and D. Stokols. 1981. Cardiovascular and Behavioral Effects of Community Noise. *American Scientist* 69(5): 528–35.

Cohn, L. F., and W. McColl. 1979. "L_{10} versus L_{eq}—A User's Perspective." In *Community Noise,* ed. R. J. Pepper and C. W. Rodman, 237–46. ASTM Special Tech. Pub. 692. Philadelphia: American Society of Testing Materials.

Cohn, L. F., and M. R. McVoy. 1982. *Environmental Analysis of Transportation Systems.* New York: Wiley.

Cowan, J. P. 1994. *Handbook of Environmental Acoustics.* New York: Van Nostrand Reinhold.

Cunniff, P. E. 1977. *Environmental Noise Pollution.* New York: Wiley.

Davis, M. L., and D. A. Cornwell. 1991. *Introduction to Environmental Engineering,* 2d ed. New York: McGraw–Hill.

EPA (U.S. Environmental Protection Agency). 1974. *Information on Levels of Environmental Noise Requisite to Protect the Public Health and Welfare with an Adequate Margin of Safety.* Technical Document 550/9-74-004. Washington, DC: EPA.

EPA (U.S. Environmental Protection Agency). 1978. *Federal Highway Administration: Noise Policy and Related Environmental Procedures.* Report EPA 550/9-77-357, Office of Noise Abatement and Control. Washington, DC: EPA.

Goldstein, J. 1979. "Description of Auditory Magnitude and Methods of Rating Community Noise." In *Community Noise,* ed. R. J. Pepper and C. W. Rodman, 38–72. ASTM Special Tech. Pub. 692. Philadelphia: American Society of Testing Materials.

Gordon, C. G., W. J. Galloway, B. A. Kugler, and D. L. Nelson. 1971. *Highway Noise: A Design Guide for Highway Engineers.* Report No. 117, National Cooperative Highway Research Program, Washington, DC.

Hothersall, D. C., and R. J. Salter. 1977. *Transportation and the Environment.* Hertfordshire, U.K.: Granada.

Kugler, B. A., et al. 1977. *Highway Noise, A Design Guide for Prediction and Control.* Report No. 174, National Cooperative Highway Research Program, Washington, DC.

Lyons, R. H. 1973. *Lectures in Transportation Noise.* Cambridge, MA: Grozier.

Magrab, E. B. 1975. *Environmental Noise Control.* New York: Wiley.

OECD (Organization for Economic Cooperation and Development). 1991. *Fighting Noise in the 1990s.* Paris: OECD.

Pallett, D. S., R. Wehrli, R. D. Kilmer, and T. L. Quindry. 1978. *Design Guide for Reducing Transportation Noise in and Around Buildings.* Building Science Series 84, U.S. Department of Commerce, National Bureau of Standards, Washington, DC.

Purdom, P. W., and S. H. Anderson. 1980. *Environmental Science, Managing the Environment.* Columbus, OH: Charles G. Merrill.

Rau, J. G., and D. C. Wooten, eds. 1980. *Environmental Impact Analysis Handbook.* New York: McGraw–Hill.

Richardson, W. J., C. R. Greene, Jr., C. I. Malme, and D. H. Thomson. 1995. *Marine Mammals and Noise.* San Diego, CA: Academic Press.

Sáenz, A. L., and R. W. B. Stephens, eds. 1986. *Noise Pollution: Effects and Control.* Chichester, U.K.: Wiley.

Schultz, T. J. 1979. "Community Annoyance with Transportation Noise." In *Community Noise,* ed. R. J. Pepper and C. W. Rodman, 87–103. ASTM Special Tech Pub. 692. Philadelphia: American Society of Testing Materials.

Sound Research Laboratories, Ltd. 1991. *Noise Control in Industry,* 3d ed. London: Chapman and Hall.

U.S. Department of Housing and Urban Development. 1985. *The Noise Guidebook.* Report No. HUD-953-CPD. Washington, DC: U.S. Government Printing Office.

U.S. Department of Transportation. 1972. *Transportation Noise and its Control.* Publication DOT P 5630.1, Washington, DC.

Chapter 22

Estimating Changes in Air Quality

Since the 1940s, much attention has been devoted to analyzing the causes and consequences of air pollution.[1] The literature on the subject is voluminous and crosses several academic disciplines. This chapter draws from that literature and introduces forecasting techniques used in formulating regulatory programs and conducting impact assessments.

THE NATURE OF AIR POLLUTION

Air pollutants are substances that, when present in ambient air, adversely affect human life and property and the natural environment. This section provides a four-part introduction to air pollutants.[2] It begins by describing contaminants widely regulated using ambient standards on air *outside* of buildings. The discussion then turns to hazardous air pollutants, which are often regulated using emission standards. Next, consideration is given to an environmental problem that is largely unregulated: air pollution *within* buildings. Finally, the section examines air quality issues of international concern: acid rain, stratospheric ozone depletion, and global climate change.

Pollutants Regulated by Ambient Standards

Many nations have introduced ambient standards on outdoor air to protect public health and the environment. While details vary from country to country, the pollutants regulated under the National Ambient Air Quality Standards (NAAQS) in the United States serve for illustrative purposes. Federal air quality laws in the United States establish national ambient standards for the following air pollutants (as of 1995): carbon monoxide (CO), ozone (O_3), nitrogen dioxide (NO_2), sulfur dioxide (SO_2), respirable particular matter, and lead (Pb).[3] Some of these

[1] Professor Lynn Hildemann of the Department of Civil Engineering at Stanford University provided a valuable critique of an early draft of this chapter.

[2] Research assistance for this section was provided by Sambhav Sankar while he was a student at Stanford University.

[3] A partial listing of the NAAQS is given in Table 8.2. See Boubel et al. (1994, p. 378) for a more complete listing. The precise definition of each pollutant in the context of implementation of the NAAQS is detailed in the *Code of Federal Regulations* (C.F.R.); see 40 C.F.R. Part 50 for details. Dozens of pages spell out the methods used to measure compliance with the standards. The pollutants included in the NAAQS are subject to change, but since the original standards were issued under the Clean Air Act of 1970, only a few changes have been made. One of the standards was renamed ("photochemical oxidants" is now referred to as ozone), another was dropped (hydrocarbons), one was modified (total suspended particulates was changed to PM-10, particulate matter with diameter <0.001 mm), and one was added (lead).

pollutants have been detected at high levels in indoor air, and damages they cause in those environments are reported later in this chapter.

Carbon monoxide, a colorless, odorless, and tasteless gas, is a product of incomplete combustion. In many places, particularly in cities, motor vehicles are the principal source of CO. Carbon monoxide combines with hemoglobin in the human bloodstream and reduces hemoglobin's ability to carry oxygen to body tissues. Depending on the degree of exposure, effects of inhaling CO range from headaches, dizziness, and nausea to unconsciousness and death. Coffin and Stokinger (1977, p. 305) report that these effects are not "to be expected at even the highest concentrations found in the ambient air. . . . However, great concern has been expressed regarding the possibility that covert effects may decrease physical or mental acuity or interfere with the function of organs already suffering oxygen deficiency, such as the heart in coronary disease." Recent research indicates that adverse physiological effects can begin at carboxyhemoglobin levels as low as about 2.5%, which corresponds to levels experienced routinely in certain occupations. For example, in a country that does not control auto emissions, driving a taxi in heavy traffic for 8 hours could bring a person's carboxyhemoglobin level to 2.5%.[4]

Volatile organic compounds (VOCs) are chemicals that contain organic carbon and vaporize readily at room temperature. Representative chemicals include benzene and carbon tetrachloride. Hydrocarbons, compounds containing only carbon and hydrogen, are an important class of VOCs. Notable sources of volatile organic compounds include the combustion exhaust from motor vehicles, and various industrial operations. For example, dry cleaning contributes large quantities of hydrocarbons to the ambient air through evaporation. Although VOCs produce some adverse effects directly, their role in chemical reactions yielding photochemical smog is of particular concern. Reaction products, referred to collectively as *photochemical oxidants,* cause a variety of ailments (such as difficulty in breathing) and damage to vegetation and property. *Ozone* is the principal reaction product and is often used as an overall indicator of photochemical oxidants.

Oxides of nitrogen (NO_x), especially nitrogen dioxide and nitric oxide, are characteristic products of high-temperature combustion processes. The most significant sources of NO_x emissions in the United States are motor vehicles and coal- and gas-fired power plants. Nitrogen oxides play an important role in producing photochemical oxidants. In addition, nitrogen dioxide has been linked to increases in upper respiratory infections.

Emissions of *sulfur oxides* (SO_x), mainly sulfur dioxide, also result from combustion processes, in this case the incomplete combustion of fuels containing sulfur impurities. Although much of the concern over SO_x emissions is directed toward coal-fired power plants, industrial facilities such as copper smelters are also an important source. Sulfur dioxide in the ambient air has been tied to human health problems, especially respiratory disorders, and to the formation of acid rain.

The term *particulates* refers to the multitude of solid and liquid particles present in the atmosphere. Somewhat more than half of airborne particulates are estimated to originate from natural sources such as dust storms and volcanic eruptions.[5] The main nonnatural sources of particulates are coal combustion and various industrial processes. Forest fires are also a major contributor. Particulates exist in a range of sizes and can absorb many chemicals present in the ambient air. High concentrations of particulates have been associated with increases in respiratory diseases and gastric cancers, poor atmospheric visibility, and the soiling of buildings and other materials. The original version of the NAAQS regulated total suspended particulates. During the 1980s, that parameter was replaced by PM-10, particulate matter with diameter $\leq$ 10 microns.[6] These particles are small enough to be breathed into the lungs. Environmentalists in the United States have called for ambi-

[4] Turiel (1985, p. 55) reports that exposure to air with 15 ppm of carbon monoxide for 10 hours is enough to reach a carboxyhemoglobin level of about 2.5%. In 1996, before federal vehicle emission controls were required in the United States, average in-vehicle CO concentrations measured within several American cities were about 32 ppm. In the 1990s, after federal controls, average in-vehicle CO emissions were less than 5 ppm (Flachsbart, 1995, p. 487). For a more complete account of adverse effects of CO, see U.S. Environmental Protection Agency (EPA, 1991).

[5] For a breakdown of sources of airborne particulates, see Perkins (1974, pp. 30–32), and Masters (1991, p. 292–94).

[6] One micron equals one thousandth of a millimeter.

ent standards on even smaller particles, those less than 2.5 microns, since high illness and mortality rates appear to be closely linked to particulates in that size range.

Lead in the atmosphere results primarily from motor vehicles fueled with leaded gasoline. A variety of human health problems, including liver and kidney damage and nervous system disorders, have been related to atmospheric lead. In the United States, concern over these problems inspired a nationwide effort to eliminate lead additives in gasoline, which caused concentrations of atmospheric lead to drop sharply during the 1980s. Notwithstanding these reductions, the exposure of children to lead continues to be a concern. Many children are exposed to old paint chips containing lead, and to soil containing lead that settled out from the atmosphere years ago.

The U.S. approach for selecting pollutants to regulate via ambient standards is far from universal. For example, some nations within the European Union have ambient standards just on sulfur dioxide and smoke, nitrogen oxide, and lead.[7] These pollutants were the subject of environmental directives passed by the European Community during the 1980s.[8] The directive establishing standards for SO_2 and smoke recognizes that these two pollutants often have synergistic effects on human health. The standard for SO_2 is such that the more smoke present, the less the amount of allowable SO_2.

Hazardous Air Pollutants

During the past few decades, regulators have given increased attention to air pollutants that are, in some sense, *hazardous.* The term hazardous requires clarification because even conventional pollutants such as carbon monoxide can be toxic in some circumstances. Given this ambiguity, how is the phrase *hazardous air pollutant* to be interpreted? In the context of environmental policy, the term hazardous is often defined in laws, and this means that the same substance can be termed hazardous in one regulatory setting and not another (even in the same country.)

Laws to control hazardous waste often include procedures that the implementing agency is to use in deciding which waste discharges are hazardous. This is demonstrated by the U.S. Clean Air Act of 1970. The act gave the administrator of the U.S. Environmental Protection Agency (EPA) responsibility for identifying air pollutants that were hazardous and thus subject to emission restrictions that would provide an ample margin of safety to protect the public health."[9] The act defined as "hazardous" an air pollutant that (1) was not among those for which a national ambient quality standard had been set, and (2) could (in the judgment of the EPA administrator) cause or contribute to "air pollution which may reasonably be anticipated to result in an increase in serious irreversible, or incapacitating reversible, illness." These provisions of the Clean Air Act of 1970 were not implemented effectively. As of 1989, only seven pollutants were listed as hazardous: arsenic, asbestos, benzene, beryllium, mercury, radioactive isotopes, and vinyl chloride.

Another approach to defining what constitutes a hazardous air pollutant is to include a list of hazardous substances in legislation. This is demonstrated by Section 112 of the U.S. Clean Air Act Amendments of 1990, which lists 189 air pollutants considered to be hazardous and subject to emission standards.[10] The 1990 amendments also include a process that the administrator of EPA is to follow in modifying the list. For example, the administrator is required to consider for inclusion substances "which are known to be, or may reasonably anticipated to be, carcinogenic, mutagenic, teratogenic, which cause reproductive dysfunction, or which are acutely or chronically toxic. . . ."[11]

[7] *Smoke* is often measured by drawing air through a filter paper so that the smoke particles in the air are retained on the paper. The density of the stain is correlated with the mass of suspended particulates.

[8] For details on ambient air quality standards used by member nations of the European Union, see Bennett (1991). Kormondy (1989) provides a digest of air quality laws in countries in Africa, Asia, Oceania, Europe, and the Americas.

[9] These provisions are set out in Section 112 of the U.S. Clean Air Act of 1970.

[10] Although most other countries have not been as aggressive in regulating hazardous air pollutants as the United States, those that do regulate toxics often rely on lists to define what substances are hazardous. For example, the Swedish government has developed a list of 13 substances that should be reduced in use or banned entirely by the year 2000; see Lee and Schneider (1994) for details.

[11] *Mutagenesis* refers to "changes, or *mutations,* in the genetic material of an organism [that] can cause cells to misfunction, leading in some cases to cell death, cancer, reproductive failure, or abnormal offspring. Chemicals that are capable of causing cancer are called *carcinogens;* chemicals that can cause birth defects are *teratogens.*" (Masters, 1991, p. 196).

TABLE 22.1 Examples of Hazardous Pollutants Released into Air in the United States in 1989[a]

Chemical	Release (millions of pounds)
Toluene	255
Methanol	200
1, 1, 1-Trichloroethane	169
Xylenes (mixed isomers)	148
Chlorine	132
Methyl ethyl ketone	128
Carbon disulfide	100
Hydrochloric acid	61
Trichloroethylene	44

[a] Based on Toxic Release Inventory data for the 15 toxic chemicals released into the air in the largest amounts in 1989 as reported by Bryner (1993, p. 64) and the list of hazardous pollutants in the Clean Air Act Amendments of 1990. Chemicals on Bryner's list that are not defined as hazardous in the 1990 amendments are acetone, ammonia, dichloromethane, freon-113, glycol ethers, and ethylene.

Table 22.1 indicates 9 air pollutants commonly released in the United States and defined as hazardous in the Clean Air Act Amendments of 1990. The table was created using a list of the 15 chemicals released into the air in the largest amounts as reported in the *Toxic Release Inventory.*[12] Of those 15 chemicals, only 9 are on the list of hazardous air pollutants in the 1990 amendments. This difference between air pollutants reported in the Toxic Release Inventory and those listed as hazardous in the Clean Air Act demonstrates that the precise meaning of terms such as hazardous and toxic can be established only by examining the laws and regulations that define them.

Indoor Air Pollution

Increasingly, indoor air pollution is being recognized as a source of significant health risks. This is reflected in the extensive news media coverage of the dangers of *environmental tobacco smoke,* a mixture of smoke from lit cigarettes, pipes and cigars, and smoke exhaled by smokers. The media have also reported widely on the incidence of fatal ailments such as *Legionnaire's disease,* which is caused by microbes found in the heating and air-conditioning systems of buildings. In addition, public health authorities have called attention to links between lung cancer and radon, a naturally occurring radioactive gas. In many places (for example, New Jersey and Pennsylvania), radon that exists naturally in the ground moves into residences through cracks in ground floor slabs and basement walls. Still another reflection of increased awareness of indoor air quality problems is the media attention given to asbestos, particularly asbestos in old heating insulation that dries out and enters the air when it is disturbed.

Table 22.2 includes examples of the more than 800 contaminants that have been detected in indoor

[12] The Toxic Release Inventory (TRI) includes emissions data from companies that manufacture, process, or otherwise use more than a threshold amount of chemicals listed as "toxic" under the U.S. Emergency Planning and Community Right to Know Act of 1986.

TABLE 22.2 Examples of Contaminants in Indoor air[a]

- Airborne biological contaminants
  - Bacteria
  - Mold
  - Spores
  - Viruses
- Asbestos
- Environmental tobacco smoke
- Heavy metals
  - Lead
  - Mercury
- Inorganic gaseous compounds
  - Carbon monoxide
  - Nitrogen dioxide
  - Ozone
- Radon
- Volatile organic compounds
  - Acetone
  - Benzene
  - Formaldehyde

[a] Based on data in Hines et al. (1993, pp. 2, 3, and 21). These authors report that over 800 different chemical compounds have been measured at least once in indoor air at office buildings. Most of the compounds were volatile organics, and not all were measured at the same time and place. At any one time in one building, there were as many as 50 or 60 volatile organic compounds.

air. Some of the items listed, particularly radon, environmental tobacco smoke, asbestos, and some volatile organic compounds, have been identified as cancer risks. Airborne biological contaminants, such as the bacterium *Legionnella pneumophilia,* have caused significant numbers of fatalities.

Problems associated with volatile organic compounds have been widespread, and they affect both residences and work environments. Sources of these gases in indoor air include carpets, furnishing, solvents, office equipment, heating insulation, and other building materials. There have been numerous reports of *sick building syndrome,* a term associated with structures whose occupants suffer from headaches, respiratory problems, and other complaints. The link between the symptoms and a building is made if the symptoms disappear when occupants are away from the building for a few days. Health officials are often able to pinpoint the source of difficulty, such as an inadequate ventilation system or a building material that is "outgassing" formaldehyde. Sometimes, however, precise causes of sick building syndrome remain elusive.

Many scientists believe that the common practice of setting ambient standards for only outdoor air pollutants is inadequate because it ignores health risks posed by indoor air. In the United States and elsewhere, many people spend over 80% of their time indoors. In addition, the ambient air concentrations of dangerous chemicals are often higher indoors than outdoors. For example, the U.S. Environmental Protection Agency demonstrated that for 18 volatile organic compounds targeted in a research study in five cities, the median air concentrations were 2 to 20 times higher in people's homes than in the outdoors. And this was for residences in cities with major chemical plants and petroleum refineries, where outdoor concentrations were expected to be notable.[13] In other EPA studies, concentrations of air pollutants indoors were over 100 times higher than for the same pollutants in nearby outdoor locations.

Researchers are attempting to more accurately gauge human health risks by developing methods to conduct *total exposure assessments.* In this context, *exposure* to a pollutant refers to time spent by an individual in an environment with a particular *concentration* of that pollutant. Exposure differs from *dose,* which is the amount of the pollutant that enters a person's body. Exposure assessments are sometimes made by having individuals wear personal monitoring devices that record pollutant concentrations they are exposed to during normal daily activities. A different technique is used to estimate exposure in microenvironments in which people are stationary for long periods (for example, office buildings and restaurants). In these situations, exposures can be calculated from measurements of ambient air quality within the microenvironments.[14]

Improved data on health risks from indoor air has informed policy debates, but regulatory actions have been slow in coming, particularly for air quality in residences.[15] Reasons include the violations of personal privacy associated with having government inspectors monitoring residences, and the enormous number of buildings that would be regulated. Many regulatory measures have been imposed in non-residential settings. For example, the U.S. Occupational Safety and Health Administration,has regulated air quality in workplace environments for decades. And restrictions on smoking in indoor public spaces have become common in many places.

Although setting ambient standards on indoor air is infeasible in many contexts, other regulatory strategies can be used. These include bans on products (such as urea-formaldehyde insulation) that release dangerous gases, and environmental standards for building materials. The legal liability of building owners for damages caused by sick buildings may, in the long run, lead to voluntary adoption of building-industry standards that improve indoor air quality.

Air Emissions of International Concern

Discharges of pollutants near national borders can be a source of international tension. Some pollutants

[13] The EPA study was one of many that have been conducted as part of the Total Exposure Assessment Methodologies (TEAM) project; see Rosenbaum (1994) for an overview of TEAM studies and other investigations to assess personal exposure to air pollutants.

[14] For more on exposure assessment, see the discussion of risk assessments in Chapter 17, and the review by Sexton and Ryan (1988) of methods for assessing human exposure to air pollution.

[15] This discussion of efforts to control indoor air quality is based on Godish (1991, pp. 381–82).

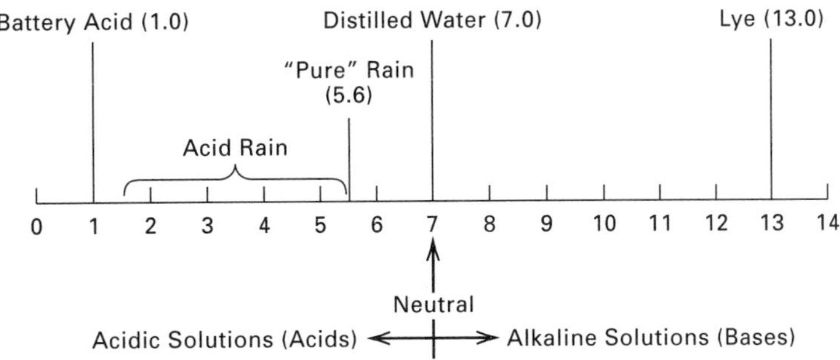

Figure 22.1 The pH scale.

are particularly significant because they can be transported over thousands of miles and affect large regions. Pollutants associated with acid rain are in this category. Other pollutants are a source of international concern because they affect the entire globe regardless of where they are discharged. Ozone depletion in the stratosphere and global climate change are air pollution problems that occur on a global scale. Pollutants associated with these problems—acid rain, ozone depletion, and climate change—are introduced here.[16]

Acid Rain

Although the term *acid rain* is commonly used, it provides an incomplete description of how lakes and other ecosystem components become acidified as a result of air emissions. *Acid deposition* is a more accurate term, because the acidification is caused by *wet deposition* (or washout) by both rain and snow, and by *dry deposition.* The latter occurs when atmospheric pollutants settle out by gravity and stick to surfaces of vegetation, soil, water, and other bodies. In keeping with conventional usage, this text employs acid rain and acid deposition interchangeably.

Acidity of precipitation is measured using pH: the negative logarithm of the hydrogen ion concentration of an aqueous solution. The pH scale varies from 0 to 14, and solutions with values less than 7 are termed *acids.* Examples include lemon juice (pH $\approx$ 2.2) and beer (pH $\approx$ 4.0). If the pH of a solution is greater than 7, it is termed a *base* or alkaline solution (see Figure 22.1).

Rain that has not been affected by pollutants is slightly acidic, owing partially to the influence of atmospheric carbon dioxide (CO_2), which dissolves in raindrops to form carbonic acid.[17] Even in areas unaffected by pollutants, the atmosphere contains small quantities of seaspray, ammonia, and other substances that affect pH. Although the pH of "pure" rain is a variable, it is common to define it at 5.6, and call rain with lower pH values acid rain. Using this convention, recorded pH values of acid rain range between 5.6 and less than 2. As an illustration, extensive areas of Europe and North America have rains with average pH values of less than 4.5.

Although environmental scientists differ over details, they generally agree that many areas are experiencing increases in acidification, that these increases are causing notable damage to some aquatic and forest ecosystems, and that escalating emissions of SO_x and NO_x contribute to the rise in acidification in some regions. Sulfur and nitrogen oxides participate in oxidation reactions that yield sulfuric acid (H_2SO_4) and nitric acid (HNO_3), respectively. SO_x NO_x, and products of chemical reactions in the atmosphere may be deposited locally or transported over

[16] Another pollutant of international concern is ionizing radiation (radionuclides). Principal anthropogenic sources include medical uses of radiation, nuclear explosions, and nuclear power plants. For an introduction to the subject, see Elsom (1992, pp. 108–31).

[17] Facts in this paragraph and the next are from Elsom (1992, pp. 82–85). He notes that the term acid rain dates as far back as 1872, when Robert Angus Smith used it "to describe the polluted air of Manchester, England, which damaged vegetables, bleached the colours of fabrics and corroded metal surfaces" (p. 85).

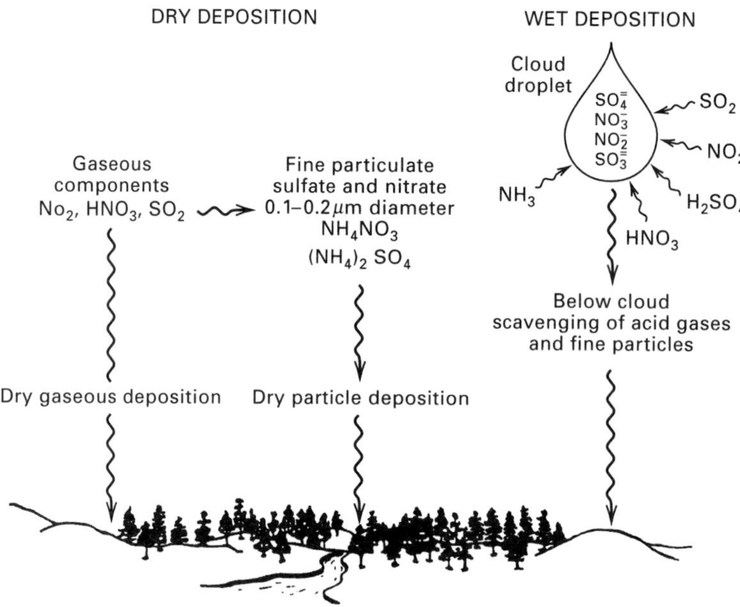

Figure 22.2 Atmospheric processes involved in acidic deposition. The two principal deposition pathways are dry deposition (nonprecipitation events) and wet deposition (precipitation events). Source: Boubel, Fox, Turner, and Stern, *Fundamentals of Air Pollution,* Third Edition, © 1994, p. 151. Reprinted with permission of Academic Press.

long distances. Examples of how these chemical compounds reach the ground by both wet and dry deposition are shown in Figure 22.2.

Approaches to reducing acid rain problems involve cutbacks in discharges of sulfur and nitrogen oxides. An example of a regulatory program designed to alleviate acid rain is the SO_2 emissions trading program that applies to coal-fired electric-generating facilities in the United States.[18]

Stratospheric Ozone Depletion

In 1985, scientists discovered a vast seasonal hole in the layer of ozone that exists normally in the *stratosphere,* the upper portion of the earth's atmosphere.[19] This discovery was significant because stratospheric ozone plays a key role in protecting humans and other organisms from dangerous ultraviolet (UV) radiation by absorbing UV energy far above the earth's surface. Recent elevations in UV energy in the lower atmosphere have been linked to reductions in stratospheric ozone and have caused numerous harmful effects, including skin cancer in humans.

Ozone is created naturally in the stratosphere by a process in which ultraviolet radiation dissociates oxygen molecules. At the same time, ozone is converted back into molecular oxygen by chemical reactions involving trace substances that exist naturally in the stratosphere. Depletion of ozone in the stratosphere (sometimes called the "destruction of the ozone layer") occurs as a consequence of human actions (for example, use of nitrogen-based agricultural fertilizers) that alter the natural cycle of ozone production and depletion.

Although emissions of nitrogen oxides and other substances have been tied to increased ozone depletion, the most important ozone-destroying chemicals linked to human activities are thought to be chlorofluorocarbons (CFCs), such as trichlorofluoromethane ($CFCl_3$, or CFC-11), and dichlorodifluoromethane (CF_2Cl_2, or CFC-12). These substances have been used extensively in aerosol sprays,

[18] For more on this SO_2 emissions trading program, see Chapter 11.

[19] Facts in this discussion are from Elsom (1992, pp. 132–44).

and as solvents and cooling agents. Applications of CFCs are of recent origin, since CFCs were invented around 1930.

Scientists believe chlorofluorocarbons diffuse upwards into the earth's stratosphere where the sun's energy is powerful enough to break the tight bonds between the atoms of CFCs to yield atomic chlorine. The chlorine atoms participate in a set of chemical reactions that destroys ozone. Because of the slow rates of diffusion of CFCs into and through the stratosphere, the concentration of CFCs in the stratosphere would continue to increase even if all emissions were stopped immediately. While all CFC releases have not been curtailed, many countries have agreed to phase out production.[20]

Global Climate Change

Not long after scientists became alarmed over stratospheric ozone depletion, a new international air pollution problem emerged: global warming. Maintenance of the earth's temperature at a steady long-term average value requires that incoming and outgoing energy be in balance. Solar energy reaches the earth as short-wavelength electromagnetic radiation, which mostly passes through the atmosphere without being absorbed. This incoming energy is radiated away from the earth's surface as long-wavelength electromagnetic radiation, and much of the reradiated energy passes outward through the atmosphere. Because some atmospheric gases, such as carbon dioxide, are strong absorbers of long-wavelength radiation, part of the reradiated long-wavelength energy does not pass into outer space. This absorption of long-wavelength energy by CO_2 and other gases is responsible for warming the earth's surface.

As carbon dioxide levels in the atmosphere increase, the balance between incoming solar energy and outgoing longer-wavelength energy radiated from the earth is disrupted. The resulting *greenhouse effect* could lead to a long-term increase in the earth's surface temperature.[21] However, average global temperatures depend on many factors besides carbon dioxide and other atmospheric gases that absorb long-wavelength electromagnetic radiation. For example, increasing particulate matter in the air may reflect incoming solar energy, thereby reducing the earth's temperature. The interactions among factors are so complex that it is difficult to predict their long-term effects on average temperature with confidence.

During the past decade or so, many scientists have become convinced that the earth's temperature is gradually increasing because of rising concentrations of gases that absorb long-wavelength radiation. Increases in these gases, collectively known as *greenhouse gases,* are shown in Table 22.3. The table also points to human activities that are sources of these gases. For example, atmospheric CO_2 is linked to the burning of carbon-based fuels such as coal, oil, natural gas, and wood, and to land use changes (particularly, deforestation of tropical rain forests) that yield net increases in atmospheric CO_2. As another example, methane releases are related to expansions in cattle populations and rice cultivation. Methane is produced by the metabolic processes of bacteria that live in oxygen-free environments such as the bottom of rice paddies and the guts of cud-chewing animals (for example, cattle and sheep).

As the data in Table 22.3 indicate, greenhouse gas concentrations have increased significantly since 1800. The key question is whether these increases have caused the average temperature of the earth's surface to rise. Atmospheric scientists have been puzzling over this question since the early 1980s, and there seems to be a consensus that global warming

---

[20] International agreements to phase out CFC production were discussed in Chapter 3. For details on these agreements, see Robinson (1995).

[21] The term *greenhouse effect* is based on the analogy between the roles of carbon dioxide in the atmosphere and the glass in a greenhouse. In a greenhouse, the glass easily allows the short-wavelength solar energy to pass through and be absorbed by objects in the interior. The warmed-up interior in turn radiates long-wavelength energy back toward the glass. The glass, however, is relatively opaque to this longer-wavelength energy, with the net effect that energy enters the greenhouse more easily than it can get back out. Therefore, it heats up. (It also heats up by reducing convection, but that is another matter.) Similarly, in the atmosphere, carbon dioxide is transparent to incoming solar energy and opaque to the long-wavelength reradiated energy from the earth. As carbon dioxide levels increase, the incoming solar energy is not affected, but the earth has a more difficult time reradiating energy back into space. The balance between the two processes is upset; more energy arrives than is lost, and the earth heats up. (This footnote was contributed by Professor Gilbert Masters of Stanford University.)

TABLE 22.3 Increases in Concentrations of Greenhouse Gases[a]

Greenhouse Gas	Atmospheric Concentrations[b] 1750–1800	1990	Illustrative Sources
Carbon dioxide	280 ppm	353 ppm	Combustion of carbon-based fuels; deforestation
Methane	0.8 ppm	1.72 ppm	Rice paddies; cud-chewing animals
Nitrous oxide	288 ppb	310 ppb	Combustion of carbon-based fuels and biomass
Tropospheric (low-level) ozone[c]	11 ppb	21 ppb	Reactions of oxides of nitrogen and volatile organic compounds in presence of sunlight
Chlorofluorocarbons			
$CFCl_3$	0	280 ppt	Propellants in aerosol sprays
CF_2Cl_2	0	484 ppt	Cooling agents in refrigerators and air conditioners

[a] Modified from Elsom, *Atmospheric Pollution: A Global Problem,* Second Edition, © 1992, Blackwell Publishers. Reprinted with permission.

[b] Units: ppm = parts per million; ppb = parts per billion; ppt = parts per trillion.

[c] The term tropospheric (or low-level) ozone is used to indicate a difference from the ozone that forms naturally in the stratosphere.

is taking place and is likely to continue.[22] However, there is no consensus on the expected amount of temperature increase or the rate at which it will take place.[23] Even though scientists disagree on these points, governments are working toward international agreements to curtail greenhouse gas production.[24] In addition, environmental professionals have argued that environmental impact assessments should stipulate how a proposed project or program will influence emissions of greenhouse gases, particularly CO_2. In practice, however, assessments of environmental impacts of proposed actions do not generally consider greenhouse gases.[25]

AIR QUALITY IMPACT ASSESSMENT PROCESS

A comprehensive assessment of a proposed project's air quality impacts involves the following steps:

1. Establish background air quality levels.
2. Identify applicable air quality regulations and criteria.
3. Forecast air pollutant *emissions* with and without the proposed action.
4. Forecast ambient air pollutant *concentrations* with and without the proposed action.

[22] Some prominent scientists do not agree that global warming is an urgent problem. For example, Richard Lindzen, a professor of meteorology at the Massachusetts Institute of Technology and a member of the National Academy of Sciences, finds "We don't have any evidence that [global climate change] is a problem." Lindzen argued that predictions of global warming are based on results from imperfect computer models, and he does not count these results as evidence. He stands with the relatively small group of atmospheric scientists who have spoken out boldly against the validity of models that predict notable long-term warming of the earth's atmosphere. This note is based on Stevens, W. K., "There is no Evidence Global Warming is a Serious Problem, Says One Scientist," *New York Times,* June 18, 1996, pp. C1 and C8.

[23] There is a vast literature on this topic; for an introduction, see Elsom (1992, pp. 145–65), and Houghton (1994).

[24] General observations on international efforts to curtail greenhouse gas emissions were given in Chapter 3. For details on these efforts, see Robinson (1995).

[25] For a discussion of efforts to introduce considerations of global warming into environmental impact assessments, see Cushman et al. (1993).

5. Compare predicted air quality with applicable standards.
6. For proposed projects and programs involving emissions of hazardous air pollutants, assess risks to human health and the environment.
7. Modify plans, if necessary, to deal with potential air quality problems.

The first step, establishing background levels of air quality, is carried out only for air quality indicators that are monitored routinely and likely to be influenced by the proposal. For example, if the proposed action consisted of constructing a new highway, the main indicators of interest would be hydrocarbons, nitrogen oxides, carbon monoxide, and photochemical oxidants. Air quality data may be obtained from government agencies. In the United States, for example, data on background air quality levels are available from the following sources: local, regional, and state air quality management agencies; state implementation plans proposed in response to federal air quality laws; and EPA's computerized data retrieval systems.

Only a small number of parameters are measured routinely by ambient air quality monitoring networks, and these are typically the indicators used in setting ambient air quality standards. An example is given by the Los Angeles area's stationary air monitoring network, which is one of the most extensive in the United States. At most of the three dozen or so monitoring sites, continuous measurements are made for only five pollutant gases, and particulate matter filter samples are collected routinely.[26]

Of the hundreds of substances present in the atmosphere, the following are most commonly used in setting ambient standards and are therefore among the parameters most commonly monitored: CO, Pb, O_3, smoke, VOCs, SO_2, NO_2, total suspended particulates, and PM-10. In regions in which acid rain is a problem, special monitoring networks have been established to measure the pH of precipitation as well as other chemicals needed to investigate acid rain, such as ammonia, sulfates, and nitrates.

As nations develop programs to control hazardous air pollutants, efforts to monitor specific organic compounds and metals (in addition to lead) will increase.[27] However, programs to monitor ambient concentrations of air toxics routinely are likely to expand slowly because of the expense and complexity of monitoring individual VOCs. For many chemical species, complete analytical procedures have not yet been developed. For example, Segall and Westin (1994, p. 166) report that as of 1992, fully validated test methods for sampling and analysis were available for less than one-quarter of the 189 hazardous pollutants regulated by the U.S. Clean Air Act Amendments of 1990.

Applicable air quality regulations and criteria are determined in the second step of the assessment process. *Criteria* relate levels of air pollutants to human health and welfare.[28] Regulations vary from place to place. For example, in the United States, applicable regulations include those establishing the national ambient air quality standards and pertinent state or local standards.

Consider, for example, the national carbon monoxide standards, which stipulate both concentration and time of exposure. The CO standards require that the average concentration for any 1-hour period be less than 35 parts of CO per million parts of ambient air. These CO measurements are said to have an *averaging time* of 1 hour. The NAAQS also require average CO concentrations during any 8-hour period to be less than 9 parts per million (ppm). Limits on CO are not to be exceeded more than once per year.

In the third step of the assessment process, emissions from the proposed action are estimated in units of weight (or mass) per time period. Procedures for estimating emissions are reviewed in the next section.

The fourth step—predicting changes in ambient *concentrations* of air quality indicators due to a new

[26] These facts about the Los Angeles area network are from Boubel et al. (1994, pp. 218–19).

[27] This increased monitoring is demonstrated by a pilot Toxics Air Monitoring System network set up by the U.S. Environmental Protection Agency for sampling ambient concentrations of volatile organic compounds (at levels of parts per billion) in three cities over a two-year period. The pilot network provided information on 13 VOCs.

[28] For example, see the air quality criteria documents that have been issued for pollutants included in the NAAQS in the United States. An illustration is given by the U.S. Environmental Protection Agency's criteria document for carbon monoxide (EPA, 1991).

discharge—is often complex. In fact, sometimes it is not carried out and the assessment considers only the increased *emissions* from the proposed action. Reasons for not estimating concentrations include (1) inadequate understanding of the underlying physical and chemical processes, and (2) constraints on time and money that preclude use of existing forecasting procedures.

The fifth and sixth steps of the assessment process concern the acceptability of proposed changes. If changes in concentrations of ambient air quality have been forecasted, they can be compared with applicable standards (step 5).

For projects in which air emissions are likely to pose significant risks to human health or the environment, a risk assessment is often called for as part of the impact assessment. As indicated in Chapter 17, risk assessments are often mandated when there is a potential for a release of dangerous quantities of hazardous substances because of accidents or engineering system failures, or because humans and other organisms will be exposed to hazardous materials at low concentrations over long time periods. Although risk assessments are often required as parts of environmental impact assessments for proposed actions, they have not been performed frequently in that context.[29]

The final step in a comprehensive impact assessment is to modify the proposed project if expected air quality degradation is unacceptable. Air pollutants are commonly reduced by changing combustion processes and using emission control devices such as scrubbers and filters.[30] However, there are many options for mitigating adverse air quality effects that do not involve control devices. Examples include reducing the scale of a facility, changing the locations of discharges, and redesigning facilities to minimize emissions and thereby prevent pollution.[31]

[29] According to Chrosotowski, proposed incinerators have been subject to more risk assessments than any other type of development project (Chrostowski, 1994, p. 161). He uses "incinerator" to include "waste to energy facilities for municipal solid waste, biohazardous waste incinerators, sewage sludge incinerators, and hazardous waste incinerators." Although Chrostowski is not explicit, the context of his observation suggests that he was referring only to the United States.

[30] Traditional technology for controlling emissions is described in detail Boubel et al. (1994).

[31] For more on strategies to minimize waste discharges, see Table 14.4 in Chapter 14.

ESTIMATING EMISSIONS OF AIR POLLUTANTS

Forecasts of air pollutant emissions and concentrations play a central role in conducting impact assessments and in designing regulatory programs. This section introduces the simplest air pollutant forecasting techniques: quick and simple methods for estimating increases in emissions. Subsequent sections introduce mathematical models to translate increases in emissions into changes in concentrations at various times and places.

Emission factors are often used to project amounts of airborne contaminants resulting from a proposed project or regulatory program. An emission factor is the estimated quantity of pollutant discharged per unit of activity. For instance, the U.S. Environmental Protection Agency (EPA, 1973) indicated that when coffee beans are processed in a direct-fired roaster, 7.6 pounds of particulates are released per ton of coffee processed. Emission factors are generally determined using emission measurements at existing sources.[32] National environmental agencies often compile data on emission factors. For example, the U.S. Environmental Protection Agency publishes a *Compilation of Air Pollutant Emission Factors* (publication AP-42 and periodic supplements) and also makes emission factors available as a computer database.[33]

The use of factors to project emissions is illustrated by EPA data for power plants fired with bitu-

[32] Emission factors are sometimes computed using engineering analysis methods. An example is the use of mass balance principles in cases where the input to and output of a production process can be quantified. The difference between input and output is assumed to be discharged to the environment. Emissions factors reported by EPA and others are highly variable in quality and it is sometimes difficult to gauge their accuracy. In commenting on this point, Boubel et al. (1994, p. 94) observe, "Some emission factors, which have been in use for years, were only rough estimates proposed by someone years ago to establish the order of magnitude of the particular source."

[33] The U.S. Environmental Protection Agency has assembled much emission factor data (including all of publication AP-42) in a computer bulletin board system available via modem hookups to EPA's *Clearinghouse for Inventories and Emission Factors* (CHIEF). It also has a compact disk read-only memory (CD-ROM) system called *Air CHIEF CD*. For details on availability, see Patrick (1994, pp. 221–223), who also describes computer databases that include emission factors for hazardous air pollutants.

minous coal. The main emissions from such plants are particulates and sulfur oxides. Other substances discharged include carbon monoxide, hydrocarbons, nitrogen oxides, and aldehydes. The EPA compilations include emission factors for each of these materials, and instructions for adjusting factors to reflect variations in the sulfur and ash content of coal. For example, EPA indicates that $16 \cdot A$ pounds of particulates are emitted per ton of coal burned where A is the percentage of ash in the coal.[34] If a proposed power plant burns bituminous coal with a 10% ash content at a rate of 200 tons/hr, the estimated emission of particulate is

$$(16 \times 10)\frac{\text{lb}}{\text{ton of coal}} \times 200\ \frac{\text{tons of coal}}{\text{hr}} = 32{,}000\ \text{lb/hr}.$$

This estimate assumes that no emission controls are employed. Using devices known as electrostatic precipitators, as much as 99.5% of the particulates can be removed.

Emission factors for mobile sources must account for variations in operating conditions. For example, carbon monoxide releases from an automobile depend significantly on the vehicle's speed and whether it has been warmed up. A car operating at 15 mph might emit twice as much CO per mile as it would at 35 mph.[35] Also, emissions of CO (per mile) during the first 3 to 5 minutes after starting a cold vehicle can be four times higher than emissions after the vehicle is warmed up.[36] Variables such as speed and whether a vehicle starts cold or hot are accounted for in EPA's computer-based procedures for estimating motor vehicle emission factors.[37] According to Cohn and McVoy (1982, p. 201), average fleet emission factors given by these procedures are especially sensitive to year of analysis, average vehicle speed, "start mode" (hot versus cold), ambient temperature, and vehicle mix (percentage of autos, motorcycles, diesel trucks, and so forth).

Once average emission factors are estimated for a fleet of vehicles, they are combined with traffic projections to yield forecasts of total emissions. For instance, suppose all vehicles on a proposed highway segment consist of passenger cars. For operating conditions of the average car in 2010, assume EPA estimating procedures indicate a CO emission of 3.4 grams per mile. Total daily CO emissions expected in 2010 equal 3.4 grams per vehicle-mile multiplied by the number of vehicle miles traveled (VMT) per day by automobiles using the highway in 2010. Projections of VMT are generally available from local or state transportation agencies.[38]

SIMPLE MODELS RELATING EMISSIONS TO CONCENTRATIONS

Most methods for predicting how emissions influence ambient air quality rely on mathematical models.[39] Rudimentary models providing rough estimates of how emissions affect concentrations will be introduced; before proceeding, however, a warning is in order. Forecasting air pollutant concentrations is a specialized endeavor that can easily be carried out incorrectly. Atmospheric pollutant transport processes are complex, especially when emissions are chemically reactive. Except for very simple cases, forecasting concentrations is best left to air quality modeling specialists.

Box Models

A *box model* assumes that a constant emission rate, P (units of mass/time), enters a volume of ambient air moving in one direction at a constant speed, U. The air in motion is confined from above by a layer

[34] This factor applies to pulverized bituminous coal in a "general-type furnace" of size greater than 10^8 British thermal units/hr. (EPA, 1973, p. 1.1–3).

[35] A graph showing variations in average CO emissions with vehicle speed is given by Papacostas (1987, p. 347).

[36] Horowitz (1982, pp. 167–171) discusses how CO emissions vary depending on whether the car is started in a cold or warmed-up state.

[37] For details on EPA's Mobile 5, a computer program for estimating vehicle emission factors, see EPA (1994).

[38] Transportation planning methods used in forecasting VMT and other aspects of travel demand are described by Papacostas (1987, chapters 8 and 9). These methods are also used to estimate the influence of proposed projects on traffic congestion.

[39] Physical models using wind tunnels are sometimes used, particularly in situations where mathematical models cannot give reasonable forecasts because of complexities such as irregular, hilly terrain.

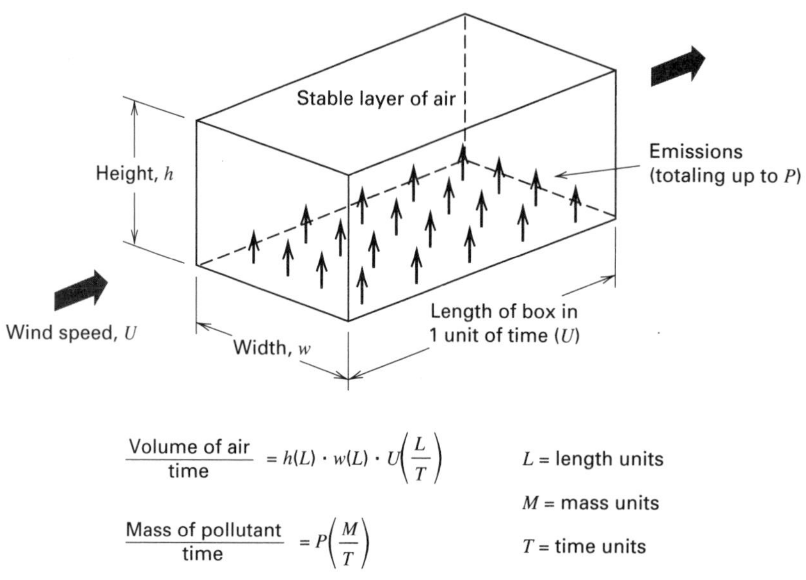

$$\frac{\text{Volume of air}}{\text{time}} = h(L) \cdot w(L) \cdot U\left(\frac{L}{T}\right)$$

$$\frac{\text{Mass of pollutant}}{\text{time}} = P\left(\frac{M}{T}\right)$$

L = length units

M = mass units

T = time units

Figure 22.3 Air available for dilution in a simple box model.

of stable air (an *inversion layer*) at an elevation, h.[40] The moving air is also confined in the direction perpendicular to the wind (see Figure 22.3). This model represents a valley, in which air passes through a zone of width, w, formed by two rows of hills. Airborne contaminants are diluted and carried from the area by motion in the direction of the wind.

The simplest box model assumes steady-state conditions: the emissions, wind speed and direction, and characteristics of air available for dilution do not vary over time. The model also assumes that contaminants mix completely and instantaneously with the air available for dilution, and that the released material is chemically stable and remains in the air. Under steady-state conditions, the air volume passing over the area in a unit of time equals the product of wind speed, U, and cross-sectional area, hw (see Figure 22.3). This relationship for volume per time is based on the law of conservation of mass applied to steady fluid motion. The concentration of the discharged material in the ambient air equals the mass of substance emitted per time divided by the corresponding air volume available for dilution. In mathematical terms,

$$c = \frac{P}{Uhw} \tag{22-1}$$

where c is the average concentration under steady-state conditions, and other terms are as previously defined. The numerator has units of mass per time, and the denominator is in volume per time. Therefore, c has units of mass of substance emitted per volume of air. Commonly employed units yield concentrations in micrograms (10^{-6} g) per cubic meter, abbreviated as $\mu g/m^3$.

Box models are used to estimate average air pollutant concentrations in regions with many small sources that are distributed uniformly under conditions where wind and emissions are steady. Kohn's (1978) application of equation 22-1 for sulfur dioxide in the St. Louis airshed during 1970 is an example.

[40] *Inversion layer* is a term used to describe a layer of atmosphere in which temperature increases with increasing altitude. In normal circumstances, temperature decreases as altitude increases. When atmospheric conditions cause an inversion layer to be formed, pollutants become trapped below it. This layer corresponds to a lid in the box model.

Average parameter values for the box model as applied to St. Louis in 1970 were estimated as

U = 15,400 m/hr, average annual wind speed
h = 1210 m, mixing height
w = 10^5 m, width of the airshed

Approximately 1375 million pounds of sulfur dioxide were discharged in the airshed during 1970. This corresponds to an emission rate of

$$P = \frac{1375 \times 10^6 \text{ lb/year} \times 454 \text{ g/lb} \times 10^6 \ \mu\text{g/g}}{8760 \text{ hr/year}}$$

$$= 7.126 \times 10^{13} \ \mu\text{g/hr}.$$

Direct substitution into Equation 22-1 yields

$$c = \frac{P}{Uhw} = \frac{7.126 \times 10^{13}}{15{,}400 \times 1210 \times 10^5} = 38 \ \mu\text{g/m}^3$$

which is equivalent to a volumetric concentration of 0.014 parts of SO_2 per million parts of ambient air.[41] During 1970, the average SO_2 concentration observed at the federal Continuous Air Monitoring Program (CAMP) station in St. Louis was 0.03 ppm.[42] The box model result is thus only about half the observed concentration. One explanation for the discrepancy is that the CAMP station was located in a zone where SO_2 concentrations were much higher than average; the box model gives only a regional average concentration.

Box models have several limitations. As illustrated by the preceding SO_2 figures, the concentrations at some times and places within a region can be very different from the average yielded by a box model. In addition, box models do not account properly for atmospheric dispersion of materials because they assume that complete mixing occurs instantaneously within the confines of the box, and that *no* pollutant dispersion occurs across the boundaries of the box. Moreover, box models have only limited applicability if pollutants are chemically reactive.[43] The models are not useful in forecasting concentrations of volatile organic compounds and other substances that contribute to photochemical smog.

Interest in box models increased with the rising concern over indoor air pollution. A simple building can be modeled as a box in which pollutants are uniformly distributed. Mass balance equations are derived to account for pollutants entering and leaving the structure and, if appropriate, decaying within the structure. Sometimes models are built by linking several boxes together. For example, predictions of concentrations of radon within a residence can be made using a two-box model, where radon concentrations are assumed to be uniform within each box. In this case, one box represents living space within a residence, and a second box represents the area under the house.[44]

Rollback Models

Rollback (or *proportional scaling*) *models* provide another simple approach for estimating how emissions influence ambient air quality. These models *assume* a linear relationship between P, the total quantity of a substance discharged in a region during a particular period, and c, the concentration of the substance at a *specific point.* In mathematical terms, the assumption is

$$c = kP + b \qquad \textbf{(22-2)}$$

where b (*background concentration*) represents the concentration when emissions are zero, and k is an empirically determined constant (see Figure 22.4). The parameters k and b are for a particular pollutant and location.

[41] The factor used by Kohn to convert from units of μg/m^3 to ppm is 0.000369. More generally, conversion from μg/m^3 to parts per million (ppm) in the air pollution field is done at a standard temperature and pressure. For example, at a temperature of 0°C and a pressure of 1 atmosphere, the ideal gas law can be used to show

$$\mu\text{g/m}^3 = \text{ppm} \times \text{M} \times 44.64$$

where M is the molecular weight of the gas (grams per mole). See Bibbero and Young (1974, p. 20) for details.

[42] The CAMP station at St. Louis is one of a network of federal monitoring stations established at various locations in the United States during the 1960s.

[43] The simplest box models assume that emissions are chemically nonreactive during the time periods used in the analysis. This assumption is often reasonable for CO, SO_2, and particulates. Box models can also be used to predict concentrations of pollutants that decay according to first-order reaction kinetics; for examples of such models, see Masters (1991, pp. 10–13).

[44] For an example demonstrating use of box models to estimate CO and radon concentrations indoors, see Masters (1991, pp. 337–41).

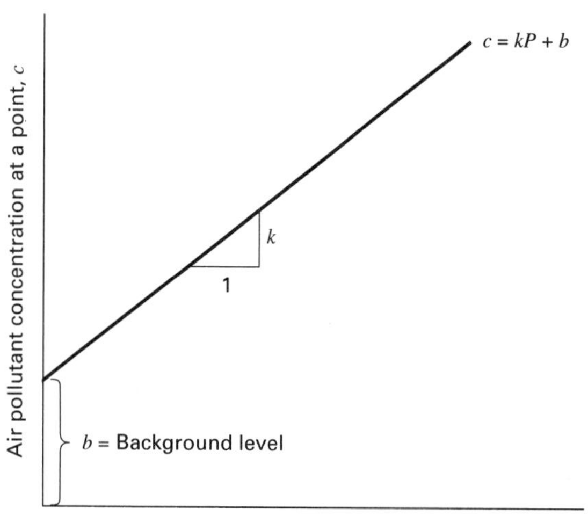

Figure 22.4 Linear relationship between emissions and concentrations in rollback models.

Kohn's (1978) application of a rollback model to calculate concentrations at the CAMP station in St. Louis is illustrative. Emission and concentration data for 1963 were used to determine k for various pollutants. For example, 300×10^6 lb of particulates were discharged in the St. Louis region in 1963, and the concentration of particulates near the CAMP station was 128 $\mu g/m^3$. Kohn estimated the background level of particulates as 31 $\mu g/m^3$. Based on this information, k was computed as

$$k = \frac{c - b}{P} = \frac{128 - 31}{300 \times 10^6} = 0.32 \times 10^{-6}.$$

This value of k was employed in calculating the particulates concentration near the CAMP station in St. Louis for 1970, a year when regional particulates emissions were 233×10^6 lb. Assuming b remained constant at 31 $\mu g/m^3$, Kohn computed the 1970 concentration using equation 22-2:

$$\begin{aligned} c &= kP + b \\ &= (0.32 \times 10^{-6})(233 \times 10^6) + 31 = 106\ \mu g/m^3. \end{aligned}$$

The measured concentration of particulates at the CAMP station in 1970 averaged 130 $\mu g/m^3$. Between 1963 and 1970, discharges of particulates decreased from 300×10^6 to 233×10^6 lb, but the average concentration at the CAMP station actually *increased* from 128 to 130 $\mu g/m^3$. Because it assumes a linear relationship between concentration and emissions, the rollback model predicted a *decrease* in concentration.

The model's failure to predict the increase in particulates concentration resulted partly because variations in meteorological conditions can cause k to change over time. For example, if average wind speeds were different in 1963 and 1970, that difference itself would cause an erroneous estimate of the 1970 particulates concentration. In addition, k can change if the locations of sources shift over time. Suppose that compared to 1963, a greater portion of the regional discharge was close to the CAMP station in 1970. The CAMP station concentration might then show an increase even though total regional emissions decreased.

Rollback models are also employed to determine discharge reductions needed to meet an ambient air quality standard, c_s. Suppose equation 22-2 is written for two time periods, a recent year in which the standard was violated, and a future year when the standard is met. Combining the two equations algebraically and then simplifying yields

$$R = 100\left(\frac{P_0 - P_s}{P_0}\right) = 100\left(\frac{c_0 - c_s}{c_0 - b}\right) \quad \textbf{(22-3)}$$

where R is the requisite percentage reduction in emissions.[45] The subscript 0 refers to when the ambient standard was violated, and s refers to the period in which the standard is met.

[45] Equation 22-3 is derived using the definition of R and the following applications of equation 22-2:

$$c_0 = k\,P_0 + b \ \ \text{(period 0), and}$$
$$c_s = k\,P_s + b \ \ \text{(period s).}$$

Combining these two equations yields

$$(c_0 - c_s) = k\,(P_0 - P_s), \text{ and}$$

Rewriting the period 0 equation gives

$$(c_0 - b) = k\,P_0$$

The last two equations, with the definition of R, provide a basis for equation 22-3.

Equation 22-3 has been used to compute required reductions in emissions from motor vehicles. This application has been disputed because the linear form of rollback models cannot account for photochemical reactions that occur among vehicular emissions of hydrocarbons and NO_x. Another objection is that rollback models have not been widely tested with observed data, and thus they can yield incorrect, misleading results. Despite these criticisms, rollback models are often used because of their simplicity. *Modified rollback models* developed by de Nevers and Morris (1975) eliminate some of the shortcomings of the models presented here.

Statistical Models

Rollback models treat the concentration of an air pollutant at a point in terms of an *average* value. Using results from Larsen's (1971) statistical analysis of air quality records, rollback models can be extended to consider fluctuations in concentration at a given location. Larsen's statistical model represents short-term variations in air pollutant concentrations at a point as *log-normal distribution.*

Log-Normal Distributions

To understand Larsen's use of log-normal distributions, consider a more common statistical form, the *normal distribution.* Figure 22.5 represents the variability of hourly temperature readings at the federal Continuous Air Monitoring Program station in Washington, DC, for a 7-year period. It is interpreted in a similar way as the noise histogram in Figure 21.4. The horizontal axis indicates hourly temperatures and the vertical axis gives the frequency of occurrence of those temperatures. In this case, the temperature intervals used for purposes of preparing the graph are so small that the results are approximated as a smooth curve instead of the collection of bars in the noise histogram.

The curve in Figure 22.5 is bell-shaped and symmetric about 57°F, the average hourly temperature observed during the 7-year period. Values above 57°F occurred with the same frequency as values below it. Temperature is a continuous variable. Its values are not limited to integer values like 60°F or 61°F. When a frequency analysis of a continuous variable yields the bell-shaped pattern in Figure 22.5, it is common to assert that the variable, in this case hourly temperature, follows a normal distribution.[46]

Larsen's analysis of air quality data for eight cities established that the frequency of occurrence of a pollutant's concentration at a particular point did *not* have a normal distribution. Consider, for example, his 7 years of data for the 1-hour average sulfur dioxide concentrations at the CAMP station in Washington, DC. The frequency analysis for SO_2 did not yield a bell-shaped curve like the one in Figure 22.5. Instead, SO_2 concentrations lower than the arithmetic average occurred much more frequently than values above the average. Based on preliminary findings, Larsen reexamined the data by taking the logarithm of each 1-hour SO_2 measurement. His frequency analysis of the transformed data indicated that variations in the logarithms of the SO_2 concentrations could be represented using a normal (bell-shaped) curve. In the language of statisticians, the original SO_2 concentrations follow a log-normal distribution. Larsen's data for CO, NO_x, hydrocarbons, and photochemical oxidants also followed log-normal distributions.

Use of log-normal probability paper simplifies the data analysis for an air quality indicator presumed to follow a log-normal distribution. The vertical axis on such paper represents the logarithm of concentration; the horizontal axis is the percentage of time a particular concentration is equaled or exceeded. Figure 22.6 shows results from using log-normal probability paper to perform a data analysis exercise suggested by Larsen. The measurements consist of average sulfur dioxide concentrations assumed to be taken in Washington, DC, for each month of 1 year. Monthly average values for January through Decem-

---

[46] As explained by Larsen (1971, p. 9), the normal (or *Gaussian*) distribution curve in Figure 22.5 is described using

$$Y = \frac{n}{s\sqrt{2\pi}} \exp\left[-\frac{1}{2}\left(\frac{t-m}{s}\right)^2\right]$$

where

Y = frequency of occurrence (hr)
t = hourly temperature (°F)
n = total number of hourly temperature readings (61,320)
m = arithmetic average of readings (57°F)
s = *standard deviation* of readings (15°F)

The curve has two parameters, m and s, and they are computed using formulas available in introductory statistics textbooks.

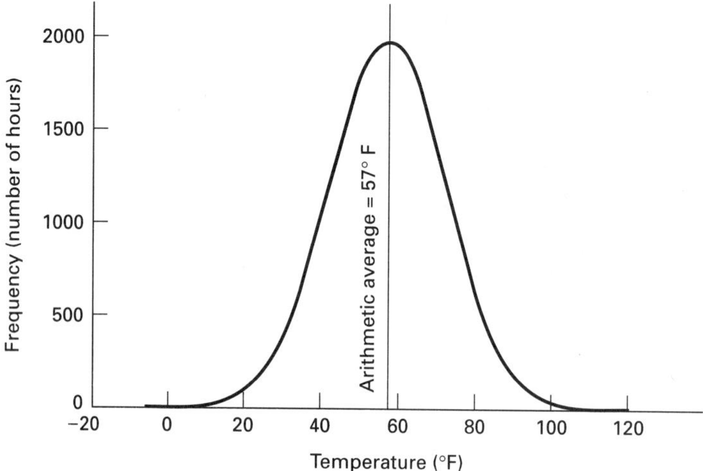

Figure 22.5 Estimated frequency of hourly temperatures at CAMP site, Washington, DC, December 1, 1961, to December 1, 1968. Source: Adapted from Larsen (1971) +C323.

ber, in micrograms per cubic meter, are 300, 250, 180, missing, 150, 60, 120, 100, 140, 160, 190, and 220, respectively.

Figure 22.6 is constructed as follows: First the 11 monthly average readings are ranked from highest to lowest as in Table 22.4. The symbol r represents the rank order of a measurement, where $r = 1$ corresponds to the maximum observed monthly SO_2 concentration. Next, f, the percentage of time a particular concentration is equaled or exceeded, is determined.

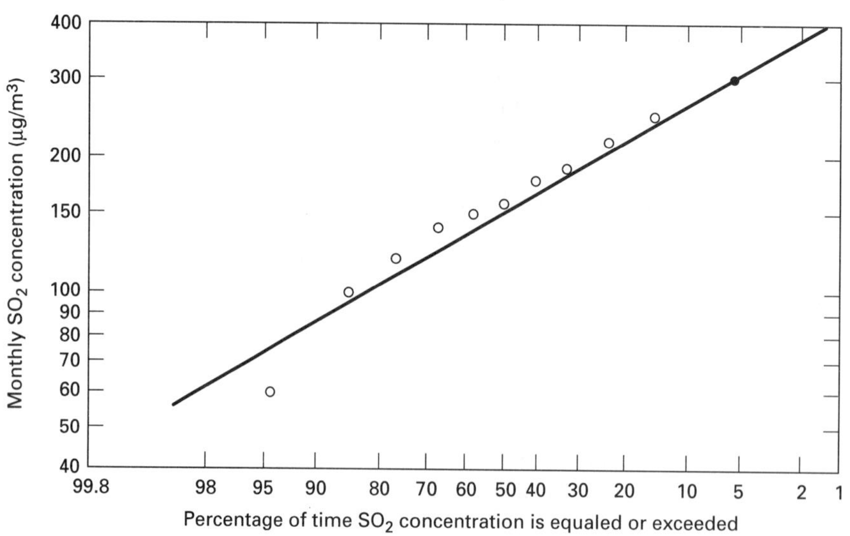

Figure 22.6 Frequency analysis of monthly average SO_2 concentrations.

TABLE 22.4 Frequency Analysis of Monthly Average SO_2 Concentrations

Month	Monthly SO_2 ($\mu g/m^3$)	Rank Order r	Percentage Equaled or Exceeded f (%)
January	300	1	5.5
February	250	2	14.5
March	180	5	41.8
April	Missing	Missing	Missing
May	150	7	58.2
June	60	11	94.5
July	120	9	76.4
August	100	10	85.5
September	140	8	67.3
October	160	6	50.0
November	190	4	32.7
December	220	3	23.6

Larsen calculated f using a formula based on the theory of log-normal distributions:

$$f = 100\%\left(\frac{r - 0.4}{n}\right) \tag{22-4}$$

where n is the number of observations ($n = 11$ in this example). Equation 22-4 is for data points higher than the *median* (or middle value) of the observations. The median (which corresponds to $r = 6$ in this case) is plotted at the 50% frequency. Values of f for points below the median are found using equation 22-4, but with 0.4 replaced by 0.6. Figure 22.6 shows f values plotted against corresponding SO_2 concentrations.

Combining Rollback and Statistical Models

Rollback models are sometimes combined with a log-normal representation of the distribution of air pollutant concentrations.[47] This is demonstrated in the U.S. Environmental Protection Agency impact assessment for a proposed wastewater treatment plant expansion in the Livermore-Amador Valley in northern California (EPA, 1975). EPA concluded that the proposed facility would allow more people to reside in the valley, and new residents would substantially increase the vehicle miles traveled during commuting hours. Because the NAAQS for photochemical oxidants were violated in the valley, EPA was very concerned about air quality impacts of the proposed expansion.

The impact assessment, which was conducted in the mid-1970s, assumed that daily maximum concentration of photochemical oxidants at a station in Livermore-Amador Valley follows a log-normal distribution. Figure 22.7 is from EPA's frequency analysis of daily high oxidant concentrations occurring in 1973, a year in which meteorological conditions were average. Point A in the figure is the *average* daily high concentration of oxidants during 1973 plotted at $f = 30\%$. For a variable that follows a log-normal distribution, the average will be equaled or exceeded only 30% of the time.

Rollback models assume that ambient air pollutant concentrations at a point are proportional to total emissions of the pollutant. Because photochemical oxidants are not discharged directly, additional assumptions were needed before the rollback concept could be applied. EPA assumed a one-to-one correspondence between average daily high concentrations of oxidants at the monitoring station and average emissions of reactive hydrocarbons (RHC) in the valley. The term *reactive* means these are hydrocarbons that take part in photochemical reactions. The complexities and nonlinearities involved in these reactions make this a tenuous assumption.

EPA predicted the frequency distribution of daily high concentrations of oxidants in Livermore-Amador Valley for 1990. Their forecasting approach required three items of information. One was the distribution of daily high concentrations of oxidants for a base year; this is the line labeled "1973" in Figure 22.7. Another was the total 1973 RHC emission in the valley, which was estimated to be 9.3 tons. The final required item was the RHC emissions expected in 1990. Based on an analysis of projected commuter travel in 1990, a total RHC discharge of 4.8 tons was predicted. The 1990 RHC figure is substantially lower than the 1973 value because of ex-

[47] For an analysis of how statistical theory can be used with rollback modeling concepts, see Ott (1995, pp. 276–93).

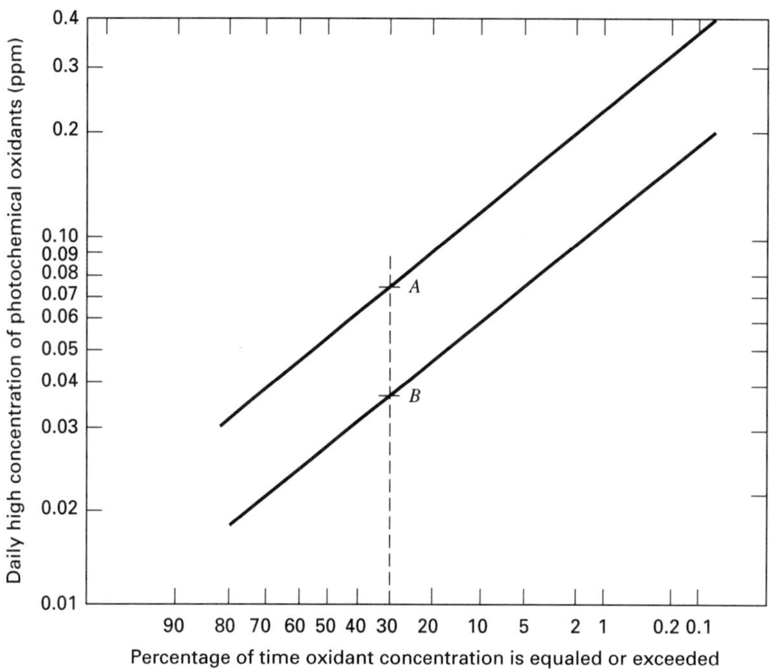

Figure 22.7 Analysis of oxidant concentration distributions in the Livermore-Amador Valley. Adapted from U.S. Environmental Protection Agency (1975).

pected improvements in motor vehicle emission controls.

EPA's application of the rollback model took the form

$$C_{f=30\%}\ (1990) = C_{f=30\%}\ (1973) \times \frac{\text{RHC emissions (1990)}}{\text{RHC emissions (1973)}} \qquad \textbf{(22-5)}$$

where C represents the concentration of photochemical oxidants. The value of $f = 30\%$ was used because the rollback model was applied for the *average* concentration of oxidants. Using the previously mentioned RHC emissions together with the 1973 average concentration (point A in Figure 22.7), equation 22-5 yields a 1990 average daily high concentration of 0.038 ppm. This is point B in the figure. To obtain the distribution of concentrations for 1990, a line parallel to the 1973 line was drawn through point B. Fluctuations in concentration, relative to the average, were assumed to be the same in 1990 as they were in 1973.

Figure 22.7 was employed to estimate the frequency with which the NAAQS for photochemical oxidants would be violated in 1990 in the Livermore-Amador Valley. This information was used to analyze alternative ways of meeting the national ambient standards for oxidants. Possible actions included reducing the proposed treatment plant expansion and adopting transportation management strategies to decrease the number of vehicle miles traveled.

POLLUTANT TRANSPORT VIA ADVECTION AND DISPERSION

Much effort has been spent analyzing how an air pollutant disperses downwind from its point of discharge. During the 1920s, the dispersion process began to be examined systematically on both theoretical and experimental levels. Mathematical models of atmospheric turbulence were used to predict the motion of particles in a puff of smoke released into the ambient air. Experimental work indicated that, under some conditions, materials downwind from an emission were distributed in a *plume,* and that the concentration distribution could be described using a bell-shaped (or normal) curve. Extensions of this early research yielded *atmospheric dispersion models* for point, line, and area sources.[48] A smokestack is a typical point source, and a highway is a common line source. An area source is illustrated by the collection of indoor fireplaces in a large suburban housing development.[49]

Point Source Gaussian Plume Models

The most widely discussed atmospheric dispersion model is the *point source Gaussian plume model.* In contrast to box and rollback models, which use average concentrations, the Gaussian model describes concentrations at numerous points in a plume of contaminants that forms downwind of a pollutant discharge. The model typically incorporates the following assumptions:

- Rate of pollutant emission is constant.
- Wind moves horizontally in a constant direction.
- Wind moves at a constant rate, the average wind speed.
- Emitted substance is chemically stable and does not settle out from ambient air.
- Area surrounding source is flat, open country.

A Gaussian plume model accounts for two types of transport: advection and dispersion. The term *advection* refers to movement of a contaminant by the motion of the wind in the horizontal direction. Advection is the dominant transport mechanism in the direction of the wind.

Dispersion is a general term describing how a contaminant is spread out by a variety of physical processes, but mainly by *turbulence,* the collection of *swirls* or *eddies* that occur in the atmosphere because of nonuniformities in the wind's velocity. The principal sources of eddies include thermal convection (bouyancy effects), and friction between the wind and objects on the earth's surface that it moves across.[50]

Atmospheric turbulence causes parcels of air within a contaminant plume to mix with unpolluted air surrounding the plume. This mixing causes the pollutant to spread out and become diluted as it is moved forward by the motion of the wind. In the plane perpendicular to the wind's direction, pollutant concentration decreases with distance from the plume's centerline.

The point source Gaussian plume model is derived by applying the law of conservation of mass to an emitted substance under the conditions outlined. The law is written as a differential equation that includes terms representing both advection and dispersion. A solution to the equation (under steady-state conditions) is the normal distribution equation used in statistics. The normal equation is also known as the Gaussian equation, which is the basis for naming the model. The Gaussian model describes the concentration of a substance downwind from a source by a series of bell-shaped curves like those in Figure 22.8. The figure's coordinate system has the x axis extending horizontally in the direction of the wind. The y axis is in the horizontal plane and perpendicular to the x axis, and the z axis extends vertically. The overall height of the plume in Figure 22.8 is somewhat greater than the source height. This occurs because emission of a gaseous material from a point source such as a

[48] Introductory textbooks on air quality management, such as Boubel et al. (1994), generally include at least one chapter on atmospheric dispersion models. For a more comprehensive treatment of the fate and transport of air pollutants, see Seinfeld (1986).

[49] Distinctions between point, line, and area sources are made by examining a source's length and width relative to the size of the area of analysis. A *point* source is small in both length and width, whereas a *line* source has one dimension (for example, a highway's length) that is large compared to the dimensions of the study area. An *area* source is relatively large in both length and width.

[50] Specialized terms are used to describe atmospheric dispersion. The main source of pollutant dispersion in the atmosphere is *eddy diffusion.* There are two sources of eddies. One is *mechanical turbulence,* which results because of variations in wind velocity that occur as winds cross over the earth's surface. Wind speed is zero at the surface and increases with elevation. The second source of eddies is *thermal turbulence,* which results because of convective currents set up as the earth's surface is heated. The air near the surface heats up (or cools down) and becomes less (or more) dense, giving rise to thermal (convective) currents.

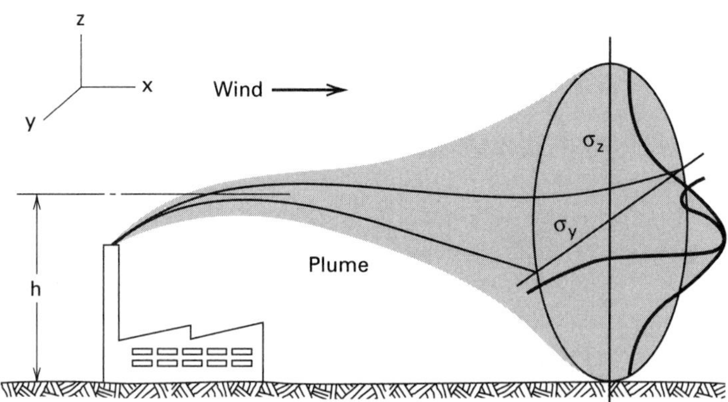

Figure 22.8 Gaussian (bell-shaped) distribution of a substance from a point source. From R. J. Bibbero, "Systems Approach Toward Nationwide Air Pollution, III, Mathematical Models," in *IEEE Spectrum,* vol. 8, no. 12, pp. 47–58. Copyright © 1971 IEEE. Reprinted with permission.

smokestack generally experiences an extra rise (due to buoyancy and momentum effects) when it hits the atmosphere. The height, h, in the figure—termed the *effective emission height*—is determined using standard equations.[51]

In its general form, the point source Gaussian plume model predicts concentration at a downwind location (x, y, z). A commonly used simplification of the model forecasts pollutant concentration at downwind points along the ground ($z = 0$):

$$c(x,y) = \frac{Q}{\pi\sigma_y\sigma_z U}\exp\left(-\frac{y^2}{2\sigma_y^2} - \frac{h^2}{\sigma_z^2}\right) \qquad \textbf{(22-6)}$$

where

Q = constant pollutant emission rate (μg/sec)
U = constant wind speed (m/sec)
h = effective emission height of the stack (m)
σ_y = horizontal dispersion coefficient (m)
σ_z = vertical dispersion coefficient (m)

Using the units shown in parentheses yields concentration in $\mu g/m^3$.

Equation 22-6 is the two-dimensional form of the Gaussian distribution equation used in statistics, and the terms σ_y and σ_z are analogous to *standard deviations* in the y and z directions. A standard deviation is a statistic related to the character of the peak in the bell-shaped curve. If the standard deviation is low, the curve has a relatively sharp peak. Values of σ_y and σ_z generally increase with x, the distance downwind from the source. This indicates that the bell-shaped curves of concentration get flatter as the distance from the source increases.

To predict the concentration at any ground location (x, y) it is first necessary to determine the dispersion coefficients, σ_y and σ_z, based on meteorological conditions to be used in making the forecast. One of several available procedures for estimating σ_y and σ_z relies on a scheme for classifying atmospheric stability based on commonly measured parameters: surface wind speed, incoming solar radiation, and nighttime cloud cover. Table 22.5 shows six stability categories (labeled A through F) defined by these factors. Once the category is established, the graphs in Figures 22.9 and 22.10 are employed to estimate σ_y and σ_z, respectively. Using these curves, only the stability category and distance downwind from the source are needed to obtain dispersion coefficients.[52] The

[51] Methods for computing effective emission height are reviewed by Perkins (1974).

[52] For more information on procedures for estimating σ_y and σ_z using Figures 22.9 and 22.10, see Turner (1994). There are other procedures for estimating dispersion parameters, and they are reviewed by Boubel et al. (1994, pp. 300–306) and Turner (1994, Chapters 2 and 4).

TABLE 22.5 Key to Stability Categories[a]

Surface Wind Speed at 10 m (m/sec)	Day: Incoming Solar Radiation			Night: Cloud Cover	
	Strong	Moderate	Slight	Thinly Overcast or ⩾4/8 Cloudiness[b]	⩽3/8 Cloudiness
<2	A	A–B	B		
2–3	A–B	B	C	E	F
3–5	B	B–C	C	D	E
5–6	C	C–D	D	D	D
>6	C	D[c]	D	D	D

[a] Source: Adapted from Turner (1970). This source should be consulted for definitions of strong, moderate, and slight incoming solar radiation.

[b] The degree of cloudiness is defined as the fraction of sky above the local apparent horizon that is covered by clouds (Bibbero and Young, 1974, p. 319).

[c] The neutral class D, should be assumed for overcast conditions during day or night. Class A is the most unstable and class F is the most stable.

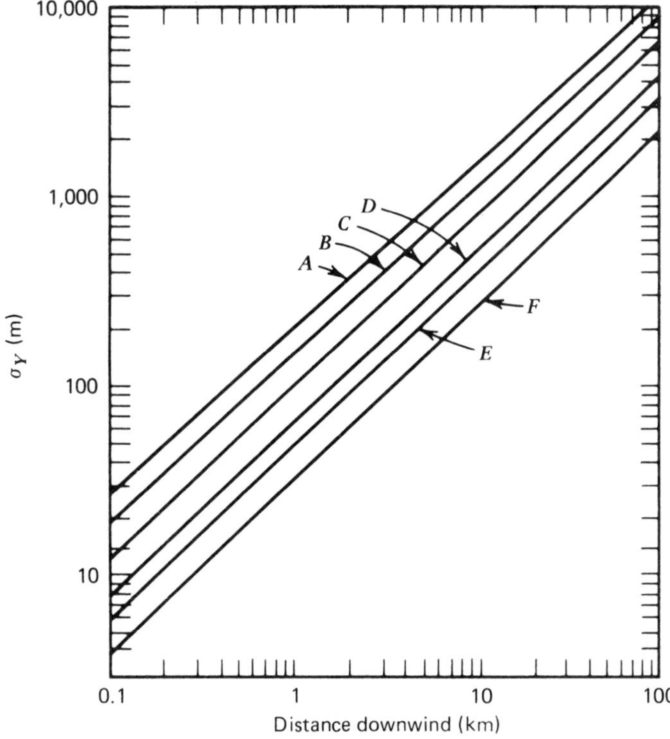

Figure 22.9 Horizontal dispersion coefficient as a function of downwind distance from the source. Adapted from Turner (1970).

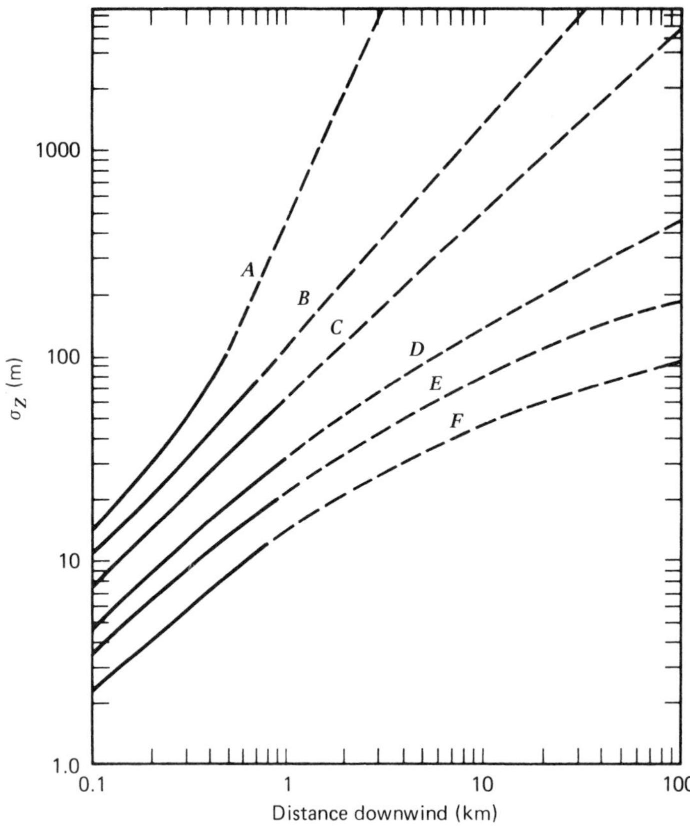

Figure 22.10 Vertical dispersion coefficient as a function of downwind distance from the source. Adapted from Turner (1970).

curves in Figure 22.9 and 22.10 were developed based on analyses of data from field experiments and they are appropriate for rural areas.[53] Dispersion coefficients developed by Briggs and tabulated by Turner (1994, Table 4.5) can be used for urban areas.

The computation of $c(x, y)$ is demonstrated with Turner's (1994) example concerning a petroleum refinery to be located in an open, unobstructed area. A smokestack at the proposed refinery would have an effective height $h = 60$ m, and emit sulfur dioxide at a rate $Q = 80 \times 10^6$ μg/sec.[54] Suppose it were necessary to forecast SO_2 concentrations at a distance 500 m downwind from the refinery stack. The conditions used in forecasting are for an overcast winter morning with a ground level wind speed $U = 6$ m/sec.

Dispersion coefficients are estimated as follows: For overcast days with wind speeds of 6 m/sec, Table 22.5 indicates D as the applicable stability category. Figures 22.9 and 22.10 show that for category D and a downwind distance of 500 m, the dispersion coefficients are $\sigma_y = 36$ m and $\sigma_z = 18.5$ m, respectively.

[53] This procedure for estimating dispersion coefficients builds on studies originally performed by F. Pasquill and then extended by F. A. Gifford. Values of σ_y and σ_z obtained using Figure 22.9 and 22.10 are sometimes termed *Pasquill-Gifford dispersion parameters.*

[54] The example assumes that SO_2 is chemically nonreactive and that any SO_2 fallout is reflected back into the ambient air when it reaches the ground.

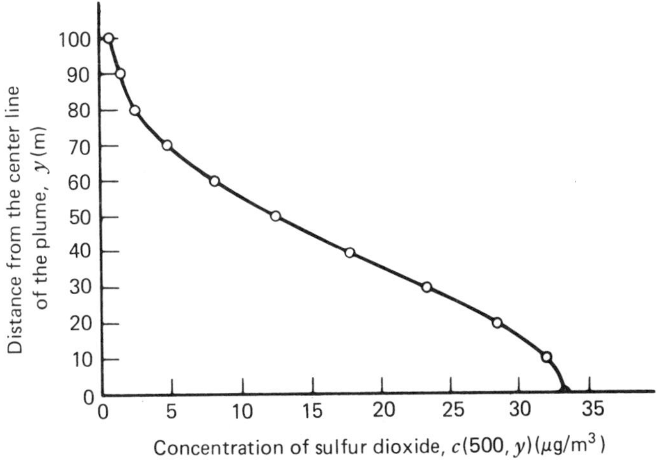

Figure 22.11 Estimated distribution of SO_2 concentration along a line 500 m downwind from a refinery smokestack.

Equation 22-6 is used to construct a curve showing the concentration distribution at 500 m downwind. This is done by setting $x = 500$ and letting y take on different values in equation 22-6. For example, at $x = 500$ and $y = 0$ (the center line of the plume), substitution of the previous values for x, y, Q, U, h, σ_y and σ_z into equation 22-6 yields

$$c(x, y) = \frac{80 \times 10^6}{(3.14)(36)(18.5)(6)} \times \exp\left[-\frac{1}{2}\left(\frac{0}{36}\right)^2 - \frac{1}{2}\left(\frac{60}{18.5}\right)^2\right]$$

$$= 33 \ \mu g/m^3.$$

Performing this computation with $y = 10$ instead of 0, and with all other values constant, the SO_2 concentration is 32 $\mu g/m^3$. The concentration distribution in Figure 22.11 is found by repeating the procedure using $y = 20$, $y = 30$, and so forth. The figure is one half of the bell-shaped distribution of SO_2 concentration at $x = 500$; the mirror image of the curve applies for negative values of y.

Many empirical investigations have been undertaken to test or *validate* the point source Gaussian model. Typically, such studies begin by estimating the model parameters, σ_y and σ_z, and appropriate values of Q, U, and h. Forecasts are then made with equation 22-6 and the results are compared with field measurements. Numerous validation studies indicate that when the point source Gaussian plume model is applied for CO, SO_2, and particulates, forecasts of concentration are reasonably close to observed values.[55] In this context, *reasonably close* means that a substantial majority of predicted values are within +/− 50% of observed values. Accuracy of forecasts may decrease significantly if there are departures from conditions used in formulating the Gaussian plume model: for example, the assumption that wind direction is constant. Because the model does not account for chemical reactions that occur between nitrogen oxides and hydrocarbons, it is not useful in predicting concentrations of photochemical oxidants.

An extension of the Gaussian point source model is used to predict effects of chemically stable discharges from line sources. Consider an example involving emissions of carbon monoxide from motor vehicles on a long, straight highway. The road is treated as an infinite number of very small point

[55] Bibbero and Young (1974, pp. 331–45) summarize results from several studies undertaken to validate the point source Gaussian model.

sources, the individual motor vehicle exhausts. For the case in which a steady wind blows perpendicular to the highway, the CO concentration resulting from the numerous point sources is

$$c(x) = \frac{2q}{\sqrt{2\pi}\,\sigma_z U} \exp - \left(\frac{h^2}{2\sigma_z^2}\right) \qquad \textbf{(22-7)}$$

where $c(x)$ is the concentration x m downwind, U, σ_z, and h are as previously defined, and q equals the *line source strength,* or pollutant emission rate per unit distance, typically given in units of μg/m-sec.

Multiple-Source Models for Chemically Stable Substances

Complex computer models have been created to forecast the concentration of chemically stable substances released from a large number of point, line, and area sources. Some of these are combinations of Gaussian plume models. Others rest on differential equations based on the law of conservation of mass. These equations are solved using numerical analysis methods to yield ambient concentrations.

In the United States, the need for models to predict ambient concentrations increased dramatically during the 1970s. Stringent federal air quality regulations in that decade led to an upsurge in modeling activity. Models were developed to treat time scales ranging from 1 hour to 1 year. In addition, numerous physical circumstances were analyzed, including hilly terrain and urban areas containing complex building configurations. Efforts were also made to model non-steady-state conditions, such as those in early morning when auto emissions gradually build to their rush hour peak.

Multiple-source models are widely available from government agencies and private vendors in many countries. For example, Elsom (1995, pp. 134–35) reports on models developed by Cambridge Environmental Research Consultants in the United Kingdom. As another example, the U.S. Environmental Protection Agency makes computer programs for several models available as part of a system called UNAMAP, the User's Network for Applied Modeling of Air Pollution.[56]

The Climatological Dispersion Model (CDM) is one of the many models available for predicting the concentration of chemically stable substances discharged from multiple sources. It relies on extensions of the Gaussian models previously described and treats both point and area sources in flat terrain. CDM forecasts concentrations of chemically stable substances under steady-state conditions with averaging times ranging from 1 month to 1 year.

A test of CDM by Prahm and Christensen (1977) indicates the model's ability to provide reasonable predictions. Their study involved a flat, urban area of 500 square kilometers that included the city of Copenhagen, Denmark. The period of analysis was the first 3 months of 1971. Information gathered for testing CDM included SO_2 discharge rates from 252 point sources and 454 area sources during the 3-month period. Measurements of average wind speed and direction were collected to estimate dispersion coefficients. These emission and meteorological data provided a basis for using CDM to calculate SO_2 concentrations at 22 monitoring stations in the Copenhagen area. Calculated values were compared to average SO_2 concentrations at the 22 stations during the 3-month analysis period.

Figure 22.12, which summarizes some of the test results, indicates that calculated SO_2 concentrations were close to measured values at many of the 22 stations.[57] Although the model's performance was good in many respects, it systematically underestimated observed concentrations. The line in Figure 22.12 intercepts the vertical axis at 39; the intercept would be at zero if there were no systematic underestimation. Prahm and Christensen explain the underestimation in terms of rural background concentrations and emis-

[56] For additional information on UNAMAP, see Turner (1979). During the late 1980s, EPA created a computer bulletin board service called TTN, Technology Transfer Network, which is based in Durham, North Carolina. Part of this bulletin board service called SCRAM, Support Center for Regulatory Air Models, contains details on models that EPA recommends for use in implementing federal air quality laws. See Boubel et al. (1994, p. 339) for details on these EPA computer systems.

[57] A statistical measure of the closeness of observed and computed concentrations is the coefficient of linear correlation, R^2. For the data in Figure 22.12, $R^2 = 0.82$. This is interpreted to mean that CDM accounts for more than 80% of the variation in the observed SO_2 concentrations during the period of analysis.

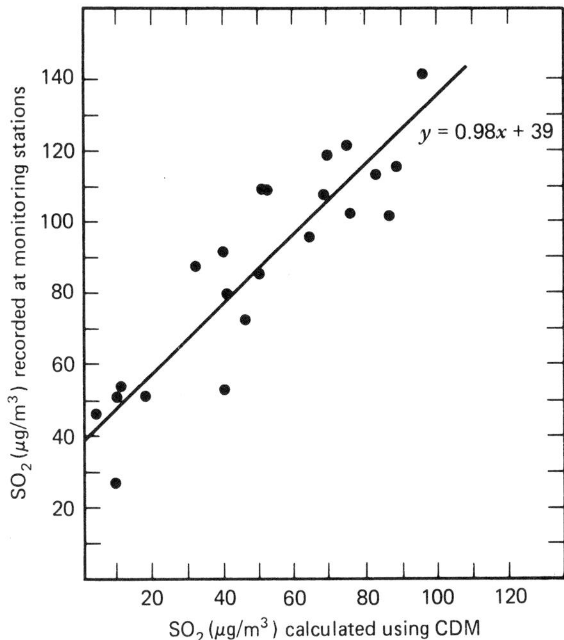

Figure 22.12 Measured and computed 3-month average SO_2 concentrations at 22 receptor points (test no. 3c). Adapted from *Atmospheric Environment,* vol. 11, Prahm and Christensen, "Validation of a Multiple Source Gaussian Air Quality Model," p. 793, © 1977, with kind permission from Elsevier Science Ltd, The Boulevard, Langford Lane, Kidlington OX51GB, UK.

sion sources not accounted for in the CDM calculations. Based on their analysis, Prahm and Christensen felt that CDM provides forecasts adequate for planning purposes when steady-state conditions and long-term averaging periods are used.[58]

CHEMICAL REACTIONS IN THE ATMOSPHERE

The dispersion models previously discussed concern chemically stable pollutants. It is much more difficult to relate discharges to ambient concentrations when the emitted material undergoes chemical transformations in the atmosphere. Sometimes the substance of interest is a reaction product and is not released directly. For example, when forecasts of ambient concentrations of ozone are made, the important emissions consist of hydrocarbons and nitrogen oxides. Ozone itself is not discharged directly. It is a product of photochemical reactions involving discharges of hydrocarbons and nitrogen oxides.

Photochemical Smog

Difficulties in modeling atmospheric dispersion for chemically reactive substances are demonstrated using photochemical smog. In this case, sunlight triggers a complex set of reactions involving numerous chemical species. The reaction products, referred to collectively as photochemical oxidants, include ozone, peroxyacetyl nitrate, and peroxybenzoyl nitrate. Photochemical oxidants have been linked to harmful effects on humans including eye irritation, coughing, and difficulty in breathing. Ozone has produced significant damage to plants. Photochemical smog has a characteristic odor and produces a haze that decreases visibility.

Some of the key reactions in the formation of photochemical oxidants involve nitrogen dioxide, nitrogen oxide (NO), molecular oxygen (O_2), and reactive hydrocarbons. When NO_2 in the atmosphere is exposed to the sun it may absorb a quantum of solar energy and dissociate:

$$NO_2 \xrightarrow{\text{sunlight}} NO + O.$$

Atomic oxygen (O) is highly reactive and may combine with molecular oxygen to form ozone:

$$O + O_2 \rightarrow O_3.$$

Ozone is capable of oxidizing NO to form NO_2 and molecular oxygen:

$$NO + O_3 \rightarrow NO_2 + O_2.$$

If ozone were the *only* agent capable of oxidizing nitric oxide, there would be no ozone buildup. Ozone created by the dissociation of NO_2 would oxidize NO

[58] As of 1990, the U.S. Environmental Protection Agency listed a later version of the Climatological Dispersion Model (CDM 2.0) as being preferred "when calculating long term (i.e., seasonal or annual) concentrations for more than a 'few' sources in an urban environment." Quoted material is from Lape (1994, p. 236), who summarizes EPA's inventory of several models preferred in different physical circumstances.

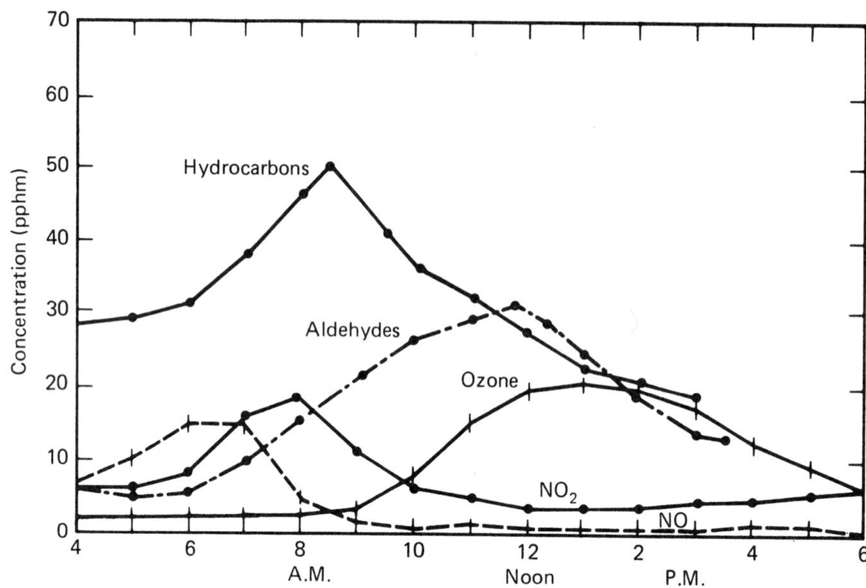

Figure 22.13 Average concentration during days of eye irritation in downtown Los Angeles, California. Hydrocarbons, aldehydes, and ozone for 1953–54. Nitric oxide and nitrogen dioxide for 1958. From Haagen-Smit and Wayne in Stern (ed.) *Air Pollution,* Third Edition, © 1976, p. 254. Reprinted with permission of Academic Press.

and create NO_2. The three reactions above would continue in a cyclic fashion.

Ozone concentrations can increase significantly on sunny days when there are reactive hydrocarbons in the atmosphere. Photochemical reactions involving hydrocarbons yield many oxidizing agents; for example, the hydroperoxyl radical. These chemical species participate in numerous reactions involving the oxidation of NO to form NO_2. And this additional NO_2 may dissociate in the presence of sunlight to produce atomic oxygen, which could yield more ozone. Analyses of the rates of these of these chemical reactions show that the concentration of ozone depends on the *ratio* of the concentrations of NO_2 and NO.[59]

Figure 22.13 shows the time pattern of pollutant formation on a typical smoggy day in Los Angeles. For the first hour or so after sunrise, there is a buildup of direct discharges of hydrocarbons and nitric oxide. During this period the concentrations of reaction products, represented by ozone, aldehydes, and nitrogen dioxide, remain low.[60] Formation of nitrogen dioxide from nitric oxide is well underway shortly after sunrise, and by about 8 A.M. the NO_2 levels peak out. By late morning, ozone concentrations begin to increase and there is a further reduction in NO. The ozone level decreases in late afternoon and reaches zero late in the night.

Relationships between photochemical oxidants and emissions of hydrocarbons and NO_x are difficult to model because of the enormous number of chemicals involved. For example, the simple process of gasoline evaporation yields hundreds of different organic compounds. After the reaction products are accounted for, the number of organic compounds in

[59] This result is demonstrated by Boubel et al. (1994, pp. 168–77).

[60] There is no direct relation between emissions of hydrocarbons and NO_x and concentrations of photochemical oxidants. For example, a 10% decrease in NO_x discharge does not imply a corresponding 10% reduction in ozone concentration. In fact, there are circumstances in which a reduction in NO_x emissions can lead to an *increase* in ozone concentration. As shown by Chock et al. (1981), this counterintuitive result can be explained using mathematical models of smog formation.

photochemical smog formation is in the thousands.[61] Moreover, the particular hydrocarbon molecules that are present influence the extent to which oxidants are formed. Some hydrocarbons (for example, tri- and tetraalkyl ethylenes) are much more chemically reactive than others.

The formation of smog involves thousands of chemical species in thousands of photochemical reactions. To cope with the multitude of chemical species and reactions, air quality modeling specialists group large numbers of different hydrocarbons into classes and then examine representative reactions for each class. There are many alternative ways of categorizing individual hydrocarbons and describing average reactions. A typical photochemical smog model involves about 100 reactions.[62]

Aerosols

Models of photochemical oxidant formation are emphasized here because they have received much attention since the 1950s. However, air quality modeling specialists have recently become concerned with the formation and transport of aerosols, especially those involving sulfates and nitrates. *Aerosols* consist of microscopic solid or liquid particles dispersed in a gas; common examples are smoke and haze. There are two reasons for the increased interest in modeling aerosols:

1. Visibility reduction caused by the light scattering properties of aerosols.
2. Acidification of watercourses due to deposition of sulfate and nitrate aerosols.

In many cases, aerosols are transported from sources several hundred miles upwind of regions experiencing visibility or acidification problems.

The role of aerosols in decreasing visibility is well known. Light scattered into an observer's line of sight by aerosols in the ambient air is perceived as a haze that obscures visible objects. Seinfeld (1986, pp. 297–98) reviewed existing information on visibility and found that from 60% to 95% of visibility reduction was attributable to light scattering by particles. The sulfate iron (SO_4^{2-}) is almost always the most important light-scattering material. Absorption of light by soot particles accounts for from 5% to 40% of the reduction in visibility.[63]

Efforts are being made to model visibility changes associated with emissions of both SO_x and NO_x. For example, an analysis of the Los Angeles air basin linked visibility reductions to nitrate aerosols caused by motor vehicle emissions, and sulfate aerosols caused by point sources of SO_x.[64]

Acid Deposition

Models have been developed to predict effects of SO_x and NO_x emissions in one region on *acidification* of water courses in distant downwind regions. These models generally include components that account for aerosols. Because the reactions involved in forming photochemical oxidants influence ambient concentrations of SO_x and NO_x, the most extensive acid deposition models also include models of oxidant formation.

An example of a model that predicts chemical species linked to acid deposition is the Regional Acid Deposition Model (RADM) developed by the National Center for Atmospheric Research (in Colorado) and the State University of New York.[65] RADM accounts for emissions of SO_2, sulfate, NO, NO_2, CO, ammonia, and 12 classes of volatile organic compounds.

The various modules that make up RADM account for the following phenomona:

- Advection and dispersion.
- Gas-phase chemical reactions.

[61] For information on the types of compounds involved, see Haagen-Smit and Wayne (1976).

[62] For example, the Bay Area Airshed Model used by the air quality management agency in the San Francisco Bay region has 81 reactions, and this model uses strategies to keep the number of reactions to a minimum. These facts were provided by Professor Lynn Hildemann of the Department of Civil Engineering at Stanford University.

[63] Seinfeld (1986, p. 22) defines soot as an "agglomeration of particles of carbon impregnated with 'tar' formed in the incomplete combustion of carbonaceous materials." Diesel-powered motor vehicles are a notable source of soot.

[64] The Los Angeles analysis is presented by White and Roberts (1977).

[65] RADM was created under EPA sponsorship. This description of RADM is based on Seigneur (1994). He also reviews and evaluates eight other models for oxidant formation, acid deposition, or both.

- Cloud formation.
- Chemical transformations in clouds.
- Aersols.
- Wet and dry acid deposition.

The model includes 124 gas-phase reactions involving 63 chemical species. And it accounts for chemical transformations in clouds by including 22 additional reactions. Before RADM is run, a meteorological model is used to predict wind speed and direction, pressure, temperature, and precipitation. RADM can be used to forecast ambient concentrations of photochemical oxidants, and wet and dry deposition of sulfate and nitrate (NO_3^-).

Contexts in Which Models Are Used

Models of acid deposition and oxidant formation are used in developing air quality regulations. Because of the significance of the outcomes, regulatory agencies and regulated firms have made substantial investments to evaluate model performance. Based on an extensive review of nine major models for predicting photochemical oxidants or acid deposition, or both, Seigneur (1994, p. 412) took the following position:

> The regional acid deposition models appear to be able to predict the wet deposition of sulfate and nitrate within a factor of two on average. Regional and urban oxidant models can generally predict ozone ambient concentrations within 30–40% on average.

Seigneur cites numerous studies providing detailed evaluations of model performance.

Despite the attention devoted to modeling chemically reactive substances, models involving reactions are not often used in routine environmental impact assessments because they are expensive to apply. Extensive field data are needed to determine a model's parameters and test the accuracy of its predictions. Because many chemical species in these models are present in very low concentrations, they are often expensive to measure.

Models for predicting the fate and transport of air pollutants are sometimes described in terms of spatial scale, a concept that provides a basis for synthesizing elements of the foregoing discussion. Modeling at a *local scale* (typically well under 100 km) is illustrated by applications of Gaussian plume models to smokestacks. Models of individual point, line, and area sources have been aggregated to create models of pollutant transport at the *urban scale;* the Climatological Dispersion Model provides an illustration. Other examples at the urban scale include models of oxidant formation. *Regional scale* models are used to predict acid deposition due to pollutant transport over about 1000 km; these models are exemplified by the Regional Acid Deposition Model. Finally, there are *global scale* models such as the *general circulation models* used to predict atmospheric temperature, wind, and rainfall at various locations.[66] Global scale models are not designed for assessing impacts of individual project proposals. Their principal use in environmental regulation has been in framing international debates on how best to respond to global environmental threats such as climate change.

Key Concepts and Terms

The Nature of Air Pollution

Carbon monoxide
Volatile organic compounds
Nitrogen oxides
Sulfur oxides
Particulates
Atmospheric lead
Photochemical smog
Photochemical oxidants
Ambient air quality standards
Air pollution and human health
Hazardous air pollutants
Indoor air pollutants
Acid rain
Stratospheric ozone depletion
Global climate change

Air Quality Impact Assessment Process

Data sources for background air quality
National ambient air quality standards

---

[66] For introductions to global scale models and their use, see Henderson-Sellers and McGuffie (1987), and Houghton, Jenkins, and Ephraums (1990).

Concentration units (ppm and $\mu g/m^3$)
Averaging time
Emissions versus ambient concentrations

Estimating Emissions of Air Pollutants
Emission factors
Stationary versus mobile sources
Vehicle miles traveled

Simple Models Relating Emissions to Concentrations
Chemically stable pollutants
Steady-state conditions
Box models
Rollback models
Frequency analysis
Log-normal distributions
Reactive hydrocarbons

Pollutant Transport Via Advection and Dispersion
Point, line, and area sources
Gaussian plume models
Turbulence
Eddy formation
Dispersion vs. advection
Effective emission height
Dispersion coefficients
Stability categories
Line source strength
Multiple-source models

Chemical Reactions in the Atmosphere
Ozone formation
Aerosols
Visibility reduction
Acid precipitation and dry deposition
Global circulation models

Discussion Questions

22-1 Consider the air quality parameters included in the NAAQS. Which of these variables are among the easiest to forecast and why?

22-2 Interpret the following statement: mathematical models to predict air quality are essential to policy formulation because they can answer "what if" questions about alternative sets of air pollution control requirements. Do you agree with this assertion?

22-3 Imagine that you work for a transportation agency that is examining several alternative highway projects in the Los Angeles metropolitan area. Because your agency does not have air quality experts on its staff, an air quality consultant will be hired to assess the air quality changes from the alternative projects. If you were on the committee interviewing prospective consultants, what questions would you ask? What types of answers would you find impressive?

22-4 What considerations are important in determining emissions from automobiles? How would you expect emission factors for the *average* auto to vary over the next decade?

22-5 Indicate the strengths and weaknesses of rollback and box models. What variables would these models have to include to reduce their shortcomings?

22-6 Suppose a metropolitan area has a mass of stable air 200 m above the ground, and surface wind speed is steady at 5 m/sec. The width of the air basin in the direction perpendicular to the wind is 16 km. Use a box model to estimate the maximum regional carbon monoxide emission (in tons per day) consistent with attaining a CO standard of 10 mg/m^3.

22-7 Consider a metropolitan area in which the average background concentration of particulates is 20 $\mu g/m^3$. Suppose the annual geometric average concentration of PM-10 in 1992 was 200 $\mu g/m^3$, and the total regional emission of particulates for that year was 400 $\times 10^6$ lb. The 1992 NAAQS for PM-10 was set at an annual geometric average concentration of 50 $\mu g/m^3$. What is the percentage by which the 1992 emissions of PM-10 would have to be "rolled back" to meet the NAAQS? How might your rollback estimate be criticized if it were used to establish an air quality management plan?

22-8 An open, burning dump in a rural area emits nitrogen oxides at the rate of 2 g/sec on an overcast night when the wind speed averages 5 m/sec. Assume the dump can be represented as a point source with an effective stack height of zero. Find the ground-level NO_x concentrations at the centerline of the plume for a point located 3 km downwind of the source. For purposes of this exercise, assume NO_x is chemically stable.[67]

22-9 Suppose a proposed rural highway is expected to accommodate a maximum of 8000 vehicles per hour during late afternoon rush hours. Overcast conditions are expected on a typical afternoon. The average emission factor for CO is estimated at 10 g/vehicle-mile, and it is assumed that a motor vehicle exhaust has an effective height of zero. Calculate the line source strength (in μg/m-sec) as the product of the emission factor, the number of vehicles per hour, and appropriate units conversion factors. Use the line source Gaussian model, equation 22-7, to predict the CO concentration 300 m downwind on a cloudy day if the wind blows perpendicular to the highway at a constant speed of 4 m/sec.

22-10 Equation 22-7 applies when wind blows perpendicular to a line source. Suppose the angle between the wind direction and the line source is Φ, where Φ is greater than or equal to about 45°. Assume the effective stack height of a motor vehicle is zero. For this case, a line source model to forecast the effects of chemically nonreactive emissions on concentrations has the form[68]

$$c(x) = \frac{2q}{\sin \Phi \sqrt{2\pi}\, \sigma_z U}$$

where all terms except Φ are as defined in the discussion of equation 22-7, and σ_z is determined at a distance x/cos (90-Φ). Using the conditions associated with question 22-9, construct a graph of CO concentration 500 m downwind of the highway versus the angle Φ (where $50° \leq \Phi \leq 90°$).

[67] This question and the next one are adapted from Turner (1994, Chap. 8).

[68] The following modeling result is presented by Horowitz (1982, pp. 339–342) and Turner (1994, p. 4–18). Horowitz indicates that the equation provides reasonable approximations for values of $\Phi > 15°$.

References

Bennett, G., ed. 1991. *Air Pollution Control in the European Community: Implementation of the EC Directives in the Twelve Member States.* London: Graham & Trotman.

Bibbero, R. J., and I. G. Young. 1974. *Systems Approach to Air Pollution Control.* New York: Wiley.

Boubel, R. W., D. L. Fox, D. B. Turner, and A. C. Stern. 1994. *Fundamentals of Air Pollution,* 3d ed. San Diego: Academic Press.

Bryner, G. C. 1993. *Blue Skies, Green Politics.* Washington, DC: CQ Press.

Chock, D. P., A. M. Dunker, S. Kumar, and C. S. Sloane. 1981. Effect of NO_x Emission Rates on Smog Formation in the California South Coast Air Basin. *Environmental Science and Technology* 15(8): 933–39.

Chrostowski, P. C. 1994. "Exposure Assessment Principles." In *Toxic Air Pollution Handbook,* ed. D. R. Patrick, 505–13. New York: Van Nostrand Reinhold.

Coffin, D. L., and H. E. Stokinger. 1977. "Biological Effects of Air Pollutants." In *Air Pollution,* 3d ed., vol. 2, ed. A. C. Stern, 231–60. New York: Academic Press.

Cohn, L. F., and G. R. McVoy. 1982. *Environmental Analysis of Transportation Systems.* New York: Wiley.

Cushman, R. M., D. B. Hunsaker, Jr., M. S. Salk, and R. M. Reed. 1993. "Global Climate Change and NEPA Analyses." In *Environmental Analysis: The NEPA Experience,* ed. S. G. Hildebrand and J. B. Cannon, 442–62. Boca Raton, FL: Lewis Publishers.

de Nevers, N., and J. R. Morris. 1975. Rollback Modeling: Basic and Modified. *Journal of the Air Pollution Control Association* 25(9): 943–47.

Elsom, D. M. 1992. *Atmospheric Pollution: A Global Problem,* 2d ed. Oxford: Blackwell.

______. 1995. "Air and Climate." In *Methods of Environmental Impact Assessment,* ed. P. Morris and R. Therivel, 120–42. London: UCL Press.

EPA (U.S. Environmental Protection Agency). 1973. *Compilation of Air Pollutant Emission Factors.*

Pub. No. AP-42, 2d ed. Research Triangle Park, NC: EPA.

———. 1975. "Draft Environmental Impact Statement, Proposed Wastewater Management Program, Livermore-Amador Valley, Alameda County, CA." EPA Region 9 Office, San Francisco, CA.

———. 1991. *Air Quality Criteria for Carbon Monoxide.* Pub. No. EPA/600/8-90/045F, Office of Health and Environmental Assessment. Washington, DC, EPA.

———. 1994. *User's Guide to Mobile 5 (Mobile Source Emission Factor Model).* Pub. No. EPA-AA-AQAB-94-01, Office of Air and Radiation. Ann Arbor, MI: EPA.

Flachsbart, P. G. 1995. Long-Term Trends in United States Highway Emissions, Ambient Concentrations, and In-Vehicle Exposure to Carbon Monoxide in Traffic. *Journal of Exposure Analysis and Environmental Epidemiology* 5(4): 473–95.

Godish, T. 1991. *Air Quality,* 2d ed. Chelsea, MI: Lewis Publishers.

Haagen-Smit, A. J., and L. G. Wayne. 1976. "Atmospheric Reactions and Scavenging Processes." In *Air Pollution,* 3d ed., vol. 1, ed. A. C. Stern, 235–88. New York: Academic Press.

Henderson-Sellers, A., and K. McGuffie. 1987. *A Climate Modeling Primer.* New York: Wiley.

Hines, A. L., T. K. Ghosh, S. K. Loyalka, and R. C. Warder, Jr. 1993. *Indoor Air: Quality and Control.* Englewood Cliffs, NJ: Prentice–Hall.

Horowitz, J. L. 1982. *Air Quality Analysis for Urban Transportation Planning.* Cambridge, MA: MIT Press.

Houghton, J. 1994. *Global Warming: The Complete Briefing.* Oxford: Lion Publishing.

Houghton, J. T., G. J. Jenkins, and J. J. Ephraums, eds. 1990. *Climate Change: The IPPC Scientific Assessment.* Cambridge, U.K.: University of Cambridge Press.

Kohn, R. E. 1978. *A Linear Programming Model for Air Pollution Control.* Cambridge, MA: MIT Press.

Kormondy, E. J., ed. 1989. *International Handbook of Pollution Control.* Westport, CN: Greenwood Press.

Lape, J. F. 1994. "Air Dispersion and Deposition Models." In *Toxic Air Pollution Handbook,* ed. D. R. Patrick, 226–40. New York: Van Nostrand Reinhold.

Larsen, R. I. 1971. *A Mathematical Model for Relating Air Quality Measurements to Air Quality Standards.* Pub. No. AP-89. Research Triangle Park, NC: EPA.

Lee, D. S., and T. Schneider. 1994. "Control of Air Toxics in Other Countries." In *Toxic Air Pollution Handbook,* ed. D. R. Patrick, 505–13. New York: Van Nostrand Reinhold.

Masters, G. M. 1991. *Introduction to Environmental Engineering and Science.* Englewood Cliffs, NJ: Prentice–Hall.

Ott, W. 1995. *Environmental Statistics and Data Analysis.* Boca Raton, FL: Lewis Publishers.

Papacostas, C. S. 1987. *Fundamentals of Transportation Engineering.* Englewood Cliffs, NJ: Prentice–Hall.

Patrick, D. R. 1994. "Emissions Estimation." In *Toxic Air Pollution Handbook,* ed. D. R. Patrick, 217–25. New York: Van Nostrand Reinhold.

Perkins, H. C. 1974. *Air Pollution.* New York: McGraw–Hill.

Prahm, L. P., and M. Christensen. 1977. Validation of a Multiple Source Gaussian Air Quality Model. *Atmospheric Environment* 11: 791–95.

Robinson, N. A. ed. *International Environmental Law and Regulation.* Charlottesville, VA: Michie.

Rosenbaum, A. S. 1994. "Population and Activity Analysis." In *Toxic Air Pollution Handbook,* ed. D. R. Patrick, 264–79. New York: Van Nostrand Reinhold.

Segall, R. R., and P. R. Westin. 1994. "Source Sampling and Analysis," In *Toxic Air Pollution Handbook,* ed. D. R. Patrick, 166–216. New York: Van Nostrand Reinhold.

Seigneur, C. 1994. "The Status of Mesoscale Air Quality Models." In *Planning and Managing Regional Air Quality: Modeling and Measurement Studies,* ed. P. A. Solomon, 403–34. Boca Raton, FL: Lewis Publishers.

Seinfeld, J. H. 1986. *Atmospheric Chemistry and Physics of Air Pollution.* New York: Wiley.

Sexton, K., and P. B. Ryan. 1988. "Assessment of Human Exposure to Air Pollution: Methods, Measurements, and Models." In *Air Pollution, the Automobile, and Public Health,* ed. A. Y. Watson, R. R. Bates, and D. Kennedy, 207–38. Washington, DC: National Academy Press.

Turiel, I. 1985. *Indoor Air Quality and Human Health.* Stanford, CA: Stanford University Press.

Turner, D. B. 1970. *Wordbook of Atmospheric Dispersion Estimates,* rev. ed. Pub. No. 999-AP-26, Public Health Service, National Center for Air Pollution Control, Cincinnati, OH.

______. 1979. Atmospheric Dispersion Modeling, A Critical Review. *Journal of the Air Pollution Control Association* 29(5): 502–19.

______. 1994. *Workbook of Atmospheric Dispersion Estimates: An Introduction to Dispersion Modeling,* 2d ed., Boca Raton, FL: Lewis Publishers.

White, W. H., and P. T. Roberts. 1977. On the Nature and Origin of Visibility-Reducing Aerosols in the Los Angeles Air Basin. *Atmospheric Environment.* 11(9): 803–12.

Chapter 23

Effects of Human Actions on Water Resources

During the past several decades, many techniques have been developed to forecast how human actions affect natural bodies of water.[1] This chapter surveys those techniques. Some of the forecasting methods predict impacts on the *quantity* of both surface water and groundwater. Others focus on changes in water *quality*. The hydrologic cycle provides a basis for understanding how water quantity and quality are inextricably related.

REPRESENTATIONS OF THE HYDROLOGIC CYCLE

Elements of the Cycle

The *hydrologic cycle* is the circulation of water from the atmosphere to the earth and back (see Figure 23.1). Water's movement from the atmosphere to earth (*precipitation*) generally consists of either rain or snow. Water molecules follow many paths after reaching the earth's surface. For example, if precipitation takes the form of snow and the weather is sufficiently cold, the snow may simply accumulate. As the weather warms, the snowmelt either moves along the land as *surface runoff* or infiltrates into the ground and percolates down to the groundwater zone. In this zone, the spaces between soil particles or rocks are completely filled with water. Groundwater itself moves very slowly in lateral directions, eventually discharging into streams, lakes, or coastal waters.

If precipitation takes the form of rain, the path it follows after reaching the earth depends on the surface it hits. If rain falls on an impermeable area such as a parking lot, it will flow across the surface until it reaches a natural or manmade channel or a depression. In the latter case, the rainwater is stored temporarily. If rain falls on a permeable area such as an open field, its path will depend on surface wetness. The rain will either infiltrate into the ground, become temporarily stored in surface depressions, or become surface runoff. If rain falls on vegetation and does not drip off, it will evaporate. This process is termed *interception* because the rain never reaches the ground.

As indicated in Figure 23.1, the return of water from the earth to the atmosphere occurs by either evaporation or transpiration. In the latter process, water diffuses out from plant cells and vaporizes into the atmosphere. Because it is hard to differentiate between evaporation and transpiration from crop-

[1] Melissa Geeslin assisted in preparing parts of the chapter concerning thermal stratification, eutrophication, Darcy's law, and the transport of pollutants in groundwater. She worked on the chapter while she was a student at Stanford University.

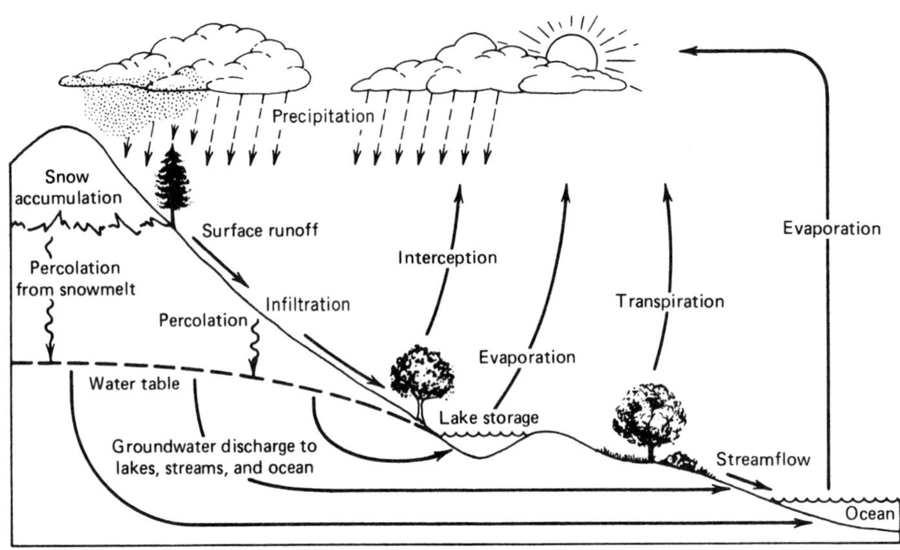

Figure 23.1 Schematic diagram of the hydrologic cycle. Adapted from *Water in Environmental Planning* by Dunne and Leopold. Copyright © 1978 by W. H. Freeman and Company. Used with permission.

lands and vegetated areas, the processes are sometimes viewed collectively as *evapotranspiration.*

Systematic analysis of parts of the hydrologic cycle has been going on for centuries.[2] Since the seventeenth century, components of the cycle have been described using equations. These relationships provide the bases for *hydrologic simulation models* that are applied in forecasting effects of human actions on elements of the hydrologic cycle.

Hydrologic Simulation Models

Computer-based models representing the land phase of the hydrologic cycle have been widely used since 1960. A typical model requires information characterizing the following:

1. The geometric configuration of the river basin, including locations of reservoirs and stream channels.
2. The land surfaces, including their potential for infiltration and evapotranspiration.
3. The soil system and physical attributes of the groundwater zone.

Once this information is specified, a simulation model can produce a forecast of streamflows and reservoir levels based on anticipated rainfall patterns. The model itself consists of equations representing processes in Figure 23.1. Some equations are based on physical principles such as the laws of conservation of mass and energy, and others rest on empirical studies. The primary model *outputs* are hydrographs of flow at selected stream locations in the basin. A *hydrograph* is a plot of how streamflow at a particular point (measured in volume of water per unit time) varies over time.

There are dozens of hydrologic simulation models.[3] One of the earliest, the Stanford Watershed Model (SWM) developed by Crawford and Linsley (1966), has been adapted for use in a variety of circumstances. Figure 23.2 delineates SWM's structure. Equations in the model provide a running account of how water entering a river basin undergoes the various processes (such as interception, infiltration, and runoff) shown in Figure 23.1. Formulas based on mass conservation keep track of all water entering and leaving the basin. Other equations route surface

[2] For a history of efforts to analyze the hydrologic cycle, see Biswas (1970).

[3] For a review of numerous hydrologic simulation models, see Overton and Meadows (1976).

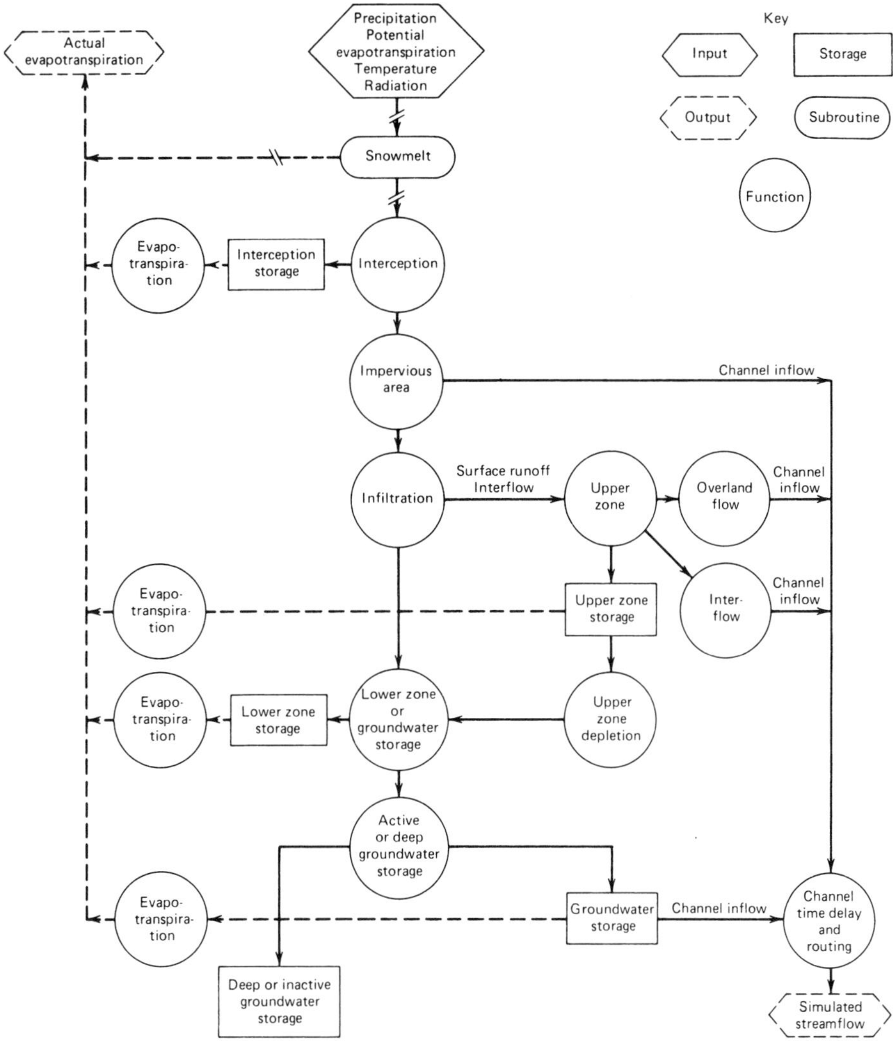

Figure 23.2 Flowchart for the Stanford Watershed Model IV. From N. H. Crawford and R. K. Linsley (1966). Reprinted with permission.

runoff and groundwater flows into stream channels. Information on streamflow is used to develop hydrographs at selected points.

The Stanford Watershed Model requires much data for *calibration,* the process of estimating model parameters. Between 20 and 40 parameters characterizing the river basin under study must be specified. Most of these can be determined from maps and hydrologic records. A small number are estimated using a trial-and-error procedure. Trial values of parameters are based on the analyst's judgment and past experience in using SWM. Once all parameter values are set, the model uses historical precipitation data to produce hydrographs at various locations. These *simulated hydrographs* are then compared with corresponding hydrographs constructed from field measurements (*observed hydrographs*). If there are significant differences between simulated and observed hydrographs, trial parameters are reestimated in an effort to reduce discrepancies. The model is then run again to produce another set of hydrographs, and a comparison with measured hydrographs is re-

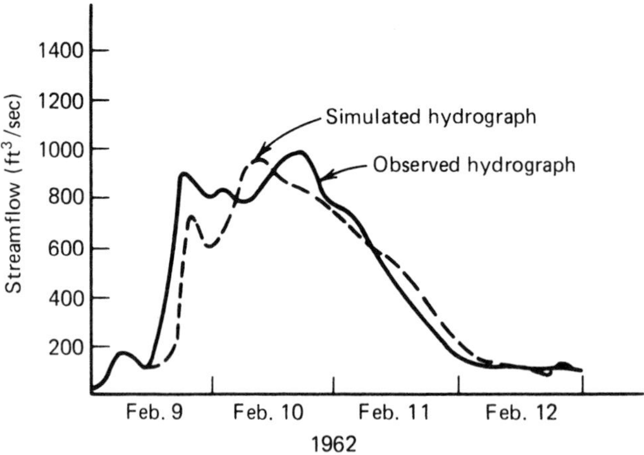

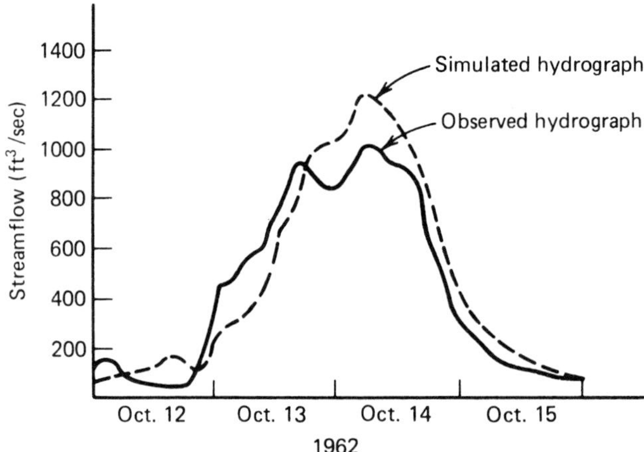

Figure 23.3 Comparisons between simulated and observed hydrographs using the Stanford Watershed Model on Morrison Creek, California. Adapted from James, *Water Resources Research,* vol. 1, no. 2, pp. 223–34, 1965. © American Geophysical Union.

peated. This iterative process continues until the model is considered calibrated.[4] Figure 23.3 demonstrates simulated and observed hydrographs obtained in calibrating SWM in a study of Morrison Creek in California. The close agreement between simulated and measured hydrographs in the figure can be expected only when much data are available for calibration. Effective calibration and use of SWM requires more than data; it requires considerable expertise. The same can be said for all of the computer-based models introduced in this chapter.

Hydrologic simulation models are used to predict how surface runoff and groundwater flows respond to water resources development projects and weather modification (for example, cloud seeding). These models can also be used to analyze land use policies and regulations. Examples of land use

[4] James (1965) presents criteria for judging whether simulated and observed hydrographs are sufficiently close to stop the trial-and-error calibration process.

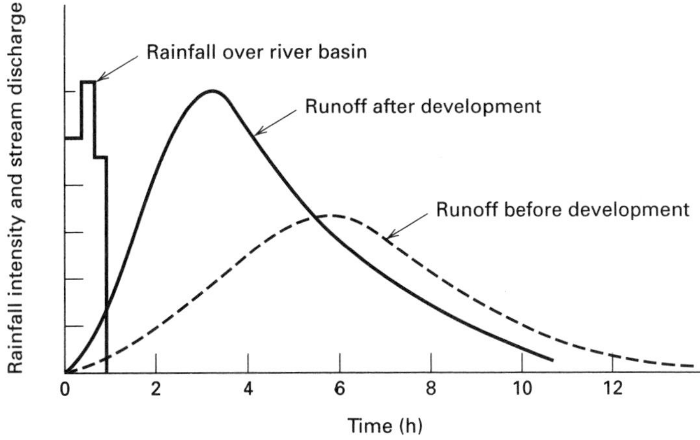

Figure 23.4 Effects of urban development on surface runoff at a particular stream location.

changes that can be modeled include suburban housing tracts that lower infiltration rates by covering permeable land, and forest harvesting practices that decrease rates of evapotranspiration.

INFLUENCE OF LAND USE CHANGES ON FLOOD FLOWS

Land use changes, especially those associated with residential, industrial, and commercial developments, can greatly influence the timing and magnitude of flood flows. This can happen if the impermeable land area is increased, or the water conveyance network is modified to remove storm water more rapidly. The latter occurs when gutters, drains, and storm sewers are installed or when natural stream channels are widened and lined with impervious materials.

Figure 23.4 depicts the impacts of urban development on a flood hydrograph. The bar graph on the left side of the figure shows the time distribution of a rainstorm over a particular river basin. Each curve is a hydrograph at a stream location that receives surface runoff produced by the storm. The dashed curve represents the basin in its natural state. The solid curve is for conditions following urban development in the basin. Because of development, the peak rate of streamflow is higher and occurs sooner. These effects are due to increased impermeability of the land surface, and conveyance system modifications using gutters and storm sewers.

The hydrographs in Figure 23.4 also demonstrate how total volumes of streamflow can be affected by land development. The area under each curve represents the volume of surface runoff associated with the rainstorm. As inspection of the figure shows, the volume of runoff is greater after development. Thus, in addition to increasing the magnitude of the peak flow at a particular stream location, a development project can contribute to downstream flooding problems by increasing total surface runoff. As demonstrated by Harbor (1994), changes in total runoff can also decrease runoff available to recharge groundwater and cause adverse effects on wetlands.[5]

To be useful in analyzing how shifts in land use influence flood flows, a hydrologic simulation model must include parameters representing river basin features affected by land development: rates of infiltration and evapotranspiration, and hydraulic characteristics of a basin's water transport network. A typical

[5] Harbor (1994, p. 103) explains the effects on wetlands: "An increase in the surface runoff, and a corresponding recharge loss, will result in wetlands that have a more eratic hydrologic regime—during storm events, the wetlands will be flooded with stormwater runoff routed from urban areas, but during dry spells the wetlands will receive less recharge from local soil and ground water." Harbor presents a "simple spreadsheet analysis" that planners can use to estimate changes in surface runoff and groundwater recharge due to proposed modifications in land use.

analysis determines how particular rainfall events are converted to surface runoff under different land development scenarios.

James's (1965) investigation of how development affected flood flows in Morrison Creek in California is illustrative. He examined two variables related to land use change: channelization and urbanization. *Channelization* was defined as the fraction of the natural stream channel system modified by either installing storm sewers or straightening, widening, or lining channels. The fraction of the study area devoted to residential, commercial, and industrial development, and to associated public services, was used to measure *urbanization.*

James employed the Stanford Watershed Model to simulate streamflows that would have occurred had the Morrison Creek basin been at each of 13 different levels of urbanization and channelization. Each level was represented by suitable values of the model's parameters. For any particular combination of urbanization and channelization, hydrographs associated with different rainfall events were computed using SWM. Since James was concerned with flood flows, he used rainfall events that occur infrequently and produce flood conditions.

Figure 23.5 is one of several graphs that summarizes the Morrison Creek study results. It indicates how the flood peak expected once in 10 years increases with rising levels of urbanization and channelization in the basin. Flood peaks are given as cubic feet per second of streamflow at a particular location on the creek. With intense urbanization and channelization, the flood peak could be three times as high as the corresponding peak under completely rural conditions. Similar outcomes were obtained for other frequencies of flood occurrence.

James's results include quantitative representations of the hydrograph changes sketched in Figure 23.4. The hydrograph shifts are major for the Morrison Creek basin, but this is not always the case. Studies by Beard and Chang (1979) indicate that when a basin's natural soils have low permeability,

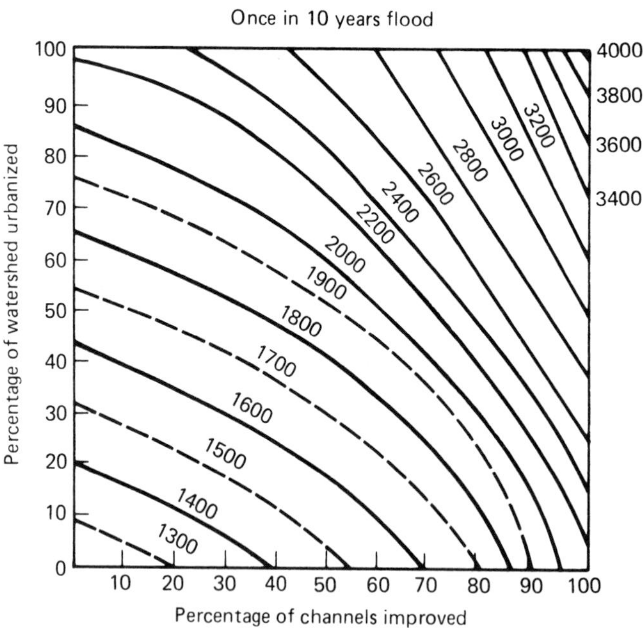

Figure 23.5 Flood peak versus urbanization and channelization (flood flows are in cubic feet per second). Adapted from James, *Water Resources Research,* vol. 1, no. 2, pp. 223–34, 1965. © American Geophysical Union.

the increase in flood peaks caused by urban development may not be large. Effects of land use changes on flood peaks depends on particular conditions in the watershed being analyzed.[6]

PARAMETERS USED TO CHARACTERIZE WATER QUALITY

There are an enormous number of constituents in water, and a correspondingly large number of water quality measures. This discussion introduces indicators commonly employed in considering how human actions affect water quality.

An especially significant aspect of water is the extent to which it contains disease-causing (*pathogenic*) organisms. Examples are bacteria causing typhoid fever and cholera, and viruses responsible for hepatitis A and poliomyelitis. Also included are protozoa such as *Giardia lamblia,* which can cause gastrointestinal disorders. Although modern methods of water disinfection have greatly reduced the threat of waterborne disease epidemics, such threats are a continuing concern. The possible contamination of water by pathogenic microbes is signaled by the presence of coliform bacteria. These bacteria, although usually not pathogenic themselves, are contained in the feces of humans and other warm-blooded animals. Although coliform bacteria are imperfect indicators of fecal contamination, they are widely used because it is impractical to test for the presence of individual disease-causing organisms on a routine basis.

Water also contains a variety of dissolved substances (*solutes*). Many solutes originate from natural processes such as the dissolution of rocks by surface runoff. Solutes are classified as organic if they contain one or more atoms of organic carbon. Otherwise the solutes are inorganic.

Inorganic solutes in water exist as chemically charged substances known as *ions.* Common negatively charged ions include carbonate, bicarbonate, sulfate, nitrate, phosphate, and chloride. Among the usual positively charged ions are calcium, magnesium, sodium, and potassium. *Total dissolved solids* (TDS) is an aggregate measure of these various ions. In recent years, much attention has been given to a group of positively charged ions in water known collectively as *heavy metals.* These metals, which are toxic to humans and other organisms at very low concentrations, include lead, mercury, and cadmium. The significance of trace quantities of heavy metals in water bodies is demonstrated by the mercury pollution in Minamata Bay in Japan between 1953 and 1961. Mercury had been discharged to the bay in wastewaters from industrial facilities. It gradually appeared in the form of methylmercury in fish and shellfish caught for human consumption from Minamata Bay. There were 121 reported cases of methylmercury poisoning caused by eating fish and shellfish from the bay. Of these, 46 cases resulted in death.[7]

It is difficult to fully characterize the *organic solutes* in water because thousands of different compounds are involved, the majority of which have not been identified. Recent developments in gas chromatography and mass spectrometry have led to remarkable advances in measuring organic solutes. They allow volatile organic compounds to be identified at concentrations as low as nanograms per liter (10^{-6} mg/l). However, many organic compounds in water are nonvolatile and not yet subject to identification. A committee of the National Research Council (1977) reviewed the health effects of 309 volatile organic solutes that had been found in drinking water supplies. A few of the substances (for example, vinyl chloride and benzene) are known or suspected causes of cancer in humans. Many others have caused cancer in laboratory animals. It was difficult for the committee to make definitive statements on cancer risks associated with human consumption of the 309 solutes. In many cases, the laboratory research needed to characterize risks of long-term exposure to trace quantities of these materials had not yet been done.

Although most nonvolatile organic solutes in water are not identified individually, they are measured collectively. A widely used aggregate measure is *biochemical oxygen demand* (BOD), the quantity

[6] Other studies on the effects of urbanization and channelization on flood flows are summarized by Dunne and Leopold (1978, pp. 324–30).

[7] This discussion of methylmercury poisoning in Minamata Bay is from the National Research Council (1977). Förstner and Wittman (1981, pp. 18–26) summarize numerous incidents of poisoning from water polluted by arsenic, cadmium, chromium, lead, and mercury.

of dissolved oxygen in a given volume of water used by bacteria to oxidize decomposable organic matter to carbon dioxide, water, nitrate-nitrogen, and other stable end products.[8] These bacteria are *aerobic;* they must have dissolved oxygen to live. BOD is widely used as a measure of potential depletions in a stream's *dissolved oxygen* (DO). High concentrations of BOD may cause decreases in dissolved oxygen and thereby harm fish and other aquatic organisms. When there is no oxygen dissolved in a stream, decomposition of organic matter often yields gases with offensive odors. In this case, decomposition is carried out by bacteria that use oxygen bound up in the organic matter itself.

Some nitrogen compounds associated with biochemical oxidation processes in natural waters are plant nutrients. Compounds of phosphorus, such as those originating in some household soaps and detergents, are also aquatic plant nutrients. The addition of nutrients to surface waters can be particularly problematic in lakes because of their effects on *eutrophication,* a process in which lakes gradually become shallower and more biologically productive due to increased nutrient inflows and sediment deposition. Young lakes are low in nutrients and have low levels of plant productivity. As an unpolluted lake ages, its nutrient and sediment contents increase from two main sources: (1) eroded materials transported to the lake by wind and runoff; and (2) algae that grow on the lake's surface and eventually decay. A lake with high plant productivity may have extensive algal blooms on its surface, and when the algae die they settle. The biochemical decay of algae uses dissolved oxygen and produces sediments. Over time, an unpolluted lake will gradually become shallower as a result of sediment buildup. When a lake receives nutrients because of pollution and other human activities (such as erosion from upstream land development), the rate of eutrophication can increase considerably.

In addition to dissolved substances, water also contains particles that remain in suspension. Land erosion and wastewater discharges are common sources of suspended matter. In flowing streams, material transported can include everything from fine clay to rocks. When a stream's velocity is reduced to zero, heavy material settles out, and the remaining particles are termed *suspended solids.* The material suspended in water causes *turbidity,* another common indicator of water quality. Turbidity is measured by viewing light through a tube of water and determining how much of the light is cut off by the suspended particles.

Suspended particles can pick up (or, more precisely, *sorb*) organic substances, heavy metals, and other chemicals dissolved in water. As explained by Hemond and Fechner (1994, p. 200), *sorption* involves two processes: a "chemical may stick to a particle surface (adsorption) or it may diffuse into the particle (absorption)." The term sorption is used because absorption and adsorption are not easily distinguished.[9]

When suspended particles settle out, the resulting *bottom sediments* become a storehouse for future releases of sorbed chemicals. This phenomenon occurred in Lake Michigan. Prior to 1970, substantial quantities of polychlorinated biphenyls (PCBs) were discharged to the lake, and some of this material accumulated in sediments on the lake's bottom. Polychlorinated biphenyls are organic solutes known to cause cancer in laboratory animals and suspected of causing cancer in humans. Direct discharges of PCBs to Lake Michigan were curtailed in 1970. For several years thereafter, the concentrations of PCBs in the lake remained nearly constant at the high, pre-1970 level because of PCBs released from sediments to the lake.[10]

During the 1970s, the number of parameters used to characterize water quality was expanded greatly owing to increased concern over hazardous substances in water. For example, the U.S. Environmental Protection Agency (EPA) issued in 1978 a

[8] By convention, BOD is often measured as oxygen consumed in biochemical decomposition of organics in a wastewater (or stream) during a *five-day* period. The term *ultimate BOD* refers to the oxygen used to convert *all* biodegradable organics to stable end products. The laboratory technique to determine BOD does not measure all nonvolatile organic solutes, since some are resistant to bacterial oxidation during the short time periods used in the test procedure. Although other aggregate measures of organic solutes have been developed to overcome its shortcomings, BOD continues to be widely used.

[9] Sorption can take place as a result of *ion exchange,* defined as "adsorption, with a charge-for-charge replacement of the ionic species on a surface by ionic species in solution. . . ." (National Research Council, 1990, p. 44).

[10] For more on this case, see Neely (1980). An introduction to the chemical and biological processes that occur in bottom sediments is given by Hemond and Fechner (1994, p. 76).

TABLE 23.1 Examples of Substances on EPA's List of 129 Priority Pollutants[a]

Purgeable organic compounds (31 total)
Chloroform
Methylene chloride
Base/neutral extractable organic compounds (46 total)
bis(2-ethylhexyl) phthalate
Di-n-butyl phthalate
Acid extractable organic compounds (11 total)
Phenol
Pentachlorophenol
Pesticides/PCBs (26 total)
Aroclor 1232
Aroclor 1242[b]
Metals (13 total)
Copper
Zinc
Miscellaneous (3 total)
Total cyanides
Asbestos (fibrous)[c]

[a] The EPA's list included information on the number of times each compound was found during tests of more than 2000 samples. In each category, the listed pollutants were the two most common for the category. Reprinted with permission from Keith and Telliard, *Environmental Science & Technology*, © 1979 American Chemical Society.

[b] In this category, b-BHC appeared in roughly the same number of samples as Aroclor 1242.

[c] Figures on frequency of occurrence were not specified for two of the three substances in the miscellaneous category: asbestos (fibrous) and total phenols.

list of 129 toxic chemicals in water. EPA called these *priority pollutants.* Table 23.1 provides a sample of the chemicals on EPA's list.[11]

Of the many other descriptors of water quality, temperature, pH, and radioactivity are notable because they are frequently influenced by human activities. The release of heated water from cooling systems of electric power plants and various industrial facilities can increase stream temperatures and harm aquatic life. The pH of natural waters is influenced by discharges from industrial facilities such as metal-plating firms, by waters draining from coal mines, and by acid deposition.[12] Acidification of surface waters is a serious concern, because potential consequences include fish kills and other damage to aquatic ecosystems. Radioactivity is another important measure of water quality. Principal nonnatural sources include fallout from nuclear weapons testing, and pollutants associated with nuclear energy production and other nonmilitary uses of radioactive materials.[13]

POINT SOURCES: MUNICIPAL AND INDUSTRIAL EFFLUENTS

Waterborne contaminants originate from both point and nonpoint sources. *Point sources* refer to discharges that can be collected and routed in pipes or other conveyance systems. Examples include municipal and industrial effluents. *Nonpoint sources* are diffuse and typically arise when contaminants enter water that flows over urban areas, farmlands, surface mines, and commercial forests. The contaminated runoff either enters surface water bodies or percolates down to the groundwater. Nonpoint sources such as runoff from city streets become point sources when runoff is diverted into pipes and channels.

Municipal wastewater discharges have been studied scientifically for over a century. Because the character of municipal wastewater is affected by industrial wastes that enter a municipal sewer system, individual municipal effluents can be quite different from each other. A substantial portion of a typical municipal discharge consists of domestic wastewater containing many organic compounds, most of which are biodegradable. The effect of municipal wastewater on water quality depends on the treatment provided and the volume of *receiving water* available to dilute the effluent.

Industrial discharges vary greatly in their physical and chemical characteristics and are generally discussed in terms of four classes of industrial water

[11] The list of 129 priority pollutants resulted from a settlement reached between EPA and the Natural Resources Defense Council and other environmental groups that sued the agency for not regulating toxics as required by applicable U.S. laws controlling pollution of surface waters.

[12] For a definition of pH and more information on acid deposition, see Chapter 22.

[13] The measurement and distribution of radioactive materials in water bodies is a complex topic; Velz (1970, pp. 254–69) provides an introductory discussion. For a review of procedures for treating radioactive waste, see Nemerow (1978, pp. 662–721).

use: sanitary, cooling, cleaning, and process.[14] The first of these, *sanitary,* consists of water used by employees for personal sanitation. Contaminants are similar to those in domestic wastewater.

The largest quantity of water used by industry is for *cooling,* as illustrated by the production of electricity. In a typical steam-electric power plant, substantial amounts of water are used for condensing steam. Water leaving a plant's condenser may be 10 to 15°F higher in temperature than the water entering. The heated water may be either discharged to a surface water body or recirculated in a loop containing a device such as a cooling tower to reduce water temperature.[15]

An example of *cleaning* water use is the washing of soft-drink bottles that have been returned for reuse. Wastewater from bottle washing contains high concentrations of BOD and suspended solids due to leftover soft drinks and debris contained in returned bottles. Because alkaline detergent baths are used in bottle cleaning, the wastewater is highly alkaline.

Process water includes water used for product formation, waste and by-product removal, and product transport. Four steps in the production of printed cotton fabric demonstrate how contaminants enter process water: desizing, scouring, dyeing, and printing. The desizing operation removes starch from cotton cloth soon after it has been woven. (Starch is added early in the production process to give new cotton thread the tensile strength it needs to be woven into cloth.) The starchy wastewater from desizing has high concentrations of BOD and suspended solids. In the scouring operation, desized woven cotton fabric is cooked in a hot, caustic solution to remove impurities. The resulting wastewater is high in BOD and suspended solids and is very alkaline. Fabric dyeing and printing that occur near the final stages of cloth production yield colored wastewater that can also be high in BOD.

[14] This discussion of industrial wastewater is based on information from Nemerow (1978); he provides detailed characteristics of wastewaters from a variety of industrial sources.

[15] Small quantities of water high in total dissolved solids are frequently discharged from power plants with recirculating cooling water systems. This *blowdown water* is released from the recirculatory cooling system to control total dissolved solids. Fresh water (low in TDS) is added to make up for the water released. Frequently, blowdown also contains chemicals used to control aquatic organisms in the cooling water system.

NONPOINT SOURCES OF WATER POLLUTION

Nonpoint sources of waterborne contaminants do not involve direct releases from a pipe or channel, and they are often difficult to describe and control. Discharges from nonpoint sources are sometimes similar in magnitude and quality to that of nearby point sources. As municipal and industrial effluents are controlled by treatment and other means, the relative importance of nonpoint sources in causing pollution generally increases. The following nonpoint sources are often significant: surface runoff from urban, agricultural, and commercial forest areas; discharges from solid waste disposal sites; wastewater from septic tanks, cesspools, and industrial lagoons; and side effects of dredging and other water resources projects.[16] Each will be discussed briefly.

Urbanization and Road Construction

Materials deposited on roads and other impervious surfaces in urban areas are carried away in surface runoff.[17] Substances accumulating on urban land include street litter, pesticides used in lawn care, animal feces, and materials deposited from the atmosphere. In addition, motor vehicles leave many contaminants, such as fibers from brake linings, and hydrocarbons and metals from engine exhausts. Pollutants also include sodium chloride and other *deicers,* chemicals used to keep roads free of ice and snow. Many drinking water supplies have had to be abandoned because they were tainted by runoff containing deicers.[18]

Soil erosion is a major source of water pollution in areas where roads and other physical facilities are being constructed. Wolman and Schick (1967) cite cases where sediment yields from erosion during construction (measured in tons per square mile per year) were 100 times higher than yields in comparable un-

[16] For evidence of the significance of these nonpoint sources in various countries, see Novotny and Olem (1994, pp. 46–55).

[17] In the United States, if runoff from a municipality is collected in a storm sewer system, the discharges from the system are regulated under federal water quality laws (see Chapter 12).

[18] Field et al. (1975) elaborate on the contamination of drinking water supplies by deicers.

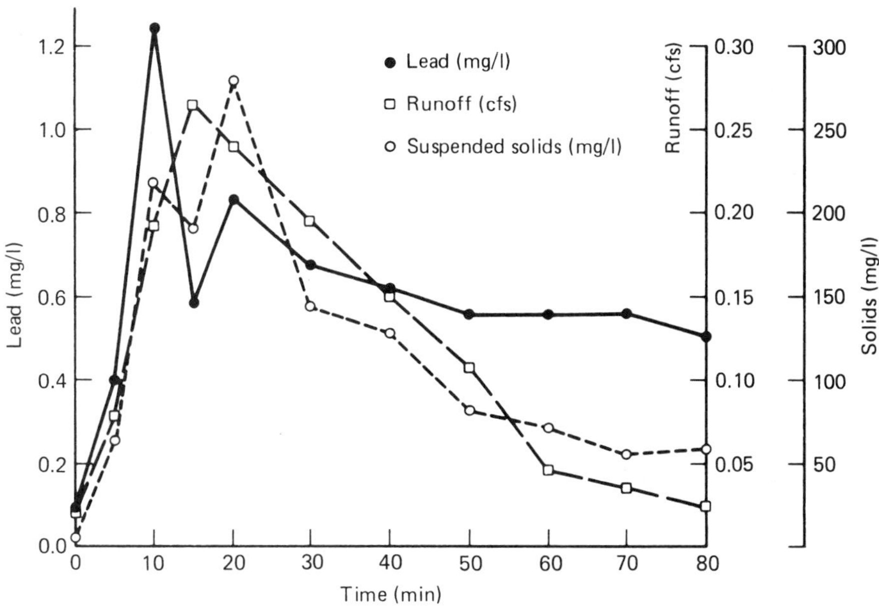

Figure 23.6 Characteristics of urban runoff at Lodi, New Jersey. From Wilber and Hunter "Contributions of Metals Resulting from Stormwater Runoff and Precipitation on Lodi, New Jersey," p. 51, in Whipple, Jr. ed., *Urbanization and Water Quality Control: Proceedings of a Symposium* © 1975 American Water Resources Association.

disturbed areas. Sediment loads affect turbidity and suspended solids in surface runoff.

Suspended solids in runoff from urban areas (*urban runoff*) are important because they often contain sorbed organic compounds and heavy metals. Loadings of suspended solids in urban runoff may be much greater than those of nearby point sources. This is illustrated by Pirner and Harms's (1978) studies of runoff from Rapid City, South Dakota. They found the amount of suspended solids in the runoff from one rainstorm in 1975 to be equal to that from the Rapid City wastewater treatment plant during 7 months. Moreover, the suspended solids yielded by this one storm were delivered to the local stream in only 3 hours.

For small areas, the time distribution of pollutants in urban runoff is often similar to that of the runoff itself. This is shown in Figure 23.6 for a 2.2-square-mile urbanized watershed at Lodi, New Jersey. The figure demonstrates a *first flush effect,* in which concentrations of contaminants (lead and suspended solids) reach their maximums shortly after runoff begins.[19] Peak concentrations result early because the initial volume of runoff carries away high loadings of contaminants that accumulated since the last rainstorm or street sweeping.

Urban runoff is of particular concern in cities that have sewers that carry both storm water and wastewater. During heavy rains, these *combined sewers* may carry flows that greatly exceed the capacity of wastewater treatment works. The resulting *combined sewer overflows* are often discharged without treatment.

Several measures can be taken to reduce water pollution caused by urban runoff. Some involve changes in operations, such as the use of more efficient devices to sweep litter from streets. Others include structural works such as detention basins to store surface runoff and allow some solids to settle out. Storage facilities can also reduce combined sewer

[19] Novotny and Olem (1994, p. 484) observe that while a first flush effect is commonly reported, there have been many cases where this effect did not occur.

overflows because collected runoff can be released slowly so as not to overload treatment facilities.

Agriculture, Forestry, and Mining

Table 23.2 indicates contaminants associated with runoff from agricultural areas (*farm runoff*). Sediments from eroding farmlands frequently carry pesticides, fertilizers, and other chemicals sorbed onto soil particles. Fertilizers often contain compounds of nitrogen and phosphorus that can stimulate the growth of algae and other aquatic plants. Nitrates in farm runoff are of special concern since they can percolate into groundwater. High concentrations of nitrates in groundwater used for drinking have been linked to methemoglobinemia, a serious disease in infants.

Another pollution problem linked to farm runoff concerns total dissolved solids, which are often high in flows draining from irrigated land. High TDS concentrations result because much water enters the atmosphere during the evapotranspiration that accompanies irrigation. Because smaller quantities of water remain to dilute solids dissolved in the water prior to irrigation, the TDS concentration increases. The solids concentration also rises because water dissolves mineral salts in the soil it irrigates. Concentrations of TDS are sometimes quadrupled due to combined effects of salt dissolution and evapotranspiration.

Other nonpoint sources associated with agriculture are animal feedlots, areas containing high densities of animals raised for slaughter. The buildup of animal manure on feedlot surfaces is often massive. When it rains, the resulting runoff is high in BOD and contains nitrogen and phosphorous compounds. Pathogenic organisms, as indicated by high concentrations of coliform bacteria, may also exist in this runoff.

Table 23.2 also lists substances in runoff from commercially harvested forests. As in the case of farmlands, runoff from commercial forests often contains high levels of sediments and pesticides. Sediment yields from eroding logging roads are especially

TABLE 23.2 Contaminants in Surface Runoff from Agriculture, Forest, and Mining Areas

Activity	Water Quality Parameters Affected by Surface Runoff
• Agriculture	
Croplands	Sediments Pesticides Compounds of phosphorous and nitrogen Total dissolved solids
Animal feedlots	Biodegradable organic matter Pathogenic organisms Compounds of phosphorous and nitrogen
• Commercial forests	Sediments Pesticides Water temperature
• Mining	Sediments Heavy metals Acidity

high. In addition, notable increases in water temperature may result when trees shading a stream are harvested.

Surface runoff from mining areas is also high in sediments. Problems can be caused by sediments from either strip mines or open pit mines because the exposed bare land is easily eroded by rainfall. The eroded soils typically contain sorbed metals and other substances associated with mining. When water comes in contact with mined materials, chemical reactions leading to acidification may take place. This occurs frequently in water draining from coal mines.

Leachates from Landfills and Wastewater Disposal Sites

When water infiltrates soil, it dissolves substances and transports them, a process called *leaching.* Water carrying the material is termed a *leachate.* Significant quantities of pollutants enter leachates passing through landfills and dumps containing solid wastes (or containerized liquid wastes), and pits and lagoons receiving industrial wastewater.

Groundwater contamination by leachates is a major problem. For example, a 1977 assessment of 50 industrial waste disposal sites in the United States found that groundwater below the sites was contaminated by leachates in nearly all cases.[20] Heavy metals were discovered in the groundwater at 49 sites, and various organic solutes (including PCBs and benzene) were measured at 40 sites. More recently, the National Research Council (1994, p. 26) reported on the most commonly detected groundwater contaminants at hazardous waste disposal sites in the United States. Six of the 15 most frequently reported hazardous substances were inorganic compounds (in order of frequency): lead, chromium, zinc, arsenic, cadmium, and manganese. The organic compounds among the 15 most commonly detected groundwater contaminants are listed in Table 23.3.

Another nonpoint source of groundwater pollution is the individual wastewater disposal system, such as a septic tank or cesspool. In 1980, nearly 20 million households in the United States relied on individual disposal systems. Groundwater contamination by septic tanks and cesspools typically consists of domestic wastes. However, problems are sometimes caused by chemicals used to clean septic tanks. These cleaners, which frequently contain trichloroethylene, benzene, or methylene chloride, have contaminated many drinking water wells.[21]

[20] For details on this assessment, see U.S. Council on Environmental Quality (CEQ, 1980, p. 89).

Water Resources Projects

Because water resources projects such as dam construction often influence water quality, they are sometimes viewed as nonpoint sources of pollution. This discussion introduces some of the many water quality issues commonly associated with reservoirs and dredging projects.[22]

Water Storage Reservoirs

Consider first the creation of a water storage reservoir. When a stream is dammed, its velocity decreases and sediments formerly held in suspension settle out. The water's turbidity is reduced and a pool of sediments is created behind the dam. Heavy metals and organic substances originally sorbed on sediments may become dissolved at the sediment-water interface. Sediment trapping also affects water quality below the dam. Because water released from the reservoir has a reduced turbidity, the downstream pattern of bank erosion and sediment deposition will be different from what it was before the dam was built.

Reservoir development also influences water quality by means of *thermal stratification,* a process in which layers of water are formed because of the way the temperature and density of stored water vary with depth. Stratification occurs seasonally in reservoirs located in temperate areas. During summer, the heat of the sun on a reservoir's surface can create a warm top layer that floats on colder water below.

[21] For examples of incidents of groundwater contamination by individual wastewater disposal systems, see U.S. Council on Environmental Quality (CEQ, 1981).

[22] In addition to reservoirs and dredging projects, channel modifications (for example, stream widening and lining) also affect water quality. A thorough examination of the literature on these subjects is given by Brookes (1988). For an overview, see Krenkel and Novotny (1980, pp. 236–39). Several case studies demonstrating the adverse ecological effects of federal water projects in the United States are presented by Hunt (1988).

TABLE 23.3 Most Frequently Detected Organic Compounds Found in Contaminated Groundwater at Hazardous Waste Sites[a]

Compound	Common Sources
Trichloroethylene	Dry cleaning; metal degreasing
Tetrachloroethylene	Dry cleaning; metal degreasing
Benzene	Gasoline; manufacturing
Toluene	Gasoline; manufacturing
Methylene chloride	Degreasing; solvents; paint removal
1,1,1-Trichloroethane	Metal and plastic cleaning
Chloroform	Solvents
1,1-Dichloroethane	Degreasing; solvents
1,2-Dichloroethane, trans-	Transformation product of 1,1,1-Trichloroethane

[a] Compounds are listed in order of frequency. The compounds are the 9 organic compounds among the 15 most frequently detected groundwater contaminants at hazardous water sites. This ranking was generated by the Agency for Toxic Substances and Disease Registry using groundwater data from the National Priorities List of sites to be cleaned up under the Comprehensive Environmental Response, Compensation, and Liability Act of 1980. The ranking is based on the number of sites at which the substance was detected in groundwater. Reprinted with permission from *Alternatives for Ground Water Cleanup.* © Copyright 1994 by the National Academy of Sciences. Courtesy of the National Academic Press, Washington, D.C.

The upper layer, known as the *epilimnion,* is often completely mixed as a result of surface winds and rainstorms. A thin layer of rapidly decreasing temperature (known as the *metalimnion*) forms below the epilimnion and separates it from the *hypolimnion,* the colder bottom layer (see Figure 23.7a). As the atmospheric temperature cools in autumn, the stratified layers become unstable, and the reservoir becomes more uniformly mixed. This is termed the *fall turnover.* Analogous stratification occurs in winter for reservoirs located in areas where temperatures drop to near or below freezing.[23]

When a reservoir stratifies during summer, it is common for algae to bloom and die in the epilimnion. Decaying organic matter settles to the hypolimnion. Because the hypolimnion cannot receive oxygen from the atmosphere, the decaying algae that settles may reduce the DO, thereby making it difficult for fish to survive. Moreover, when a reservoir stratifies, water released from the hypolimnion may have low concentrations of dissolved oxygen and high concentrations of iron, manganese, and other substances.[24]

Figure 23.7b, which is for a small impoundment in northern California, demonstrates that dissolved oxygen can approach zero in the hypolimnion during summer stratification. Such variations in DO affect reservoir ecosystems, and they may also cause dramatic changes downstream. This might occur, for example, if water were withdrawn from a depth of greater than about 15 feet for the impoundment in Figure 23.7b.

Navigation Channels

Dredging to develop navigation channels is another type of project influencing water quality. Coastal dredging in highly populated regions is especially problematic because the water bodies frequently con-

[23] Information on thermal stratification is from Masters (1991, p. 139–41). He also describes the process of stratification during winter.

[24] These conditions result when dissolved oxygen in the hypolimnion drops to zero, and decomposition of organic matter is carried out by anaerobic bacteria. In the absence of DO, chemical reactions at the sediment-water interface include the release of iron and manganese from the sediments into solution. Substances having high concentrations in water releases from the hypolimnion may also include sulfide, methane, ammonium, and inorganic phosphorous (Canter, 1996, p. 233).

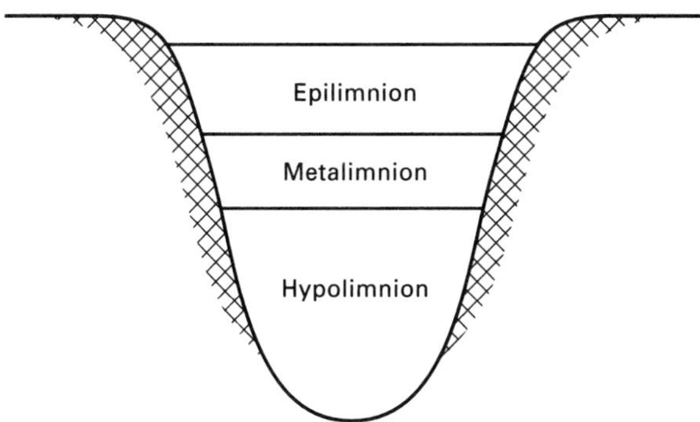

(*a*) Thermal stratification of a reservoir in summer. Tchobanoglous and Schroeder, Water Quality. ©1985 Addison-Wesley Publishing Company, Inc. Reprinted by permission of Addison-Wesley Publishing Company, Inc.

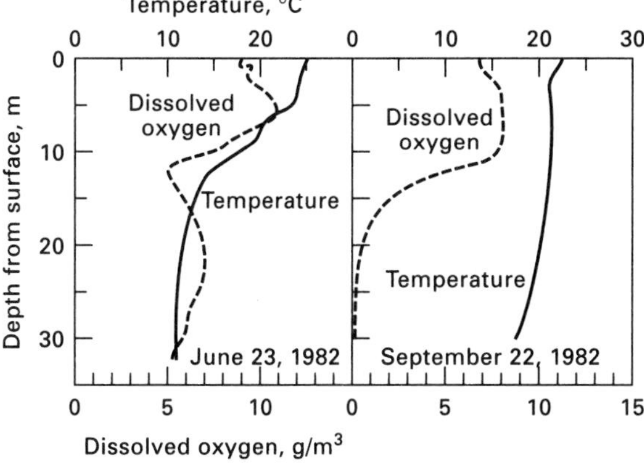

(*b*) Typical temperature and dissolved oxygen profiles for a small thermally stratified impoundment—Lake Mendocino near Ukiah, California. Source: Data from U.S. Geological Survey as it appears in Tchobanoglous and Schroeder, Water Quality. ®1985 Addison-Wesley Publishing Co., Inc. Reprinted by permission of Addison-Wesley Publishing Company, Inc.

Figure 23.7 Reservoir stratification—layering and profiles of temperature and dissolved oxygen.

tain bottom sediments that have sorbed heavy metals and synthetic organic compounds. Dredging exposes a large surface area of bottom sediments to water, and this can lead to many chemical and biological changes. For example, resuspension of bottom muds having a high BOD often depresses dissolved oxygen concentrations in the dredging area. In addition, complex chemical interactions at newly created sediment-water interfaces can increase the concentration of many solutes, some of which may be toxic to aquatic life.

FORECASTING CHANGES IN SURFACE WATER QUALITY

Mathematical models are often used to forecast how proposed projects influence surface water quality. The simplest cases involve substances such as chlorides, which are *conservative* in that they are not modified by chemical or biological reactions in natural waters. Materials that undergo biochemical transformations (for example, nitrogen compounds) are more difficult to model. For some substances, the transformations that occur in natural waters are not understood well enough to be described with equations.

The degree of difficulty in modeling water quality impacts depends on whether the water body is a stream, lake, or estuary. The simplest forecasting exercise is for a stream where the transport of contaminants is in one direction and due almost entirely to the stream's average motion (*advection*). In this case, *dispersion* of pollutants—mixing due to factors such as molecular diffusion, turbulence, and density differences—can be ignored. In comparison, analyses of lakes and estuaries are often more complex because they typically include pollutant transport due to both advection and dispersion, and they frequently account for flows in more than one direction.[25] For the most part, the following discussion considers only streams.[26]

Predicting Effects of Point Source Discharges

Mathematical models have been used to predict effects of point sources on water quality since the early 1900s. Breakthroughs in computer technology since the 1960s have been accompanied by advances in model development and use. Currently, widely available computer-based models that are maintained by

[25] There are exceptions: for example, lakes are sometimes modeled as uniformly mixed systems.

[26] For information on models to predict water quality in reservoirs, lakes, and estuaries, see Krenkel and Novotny (1980), Thomann and Mueller (1987), and James (1993). These authors also introduce models used in analyzing non-steady-state conditions, a subject not treated in this text.

government agencies are used extensively. This section introduces one such model, QUAL 2e, which is kept up-to-date by the U.S. Environmental Protection Agency's Environmental Research Laboratory in Athens, Georgia. To provide a context for introducing QUAL 2e, this section begins by considering basic equations used in modeling simple systems.

Simple Models: First-Order Reaction Kinetics

The law of conservation of mass is the centerpiece of most mathematical models used in predicting changes in surface water quality. The simplest application involves a conservative pollutant, such as chlorides, and steady-state conditions in a stream. As shown in Chapter 16, a *mass balance analysis* for conservative substances under steady-state conditions yields C, the concentration of pollutant immediately downstream of the discharge point:

$$C = \frac{Q_1 C_1 + Q_2 C_2}{Q_1 + Q_2} \qquad \textbf{(23-1)}$$

where

Q_1 = flowrate just upstream of the discharge (L^3/T)
Q_2 = rate of wastewater flow (L^3/T)
C_1 = pollutant concentration just upstream of the discharge (M/L^3)
C_2 = pollutant concentration of the wastewater (M/L^3)

and M, L, and T indicate units of mass, length, and time, respectively. (This use of M, L, and T is employed throughout the chapter.) In the United States, illustrative units are ft^3/sec for flowrate and mg/l for concentration. In countries that employ metric units, typical units for flowrate are m^3/sec and km^3/yr.

Because the pollutant in equation 23-1 is conservative and streamflow is constant, the increase in concentration caused by the discharge is independent of the distance downstream. This analysis assumes that the effluent is completely mixed once it enters the stream, and that the concentration of pollutant is uniform throughout any stream cross section. An illustrative application of equation 23-1 is in Chapter 7, where the equation was used to determine how chloride reduction by the Margarita Salt Company influences the concentration of chlorides in the Cedro River.

The next level of modeling complexity is for a substance that decreases in concentration in accordance with *first-order reaction kinetics.* Underlying physical or biochemical causes of the decrease are not treated explicitly. At any instant, the rate of decay is assumed to be proportional to the amount of substance present; the proportionality factor is called the *rate constant.* This modeling situation was introduced in Chapter 16. As indicated in the discussion of equation 16-2, coliform bacteria are predicted using a first-order reaction model. More generally, a model based on first-order reaction kinetics has the form

$$C(x) = C_0 e^{-k(x/u)} \qquad \textbf{(23-2)}$$

where

x = distance downstream of the discharge (L)
$C(x)$ = pollutant concentration at downstream distance x (M/L^3)
C_0 = pollutant concentration at $x = 0$ (M/L^3)
u = velocity of streamflow (L/T)
k = empirically determined rate constant (T^{-1})

The concentration at $x = 0$ can be calculated using equation 23-1. The value of C_0 is the total mass of pollutant per time immediately below the discharge point divided by the volume of streamflow per time at that point. Because stream velocity is assumed constant, (x/u) represents the *time of travel* below the discharge location.

Chapter 16 shows how k can be estimated using observed data. It also demonstrates how the first-order reaction model predicts reductions in downstream coliform concentrations due to wastewater treatment. BOD and total phosphorous are also modeled by assuming they decrease according to a first-order reaction.

Sometimes the water quality variable of interest interacts with other water quality measures. This holds for dissolved oxygen, which depends on the concentration of biochemical oxygen demand. The simplest mass balance analysis of DO considers two effects: loss of oxygen caused by the biochemical oxidation of organic solutes, and gain of oxygen due to transfers across the air-water interface (*atmo-*

spheric reaeration). One mass balance equation is written for DO and another for BOD, but the equations are linked. The dissolved oxygen equation includes a BOD term, since oxygen is consumed as bacteria metabolize organic matter. Mass balance equations for dissolved oxygen and biochemical oxygen demand are solved simultaneously to yield the *oxygen sag equation*:[27]

$$C(x) = C_s(x) - \left[\frac{k_1 L_0}{k_2 - k_1} \left(10^{-(k_1 x/u)} - 10^{-(k_2 x/u)} \right) + D_0\, 10^{-(k_2 x/u)} \right] \quad \textbf{(23-3)}$$

where

x = distance downstream from BOD discharge (L)
$C(x)$ = concentration of DO at location x (M/L^3)
$C_s(x)$ = *saturation concentration* of DO at location $x(M/L^3)$
D_0 = *initial DO deficit*, $C_s(x) - C(x)$, at x = 0 (M/L^3)
L_0 = *ultimate BOD* at $x = 0$ (M/L^3)
k_1 = coefficient of deoxygenation (T^{-1})
k_2 = coefficient of reaeration (T^{-1})
u = velocity of streamflow (L/T)

An illustrative representation of BOD decay and dissolved oxygen sag is shown in Figure 23.8. The oxygen sag equation holds only if the oxygen-consuming organic solutes are *not* so plentiful that they drive the DO concentration to zero.

Several terms in equation 23-3 require elaboration. The coefficient of deoxygenation, k_1, is associated with the biochemical oxidation of organic matter. Because BOD decay is assumed to follow a first-order reaction, k_1 is estimated using the previously discussed method for finding decay rates for coliform (see Figure 16.9). The coefficient of reoxygenation, k_2, is computed based on stream velocity and average stream channel depth.[28] The saturation value of dissolved oxygen, $C_s(x)$, is the maximum amount of oxygen that can be dissolved per unit volume of water. Tables for determining $C_s(x)$ based on water temperature are available in water quality textbooks.[29] The ultimate biochemical oxygen demand, L_0, is the amount of dissolved oxygen required to convert all the biodegradable organic matter (per unit volume) to carbon dioxide, water, and other stable end products.

If a wastewater discharge contains substantial amounts of organic nitrogen, the oxygen sag analysis can be modified to distinguish between *carbonaceous* BOD and *nitrogenous* BOD. The former is due to organic compounds consisting only of carbon, oxygen, and hydrogen, the latter is due to the oxidation of ammonia-nitrogen.[30] Equation 23-3 has also been extended to account for other factors affecting DO, such as photosynthetic activities of aquatic plants and biochemical oxidation of organic sediments.

Some water quality experts prefer to analyze DO by maintaining separate accounts for deoxygenation and reoxygenation instead of combining them as is done in the oxygen sag equation. This approach, due to Velz (1970), is referred to as the "rational method of stream analysis." Figure 23.9 illustrates results from using Velz's approach to analyze DO in a reach of the Upper Chattahoochee River in Georgia. Several municipal wastewater discharges were accounted for.

Figure 23.9 indicates a close correspondence between observed DO and forecasts obtained using the rational method. Dissolved oxygen decreases immediately below the principal waste sources (river miles 300 to 275) and then gradually increases as the oxygen required in stabilizing organic matter is reduced. If there were no additional effluents downstream, atmospheric reaeration would eventually raise the river's dissolved oxygen to near its saturation value.

Computer-based Models as Illustrated by QUAL 2e

For all but the simplest applications of water quality modeling, it is convenient to rely on computer

[27] Equation 23-3 is often called the *Streeter-Phelps equation* in honor of the researchers who formulated it several decades ago: N. W. Streeter and E. B. Phelps.

[28] Numerous methods for calculating k_2 are presented by McCutcheon (1989, pp. 217–39).

[29] For example, saturation values of DO are tabulated by Velz (1970, p. 194). Because all streams have some background BOD, the dissolved oxygen level of an "unpolluted" stream is often lower than the saturation value given in the tables.

[30] The distinction between carbonaceous and nitrogenous BOD is discussed by O'Connor, Thomann, and Di Toro (1976).

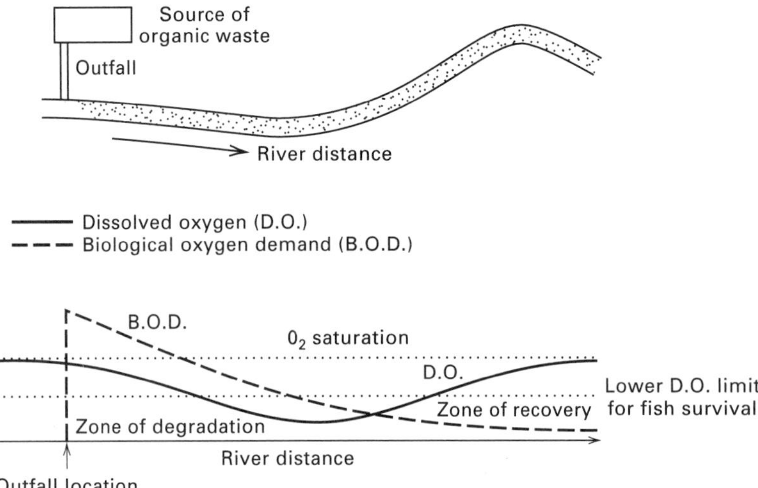

Figure 23.8 "DO sag" induced in a river by inputs of organic waste such as sewage. In the zone of degradation, oxygen is consumed more rapidly by biodegrading microorganisms than it can be replenished from the atmosphere. Recovery begins downstream after much of the organic waste is degraded, and the river becomes reaerated. Source: Hemond and Fechner, *Chemical Fate and Transport in the Environment,* 1994, p. 109. Reprinted with permission of Academic Press.

programs, many of which are maintained and made available by government agencies. One of the U.S. Environmental Protection Agency's water quality models, QUAL 2e, provides an illustration. According to McCutcheon (1989, p. 50), of the many surface water quality computer models developed in the United States, only nine have been evaluated "to some extent," and EPA's QUAL 2e is one of them. Based on his review of available model evaluation studies, McCutcheon (1989, p. 68) concludes that QUAL 2e is among the three "one-dimensional, steady state-state models [developed in the U.S. that] have been completely peer reviewed and evaluated."[31]

The simplest applications of QUAL 2e involve water quality variables that are not significantly influenced by concentrations of other variables. Water quality parameters of this type include temperature and conservative solutes such as chlorides.[32] In addition, the transport of coliform bacteria and other parameters that decay with first-order reaction kinetics is easily modeled. More complex applications of QUAL 2e involve dissolved oxygen and variables that affect oxygen levels (for example, BOD). Figure 23.10 shows the components of QUAL 2e used in modeling dissolved oxygen. As the figure shows, QUAL 2e accounts for much more than just atmospheric reaeration and carbonaceous BOD. The effect of plant photosynthetic activity on the dissolved oxygen budget of a watercourse is also considered (see box labelled *chlorophyll a, algae* in Figure 23.10). The influence on dissolved oxygen of biochemical degradation of organic matter in bottom sediments is also treated. This effect is represented in Figure 23.10 by the box labelled *benthic demand,* oxygen used by

[31] McCutcheon (1989, p. 68), goes on to argue that because QUAL 2e is easy to use and has been employed extensively, its use is justified "over any other simpler model." The evaluations that McCutcheon analyzed include documented, independent checks on computer codes as well as conventional model calibration and validation studies.

[32] Temperature, which is modeled using the law of conservation of energy, affects other water quality variables because many rates of reaction (such as k_1 and k_2 in equation 23-3) are influenced by temperature. For an introduction to the effect of temperature on reaction kinetics and growth rates of aquatic organisms, see McCutcheon (1989, p. 171–209). He also describes models used to predict temperature in surface waters.

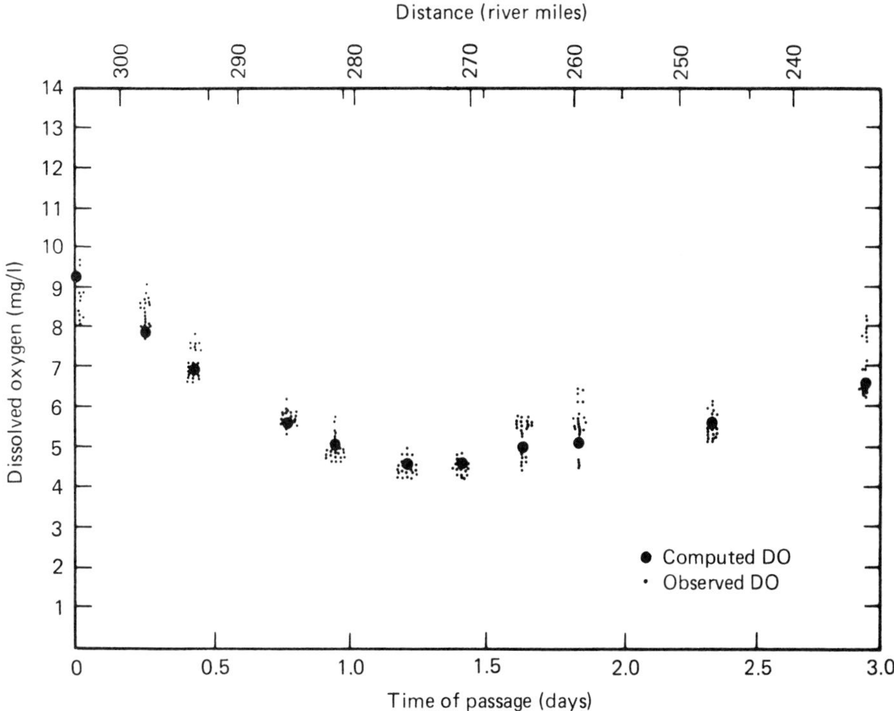

Figure 23.9 Computed and observed dissolved oxygen concentrations in the Atlanta to Franklin reach of the upper Chattahoochee River in Georgia. From Cherry et al. "An Interim Report on the Intensive River-Quality Assessment of the Upper Chattahoochee River Basin, Georgia," p. 36, in Greeson ed., *Urbanized and Water Quality Control Proceedings of a Symposium,* © 1977 American Water Resources Association.

bacteria in bottom sediments while stabilizing organic matter.[33] Bottom sediments may contain organic material in the form of settleable solids from wastewater discharges, as well as aquatic plants and leaves in varying states of decay. The surface layer of bottom sediments, which is in direct contact with water, typically removes oxygen from the overlying water in the course of biochemical oxidation.

The QUAL 2e model includes separate mass balance equations for different forms of nitrogen: organic nitrogen, ammonia-nitrogen, nitrite-nitrogen, and nitrate-nitrogen. Figure 23.11 indicates how these various forms of nitrogen (together with plant and animal nitrogen) are linked in the *nitrogen cycle* for aquatic systems. Phosphorous cycles also exist in aquatic systems and QUAL 2e accounts for them (see Figure 23.10). Mass balance equations within QUAL 2e are linked to reflect the interactions among water quality variables. Running the QUAL 2e model consists of solving sets of mass balance equations simultaneously. Outputs from QUAL 2e can include steady state predictions of DO, carbonaceous BOD, chlorophyll a, and the numerous other parameters shown in Figure 23.10.[34]

An extension of QUAL 2e called TWQM (tailwater quality model) can be used to estimate the quality of waters released from the hypolimnia of stratified reservoirs. In addition to treating the parameters included in QUAL 2e, the tailwater quality model con-

[33] For additional information on benthic demand and the influence of photosynthesis on the oxygen balance of surface waters, see Thomann and Mueller (1987).

[34] QUAL 2e is also capable of accounting for some variations over time. The model has an option "that bases water-quality calculations on the specification of constant flows and loads, but allows specification of time varying meteorological conditions and simulates the effect of variable sunlight, air temperature and wind speed on water quality conditions" (McCutcheon, 1989, p. 12).

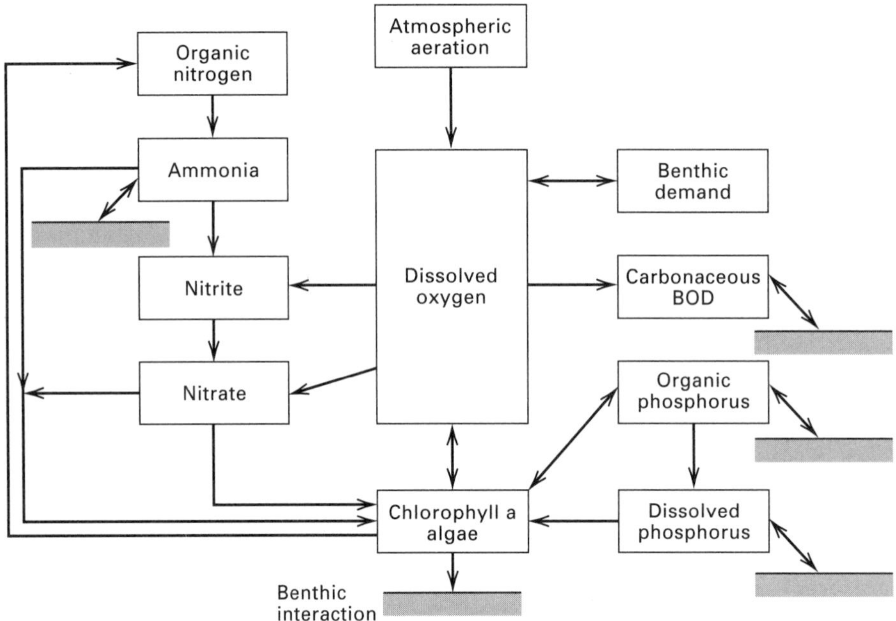

Figure 23.10 QUAL 2e model components related to dissolved oxygen. Reprinted with permission from McCutcheon, *Water Quality Modeling: Vol. 1, Transport and Surface Exchange in Rivers.* Copyright 1989 CRC Press, Boca Raton, FL.

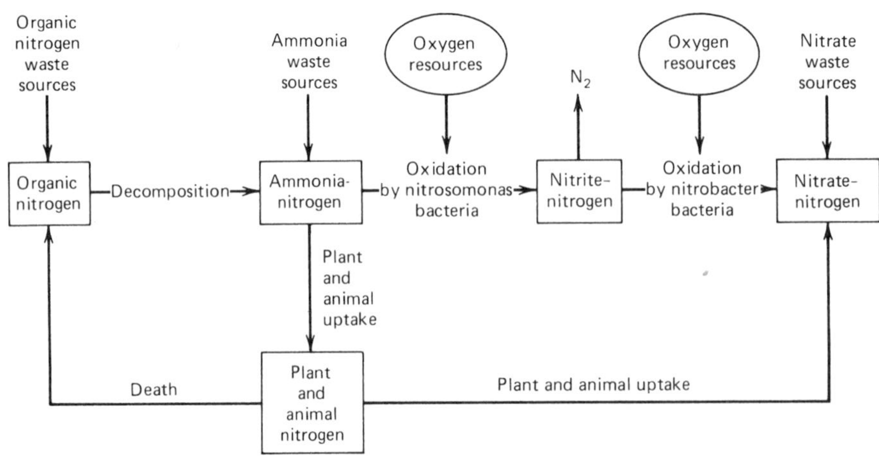

Figure 23.11 Major features of the nitrogen cycle in waters containing dissolved oxygen. From Loucks, Stedinger, and Haith, *Water Resource Systems Planning and Analysis,* p. 451, copyright © 1981 Prentice-Hall. Reprinted with permission.

tains equations representing chemical reactions at the sediment-water interface. These reactions involve iron, manganese, and hydrogen sulfide, among others.[35]

Although water quality models provide a basis for setting wastewater discharge standards and predicting environmental impacts, time and resource constraints limit their use. Gathering the data needed to calibrate model parameters can be expensive and time consuming. In addition, the calibration process itself can be complex, especially for models that include phosphorous and nitrogen cycles, or models that involve physical systems where flows and pollutant dispersion patterns are poorly understood. In any case, use of computer-based water quality models requires specialists who have the training and experience to know which models to use, how to calibrate them, and how to interpret model outputs.

Forecasting Effects of Nonpoint Sources

Estimating water quality impacts of nonpoint sources involves predicting how contaminants on land are transported overland by surface runoff and into the ground by infiltration (see Figure 23.1). Water entering the ground may either percolate down to the groundwater table or move laterally, a process known as *interflow.*

A key step in forecasting effects of nonpoint source pollution is determining the source *loadings,* the amounts of pollutants on land that are transported during a precipitation event. Once loadings have been estimated, hydrologic simulation tools (such as the Stanford Watershed Model) can be used to route the resulting flow and pollutants to the groundwater and various streams, lakes, and coastal waters. After estimates are made of pollutants arriving at surface waters, extensions of the previously mentioned point source water quality models can be used to compute pollutant concentrations at downstream locations.

Nonpoint Source Pollution from Urban Areas

Figure 23.12 delineates processes that affect how nonpoint source pollutants in urban areas are accumulated during dry weather and transported during wet weather. In cities where storm drainage facilities are separate from wastewater collection networks, the main contaminants from surface runoff arrive at surface waters (that is, *receiving water bodies*) via two routes shown at the bottom of Figure 23.12: storm sewers, and shallow groundwater aquifers (*base flow discharges*).[36] In areas where storm runoff is carried in the same pipe network as municipal sewage, contaminants also reach receiving water bodies via overflows of combined sewer systems. Portions of Figure 23.12 concerning infiltration of runoff into aquifers are ignored here but subsequently considered in the discussion of runoff from nonurban areas.

As shown in Figure 23.12, pollutants that build up on impervious surfaces are carried away during precipitation events (*washoff*). These pollutants accumulate from a variety of sources: motor vehicles (for example, vehicle oil leaks); litter;[37] atmospheric fallout (dry deposition); and fallen leaves.

One approach to characterizing pollution loadings in washoff involves estimating (1) the weight of material accumulated on impervious surfaces since the last precipitation event, and (2) how much of the accumulated material will be washed away (either as solids or dissolved substances) by a subsequent precipitation event. Monitoring studies conducted in many cities provide a basis for developing empirical equations to carry out these steps.[38]

An example of how monitoring data have been used to develop equations for predicting accumulation and washout is given by Sartor and Boyd (1975). They gathered data from several cities on the solids accumulating between periods of rainfall or street sweeping. Their data, summarized in Figure

---

[35] For more on TWQM, see Canter (1996, pp. 233 and 237) and references cited therein.

[36] *Aquifers* are layers of the subsurface environment that consist of permeable materials that are fully saturated with water. These layers are capable of yielding significant quantities of water to wells or springs. "During prolonged dry periods, most of the natural flow in surface waters, as well as infiltration into deep sewers, originates from gound-water systems and is referred to as *base flow* or *gound-water runoff*" (Novotny and Olem, 1994, pp. 171–72). In contrast, *aquitards* are subsurface formations that are relatively impermeable and not suitable for development as water supplies.

[37] The term *litter* refers to items such as cans, broken glass, plastic, paper, garbage, dead animals, animal excreta, and the like (Novotny and 0Olem, 1994, p. 454).

[38] This discussion includes an illustrative equation to estimate accumulated solids. For other examples and citations for the data used to derive these empirical equations, see Novotny and Olem (1994, pp. 461–83).

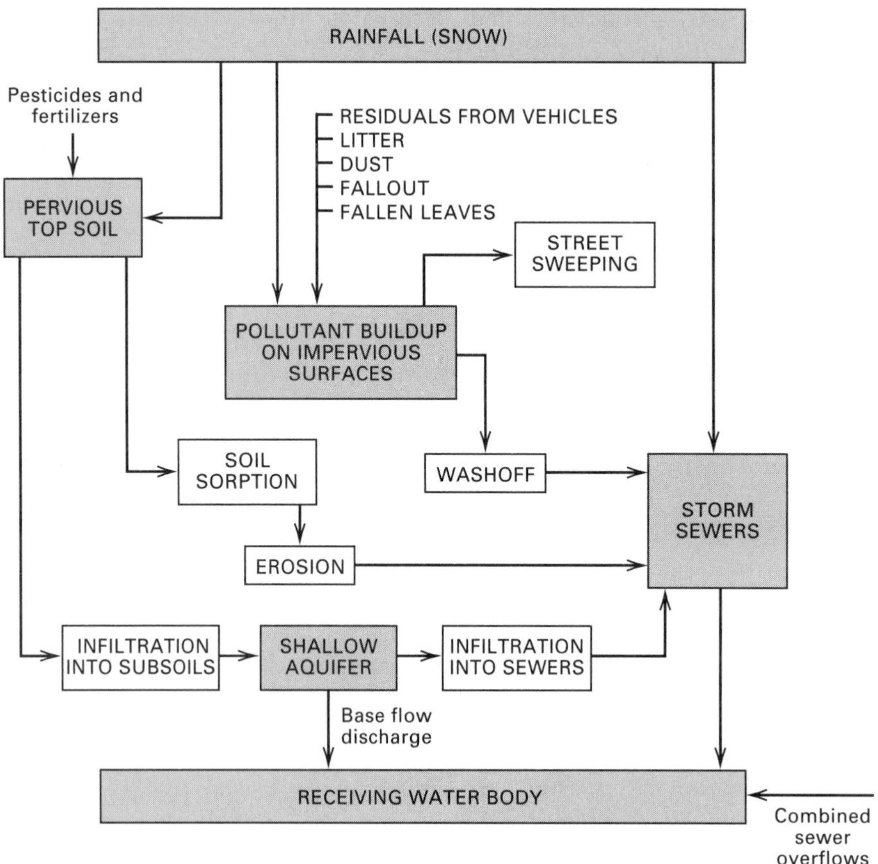

Figure 23.12 Sources of pollution linked to precipitation in urban areas. Adapted from Novotny and Olem, *Water Quality: Prevention, Identification, and Management of Diffuse Pollution,* p. 445, © 1994 Van Nostrand Reinhold. All Rights Reserved.

23.13, were used to derive an equation for estimating solids accumulation on streets per unit length of curb:

$$F_t = \frac{G}{k}(1 - e^{-kt}) \tag{23-4}$$

where

F_t = loading of solids on street surfaces (M/L)
G = constant rate of solids deposition ($M/L\text{-}T$)
k = empirically determined rate constant (T^{-1})
t = time elapsed since last rainfall or street sweeping (T)

Estimates of the parameters k and G have been made for many cities.

Equation 23-4 is demonstrated by a case presented by Overton and Meadows (1976) in which G = 100 lb of solids/curb-mile-day and k = 0.40/day. The estimated quantity of solids accumulating in 2 days after the last rainfall or street sweeping is

$$F_2 = \frac{100}{0.4}(1 - e^{-0.4(2)}) = 138 \text{ lb/curb-mile}$$

As in Figure 23.13, this equation shows the solids buildup reaching a maximum as t, the time elapsed,

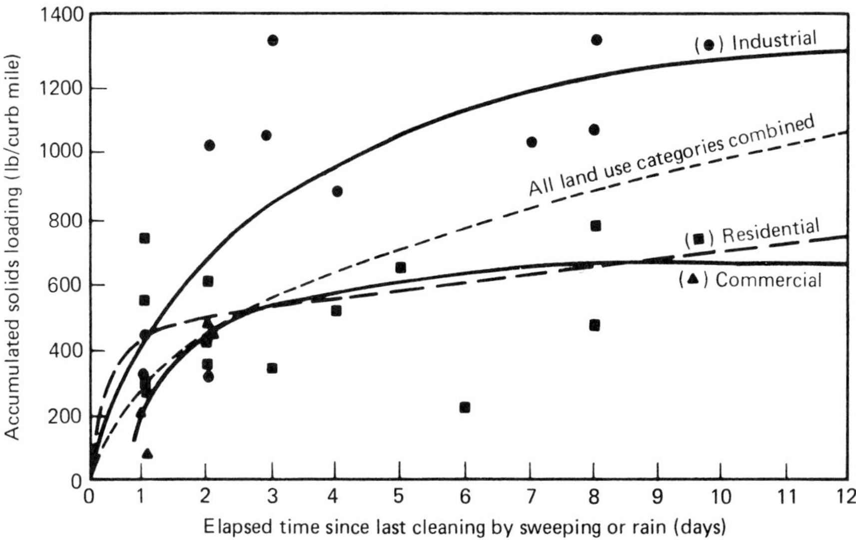

Figure 23.13 Solids accumulation for different land uses. Results of a study for eight U.S. cities. Reprinted from Sartor and Boyd, in Jewell and Swan (eds.), *Water Pollution Control in Low Density Areas,* by permission of University Press of New England. Copyright © 1975 by Trustees of the University of Vermont.

becomes very large. The maximum value, G/k, is 250 lb/curb-mile in this example.

To predict constituents in urban runoff, substances associated with solids accumulating on streets must be identified. Data indicating the amount of a substance per unit of solids for different pollutants and land use categories are used for this purpose. For example, the phosphate in solids accumulating on residential neighborhood streets has been estimated as 50 mg/kg of solids.[39] The mass of phosphate accumulating in such an area is calculated by multiplying 50 mg/kg by the total solids accumulation (in kg).

After individual contaminants in the solids accumulating on streets are determined, it is necessary to estimate how much material will be washed away by surface runoff. Loucks, Stedinger, and Haith (1981) determined this quantity by assuming there is a runoff rate that, if equaled or exceeded, will flush away all accumulated solids. If the surface runoff is less than this rate, the fraction of solids washed off is assumed equal to the actual surface runoff divided by the minimum required for a complete washout.[40]

A remaining question concerns how much contaminant sorbed on the solids goes into solution when the solids are flushed. This can be estimated using a *partition coefficient,* defined as the (constant) ratio of the steady state concentration of a material in one phase (or environmental medium) to its steady state concentration in a second phase.[41] In this case, the phases are accumulated solids and water. Loucks, Stedinger, and Haith (1981) demonstrate how partition coefficients are used to calculate pollutant concentrations in surface runoff.

Procedures to estimate pollutant loadings based on mathematical models of accumulation and washout require monitoring data for the urban area under

[39] This value, which is for neighborhoods with single-family homes, is from Loucks, Stedinger, and Haith (1981, p. 481). Other ways of characterizing the quantities of substances in urban runoff are given by Whipple et al. (1983, pp. 84–90), and Novotny and Olem (1994, chap. 8).

[40] More-complicated approaches to estimating the extent of washout are discussed by Overton and Meadows (1976, pp. 318–28), and Whipple et al. (1983, pp. 104–107).

[41] An introduction to partition coefficients was given in Chapter 16.

study. In the absence of such data, rough estimates of loadings can be made using monitoring results from other urban areas.[42]

There are many computer-based models that use information on mass of pollutants generated by a storm event together with hydrologic simulation procedures to estimate pollution loadings delivered to surface waters.[43] One of the oldest and most widely used is EPA's Stormwater Management Model (SWMM). This model, which is maintained by EPA's Environmental Research Laboratory in Athens, Georgia, has been applied in many locations throughout the world. As in the case of hydrologic simulation models, use of models to predict pollution from urban runoff requires a large data set to calibrate parameters, and extensive experience in model selection, calibration, and output interpretation.

Runoff Outside of Urban Areas

In nonurban settings, the amount of soil eroded by surface runoff is frequently the key factor in determining water quality effects of nonpoint sources. Erosion is often significant for activities related to construction, farming, open pit and strip mining, and forest harvesting.[44]

[42] As an example, the U.S. Environmental Protection Agency (EPA, 1983, chap. 6) developed a method for estimating pollutant loadings at unmonitored sites using data from its Nationwide Urban Runoff Project. The EPA approach is based on a statistical analysis of data from 2300 separate storm events monitored at a total of 81 sites within 22 cities throughout the United States. The pollutant-loading estimation procedure based on EPA's statistical analysis results is demonstrated by Novotny and Olem (1994, pp. 491–95).

[43] More than a dozen computer simulation models have been developed to predict pollutant concentrations in urban runoff, and several of them are reviewed by Overton and Meadows (1976), Whipple et al. (1983), and Novotny and Olem (1994). The models generally assume that no chemical or biological transformations of pollutants take place in the time during which the runoff is routed to a stream, lake, or estuary. In addition, contaminants transferred from surface runoff to the ground via infiltration are often considered negligible in these models.

[44] For more information on soil erosion, see Novotny and Olem (1994, chap 5). In addition to affecting concentrations of contaminants, soil eroded from land surfaces can greatly influence the shape of streams. The increase in suspended sediments delivered to a stream alters the downstream pattern of bank erosion and sediment deposition. These effects are difficult to forecast because sediment transport processes are complex and not fully understood. An introduction to sediment transport modeling is given by Dunne and Leopold (1978, pp. 672–85).

Many relationships exist for predicting soil erosion due to surface runoff. One of the most widely used is the *universal soil loss equation:*

$$A = (R)\,(K)\,(L\ S)\,(C)\,(P) \qquad \textbf{(23-5)}$$

where

A = average soil loss per unit area ($M/T\text{-}L^2$)
R = local rainfall factor (T^{-1})
K = soil erodibility factor
$L\ S$ = slope-length factor
C = cropping management factor
P = erosion-control-practice factor

The equation was developed by applying statistical analysis procedures to more than forty years of field data gathered by the U.S. Department of Agriculture. Although equation 23-5 was originally created to analyze soil conservation activities on farms, it has been adapted for use in nonagricultural settings. Factors on the rightside of equation 23-5 are determined using empirical relationships such as those summarized by Novotny and Olem (1994, pp. 254–68).

The universal soil loss equation can be applied for each of several time periods, depending on how R is selected. Values of R have been developed to yield average annual quantities of eroded soil. They have also been developed to correspond to different frequencies of occurrence for storm events. For example, depending on how R is selected, soil loss can be estimated for a storm whose magnitude is normally exceeded once in 10 years, once in 20 years, and so forth.[45]

After soil loss is determined using the universal soil loss equation or some other model, it is necessary to examine a question related to desorption: How will substances, such as pesticides and heavy metals, sorbed on the eroded soil be distributed between the soil and the runoff that carries the soil. Available analysis methods rely on partition coefficients and other empirical equations to calculate how much sorbed material will be desorbed.[46]

[45] Overton and Meadows (1976, pp. 311–18) present several applications of the universal soil loss equation.

[46] Approaches for determining the disposition of substances sorbed on eroded soil are discussed by Novotny and Olem (1994, chap. 6), and Ghadiri and Rose (1992, pp. 58–65). Many of the available methods focus on desorption of nitrogen, phosphorous, heavy metals and pesticides.

Models that predict surface water quality impacts of nonpoint sources from nonurban areas combine procedures for analyzing desorption of chemicals with hydrologic simulation methods. Some models emphasize the overland transport of eroded materials in surface runoff, and others treat both surface runoff and infiltration into soils with subsequent effects on groundwater quality.

An example of a model for analyzing impacts of nonpoint source pollution from nonurban areas is Chemicals, Runoff, and Erosion from Agricultural Management Systems (CREAMS) developed by the U. S. Department of Agriculture. The erosion component of CREAMS uses a modified version of the universal soil loss equation. The model also includes a component to account for chemicals sorbed on soils going into solution when soils are eroded by surface runoff. CREAMS (and an extension of it that treats transport of chemicals into groundwater) is available from the Agricultural Research Service of the U.S. Department of Agriculture in Tifton, Georgia.[47]

FORECASTING CHANGES IN GROUNDWATER QUALITY

The subsurface environment is represented in Figure 23.14. Porous, granular material (*porous media*) lies below the land surface and overlays solid rock (*bedrock*). Material in the first few feet below the surface is called *soil,* which is a zone characterized by intense chemical and biological activity. The soil and material below it that are not saturated with water make up the *unsaturated* (or *vadose*) *zone*. While there is water in this zone, it does not completely fill the *pores;* that is, the spaces between particles. In the *saturated zone,* all pore spaces are completely filled with water. The water level in a well (that is not being pumped) defines the elevation of the *water table* at the location of the well.

[47] The extension of CREAMS is called Groundwater Loading Effects of Agricultural Management Systems (GLEAMS). The name is misleading, since GLEAMS is mainly concerned with surface runoff and erosion processes. It is similar to CREAMS but includes a component for estimating the movement of pollutants to groundwater (Ghadiri and Rose, 1992, p. 73). For an introduction to these and other, similar models, see Novotny and Olem (1994, chap. 9), and Ghadiri and Rose (1992).

Figure 23.14 illustrates how both water and contaminants from a waste disposal area might flow in an *aquifer,* a geologic layer of porous media permeable enough to allow water to flow through it.[48] The arrows indicate the direction of flow. Because groundwater flows at rates between a few inches and a few feet per day, it can take many years before contaminants leached from a waste disposal area are detected in a well or surface water body.

Human actions affect the quality of groundwater in numerous ways; many of them have been mentioned in the previous discussions of nonpoint sources of pollution (for example, runoff carrying pesticides from farmlands). During the past few decades, environmental agencies have focused much attention on groundwater contamination by leachates from hazardous waste landfills and wastewater lagoons, and by leaks from underground tanks storing gasoline, solvents, and other chemicals.

Forecasting how human actions influence groundwater quality is generally more difficult than predicting effects on surface water quality. One reason is that test wells and other equipment are needed to delineate the groundwater zone and determine hydraulic properties of material in that zone. This type of data gathering is costly. Heterogeneity in the porous media through which groundwater flows is another source of difficulty. Hydraulic properties of subsurface material can vary greatly, even over small areas. And these variations sometimes make it necessary to analyze water movement in all three spatial dimensions. Further complications in predicting groundwater quality are the chemical and biological transformations that occur at the interface of water and subsurface material.

Darcy's Law

Models that estimate how pollutants are transported in groundwater include a component that accounts for the movement of water itself. Groundwater moves because of differences in *hydraulic head* (or simply "head"), a term engineers use to refer to the energy associated with a unit volume of water. When groundwater moves slowly, as it often does,

[48] While water flows fairly easily through an aquifer, it does not pass easily through an aquitard, a layer of clay or other material with low permeability.

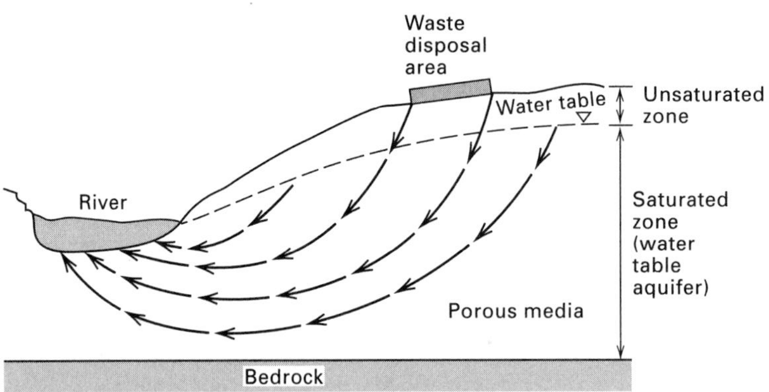

Figure 23.14 Flow of water and contaminants in a water table aquifer. Adapted from Miller in Ward, Giber, McCarty, eds., *Chemical Contamination of Ground Water,* copyright © 1985 John Wiley & Sons, Inc. Reprinted by permission of John Wiley & Sons, Inc.

the kinetic energy associated with water velocity can be ignored. In this case, the hydraulic head of a unit volume of water is calculated as the sum of (1) the elevation of the volume unit above an arbitrary reference level, and (2) the pressure (above atmospheric pressure) of water in the volume unit.

Hydraulic head is clarified by Figure 23.15, which shows a laboratory experiment involving a section of a pipe packed with sand. Water under pressure moves through the sand at a constant flowrate Q (which has units of volume/time). Monitoring devices (called *piezometers*) are at either end of the sand section, and the height that the water rises in these devices measures the pressure (above atmospheric pressure) of water in the spaces between the sand particles. The hydraulic head (h_1) at location 1 can be shown to be[49]

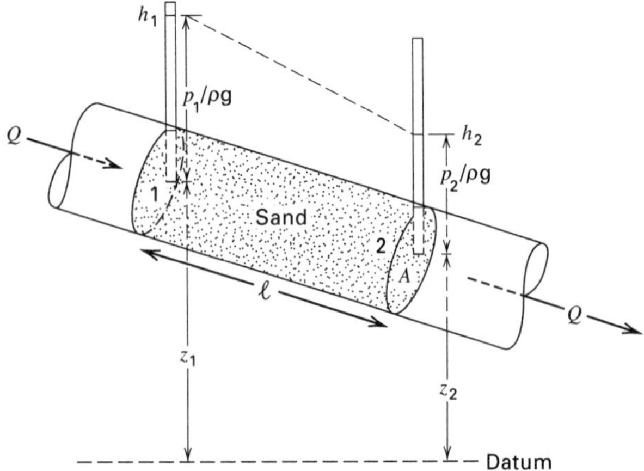

Figure 23.15 Sand-packed pipe section carrying a flow Q. Zheng and Bennett, *Applied Contaminant Transport Modeling: Theory and Practice,* 1995, p. 4, © Van Nostrand Reinhold. All Rights Reserved.

$$h_1 = Z_1 + \frac{p_1}{\rho g} \tag{23-6}$$

where

Z_1 = height of water at location 1 above the reference location (L)
p_1 = pressure of water in the pore spaces at location 1 ($M/L\text{-}T^2$)[50]
ρ = mass density of water (M/L^3)
g = acceleration due to gravity (L/T^2)

[49] For the derivation of equation 23-6, see introductory textbooks on fluid mechanics, such as Street, Watters, and Vennard (1996, chap. 2).

[50] Pressure is a force per unit area, and by tradition, it is expressed in different units in different fields of science and engineering. Using one of Newton's laws of motion, force equals mass times acceleration, and has units of $(M\text{-}L)/T^2$. Thus dividing force by area (L^2) gives pressure with units of $M/L\text{-}T^2$. Hydraulic engineers often measure pressure with piezometers, and it is common to express head associated with pressure, such as $p_1/(\rho g)$, in units of L.

The expression $p_1/(\rho g)$ has units of length and represents the height at which water rises in the piezometer at location 1 (see Figure 23.15). The hydraulic head at location 2 is computed in a similar fashion.

The line connecting h_1 and h_2 is called the *hydraulic grade line,* and its slope, $(h_1 - h_2)\ \ell$ (where ℓ is the length of the sand section), is termed the *hydraulic gradient.* The existence of a non-zero gradient is what causes groundwater to flow. Since water will lose energy as it overcomes friction in moving around particles, the drop in the hydraulic grade line is in the direction of flow.

The laboratory experiment in Figure 23.15 is employed here to introduce a fundamental equation used in modeling groundwater flow: *Darcy's law,* named for Henry Darcy, a nineteenth-century French hydraulic engineer who developed groundwater flow relationships based on empirical studies.[51] Darcy's law can be used to calculate flowrate, Q, as

$$Q = KA\left(\frac{h_2 - h_1}{\ell}\right) \tag{23-7}$$

where

h_i = hydraulic head at location i, for $i = 1, 2$ (L)
A = cross-sectional area of pipe (L^2)
ℓ = length of sand section (L)
K = hydraulic conductivity of the sand (L/T)

Hydraulic conductivity is an empirically derived constant, and it depends on the characteristics of the porous media through which the water flows. An illustrative characteristic is the range of particle sizes. For instance, a uniformly sized gravel will have a higher hydraulic conductivity than a gravel with many irregular-sized particles. In the latter case, smaller particles fill spaces between larger particles and make it more difficult for water to flow.

A basic relationship used in analyzing the flow of moving fluids is[52]

$$Q = VA \tag{23-8}$$

where Q and A are as previously defined and V is velocity. Using this expression, Darcy's law can be reformulated in terms of velocity. Dividing equation 23-7 by A gives what is sometimes called the *Darcy velocity:*

$$v = \frac{Q}{A} = -K\left(\frac{h_2 - h_1}{\ell}\right). \tag{23-9}$$

Equation 23-9 indicates that the Darcy velocity (v) depends on hydraulic conductivity and the hydraulic gradient.[53]

The velocity of water moving through the pore spaces is termed *seepage velocity.* It is not the same as the Darcy velocity because the latter is calculated assuming that the water flows through area A, the *entire* cross-sectional area perpendicular to the water's movement. However, the area available to accommodate flow is smaller than A because the water moves through the spaces between particles. Irregularities in the size of individual particles and pore spaces make it impossible to calculate anything other than an average of the seepage velocity. This average is computed by accounting for porosity (η), defined as the fraction of the total volume of porous media occupied by pore space (see Figure 23.16). It can be shown that V, the average seepage velocity of groundwater, is[54]

$$V = \frac{v}{\eta}. \tag{23-10}$$

Seepage velocity is greater than Darcy velocity because flow (Q) moves through an area smaller than the bulk cross section (A).

Advection and Dispersion of Conservative Solutes

As in the case of surface waters, the simplest procedures for predicting effects of human actions on

---

[51] Darcy based his original formulation for calculating flow through a groundwater aquifer on field studies he conducted in the course of developing a water supply for his native city, Dijon, France (Biswas, 1970, p. 308).

[52] This relationship, called the *continuity equation,* is based on the conservation of mass. It is derived in fluid dynamics textbooks such as Street, Watters, and Vennard (1996, pp. 108–11).

[53] The Darcy velocity, which is sometimes termed *specific discharge,* is actually a discharge per bulk area; the latter is illustrated by A, the bulk area of sand face in Figure 23.15.

[54] This relationship is derived in textbooks introducing the mechanics of groundwater flow; see, for example, Zheng and Bennett (1995, pp. 3–7).

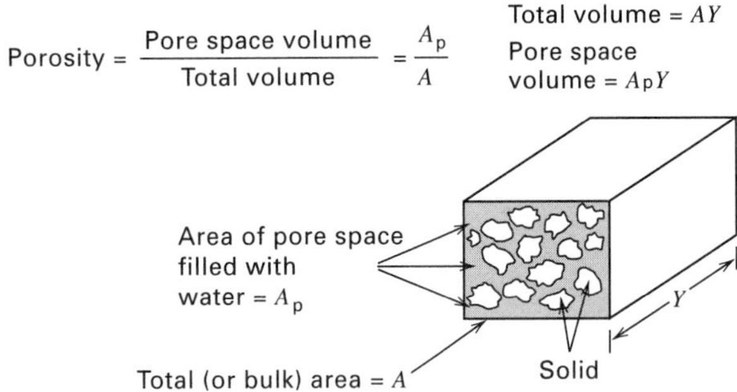

(*a*) Representation of pore spaces and definition of porosity.

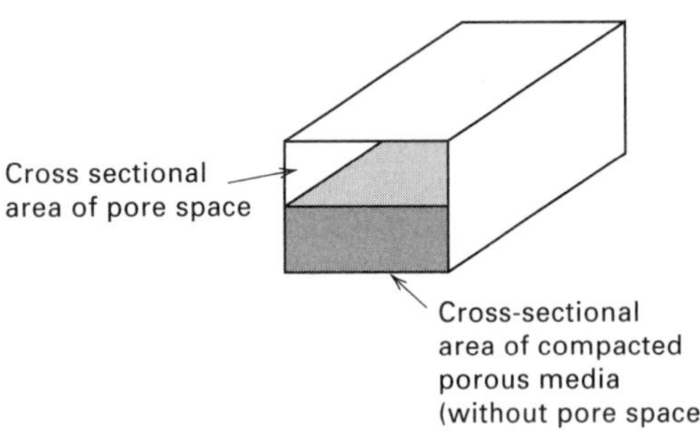

(*b*) Area available for groundwater flow.

Figure 23.16 Pore space area versus total cross-sectional area. From *Introduction to Environmental Engineering and Science* by Masters and Gilbert, © 1991. Adapted by permission of Prentice-Hall, Inc., Upper Saddle River, NJ.

groundwater quality are those involving conservative solutes. Forecasting procedures for conservative solutes generally rest on the following: (1) a hydraulic analysis to estimate seepage velocity, including both speed and direction; and (2) a mass balance analysis to determine solute concentrations at various locations.

In developing forecasting methods, the behavior of a conservative substance is assumed to be affected by only two processes, advection and dispersion. In this context, *advection* is solute transport by the average motion of groundwater flow.[55] It is the main process responsible for pollutant transport in groundwater. A contaminant is transported simply because the groundwater in which it is dissolved is moving. Under most circumstances, the mass of dissolved material is assumed to move with the same average speed and in the same direction as the groundwater itself.

Dispersion refers to the way contaminants spread out as groundwater moves. Because groundwater often moves slowly, turbulence is not an important cause of dispersion. Contaminants spread out mainly because of nonuniformities in velocity caused by heterogeneities in the porous media within the subsurface environment.[56] Figure 23.17 illustrates heterogeneities in an individual pore space and within a small assemblage of pores. As shown in Figure 23.17a, for any parcel of water within a pore, velocity varies from a maximum along the centerline of the pore to zero along the pore walls. Figure 23.17b shows the tortuous paths followed by individual water parcels. As individual parcels wind around soil particles, they vary in velocity.

In field settings, nonuniformities in velocity due to pore level irregularities are small compared to those caused by macroscopic changes in aquifer properties, such as hydraulic conductivity and porosity. Figure 23.18 shows results from an experiment involving a sandbox a few feet long that contained porous materials with two different levels of hydraulic conductivity. The sandbox was injected with water containing a *tracer,* an easily detectable chemical used to mark the movement of dissolved material. The figure illustrates how the tracer spread out, largely as a result of variations in hydraulic conductivity.

Differences between advection and dispersion are clarified by a laboratory experiment stimulating the flow of a conservative substance in groundwater. Figure 23.19a shows a cylindrical column packed with clean sand through which ordinary tap water flows at a constant average velocity. Assume that a nonreactive tracer material is injected into the tap water at time t_0, and the concentration of tracer in the tap water prior to time t_0 is zero.

After time t_0, the tracer concentration in the column inflow is C_0. A short time later, the concentration of tracer in the column *outflow* jumps from zero to C_0. If advection were the only transport

[55] The remainder of this paragraph is based on National Research Council (1990, p. 37).

[56] A minor source of dispersion in groundwater is molecular diffusion, described by Zheng and Bennett (1995, p. 31) as follows: "The classical diffusion model assumes that all ions or molecules within a fluid system are subject to random movement. . . . Where this movement occurs in the presence of a concentration gradient, it results in a net flux or transport of that substance."

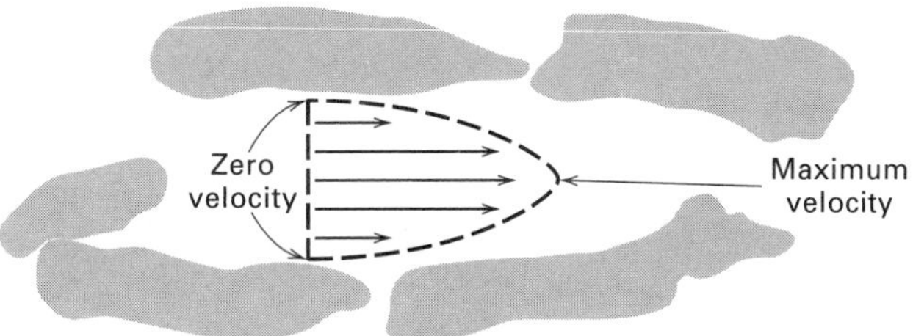

(*a*) Theoretical variation of water velocity within an individual pore.

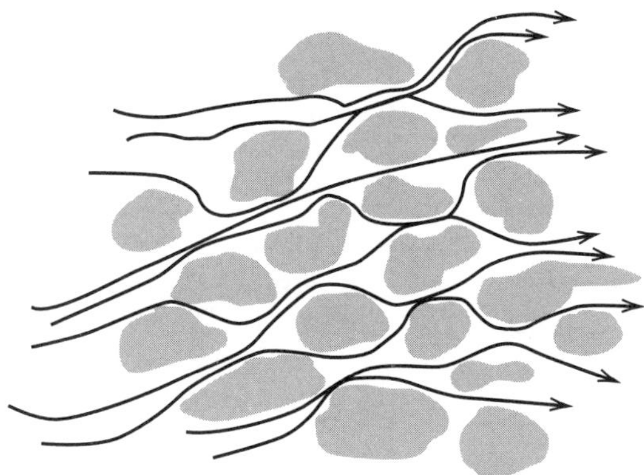

(*b*) Tortuous branching and reconnection of microscopic waterflow paths in a porous medium.

Figure 23.17 Sources of nonuniformities in water velocity within individual pores and small assemblages of pores. From Zheng and Bennett, *Applied Contaminant Transport Modeling: Theory and Practice,* 1995, pp. 25–6, © Van Nostrand Reinhold. All Rights Reserved.

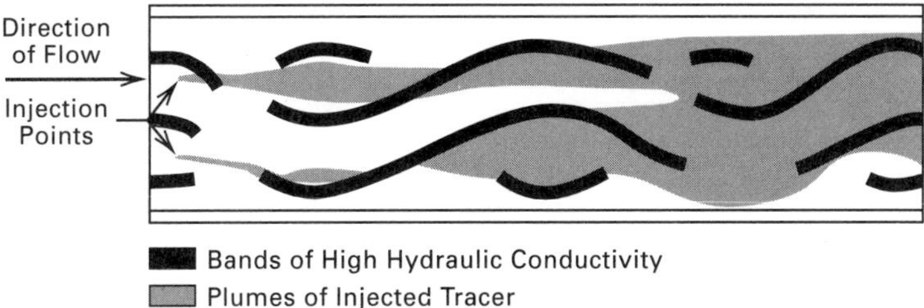

Figure 23.18 Results of a laboratory experiment to determine the effects of macroscopic heterogeneity on a tracer. From Zheng and Bennett, *Applied Contaminant Transport Modeling: Theory and Practice,* 1995, p. 49, © Van Nostrand Reinhold. All Rights Reserved.

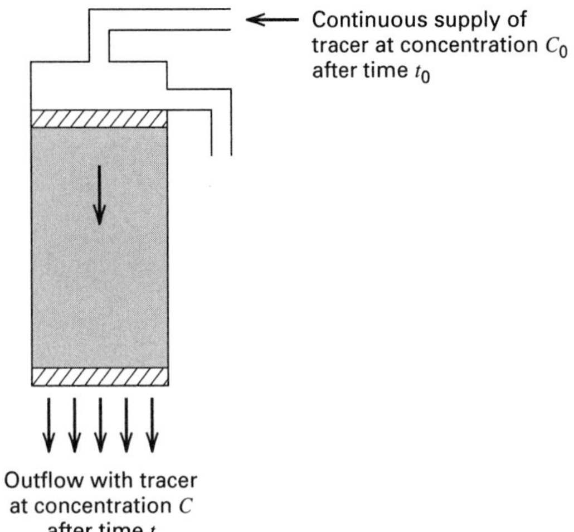

(*a*) Column with uniform flow and continuous supply of tracer after time t_0

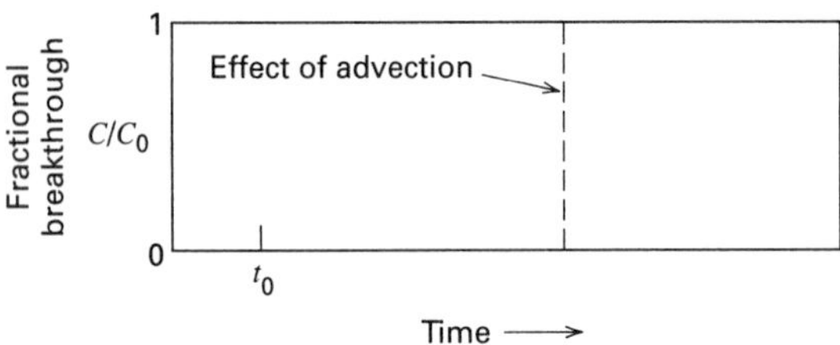

(*b*) Concentration of tracer in outflow if transport is due only to advection (hypothetical condition).

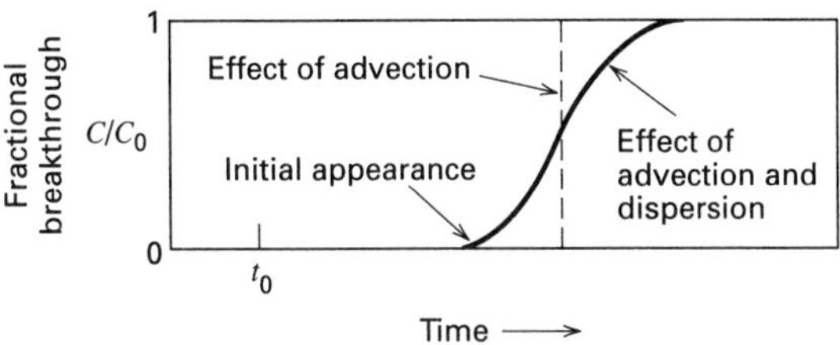

(*c*) Concentration of tracer in outflow due to combined effects of advection and dispersion. Source: Adapted from R. Allan Freeze and John A. Cherry, *Groundwater*, copyright © 1979, p. 390. Adapted by permission of Prentice-Hall, Englewood Cliffs, N.J.

Figure 23.19 Dispersion of conservative tracer passing through a column of clean sand. From *Groundwater* by Freeze and Cherry, © 1979. Adapted by permission of Prentice-Hall, Inc., Upper Saddle River, NJ.

process, the increase in outflow concentration from zero to C_0 would occur in an instant. The tracer injected at time t_0 would advance through the column with a uniform velocity equal to the average seepage velocity. Figure 23.19b, which assumes that advection is the only transport mechanism, shows tracer concentration in the column outflow versus time. By convention, the graph's vertical axis is C/C_0, the fraction of the original solute concentration that appears in the column outflow. Use of this ratio, termed the *fractional breakthrough,* makes it possible to compare the transport of different solutes on a single graph.

When the tracer experiment is actually conducted, the graph of outflow concentration versus time does not have the form in Figure 23.19b. Instead, the outcome is more like the S-shaped curve in Figure 23.19c. Some of the initial tracer injection arrives at the outlet sooner than would be the case if transport were due only to advection, and some of it arrives later. The spreading of the initial tracer injection is attributed to dispersion.

Mass balance analyses have been used to forecast how the outflow concentration of a nonreactive tracer varies with time in the foregoing experiment. The parameters in the resulting mathematical model include average path length of a unit volume of water, average velocity of water, and measures of the dispersion characteristics of the porous material in the column.[57]

For the simple experimental system in Figure 23.19, the concentrations of conservative solutes predicted with mathematical models are often close to observed values. However, real groundwater aquifers are far more complex than a laboratory column, and the accuracy of forecasted concentrations in real aquifers may be quite low, particularly if data used to determine the hydraulic properties of aquifers are incomplete. Complications result because of heterogeneities in the porous media, and the need to consider more than one dimension when analyzing paths of flowing groundwater.

Transport of Sorbing and Reactive Solutes

Processes influencing nonconservative solutes in groundwater include sorption, chemical precipitation, acid-base reactions, oxidation-reduction reac-

[57] Mathematical models based on mass balance analyses for conservative solutes are presented by Freeze and Cherry (1979, pp. 388–97).

tions, and microbial cell synthesis.[58] For many solutes there is not sufficient information to model transformations mathematically.

Studies by Roberts and associates (1980) illustrate the complexities in predicting solute concentrations when biological and chemical reactions influence transport processes. Their investigation concerned the suitability of injecting highly treated municipal wastewater into an aquifer as a means of replenishing the groundwater in Palo Alto, California. This discussion focuses on the part of their study concerning chlorides and two reactive organic solutes, chlorobenzene and naphthalene.

Roberts and his colleagues measured solute concentrations at two locations, an injection well and a test (or observation) well. The quantity of wastewater pumped into the injection well was also measured. Observations were presented on graphs of fractional breakthrough (C/C_0) versus volume of wastewater pumped into the aquifer at the injection well (*cumulative volume injected*). For any solute, the concentrations are C_0 at the injection well and C at the observation well.

Figure 23.20a summarizes results for the transport of chlorobenzene and chlorides. Compared to chlorides, a much greater volume of wastewater had to be injected before any chlorobenzene was detected at the observation well. The reasoning used to explain the differences in transport patterns was that for chlorides, transport between the injection and test wells was due only to advection and dispersion. The average time of travel for chlorides was identical to that of water.

For chlorobenzene, however, transport was affected by sorption in addition to advection and dispersion. Roberts and his colleagues postulated that in the early stages of pumping, much of the chlorobenzene was sorbed on the porous media. Eventually, the sorption capacity of the porous media in the path between the two wells was exhausted. When this occurred, values of C/C_0 were equal to unity for chlorobenzene. Thereafter, because sorption was no longer significant, only advection and dispersion played a role in the transport process, and the chlorobenzene concentration at the test well remained at C_0.

The movement of naphthalene, shown in Figure 23.20b, is quite different from the movement of either chlorides or chlorobenzene. In the early stages of pumping, the naphthalene concentration at the test well increased, but then it decreased without reaching the concentration at the injection well. Roberts and his colleagues hypothesized that naphthalene transport was influenced by biochemical oxidation in addition to advection and dispersion. In the first stages of pumping, the naphthalene concentration increased because bacteria were insufficient to accomplish the biochemical oxidation of naphthalene. Once these bacteria were established, the biochemical oxidation rate increased and there was a sharp decrease in naphthalene at the test well. Finally, after the biological oxidation process reached a steady state, the naphthalene concentration stabilized at less than one-tenth of the concentration at the injection well. This type of field research has provided a basis for extending mass balance analyses for conservative pollutants to include solute transformations such as sorption and biochemical oxidation.

Transport of Nonaqueous-Phase Liquids

Chemical and biological transformations are not the only complications in modeling the fate and transport of contaminants in groundwater. Some liquids, such as oil, don't mix with water and sometimes become trapped as small, immobile globules. Two categories of these *nonaqueous-phase liquids* (NAPLs) are associated with contaminated groundwater: light NAPLs, such as gasoline, which are less dense than water and tend to float on the surface of groundwater; and dense NAPLs, such as trichloroethylene, which are more dense than water and tend to migrate deep underground where they often remain undetected in pools and dissolve slowly to contaminate groundwater.[59]

Figure 23.21 illustrates complexities in modeling NAPLs by showing how a dense nonaqueous-phase liquid (DNAPL) can spread. In the unsaturated zone, portions of a DNAPL can vaporize, and the contaminant may be transported laterally in the unsaturated zone as a gas. The liquid phase of a DNAPL will move vertically downward, however, because gravitational

[58] For an introduction to each of these processes and their role in the transport of solutes in groundwater, see National Research Council (1990, pp. 38–52).

[59] For information on the difficulties in predicting the fate and transport of NAPLs in groundwater, see National Research Council (1994, pp. 32–37).

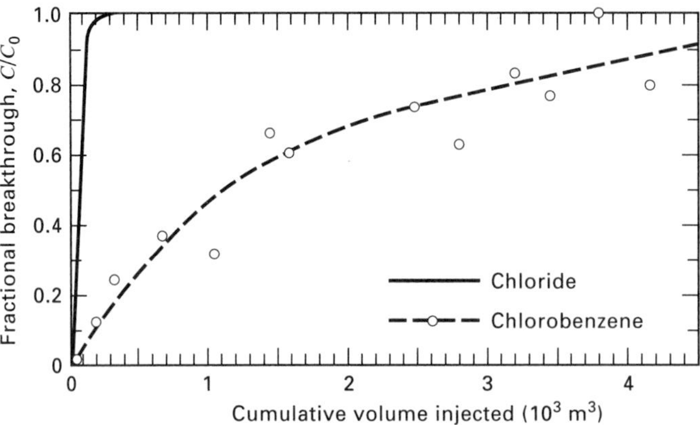

(*a*) Concentrations of chlorobenzene and chloride.

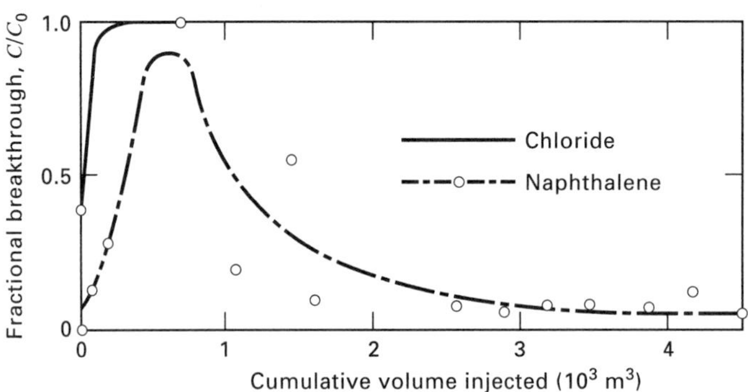

(*b*) Concentrations of naphthalene and chloride.

Figure 23.20 Concentrations of organic solutes and chloride at the observation well. From P. V. Roberts et al., *Journal of Water Pollution Control Federation,* vol. 52, no. 1, pp. 161–72. Water Pollution Control Federation. Copyright © 1980.

forces largely determine its transport in the unsaturated zone.

In the saturated zone, there are two principal mechanisms of transport. One involves portions of the DNAPL that dissolve in the flowing groundwater. This material becomes a contaminant plume, and its movement is governed by advection, dispersion, sorption, and various reactive processes. A second transport mechanism involves the immiscible fluid itself. Because of its density, this material will sink to either bedrock or other relatively impenetrable material (an *aquitard*), where it may spread laterally to form a pool. The vertical transport is essentially uncoupled from the advective-dispersive transport associated with the groundwater's horizontal flow. In addition, some of the dense NAPL may remain as globules entrapped between soil particles (*residual DNAPL*).

Computer-based Groundwater Models

There are many computer-based models that describe flow and solute transport in groundwater, and they vary extensively in terms of what they can accomplish and how well they have been tested.[60] The

[60] Computer-based groundwater models are available from government agencies, universities, and commercial vendors. For an illustrative listing of sources, see Zheng and Bennett (1995, pp. 400–401). An overview of dozens of computer-based models of contaminant transport in groundwater is given by Barry (1992).

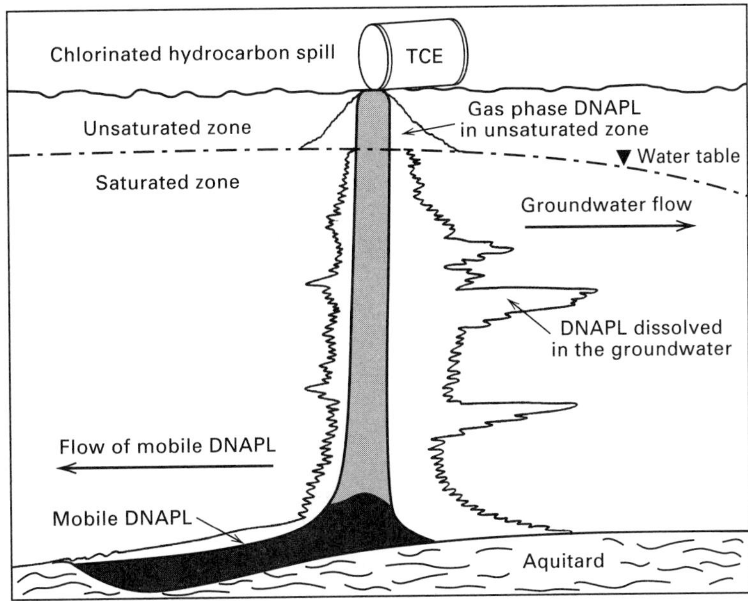

Figure 23.21 Distribution of a dense nonaqueous-phase liquid in the vadose and saturated zone. TCE, the common name for trichloroethylene, is an example of a DNAPL. Source: Fetter (1993, p. 236).

following discussion gives an indication of what types of water quality changes can be predicted using groundwater models.

The simplest groundwater models analyze only flow. These models are of fundamental importance, because it is impossible to understand anything about pollutant transport in groundwater without knowing the speed and direction of subsurface flows. Models of flow were the first groundwater models to be developed, and they date back to Darcy's studies in the 1850s. Over the next hundred years, groundwater specialists and petroleum engineers developed much of what constitutes the contemporary theory of flow through porous media.[61] Because of this long history, well-documented models (based on extensions of Darcy's law under steady state conditions) are available to yield good predictions of hydraulic head and flow in a variety of practical field settings.[62]

[61] Historical observations in this section are from Zheng and Bennett (1995, pp. xv–xxi).

[62] Theoretical and computational difficulties arise in dealing with unsteady-state flow conditions in the unsaturated zone, and in treating flow in fractured subsurface formations. Example of fractures are conduits in cracked subsurface rock formations.

The main difficulty in using computer-based models of groundwater flow is a problem faced in all types of groundwater modeling: the time and expense of gathering data to characterize the geologic environment. The most satisfying results in applying groundwater flow models have been at field sites where large numbers of monitoring wells have been set up over several years to develop a thorough set of data for calibrating models and validating predictions.[63] In settings where decisions must be made in the absence of extensive data on hydraulic conductivity, porosity, and other subsurface characteristics, modeling results can be misleading, even when only flow is being modeled.

The most basic models of pollutant transport in the subsurface treat only conservative solutes, such as chlorides. These models have a relatively long

[63] One such field site is an abandoned landfill at Canadian Forces Base in Borden, Ontario. Experiments in the 1980s used data from 275 multilevel groundwater samplers, each of which had from 14 to 18 sampling ports (Fetter, 1993, p. 150). Since about 1980, over a dozen groundwater modeling investigations have been conducted at the Borden landfill site, and they have involved a variety of modeling approaches. A number of these studies are described by Zheng and Bennett (1995, pp. 312–37).

history, starting with the computer-based models developed during the 1960s to study seawater intrusion problems. In additon to modeling conservative solutes, groundwater specialists have modeled chemicals that decay according to first-order reaction kinetics. Although sophisticated computer-based models of chemical reactions in groundwater have been developed by geochemists, they have not been validated extensively using experimental data.[64]

Computer-based models that account for sorption as a phenomenon affecting contaminant transport have also been developed. Here the main problem (in additon to the usual difficulties in characterizing the subsurface environment) is in obtaining reliable data for partition coefficients and other parameters needed to describe sorption and desorption in mathematical terms.[65]

After reviewing the various ways in which contaminants are transported in groundwater, a committee of the National Research Council (1994, pp 73–74) noted:

> Most investigations of contaminant fate in the subsurface have been carried out under ideal conditions such as in homogeneous aquifers with single contaminants. . . . [H]owever, the subsurface is neither chemically, biologically, nor physically homogeneous. Unfortunately, relatively little is understood about the impact of heterogeneities on processes that control the fate and transport of contaminants, including sorption, abiotic and biotic reactions, and residual entrapment and dissolution.

Much additional research is needed before the fate of nonaqueous-phase liquids and reactive substances in groundwater can be predicted with confidence. Regardless of the solute type, the difficulty and expense of gathering field data to characterize the subsurface environment and estimate model parameters will constrain the use of these models.

[64] This is the view of a committee of the National Research Council (1990, pp. 129–30). The committee argues, "So far, researchers have validated only sections or portions of geochemical [computer] codes against field or laboratory data. . . ." According to the committee, this lack of model testing results from a division of interest among groundwater specialists: "modelers tend to go their own way, building impressive computer codes, while experimentalists tend to gather data for purposes other than evaluating models."

[65] For an example of difficulties in obtaining suitable partition coefficients, see the modeling investigation summarized by Fetter (1993, pp. 150–56).

Key Concepts and Terms

Representation of the Hydrologic Cycle
Surface runoff
Infiltration
Percolation
Evapotranspiration
Hydrograph
Hydrologic simulation model
Model calibration

Influence of Land Use Changes on Flood Flows
Flood frequencies
Urbanization
Channelization

Parameters Used to Characterize Water Quality
Coliform bacteria
Suspended solids
Bottom sediments
Inorganic and organic solutes
Volatile and nonvolatile solutes
Total dissolved solids
Biochemical oxygen demand
Dissolved oxygen
Heavy metals
Priority pollutants

Point Sources: Municipal and Industrial Effluents
Domestic wastewater
Industrial cooling water
Cleaning and process water

Nonpoint Sources of Water Pollution
Urban runoff
First flush effect
Farm runoff
Contaminated leachates
Hazardous compounds in groundwater
Thermal stratification
Hypolimnion and epilimnion

Forecasting Changes in Surface Water Quality
Mass balance analysis
Conservative solute

Biochemical oxidation
First-order reaction kinetics
Oxygen sag equation
Nitrogen cycle
Benthic demand
Nonpoint source loadings
Accumulation and washoff
Sorption and desorption
Partition coefficient
Universal soil loss equation

Forecasting Changes in Groundwater Quality
Heterogeneity of porous media
Unsaturated versus saturated zone
Water table
Aquifer
Hydraulic head
Hydraulic gradient
Darcy's law
Darcy velocity vs. seepage velocity
Advection
Dispersion
Tracer
Reactive solutes
Nonaqueous-phase liquids
Model validation

Discussion Questions

23-1 Consider a proposed reservoir project that would supply water for irrigation and hydroelectric power generation. How might the project influence water flows in various components of the hydrologic cycle? Indicate which hydrologic impacts are likely to be the most significant.

23-2 When stream channels are cleared and widened to reduce flooding, there is sometimes an increase in flooding downstream. Give two possible explanations for this.

23-3 Provide examples demonstrating how sorption and desorption phenomena affect the quality of both surface water and groundwater.

23-4 Explain the difference between advection and dispersion in surface water. How would you modify your statement to explain the difference between advection and dispersion in groundwater?

23-5 Benzene (C_6H_6) in a river is subject to a biodegradation process that can be approximated using first-order reaction kinetics. Assume that the decay coefficient is 0.11 day^{-1}. Suppose spilled benzene dissolves into a river flowing at an average velocity of 0.3 m/sec. Let C_0 = the concentration of benzene in the river at the point of discharge. How many kilometers will the river have to flow before the concentration of benzene is cut in half as a result of biodegradation?

23-6 Consider a situation in which the following oxygen sag equation parameters are given:

$$k_1 = 0.2/\text{day}$$
$$k_2 = 0.4/\text{day}$$
$$D_0 = 1 \text{ mg/l}$$
$$C_s(x) = 8 \text{ mg/l for all } x$$
$$L_0 = 15 \text{ mg/l}$$

a. Plot the oxygen sag curve for $t = 1, 2, \ldots, 5$, where t, the time of travel, equals (x/u) in equation 23-3. Streamflow velocity is constant for this steady state analysis. How much travel time passes before the DO deficit is at a maximum?
b. Replot the sag curve for the case where wastewater treatment reduces L_0 to 7.5 mg/l. Use the same graph paper as in (*a*).
c. Employ the results of (*a*) and (*b*) to characterize the effects of wastewater treatment on stream dissolved oxygen.

23-7 Consider an aquifer with an average hydraulic conductivity of 15 m/day. Suppose the slope of the hydraulic grade line is 5 m/1000 m. What is the Darcy velocity? If the average porosity is 0.3, what is the seepage velocity? Explain why the seepage velocity must always be greater than the Darcy velocity.

23-8 Of all the models introduced in this chapter, select the one that would be easiest to calibrate and validate. For the model you selected, describe how you might design an exercise to validate the model.

23-9 Your office is responsible for making a decision on a proposed irrigation project that will contribute significantly to nonpoint source pollution of both surface and groundwater flows. Several consultants have been asked to present their qualifications to do modeling studies to assist in quantifying the impacts. Prepare three questions to ask each consultant so that you can help decide which one to select.

References

Barry, D. A. 1992. "Modeling Contaminant Transport in Subsurface: Theory and Computer Programs." In *Modeling Chemical Transport in Soils: Natural and Applied Contaminants,* ed. H. Ghadiri and C. W. Rose, 105–44. Boca Raton, FL: Lewis Publishers.

Beard, L. R., and S. Chang. 1979. Urbanization Impact on Streamflow. *Journal of the Hydraulics Division,* Proceedings of the American Society of Civil Engineers, 105 (HY6): 647–59.

Biswas, A. K. 1970. *History of Hydrology.* Amsterdam: North Holland.

Brookes, A. 1988. *Channelized Rivers: Perspectives for Environmental Management.* Chichester, U.K.: Wiley.

Canter, L. W. 1996. *Environmental Impact Assessment,* 2d ed. New York: McGraw–Hill.

CEQ (U.S. Council on Environmental Quality). 1980. *Eleventh Annual Report.* Washington, DC: Council on Environmental Quality.

———. 1981. *Contamination of Ground Water by Toxic Organic Chemicals.* Washington, DC: Council on Environmental Quality.

Crawford, N. H., and R. K. Linsley. 1966. *Digital Simulation in Hydrology: The Stanford Watershed Model IV.* Technical Report No. 39, Department of Civil Engineering. Stanford, CA: Stanford University.

Dunne, T., and L. B. Leopold. 1978. *Water in Environmental Planning.* San Francisco: Freeman.

EPA (U.S. Environmental Protection Agency) 1983. *Results of the Nationwide Urban Runoff Program,* vol. 1 (Final Report), Water Planning Division. Washington, DC: EPA.

Fetter, C. W. 1993. *Contaminant Hydrogeology.* New York: Macmillan.

Field, R., E. J. Struzeski, H. E. Masters, and A. N. Tafuri. 1975. "Water Pollution and Associated Effects from Street Salting." In *Water Pollution in Low Density Areas: Proceedings of a Rural Environmental Engineering Conference,* ed. W. J. Jewell and R. Swan, 317–40. Hanover, NH: University Press of New England.

Förstner, V., and G. T. W. Wittman. 1981. *Metal Pollution in the Aquatic Environment,* 2d rev. ed. Berlin: Springer-Verlag.

Freeze, R. A., and J. A. Cherry. 1979. *Groundwater.* Englewood Cliffs, NJ: Prentice–Hall.

Ghadiri, H., and C. W. Rose. 1992. "Sorbed Chemical Transport Modeling." In H. Ghadiri and C. R. Rose (eds.), *Modeling Chemical Transport in Soils: Natural and Applied Contaminants,* ed. H. Ghadiri and C. R. Rose. Boca Raton, FL: Lewis Publishers.

Harbor, J. M. 1994. A Practical Method for Estimating the Impact of Land Use Change on Surface Runoff, Groundwater Recharge and Wetland Hydrology. *Journal of the American Planning Association* 60(1): 95–108.

Hemond H. F. and E. J. Fechner. 1994. *Chemical Fate and Transport in the Environment.* San Diego, CA: Academic Press.

Hunt, C. E. 1988. *Down by the River: The Impact of Federal Water Projects and Policies on Biological Diversity.* Washington, D.C.: Island Press.

James, A., ed. 1993. *An Introduction to Water Quality Modeling,* 2d ed. New York: Wiley.

James, L. D. 1965. Using a Digital Computer to Estimate the Effects of Urban Development on Flood Peaks. *Water Resources Research* 1(2): 223–34.

Keith, L. H., and W. A. Telliard. 1979. Priority Pollutants: I-A Perspective View. *Environmental Science & Technology* 13(4): 416–23.

Krenkel, P. A., and V. Novotny. 1980. *Water Quality Management.* New York: Academic Press.

Loucks, D. P., J. R. Stedinger, and D. A. Haith. 1981. *Water Resources Systems Planning and Analysis.* Englewood Cliffs, NJ: Prentice–Hall.

Masters, G. 1991. *Introduction to Environmental Engineering and Science.* Englewood Cliffs, NJ: Prentice–Hall.

McCutcheon, S. C. 1989. *Water Quality Modeling: Vol. I, Transport and Surface Exchange in Rivers.* Boca Raton, FL: CRC Press.

Miller, D. 1985. "Chemical Contamination of Ground Water." In *Ground Water Quality,* ed. C. H. Ward, W. Giber, and P. L. McCarty, 39–52. New York: Wiley.

National Research Council (U.S.). 1977. *Drinking Water and Health.* Washington, DC: National Academy of Sciences.

———. 1990. *Ground Water Models: Scientific and Regulatory Applications.* Washington, DC: National Academy Press.

———. 1994. *Alternatives for Ground Water Cleanup.* Washington, DC: National Academy Press.

Neely, W. B. 1980. *Chemicals in the Environment: Distribution Transport, Fate, Analysis.* New York: Dekker.

Nemerow, N. L. 1978. *Industrial Water Pollution: Origins, Characteristics, and Treatment.* Reading, MA: Addison–Wesley.

Novotny, V., and H. Olem. 1994. *Water Quality: Prevention, Identification, and Management of Diffuse Pollution.* New York: Van Nostrand Reinhold.

O'Connor, D. J., R. V. Thomann, and D. M. Di Toro. 1976. "Ecologic Models." In *Systems Approach to Water Management,* ed. A. K. Biswas, 294–334. New York: McGraw–Hill.

Overton, D. E., and M. E. Meadows. 1976. *Stormwater Modeling.* New York: Academic Press.

Pirner, S. M., and L. L. Harms. 1978. Rapid City Combats the Effect of Urban Runoff on Surface Water. *Water and Sewage Works* 125(2): 48–53.

Roberts, P. V., P. L. McCarty, M. Reinhard, and J. Schreiner. 1980. Organic Contaminant Behavior, During Groundwater Recharge. *Journal of the Water Pollution Control Federation* 52(1): 161–72.

Sartor, J. D., and G. B. Boyd. 1975. "Water Quality Improvement Through Control of Road Surface Runoff." In *Water Pollution Control in Low Density Areas, Proceedings of a Rural Environmental Engineering Conference,* ed. W. J. Jewell and R. Swan, 301–16. Hanover, NH: University Press of New England.

Street, R. L., G. Z. Watters, and J. K. Vennard. 1996. *Elementary Fluid Mechanics,* 7th ed. New York: Wiley.

Tchobanoglous, G., and E. D. Schroeder. 1987. *Water Quality; Characteristics, Modeling, Modification.* Menlo Park, CA: Addison-Wesley.

Thomann, R. V., and J. A. Mueller. 1987. *Principles of Surface Water Quality Modeling and Control.* New York: Harper & Row.

Velz, C. V. 1970. *Applied Stream Sanitation.* New York: Wiley.

Whipple, W., N. S. Grigg, T. Grizzard, C. W. Randall, R. P. Shubinski, and L. S. Tucker. 1983. *Stormwater Management in Urbanizing Areas.* Englewood Cliffs, NJ: Prentice–Hall.

Wolman, M. G., and A. P. Shick. 1967. Effects of Construction on Fluvial Sediment, Urban and Suburban Areas of Maryland. *Water Resources Research* 3(2): 451–64.

Zheng, C., and G. D. Bennett. 1995. *Applied Contaminent Transport Modeling: Theory and Practice.* New York: Van Nostrand Reinhold.

Index

Terms followed by lowercase roman t indicate material in tables. Terms followed by italic *n* indicate material in footnotes or end of chapter problems. United States, or U.S., is *not* used as a modifier. Thus look for "Army Corps of Engineers" rather than U.S. Army Corps of Engineers.